Brief Contents

PART I THE MICROBIAL WORLD

PART II MICROBES AND HUMAN AFFAIRS

Courtesy of Dr. Rodney M. Donlan and Janice Carr/CDC

Contents

© Linda Armstrong/ShutterStock, Inc.

© Aerial Archives/Alamy Images

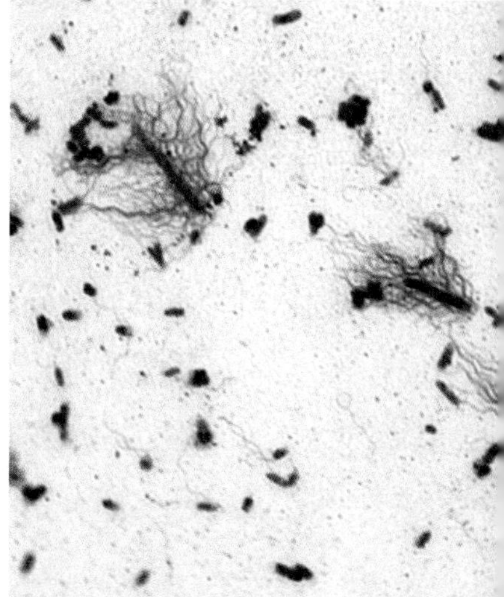
Courtesy of Dr. Jeffrey Pommerville

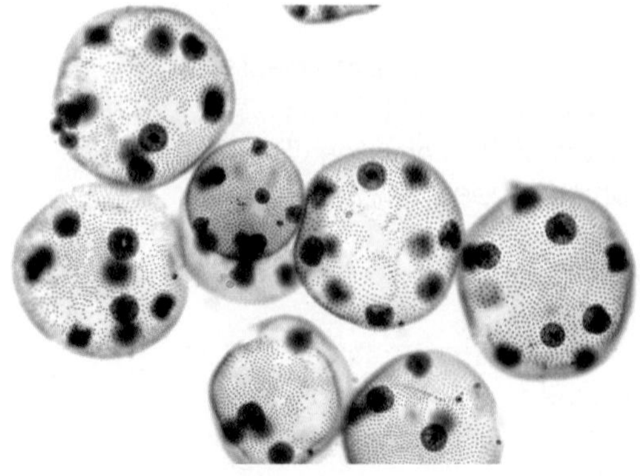

Courtesy of Dr. Jeffrey Pommerville

© Izatul Lail bin Mohd Yasar/ShutterStock, Inc.

© Photodisc/age fotostock

Preface

When I was approached by Jones & Bartlett Learning to become the new author of *Alcamo's Microbes and Society, Fourth Edition*, I was delighted. Why? It certainly wasn't because I was looking for more to do. With authoring other college textbooks and having a full-time faculty teaching position at Glendale Community College, I am plenty busy. The actual reason was that I saw this as an opportunity to convey the excitement and wonder of microbes to college students not majoring in the sciences, and to the public.

Today, hardly a day goes by when a story about microorganisms (microbes) does not appear in a newspaper, magazine, or on the TV news. When we see or hear a report on a disease outbreak, the discovery of some new and often exotic microorganism, or a news story about how microorganisms influence our good health and the environment, many students and the public in general often do not have the background to properly evaluate many of the statements or claims made. After reading *Microbes and Society*, all that will change.

It is an exciting time in microbiology, so I see my charge as one to capture that excitement, and transform the reader into a knowledgeable and scientifically literate citizen in today's society. This is not to make you into a science nerd, although there is nothing wrong with that, but rather to illustrate how the science of microbiology affects all of us in tremendous ways. As citizens of the world, we need to understand how the science operates, the challenges it faces, and the discoveries being made. Today, microbes are important to our lives, our good health, our society, and the daily operation of the world around us.

So, welcome to the world of microbiology! I hope you find the journey exciting, enriching, informative, and fascinating.

Audience

Alcamo's Microbes and Society, Fourth Edition is written to the nonscience undergraduate and inquiring citizen of the 21st century. It discusses such topics as the place of microbes in ecology and the environment, the use of microbes in biotechnology, the role of microbes in food production, and the numerous other ways that microbes contribute to the quality of our lives and our wellbeing. The book also examines the problem of antibiotic resistance, discusses the problems of nosocomial infections, and surveys several microbial diseases of history and contemporary times. Students will find that understanding microbes will help them do well in such fields as business, sociology, food science, pharmaceutical and health sciences, economics, and agriculture. The book assumes little or no science background, and it should accommodate one-quarter or one-semester courses.

Objectives

The 21st century is destined to be the Century of Biology. In future decades we can anticipate new products of biotechnology, new ways of preserving and protecting our environment, new methods in agriculture, new practices to maintain human health, and new technologies not yet even in the idea stage. And importantly, microbes are at the center of all of these. They are the hammers and nails of genetic engineering, the worker bees for purifying polluted water, the sources of imaginative insecticides and pesticides, our internal guardians protecting us from disease and helping keep us healthy, and the jumping off points for futuristic technologies. Knowing the microbes is essential to knowing the future. Helping you to know the microbes is the first major objective of this book.

What of today? Rarely does a day go by when we do not enjoy a "microbial food;" each time we put out the garbage, we assume that microbes will break it down; whenever we take a breath, we inhale oxygen that microbes have put into the atmosphere; and each time we cover a sneeze, we try to stop the spread of microbes. Helping you to understand the places that microbes occupy in our day-to-day existence is this book's second major objective.

But what would the present and future be without the past? The third major objective is to show you how microbes have had a significant impact on history. We shall study, for example, how microbes changed the course of Western civilization, how microbes stopped Napoleon's conquest of Russia, how microbes influenced the way cultures arose, and how microbes made much of the current work in biotechnology possible. Few groups of organisms have played such a rich and powerful role in history.

I hope you will enjoy your education in microbiology and come to understand the influence of microbes on our society today, in the past, and in the future.

Organization

Alcamo's Microbes and Society, Fourth Edition contains two parts. Part I introduces the microbial world over the span of 11 chapters. Individual chapters explore the bacteria, viruses, fungi, protists, and other microbes; other chapters describe how these microbes grow and reproduce, the unique genetic patterns they display, and the methods used to control them.

Part II moves to the practical applications of microbiology. We visit a restaurant for a microbial meal, we wander through a research facility and see microbes at work, we stop at various locations in the environment and observe microbes acting on our behalf, and we examine their place in disease. The bottom line is that microbes are relevant.

Must the chapters be studied in sequence? Absolutely not. Time constraints often prevent courses from using the entire book, so instructors and students are invited to "cherry pick" those topics that fit best. To encourage flexibility, each chapter has been written independently of the others, and each section in a chapter stands alone. Instructors may, therefore, design their own approach to microbiology according to their students' needs.

The Student Experience

Approaching a course in microbiology can be an anxious experience. There are new insights to learn, new concepts to master, and an entirely new vocabulary to memorize. To help smooth over the bumps, this book incorporates several features that should help increase students' comfort level as they read *Alcamo's Microbes and Society, Fourth Edition.*

Each chapter begins with a section titled "**Looking Ahead**" to let students know what they should take away from the chapter. The reading then opens with an engaging story to set a tone for the pages that follow. Key terms in the chapter appear in bold to draw the attention of readers, and organism pronunciations (sometimes appearing to be tongue twisting names) appear in the chapter margins along with the definitions of some more specialized terms (see margin). In addition, a pronunciation guide has been added as an Appendix.

12 Microbes and Food: A Menu of Microbial Delights

Many food and beverage products we consume and enjoy every day are either created by certain species of microbes or they are part of the production process in making the commercial product.

© Roman Sigaev/ShutterStock, Inc.

Looking Ahead

Although we often think of microbes as food contaminants, many species actually play vital roles in producing the foods we enjoy, as we will discover in this chapter.

On completing this chapter, you should be capable of:

- Describing the process of wine fermentation.
- Comparing the roles of microbes in the fermentation processes for olives and cheese making.
- Explaining the role of yeast in bread making.
- ining the fermentation process for sausages and sauerkraut.
- ussing the process of brewing beer.

he decades, many food-related microbes have received ... y have been umerous disease outbreaks associated ... en received although no ... this nega- ... microbes ... associated ... ondemn-

■ *Bacillus* bä -cil′lus

■ *Clostridium* klos-tri′dē-um

■ rem (Roentgen Equivalent in Man): A measure of radiation dose related to biological effect.

Endospores

A few bacterial species, particularly members of the genera *Bacillus* and *Clostridium*, have the ability to produce an extraordinarily resistant structure called the **endospore** (FIGURE 5.13). This structure is usually formed when nutrient stress is encountered in the environment. The spore is formed during an involved process during which the mother cell produces an internal spore (i.e., endospore) that contains a complete set of genetic information surrounded by a very thick peptidoglycan set of walls. Once formed and mature, the dormant spore is released with the death of the mother...

Endospores are probably the most resistant biological st... Desiccation has little effect on the spore. By containin... are heat resistant and undergo very few chemic... them difficult to eliminate from contaminated m... For example, endospores can remain viable in ... When placed in 70% ethyl alcohol, endospores h... can barely withstand 500 rems of radiation, but er... rems. In this dormant condition, endospores ca... centuries.

240 CHAPTER 11 Controlling Microbes: From Outside and Within the Body

A CLOSER LOOK 11.4
Hiding a Treasure

Their timing could not have been worse. Howard Florey, Ernst Chain, Norman Heatley, and others of the team had rediscovered penicillin, purified it, and proved it useful in infected patients. But it was 1939, and German bombs were falling on London. This was a dangerous time to be doing research into new drugs and medicines. What would they do if there was a German invasion of England? If the enemy were to learn the secret of penicillin, the team would have to destroy all their work. So, how could they preserve the vital fungus yet keep it from falling into enemy hands?

Heatley made a suggestion. Each team member would rub the mold on the inside lining of his coat. The *Penicillium* mold spores would cling to the rough coat surface where the spores could survive for years (if necessary) in a dormant form. If an invasion did occur, hopefully at least one team member would make it to safety along with his "moldy coat." Then, in a safe country, the spores would be used to start new cultures and the research could continue. Of course, a German invasion of England did not occur, but the plan was an ingenious way to hide the ... organism. ... told in *The Mold in* Dr. ...

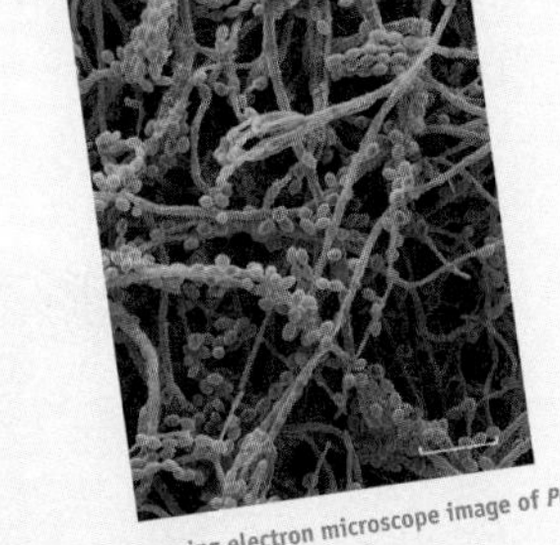

False-color scanning electron microscope image of *Penicillium* spores and hyphae. (Bar = 10 um.)

© RGB Ventures LLC dba SuperStock/Alamy

Special topics boxes in each chapter ("**A Closer Look**") encourage a moment of relief from the rigors of study, and present a historical insight, an interesting aside, or a health issue. Most figures are presented in full color, and special attention has been given to setting them close to their text reference.

The chapter concludes with **A Final Thought**, a set of **Questions to Consider** that provide challenging opportunities to apply what has been learned, and a **Key Terms** list for review. Students may note that all chapters are about the same length. This was done purposefully to provide a symmetrical framework in which students can learn. Each chapter has several sections and numerous smaller subsections to accommodate limited study times.

A Final Thought

When one of us (J.P.) was a young assistant professor of biology, the question of whether viruses are "alive" came up for discussion at an informal social gathering. As several people had had some beer to drink, the conversation became quite animated supporting one or the other viewpoint on viruses being alive. Perhaps one of the best answer was, "Who cares! We treat viral diseases the same way regardless of whether they are alive or not."

But the question has continued to pique my interest philosophically to this day. And now I put it to you. Are viruses alive? Although this textbook often refers to viruses as microbes for the sake of convenience, we have avoided references to "live" or "dead" viruses. Instead, we have used the words "active" for replicating viruses and "inactive" for viruses not replicating. Perhaps we should consider viruses to be inert chemical molecules with at least two properties of living organisms—the ability to replicate and adapt. Thus, viruses are neither totally inert nor totally alive, but somewhere on the threshold of life. Your thoughts?

Even many of the paragraphs are about the same size (there should be a rhythm in reading). The ultimate goal has been to provide a thorough and balanced presentation of microbiology within an enjoyable context.

What's New in This Edition?

Microbiology is a dynamic science and so this fourth edition reflects many of these advances as they pertain to humans, the environment, and society. The textbook also features new, and improved schematics to represent essential concepts and paradigms of thought in a more accurate and approachable way.

Each chapter has been revised based on the most current and significant knowledge available. New to this edition is:

- Updated coverage of microbes and new information on the human microbiome (Chapters 1 and 2).
- Revised coverage of microbial molecules (Chapter 3).
- Clearer descriptions of genes and the science of genomics (Chapter 4).
- New information on viruses and all microbial groups (Chapters 5–8).
- Revised descriptions of bacterial growth and metabolism (Chapter 9).
- Current applications of genetic engineering and biotechnology (Chapters 10 and 14).
- New discussions on bacterial resistance to antibiotics (Chapter 11).
- Expanded coverage of food microbiology, safety, and prevention. (Chapters 12 and 13).
- Expanded coverage of vaccines and their importance to human health (Chapter 14).
- New data on microbes in agriculture and the environment (Chapters 15 and 16).
- Clearer discussions of the immune system function (Chapter 17).
- More concise descriptions of viral and bacterial diseases (Chapters 18 and 19).

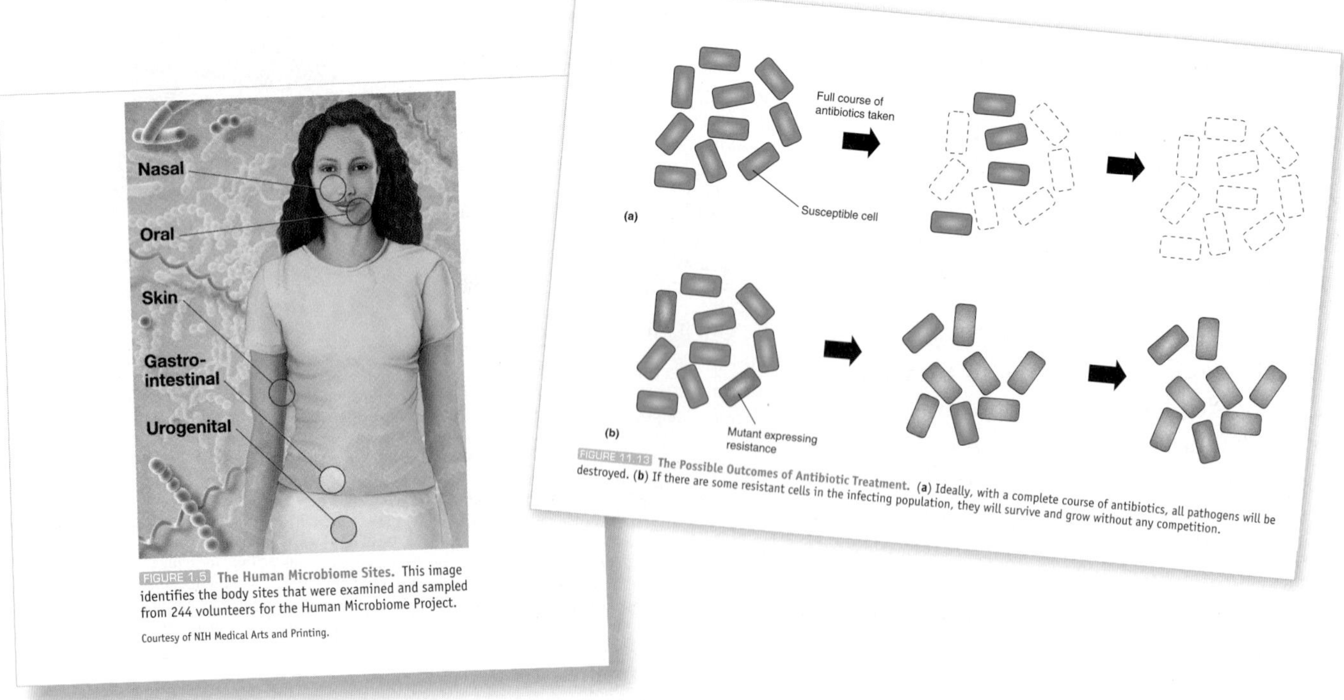

Nasal
Oral
Skin
Gastro-
intestinal
Urogenital
FIGURE 1.5 The Human Microbiome Sites. This image identifies the body sites that were examined and sampled from 244 volunteers for the Human Microbiome Project.
Courtesy of NIH Medical Arts and Printing.
Full course of
antibiotics taken
Susceptible cell
(a)
(b)
Mutant expressing
resistance
FIGURE 11.13 The Possible Outcomes of Antibiotic Treatment. (a) Ideally, with a complete course of antibiotics, all pathogens will be destroyed. (b) If there are some resistant cells in the infecting population, they will survive and grow without any competition.

All chapters have a small number of special topics boxes that are interesting, intriguing, and often important to us, and society as a whole. Some of these boxes have been updated or replaced. Among the new boxes are:

1.3 Antibiotics in the Feedlots	The abuse of antibiotics in fattening up livestock
2.4 Size Matters	Cell size is critical for proper nutrient transport
3.3 Microbes to the Rescue!	Microbes help clean up the *Deepwater Horizon* oil spill
4.2 The Tortoise and the Hare	Personalities affect the search to identify the structure of DNA
5.4 The Blood of History	Microbes have caused some "bloody" miracles in history
6.4 The Power of the Virus	Engineered viruses help to cure genetic diseases
7.4 "Black '47"	A microbial plant disease changes history
8.3 The Work of the Devil	Fungal hallucinations were behind the Salem witchcraft trials
9.3 Microbes "Raise a Stink"	Microbes compete successfully by being offensive to others
10.2 Gene Swapping in the World's Oceans	Microbes undergo an amazing amount of gene recombination
11.3 Are Antibacterial Soaps Worth the Money?	Investigating the reported usefulness of antibacterial soaps
12.3 This hqt's for You!	Ancient beer making by the Egyptians
13.3 Keeping Microorganisms Under Control	Quizzing your food safety precautions at home
14.1 Tag Team Microbes	Microbes working together can produce successful results
15.2 Stilton Cheese—Slicing through a Microbial Community	Some bacteria and fungi are essential to making many cheeses
16.2 The Great Sanitary Movement	Edwin Chadwick and the birth of modern sanitation
17.1 Probiotics and Your Health	Bacteria, probiotics, and their role in human health
18.4 The HIV Hideouts	Where the AIDS virus can hide out in the body
19.1 The Killer of Children	Pneumonia is deadly to children in many developing nations

Teaching Tools

The following resources are available to instructors to assist with course preparation:

- Sample Syllabus
- Lecture Outlines
- In-depth Lecture Presentations in PowerPoint format that feature art from the book
- Test Banks

- Image Bank of figures from the book
- Answers to the end-of-chapter questions from the book

Qualified instructors can obtain this material by contacting their Jones & Bartlett Learning Account Specialist or by going to www.jblearning.com.

Jones & Bartlett Learning also publishes the following titles that may be of use for your microbiology course:

- Infectious Diseases: *The Guide to Infectious Diseases by Body System, Second Edition* is an excellent ancillary tool for learning about microbial diseases. Each of the fifteen body systems units presents a brief introduction to the anatomical system and the bacterial, viral, fungal, or parasitic organism infecting the system.
- Encounters in Microbiology: *Encounters with Microbiology, Volume I, Second Edition* and *Volume II* bring together "Vital Signs" articles from *Discover* magazine in which health professionals use their knowledge of microbiology in their medical cases.

Note to Instructors

Microbiology embodies the beautiful and ugly, the simple and complex, and the big and the small of life, and in this regard is a fascinating, useful, and approachable topic for non-science majors to learn. Because of the real value of microbes to the quality of human life, the environment, and society, as well as their ability to cause disease, I saw the need for a course in microbiology for the non-science major. *Alcamo's Microbes and Society, Fourth Edition* is written and designed to support such a course; I hope you find it to be a useful tool in your pedagogical mission.

Please feel free to email me (jeffrey.pommerville@gccaz.edu) anytime with questions, comments, ideas, and/or suggestions that you believe could strengthen this text and make it an even more exciting learning experience for students.

Additional Microbiology Resources from Jones & Bartlett Learning

The following additional products are available to supplement your microbiology course. Our full catalog can be browsed at go.jblearning.com/Microbiology

- *Alcamo's Laboratory Fundamentals of Microbiology* is a series of 30 multipart laboratory exercises that provide basic training in the handling of microorganisms and help students understand the properties and uses of microbes.
- A two-volume anthology called *Encounters in Microbiology* brings together "Vital Signs" articles from *Discover* magazine in which health professionals describe in layman's terms their experiences with microbes and their knowledge of microbiology to solve medical mysteries.
- *Guide to Infectious Diseases by Body System, Second Edition* is an excellent tool for learning about microbial diseases. Each of the 15 body system units in the booklet presents a brief introduction to the anatomical system and a sampling of the bacterial, viral, fungal, or parasitic organisms that can infect the system.

Acknowledgments

Putting together a new edition of a textbook always requires the input of a whole team and so I must recognize everyone at Jones & Bartlett Learning who helped put together this new edition of *Alcamo's Microbes and Society*. I want to thank my editors Erin O'Connor and Matt Kane for giving me the opportunity to revise and update *Alcamo's Microbes and Society, Fourth Edition*, and for their support and encouragement throughout the process; and Audrey Schwinn for managing the process expertly, guiding the revisions, and keeping me on a timetable. I also want to thank Leah Corrigan for her expert work at the production end of the publication process, and to photo researcher Lauren Miller. Thanks also to Shellie Newell for copyediting the manuscript and to Jan Cocker for proofreading the text. It takes a team of talented and dedicated professionals to put together a text and the *Alcamo's Microbes and Society* team is stellar. I salute you all and am honored to work with you.

The author and the publisher would also like to thank the following individuals for their services as reviewers of this edition.

James C. Aumer, MS, C(ASCP), Rochester Institute of Technology
Marisa Barbknecht, MS, University of Wisconsin-La Crosse
Janet M. Cullen, MD, North Park University
Cynthia A. Fuller, PhD, Henderson University
Joanna Miller, PhD, Drew University
Robert Charles Osgood, PhD, Rochester Institute of Technology
Karen A. Palin, PhD, Bates College
Ekaterina Vorotnikova, PhD, University of Massachusetts, Lowell

About the Author

Courtesy of Dr. Jeffrey Pommerville

Today, I am a microbiologist, researcher, and science educator. My plans did not start with that intent. While in high school in Santa Barbara, California, I wanted to play professional baseball, study the stars, and own a '66 Corvette. None of these desires would come true—my batting average was miserable (but I was a good defensive third baseman), I hated the astronomy correspondence course I took, and I have yet to buy that Corvette.

I found an interest in biology at Santa Barbara City College. After squeaking through college calculus, I transferred to the University of California at Santa Barbara (UCSB), where I received a BS in biology and stayed on to pursue a PhD degree studying cell communication and sexual pheromones in a water fungus in the lab of Ian Ross. After receiving my doctorate in cell and organismal biology, my graduation was written up in the local newspaper as a native son who was a "fungal sex biologist"—an image that was not lost on my three older brothers!

While in graduate school at UCSB, I rescued a secretary in distress from being licked to death by a German shepherd. Within a year, we were married (the secretary and I). When I finished my doctoral thesis, I spent several years as a postdoctoral fellow at the University of Georgia. Worried that I was involved in too many research projects, a faculty member told me something I will never forget. He said, "Jeff, it's when you can't think of a project or what to do that you need to worry." Well, I have never had to worry!

I then moved on to Texas A&M University, where I spent 8 years in teaching and research—and telling Aggie jokes. Toward the end of this time, after publishing over 30 peer-reviewed papers in national and international research journals, I realized I had a real interest in teaching and education. Leaving the sex biologist career behind, I headed farther west to Arizona to join the biology faculty at Glendale Community College, where I continue to teach introductory biology and microbiology.

I have been lucky to be part of several educational research projects and have been honored, with two of my colleagues, with a Team Innovation of the Year Award by the League of Innovation in the Community Colleges. In 2000, I became project director and lead principal investigator for a National Science Foundation grant to improve student outcomes in science through changes in curriculum and pedagogy. I had a fascinating three years coordinating more than 60 science faculty members (who at times were harder to manage than students) in designing and field testing

18 interdisciplinary science units. This culminated with me being honored in 2003 with the Gustav Ohaus Award (College Division) for Innovations in Science Teaching from the National Science Teachers Association.

I am the Perspectives Editor for the *Journal of Microbiology and Biology Education*, the science education research journal of the American Society for Microbiology (ASM) and in 2004 was co-chair for the ASM Conference for Undergraduate Educators. From 2006 to 2007, I was the chair of Undergraduate Education Division of ASM. In 2006, I was selected as one of four outstanding instructors at Glendale Community College. The culmination of my teaching career came in 2008 when I was nationally recognized by being awarded the Carski Foundation Distinguished Undergraduate Teaching Award for distinguished teaching of microbiology to undergraduate students and encouraging them to subsequent achievement.

I mention all this not to impress but to show how the road of life sometimes offers opportunities in unexpected and unplanned ways. The key though is keeping your "hands on the wheel and your eyes on the prize"—then unlimited opportunities will come your way. And, hey, who knows—maybe that '66 Corvette could be in my garage yet.

Courtesy of Dr. Rodney M. Donlan and Janice Carr/CDC

To the Student—Study and Read Smart

When I was an undergraduate student, I hardly ever read the "To the Student" section (if indeed one existed) in my textbooks because the section rarely contained any information of importance.

This one does, so please read on.

In college, I was a mediocre student until my junior year. Why? Mainly because I did not know how to study properly, and, important here, I did not know how to read a textbook effectively. My textbooks were filled with underlined sentences (highlighters hadn't been invented yet!) without any plan on how I would use this "emphasized" information. In fact, most textbooks *assume* you know how to read a textbook properly. It is not like reading a fictional novel.

Reading a textbook is difficult if you are not properly prepared. So that you can take advantage of what I learned as a student and have learned from instructing thousands of students, I have worked hard to make this text user friendly with a reading style that is not threatening or complicated. Still, there is a substantial amount of information to learn and understand, so having the appropriate reading and comprehension skills is critical. Therefore, I encourage you to spend 20 minutes reading this section, as I am going to give you several tips and suggestions for acquiring those skills. Let me show you how to be an active reader.

Be a Prepared Reader

Before you jump into reading a section of a chapter in this text, prepare yourself by finding the place and time and having the tools for study.

Place. Where are you right now as you read these lines? Are you in a quiet library or at home? If at home, are there any distractions, such as loud music, a blaring television, or screaming kids? Is the lighting adequate to read? Are you sitting at a desk or lounging on the living room sofa? Get where I am going? When you read for an educational purpose—that is, to learn and understand something—you need to maximize the environment for reading. Yes, it should be comfortable but not to the point that you will doze off.

Time. All of us have different times during the day when we perform some skill the best, be it exercising or studying. The last thing you want to do is read when you are tired or simply not "in the zone" for the job that needs to be done. You cannot learn and understand the information if you fall asleep or lack a positive attitude. I have kept the chapters in this text to about the same length so you can estimate the time necessary for each and plan your reading accordingly. If you have done your preliminary survey of the chapter or chapter section, you can determine about how much time you will need. If 40 minutes is needed to read—and comprehend (see below)—a section of a chapter, find the place and time that will give you 40 minutes of uninterrupted study.

Brain research suggests that most people's brains cannot spend more than 45 minutes in concentrated, technical reading. Therefore, I have avoided lengthy presentations and instead have focused on smaller sections, each with its own heading. These should accommodate shorter reading periods.

Reading Tools. Lastly, as you read this, what study tools do you have at your side? Do you have a highlighter or pen for emphasizing or underlining important words or phrases? Notice, the text has wide margins, which gives you the space to make notes or to indicate something that needs further clarification. Do you have a pencil or pen handy to make these notes? Lastly, some students find having a ruler is useful to prevent your eyes from wandering on the page and to read each line without distraction.

Be an Explorer Before You Read

When you sit down to read a section of a chapter, do some preliminary exploring. Look at the section head and subheadings to get an idea of what is discussed. Preview any diagrams, figures, tables, graphs, or other visuals used. They give you a better idea of what is going to occur. We have used a good deal of space in the text for these features, so use them to your advantage. They will help you learn the written information and comprehend its meaning. Do not try to understand all the visuals, but try to generate a mental "big picture" of what is to come. Familiarize yourself with any symbols or technical jargon that might be used in the visuals.

Be a Detective as You Read

Reading a section of a textbook requires you to discover the important information (the terms and concepts) from the forest of words on the page. So, the first thing to do is read the complete paragraph. When you have determined the main ideas, highlight or underline them. However, I have seen students highlighting the entire paragraph in yellow, including every *a*, *the*, and *and*. This is an example of highlighting before knowing what is important. So, I have helped you out somewhat. Important terms and concepts in the textbook are in **bold face** followed by the definition (or the definition might be in the page margin). So, in many cases, you should only need to highlight or underline essential ideas and key phrases—not complete sentences. By the way, the important microbiological terms and major concepts also are in the **Glossary** at the back of the text.

What if a paragraph or section has no boldfaced words? How do you find what is important here? From an English course, you may know that often the most important information is mentioned first in the paragraph. If it is followed by one or more examples, then you can backtrack and know what was important in the paragraph.

Say It in Your Own Words

Brain research has shown that each individual can only hold so much information in short-term memory. If you try to hold more, then something else needs to be removed—sort of like having a full computer disk. So that you do not lose any of this important information, you need to transfer it to long-term memory—to the hard drive if you will. In reading and studying, this means retaining the term or concept; so, write it out in your notebook *using your own words*. Memorizing a term does not mean you have learned the term or understood the concept. By actively writing it out in your own words, you are forced to think and actively interact with the information. This repetition reinforces your learning.

Be a Patient Student

In textbooks, you cannot read at the speed that you read your text messages, email, or a magazine story. There are unfamiliar details to be learned and understood in a textbook—and this requires being a patient, slower reader. Actually, if you are like me and not a fast reader to begin with, it may be an advantage in your learning process. Identifying the important information from a textbook chapter requires you to *slow down* your reading speed. Speed-reading is of no value here.

Know the What, Why, and How

Have you ever read something only to say, "I have no idea what I read!" As I've already mentioned, reading a microbiology text is not the same as reading *Sports Illustrated* or *People* magazine. In these entertainment magazines, you read passively for leisure or perhaps amusement. In *Alcamo's Microbes and Society, Fourth Edition,* or any other textbook, you must read actively for learning and understanding—that is, for *comprehension*. This can quickly lead to boredom unless you engage your brain as you read—that is, be an active reader. Do this by knowing the *what, why,* and *how* of your reading.

- *What* is the general topic or idea being discussed? This often is easy to determine because the section heading might tell you. If not, then it will appear in the first sentence or beginning part of the paragraph.
- *Why* is this information important? If I have done my job, the text section will tell you why it is important or the examples provided will drive the importance home. These surrounding clues further explain why the main idea was important.
- *How* do I "mine" the information presented? This was discussed under being a detective.

A Marked Up Reading Example

So, let's put words into action. Below is a passage from *Alcamo's Microbes and Society, Fourth Edition*. I have marked up (highlighted) the passage as if I were a student

Domain Bacteria

The Bacteria have adapted to the diverse environments on Earth, inhabiting the air, soil, and water, and they exist in enormous numbers on the surfaces of virtually all plants and animals. They can be isolated from Arctic ice, thermal hot springs, the fringes of space, and the tissues of animals. Bacterial species, along with their archaeal relatives, have so completely colonized every part of the Earth that their mass is estimated to outweigh the mass of all plants and animals combined. Let's look briefly at some of the more common phyla and other groups, which give us a sense of the bewildering diversity that actually exists.

■ Cyanobacteria sī′-an-ō-bak-tėr′-ē-ä

■ bloom: A sudden increase in the number of cells of an organism in an environment.

Photosynthetic Bacteria. The members of the **Cyanobacteria** thrive in freshwater ponds and in the oceans and exist as unicellular, filamentous, or colonial forms. The pigments they contain give a blue-green, yellow, or red color to the organisms and the periodic redness of the Red Sea is due to blooms of those cyanobacterial species containing large amounts of red pigment.

The phylum Cyanobacteria is unique among bacterial groups because its members carry out photosynthesis similar to unicellular algae and plants using the light-trapping pigment chlorophyll. Their evolution on Earth was responsible for the "great oxidation event" some 2.5 billion years ago, the consequences of which transformed life on the young planet (see Figure 5.1). Such photosynthetic species are known as **autotrophic** microbes because they synthesize their own food (*auto* = "self;" *troph* = "feeder"; hence, "self-feeder"). By providing organic matter at the base of the food chain, the cyanobacterial species occupy a key position in the nutritional patterns of nature. In addition, chloroplasts probably evolved from a free-living cyanobacterial ancestor that took up residence in an evolving eukaryotic cell millions of years ago.

In addition to photosynthesis, some cyanobacterial species carry out **nitrogen fixation**. In this process, specialized cells called **heterocysts** within the cyanobacterial filament take up nitrogen from the atmosphere and convert it to ammonia and other nitrogen-containing substances used within the organism as well as within marine and aquatic plants during the synthesis of organic molecules. Because cyanobacterial species carry out photosynthesis as well as nitrogen fixation, they are among the most independent organisms on Earth.

Heterotrophic Bacteria. We now turn to bacterial species representing **heterotrophic** microbes (*hetero* = "other"), meaning they obtain their organic food molecules from other sources. Many do this by playing key roles as **decomposers**,

reading it for the first time. It uses many of the hints and suggestions I have provided. Remember, it is important to read the passage slowly, and concentrate on the main idea (concept) and the special terms that apply.

Have a Debriefing Strategy

After reading the material, be ready to debrief. Verbally summarize what you have learned. This will start moving the short-term information into the long-term memory storage—that is, *retention*. Any notes you made concerning confusing material should be discussed as soon as possible with your instructor. For microbiology, allow time to draw out diagrams. Again, repetition makes for easier learning and better retention.

In many professions, such as sports or the theater, the name of the game is practice, practice, practice. The hints and suggestions I have given you form a skill that requires practice to achieve and use efficiently. Change will not happen overnight; perseverance and willingness though will pay off with practice. You might also check with your college or university academic resource center or center for learning. These folks will have more ways to help you to read a textbook better and to study well.

Send Me a Note

In closing, I would like to invite you to write or email me. Let me know what is good about this textbook so I can build on it and what may need improvement so I can revise it. Don't be shy; let me know your thoughts. Also, I would be pleased to hear about any news of microbiology in your community, and I'd be happy to help you locate any information not covered in the text.

I wish you great success in your microbiology course. Welcome! Let's now plunge into the wonderful and sometimes awesome world of microorganisms.

—Dr. P

Email: jeffrey.pommerville@gccaz.edu
Website: http://web.gccaz.edu/~jpommerv/

Courtesy of Dr. Rodney M. Donlan and Janice Carr/CDC

A Tribute to I. Edward Alcamo

Courtesy of I. Edward Alcamo

Dr. Ignazio Edward Alcamo was a long-time Professor of Microbiology at the State University of New York at Farmingdale and the author of numerous textbooks, lab kits, and educational materials. He was the 2000 recipient of the Carski Foundation Distinguished Undergraduate Teaching Award, the highest honor bestowed upon microbiology educators by the American Society for Microbiology.

Ed Alcamo was educated at Iona College and St. John's University and held a deep belief in the partnership between research scientists and allied health educators. He sought to teach the scientific basis of microbiology in an accessible manner as well as to inspire students with a sense of topical relevance. Michael Vinciguerra, Provost at the SUNY Farmingdale wrote, "In 1970, when I joined the faculty as a chemistry professor, Ed's reputation as an excellent biology educator was already well known."

A prolific author, Dr. Alcamo produced a broad array of publications including several learning guides and textbooks—*Fundamentals of Microbiology* is now in its tenth edition. He also prepared the *Encarta* encyclopedia entry entitled "Procaryotes," as well as *The Microbiology Coloring Book*, and *Schaum's Outline of Microbiology*. His other books published within the past several years include *AIDS: The Biological Basis, DNA Technology: The Awesome Skill, The Biology Coloring Workbook,* and *Anatomy and Physiology the Easy Way*. In December 2002, after a six-month illness, Dr. Alcamo died of acute myeloid leukemia.

Dr. Alcamo's teaching career was dedicated to the proposition that emphasizing quality in education is central to turning back the tide of fear and uncertainty and enabling doctors to find cures for disease. In the early 1980s, when the early cases of an unknown acquired immunodeficiency syndrome were turning into a mysterious and intractable epidemic, Dr. Alcamo told this to his class:

One afternoon, about 350 years ago, in the countryside near London, a clergyman happened to meet Plague.
"Where are you going?" asked the clergyman.
"To London," responded Plague, "to kill a thousand."
They chatted for another few moments, and each went his separate way.
Some time later, they chanced to meet again. The clergyman said, "I see you decided to show no mercy in London. I heard that 10,000 died there."
"Ah, yes," Plague replied, "but I only killed a thousand. Fear killed the rest."

I

The Microbial World

Courtesy of Dr. Rodney M. Donlan and Janice Carr/CDC

In areas of Great Britain, milkmen still visit homes regularly and deliver bottles of fresh, pasteurized milk. Unfortunately, magpies, crows, and other birds arrive soon thereafter, use their strong beaks to peck through the foil caps covering the bottles, and then help themselves to the milk. In doing so, they often transmit species of bacteria capable of causing intestinal illness, and the unsuspecting families soon suffer the discomfort of diarrhea, cramps, and nausea. One British scientist has whimsically suggested: "Battered Brits better beware bacteria-bearing birds."

But we should not blame all the microbes for the ills a few species cause. In the same home refrigerators where the milk is stored is a variety of other dairy products that owe their existence to microbes. For instance, both the sour cream and buttermilk have been produced by harmless bacterial species. The cartons of yogurt would remain condensed milk without the help of two types of bacteria. And those blue streaks in the Roquefort cheese are harmless fungi purposely added to turn milk curds into a protein-rich and perfectly safe cheese we can sprinkle on a salad.

Microbes are a fascinating group of living organisms. Unappreciated by the vast majority of people, microbes bring about changes both mind-boggling and awe-inspiring. For example, some bacterial species and algae are responsible for putting much of the oxygen we breathe into the atmosphere; other bacterial species convert the nitrogen gas in the atmosphere into forms used by all organisms to make proteins; numerous bacteria and fungi are waste decomposers par excellence; and many microorganisms are used in industry to make such diverse products as perfumes, growth supplements, and even chocolate-covered cherries. Unfortunately, a few are also responsible for the profound changes that can occur during the course of infectious disease. In fact, diseases like the plague (Black Death) in the 1350s were so powerful, they changed European society.

All these microbes live in a world we can scarcely comprehend. Many species survive quite well in oxygen-free environments; others live six miles below the ocean surface and in solid rock; and still others thrive in acidic conditions that would dissolve most metals. Moreover, many microbes have the ability to pick up genes from their environment and incorporate those genes into their hereditary material. These activities serve to point out some of the extraordinary processes taking place in microbes, processes that set them apart from other living organisms and make knowledge of microbiology almost indispensable to all segments of society.

And so in Part I of *Microbes and Society*, we get to know the microbes and become familiar with their everyday activities. We explore how their world was discovered (Chapter 1) and see how they fit into the scheme of the living world (Chapter 2). In Chapters 3 and 4, we review the basic structure of microbial cells and highlight the roles of microbes in the discoveries surrounding DNA. Next, we look at their remarkable diversity, examine some of the unique characteristics of different microbial groups, and discover the influence they exert on society (Chapters 5 to 8). Then, in the last three chapters of Part I, we see how microbes grow (Chapter 9), we survey their extraordinary genetics (Chapter 10), and we study some methods by which they can be controlled (Chapter 11). The pages of Part I contain some eye-opening concepts that will tune you into a microbial world unknown to and unappreciated by most of your friends and family.

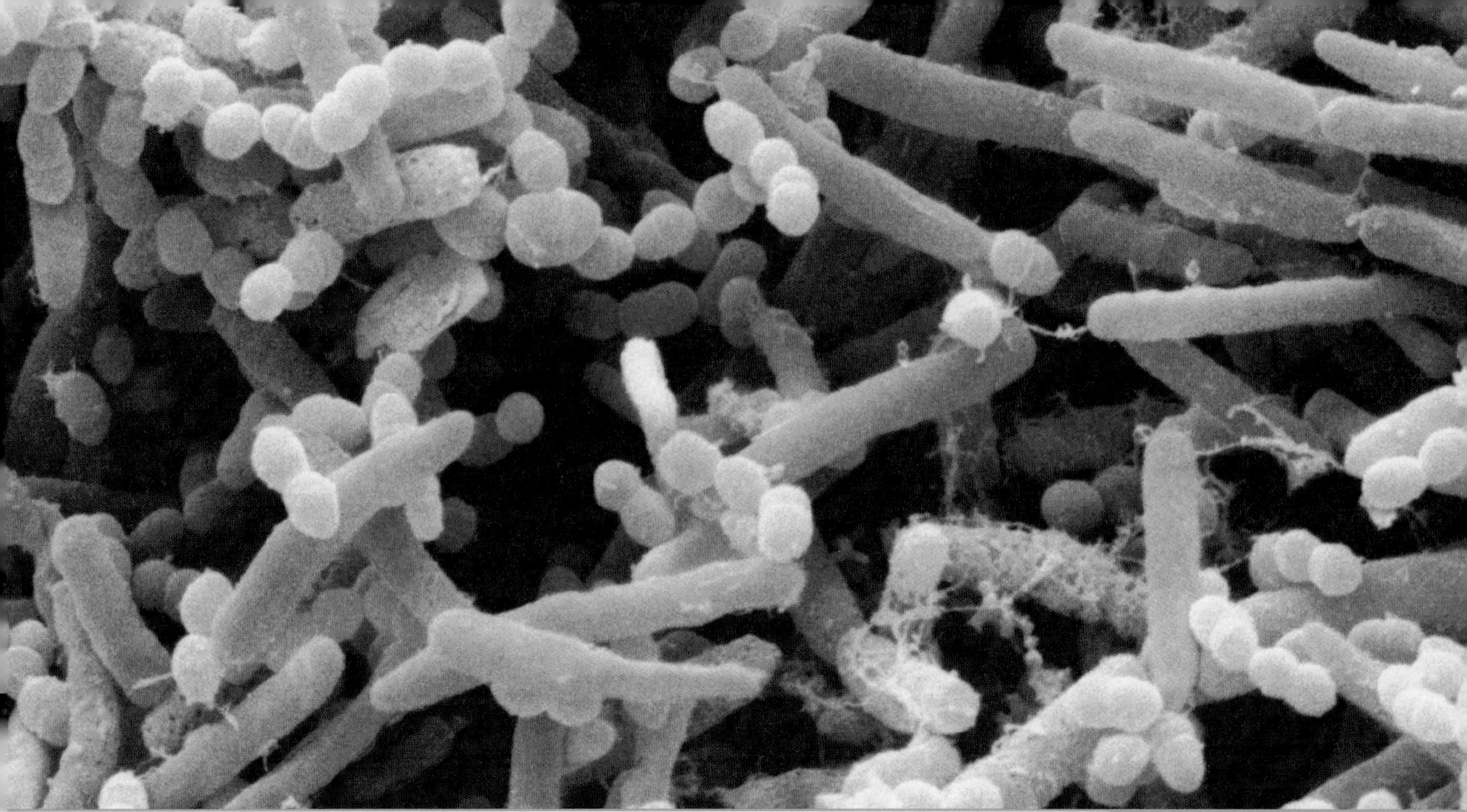

False-color scanning electron microscope image of bacteria found in compost. Microbes, including many bacterial and fungal species, begin the process of decay by breaking down plant tissue and converting it into compost, which is used as fertilizer and a soil additive.

The Microbial World: Surprising and Awesome

1

Looking Ahead

We share our world with thousands of species of plants and animals we can see (macroscopic) and thousands of species of microbes we cannot see (microscopic). This microbial world is surprising and awesome—surprising because it contains such a wealth of different forms of life, and awesome because we are learning just how essential microbes are and how they affect our world. This chapter begins our trek into the microbial world with some insights into this invisible realm.

After reading and completing this chapter, you should be capable of:

- Describing at least five ways microbes matter in the environment and to society.
- Evaluating the contributions made by the four pioneers of microbiology.
- Identifying contributions made during the Golden Age of microbiology.
- Comparing and contrasting eukaryotic and prokaryotic cell structures and distinguishing the four microbial groups.
- Identifying three challenges facing microbiology today.

Each year, a group of pilgrims gather in the English countryside outside the village of Eyam (about 160 miles north of London) and pay homage to the townsfolk, who 350 years earlier gave their lives so others might live. The pilgrims bow their heads and remember what happened that fateful year.

In the summer of 1665, a village tailor in Eyam received a shipment of cloth from London where plague was epidemic. Unknown to the cloth merchant, the shipment harbored fleas, which today we know can carry the plague bacterium. Within one week, the merchant was dead, and by late fall 28 citizens of Eyam had died of the disease. With the coming spring, plague erupted in Eyam and many had fled to the countryside. But many of the townsfolk realized that by doing so, they would probably spread the disease to nearby villages. After much soul-searching, they reluctantly decided to stay and take their chances. They marked off the village limits with a circle of boundary stones, where the neighboring villagers set food and other supplies for the self-quarantined group. By late in the year, the plague subsided but not before taking the lives of 260 of Eyam's 350 remaining citizens. And in London, where the Great Plague started, some 100,000 people died, representing 15% of the population.

Disease has always left people thunderstruck with terror. Before the late 1800s, however, such fear was compounded by ignorance because no one knew what caused disease, much less how to deal with it. As we shall see later in this chapter, the veil of ignorance would not lift until the late 1800s, when Louis Pasteur and Robert Koch confirmed the link between infectious disease and microscopic forms of life.

These life forms include bacteria (*sing.* bacterium), protozoa and algae, fungi (*sing.* fungus), and viruses, most all invisible to the unaided eye (FIGURE 1.1). Scientists refer to them as **microorganisms**, but we shall use the equally acceptable and more simplified term **microbes**.

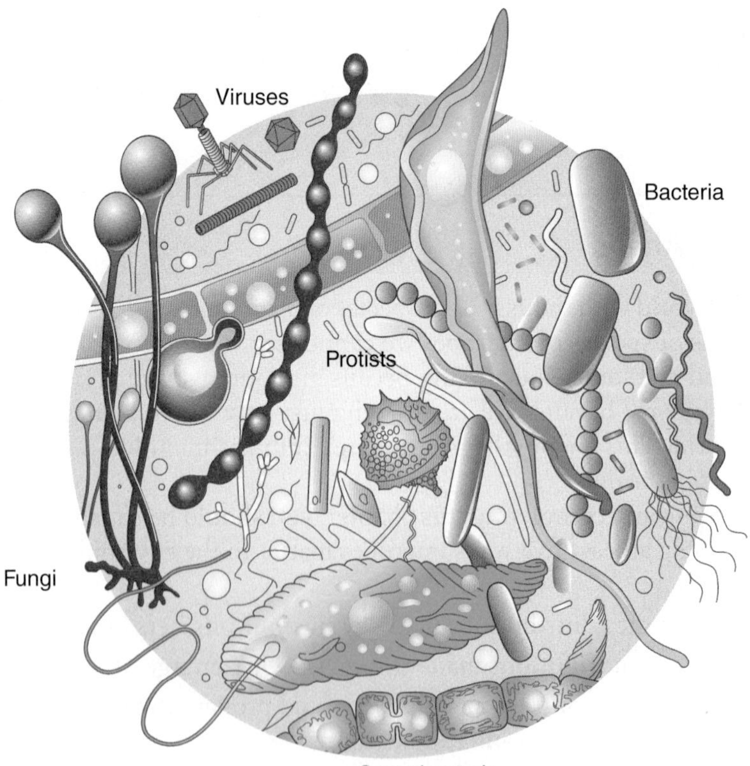

FIGURE 1.1 **A Microbial Menagerie.** The microbes consist of the bacteria, which includes the cyanobacteria, as well as the protists, fungi, and viruses. Many of these microbes would be present in a pinch of rich soil or in a drop of seawater.

1.1 Why Microbes Matter: Some Examples

Most of us are inclined to think of disease when we think of microbes. And this negative connection is probably justified because some microbes do cause much misery and pain to both individuals and to society, as the citizens of Eyam discovered. In addition, public news media are inclined to report on only those microbes causing human disease, and consumer advertising tends to emphasize microbe-destroying products, such as soaps and household cleaning products. But the majority of microbes are responsible for much of what adds quality to our lives. Microbes break down the remains of everything that dies and recycle the essential chemical elements so that vital nutrients can be renewed and reused by other living organisms. Moreover, about half the oxygen gas we breathe and many other organisms use is a product of microbial **photosynthesis**, the process by which light energy is converted to chemical energy.

Microbes are also responsible for the final forms of many foods we eat, including dairy products such as yogurt, buttermilk, and sour cream. They are the active partners in wine and beer production and often are the flavorful agents in cheese products.

Closer to home, there are hundreds of microbial species normally found on our skin and in our mouth, intestines, respiratory tract, and portions of the urogenital tract. This community of resident microbes, called the **human** microbiome, not only helps us fight off potentially disease-causing microbes, but also plays a critical role in maintaining our good health. In fact, a healthy adult human is host to some 100 trillion bacterial cells in the gut alone, a number ten times greater than the number of human cells building the entire human body.

microbiome: A specific environment characterized by a distinctive microbial community and its collective genetic material.

Huge communities of microbes also exist in the soil and the world's oceans. If you take a pinch of rich soil and place it in the palm of your hand, you will be holding an estimated billion microbes. If you examined one drop of seawater, you would discover some 10 million viruses. If you now piled up all the viruses from all the oceans end-to-end, they would extend out from the Earth 100 million light years (the diameter of our Milky Way galaxy is only 100,000 light years).

Fortunately, we are not here to count microbes in our bodies, the soil, or the oceans. Rather, we are here to study them, and so we should ask: What are the microbes doing in our bodies, the soil, and the oceans? And why are they there? To be sure, a few are **pathogens**—disease-causing microbes—probably in transit from one living organism to another. However, most of our microbiome is quite benign. By their sheer numbers, they control the pathogens and maintain nature's balance in our bodies. In the soil and especially in the oceans, many species capture energy from the sun and, through photosynthesis, store that energy in the form of sugar molecules while supplying 50% of the oxygen to Earth's atmosphere (plants supply the other 50%). Other soil and marine species are **decomposers** that break down dead plant and animal matter and recycle the carbon, nitrogen, sulfur, and other elements back to the atmosphere so the elements can be used to build the chemical molecules essential to maintaining life.

Here are a few other examples illustrating microbial diversity and the roles of those microbes.

Down on the Farm

A cow grazes peacefully on a sunlit day in an open field and chews its cud (FIGURE 1.2). But in the gut of the cow, microbes are busily at work because cows and other ruminants cannot digest grasses. When ingested, the grasses mix with resident

FIGURE 1.2 **Microbial Factories.** These cows lazily enjoying the afternoon sun are in fact microbial factories. Within a cow's first stomach (the rumen), various species of microbes (primarily protists and bacteria) break down ingested grasses and other plant materials.

microbes in the first stomach of the cow (called the rumen). There the microbes begin the digestion process, a transformation that will provide nutrients not only for the microbes but also provide energy and the building blocks the cow needs to build muscle and provide us with milk.

In other areas of agriculture, humans have put microbes to work as insecticides and pesticides. One bacterium, called *Bacillus thuringiensis*, is sprayed onto plants, where its poisonous toxins kill the caterpillar (or larval) stage of many agricultural pests. Scientists have even extracted from these bacteria the genes that encode the toxins and genetically engineered those genes into plant cells. Remarkably, these transgenic plants then produce the insecticidal toxins themselves as they grow in the field. Most of the soybean plants currently in American fields contain these bacterial genes.

■ *Bacillus thuringiensis* bä-sil′lus thur-in-jē-en′sis

■ transgenic: Referring to an organism containing a gene or genes from another organism.

In Our Foods

The food flavor market is a multibillion-dollar industry. On this industrial scale, microbes are grown in huge batches in building-sized tanks, where the microbes produce useful ingredients for many foods and beverages. Citric acid, for example, is produced by some molds and is typically used to enhance flavor in fruit juices, soda drinks, and candies. Lactic acid, produced by certain bacteria, is an emulsifier and, like citric acid, is also a food and beverage preservative. In addition, bacterial proteins produced in industrial quantities are used in baking, to tenderize meat, to clarify fruit juices, and even to soften the centers of chocolate-covered cherries. Working like miniature chemical factories, microbes can churn out industrial quantities of vitamins (especially B vitamins) that are then placed in foods and incorporated into the vitamin supplements many of us take.

Microbes also play an important role in the natural flavoring and aroma of many foods. Many peach, banana, pear, and coconut flavor enhancers are substances derived from fungal species. The most common food flavor enhancer, monosodium glutamate (MSG), is derived from the production of glutamic acid by specific bacteria. Food thickeners and stabilizers, such as xanthan gum and alginate, are bacterial and algal products, respectively. Even the artificial sweetener aspartame (NutriSweet®) is derived from an amino acid produced by specific types of bacteria.

With Most Meals

Rarely does a meal go by when we do not rely on microbes for something on our plate or in our glass, as FIGURE 1.3 illustrates. For example, the tangy taste in sausages and the specific tastes of sauerkraut, pickles, and vinegar are due to microbial activity. Many dairy products and cheeses result from microbial action on milk while virtually all types of bread depend on microbes for their taste and spongy textures. And certainly, wine and beer at the hub of the fermentation industries could not be produced without microbes. Even the coffee we drink depends, in part, on microbes for its flavor and aroma.

FIGURE 1.3 **An Selection of Foods and Beverages Produced by Microbes.** Society is dependent on microbes for many of its foods. Fermented foods and dairy products are examples.

In the Environment

Besides playing important roles as decomposers in the soil, microbes can and have been used effectively to lessen pollution. This immensely appealing way of putting microbes to work for society is called **bioremediation**, the application of bacteria and fungi to degrade toxic compounds in the environment. The process has been used to eliminate environmental pollutants, such as the waste products of explosives, as well as cleaning agents and radioactive compounds. When an oil spill occurs, technologists may add mineral nutrients to the water to encourage microbes to grow so they can gorge themselves on the petroleum. More impressive is the recent action of natural oil-eating microbes. Following the April 2010 explosion on the *Deepwater Horizon* oil rig in the Gulf of Mexico, tons of oil spilled into the seawater, affecting the shoreline of the Gulf Coast. For five months following the capping of the spill, oil-eating bacterial species multiplied and gobbled up at least 200,000 tons of oil trapped in underwater layers.

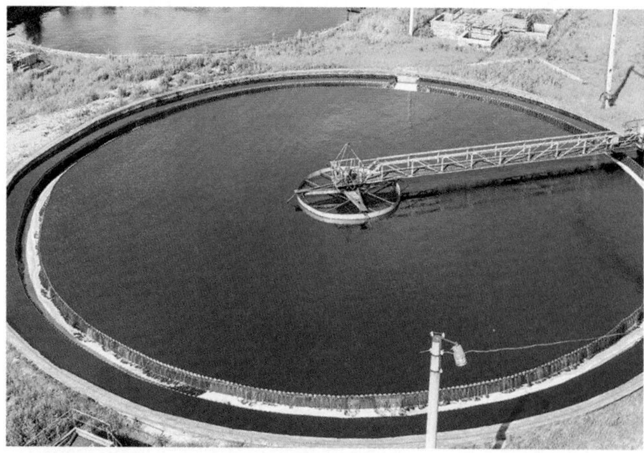

FIGURE 1.4 **Oxidation Lagoons.** This aerial view of a wastewater treatment plant shows an oxidation lagoon. Exposure to sunlight and oxygen allow the bacteria to degrade organic matter in the sewage.

Microbes also remain the prime factor in sewage treatment. In some municipalities, waste is piped into so-called oxidation lagoons (FIGURE 1.4), where the microbial community digests the organic matter and completely converts the complex compounds into simple ones that can be recycled. Waste treatment plants also rely on microbial chemistry to handle the massive amounts of sewage and garbage generated daily. Thanks to microbes, the garbage is converted to chemical fertilizers used as soil enhancers for growing crops.

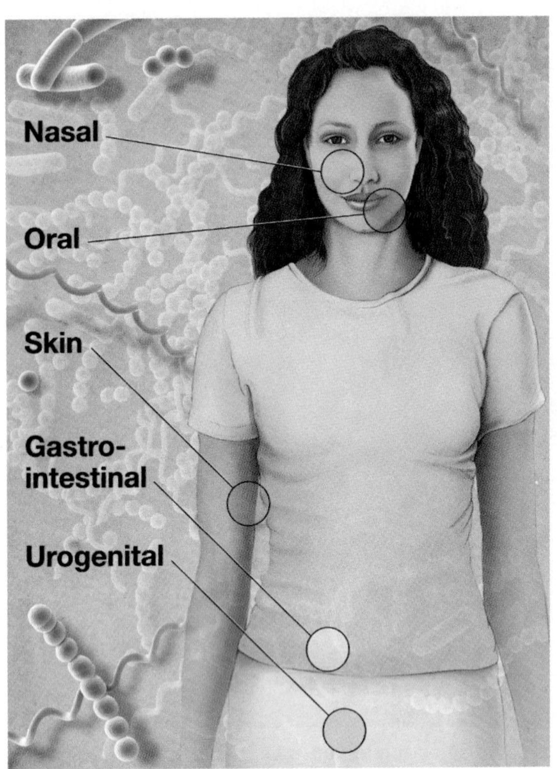

FIGURE 1.5 **The Human Microbiome Sites.** This image identifies the body sites that were examined and sampled from 244 volunteers for the Human Microbiome Project.

Courtesy of NIH Medical Arts and Printing.

In the Pharmaceutical/Biotechnology Industries

Pharmaceutical scientists are hard at work searching for effective medicines, antimicrobial drugs, and vaccines to combat the infectious diseases of our era. One of the approaches for innovative therapies is **biotechnology**, the use of microbes and their chemistry to manufacture products intended to improve the quality of human life.

Today bacterial species and fungal yeasts are being manipulated to act as living factories producing useful products, such as human insulin for diabetics, blood-clotting factors for hemophiliacs, and human growth hormone for children suffering dwarfism. Viruses have been genetically altered to increase disease resistance in plants, to kill insect pests in the environment, and to deliver new genes into cells. Some fungal yeasts have been modified to produce viral proteins used in vaccines, such as the one for hepatitis B.

In the Human Body

Humans are essentially free of any microbes until birth, but during and after birth, body surfaces (skin, mouth, gut) on and in the newborn become host to an enormous variety of microbes, our so-called human microbiome. Although scientists have known for a long time that the human microbiome influences our health by providing a cellular barrier to pathogen invasion, only recently have scientists realized the breadth of the microbiome's involvement (FIGURE 1.5).

Should these microbes become unbalanced in an individual, the person may develop medical conditions ranging from inflammatory bowel disease (Crohn's disease and ulcerative colitis) to obesity. Babies born through Cesarean section have an altered intestinal microbiome, which may later in life lead to a predisposition to develop allergies or asthma. From a societal perspective, the human microbiome will be a key component of future personalized medicine. This means some "societal diseases," such as obesity, diabetes, allergies, acne, Crohn's disease, some cancers, and even depression, may be personally treated based, in part, on one's "microbiome fingerprint."

To be sure, it is an awesome time to be studying microbiology. However, we must pause before we get too deeply into the study of microbes, for, as an old proverb states, "*To understand where you are going, you must know where you have been.*" And so before we launch into our study of microbiology, we must briefly examine its origins.

1.2 The Roots of Microbiology: The Pioneers

Many people mistakenly perceive science as nothing more than a collection of facts and data, and scientific principles and theories as something one simply "plucks from a tree." Rather, **science** is a human effort that uses observation and experimentation to understand, or to understand better, how the natural world works. It is only through human endeavor and creativity that newly found knowledge is built into important and logical frameworks comprising those principles and theories. This is made no clearer than by examining the historical observations on and experiments with microbes carried out by some very inquisitive individuals.

Robert Hooke (1635–1703)

The microbial world was virtually unknown until the mid-1600s, when an English scientist named Robert Hooke became fascinated with a newly developed instrument, the microscope, and wrote about his observations. In a book he published, called *Micrographia*, Hooke recorded his descriptions of minute compartments in slices of cork (which he named "cells"). He also reported the microscopic details of threadlike fungi he found growing on the sheepskin cover of a book. Hooke's curiosity led him to study the fungus, and his stunning illustrations of the mold were among the first visual descriptions of an invisible world.

Meanwhile, across the North Sea in Delft, Holland, more astonishing discoveries of the microbial world were being made.

Antony van Leeuwenhoek (1632–1723)

The individual who first brought the diversity of microbes to the world's attention was a Dutch merchant named Antony van Leeuwenhoek. In the 1670s, Leeuwenhoek developed the skill of grinding lenses for the purpose of magnifying and inspecting cloth. Knowing of Hooke's *Micrographia*, Leeuwenhoek began using his own lenses to satisfy his curiosity about the microscopic world. Leeuwenhoek studied the eye of an insect, the scales of a frog's skin, and the intricate details of muscle cells. Then, in 1673, while peering with his microscope into a drop of pond water, he came upon microscopic forms of life, darting back and forth and rolling and tumbling when observed with his simple microscope (FIGURE 1.6a). He dubbed the organisms **animalcules** (he assumed they were tiny animals). They at first delighted him, then amazed him with their variety, and finally perplexed him as he pondered their meaning. Today, we are fairly sure the organisms he saw were protists (protozoa and algae).

Leeuwenhoek
lā-ven-hōk

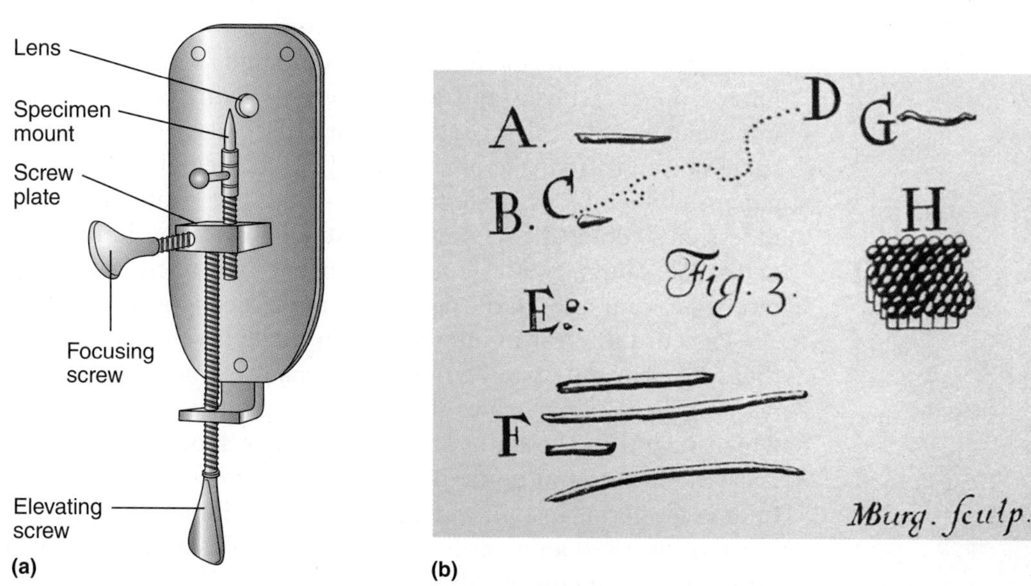

FIGURE 1.6 **Leeuwenhoek's Microscope and Drawings of Bacteria.** (**a**) To view his animalcules, Leeuwenhoek placed his sample on the tip of the specimen mount that was attached to a screw plate. An elevating screw moved the specimen up and down, while the focusing screw pushed against the metal plate, moving the specimen toward and away from the lens. (**b**) With such an instrument, Leeuwenhoek drew the animalcules (bacteria in this drawing) he saw.

Leeuwenhoek excitedly communicated his findings to a group of English scientists called the Royal Society. Over the next 40 years, he wrote many letters describing the new microscopic forms he observed. His letter dated September 17, 1683, is particularly noteworthy because it contains the first known descriptions and drawings of bacterial cells (FIGURE 1.6b). Leeuwenhoek, in fact, saw all the organisms we today know as microbes (except the viruses). The observations and descriptions made by Hooke and Leeuwenhoek introduced the world to the microbes.

Louis Pasteur (1822–1895)

After the death of Leeuwenhoek, interest in animalcules gradually waned as most people believed the animalcules were mere curiosities with little or no influence on society. In the 1850s, however, Louis Pasteur, the renowned French chemist and scientist, called attention to the animalcules as possible agents of infectious disease.

Pasteur (FIGURE 1.7) believed the discoveries of science should have practical applications to society, so in 1857 he seized the opportunity to unravel the mystery of why some French wines were turning sour. The prevailing theory held that wine fermentation resulted from the purely chemical breakdown of grape juice to alcohol; no living organisms were involved. But Pasteur consistently saw large numbers of tiny yeast cells in the wine. Moreover, he noticed sour wines also contained populations of the barely visible bacteria similar to some described by Leeuwenhoek. In a classic series of experiments, Pasteur boiled several flasks of grape juice and removed all traces of yeast cells from the flasks; he then set the juice aside to see if it would ferment. Nothing happened. Next, he carefully added pure yeast cells back into the flasks, and soon the fermentation was proceeding normally. Moreover, he found that when he used heat to remove all bacteria from the grape juice, the wine would not turn sour; it would not "get sick." Pasteur's recommendation of using heat to control bacterial contamination led to the process that today we call **pasteurization**.

FIGURE 1.7 **Louis Pasteur.** Pasteur developed pasteurization and proposed the germ theory of disease.

Courtesy of the National Library of Medicine

Pasteur's work shook the scientific community because it revealed microscopic yeast cells and bacteria as tiny, living factories where important chemical changes take place; indeed, Pasteur wondered if bacteria could also make people sick. In 1857, he published a short paper in which he suggested microbes might be related to human illness. In so doing, he set down the foundation for the **germ theory of disease**, a fundamental tenet holding that some microbes (germs) play significant roles in the development of infectious disease.

Pasteur's interest in microbes grew as he learned more about them. Prior to and up to Pasteur's time, many learned and common people believed animalcules arose through **spontaneous generation**, which suggested these tiny forms could arise spontaneously from decaying matter and nonliving substances. However, Pasteur was finding his germs in soil, water, and air, and—importantly—in the blood of diseased animals. So, he went about designing an elegant series of experiments to validate his belief and refute the spontaneous generation of germs.

Pasteur prepared a nutrient-rich solution (called a broth) in a series of swan-neck flasks (so named because their S-shaped necks resembled those of swans), as explained in **A Closer Look 1.1**. Pasteur boiled the broth in the flasks, thereby destroying all microbes; then he left the flasks open to the air. However, the S-shaped neck trapped any microbes in the air and prevented their entry into the flasks. Thus, when the flasks were set aside to incubate in a warm environment, no microbes appeared in the broth. However, Pasteur reasoned that if the neck was cut off a flask, microbes, if present in the air, should enter the broth and grow. And grow they did, becoming so dense that the liquid soon became cloudy with microbes. These experiments clearly showed microbes were in the air and probably in other environments as well.

A CLOSER LOOK 1.1

Experimentation and Scientific Inquiry

Science certainly is a body of knowledge, as you can see from the pages in this text! However, science also is a process—a way of learning. Often we accept and integrate into our understanding new information because it appears consistent with what we believe is true. But, are we confident our beliefs are always in line with what is actually accurate? To test or challenge current beliefs, scientists must present logical arguments supported by well-designed and carefully executed experiments.

The Components of Scientific Inquiry

There are many ways of finding out the answer to a problem. In science, scientific inquiry—or what has been called the "scientific method"—is the way problems are investigated. Let's understand how scientific inquiry works by following the logic of the experiments Louis Pasteur published in 1861 to refute the idea of spontaneous generation; that is, that life could arise from nonliving materials.

When studying a problem, the inquiry process usually begins with "observations." For Pasteur, his earlier observations suggested organisms do not appear from nonliving matter but rather are in the air.

Next comes the "question," which can be asked in many ways but usually as a "what," "why," or "how" question. For example, "What accounts for the growth of microorganisms in the animal broth?"

From the question, various hypotheses are proposed to answer the question. A "hypothesis" is a provisional but testable explanation for an observed phenomenon. Pasteur's previous work suggested life arising spontaneously in the broths was, in reality, a case of airborne microorganisms in dust landing on a suitable substance and then multiplying in such profusion they could be seen as a cloudy liquid.

Pasteur's Experiments

Pasteur set up a series of experiments to test the hypothesis that "Life only arises from other life" (see facing page).

In experiment 1, Pasteur sterilized animal broths in glass flasks by heating. He then either left the neck open to the air (a) or sealed the glass neck (b). Organisms only appeared (turned the broth cloudy) in the open flask.

In experiment 2, Pasteur sterilized a broth in swan-necked flasks (a). No organisms appeared, even after many days. However, if the neck was snapped off or the broth tipped to come in contact with the neck (b), organisms (cloudy broth) soon appeared.

Analysis of Pasteur's Experiments

Let's analyze the experiments. Pasteur had a preconceived notion of the truth and designed experiments to test his hypothesis. In his experiments, only one "variable" (an adjustable condition) changed. In experiment 1, the flask was open or sealed; in experiment 2, the neck was left intact or exposed to the air. Pasteur kept all other factors the same; that is, the broth was the same in each experiment; it was heated the same length of time; and similar flasks were used. Thus, the experiments had rigorous "controls" (the comparative condition). For example, in experiment 1, the control was the flask left open. Such controls are pivotal when explaining an experimental result. Pasteur's finding that no life appeared in the sealed swan-necked flask is interesting but tells us very little by itself. Its significance comes by comparing this to the broken neck (or tipped flask) where microbes quickly appeared.

Hypothesis and Theory

When does a hypothesis become a theory? Well, there is no set time or amount of evidence specifying the change from hypothesis to theory.

A "theory" is defined as a hypothesis that has been tested and shown to be correct every time by many separate investigators. So, at some point, if validated, sufficient evidence exists to say a hypothesis is now a theory. However, theories are not written in stone. They are open to further experimentation and so can be refuted.

As a side note, today a theory often is used incorrectly in everyday speech and in the news media. In these cases, a theory is equated incorrectly with a hunch or belief—whether or not there is evidence to support it. In science, a theory is a general set of principles supported by large amounts of experimental evidence.

(continued)

A CLOSER LOOK 1.1

Experimentation and Scientific . . . (continued)

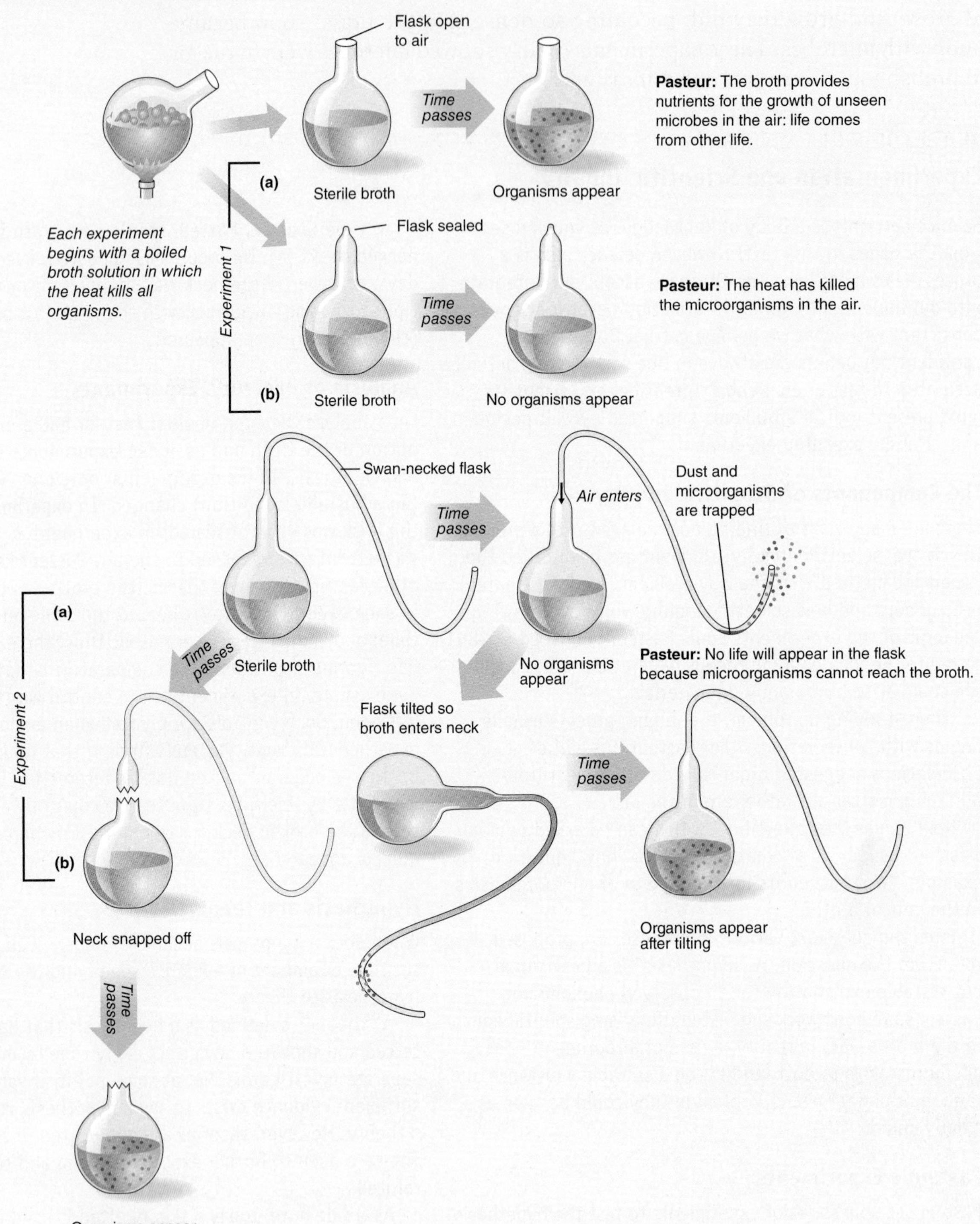
Flask open
to air
Time
passes
Pasteur: The broth provides nutrients for the growth of unseen microbes in the air: life comes from other life.
(a)
Sterile broth
Organisms appear
Each experiment begins with a boiled broth solution in which the heat kills all organisms.
Experiment 1
Flask sealed
Time
passes
Pasteur: The heat has killed the microorganisms in the air.
(b)
Sterile broth
No organisms appear
Swan-necked flask
Time
passes
Air enters
Dust and microorganisms are trapped
(a)
Time
passes
Sterile broth
No organisms appear
Pasteur: No life will appear in the flask because microorganisms cannot reach the broth.
Experiment 2
Flask tilted so broth enters neck
Time
passes
(b)
Neck snapped off
Organisms appear after tilting
Time
passes
Organisms appear

Pasteur and the Spontaneous Generation Controversy. (**1a**) When a flask of sterilized broth is left open to the air, organisms appear. (**1b**) When a flask of broth is boiled and sealed, no organisms appear. (**2a**) Broth sterilized in a swan-neck flask is left open to the air. The curvature of the neck traps dust particles and microorganisms, preventing them from reaching the broth. (**2b**) If the neck is snapped off to allow in air or the flask is tipped so broth enters the neck, organisms come in contact with the broth and grow.

Although Pasteur's work suggested some microbes (germs) could cause disease, he was stymied by his inability to prove that a particular germ caused a specific disease. Adding to his frustration was the death of three of his five children, two dying from typhoid fever, a germ-caused disease.

Robert Koch (1843–1910)

Although Pasteur failed to prove a specific organism caused a specific human disease, his work stimulated others to investigate the association of microbes with human disease. Among them was Robert Koch (FIGURE 1.8a), a German country doctor. Koch's primary interest was anthrax, a deadly blood disease in cattle and sheep. Koch injected mice with the blood of sheep that had died from anthrax. The mice soon died and, using his microscope, Koch observed rod-shaped bacteria, all of the same type as in the blood sample from the sheep. He watched for hours as the bacterial cells elongated, multiplied, formed long filaments, and produced highly resistant bodies called spores. At this point, Koch took some spores on a sliver of wood and injected them into healthy mice. Several hours later, the symptoms of anthrax appeared. Koch excitedly autopsied the mice and found their blood swarming with the rod-shaped anthrax bacteria. The cycle was complete—a specific microbe was shown to cause a specific human disease.

Here was the verification of the germ theory of disease that had escaped Pasteur. Koch's procedures, which became known as **Koch's postulates**, are illustrated in

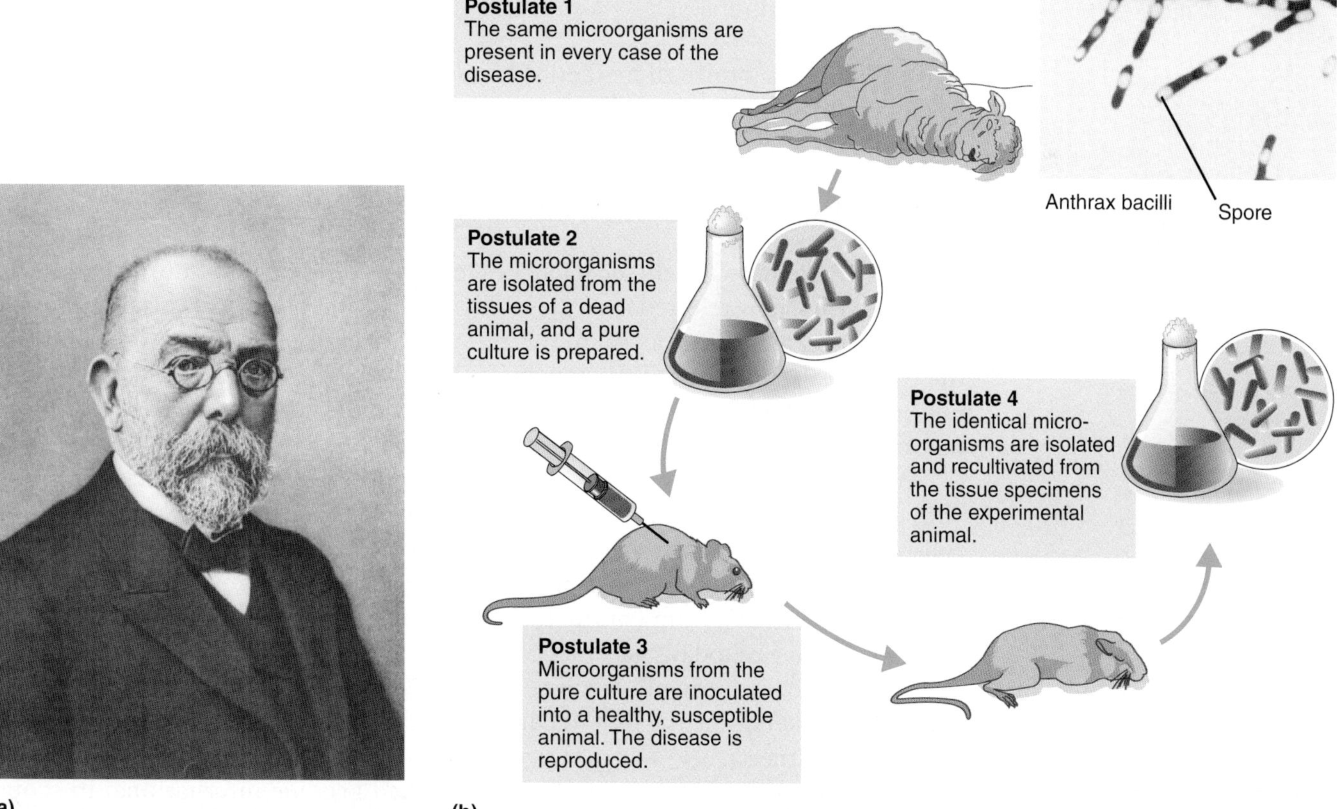

Postulate 1
The same microorganisms are present in every case of the disease.
Anthrax bacilli
Spore
Postulate 2
The microorganisms are isolated from the tissues of a dead animal, and a pure culture is prepared.
Postulate 4
The identical micro-organisms are isolated and recultivated from the tissue specimens of the experimental animal.
Postulate 3
Microorganisms from the pure culture are inoculated into a healthy, susceptible animal. The disease is reproduced.
(a)
(b)

FIGURE 1.8 **Koch's Postulates.** (**a**) Robert Koch was the first to verify Pasteur's germ theory by using a set of postulates (**b**) to relate a single microbe to a single disease. The photo (inset) shows the rods of the anthrax bacillus as Koch observed them. Many rods are swollen with spores.

(a) © National Library of Medicine. (b) Courtesy of CDC

A CLOSER LOOK 1.2

Jams, Jellies, and Microorganisms

One of the major developments in microbiology was Robert Koch's use of a solid culture surface on which bacteria would grow. He accomplished this by solidifying beef broth with gelatin. When inoculated onto the surface of the nutritious medium, bacterial cells grew vigorously at room temperature and produced discrete, visible clumps of cells called "colonies."

On occasion, Koch grew bacteria at temperatures above room temperature. In many cases, at these slightly higher temperatures, the gelatin turned to liquid. Moreover, some bacterial species could digest the gelatin. Walther Hesse, an associate of Koch's, mentioned the problem to his wife and laboratory assistant, Fanny Hesse. She had a possible solution. For years, she had been using a seaweed-derived powder called agar (pronounced ah'gar) to solidify her jams and jellies. Agar was valuable because it mixed easily with most liquids and once gelled, it did not liquefy, even at the warmer temperatures Koch used.

In 1880, Hesse was sufficiently impressed to recommend agar to Koch. Soon Koch was using it routinely to grow pure cultures of bacteria. It is noteworthy that Fanny Hesse may have been among the first Americans (she was originally from New Jersey) to make a significant contribution to microbiology.

Another key development, the common petri dish (plate), also was invented about this time (1887) by Julius Petri, one of Koch's former assistants.

Fanny Hesse.

FIGURE 1.8b. With improved methods for culturing bacteria (see **A Closer Look 1.2**), Koch's postulates were adopted as the guide for relating a specific microbe to a specific disease, and these postulates are still used today.

1.3 The Roots of Microbiology: The Golden Age

Thanks to Pasteur and Koch, the science of microbiology blossomed during a period of about 60 years, now referred to as the Golden Age of microbiology. This period began in 1857 with Pasteur's proposal of the germ theory of disease, and it continued into the twentieth century, until the advent of World War I. During these years, numerous branches of microbiology were established, and the foundations were laid for the maturing process that has led to the modern science.

During the Golden Age, Pasteur found he could weaken the bacteria associated with chicken cholera and inject the weakened cells into healthy chickens, where the cells protected the animals against a lethal dose of the pathogenic bacteria. He also showed that a rabies vaccine he and coworkers developed prevented a young boy bitten by a rabid dog from contracting rabies. These vaccines became the basis for developing some of today's vaccines.

Meanwhile, in Germany, Koch soon isolated the microbe causing tuberculosis, and his coworkers were the first to cultivate the bacterial organisms causing typhoid fever and diphtheria.

By the turn of the century, microbiology had become international in scope, moving far beyond France and Germany. For example:

- Ronald Ross, an English physician working in the Far East, proved that mosquitoes transmit the microbes causing malaria;
- David Bruce, another Englishman, identified tsetse flies as the insect transmitting sleeping sickness;

- Masaki Ogata, a Japanese investigator, reported rat fleas could transmit bubonic plague;
- American microbiologist Howard Taylor Ricketts located the agent of Rocky Mountain spotted fever in the human bloodstream and demonstrated its transmission via ticks;
- Walter Reed, another American, pinpointed mosquitoes as the insects transmitting yellow fever.

Amid the growing interest in microbes, other scientists studied the environmental importance of microbes. The Russian scientist Sergei Winogradsky discovered how certain bacteria use carbon dioxide to produce sugars. And Martinus Beijerinck, a Dutch investigator, isolated bacteria that trap nitrogen in the soil and make it available to plants for use in constructing amino acids and proteins. An interest in microbes was reaching far beyond their role and importance in medicine.

1.4 The Microbial World: The Microbes

With the advent of World War I, scientists turned to research on blood products and vaccines to treat and prevent war-related infections. Importantly, antibiotics such as penicillin were discovered. Many of these drugs are naturally produced by bacteria and fungi, so physicians were presented with previously unimagined therapies to cure war-related infections as well as other infectious diseases afflicting society.

Types of Cellular Organization

The understanding of microbes expanded greatly during the 1930s and 1940s with the invention of the electron microscope. An ordinary light microscope permits magnifications of 1,000 times (1,000×), while the electron microscope permits magnifications of 200,000×. With the electron microscope, for the first time bacterial cells were seen as being cellular like all other microbes, plants, and animals. In fact, biologists now saw the cell as the basic unit of structure and function in biology. However, the bacterial cells were organized in a structurally different way from other microbes and organisms.

It was known that animal and plant cells contained a cell nucleus that houses the genetic instructions in the form of chromosomes and was separated physically from other cell structures by a membrane envelope (FIGURE 1.9a). This type of organization defines the **eukaryotic cell** (*eu* = "true"; *karyon* = "nucleus"). Microscope observations of the protozoa, algae, and fungi also revealed these organisms had a eukaryotic organization. Thus, not only are all plants and animals "eukaryotes," so are the microorganisms that comprise the fungi and protists.

Studies with the electron microscope revealed that bacterial cells had few of the membranous compartments typical of eukaryotic cells. They lacked a cell nucleus, indicating the bacterial chromosome was not surrounded by a membrane envelope (FIGURE 1.9b). Therefore, a bacterial cell represents a **prokaryotic cell** (*pro* = "before") and as a group are "prokaryotes."

In addition, the mysterious viruses could now be visualized for the first time, and an intense period of virus research ensued. Viruses lacked a cellular organization; they are neither prokaryotes nor eukaryotes.

The Microbial Menagerie Revisited

In the period after World War II, large sums of money became available for biochemical research, and scientists worked out the key processes by which microbes synthesize proteins. These processes are centered in the hereditary material, **deoxyribonucleic acid (DNA)**, and soon the discipline of microbial genetics was in full flower. Elegant applications of this research were realized in the 1970s, with the advent of genetic

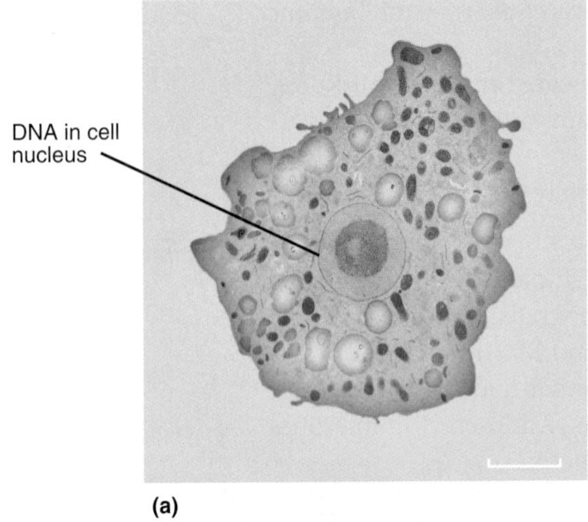

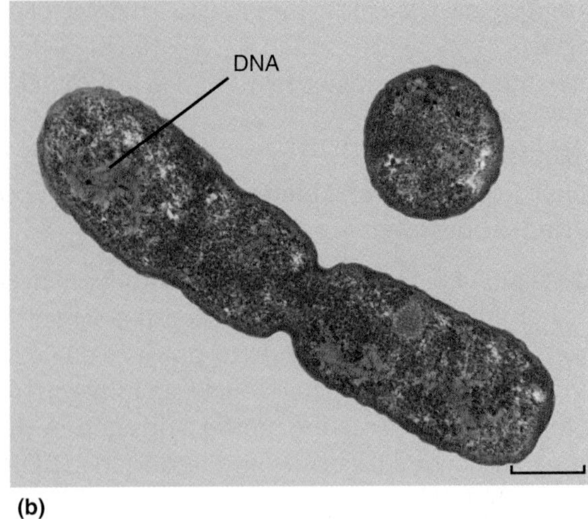

FIGURE 1.9 **False Color Images of Eukaryotic and Prokaryotic Cells.** (**a**) An electron microscope image of a protozoan cell. All eukaryotes, including the protists and fungi, have their DNA enclosed in a cell nucleus. (Bar = 3 μm. (**b**) An electron microscope image of a dividing *Escherichia coli* cell. (Bar = 0.5 μm.)

engineering and biotechnology. Scientists learned to isolate genes, manipulate and splice them, and control the biosynthetic processes that genes control in microbial cells.

For many scientists, microorganisms were the workhorses they used to make all the brilliant discoveries and advances. Therefore, let's briefly examine the members of the microbial world. This includes those shown back in Figure 1.1 and represent the prokaryotes, most of the eukaryotes, and the viruses.

Bacteria and Archaea. Among the best known of all microbes are the **Bacteria**—small, single-celled organisms that lack much in the way of visible internal structure (FIGURE 1.10a). Another group of similarly structured single-celled organisms is the **Archaea**. Although the archaeal cells look like bacterial cells, they have a different chemical structure for the cell wall and lack the metabolic diversity exhibited by the bacterial organisms.

■ Archaea
ar-kē′a

Prokaryotes on Earth first appeared approximately 3.8 billion years ago and, over this extraordinarily long period of time, they have evolved to occupy every conceivable place on Earth. For instance, various species live in the outer reaches of the atmosphere, at the bottoms of the oceans, in the frigid valleys of the Antarctic continent, in the Great Salt Lake, and in the scalding hot deserts of North Africa. No other organisms of any kind (including humans) have adapted so thoroughly to Earth's varied conditions.

Many species acquire their food from organic matter while other members, such as the **cyanobacteria**, synthesize their own food similar to algae and green plants (FIGURE 1.10b). Many species use oxygen in their chemistry, while others exist in oxygen-poor environments. We have already identified some of the roles bacterial species play in the food industry, in the industrial plant, on the farm, and in the biotechnology industry. Although there are no known archaeal species that cause disease in humans, other animals, or plants, a variety of bacterial species can cause disease. Using their considerable powers of reproduction and their ability to overcome body defenses, bacterial pathogens can infect vital tissues or organ systems and bring on illness and disease. We will study bacterial diseases in another chapter.

Cells of the protists and fungi have a very different structural organization and compose a large part of the eukaryotic group of organisms.

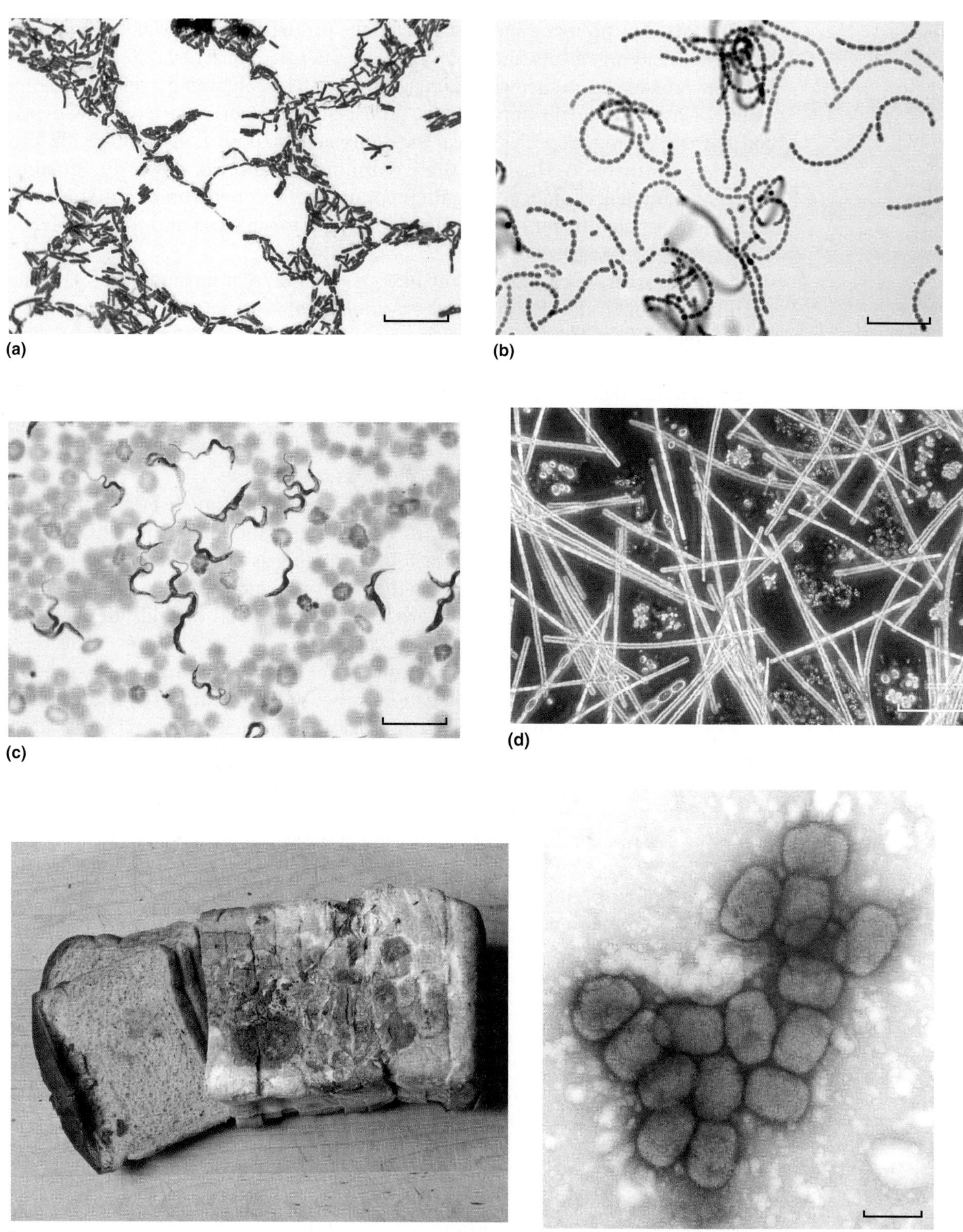

FIGURE 1.10 **A Gallery of Microbes.** (**a**) A stained microscope preparation showing rod-shaped bacterial cells typically found in the soil. (Bar = 10 µm.) (**b**) Filamentous strands of a cyanobacterium that carries out photosynthesis. (Bar = 100 µm.) (**c**) The ribbon-like cells of the protist *Trypanosoma,* the causative agent of African sleeping sickness. (Bar = 10 µm.) (**d**) A sample of phytoplankton (Bar = 10 µm.) (**e**) A typical blue-gray mold growing on a loaf of bread. (**f**) False-color image of smallpox viruses. (Bar = 100 nm.)

(a–c) Courtesy of Dr. Jeffrey Pommerville. (d) © M I (Spike) Walker/Alamy. (e) © Jones & Bartlett Learning. Photographed by Kimberly Potvin. (f) Courtesy of Dr. Fred Murphy/CDC.

Protists. The protozoa and algae, called the **protists**, are a large and diverse group of single-celled organisms, most of which are microscopic. They differ substantially from the prokaryotes in their structural composition by containing a cell nucleus and a variety of interior cellular compartments. Protists share certain characteristics with plants and animals, and some species appear to be ancestors of those more complex life forms.

Many **protozoa** live freely in the environment and break down the remains of plants and animals while recycling their components. Others can be pathogens, perhaps the most significant ones being responsible for malaria and African sleeping sickness (FIGURE 1.10c).

Other protists, such as the plant-like green algae, diatoms, and dinoflagellates, are photosynthetic and, along with the cyanobacteria, are part of the **phytoplankton** inhabiting the ocean surfaces in astronomical numbers (FIGURE 1.10d). As such, they form the base of many food chains, serving as food sources for other organisms in Earth's waters. Algae do not infect humans; however, they can cause serious poisoning. In addition, so-called **algal blooms**, representing enormous growths of an algal population, often cause massive fish kills by removing much of the oxygen dissolved in the water. Such areas often are referred to as "dead zones."

Fungi. The fungi (sing. fungus) includes the molds, mushrooms, and yeasts. The molds and mushrooms are formed from long, branching chains (filaments) of cells, while the yeasts typically can be found growing as single cells. In their feeding patterns, all fungi secrete enzymes into the environment and break down the nearby organic matter, which they then absorb. In fact, many of the molds are among the major decomposers of organic matter on Earth.

Many fungi benefit society. For example, some antibiotics are naturally produced by molds. Another mold, called *Aspergillus*, is used to produce soy sauce and the alcoholic rice beverage called sake. Yet another mold named *Claviceps* is cultivated to produce drugs that help relieve migraine headaches and induce uterine contractions during labor. And the yeasts are involved in wine and beer production.

■ *Aspergillus* a-spėr-jil'lus

■ *Claviceps* kla'vi-seps

Unfortunately, some fungal species also cause food spoilage (FIGURE 1.10e) while others are serious agricultural pests, causing grave economic damage. Although they are not dominant agents of human disease, a few fungi may cause disease if they infect the skin, respiratory tract, or urinary tract.

Viruses. Although most people are familiar with the word "virus," few are aware of their characteristics. **Viruses** are not cellular. Rather, they consist of genetic information surrounded by a coat of protein. In some cases, a membranous envelope encloses the protein coat (FIGURE 1.10f). Viruses do not grow; they produce no waste products; they display none of the chemical reactions we associate with living organisms; and they cannot reproduce independently.

However, viruses do reproduce (replicate) actively within appropriate host cells, and in doing so, they use their genetic information and the chemical machinery of the host cells for their own purposes. Some minutes or hours later, hundreds of new viruses exit from each cell, often leaving damaged or dead cells in their wake. As this wave of cell destruction spreads, the tissue suffers damage and the symptoms of disease ensue.

■ host: An organism infected by a pathogen.

Despite their role in causing infectious disease (e.g., hepatitis, influenza, AIDS), viruses perform some valuable services to medical science. For example, specific viruses can be genetically altered to make them less infectious and then customized to carry "good" genes capable of relieving the symptoms of a genetic disorder, such as cystic fibrosis or hemophilia. Viruses are also being used as agents to kill certain types of human cancers and to manufacture vaccines.

TABLE 1.1 summarizes some of the characteristics of the prokaryotes, eukaryotes, and viruses.

TABLE 1.1 Some Characteristics of Microbes and Viruses

Group	Cellular organization	Presence of a cell wall	Type of genetic information	Presence of internal compartments	Cause infectious disease in humans
Bacteria	Prokaryotic	Most; made of peptidoglycan	Single, circular DNA chromosome	Few	Some
Archaea	Prokaryotic	Yes; different from Bacteria	Single, circular DNA chromosome	Few	None known
Protists	Eukaryotic	Algae; made of cellulose	Several linear DNA chromosomes	Many	Some
Fungi	Eukaryotic	Yes; made of chitin	Several linear DNA chromosomes	Many	Some
Viruses	Acellular	No	DNA or RNA[1]	None	Some

[1] RNA = ribonucleic acid

1.5 Microbiology Today: Challenges Remain

Microbiology finds itself on the world stage again, in part from the biotechnology advances made in the latter part of the 20th century. However, the science also faces several challenges, many of which still concern infectious diseases, which today are responsible for 26% of all deaths globally and are spreading geographically faster than at any time in history (FIGURE 1.11). Even in the United States, more than 100,000 people die each year from bacterial infections, making them the fourth leading cause of death. Here are some of the challenges facing microbiology today.

New Disease Outbreaks

Experts estimate that more than 5 billion people traveled by air in 2013, making a disease outbreak or epidemic in one part of the world only a few airline hours away from becoming a potentially dangerous threat in another region of the globe. It is a sobering thought to realize that since 2002, the World Health Organization (WHO) has verified more than 1,100 epidemic events worldwide. So, unlike past generations, today's highly mobile, interdependent, and interconnected world provides potential opportunities for the rapid spread of infectious diseases.

■ epidemic: The occurrence of more cases of a disease than expected in individuals within a geographic area.

There are now nearly 40 infectious diseases that were unknown or rare a generation ago. These include **emerging infectious diseases**, which are those having recently surfaced for the first time in a population of a given area. Among the more newsworthy have been AIDS, severe acute respiratory syndrome (SARS), Lyme disease, mad cow disease, and most recently swine and bird flu. There is no cure for any of these and undoubtedly there are more such agents ready to emerge. **Reemerging infectious diseases** are ones that have existed in the past but are now showing a resurgence in resistant forms or a spread in geographic range. Among the more prominent reemerging diseases are drug-resistant tuberculosis, cholera, dengue fever, and, for the first time in the Western Hemisphere, West Nile fever. Therefore, emerging and reemerging diseases will remain as perpetual challenges to public health and society.

Drug-Resistant Microbes

Another challenge concerns our increasing inability to fight infectious disease because many pathogens are now resistant to one or more antibiotics and such antibiotic

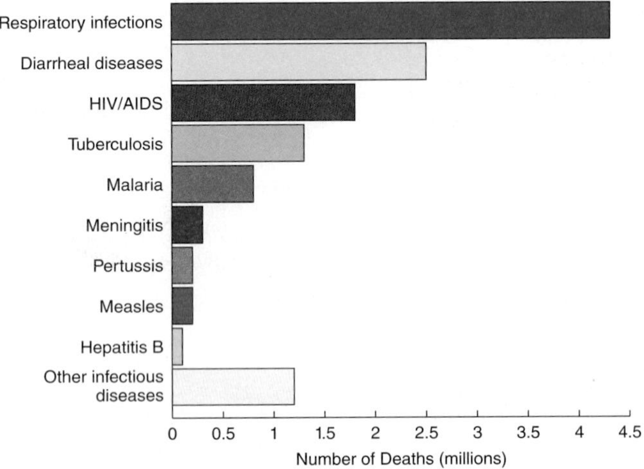

FIGURE 1.11 **Global Mortality from Infectious Diseases.** On a global scale, some 15 million annual deaths are caused by infectious diseases. Noninfectious causes include chronic diseases, injuries, nutritional deficiencies, and maternal and perinatal conditions.

Data from *World Health Statistics 2011*, World Health Organization.

resistance is developing faster than new antibiotics are being discovered. Ever since it was recognized that many microbes could mutate into multidrug resistant forms, informally called **superbugs**, a crusade has been waged to restrain the inappropriate use of these antimicrobial drugs by doctors and to educate the public not to demand them in uncalled-for situations. If actions are not taken to contain and reverse resistance, the world could be faced with previously treatable diseases that have again become untreatable, as in the days before antibiotics were developed. One of the biggest challenges is in the American feedlots as **A Closer Look 1.3** reports.

Climate Change

A very controversial issue is how **climate change** (including warming temperatures and altering rainfall patterns) may affect the frequency and distribution of infectious diseases around the world. Many scientists believe that as temperatures rise in various regions of the world, mosquitoes transmitting diseases like malaria and dengue fever will broaden their range, especially into more temperate climates such as North America. Therefore, microbiologists, health experts, climate scientists, and many others are studying new strategies to limit potential pandemics before they can get started and to understand how such climatic changes could affect the health of our society, as well as livestock, plants, and wildlife.

A CLOSER LOOK 1.3

Antibiotics in the Feedlots

For decades, many farmers in developed nations have used animal feeds mixed with low amounts of antibiotics. Why the antibiotics? The bottom line is that these doses of antibiotics fatten up livestock faster and get the animals to market sooner.

Many of the feed antibiotics are the same ones used to treat human infectious disease. In fact, every year in the United States 50% to 80% of antibiotics produced are dispensed on large commercial feedlots to promote the faster growth of healthy cattle, swine, and poultry. By using these antibiotics, many feedlots are becoming breeding grounds for the evolution of antibiotic-resistant bacterial species. Numerous scientific studies have shown that using subtherapeutic levels of antibiotics in animals kills many bacterial species but not all. Those that survive are the resistant ones, which, as they reproduce, will spread their resistance genes to many similar and dissimilar species, including any pathogens that happen to pass by. Eventually these antibiotic-resistant strains can end up in people.

Some countries, like Denmark, have enacted tough rules for the use of subtherapeutic antibiotics in livestock, including pigs. The result: animal health on the large and intensively run farms (similar to those in the United States) was not harmed, pigs gained as much weight as with antibiotics, and the litter size actually increased. The only necessity was to maintain high hygiene practices and reduce overcrowding in the feedlots. On the flip side, antibiotic resistant samples in pigs decreased from 20% to 6%. This all culminated in 2006 with the European Union banning antibiotics as growth promoters in farm animal feed.

In 1977, the U.S. Food and Drug Administration (FDA) stated that the overuse of antibiotics in livestock could generate resistance and make these drugs less effective in humans. So, the agency issued an order to ban nonmedical use of penicillin and tetracycline in farm animals. The rule, however, has never been applied. Then in 2005, the FDA prohibited the use of another group of antibiotics, the fluoroquinolones, in poultry in the hopes of reducing fluoroquinolone resistance; and, in January 2012, the agency prohibited the use of certain cephalosporins for use in livestock. In March 2012, a federal court judge ordered the FDA to take the necessary steps to ban the two antibiotics for routine use in animal feed. It remains to be seen if those steps actually occur.

■ pandemic: Refers to a disease occurring over a wide geographic area (worldwide) and affecting a substantial proportion of the global population.

Microbiology is a very hot topic in today's world, and you are fortunate to be studying this discipline of biology. The science penetrates virtually every aspect of our existence, and despite the challenges mentioned above, as we shall see time and time again, microbes are essential elements for improving the quality of our lives, our health, and our society.

A Final Thought

It should be clear from this chapter that microbes, despite their involvement in disease, contribute substantially to the quality of life in our society. Rather than gush with enthusiasm on the positive roles they play, perhaps it is best to paraphrase several concepts of applied microbiology set down by the late industrial microbiologist David Perlman of the University of Wisconsin. Perlman wrote:

1. The microbe is always right, your friend, and a sensitive partner.
2. There are no stupid microbes.
3. Microbes can and will do anything.
4. Microbes are smarter, wiser, and more energetic than chemists, engineers, and others.
5. If you take care of your microbial friends, they will take care of you.

Questions to Consider

1. Pasteur's work on the fermentation of wine has been referred to as "the birth certificate of microbiology." Why do you suppose that is so? Further, because of the "Pasteurian revolution," medicine could no longer do without science, and hospitals could no longer be mere hospices; that is, places to die. Why is that so?
2. "Microbes? All they do is make you sick!" From your introduction to microbes in this chapter, how might you counter this argument?
3. Every now and then in science, a seminal experiment sets off a barrage of studies that lead to the discovery of an important principle. Which experiment do you believe was the spark that ultimately led to our understanding of the germ theory of disease?
4. If you were to ask someone to describe a microbe, he or she might think of a dot under a microscope. However, the microbial world is quite varied, and each of its members is unique. Although your experience in microbiology is somewhat limited at this juncture, the information in this chapter should give you some insight into the microbial world. How, then, would you now describe a microbe?
5. This chapter noted two reasons why interest in microbes ebbed after the death of Leeuwenhoek. Can you think of any other reasons?
6. The poet John Donne once wrote: "No man is an island, entire of itself; every man is a piece of the continent." This maxim applies not only to humans, but to all living things in the natural world. What are some roles microbes play in the interrelationships among living things?
7. Our world is somewhat "germ-phobic." The media cover new outbreaks of disease, we eagerly await new antibacterial medicines, and we hear of new ways to "fight germs." But suppose there were no microbes to contend with. What do you suppose life would be like?
8. If you could only select one individual, who would you choose as the "first microbiologist?" Justify your choice.

Key Terms

Informative facts are necessary for the expression of every concept, and the information for a concept is founded in a set of key terms. The following terms form the basis for the concepts of this chapter. On completing the chapter, you should be able to explain and/or define each one.

alga (pl. algae)
algal bloom
animalcule
Archaea
Bacteria
bacterium (pl. bacteria)
bioremediation
biotechnology
climate change
cyanobacteria
decomposer
deoxyribonucleic acid (DNA)
emerging infectious disease
eukaryotic cell
fungus (pl. fungi)
germ theory of disease
human microbiome
Koch's postulates
microorganism (microbe)
pasteurization
pathogen
photosynthesis
phytoplankton
prokaryotic cell
protist
protozoan (pl. protozoa)
reemerging infectious disease
science
spontaneous generation
superbug
virus

Microbes in Perspective: Of Collectors, Classifiers, and Microscopists

Looking Ahead

Despite their incredibly small size, microbes occupy extensive, well-established, and integral places in the living world. Furthermore, their names, chemical makeups, and other characteristics conform to the principles that apply to all life forms, as we shall note in these pages.

On completing this chapter, you should be capable of:

- Explaining how microbes are named.
- Illustrating how organisms are cataloged in the tree of life.
- Distinguishing between the different forms of microscopy.
- Constructing and labeling a typical eukaryotic and prokaryotic microbial cell.

The diversity of life is enormous, coming in many forms and variations. Here, you see part of the Spectrum of Life exhibit in the Hall of Biodiversity at the American Museum of Natural History in New York City. The exhibit displays more than 1,500 specimens, from microorganisms to terrestrial and aquatic plants and animals—but this only represents a small fraction of all known living organisms on this planet.

Carolus Linnaeus (FIGURE 2.1) was in trouble. The ship from Africa was pulling into port, and soon the Swedish botanist would be confronted with new plants, new animals, and new problems. "What shall we call this one?" he would be asked. "Or this one?" "Or that one?" The museums were full of newly discovered organisms (such as penguins, manatees, kangaroos, tobacco plants, bananas, and potatoes) arriving from distant corners of the globe and people were

FIGURE 2.1 **Carolus Linnaeus.** Karl von Linne, the Swedish botanist known in scientific history as Carolus Linnaeus, took on the daunting task of classifying the known plants and animals of the biological world and giving them scientific names.

bringing new specimens to him almost daily (he soon learned to dread the sight of arriving ships). Linnaeus even added to this frenzy: he inspired an unprecedented worldwide program of specimen hunting by sending his students around the globe in search of new and unknown plants and animals. This need to apply a scientific name to all organisms, called "nomenclature," had gotten thoroughly out of hand. To be sure, the biological world needed some order, but who had appointed Linnaeus king of nomenclature? It was exasperating, to say the least!

Linnaeus is just one of the collectors and classifiers we will encounter in this chapter as we fit microbes into the same scheme to which all living organisms belong and see how they are related to members of the visible world. Linnaeus could hardly anticipate what the future would hold.

2.1 Naming Microbes: What's in a Name?

Although at times it was exasperating, Linnaeus performed a valuable service. In a 1753 book on plants, he supplied scientific names for some 6,000 different plants known at that time. In his tenth edition of *Systema Naturae* (1759), Linnaeus extended the nomenclature scheme to thousands of animals. In the end, his scientific names were widely adopted by European scientists and were introduced around the world.

Escherichia coli
esh-ėr-ē kē-ä kō'lī (or kō'lē)

Binomial and Common Names

Because all Linnaeus' assigned scientific names were derived from Latin, he had a monumental task of devising new Latin names and word endings for the thousands of new scientific names. So what is this naming system Linnaeus devised and which is still used today?

Linnaeus' system is called **binomial nomenclature**; that is, each organism is assigned a two-word name, the binomial, derived from Latin. The first word is called the **genus** and it is followed by a second word, called the species modifier or the specific epithet. The binomial indicates the **species** of organism. For instance, a bacterial species normally found in the human colon is called *Escherichia coli* (FIGURE 2.2). The first part of the binomial, *Escherichia*, is the genus name to which the organism belongs. It is derived from the name of the scientist, Theodor Escherich, who first identified the microbe in 1888. The second word of the binomial, *coli*, is the specific epithet derived from the word "colon," which is where the bacterium was first found by Escherich. [The same rules apply to humans, who are scientifically known as *Homo sapiens* (*Homo* = "man"; *sapiens* = "wise")].

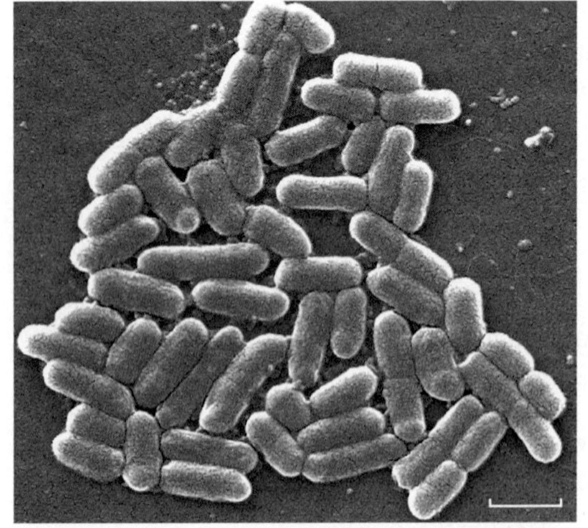

FIGURE 2.2 *Escherichia coli.* A false-color electron microscope image of *E. coli* cells. Most strains of the species are harmless and live in the intestines of healthy humans and animals. However, some strains such as this one (0157:H7) produce a powerful toxin that can cause severe illness. (Bar = 2 μm.)

In Linnaeus' time and even today, when a new species is discovered the binomial name might reflect the discoverer's name, the organism's manner of growth, a location where the organism was first found, or even everyday names. **A Closer Look 2.1** examines several other microbial species names and their origins.

The correct way to write the species name for a microbe (or for any organism) is to capitalize the first letter of the genus name and to write the remainder of the genus name and the specific epithet in lowercase letters. The binomial should always be italicized (if this is not possible, it should be underlined). After the full species name has been introduced in a piece of writing, the name can be abbreviated by using the first letter of the genus name and the full specific epithet. Thus, *Escherichia coli* is abbreviated *E. coli*. Unfortunately, in today's newsprint and Internet sites, an organism's binomial often is written in normal text; for example, you might see Escherichia coli.

A microbial species may have various strains or subspecies whose identifiers are added to the binomial. An example is *Escherichia coli* O157:H7, a strain that does not occur normally in the human gut. The combination of letters and

A CLOSER LOOK 2.1

"What's in a Name?"

In Shakespeare's *Romeo and Juliet*, Juliet tells Romeo a name is an artificial and meaningless convention. Perhaps from her perspective in convincing Romeo that she is in love with the man called Montegue and not the family Montegue, there is a love-based reason for "What's in a name." As you read this book, you have and will come across many scientific names for microbes. Not only are many of these names tongue twisting to pronounce (many are listed with their pronunciation in the text side margins and inside the front and back covers), but how in the world did the organisms get those names? Most are derived from Greek or Roman word roots. Here are a few examples.

Species	Meaning of Name
Genera Named After Individuals	
Bordetella pertussis (bor-de-tel'lä pėr-tus'sis)	Named after the Belgian Jules Bordet who in 1906 identified the small bacterium (*ella* means "small") responsible for pertussis (whooping cough).
Neisseria gonorrhoeae (nī-se'rē-ä go-nôr-rē' ī)	Named after Albert Neisser who discovered the bacterial organism in 1879. As the specific epithet points out, the disease it causes is gonorrhea.
Genera Named for a Microbe's Shape	
Vibrio cholerae (vib'rē-ō kol'ėr-ī)	*Vibrio* means "comma-shaped," which describes the shape of the bacterial cells causing cholera.
Staphylococcus epidermidis (staf-i-lō-kok kus e-pi-der'mi-dis)	The stem *staphylo* means "cluster" and *coccus* means "spheres." So, these bacterial cells form clusters of spheres found on the skin surface (epidermis).
Genera Named After an Attribute of the Microbe	
Saccharomyces cerevisiae (sak-ä-rō-mī'sēs se-ri-vis'ē-ī)	In 1837, Theodor Schwann observed yeast cells and called them *Saccharomyces* (*saccharo* = "sugar"; *myce* = "fungus") because the yeast converted grape juice (sugar) into alcohol; *cerevisiae* (*Ceres* was the Roman goddess of agriculture) refers to the use of yeast since ancient times to make beer.
Myxococcus xanthus (micks-ō-kok'kus zan'thus)	The stem *myxo* means "slime," so these are slime-producing spheres that appear as a yellow (*xantho* = "yellow") growth in culture.

Lastly, in keeping with Shakespeare's poetic style, there is the organism *Thiomargarita namibiensis*. This bacterial species was first isolated in 1997 from sediment samples in the Atlantic Ocean off the coast of Namibia, a country in southwestern Africa (*ensis* = "belonging to"). These spherical-shaped bacterial cells accumulate sulfur (*thio* = "sulfur") so when they are observed with the microscope the cells appear white and look like a microscopic string of pearls (*margarit* = "pearl"; see A Closer Look 2.3). Thus, we have *Thiomargarita namibiensis* (thī'ō-mär-gä-rē-tä na'mi-bē-en-sis)—the "Sulfur Pearl of Namibia." Juliet would be impressed!

numbers for the strain refers to specific markers found on the cell's surface, which distinguish it from other types of *E. coli*. In this case, the O157:H7 strain produces a powerful toxin, which if ingested in contaminated food, can cause intestinal hemorrhaging.

Finally, besides the binomial name, many microbes have a common name often used in conversation. For instance, one of the bacterial species causing bacterial pneumonia, *Streptococcus pneumoniae*, is commonly referred to as the pneumococcus; and a bacterial species causing bacterial meningitis, *Neisseria meningitidis*, is often called the meningococcus.

- *Streptococcus pneumoniae* strep-tō-kok kus nü-mō nē-ī
- *Neisseria meningitidis* nī-se rē-ä me-nin ji'ti-dis
- meningitis: An inflammation of the membranes surrounding and protecting the brain and spinal cord, often resulting from a bacterial or viral infection.

2.2 Taxonomy: Cataloging Life

In addition to providing the names to many organisms, Linnaeus also established the ground rules of **taxonomy**, which involves the classification of organisms into hierarchical groups that indicate natural relationships. Thus, the taxonomic rules bring order to the living world by placing all organisms into related categories that can be more easily studied and related to other microscopic and macroscopic forms of life.

The Hierarchical Groups

As outlined in TABLE 2.1, the least inclusive, most fundamental group is the species (pl. species). For most animals and plants, a species is usually defined as a group of organisms that interbreed sexually with one another in nature, produce offspring similar to the parents, and are fertile. This definition also holds true for the eukaryotic microbes, most of which also reproduce sexually. However, problems exist in classifying prokaryotic organisms because these organisms do not interbreed sexually. Therefore, one cannot define a prokaryotic species based on sexual reproductive patterns and a valid definition remains controversial. For our purposes, a group of bacteria belong to the same species if the members share many stable physiological, biochemical, and genetic properties that are absent from other prokaryotic groups.

In this Linnaean classification, a group of species closely related are gathered together to form a genus (pl. genera). For example, familiar to us are lions, tigers, and leopards. They all belong to separate species, but they are classified together in the same genus

TABLE 2.1 Taxonomic Classification of Humans, Brewer's Yeast, and a Common Bacterium

	Humans	Brewer's Yeast	*Escherichia coli*
Domain	Eukarya	Eukarya	Bacteria
Kingdom	Animalia	Fungi	
Phylum	Chordata	Ascomycota	Proteobacteria
Class	Mammalia	Saccharomycotina	Gammaproteobacteria
Order	Primates	Saccharomycetales	Enterobacteriales
Family	Hominidae	Saccharomycetaceae	Enterobacteriaceae
Genus	*Homo*	*Saccharomyces*	*Escherichia*
Species	*H. sapiens*	*S. cerevisiae*	*E. coli*

Panthera because they have similar big cat-like features. Likewise , a group of similar genera comprise a **family**. For instance, lions, tigers, and leopards (genus *Panthera*) and domestic housecats (genus *Felis*) are categorized in the same family Felidae.

Taxonomists then use progressively more inclusive categories of classification. Related families are organized into an **order**, and orders are brought together in a **class**. Various classes comprise a **phylum** (pl. phyla) and all phyla are grouped together in a **kingdom**, except for the prokaryotes where there is no designed kingdom (see below). Notice in Table 2.1 that above the genus level, broader categories are not italicized but are capitalized.

The Domains of Life

Before the invention of the microscope, people believed the living world consisted of plants and animals. Therefore, they had little difficulty classifying life—after all, animals moved about, while plants were rigid and immobile. However, soon after Leeuwenhoek reported the existence of animalcules, it became clear that these organisms should be incorporated into the biological kingdoms. But scientists were unsure exactly where the microbes belonged. Even Linnaeus was unsure and therefore grouped them apart from plants and animals under the heading *Vermes* (as in "vermin") in a category he called *Chaos* (as in "confusion").

The classification of microbes remained somewhat chaotic until 1866, when the German naturalist Ernst H. Haeckel proposed a new kingdom called the Protista (the Greek word *protist* means "the very first"). This new third kingdom soon came to include virtually all microbes (bacteria, protozoa, algae, and fungi).

As the twentieth century progressed, advances in cell biology and microscopy identified structural differences between cells—the prokaryotes and eukaryotes. This and an interest in evolutionary biology led scientists to question the three-kingdom classification scheme. As a result, two more kingdoms were added to account for the increasing number of microorganisms being discovered and studied. All the fungi were placed in the kingdom Fungi and all the bacteria were placed in the kingdom Monera (FIGURE 2.3). The kingdom Protista now included any living organism that was single-celled and not in the other kingdoms; namely, the unicellular algae and protozoa.

By the 1980s, better and more detailed research techniques were used to study living organisms. These techniques in biochemistry and molecular biology brought about another, and perhaps the most dramatic, change to the classification of life. First developed in the 1980s by Carl Woese and his coworkers at the University of Illinois, these scientists compared organisms based on specific genes in their DNA, the genetic information found in all organisms. Their comparisons from prokaryotes and eukaryotes showed all these genes could be lumped into three separate groups. Therefore, they suggested life should be cataloged into three groups called **domains**, which are the most inclusive and encompass all five original kingdoms. Today, this classification system is called the **three-domain system** (FIGURE 2.4). It is more commonly referred to as the "tree of life."

From the comparison of genes, molecular composition of the cell membrane, and other structural and biochemical traits, it was clear the prokaryotes in the kingdom Monera actually had two different evolutionary histories. Therefore, they represent two very different branches of prokaryotes and so were split into separate domains in the tree of life. The domain **Bacteria** includes most of the members with one set of similar genes, such as *E. coli*. The remaining prokaryotes with a different set of similar genes were assigned to the domain **Archaea**. Although many live at ordinary temperatures and conditions, such as floating in the open oceans and even living in the human mouth and gut, some members of the domain prefer more extreme environments, as **A Closer Look 2.2** highlights. The third domain, which includes all the eukaryotic organisms (protists, fungi, plants, animals), had a third set of similar genes and appropriately is called the domain **Eukarya**.

Archaea
ar-kē′-a

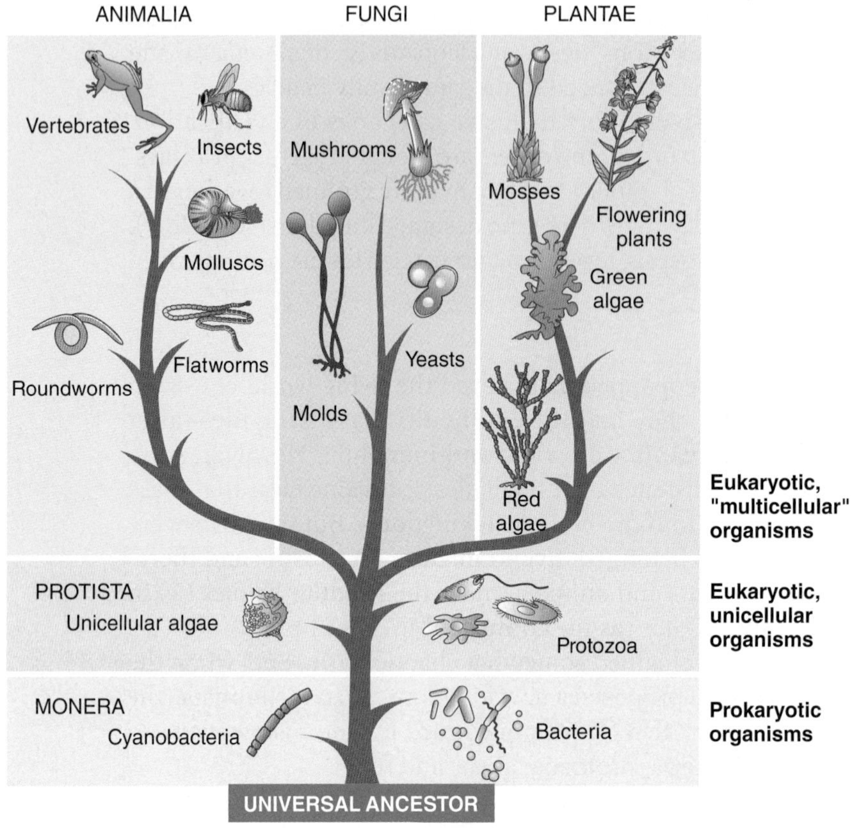

FIGURE 2.3 **The Five-Kingdom System.** This system of classification implies an evolutionary lineage, beginning with the Monera and extending to the Protista. Certain of the Protista were believed to be ancestors of the Plantae, Fungi, and Animalia. Divergence at each level was based on the mode of nutrition: photosynthesis, absorption, or ingestion. Unicellular or multicellular organization was also a key feature in the system.

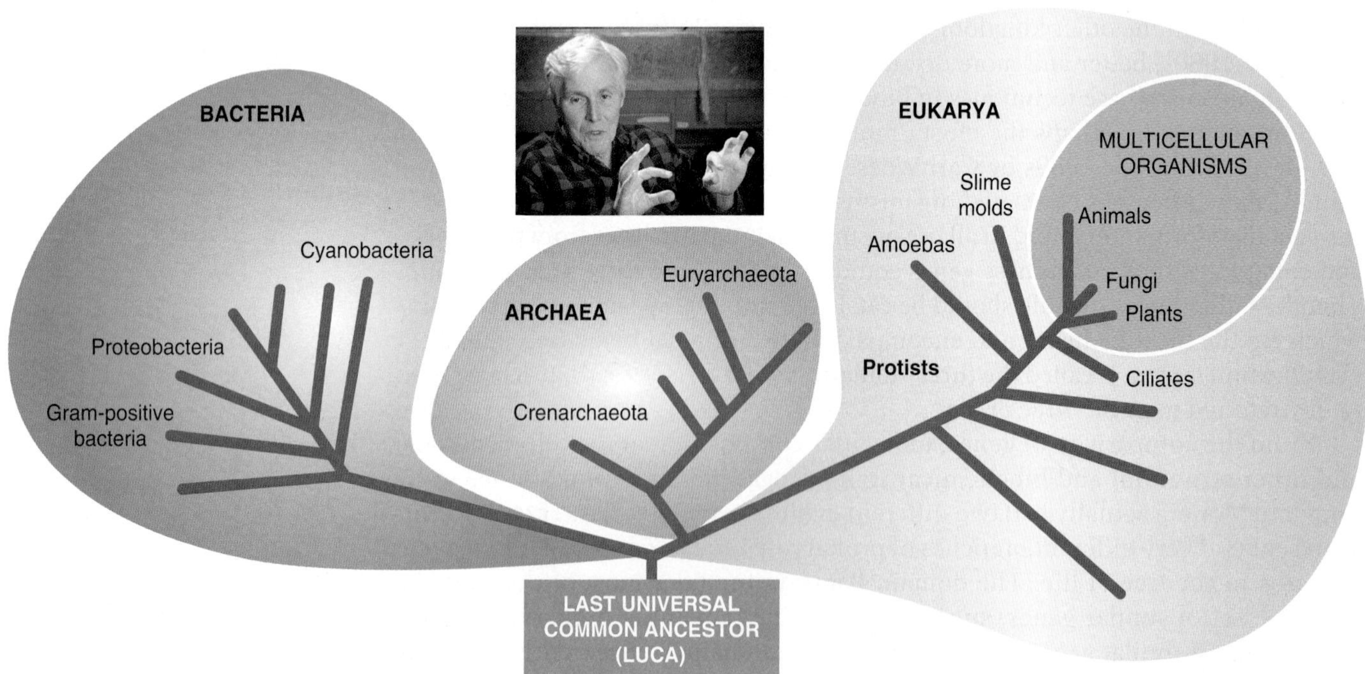

FIGURE 2.4 **The Tree of Life.** Carl Woese (inset) developed the tree of life and the three domain system to illustrate the relatedness of all living organisms. All organisms in the yellow portion of the domain Eukarya represent protists and all organisms in the three domains are single-celled, except for the plants, animals, and some fungi.

A CLOSER LOOK 2.2

Prepare for Landing

"Ladies and gentlemen, as we start our descent to the San Francisco international airport, please make sure your seat backs and tray tables are in their full upright position. Make sure your seat belt is securely fastened and all carry-on luggage is stowed underneath the seat in front of you or in the overhead bins. Please turn off all electronic devices until we are safely parked at the gate. Thank you."—and, oh yes, look out your cabin window and marvel at the red-colored salt evaporation ponds below (see photo).

An aerial view of evaporation salt ponds.

Many members of the Archaea are truly startling in their ability to thrive in extremely hostile environments. Some of these so-called "extremophiles" can grow at extremely high temperatures, even above the boiling point of water. Others are quite happy growing in very acidic environments such as acid-laden streams around old mines. Yet other archaeal species prefer high concentrations of salt.

Salt evaporation ponds, also called salterns, are shallow, man-made ponds filled with seawater. As the water evaporates, the salt [primarily sodium chloride (NaCl) or table salt] concentration rises. When the salt concentration goes above about 26 percent, NaCl starts to crystallize at the bottom of the ponds and it can be harvested.

Several microbial species found in these ponds are the halophiles (*halo* = "salt", *phile* = "loving"), which are microbial species that prefer growing in salty environments such as a saltern. Therefore, as the water in the salt ponds evaporates, a succession of microbes will develop as the salt concentration increases.

The salterns with the lowest concentration of salt appear green because they contain an abundance of halophilic green algae. Then, in ponds with somewhat higher salt concentrations, another green alga, *Dunaliella salina*, predominates. The green algae in such ponds not only provide pretty colors, they also speed up the rate of evaporation by absorbing sunlight (heat).

Then, as evaporation continues, ponds become too salty for the growth and survival of green algae and most microbes. Such high salt concentrations, however, become a perfect environment for the growth of the extreme halophiles—members of the Archaea. In high-salinity ponds, archaeal species, like *Halobacterium salinarum*, predominate and shift the color of the evaporation ponds from algal green to archaeal pink, orange, red, and even purple. Note: even though the genus *Halobacterium* sounds like it is in the domain Bacteria, it actually is a member of the Archaea.

Thus, the succession of microbes that develops in the various evaporation ponds not only provide vivid colors to the observer but also indicate the relative salinity of the pond. So, the next time you fly into San Francisco, look at the very colorful evaporation ponds filled with green algae or the extremely halophilic Archaea.

"On behalf of our airline and the entire crew, I'd like to thank you for joining us on this flight and we are looking forward to seeing you on board again in the near future. And weren't the microbes awesome?"

In the tree of life, the tips of the branches represent currently living organisms, be it a species, genus, phylum (e.g., Proteobacteria) or a common biological group (e.g., amoebas). They represent organisms that we actually know something about. Based on DNA sequence analysis, it is possible to trace evolutionary history (the red lines in Figure 2.4) back to hypothetical common ancestors of today's organisms, which are represented by the branch points called nodes (e.g., the node at the center right represents the ancestor that gave rise to the organisms in the domain Archaea and Eukarya). The root of the tree (e.g., the last common universal ancestor; LUCA) represents a hypothetical common ancestor that about 3.8 billion years ago gave rise to two lineages, one of which would evolve into the Bacteria. The other lineage in another 500 million years would split into two lines forming the beginnings of the Archaea and Eukarya.

Dunaliella salina
dun-al-ē-el′-a sa-lī′-na

Halobacterium salinarum
hā′lo-bak-tēr′ē-um sal-i-nar′-um

Notice the viruses are not included in the tree of life. This omission is intentional because the viruses do not have the cellular organization characteristic of the prokaryotes and eukaryotes.

2.3 Microscopy: Seeing the Unseen

As we have learned, the existence of microbes was largely a matter of speculation until the 1600s, for the simple reason that seeing them required a microscope and the magnifying lenses of the time lacked the quality needed to see tiny objects. Microscopes did not come into existence until the early 1600s, when a spectacle maker named Zacharius Janssen placed two lenses together to make a crude microscope. Galileo Galilei, the great astronomer, perfected the microscope in the 1620s, and Robert Hooke, the imaginative British microscopist, used the microscope to describe cells and other objects in the 1660s.

The stage now was set for discovering the microbial world. As mentioned in another chapter, Leeuwenhoek was the first to provide detailed descriptions of microbes, including protozoa, algae, and bacterial cells, because the lenses he made could magnify specimens over 200× and, by some accounts, up to about 400×. Hooke's microscope only magnified specimens about 20×.

Microbial Measurements and Cell Size

One of the defining features of microbes is their extremely small size. Therefore, they are not measured in inches, millimeters, or other well-known units, but in much smaller and less familiar units. The unit most often used for cellular measurements is the **micrometer** (sometimes referred to as the micron). A micrometer is a millionth of a meter. The abbreviation for a micrometer is expressed by using the Greek letter mu (written as μ) together with the letter m. Thus, the length of a typical bacterial cell would be expressed as 5 micrometers, or 5 μm.

To conceptualize the extraordinarily small size of a micrometer, take a look at FIGURE 2.5. There are 1,000 μm in one millimeter (mm). Put another way, the cells of the bacterial species *Staphylococcus aureus* (commonly called "staph") have a diameter of roughly 1 μm. Therefore, 1 million staph cells lying side-by-side would occupy the space taken up by a single millimeter.

■ *Staphylococcus aureus*
staf-i-lō-kok′ kus ô-rē-us

Most of the eukaryotic species of microbes are also measured in micrometers. For example, yeast cells are approximately 5 μm in diameter and the cells of molds may be 25 μm long or longer, with varying widths. Some protists may be as large as 100 μm (0.1 mm), or about the size of the period at the end of this sentence.

At the opposite end of the microbial scale are the viruses, which are measured in **nanometers**. A nanometer (nm) is equivalent to one-thousandth of a micrometer (see Figure 2.5). Therefore, about 10 flu viruses (100 nm in diameter) would occupy the space of one micrometer.

Many cell structures are also measured in nanometers. For example, the bacterial ribosome, which manufactures proteins, is about 20 nm in diameter, so you could line up 50 ribosomes in the space of one micrometer.

FIGURE 2.6 illustrates the broad spectrum of microbial sizes, from the incredibly tiny viruses to the near visible protists. In fact, some filamentous and colonial algae and fungal molds are visible to the unaided eye. However, there are some notable size exceptions, as **A Closer Look 2.3** describes.

Light Microscopy

Since Leeuwenhoek's time, great strides have been made in the construction of **light microscopes**, and today's instruments routinely achieve magnifications of 1,000× (FIGURE 2.7a). The component parts of the compound microscope are the ocular lens (or eyepieces), the objective lenses (closest to the object), and the substage condenser, which concentrates light coming from the light source on the object.

Most microscopes have a revolving nosepiece with three or more objective lenses: the low-power lens (10×), the high-power lens (40×), and the oil-immersion lens (100×).

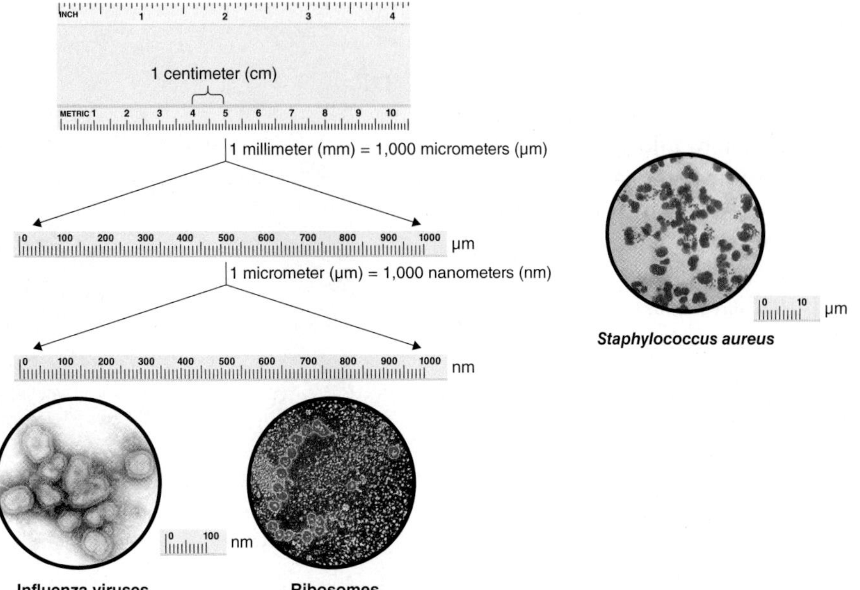

FIGURE 2.5 **Measurement of Size.** In the sciences, and in over two-thirds of the world's nations, the metric system is used to measure volumes, weights, and lengths (distance). The advantages are its accuracy and simplicity, as seen for the measurement of cell and virus size.

(a) Courtesy of F. A. Murphy/CDC (b) © Science Source (c) Courtesy of Dr. Norman Jacobs/CDC

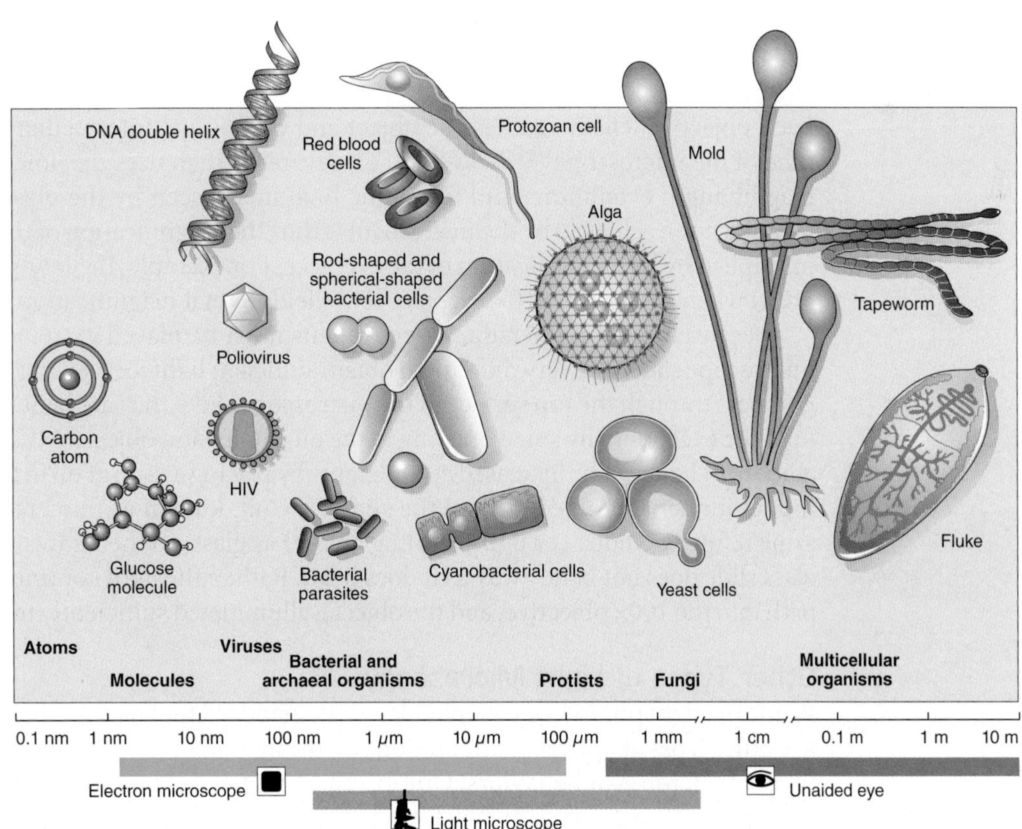

FIGURE 2.6 **Size Comparisons among Atoms, Molecules, Viruses, and Microbes (not drawn to scale).** Although tapeworms and flukes usually are macroscopic, they are often studied by microbiologists because of their disease potential.

A CLOSER LOOK 2.3

Exotic and Extreme

Copper mines can be acidic caldrons. Take, for example, the Richmond Mine at Iron Mountain near Redding, California. It is the source of the most acidic water naturally found on Earth. And believe it or not, there are acid-loving prokaryotes present that form a pink, floating film several millimeters thick on the surface of the hot, toxic water, which has a pH of 0.8. If that isn't amazing enough, in 2006 University of California scientists identified one of these microbes, a member of the domain Archaea, as the smallest living organism yet discovered. Called ARMAN (Archaeal Richmond Mine Acidophilic Nanoorganism), it is only 0.2 to 0.4 µm in diameter (about the size of a large virus). By comparison, an *Escherichia coli* cell is three times this diameter and has up to 100 times the cell volume. ARMAN cells are free living, they contain few ribosomes, and possess a relatively small number of genes.

At the other extreme, while on an expedition off the coast of Namibia (western coast of southern Africa) in 1997, scientists from the Max Planck Institute for Marine Microbiology in Bremen, Germany, found a bacterial monster in sediment samples from the sea floor. Named *Thiomargarita namibiensis* (see A Closer Look 2.1), these chains of spherical cells (see figure) were 100 µm to 300 µm in diameter—but some as large as 750 µm—about the diameter of the period in this sentence. Their volume is about 3 million times greater than that of *E. coli*. Another closely related strain was discovered in the Gulf of Mexico in 2005.

Yes, the vast majority of microorganisms are of typical microscopic size, but exceptions have been found in some extremely exotic places.

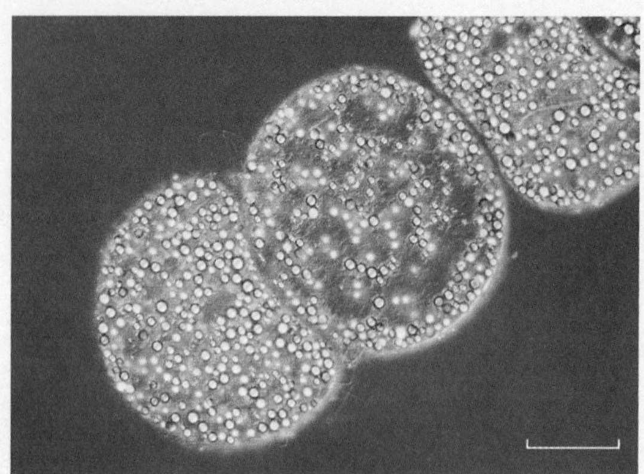

Light microscope image of *Thiomargarita namibiensis* cells. (Bar = 100 µm.)

Courtesy of Heide Schulz-Vogt, Max Planck Institute of Marine Microbiology, Germany

Each objective lens magnifies the object and creates an "intermediate image" in the tube of the microscope (FIGURE 2.7b). The eyepiece then uses this image as an object, magnifying it even more, and forms the final image seen by the observer. The total magnification achieved by the instrument is thus the magnification of the objective lens multiplied by the magnification of the eyepiece. For example, the low-power objective (10×) when used with a 10× eyepiece lens yields a total magnification of 100×.

Because the oil-immersion objective lens must be placed extremely close to the microscope slide, it is very difficult to obtain sufficient light for viewing because as light rays pass through the top surface of the microscope slide, they are bent by the glass and miss the exceptionally small opening in the oil-immersion objective (FIGURE 2.7c). The amount of light can be increased considerably by placing a drop of oil in the gap between the oil-immersion objective and the slide. The oil, known as immersion oil, has the same refractive index (or light-bending ability) as glass, so the light coming out of the glass slide does not bend away as it does in air. Rather, the light continues on a straight path into the 100× objective, and the object is illuminated sufficiently to be seen clearly.

Other Types of Light Microscopy

While the light microscope is considered the standard tool of microbiology, scientists use other types of optical configurations on the light microscope for viewing particular types of microbes or behaviors. One example is **dark-field microscopy** that produces a white image on a dark or black background. The light microscope uses a special condenser system to illuminate objects from the sides rather than from the bottom. The effect is somewhat like seeing dust particles illuminated when a beam of sunlight

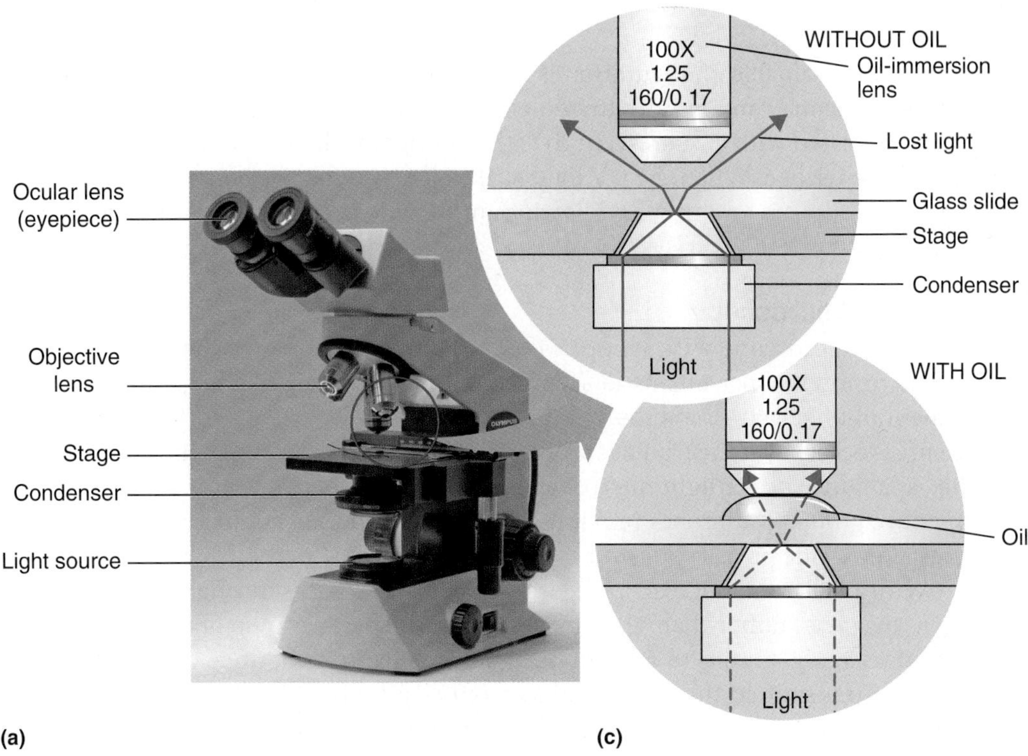

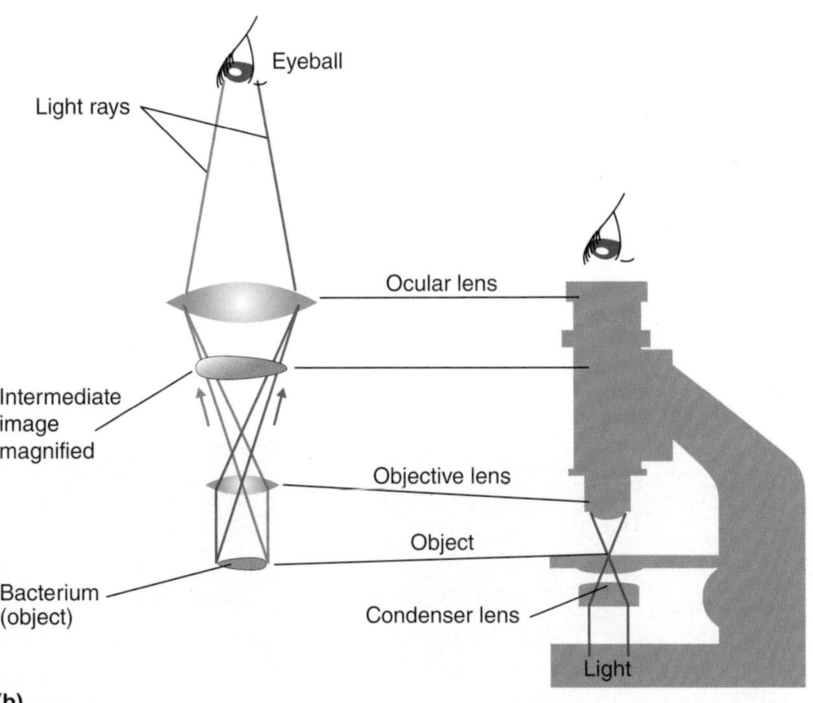

FIGURE 2.7 **Light Microscopy.** (**a**) This is a familiar light microscope used in many instructional and clinical laboratories. Note the important features of the microscope that contribute to the visualization of the object. (**b**) Image formation with the light microscope requires the light to pass through the objective lens, forming an intermediate image. (**c**) When using the oil-immersion (100×) lens, light rays enter the air and bend (solid arrows), missing the objective lens. However, they remain on a straighter line (dashed arrows) when oil is placed between lens and slide.

(a) Courtesy of Dr. Jeffrey Pommerville

passes through a darkened room. Spiral-shaped bacterial cells can be seen clearly with this instrument (FIGURE 2.8a).

Another valuable configuration is **fluorescence microscopy**. For this set up, the light microscope contains an ultraviolet (UV) light source. After the slide specimen has been coated with a fluorescent dye, the UV light is directed at the specimen now on the microscope. When the UV light strikes the dye, the dye emits visible light and the object appears as a brightly glowing image whose color varies with the type of fluorescent dye used (FIGURE 2.8b).

Electron Microscopy

The light microscope with its optical configurations increases the lower limits of human vision and permits us to see most microbial cells. However, with the development of the electron microscope in the 1940s, a whole new world opened up to scientists because this instrument represented a quantum leap in magnification beyond the capabilities of the light microscope. Microbiologists could now see the viruses, an entire group of microbes previously invisible, and they could visualize the finer structures inside prokaryotic and especially eukaryotic microbial cells.

With an electron microscope, a beam of electrons passes through a vacuum tube. Then, magnets, rather than glass lenses, focus the beam on the object prepared for viewing. Acting similar to a beam of light, the electrons bounce off, are absorbed by, or are transmitted through the object and create a final image. The image can be viewed on a microscope screen or television monitor, or captured in digital format for a permanent record.

Two types of electron microscopes are in widespread use today: the **transmission electron microscope** (**TEM**) and the **scanning electron microscope** (**SEM**). The TEM (FIGURE 2.9) produces images of a specimen that previously had been cut into thin slices (100 nm thick), while the SEM uses whole cells and permits us to see the surfaces of cells in three dimensions. The final magnification possible with the TEM is approximately 200,000× and the microscope can see objects as small as 2 nm. The SEM produces a final magnification of about 20,000× and will resolve objects as small as about 7 nm.

Pseudomonas sū-dō-mō näs

FIGURE 2.10 shows images of the bacterium *Pseudomonas* taken with the light microscope, and with the TEM and SEM.

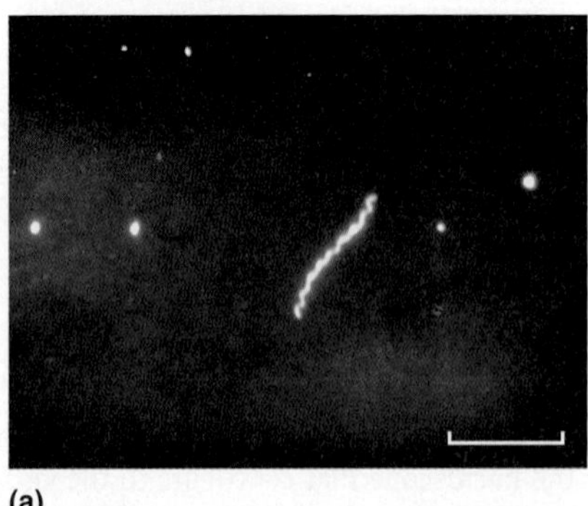

(a)

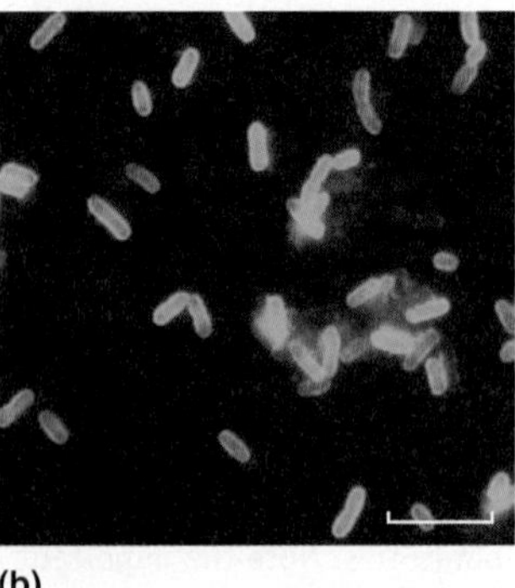

(b)

FIGURE 2.8 **Observing Cells with Other Types of Light Microscopy.** The two images show some common techniques ([a] dark field; [b] fluorescence) for contrasting bacterial cells. These methods require special optical configurations added onto the light microscope. (**a** and **b**, Bar = 10 μm.)

(a) Courtesy of Schwartz/CDC. (b) Courtesy of Larry Stauffer, Oregon State Public Health Laboratory/CDC.

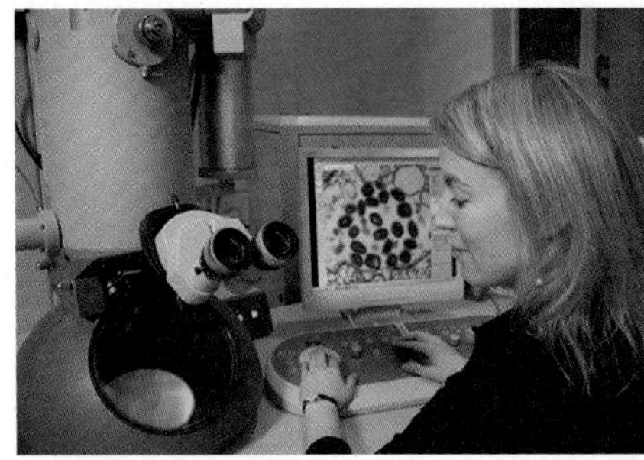

(a)

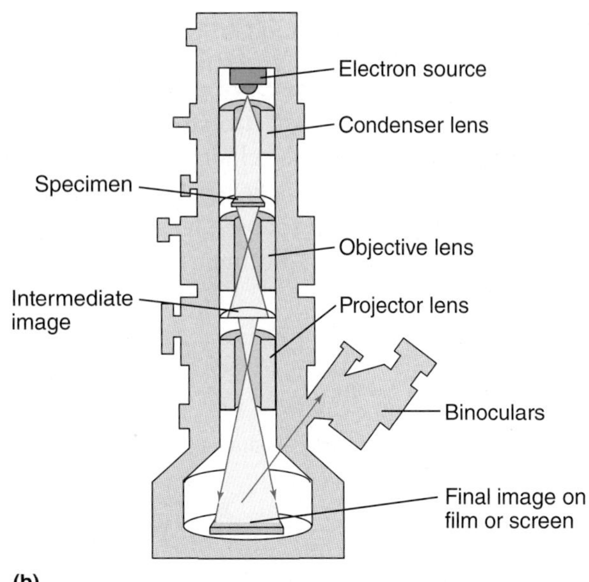

(b)

FIGURE 2.9 **The Electron Microscope.** (**a**) A transmission electron microscope (TEM). (**b**) A schematic of the vacuum tube. A beam of electrons is emitted from the electron source and electromagnets function as lenses to focus the beam on the specimen. The image is magnified by objective and projector lenses.

(a) Courtesy of Cynthia Goldsmith/James Gathany/CDC

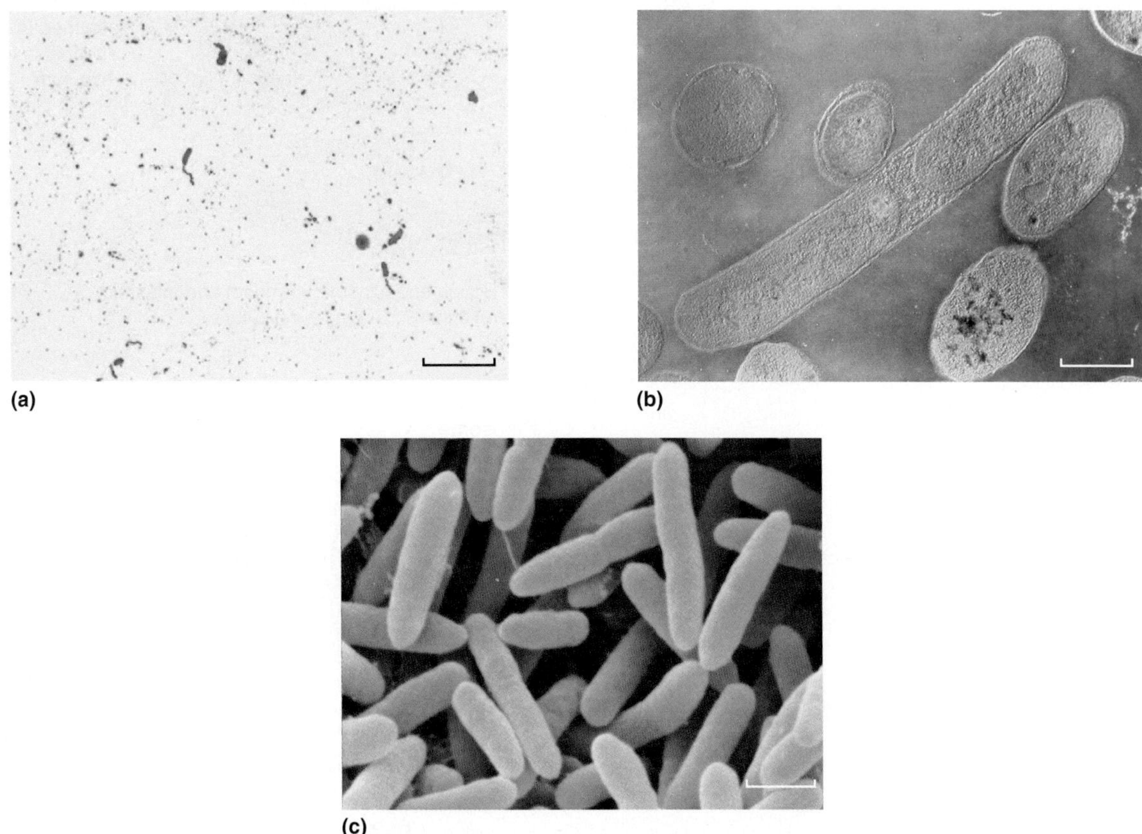

FIGURE 2.10 **Light, Transmission, and Scanning Electron Microscopy Compared.** Three false-color images of the bacterial genus *Pseudomonas* as seen with three types of microscopy. (**a**) A photograph of stained cells as seen with the light microscope. (Bar = 5.0 µm.) (**b**) A view of sectioned cells taken with a transmission electron microscope. (Bar = 1.0 µm.) (**c**) A view of whole cells taken with a scanning electron microscope. (Bar = 2.0 µm.) The difference in perspective between the three microscope images is clear.

(a) Courtesy of Dr. William A. Clark/CDC. (b) © CNRI/Science Source. (c) © SciMAT/Science Source.

2.4 Cell Ultrastructure: Comparing Eukaryotic and Prokaryotic Cells

With the use of the electron microscopes, especially the TEM, a much clearer and more detailed view inside cells is possible. Today, we have a fairly good idea of the finer structural details inside a cell, which is often referred to as the **ultrastructure** of the cell. To finish this chapter, let's see what's inside a typical eukaryotic and prokaryotic cell (FIGURE 2.11). Another chapter will discuss the bacterial structures in more detail.

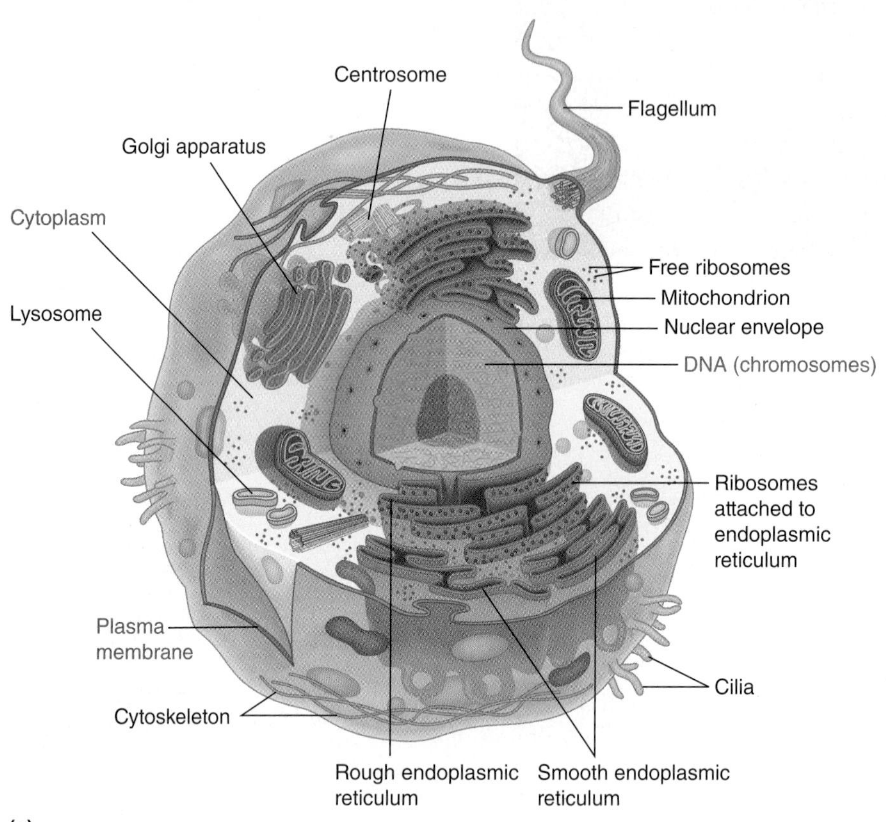

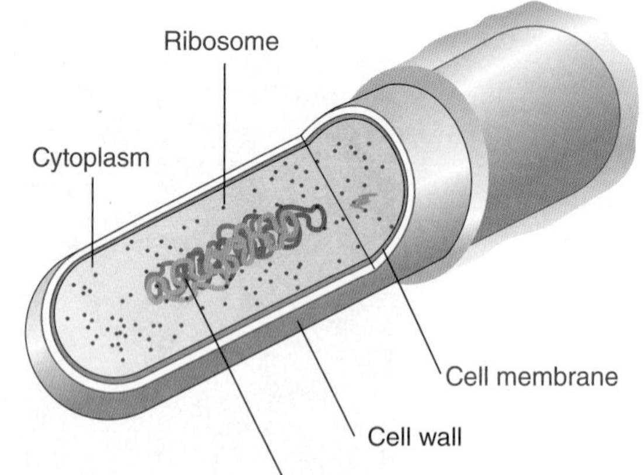

FIGURE 2.11 **A Stylized Comparison of a Eukaryotic and Prokaryotic Cell.** (**a**) A protistan cell represents a typical eukaryotic cell. Note the variety of the cellular features, some of which are noted in the text. (**b**) A prokaryotic cell, typical of *Escherichia coli*. Relatively few structures are seen with light microscopy, the presence of a cell nucleus is the primary feature distinguishing the eukaryotic cell from the prokaryotic cell. Universal structures common to all cells are indicated in red.

Cell Structure

All cells are surrounded by a selective barrier called the **cell** or **plasma membrane** that separates the environment from the semifluid **cytoplasm** inside the cell. Being a selective barrier means the membrane can control what enters and leaves the cell. The cytoplasm has the consistency of olive oil and contains many nutrients like sugars and salts. Suspended in the cytoplasm are various cellular components.

Notice in Figure 2.11 that both prokaryotic and eukaryotic cells contain **chromosomes**, which carry the genetic information (genes) in the form of DNA. One major difference between the two cell types is the way the DNA is organized. In eukaryotic cells, the chromosomes are surrounded by a membrane (nuclear) envelope forming the **cell nucleus**. In a prokaryotic cell, the DNA chromosome is not surrounded by any membranes and lies "bare" in the cytoplasm.

Also inside the eukaryotic cell are several complex, membrane-enclosed cellular compartments, called **organelles** having specific functions. Among the more prominent organelles are the cell nucleus, an "endomembrane system" and mitochondria (and chloroplasts in plants and algal cells). Bacterial and archaeal cells have no such organelles, but both prokaryotes and eukaryotes have **ribosomes**, the tiny bodies in the cytoplasm where proteins are constructed based on information received from the genes. In fact, it was one of the genes needed to produce ribosomes that Woese used to construct the three domains and the tree of life. TABLE 2.2 lists many of the cell structures and organelles with their functions.

Throughout this discussion of cells and sizes, have you wondered why most microbial cells stay so small compared to their eukaryotic relatives? **A Closer Look 2.4** provides the answer.

A CLOSER LOOK 2.4

Size Matters

Take a look back at A Closer Look 2.3. The *Thiomargarita namibiensis* cells shown in the photo are huge! Most bacterial cells are at least 100-times smaller; that is, not the 100-300 μm but rather more like 2 μm typical of an *Escherichia coli* cell. How can some bacterial species like *Thiomargarita*, though rare, survive as such large cells?

The basics of cell size can be simply stated as follows: size matters. If a cell gets too large, nutrients needed for metabolism cannot pass across the cell membrane fast enough to accommodate the metabolic demands of the cell. In most cases, such cells would die either from "starvation" or from the buildup of toxic waste products in the cytoplasm that cannot be eliminated fast enough.

So, for species like *E. coli*, when the cell reaches a critical size, it will divide into two smaller cells, making the transport of nutrients and waste products much easier and highly efficient.

But what about *T. namibiensis* and similar "bacterial giants?" The cells of *T. namibiensis* have solved the size problem by "pushing" all the cytoplasm out toward the edge of the cell (see figure). In other words, the whole central part of the cell is one big cavity – well it is not really a cavity but rather a space filled up with essential nutrients to "feed" the thin layer (0.5-2 μm thick) of cytoplasm. In fact, this thickness is identical to the cytoplasmic thickness of an *E. coli* cell.

By the way, the larger eukaryotic cells have specialized transport systems (endomembrane and cytoskeleton) to more rapidly transport essential materials throughout the cytoplasm. And the viruses have no independent metabolism or need for nutrients, so they can be exceedingly small – as most are.

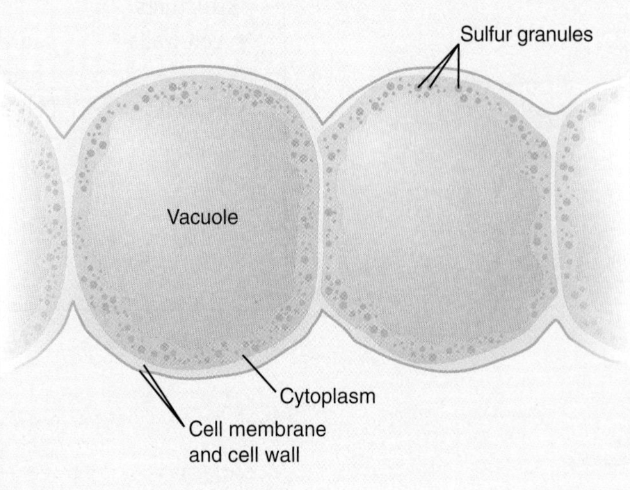

TABLE
2.2 A Comparison of Eukaryotic and Prokaryotic Cell Structures/Processes

Cell Structure/ Process	Function	Prokaryotes	Eukaryotes
Cell/plasma membrane	Semipermeable barrier separating environment from cell cytoplasm	Yes	Yes
Cell nucleus	Houses the genetic information needed for growth and metabolism	No structure	Yes
	• Presence of DNA	Yes, as a single circular chromosome in cytoplasm	Yes, as multiple, linear chromosomes surrounded by a double membrane
Cytoplasm	Fluid and contents that fill the cell and in which most metabolism occurs	Yes	Yes
• Endomembrane system	Membranes that divide the cell into functional and structural compartments and regulates protein traffic	No	Yes
• Cytoskeleton	Cytoplasmic cellular scaffolding for transport and cell division	Yes (organization unique from eukaryotes)	Yes
• Ribosomes	Site of protein manufacture	Yes	Yes
• Mitochondria	Conversion of chemical energy to cellular energy (ATP)	No structure	Most
	• Make ATP	Yes (on cell membrane)	Yes
• Chloroplasts	Conversion of light energy to chemical energy (photosynthesis)	No structure	Algae and plants only
	• Carry out photosynthesis	Some (on cell membrane)	Algae and plants
Exterior structures			
• Cell walls	Cell structure and water balance	Most	Algae, fungi, and plants
• Flagella	Cell movement (motility)	Some (structurally unique from eukaryotes)	Some
• Cilia	Cell movement (motility)	No	Some protists and animals only

A Final Thought

The diversity of life forms has astounded scientists for centuries, and the challenge of characterizing and categorizing organisms (taxonomy) has been the focus of many scientists. The challenge of naming and categorizing this vast diversity is nowhere more complicated than it is with the microbial world. When it comes to microscopic life forms, which are often unicellular, the scientist is dependent on different forms of microscopy and various biochemical, genetic, and molecular means for detecting cellular activity. The fruits of this labor are shown in the three domain system, which visibly describes the various forms of microscopic life and which has challenged our thinking about the basic nature of life itself. The history of taxonomy, from Linnaeus to Woese, is a magnificent study of life forms, and we encourage you in this Chapter to gain an appreciation of the diversity of microorganisms, how they can be viewed, and the fundamental differences in cell architecture between domains.

Questions to Consider

1. A student is asked on an examination to write a description of the fungi. She blanks out. However, she remembers that fungi are eukaryotes, and she recalls the properties of eukaryotes. What information about eukaryotes from this chapter can she use to answer the question?
2. A local newspaper once contained an article about "the famous bacterial genus ecoli." How many errors can you find in this phrase? Rewrite the phrase with the mistakes corrected.
3. In 1987, in a respected scientific journal, an author wrote, "Linnaeus gave each life form two Latin names, the first denoting its genus and the second its species." A few lines later, the author wrote, "Man was given his own genus and species *Homo sapiens*." What is conceptually and technically wrong with both statements?
4. Biologists tend to be collectors and compulsive classifiers. Why do you think this is so? Also, which classifier mentioned in this chapter do you think had the most impact on the science of his day?
5. Microbes have been described as the most chemically diverse, the most adaptable, and the most ubiquitous organisms on Earth. From this chapter, what can you add to this list of "mosts"?
6. Prokaryotic cells lack the extensive group of organelles found in most eukaryotic cells. Provide a reason for the structural difference. (Hint: remember size matters!)
7. A classmate who missed last week's first lab on microscopy observes a lab partner in this week's lab using oil immersion with the 100× objective lens. She asks her partner why he used it. "To increase the magnification of the microscope" was his answer. Do you agree or disagree? Why?
8. Every state has an official animal, flower, and/or tree, but one state has an official bacterial species named in its honor: *Methanohalophilus oregonense*. What's the state and decipher the meaning of the genus name. (Note: *ense* = "belonging to")

Key Terms

Informative facts are necessary for the expression of every concept, and the information for a concept is founded in a set of key terms. The following terms form the basis for the concepts of this chapter. On completing the chapter, you should be able to explain and/or define each one.

Archaea
Bacteria
binomial nomenclature
cell (plasma) membrane
cell nucleus
chromosome
class
cytoplasm
dark-field microscopy
domain
Eukarya
eukaryote
family
fluorescence microscopy
genus (pl. genera)
kingdom
light microscope
micrometer (μm)
nanometer (nm)
order
organelles
phylum (pl. phyla)
prokaryote
ribosome
scanning electron microscope (SEM)
species
taxonomy
three-domain system
transmission electron microscope (TEM)
ultrastructure

Molecules of the Cell: The Building Blocks of Life

Many ideas have been put forward to explain how life began on planet Earth. Here Dr. Stanley Miller is shown recreating his famous experiment performed in 1953. A spark was sent through a flask containing gases believed present in the atmosphere of the Earth billions of years. The experiment produced several amino acids, one of the building blocks of life.

© Roger Ressmeyer/CORBIS

Looking Ahead

Having examined the basic structure of microbial cells, it is important to consider the molecules building cells and understand the roles these molecules play in cell function.

On completing this chapter, you should be capable of:

- Drawing the structure of an atom.
- Describing the various simple sugars and polysaccharides.
- Comparing and contrasting fats, phospholipids, and sterols.
- Drawing an amino acid and contrasting the four levels of protein structure.
- Identifying the parts of a nucleotide and distinguishing DNA from RNA.

Stanley Miller was intrigued. As a young graduate student at the University of Chicago, he became fascinated with the ideas presented in a lecture presented by Professor Harold Urey. Urey told the audience he believed the atmosphere of primitive Earth was very different from today's atmosphere and likely consisted of several gases, including methane, ammonia, hydrogen sulfide, and hydrogen. Urey further suggested that within such an atmosphere it might be possible to synthesize some of the chemical building blocks needed to form the raw materials for the emergence of life.

■ primordial soup: A pond or body of water rich in substances that could provide favorable conditions for the emergence of life.

Life on Earth began around 3.8 billion years ago, evolving from some form of "primordial soup" into the incredible diversity of microbes and life we see today. So, where did the very first molecules of life on Earth come from?

In 1952, Miller took a research position in Urey's lab and asked to test Urey's hypothesis that a primitive gas mixture could form some of the building blocks of life. After much argument and Miller's persistence, Urey gave his approval to the experiment. After three months of planning, Miller's experiment would attempt to simulate the primitive Earth atmosphere by putting methane, ammonia, hydrogen gas, and water vapor together in a closed reaction vessel (FIGURE 3.1). But how would he get these gases to interact with one another. There needed to be a spark.

Because lightning storms would have been very common in the atmosphere of early Earth, Miller had the brilliant idea of putting an electric charge through the gas mixture to simulate lightning going through the atmosphere. So, using electrical sparks in the reaction vessel, Miller allowed the reaction to continue for several days. As the days passed, he noticed the accumulation of a brown slime on the reaction vessel and a yellow-brown color in the water. When analyzed, Miller detected five amino acids, which are some of the building blocks of proteins.

The Miller-Urey experiment showed for the first time that some of the building blocks of life, like amino acids, could be made under conditions simulating early Earth. Today, we know Earth's primitive atmosphere did not have the exact composition Urey proposed. Still, Miller's work represents a landmark experiment. In fact, in 2008, a year after Miller died, more modern techniques of analysis turned up more amino acids and other substances of interest from the original reaction vessels Miller used.

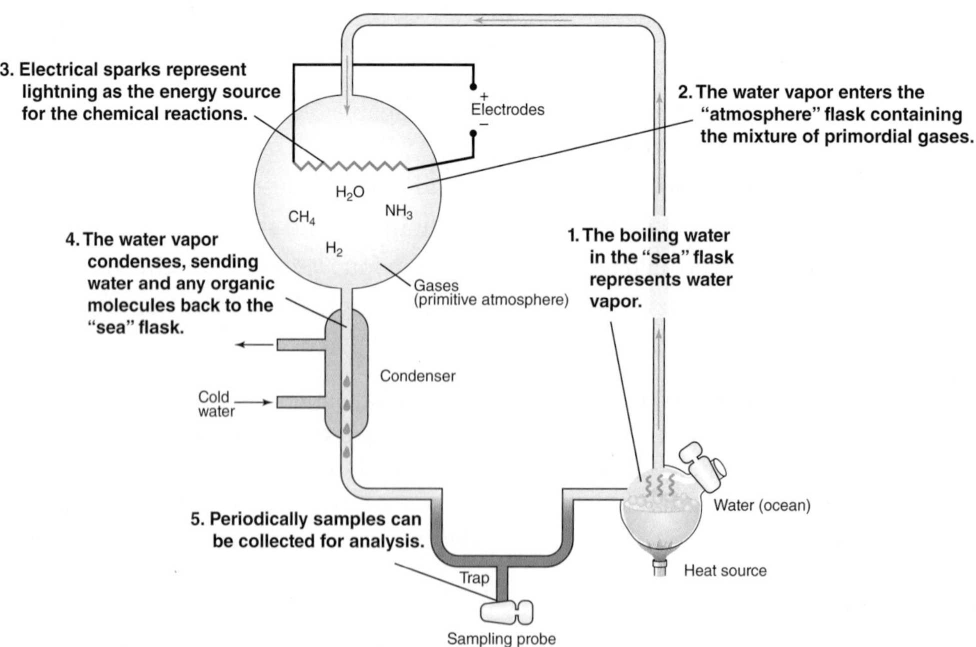

FIGURE 3.1 **The Miller-Urey Experiment.** The experiment consisted of a reaction vessel containing Earth's primordial gases in which an electrical spark provided the catalyst for the chemical reactions.

Today, we cannot answer for certain how life began, but we can study the building blocks and other large molecules forming the structure and function of microbes, viruses, and all life. These are the carbohydrates, lipids, proteins, and nucleic acids—and they are the topic in the coming pages.

3.1 Chemistry Basics: Atoms, Bonds, and Molecules

In the primordial soup, the capture of organic molecules, the building blocks of life, into a concentrated area, within a membrane bound compartment, permitted the chemical reactions of life (metabolism) to take place at a reasonable rate, something that would not have happened with the molecules floating freely and randomly in the environment. Of course, when the membrane of the cell formed, water was also captured within the cell and would be the substance in which metabolism evolved.

Atoms and Bonding

The basic units of matter, called **atoms**, have three basic components: negatively charged **electrons**, positively charged **protons**, and uncharged **neutrons**. Protons and neutrons possess most of the mass of an atom and form the core or atomic nucleus, while the electrons orbit the nucleus in regions called "shells" (FIGURE 3.2a). There are some 92 different naturally occurring **elements**, all of which differ from one another based on the number of protons and neutrons in the atomic nucleus and the number of electrons orbiting the atomic nucleus (FIGURE 3.2b). Of these, about 25 are commonly found in microbes and other organisms. The four most abundant are hydrogen (H), carbon (C), nitrogen (N), and oxygen (O).

In a very real way, the atoms are like a box of Legos® in that atoms of different sizes and configurations can "snap together," provided the atoms fit, to form different structures. When this happens, the electrons of the interacting atoms are held together by **chemical bonds** formed between electrons in the interacting atoms. So, **molecules** are simply two or more atoms bonded together by their electrons. For example, the gases used in the Miller-Urey experiment included methane and ammonia. Methane consists of one carbon atom bonded to four hydrogen atoms

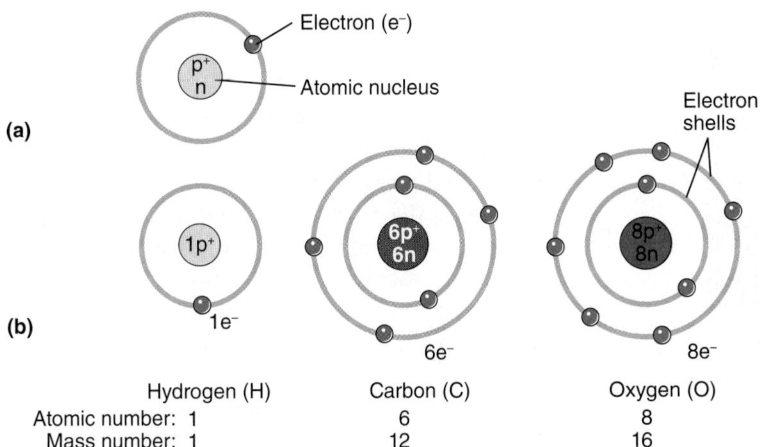

FIGURE 3.2 **The Structure of Atoms.** (**a**) An atom consists of protons and neutrons in the atomic nucleus and surrounding shells of electrons. (**b**) Different atoms have different numbers of protons, neutrons, and electrons.

Methane

Ammonia

Water

and is written as CH_4 (Note: if there is just one atom in a molecule, there is no subscript number used); ammonia is written as NH_3. In addition, many of the molecules contain carbon and one refers to such carbon-containing chemicals as **organic** molecules.

Water

Another very important molecule for life is water, a molecule consisting of three atoms—two hydrogen (H) and one oxygen (O) and is written as H_2O. Approximately 70% of the mass of a cell is water, demonstrating that no organism can survive and grow without water, and many organisms, including microbes, live in water.

Liquid water is the medium in which all cellular chemical reactions occur because water acts as the universal solvent in cells. Take for example what happens when you put a solute like salt or sugar in water. The salt and sugar dissolve, forming an **aqueous solution**; that is, one or more solutes dissolved in water. Water molecules also are part of many chemical reactions, as we will see later in this chapter.

Besides its solvent properties, water also has other characteristics that make it an ideal molecule for life, as pointed out in **A Closer Look** 3.1.

solvent: The substance doing the dissolving in a solution.

solute: The substance dissolved in a solution.

A CLOSER LOOK 3.1
Water and Life

Life on Earth certainly evolved in a watery primordial soup some 3.8 billion years ago and remained in a water environment for some 3 billion years before spreading to land where survival still depended on water. Without water, the human body would be unable to survive for more than one week. In fact, water is such a necessity for life as we know it, that when scientists send spacecraft and rovers in search for traces of microbial life on other worlds, such as Mars, water is one of the most important molecules they look for (see figure).

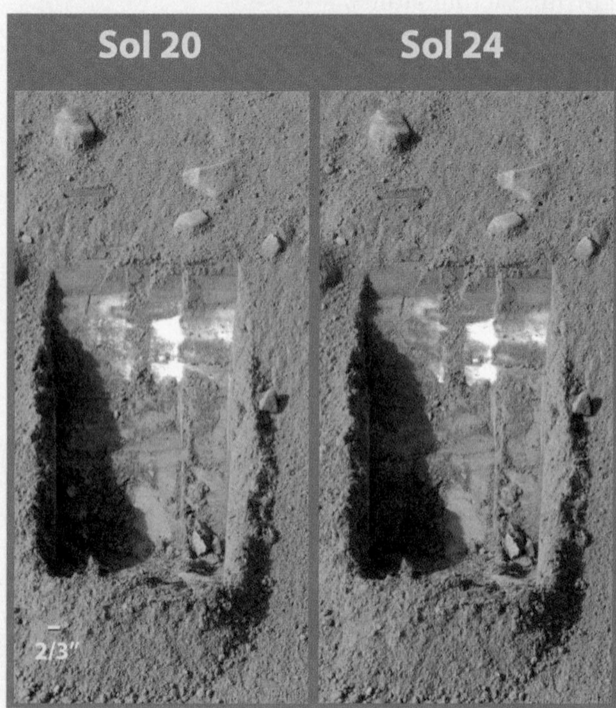

The white bits in this photo are Martian water ice.

Courtesy of JPL-Caltech/University of Arizona/Texas A&M University/NASA.

Besides its property as the solvent for life, what are water's other life-supporting properties? We can identify three.

- **Cohesion.** Water molecules will stick together due to weak bonding. The sticking together is called cohesion. For example, it is the cohesion between water molecules that allows water to be transported up tress from the roots to the leaves.

  Also, by sticking together, water has a high surface tension; it is hard to break water molecules apart. That means small insects can "walk on water" as if there was a film on the water surface.
- **Temperature moderation.** It takes a lot of energy (heat) to raise the temperature of water. For instance, if you have ever accidently touched a metal pot full of water being heated, the metal heats up much faster than the water as a misdirected finger will attest. The bonding between water molecules gives water a stronger resistance to temperature changes than occurs with most other substances. Likewise, coastal areas, bordering an ocean, such as San Diego, in the summertime tend to have more moderate air temperatures than deserts, such as around Phoenix, because of the higher humidity (water content) in the atmosphere. There is more water to absorb the heat along the ocean coast.

  In the human body, water also moderates body temperature by evaporative cooling. When water evaporates, the water left behind cools down because the water molecules with the greatest energy (the

A CLOSER LOOK 3.1

Water and Life . . . (continued)

"hottest" ones) vaporize first. That's why sweating is important—it helps prevent an individual from overheating.

- **Insulation.** As you know, frozen water (ice) floats, unlike most substances that will sink on freezing. These substances sink because the atoms move closer together and become denser than the surrounding liquid when frozen. Due to bonding between water molecules, these molecules move farther apart, making the ice less dense than the surrounding liquid water and the ice therefore floats.

By ice floating it acts as insulation. If frozen water was denser than liquid water, it would sink and in the winter all ponds and lakes (and perhaps oceans too) would eventually freeze solid, freezing any living creatures (including microbes). However, because ice floats, it forms an insulating "blanket" over the body of water in winter, allowing the water underneath to remain liquid—and life to survive.

So drink up and stay hydrated. And that goes for microbes too!!

The Molecules of Microbes

Most of the molecules in a cell involved in structure and cell function—including metabolism, cell movement, and cell growth—are significantly larger than simple molecules like methane and water. Indeed, the molecules of life often are hundreds to billions of atoms in size. For this reason, they are called "macromolecules" (*macro* = "large") and form the carbohydrates, lipids, proteins, and nucleic acids. Their structural differences will be in the way the atoms in each macromolecule are bonded together.

3.2 Carbohydrates: Simple Sugars and Polysaccharides

Carbohydrates are organic molecules containing carbon, hydrogen, and oxygen, generally in a ratio of one atom of carbon to two atoms of hydrogen to one atom of oxygen. Thus, the basic formula unit for a carbohydrate is CH_2O.

Carbohydrates vary from relatively small, simple molecules to extremely large, complex macromolecules. The smallest carbohydrates are the simple sugars, which include the **monosaccharides** ("single sugars") and **disaccharides** ("double sugars"), both names derived from the Latin word *saccharo* meaning "sugar." Slightly larger molecules are referred to as "oligosaccharides" (*oligo* = "few"), while those composed of thousands of monosaccharides are called **polysaccharides** (*poly* = "many").

Some carbohydrates, such as the sugars, serve as energy sources in cells while others, specifically the polysaccharides, either store energy (starches) or build the cell walls of most all prokaryotes, and those of the eukaryotic algae and fungi (and plants of course).

Simple Sugars

A monosaccharide may contain three to seven carbon atoms. Among the most significant monosaccharides are the five-carbon sugars (pentoses), such as ribose and deoxyribose and the six-carbon sugars (hexoses), including glucose, fructose, and galactose. These three hexoses have the same numbers of carbon, hydrogen, and oxygen atoms, and they all have the same chemical formula: $C_6H_{12}O_6$. However, their atoms are snapped together in a different order and are called isomers of one another.

isomer: A substance that has the same number and types of atoms as another substance but where the atoms are arranged differently.

Monosaccharides, especially glucose, act as the fundamental building blocks for larger carbohydrate molecules and as sources of energy for cellular processes. These monosaccharides can be bonded together in a chemical reaction that involves

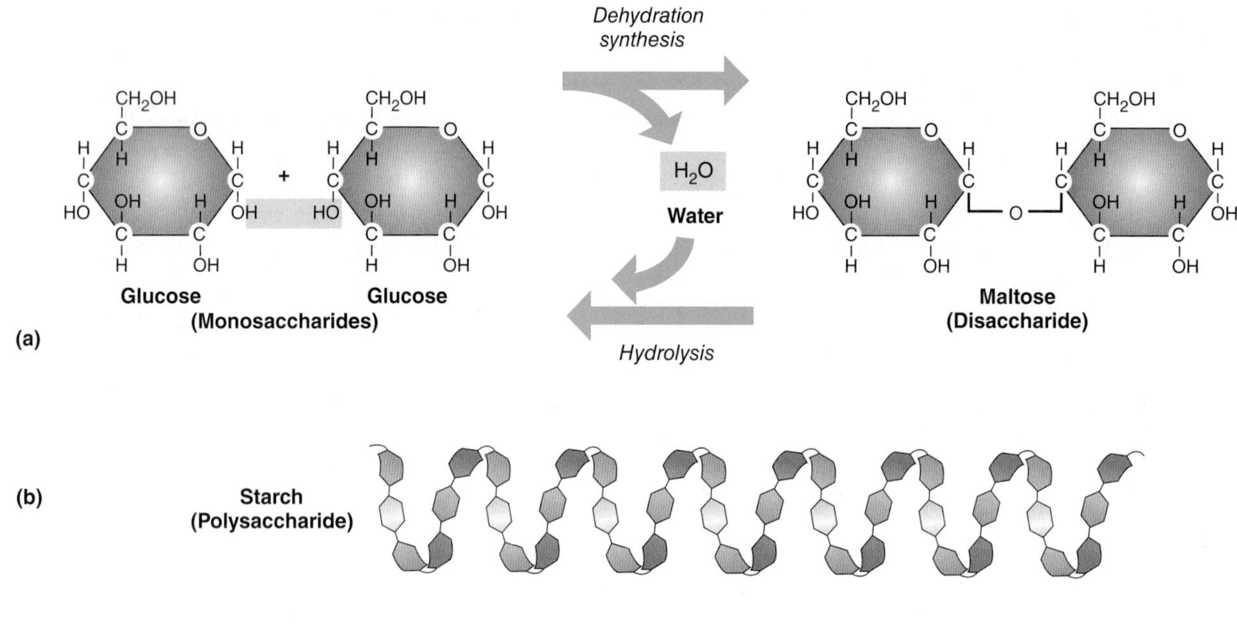

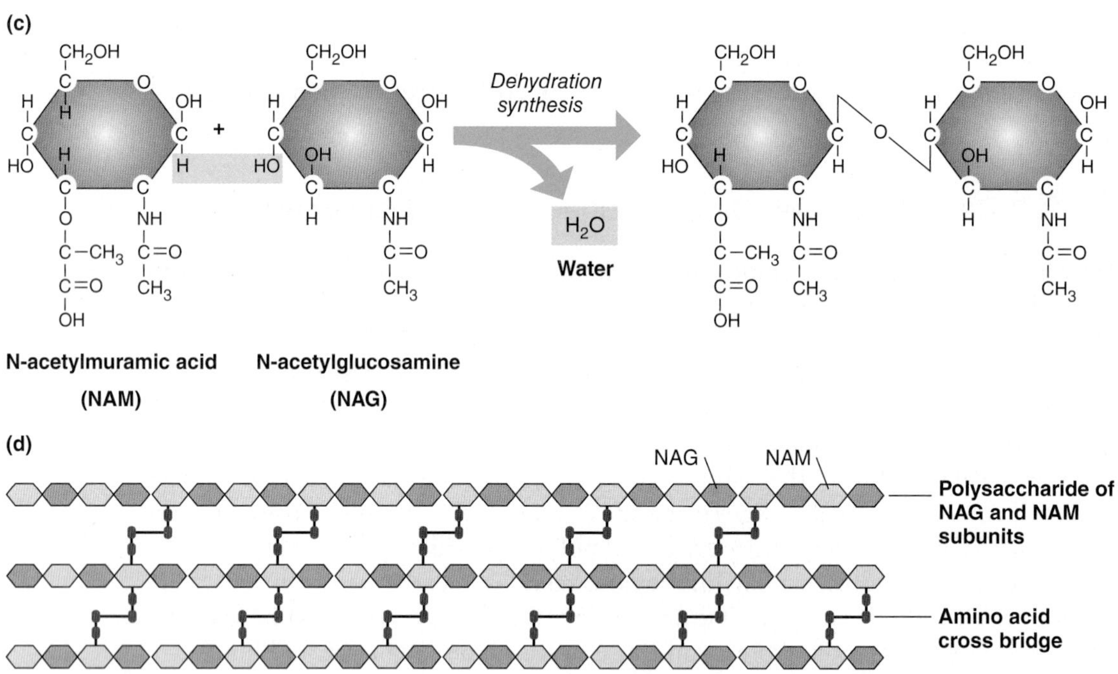

FIGURE 3.3 **Carbohydrates Consist of the Monosaccharides, Disaccharides, and Polysaccharides.** (**a**) Glucose is a monosaccharide that can bond with another glucose in a dehydration synthesis reaction to form maltose, a disaccharide. (**b**) The bonding of additional glucose molecules leads to the formation of a polysaccharide, such as starch. Note that each little hexagon is glucose. (**c**) N-acetylmuramic acid (NAM) and N-acetyl glucosamine (NAG) are modified simple sugars that can bond together to form the bacterial cell wall peptidoglycan. (**d**) Peptidoglycans can be held together by side chains of amino acids.

the removal of H_2O from the sugars. When water is lost during the synthesis of a molecule, it is known as a **dehydration synthesis reaction** (FIGURE 3.3a). These reactions involve the action of cellular **enzymes**, protein molecules capable of rearranging the components of organic substances while themselves remaining unchanged.

The linking together of two monosaccharide molecules through a dehydration synthesis reaction forms a disaccharide (see FIGURE 3.3a). Among the commonly encountered disaccharides is maltose, also known as malt sugar. This disaccharide is found in cereal grains such as barley, where the sugar is fermented by yeast cells in the absence of oxygen to produce the alcohol in beer. Another well-known disaccharide is lactose, the principal carbohydrate in milk. This carbohydrate is a combination of a glucose molecule and a galactose molecule. Lactose can be chemically changed to lactic acid by certain species of bacteria. The acid causes milk to become sour. However, the reaction can be controlled and used in the dairy industry to produce yogurt, buttermilk, and sour cream (depending on the starting material). A third disaccharide, sucrose, is a combination of a glucose molecule and a fructose molecule. It is commonly known as table sugar.

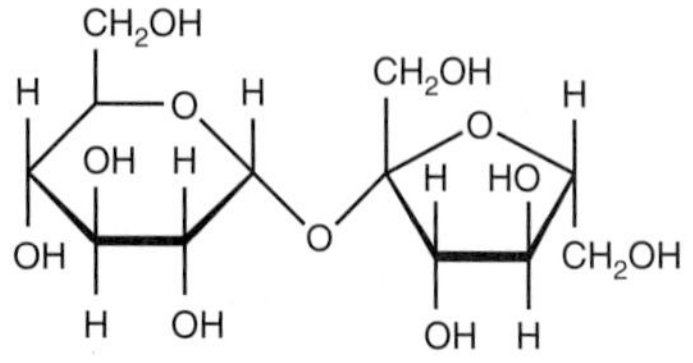

Lactose

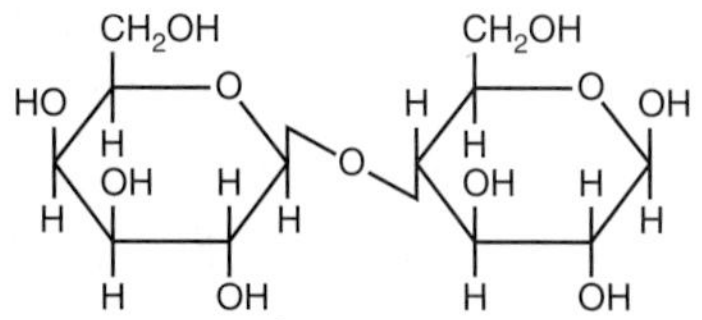

Sucrose

Polysaccharides

The polysaccharides are extremely large and complex carbohydrate molecules. A single polysaccharide molecule may contain hundreds or thousands of monosaccharide subunits bonded together through dehydration synthesis reactions. One example of an "energy polysaccharide" is **starch**, which is composed exclusively of glucose molecules (FIGURE 3.3b). Starch is typically found in many plants like potatoes and corn. Organisms from microbes to humans can break down starch to obtain the constituent glucose monosaccharides that are then used for energy production. This tearing apart of a molecule is known as a **hydrolysis reaction** and it involves enzymes along with the addition of water molecules.

Other large carbohydrate polymers are "structural polysaccharides." For example, **cellulose**, a major part of plant cell walls, is also composed of long fibers made of glucose. However, the glucose molecules in cellulose are linked together differently than in starch and humans lack the necessary enzyme to break these linkages. Therefore, we cannot digest the cellulose present in the plant foods we eat—and we therefore excrete the cellulose as roughage or so-called "dietary fiber." Unlike humans, termites and herbivores primarily ingest cellulose (e.g., wood, grasses, and hay) for energy. How can they break down cellulose but we can't? **A Closer Look 3.2** provides the microbial solution.

The bacterial cell wall also is a structural polysaccharide made from sugars that have been modified to contain nitrogen (FIGURE 3.3c). These sugars, N-acetylmuramic acid (NAM) and N-acetylglucosamine (NAG), form through dehydration synthesis reactions into long fibers called **peptidoglycan**. In many cases, multiple layers of peptidoglycan build the cell wall and the adjacent chains are stabilized and held together by cross bridges made of amino acids.

3.3 Lipids: Fats, Phospholipids, and Sterols

We are all familiar with some types of **lipids**, such as the animal fats and plant oils in our diet. Like the carbohydrates, these lipids are used by some microbes and many other organisms for energy and energy storage. Lipids are built from carbon, hydrogen, and oxygen atoms, but there are more carbon-hydrogen bonds in a lipid than in a carbohydrate. As a result, there is more energy that can be harvested from a fat or oil than from a similar weight carbohydrate.

A CLOSER LOOK 3.2

What's in Your Gut?

Due to their wood-eating habits, termites cause more damage to American homes than do tornadoes, fires, and earthquakes combined. The cost is estimated to be over $5 billion annually. On the positive side, around the world, more than 2,600 species of termites are busily breaking down tons of fallen, dead trees and other woody plants, digesting the wood (primarily cellulose) that is the main component of plant cell walls. Ecologically, termites are critical to the recycling of the forests nutrients. But without microbes, the digestion of these polysaccharides (for good or bad) by termites would be impossible.

Within a termite's gut is one of nature's most efficient bioreactors—a microbial community capable of converting 95% of cellulose into simple sugars within 24 hours. This microbial community consists of more than 200 bacterial, archaeal, and protist species—many species not found anywhere else on Earth—that produce a variety of wood-digesting enzymes. So, without the wood-eating microbes, a termite could not extract needed nutrients and energy from the wood and without the termite to initially grind the wood into tiny pieces, the microbes could not survive in the termite's gut. This "living together," called a symbiosis, between termite and gut microbes creates the energy-producing end-products both the termite and microbes need (see figure).

A similar symbiotic scenario plays out between cows (and other ruminants) and their home-dwelling microbes. For example, cows re-chew their food [primarily grass in summer and silage (preserved grass or corn) in winter] to increase the surface area for digestion. However, that digestion is not primarily done by cow enzymes. Of the cow's four so-called "stomachs," the rumen is especially important because it contains billions of microbes (bacterial, archaeal, protist, and fungal species) that help the bovines digest cellulose and other similar polysaccharides. And like those in the termite gut, the bovine microorganisms all depend on each other for survival, the by-products from one species being used by another.

So why can't we digest cellulose and most other plant polysaccharides? After all, we also have billions of microbes in our gut. Unfortunately we do not have the microbes capable of producing the massive numbers of cellulose-busting enzymes. Here's a science fiction thought for discussion: Suppose an "experiment" was carried out that supplied our gut with a resident population of microbes capable of digesting cellulose. How would this change affect our food consumption, both personally and on a global scale?

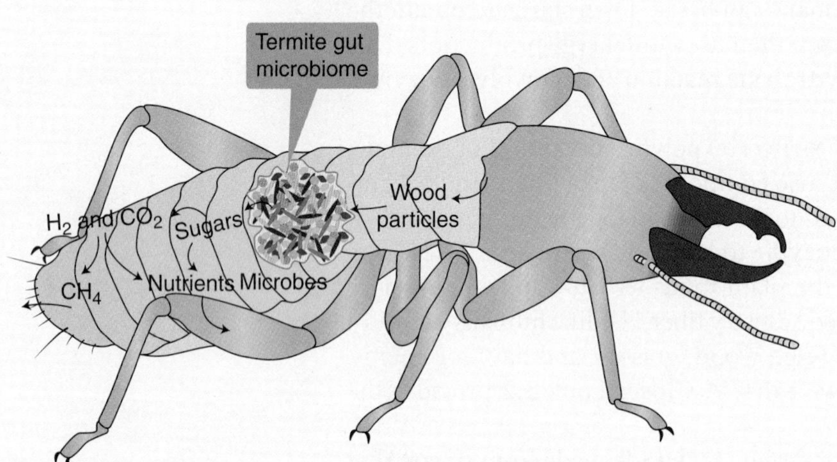

A drawing illustrating the microbial products of cellulose digestion in the termite gut.

As you know, if you shake a mixture of water and oil, the two liquids will separate from one another because lipids do not dissolve in water. We therefore say lipids are **hydrophobic** (*hydro* = "water;" *phobic* = "fearing") whereas simple sugars like glucose and sucrose are **hydrophilic** (*philic* = "loving") and dissolve in water. Because lipids do not dissolve in water, environmental cleanup of major oil spills is a challenging task to which microbes can be of great assistance, as revealed in **A Closer Look 3.3**.

A CLOSER LOOK 3.3

Microbes to the Rescue!

On April 20, 2010, the *Deepwater Horizon* drilling rig that was working on a well for the British Petroleum oil company blew up in the Gulf of Mexico. Four days later, engineers discovered the wellhead was damaged and was leaking oil and methane gas into the Gulf. For 3 months, oil spilled into the Gulf and contaminated nearby shores and wetlands (see figure), making it the largest accidental oil spill in history. According to federal government estimates, some 5 million barrels, or 780 million liters, were spilled, and it was not until September that the well was declared sealed. Then, within weeks, many parts of the Gulf were almost oil free. Where did all the spilled oil go?

An oil well, such as the *Deepwater Horizon*, produced primarily crude oil (petroleum), which is "unprocessed" oil. For decades, scientists have tried to use bioremediation—the breakdown (biodegradation) of contaminating compounds using microorganisms—as a natural method for cleaning up some of the environment's worst chemical hazards, including oil spills.

But what about all the crude oil released and dispersed from the ruptured *Deepwater Horizon* well? Could local bacterial species present in the Gulf act as natural bioremediation agents? It appears they did!

The Gulf has many natural oil seeps and resident bacterial species have evolved the metabolism needed to break down the oil. So, when the Deepwater Horizon spill occurred, oil-hungry bacterial cells were ready to multiply and able to consume much of the oil. It remains to be determined just how efficient these oil denizens were in mopping up the oil, but it is clear they were intimately involved in the cleanup. The spill might have been even worse if it was not for the bacterial species present in the Gulf waters. Still, microbes cannot eliminate or digest all the oil and it may be several years before the Gulf and the wetlands are completely recovered.

Sunlight is reflected off the Deepwater Horizon oil spill in the Gulf of Mexico on May 24, 2010. The image was taken by NASA's Terra satellite.

Courtesy of NASA/GSFC, MODIS Rapid Response.

Based on their chemical composition, lipids may be subdivided into three different groups: fats and oils, phospholipids, and sterols.

Fats and Oils

Fats and **oils** contain two components, a three-carbon molecule called glycerol and three long chains of carbon atoms called fatty acid molecules (FIGURE 3.4). The synthesis of a fat or oil is the result of each fatty acid being joined to the glycerol through a dehydration synthesis reaction. A fatty acid is considered to be **saturated** if it contains the maximum number of hydrogen atoms extending from the carbon backbone, whereas a fatty acid is **unsaturated** if it contains less than the maximum hydrogen atoms. Notice in FIGURE 3.4 that two of the chains are saturated and straight while the third is unsaturated and bent. Importantly, saturated fatty acids tend to make the fat solid at room temperature (e.g., butter) while unsaturated fatty acids tend to make the fat liquid at room temperature and would be then be called an oil (e.g., vegetable and fish oils).

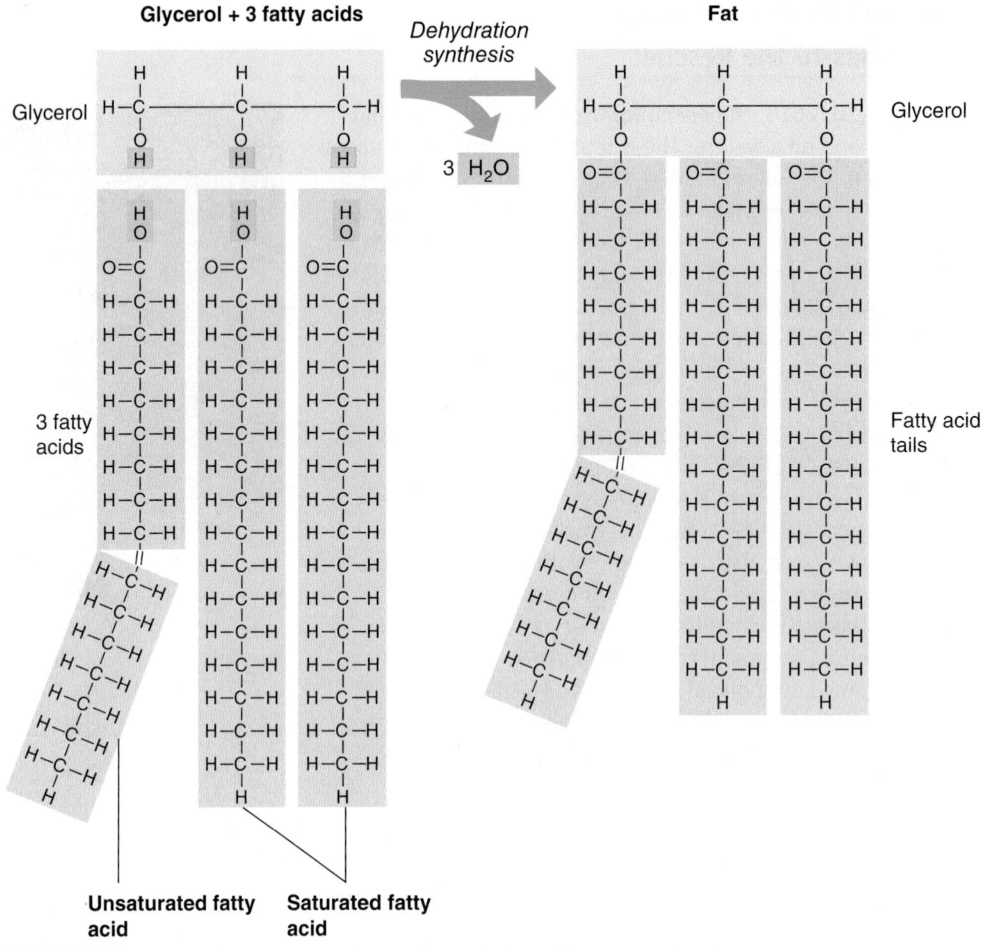

FIGURE 3.4 **Glycerol and Fatty Acids Combine to Form a Fat.** Glycerol is a three-carbon molecule and fatty acids are long carbon–hydrogen chains that can be saturated or unsaturated. Three fatty acid chains can combine with each glycerol through dehydration synthesis reactions to form fat.

Phospholipids

Phospholipids are phosphorus-containing lipids that possess a phosphate group (PO_4) in the place of one fatty acid chain (FIGURE 3.5a). The phosphate group makes the "head" end of the phospholipid hydrophilic, while the fatty acid tails are hydrophobic. This gives the phospholipid the property of being "amphipathic," meaning one portion is hydrophobic (the tails) and another portion is hydrophilic (the head) (FIGURE 3.5b). The amphipathic property of the phospholipid is the basis on which the cell forms a membrane in a watery environment. By organizing as a double-layer (or bilayer), phospholipids can accommodate a watery exterior with hydrophilic head groups toward the cell environment, and accommodate the watery interior of the cell with hydrophilic head groups toward the cytoplasm (FIGURE 3.5c). Such a configuration also allows the hydrophobic fatty acid tails to associate with one another and not be exposed to water.

Sterols

Other types of lipids include the **sterols**, which are very different from fats and phospholipids, and are included with lipids solely because they too are hydrophobic molecules. Sterols play structural roles in microbes by stabilizing cell membranes of a few bacterial species as well as the membranes of protists and fungi. The sterol cholesterol is found in the plasma membranes of human cells.

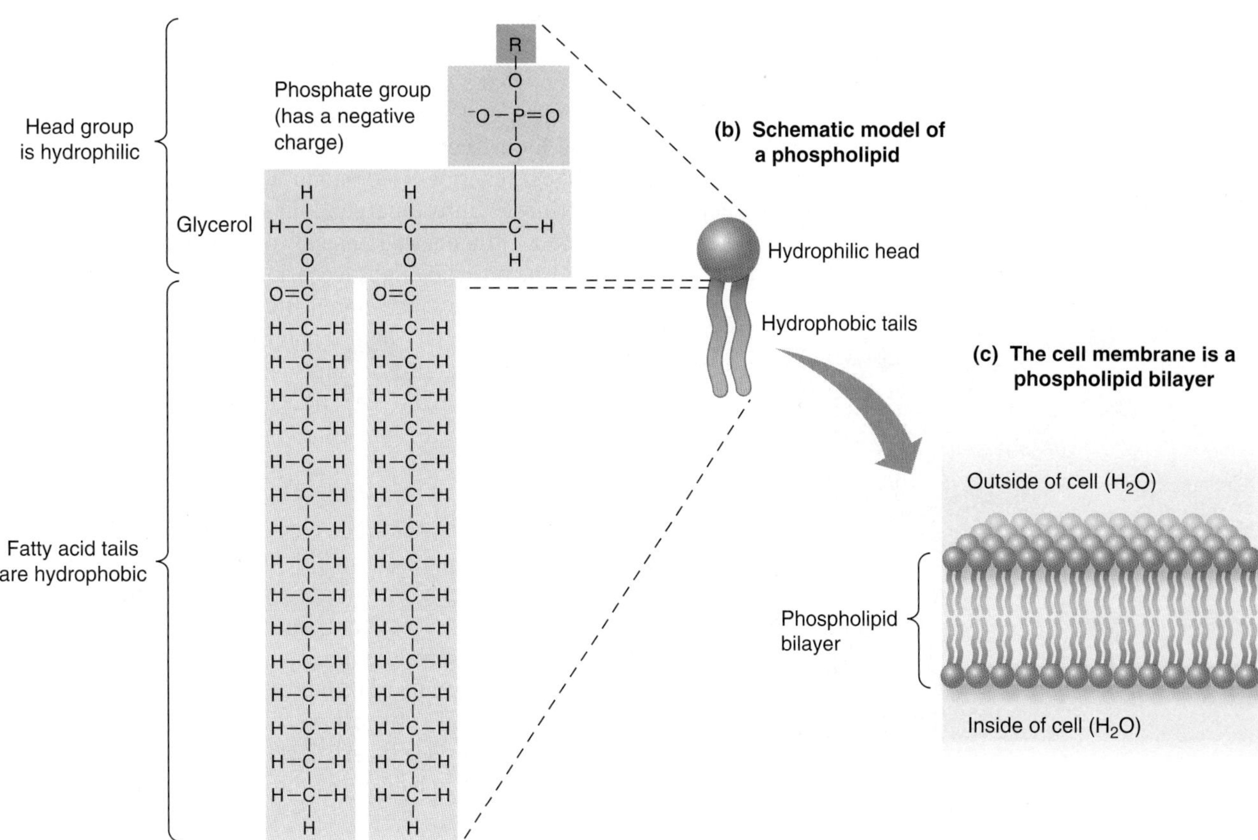

FIGURE 3.5 **Phospholipid, the Lipid of Cell Membranes.** (**a**) Phospholipids are composed of glycerol and two fatty acid tails and a charged phosphate group. The charge makes this region of the phospholipid hydrophilic. (**b**) A schematic drawing of a phospholipid with the glycerol and head group shown as a circle with the fatty acid tails extending downward. (**c**) The membrane of the cell is a phospholipid bilayer. This allows the hydrophilic head groups to associate with the watery exterior and interior of the cell.

3.4 Proteins: Amino Acids and Polypeptides

Proteins are the most abundant large molecules in all living organisms, including microorganisms. Composed of carbon, hydrogen, oxygen, nitrogen, and, usually, sulfur atoms, proteins make up about 60% of a microbial cell's dry weight, this high percentage indicating the essential and diverse roles of proteins. Many proteins function as structural components of cells and cell walls, and as transport agents in membranes. A large number of proteins also serve as enzymes that catalyze all the chemical reactions of metabolism, although sometimes we need to help metabolism. **A Closer Look 3.4** describes one embarrassing situation where microbial enzymes come to the rescue. In addition, viruses are composed of genetic information enclosed in a covering of protein.

■ dry weight: The weight of the materials in a cell after all the water is removed.

Amino Acids

All proteins are built from subunits called **amino acids**, whose fundamental structure is shown in FIGURE 3.6a. This structure includes one amino group (NH_2) and one carboxyl group (COOH) bonded together through a central carbon atom. The amino acids vary from one another on the basis of what atoms are attached as a side group to this central carbon. This side-group of atoms, known as the "R-group," can be as simple as a hydrogen in the case of glycine, or involve other combinations of atoms, some of which are shown

A CLOSER LOOK 3.4

"Not Without My Beano®!"

Some people would not dare sit down to a meal of corned beef and cabbage without a knife, fork, soda bread—and, of course, their Beano®. Nor would they have Brussels sprouts with their steak or broccoli with their fried chicken unless they were sure their Beano was nearby. Eating pasta e fagioli ("pasta and beans") without Beano? Some say—"No thanks!"

Beans are one source of raffinose carbohydrate.

Courtesy of Debora Cartagena/CDC.

So what's the problem? Cabbage, broccoli, Brussels sprouts, and other "strong tasting" vegetables, and high-fiber foods like beans, contain a family of oligosaccharides (3–5 monosaccharides in length) called raffinose. Unfortunately, humans lack the ability to produce the enzyme needed to break down raffinose. That means the oligosaccharides pass undigested into the large intestine where resident bacteria of our human microbiome have the enzyme to digest raffinose, but in the process also produce gas (methane, carbon dioxide, and hydrogen). Now unfortunate individuals pay a heavy price: gas (flatulence), bloating, embarrassment—and an unwillingness to go back for seconds.

Enter Beano, the trade name for an enzyme preparation produced from the mold *Aspergillus niger*. The enzyme breaks down raffinose into the simple sugars galactose, fructose, and glucose, and leaves nothing for the intestinal bacteria to work on. So, all it takes is a couple of Beano tablets just before eating the first bites of food (it tastes somewhat like soy sauce) and the fungal enzyme breaks down the raffinose, leaving a happy memory of the meal (or so says the manufacturer).

Aspergillus niger
a-spėr-jil'lus nī'jer

in FIGURE 3.5b for the amino acids alanine, valine, and lysine. Notice cysteine is one of the sulfur-containing amino acids. In all, up to 20 different combinations of atoms form the R-groups, which means there are 20 different amino acids available to build proteins.

Similar to the way polysaccharides are built from monosaccharide subunits, proteins are built from amino acids by means of dehydration synthesis reactions (FIGURE 3.7). The bond linking two amino acids is referred to as a **peptide bond**. By forming successive peptide bonds, more and more amino acids can be added to the growing chain. The final number of amino acids in a chain may vary from a very few (in which case, the small protein is called a "peptide") to thousands making a **polypeptide**. A protein then is composed of one or more polypeptides and an extraordinary variety of proteins can be formed from the 20 available amino acids.

Polypeptides and Protein Shape

Because proteins play diverse roles and carry out most of the metabolic activities of the cell, they have tremendous differences in shape. Protein shape can be defined on three or four levels of structure.

Primary Structure. The specific sequence of amino acids in a polypeptide is referred to as the **primary structure** (FIGURE 3.8a). This sequence is unique to each polypeptide and is determined by the genetic information in each gene coding for a

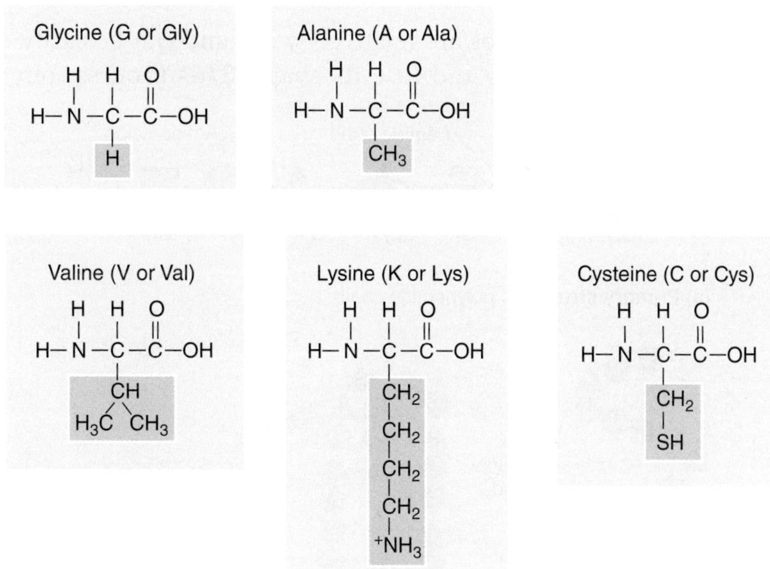

FIGURE 3.6 **The Structure of Amino Acids.** (**a**) All amino acids have the same basic structure, but each varies by the R-group, which is a set of atoms attached to the central carbon. (**b**) Five of the 20 amino acids are shown with their unique R-group.

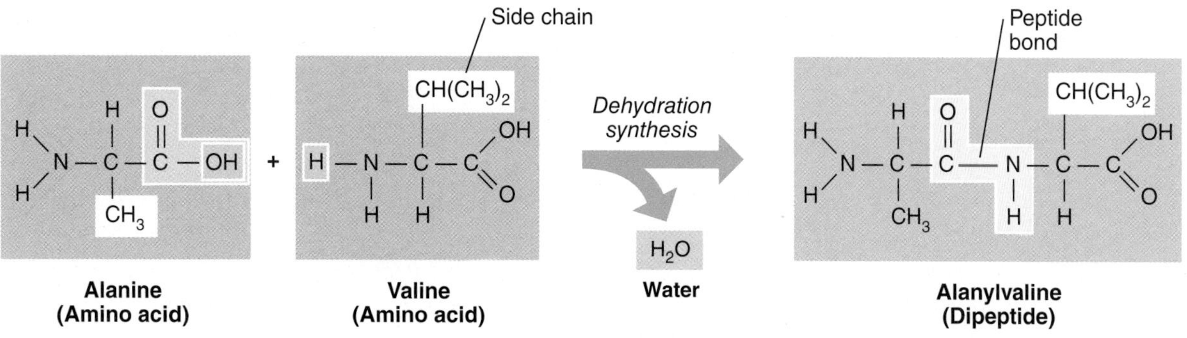

FIGURE 3.7 **Formation of a Dipeptide.** The amino acids alanine and valine are shown. The OH group from the acid group of alanine combines with the H from the amino group of valine to form water. The carbon atom of alanine and the nitrogen atom of valine then link together, yielding a peptide bond. Continued dehydration synthesis reactions will form a polypeptide.

unique sequence of amino acids. However, a long chain of amino acids does not, in itself, form the final shape of the protein.

Secondary Structure. The amino acids in the primary sequence can interact with one another, forming the polypeptide's **secondary structure** (FIGURE 3.8b).

Often this secondary structure takes on the form of a helix (coil) or a folded sheet-like structure.

Tertiary and Quaternary Structure. Most polypeptides can further fold to form the **tertiary structure** (FIGURE 3.8c). This three-dimensional shape depends on the interactions between R-groups of various amino acids in different regions of the polypeptide, giving the polypeptide a globular or fibrous structure.

In some cases, two or more polypeptides bond together to form the final functional protein and this represents the **quaternary structure**. Examples include the red blood cell protein hemoglobin and antibodies, both of which consist of four polypeptides.

The bonds between R-groups are relatively weak and can be easily disrupted, causing the protein to unravel and lose its shape. This process, referred to as

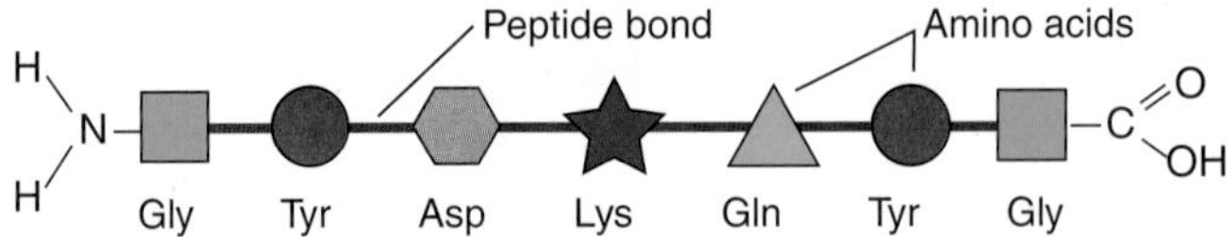

(a) Primary structure: polypeptide chain

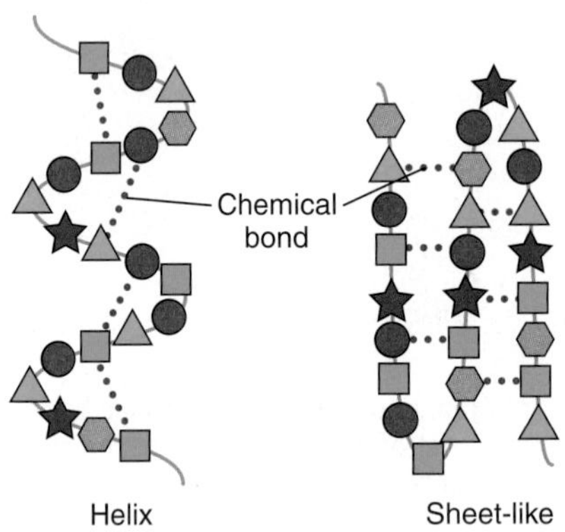

(b) Secondary structure

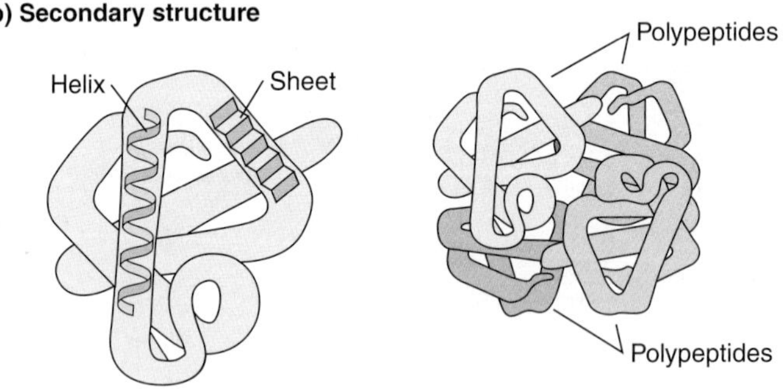

(c) Tertiary and quaternary structure

FIGURE 3.8 **Protein Structure.** (**a**) The primary structure refers to the sequence of amino acids. (**b**) Interactions between amino acids can cause changes in shape of the polypeptide. Called the secondary structure, the shape is usually a helix or sheet-like structure. (**c**) A polypeptide can further fold back on itself through interactions between R-groups, forming the so-called tertiary structure. In addition, some proteins require more than one polypeptide to function, and the polypeptide configuration of these proteins is known as the quaternary structure.

denaturation, is brought about by heat, antiseptics, disinfectants, or other agents that alter the protein's surrounding environment. Loss of protein function may kill microbes or inhibit their growth. Also, a high fever (104°C or higher) can be fatal to an individual because blood proteins and enzymes are subject to denaturation and loss of function at high body temperatures.

3.5 Nucleic Acids: DNA and RNA

Nucleic acids are the fourth major group of large molecules found in all organisms and they are composed of carbon, hydrogen, oxygen, and nitrogen atoms. The two important nucleic acids are **deoxyribonucleic acid (DNA)** and **ribonucleic acid (RNA)**. DNA is the nucleic acid of which chromosomes are composed and RNA is the nucleic acid involved in converting the information in a gene into a polypeptide.

Nucleic acids are built from subunits called **nucleotides** (FIGURE 3.9a). Each nucleotide is composed of three parts: a pentose sugar, a phosphate group (PO_4), and a nitrogen-containing molecule called a nitrogenous base, or simply, a **nucleobase**. The pentose sugar in DNA is deoxyribose, while in RNA it is ribose.

The four nucleobases in DNA are adenine, guanine, cytosine, and thymine; in RNA, they are adenine, guanine, cytosine, and uracil. (Note that DNA has thymine but no uracil, and RNA has uracil but no thymine.) Adenine and guanine are double-ring molecules called "purines," while cytosine, thymine, and uracil are single-ring molecules called "pyrimidines."

The phosphate group found in nucleic acids links the sugars to one another in both DNA and RNA (FIGURE 3.9b). The chain of alternating sugar and phosphate subunits forms the so-called sugar-phosphate "backbone."

To visualize a DNA molecule, picture a ladder. In the molecule, two sugar-phosphate backbones make up the sides of the ladder, and the rungs (the steps) of the ladder are composed of the nucleobases (FIGURE 3.9c). On one side of each rung is a purine molecule, and on the other side is a pyrimidine molecule. Thus, an adenine molecule always bonds with a thymine molecule (and vice versa), and a guanine molecule with a cytosine molecule (and vice versa). The ladder is then twisted into a spiral staircase-like structure called a **double helix**. Microbes and all other living organisms have their DNA in this form.

RNA is a single-stranded molecule with a single sugar-phosphate backbone from which protrudes the nucleobases. TABLE 3.1 summarizes the differences between DNA and RNA.

Like proteins, nucleic acids cannot be denatured without injuring the microbe or killing it. For example, ultraviolet (UV) and gamma ray radiations damage or break DNA and can be used to lower the microbial population on an environmental surface or sterilize food products. Chemicals such as formaldehyde alter the nucleic acids of viruses and can be used to produce some viral vaccines. Moreover, certain antibiotics interfere with nucleic acid activity, thereby killing bacteria. We shall encounter many other instances where tampering with nucleic acids or with the other key organic molecules of a microbe leads to its destruction.

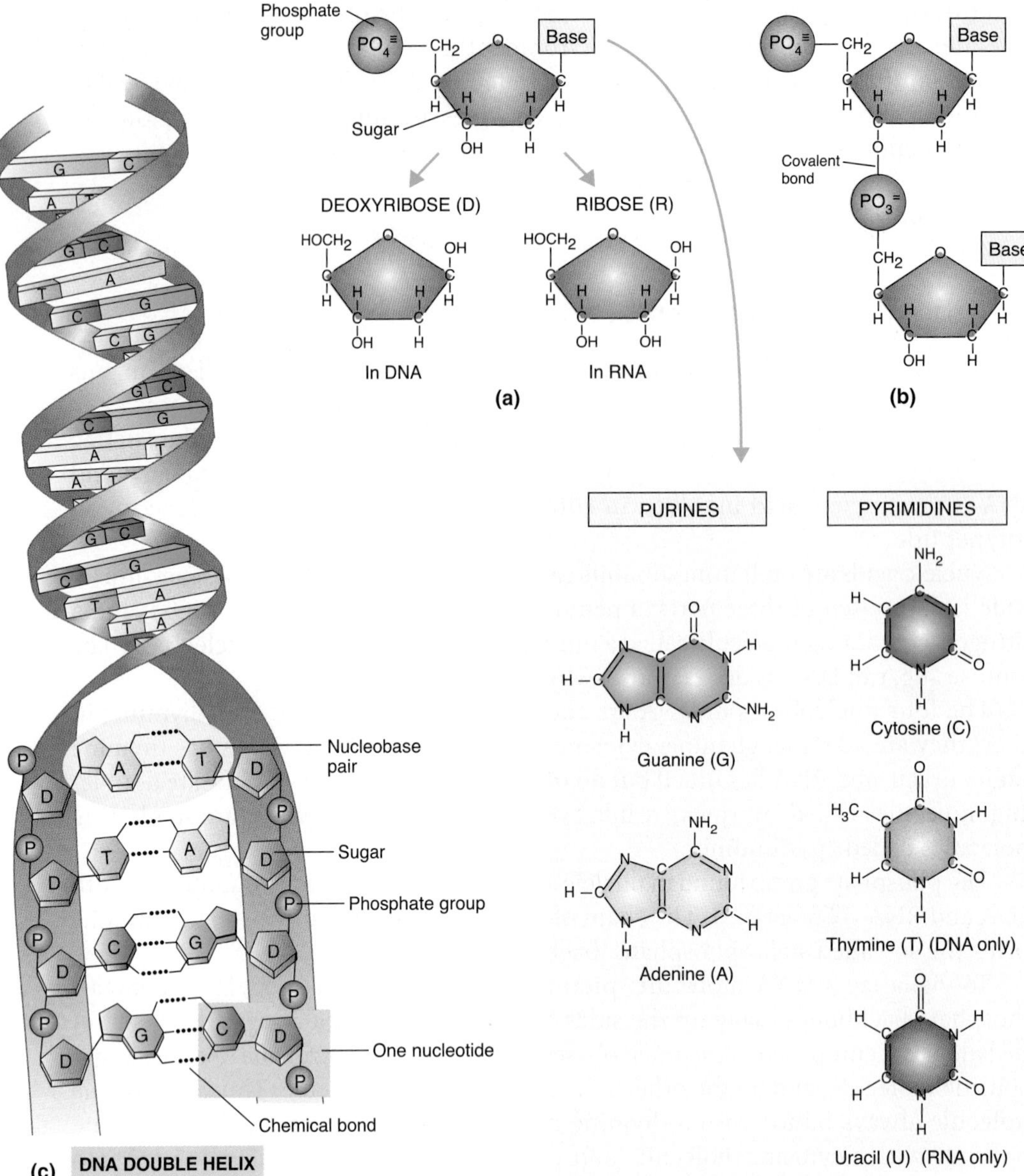

FIGURE 3.9 **The Molecular Structures of Nucleotide Components and the Construction of DNA.** (**a**) The sugars in nucleotides are ribose and deoxyribose, which are identical except for one additional oxygen atom in RNA. The nucleobases include the purines adenine and guanine and the pyrimidines thymine, cytosine, and uracil. (**b**) Nucleotides are bonded together by dehydration synthesis reactions. (**c**) The two polynucleotides of DNA are held together by chemical bonds between adenine (A) and thymine (T) or guanine (G) and cytosine (C) to form a double helix.

TABLE 3.1 A Comparison of DNA and RNA

Property	DNA	RNA
Pentose sugar	Deoxyribose	Ribose
Nucleobases	Adenine (A), guanine (G), cytosine (C), thymine (T)	Adenine (A), guanine (G), cytosine (C), uracil (U)
Number of polynucleotides	Two (double helix)	One

A Final Thought

We could probably talk about microbes without talking about their chemistry, but it would be like trying to describe a Big Mac® without knowing what is in the hamburger. We realize that to some people, the word "chemistry" is equivalent to "root canal," but we also know chemical molecules are the nuts and bolts of all living organisms—and viruses, too.

In other chapters, we shall discuss milk products containing "carbohydrates," membranes composed of "lipids," antibodies consisting of "proteins," genes composed of "nucleic acids," and a host of other concepts that include a smattering of chemistry. To understand how yeasts cause bread to rise, we must understand the chemistry of the process; and to explain genetic engineering to our friends, we must know a bit of the chemistry behind the process. Viral replication is centered in chemistry; the production of yogurt is a chemical process; and the process of disinfection is based in chemistry.

We recommend you give this chapter on chemistry a careful reading. In succeeding chapters, you will find that your investment of time was worthwhile.

Questions to Consider

1. Polysaccharides are important sources of sugar for energy; however, polysaccharides are also important structural components of the cell. Give an example of at least two structural polysaccharides. Which sugars are linked together to form these polysaccharides?
2. The cell membrane is made of phospholipids. What unique properties of phospholipids make this molecule uniquely adept at forming cell membranes?
3. Suppose you had the option of destroying one type of organic molecule in a bacterial species as a way of eliminating the microbe. Which type of molecule would you choose and why?
4. If proteins are all long chains of amino acids, then how can different proteins have different shapes and take on different functions?
5. Oxygen comprises about 65% of the weight of a living organism. This means a 120-pound person contains 78 pounds of oxygen. How can this be?
6. The toxin associated with the foodborne disease botulism is a protein. To avoid botulism, home canners are advised to heat preserved foods to boiling for at least 12 minutes. How does the heat help?

Key Terms

Informative facts are necessary for the expression of every concept, and the information for a concept is founded in a set of key terms. The following terms form the basis for the concepts of this chapter. On completing the chapter, you should be able to explain and/or define each one.

amino acid
aqueous solution
atom
carbohydrate
cellulose
chemical bond
dehydration synthesis reaction
denaturation
deoxyribonucleic acid (DNA)
disaccharide
double helix
electron
element
enzyme

fat
hydrolysis reaction
hydrophilic
hydrophobic
lipid
molecule
monosaccharide
neutron
nucleic acid
nucleobase
nucleotide
oil
organic
peptide bond
peptidoglycan
phospholipid
polypeptide
polysaccharide
primary structure
protein
proton
quaternary structure
ribonucleic acid (RNA)
saturated
secondary structure
starch
sterol
tertiary structure
unsaturated

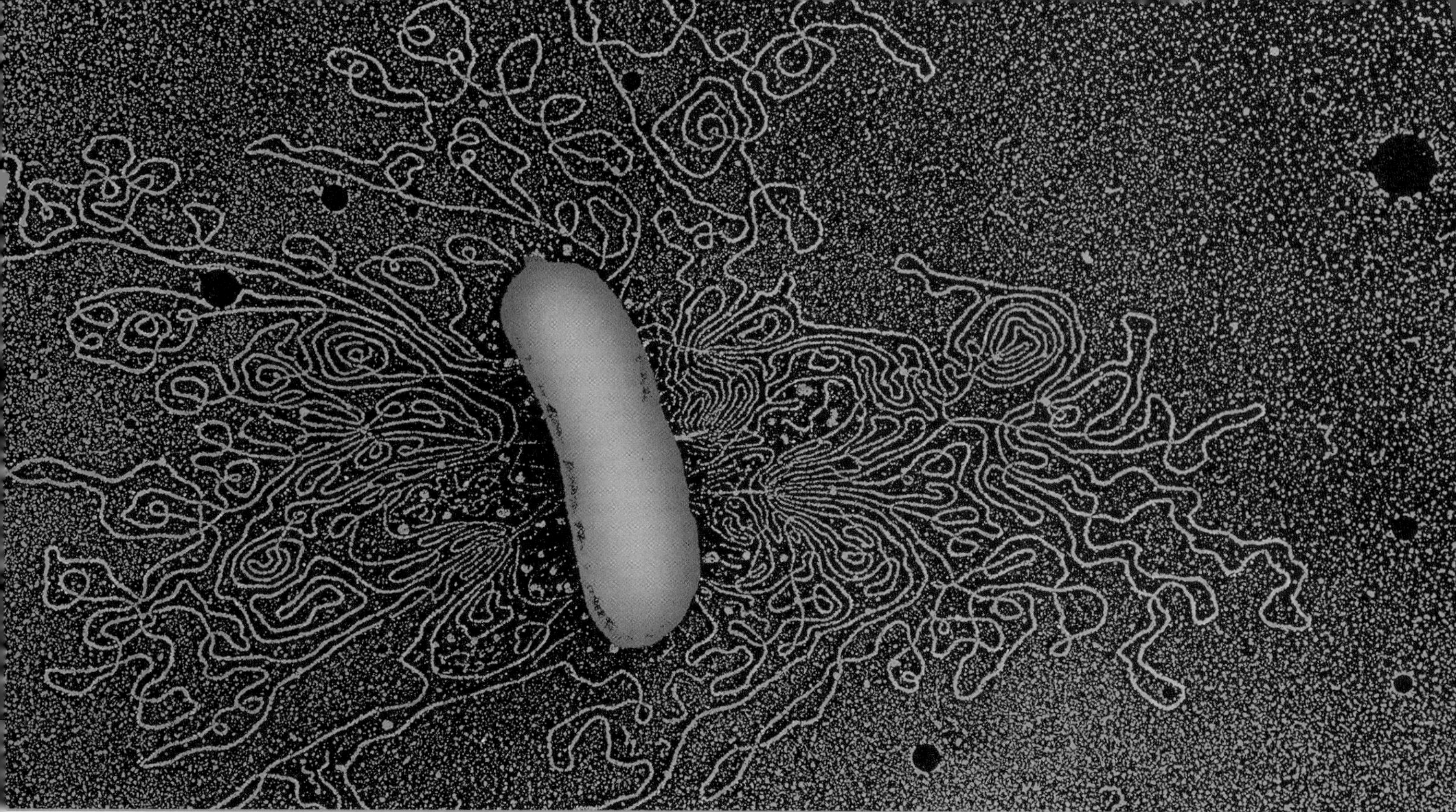

The DNA Story: Chromosomes, Genes, and Genomics

4

DNA is the molecule that codes the genetic information of all living organisms, including microbes, and some viruses. Here, we see an electron microscope image of an *Escherichia coli* cell that has burst open releasing its DNA. The DNA fiber you see represents a single, circular molecule.

Looking Ahead

Microbes have made vital contributions to our understanding of all life by serving as the nuts and bolts of biochemical research. Nowhere is this more evident than in the remarkable advances in genetic engineering and biotechnology. This chapter explores the background of these modern technologies.

On completing this chapter, you should be capable of:

- Describing the experiments identifying DNA as the hereditary material in cells.
- Discussing the evidence that led to the discovery of the structure of DNA.
- Distinguishing between transcription and translation.
- Listing examples illustrating the "microbial nature" of the human genome and discussing the Human Microbiome Project.

In 1853, Gregor Mendel, an Austrian monk in the Augustinian order at the St. Thomas Monastery in Brünn (now Brno in the Czech Republic), had been put in charge of the monastery's experimental garden (FIGURE 4.1). Being the son of a farmer, Mendel had a special liking for natural science and agriculture. As such, while tending to the garden, he was curious as to how plants acquired their characteristics. He knew a plant's respective offspring retained the essential traits of the parents and were not influenced by the environment. But, how were these traits inherited?

At this time, it was customary to explain inheritance by saying, "It's all in the blood." People believed something in the blood was responsible for the blending of parental characteristics. Such references are still heard today, in phrases such as "blood relations" and "blood lines." However, the fact that semen contains no blood made it difficult to explain how blood could be the inheritance "factor." But if blood was not the substance of heredity, then what was?

■ semen: The thick white fluid containing sperm that a male ejaculates.

In 1854, Mendel set out to test his ideas of inheritance using the pea plants growing in the monastery's experimental garden. He studied seven traits in these plants (e.g., plant height, flower color, seed color) during an extensive series of breeding experiments that took place over eight years. His experiments led him to suspect the plant's reproductive cells had some type of "factors" controlling heredity. Further, these factors occur in pairs and express themselves as traits in the plants and their offspring. From his work, Mendel developed a theory of inheritance completely at odds with a blood origin. He reported on his plant hybrid experiments to the Brno Natural Science Society in 1865 and the society published his paper in 1866. Mendel's work provided the first inkling for inherited factors as discrete units transmitted from parent to offspring.

FIGURE 4.1 **At the Roots of DNA and Genetics.** Gregor Mendel, the Austrian monk who established the principles of genetics through meticulous experiments with pea plants.

© Pictorial Press Ltd/Alamy.

Mendel's work on inheritance in pea plants was not appreciated in his lifetime. In fact, it was not until 1900, 16 years after his death, that three agricultural scientists discovered his 1866 paper. Verifying his experiments, the "age of genetics" was born and in the next several decades, scientists would learn more about these inherited factors and the special substance called **deoxyribonucleic acid** (**DNA**) that carried each living organism's specific traits. Mendel's theory of inheritance ranks with the germ theory of disease, the cell theory, and the theory of evolution in changing the way we and society perceive the biological world.

To understand the great revolution taking place in genetics today, we must examine the role of DNA and its functioning inheritance factors, the genes. Then, we will explore how the genetic message in DNA yields the biochemical message in protein and see how scientists have used their understanding of DNA to develop the new discipline of genomics.

And at the center of the wave of discoveries about DNA were the microbes, which often were used to test the new ideas in genetics because they provided a simple way to verify or refute newly discovered genetic processes. Furthermore, microbes were the sources of the chemical "nuts and bolts" that made it possible to replicate the genetic processes in test tubes. As models for other forms of life, microbes provided many of the answers about the secrets of **heredity**, that is, how genetic traits are transmitted from one generation to another. It is a thrilling story of science and society, and one that is current as today's headlines.

4.1 The Roots of DNA Research: Peas, Fruit Flies, and Microbes

Today's students discuss DNA as if scientists always knew about it. Over the years, the term DNA has become part of everyday speech, and DNA is as familiar as "Big Mac®." Of course, it was not always that way. Scientists spent decades figuring out the significance of DNA and the processes it controls. In this section, we shall examine some of their work, especially focusing on the microbial contribution.

Morgan's Fruit Flies

As mentioned above, in 1900 Mendel's work on inheritance with pea plants came to light and was verified. Within a few short years, scientists tentatively connected Mendel's inheritance factors to a cell's **chromosomes**, the thread-like strands of DNA occurring in all cells. However, scientists were uncertain whether chromosomes contained the units of heredity until Thomas Hunt Morgan performed landmark experiments with fruit flies in 1910. Morgan determined that among the four pairs of fruit fly chromosomes is the information for gender, eye color, and many traits. Morgan's work provided the first evidence relating traits to chromosomes, placing the theory of inheritance on a firmer footing.

At Morgan's time, geneticists believed only a part of a chromosome specified a trait because if one chromosome carried but one trait there simply were not enough chromosomes in cells to account for the hundreds of traits of any one individual—even in microbes. Therefore, each chromosome must carry multiple traits, like beads on a string, and before long the concept of the gene emerged. **Genes** were seen as being the hereditary units on a chromosome and represented separate segments of a chromosome each containing the information for one trait, be it height of pea plants or eye color in fruit flies. But chromosomes were known to be made of DNA and protein. Which was the source for the hereditary information? Microbes would answer the question.

Microbes as Experimental Systems

In 1928, the English bacteriologist Frederick Griffith reported some puzzling results of experiments with bacteria. Griffith was working with the pneumococcus *Streptococcus pneumoniae*. One strain produced a thick, carbohydrate coat called a capsule. When Griffith injected this smooth-coated strain (S strain) into mice, the animals soon died of pneumonia. Another strain without a capsule (rough or R strain) was harmless when injected into mice (FIGURE 4.2). In the most interesting experiment, Griffith took the cellular debris from the dead pathogenic strain and mixed it with the live harmless strain. When injected into mice, the mice soon died as the harmless strain had been transformed into a capsule containing, pathogenic strain. Somehow the genetic information to make a capsule was transferred to (inherited by) the harmless strain, transforming the bacterial cells into a deadly form. But Griffith was unable to isolate the transforming factor. Years later, in 1944, Oswald Avery and his research group identified the transforming substance as DNA. When he used enzymes that digest DNA, the effect of the transforming substance was lost.

■ *Streptococcus pneumoniae*
strep-tō-kok′kus nü-mō′nē-ī

Finally, to pin down DNA's role, Alfred Hershey and Martha Chase worked with the bacterium *Escherichia coli* and a virus called a **bacteriophage** (or **phage** for short) that attacks and infects the bacterial cells. Scientists knew phages use the infected bacterial cells as chemical factories for producing more viruses. They also knew phages consist of nothing more than a core of DNA enclosed in a coat of protein. What they did not know was whether the DNA or the protein was the hereditary information to produce new viruses. **A Closer Look 4.1** looks at their experiments.

■ *Escherichia coli*
esh-e˙r-ē′kē-ä kō′lī (or kō′lē)

Certain experiments stand out as tipping points in scientific history, and the Hershey–Chase experiments are an example. The results obtained in 1952 had great influence on the biochemical thinking of the era because their work solidified the understanding of DNA's role in heredity.

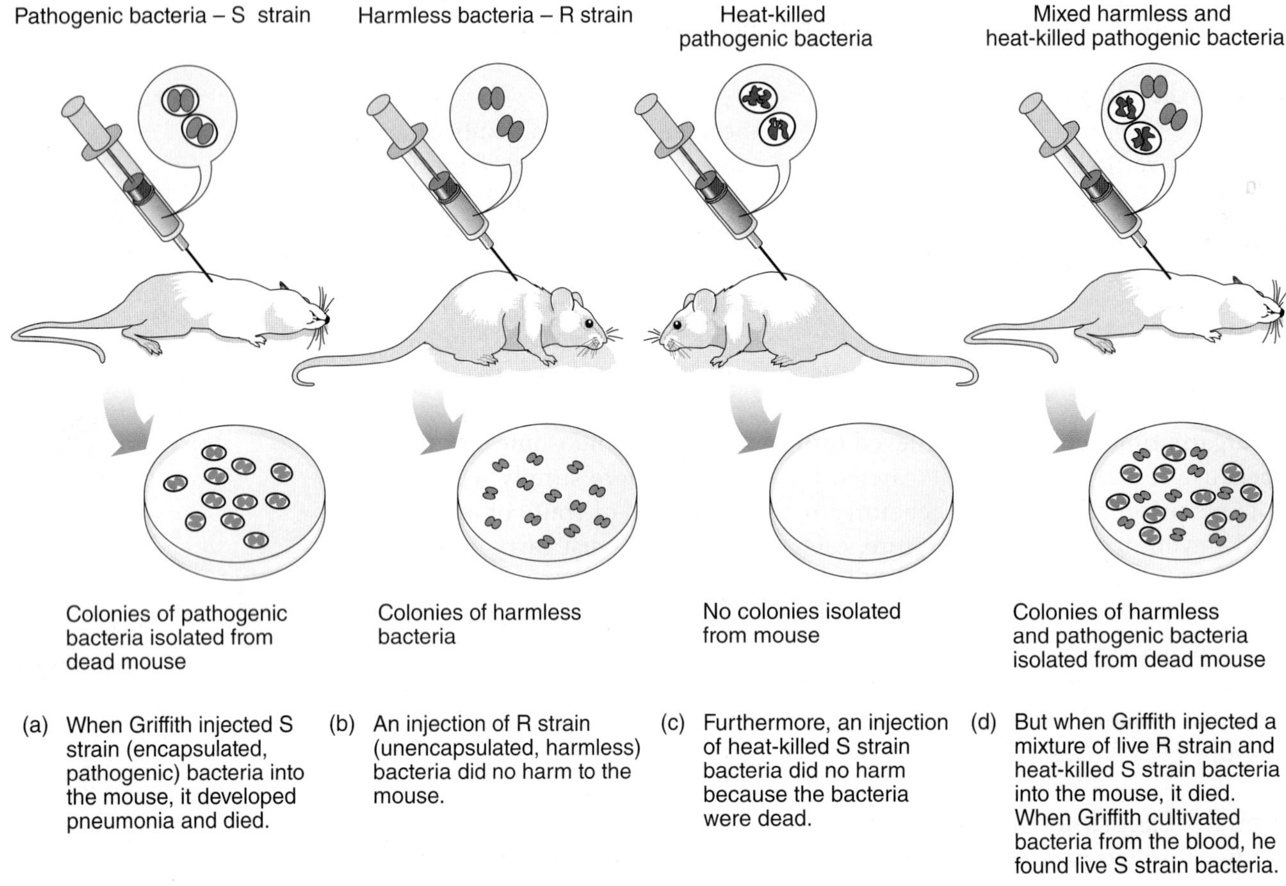

Pathogenic bacteria – S strain
Harmless bacteria – R strain
Heat-killed pathogenic bacteria
Mixed harmless and heat-killed pathogenic bacteria
Colonies of pathogenic bacteria isolated from dead mouse
Colonies of harmless bacteria
No colonies isolated from mouse
Colonies of harmless and pathogenic bacteria isolated from dead mouse
(a) When Griffith injected S strain (encapsulated, pathogenic) bacteria into the mouse, it developed pneumonia and died.
(b) An injection of R strain (unencapsulated, harmless) bacteria did no harm to the mouse.
(c) Furthermore, an injection of heat-killed S strain bacteria did no harm because the bacteria were dead.
(d) But when Griffith injected a mixture of live R strain and heat-killed S strain bacteria into the mouse, it died. When Griffith cultivated bacteria from the blood, he found live S strain bacteria.

FIGURE 4.2 **The Transformation Experiments of Griffith.** Griffith's experiments were among the first to demonstrate the transfer and inheritance of genetic information using a bacterial species.

4.2 DNA: Its Structure and Duplication

Once it was certain DNA was the molecule of heredity, scientists needed to discover its chemical structure because knowing the structure might explain how DNA directs the hereditary activities of the cell.

The Double Helix

Since the 1920s, DNA was known to consist of three parts: a five-carbon sugar (deoxyribose); a number of phosphate groups (PO_4); and a series of molecules called nitrogenous bases (called nucleobases, or just bases for short). The bases were adenine (A), thymine (T), guanine (G), and cytosine (C); adenine and guanine are double-ring molecules known as purines; cytosine and thymine are single-ring molecules called pyrimidines (FIGURE 4.3a).

DNA also was known to contain roughly equal proportions of phosphate groups, deoxyribose molecules, and bases. Therefore, scientists correctly concluded that DNA must be composed of these three components joined to one another as a unit. The unit was known as a **nucleotide** with the base distinguishing one nucleotide from another (FIGURE 4.4b).

In the early 1950s, almost nothing was known about the spatial arrangement of the nucleotides in DNA. About this time, a new technique called X-ray diffraction became available. In X-ray diffraction, crystals of a chemical substance are bombarded with X rays, which are scattered (or diffracted) by electrons in the chemical substance,

A CLOSER LOOK 4.1

Eureka! It Is DNA

Up until about 1952, scientists debated whether DNA or protein was the genetic material. Even though Avery's experiment strongly pointed to DNA, publication of the work had little impact because it was not clear yet that bacterial cells even had genes. More importantly, how could the chemical complexity of any organism be determined by just the four nucleotides (letters) present in DNA when proteins have 20 amino acids (letters)? By analogy, it would be hard to build the entire English language from just four letters rather than the 26 we actually have in the English alphabet.

With the realization that bacterial cells (and viruses) do have genes just like pea plants and fruit flies, Alfred Hershey and Martha Chase of the Department of Genetics at Cold Spring Harbor on Long Island set out in 1952 to determine whether DNA or protein was the genetic material. They used *Escherichia coli* and a virus (phage) that infects the bacterial cells and then uses those cells as factories to make more viruses. Importantly, Hershey and Chase made use of the fact that the DNA in the core of the phage contains phosphorus but no sulfur atoms and protein in the phage covering or coat has sulfur but no phosphorus atoms. They then produced phages that had radioactive phosphorus (^{32}P) in their DNA and radioactive sulfur (^{35}S) in their protein coat. Radioactive phages were then mixed with *E. coli* cells and the microbiologists waited just long enough for new phages to be made in the infected cells (see figure). At this point, they removed any remaining phages and debris from the surface of the bacterial cells.

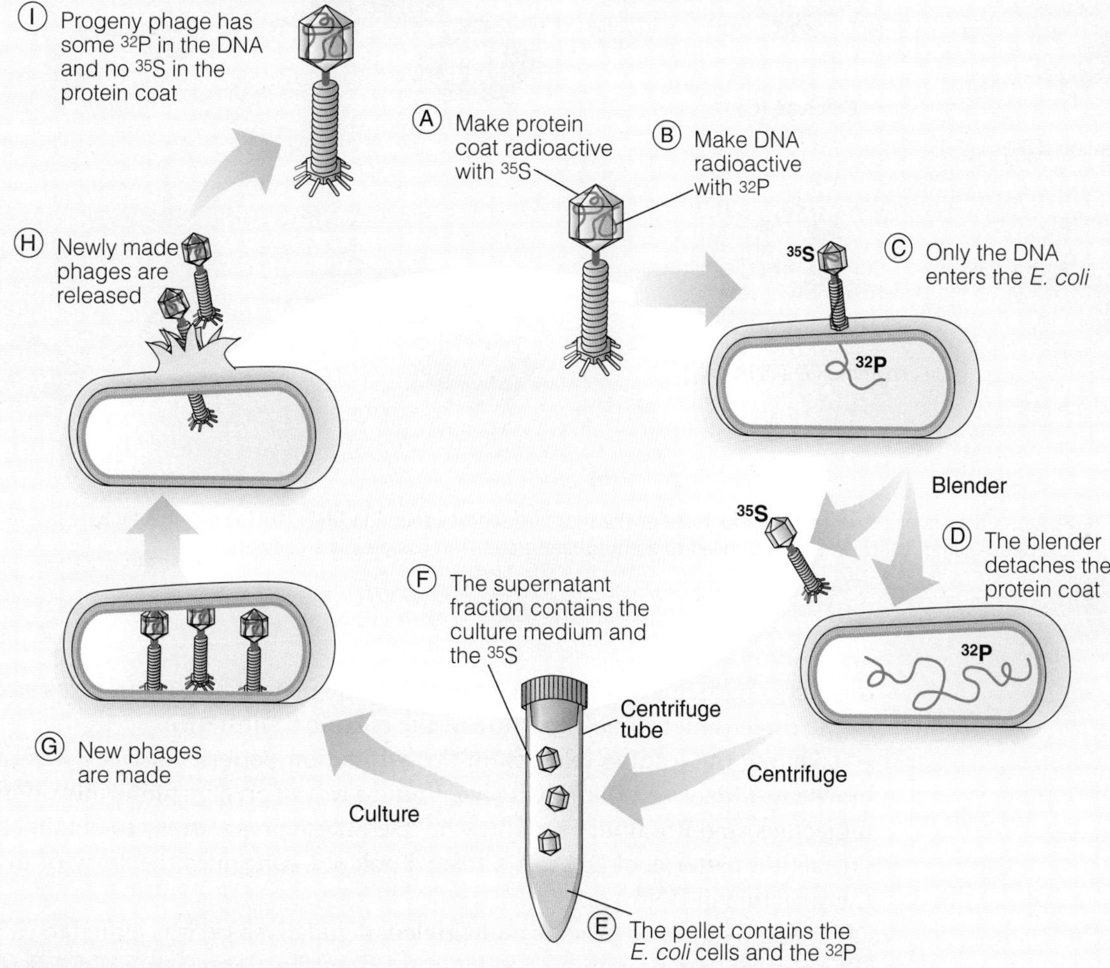

The Hershey-Chase experiment with viruses and their host bacterium *Escherichia coli*.

Now Hershey and Chase examined the *E. coli* cells and the surrounding fluid to find out where the radioactive substances were located. They discovered that most of the radioactive phosphorus (i.e., the DNA) was in the cytoplasm of the *E. coli* cells and most of the radioactive sulfur (i.e., the protein) was in the virus coats in the surrounding fluid. Thus, their observations indicated that only DNA entered the bacterial cells while the protein coat of the phage remained outside. Because the genetic information to make new phages would have to get into the cells' cytoplasm, Hershey and Chase concluded that virus DNA—and only DNA—is the substance responsible for producing more phages; protein apparently has no role in the process.

The question was answered: the hereditary information in cells is DNA.

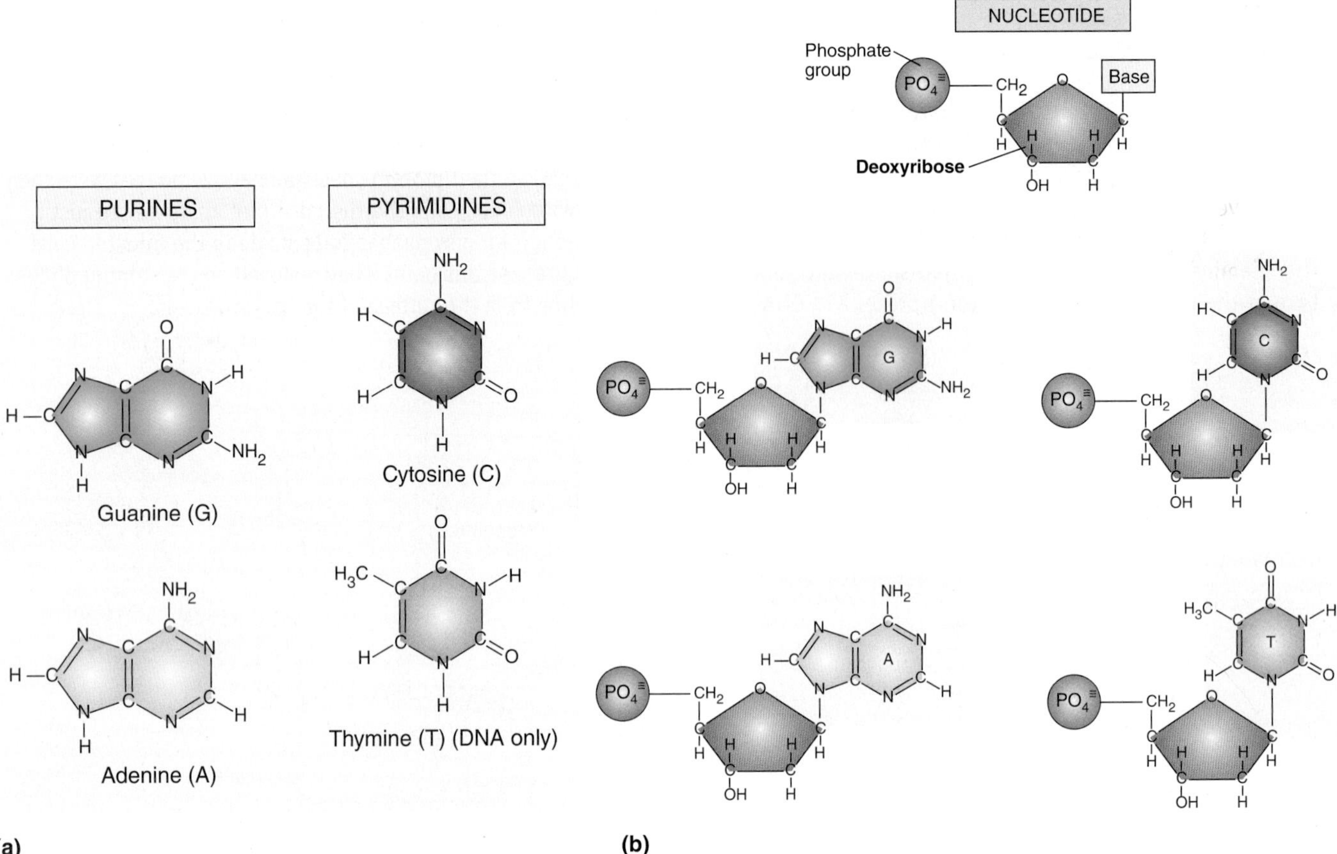

FIGURE 4.3 **The Building Blocks of DNA.** (**a**) The structures of the four nucleobases found in DNA. (**b**) The DNA nucleobases are bonded to a deoxyribose sugar, which in turn is bonded to a phosphate group. The complex is a nucleotide.

forming a particular pattern on a photographic plate. The pattern gives strong clues to the three-dimensional structure of the chemical substance.

Among the leading experts on the diffraction patterns of DNA was the British biochemist Maurice Wilkins who had found a way to prepare uniformly oriented DNA molecules, and Rosalind Franklin who used these preparations to obtain clear X-ray diffraction patterns of DNA. **A Closer Look 4.2** continues the story of discovering the structure of DNA.

Using X-ray data, Watson and Crick concluded a base was joined to each deoxyribose sugar, hanging out as a side group from the chain, as FIGURE 4.4a illustrates. Next, they knew Franklin's X-ray diffraction data suggested DNA was a **double helix**, so somehow the bases must interact.

At this point, the observations made years before by biochemist Erwin Chargaff came into the picture. Chargaff reported that in DNA, the amounts of adenine and thymine are nearly identical, as are the amounts of cytosine and guanine. Watson and Crick thus envisioned that for every adenine molecule on one helix of DNA, there must be a thymine molecule on the opposing helix (and vice versa). This would be the "complementary" base. Similarly, for every guanine molecule on one helix of the DNA molecule, there must be a complementary cytosine molecule on the other (and vice versa). The weak chemical bonds formed between the opposing bases held the two helices together. The model was complete (FIGURE 4.4b, c).

A CLOSER LOOK 4.2

The Tortoise and the Hare

We all remember the children's fable of the tortoise and the hare. The moral of the story was those who move along slowly and methodically (the tortoise) will win the race over those who are speedy and impetuous (the hare). The race to discover the structure of DNA is a story of collaboration and competition—a science "tortoise and the hare."

Rosalind Franklin (the tortoise) was 31 when she arrived at King's College in London in 1951 to work in J. T. Randell's lab. Having received a Ph.D. in physical chemistry from Cambridge University, she moved to Paris where she learned the art of X-ray crystallography (see figure). At King's College, Franklin was part of Maurice Wilkins's group and she was assigned the job of using X-ray crystallography to work out the structure of DNA fibers. Her training and constant pursuit of excellence allowed her to produce superb, high-resolution X-ray images of DNA.

Rosalind Franklin.

Meanwhile, at the Cavendish Laboratory in Cambridge, James Watson (the hare), an American post-doctoral student, was working with a British graduate student, Francis Crick on the structure of DNA. Watson, who was in a rush for honor and greatness that could be gained by figuring out the structure of DNA, had a brash "bull in a china shop" attitude. This was in sharp contrast to Franklin's philosophy where you don't make conclusions until all of the experimental facts have been analyzed. Therefore, until she had all the facts, Franklin was reluctant to share her data with Wilkins—or anyone else.

Perhaps feeling left out, Wilkins was more than willing to help Watson and Crick. Because Watson thought Franklin was "incompetent in interpreting X-ray photographs" and he was better able to use the data, Wilkins shared with Watson an X-ray image and report that Franklin had filed. From these materials, it was clear that DNA was a helical molecule, something that Franklin knew but, perhaps being a physical chemist, she did not grasp its importance because she was concerned with getting all the facts first and making sure they were absolutely correct. But, looking through the report that Wilkins shared, the proverbial "light bulb" went on when Crick saw what Franklin had missed; that the two DNA strands formed a double helix. This knowledge, together with Watson's ability to work out the base pairing, led Watson and Crick to their "leap of imagination" and the structure of DNA.

In her book entitled, *Rosalind Franklin: The Dark Lady of DNA* (HarperCollins, 2002), author Brenda Maddox suggests it is uncertain if Franklin could have made that leap as it was not in her character to jump beyond the data in hand. In this case, the leap of intuition won out over the methodical data collecting in research—the hare beat the tortoise this time. However, it cannot be denied that Franklin's data provided an important key from which Watson and Crick made the historical discovery.

In 1962, Watson, Crick, and Wilkins received the Nobel Prize in Physiology or Medicine for their work on deciphering the structure of DNA. Should Franklin have been included? The Nobel Prize committee does not make awards posthumously and Franklin had died 4 years earlier from ovarian cancer. So, if she had lived, did Rosalind Franklin deserve to be included in the award? But then who would have been left out, as the Nobel Prize cannot have more than three recipients for any one award?

Watson and Crick expressed their hypothesis in a seminal paper published in 1953 in the prominent journal *Nature* (FIGURE 4.5). The scientific community hailed their work as a great scientific leap forward. It made it possible for biochemists to postulate how DNA replicates and passes the genetic information onto the next generation.

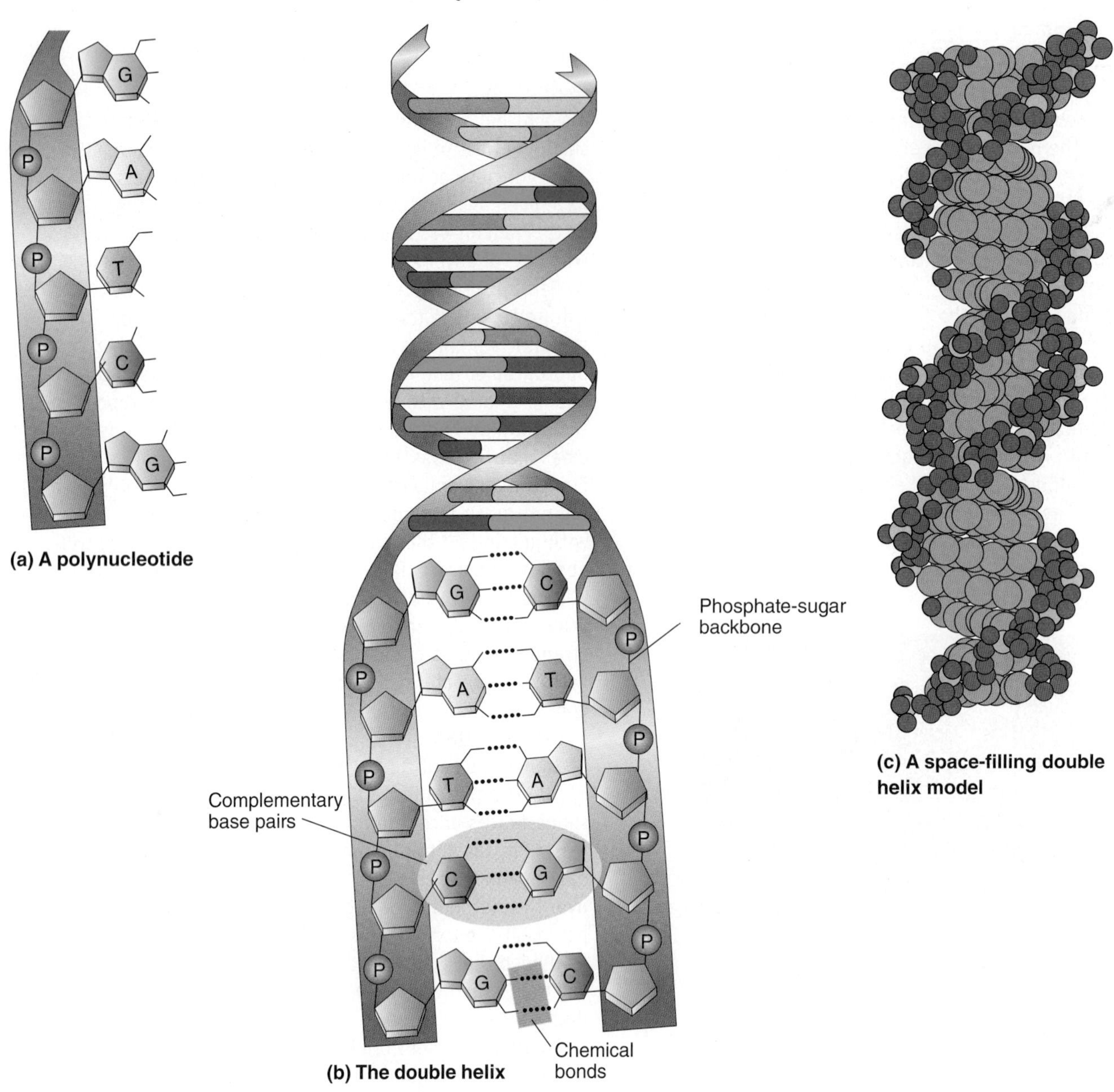

FIGURE 4.4 **The DNA Molecule.** (**a**) The nucleotides are bonded together into a polynucleotide chain. (**b**) A stylized model of DNA. The molecule consists of a double helix of two intertwined polynucleotide strands. Complementary bases extend out and toward one another by forming A-T and G-C base pairs. (**c**) A space-filling model of the DNA molecule.

DNA Replication

Toward the end of their article in *Nature* magazine, Watson and Crick pointed out that the structure of DNA might provide insight into how the molecule copies (replicates) itself. They implied that if the two polynucleotides of the double helix were separated, each could provide the information (template) necessary to synthesize a new complementary polynucleotide strand (FIGURE 4.6a). For example, an adenine molecule always pairs with a thymine molecule, so the adenine on one strand would specify a complementary thymine on a new strand.

In 1956, Arthur Kornberg discovered the enzyme that participates in the copying of the DNA molecule. Called **DNA polymerase**, an important question to be answered was whether the two DNA strands serve as templates for new strands of DNA and then combine with the new strands, or recombine back with each other after serving as templates for new strands (FIGURE 4.6b). The answer to this question was found in

FIGURE 4.5 **Elucidating the Structure of DNA.** Pictured in this 1952 photo are James Watson (left) and Francis Crick (right) who correctly deduced the structure of DNA.

© A. Barrington Brown/Science Source.

1958 in experiments performed by Matthew Meselson and Franklin Stahl using the workhorse microbe *E. coli*.

When bacterial cells divide, each of the two daughter cells receives a complete copy of the chromosome. Meselson and Stahl analyzed the newly forming bacterial cells to determine which of the two possible forms of replication was actually found in the daughter cells. The analysis indicated the double helix in each daughter cell had one completely old strand and one completely new strand. This mechanism of DNA replication has been called **semiconservative replication** because one strand of the original DNA is "conserved" and represents half (*semi* = "half") of the newly formed double helix.

By 1960, the structure and method of replication of DNA were finally understood. Not only did most of this work use microbes and viruses, the insights stimulated a series of experiments that would solve the mystery of how DNA acts as the hereditary substance.

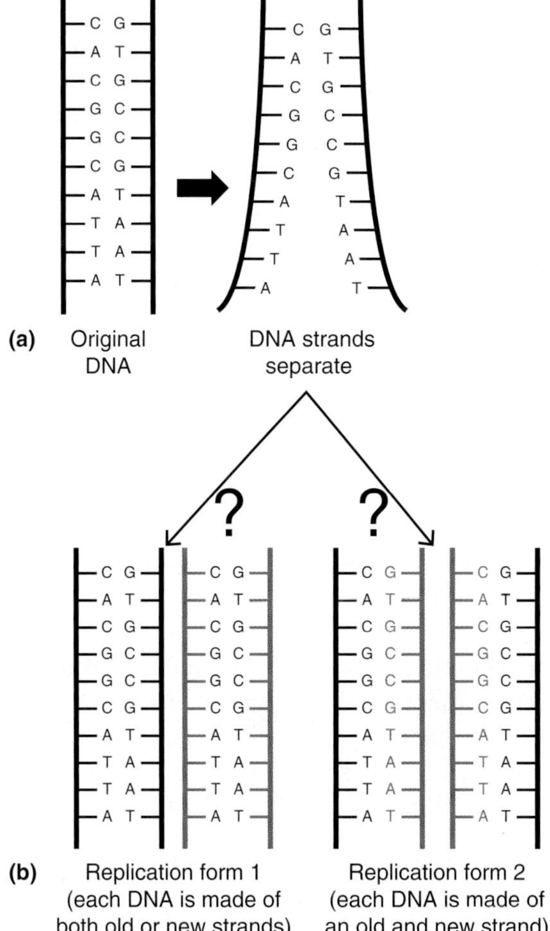

FIGURE 4.6 **How DNA Replicates.** The complementary bases found in the two polynucleotide strands of DNA suggested that once the two strands separate (**a**), each strand could be a template for a complementary strand. (**b**) The question was how did the two strands form the complementary strands? Two potential replication forms are shown. The work of Meselson and Stahl supported replication form 2, which is an example of semiconservative replication.

4.3 Gene Expression: DNA to Protein

Proteins are the working and structural components of all cells. Chemically, they are composed of building blocks called amino acids (much as the building blocks of nucleic acids are nucleotides). Only 20 different amino acids yield the countless combinations found in the proteins of cells. Biochemists estimate there are roughly 10,000 different enzymes operating in microbial cells, which implies there are at least 10,000 unique combinations of the 20 amino acids.

What makes one protein different from another protein is its sequence of amino acids. For example, two proteins may each consist of 33 amino acids linked together, but if the sequences of the amino acids are different, then the proteins are different. Consider, for a moment, how many words can be composed from our 26-letter alphabet. For proteins, the alphabet is composed of 20 amino acids.

If DNA molecules specify amino acid sequences in proteins, it follows that some message in the DNA molecule must specify an amino acid sequence. The contemporaries of Watson and Crick suggested the sequence of bases in DNA might constitute a code expressing such a message. The questions were: What was the code? How does the code get translated from the language of nucleotides to the language of amino acids? In other words, how does **gene expression**— the process by which information in a gene is used (expressed) to make a protein—occur?

Ribonucleic Acid and the Genetic Code

One of the first points biochemists had to work out was whether the biochemical information in DNA is used directly to make protein or some intermediary molecule is employed. The assembly of amino acids into proteins was known to take place in the cell's cytoplasm, but DNA is largely restricted to the cell's nucleus in eukaryotic organisms and to the area of the nucleoid in prokaryotic organisms. A direct passage of information from DNA to protein seemed unlikely.

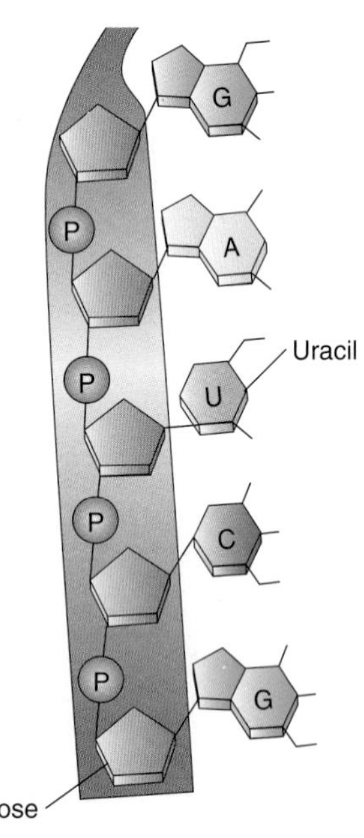

FIGURE 4.7 **The RNA Molecule.** Ribonucleic acid (RNA) is a single-stranded molecule consisting of nucleotides containing a ribose sugar and the nucleobases adenine, guanine, cytosine, and uracil.

In the 1940s, biochemists reported that when *E. coli* cells are synthesizing proteins, they contain an unusually large amount of the second type of nucleic acid, **ribonucleic acid (RNA)**. RNA is different from DNA in that RNA contains the sugar ribose, has the nucleobase uracil (U) rather than thymine, and is a single-stranded molecule (FIGURE 4.7). In eukaryotic cells, the RNA molecules move from the cell nucleus to the cytoplasm, so attention centered on RNA as the possible intermediary molecule. Moreover, it was easy to see how the sequence of bases in DNA could determine the complementary sequence of bases in RNA. The base sequence of DNA would serve as a template for synthesizing a molecule of RNA with complementary bases (except uracil replaces thymine in the RNA molecule). The process would be similar to DNA replication, except single-stranded RNA molecules would result.

So, how does the genetic message in DNA get copied to RNA, and what is the genetic code? Answers to these questions came in a series of experiments performed on *E. coli* in 1961 by Crick and his colleagues. Crick's group reasoned that the genetic code in DNA is in a series of blocks of information, each block composed of three bases. A single block of three bases could specify a single amino acid in the protein to be synthesized.

A bit of simple mathematics pointed to the three-base code. Virtually all proteins are made of just 20 amino acids. If a single base encoded a single amino acid, then there would be only enough codes for four amino acids (remember, there are only four bases). If two bases (e.g., AG, TC, and so on) encoded a single amino acid, then only 16 codes would exist and only 16 amino acids could be encoded. However, if three bases comprised a code unit, then 64 possible codes could exist in DNA (for example, ATA, GAT, TCG, CAT, and so on). Sixty-four codes were more than enough to specify the 20 amino acids. Experiments verified the hypothesis and indicated a block of three bases, called a **codon**, specifies a single amino acid in a protein.

Scientists then determined the **genetic code**, the base codes for all 20 amino acids (TABLE 4.1). In fact, the genetic code is nearly universal because the same three-base codes specify the same amino acids regardless of whether the organism is a bacterium, a mushroom, or a human. Also note in Table 4.1 that there is a redundancy in the genetic code, as often there is more than one codon specifying the same amino acid. For example, codons UCU, UCC, UCA, and UCG all code for the amino acid serine.

FIGURE 4.8 summarizes our discussion. Each gene in an organism's DNA supplies the information for a specific genetic code found in RNA, with different genes supplying different sequences of bases comprising the code. This process is called **transcription**. The sequence of bases in RNA specifies a specific sequence of amino acids in a protein. This process is called **translation**. This series of steps in gene expression is called the **central dogma** of genetics.

TABLE

4.1 The Genetic Code Decoder

The genetic code embedded in an mRNA is decoded by knowing which codon specifies which amino acid. On the far left column, find the first letter of the codon; then find the second letter from the top row; finally read up or down from the right-most column to find the third letter. The three-letter abbreviations for the amino acids are given.

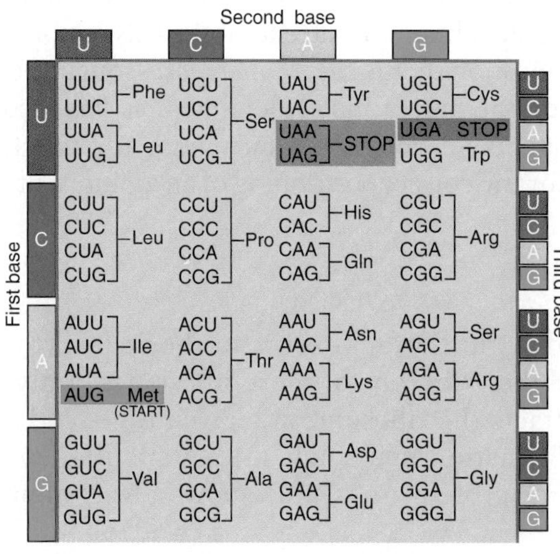

First base	Second base U	Second base C	Second base A	Second base G	Third base
U	UUU Phe	UCU Ser	UAU Tyr	UGU Cys	U
U	UUC Phe	UCC Ser	UAC Tyr	UGC Cys	C
U	UUA Leu	UCA Ser	UAA STOP	UGA STOP	A
U	UUG Leu	UCG Ser	UAG STOP	UGG Trp	G
C	CUU Leu	CCU Pro	CAU His	CGU Arg	U
C	CUC Leu	CCC Pro	CAC His	CGC Arg	C
C	CUA Leu	CCA Pro	CAA Gln	CGA Arg	A
C	CUG Leu	CCG Pro	CAG Gln	CGG Arg	G
A	AUU Ile	ACU Thr	AAU Asn	AGU Ser	U
A	AUC Ile	ACC Thr	AAC Asn	AGC Ser	C
A	AUA Ile	ACA Thr	AAA Lys	AGA Arg	A
A	AUG Met (START)	ACG Thr	AAG Lys	AGG Arg	G
G	GUU Val	GCU Ala	GAU Asp	GGU Gly	U
G	GUC Val	GCC Ala	GAC Asp	GGC Gly	C
G	GUA Val	GCA Ala	GAA Glu	GGA Gly	A
G	GUG Val	GCG Ala	GAG Glu	GGG Gly	G

Key

Ala = Alanine
Arg = Arginine
Asn = Asparagine
Asp = Aspartic acid
Cys = Cysteine
Gln = Glutamine
Glu = Glutamic acid
Gly = Glycine
His = Histidine
Ile = Isoleucine
Leu = Leucine
Lys = Lysine
Met = Methionine
Phe = Phenylalanine
Pro = Proline
Ser = Serine
Thr = Threonine
Trp = Tryptophan
Tyr = Tyrosine
Val = Valine

Transcription

As evidence about the nature of the genetic code continued to accumulate, a persuasive body of data was being generated on how DNA transfers its information to RNA—and microbes were at its core.

The process of gene expression begins with an uncoiling of the DNA double helix and a separation of the two DNA strands within a gene (FIGURE 4.9). This separation exposes the nucleotide bases. Using the enzyme **RNA polymerase**, cells synthesize a single-stranded molecule of RNA having bases complementary to the bases along one strand of the gene. The synthesis of RNA by RNA polymerase is specified by a segment of DNA bases called the **promoter site**. For a given gene, the promoter site exists on one DNA strand but not on the other, which is why one strand transcribes its message to the RNA and the other does not. So, as a result of transcription, the base code in the gene is copied as a base code in the RNA molecule. The RNA fragment so constructed is known as **messenger RNA (mRNA)**. In the cell's cytoplasm, each single-stranded mRNA molecule will specify a unique sequence of amino acids.

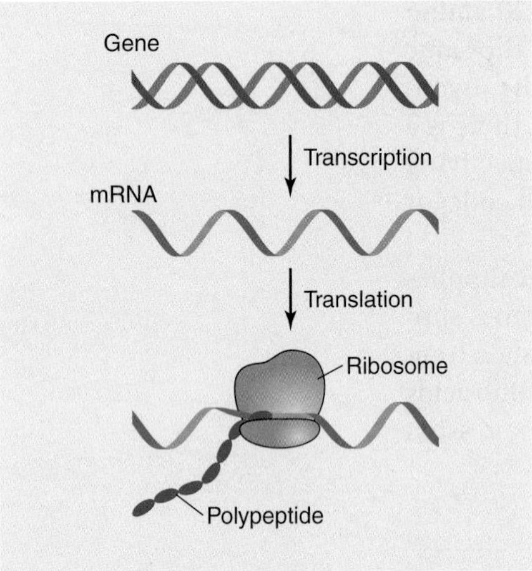

FIGURE 4.8 **The Central Dogma.** The flow of genetic information proceeds from DNA to RNA by transcription and from RNA to protein (polypeptide) by translation.

In the production of the final mRNA molecule, some modification of the RNA may take place. In eukaryotic microbes, such as fungi and protists, and in other eukaryotic organisms as well, bits of RNA are snipped out of the preliminary mRNA molecule. These fragments of RNA are referred to as **introns** because they are intervening codons within a gene. The sections of remaining mRNA are known as **exons** because their codes will be expressed as the final mRNA. In bacterial organisms, the preliminary mRNA is the final mRNA molecule—there are no introns.

Two other types of RNA are also encoded by DNA and transcribed into RNA (FIGURE 4.10). One type is **ribosomal RNA (rRNA)**, which combines with protein to form **ribosomes**, the structures in the cell cytoplasm of all organisms where amino acids are assembled into proteins according to the instructions delivered by mRNA molecules. The other type is **transfer RNA (tRNA)**, which, in the cytoplasm, binds to an amino acid and delivers it to a ribosome for use in translation. There are about 30 different tRNA molecules to handle all the codons and redundancies found in the genetic code. As a result, the amino acid binds at one end of the tRNA molecule, as shown in Figure 4.10. At another location on the tRNA, there is a sequence of three bases that complements a codon on the mRNA molecule; this three-base sequence is known as an **anticodon**. The matching of complementary codon (on mRNA) and anticodon (on tRNA) is essential for the correct positioning of an amino acid during translation.

Translation

The final step of gene expression, translation, occurs in the cell's cytoplasm, where the mRNA molecule combines with one or more ribosomes. Only a portion of the mRNA molecule contacts the ribosome at a given time, and several ribosomes may be using the information in a single mRNA simultaneously to form several molecules of the same protein. FIGURE 4.11 shows a brief moment in the interaction.

While the mRNA-ribosome complex is forming, amino acids are binding to their specific tRNA molecules in the cytoplasm. Then, the tRNA molecules transport the amino acids to the ribosome where mRNA is stationed. Next, one codon of the mRNA

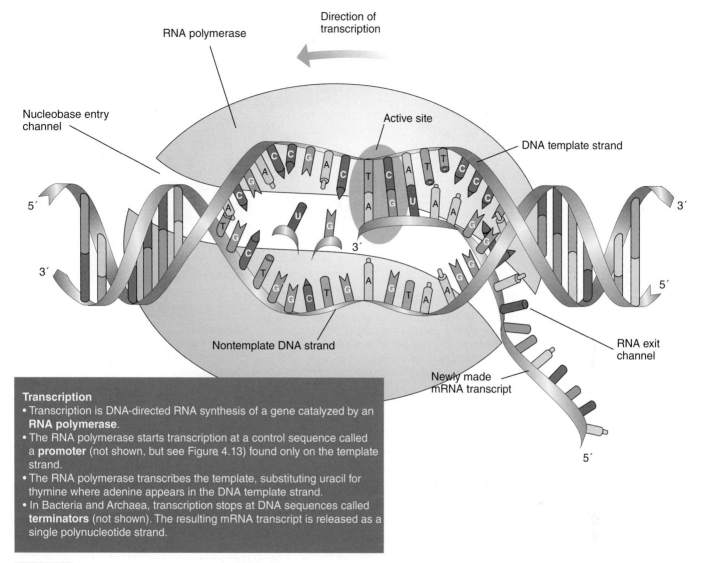

Direction of transcription

RNA polymerase

Nucleobase entry channel

Active site

DNA template strand

5´ 3´ 3´ 5´ 3´

Nontemplate DNA strand

RNA exit channel

Newly made mRNA transcript

5´

Transcription

- Transcription is DNA-directed RNA synthesis of a gene catalyzed by an **RNA polymerase**.
- The RNA polymerase starts transcription at a control sequence called a **promoter** (not shown, but see Figure 4.13) found only on the template strand.
- The RNA polymerase transcribes the template, substituting uracil for thymine where adenine appears in the DNA template strand.
- In Bacteria and Archaea, transcription stops at DNA sequences called **terminators** (not shown). The resulting mRNA transcript is released as a single polynucleotide strand.

FIGURE 4.9 **The Transcription Process.** The enzyme RNA polymerase moves along the template strand of the DNA and synthesizes a complementary molecule of RNA using the base code of DNA as a template. The mRNA will carry the genetic message into the cytoplasm, where translation occurs. Note that the nontemplate DNA strand of the gene is not transcribed.

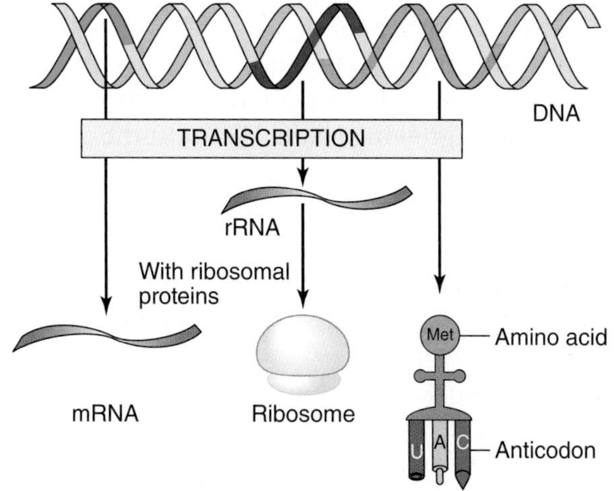

FIGURE 4.10 **The Transcription of the Three Types of RNA.** Different genes in the DNA contain the information to produce mRNAs, rRNAs that along with protein builds the ribosomes, and tRNAs that carry one of the 20 amino acids needed for translation.

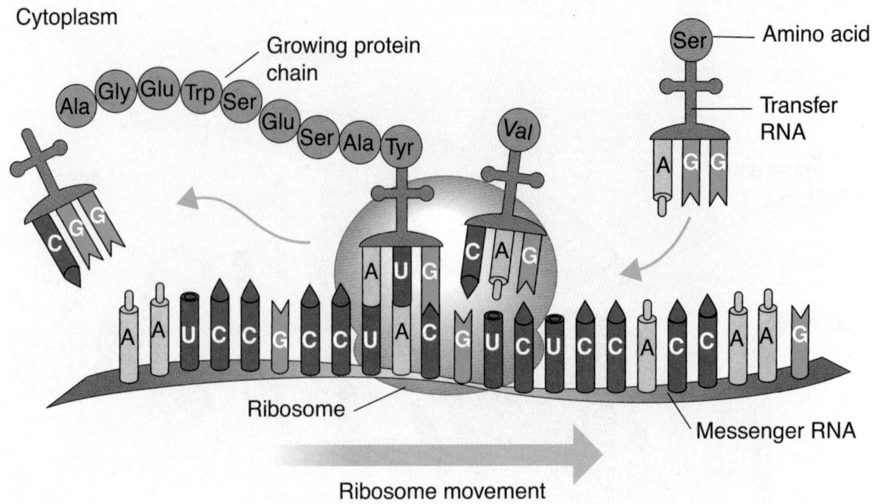

FIGURE 4.11 **The Translation Process.** The messenger RNA moves to the ribosome, where it is met by transfer RNA molecules bonded to different amino acids. The tRNA molecules align themselves opposite the mRNA molecule and bring the amino acids into position. A peptide bond forms between adjacent amino acids on the growing protein chain, after which the amino acid leaves the tRNA. The tRNA returns to the cytoplasm to bond with another amino acid molecule.

molecule attracts its complementary anticodon on a tRNA molecule, and the pairing brings a certain amino acid into position. Within the ribosome, a second tRNA molecule (holding a second amino acid) pairs its anticodon with the second codon. Two tRNA molecules have thus positioned their amino acids next to each other. Next, an enzyme joins the two amino acids to form a dipeptide (two amino acids joined together). The first tRNA molecule then is released from its amino acid and moves back into the cytoplasm to bind to another identical amino acid.

Now, the ribosome moves to the next codon of the mRNA molecule. The codon attracts its complementary anticodon on a tRNA molecule, which brings a third amino acid into position. As before, an enzyme chemically joins the first two amino acids to the newest amino acid to form a tripeptide (three amino acids in a chain). The second tRNA is then released back into the cytoplasm. And so it goes: Codon after codon is paired with its anticodon, and amino acid after amino acid is positioned and joined to the growing peptide chain. The genetic message of DNA is being translated via mRNA into an amino acid sequence in a protein. Translation is in full swing and will continue until hundreds or thousands of amino acids have been added one by one to the growing chain. We are at the crux of one of life's most fundamental processes.

The final codons of the mRNA molecule are chain terminators, or "stop" signals. When the ribosome reaches one of these codons (UAA, UAG, or UGA; see Table 4.1), no complementary tRNA molecules exist and thus no amino acids are added to the chain. The stop signals activate release factors that free the amino acid chain from the ribosome, and translation comes to a conclusion.

With the conclusion of the process, the genetic message of DNA has been expressed as the amino acid sequence in a protein. This genetic message will manifest itself through the activity of an enzyme, a structural protein of the cell, a transport protein, or any of numerous other protein-based molecules. This extraordinary process, summarized in FIGURE 4.12, is one of the key underpinnings of microbial life—indeed, of all life.

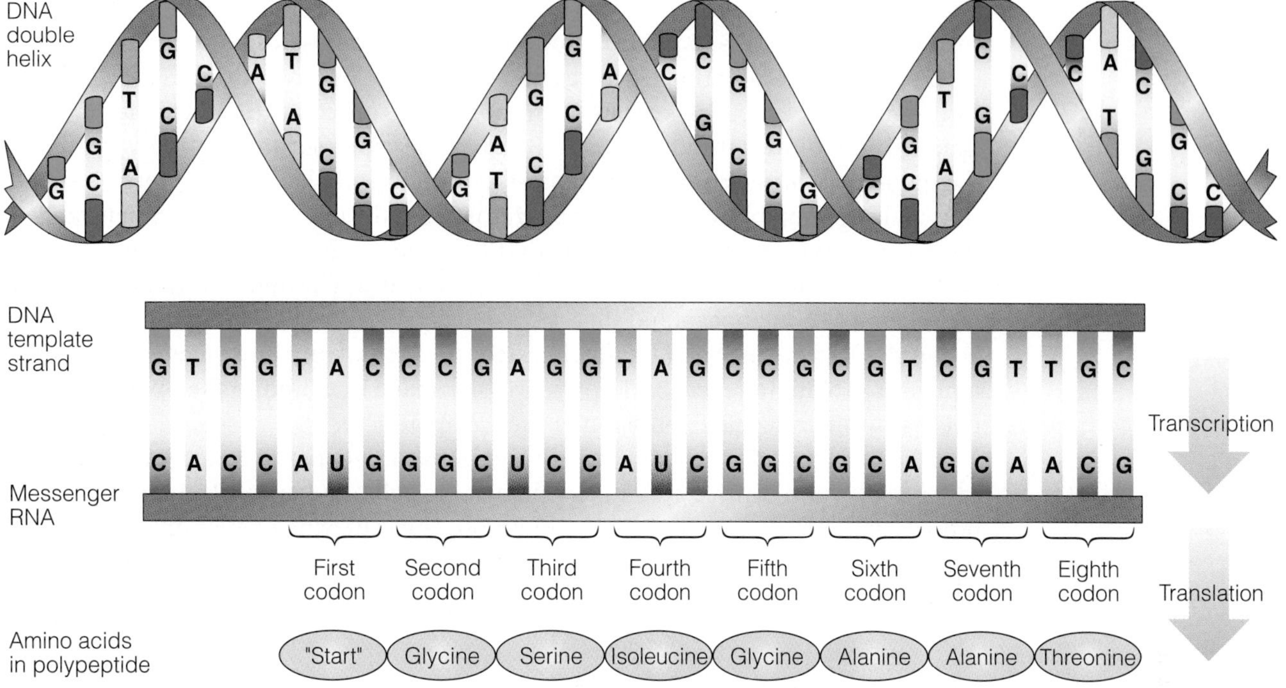

FIGURE 4.12 **A Summary View of Protein Synthesis.** The DNA molecule unwinds, and one strand is transcribed as a molecule of messenger RNA. The mRNA then operates as a series of codons, each codon having three bases. During translation, a codon specifies a specific amino acid for placement in the growing polypeptide. Note that codons 6 and 7 are identical and specify the same amino acid, alanine. Similarly, codons 2 and 5 are identical and encode glycine.

Gene Regulation

As biochemical knowledge increased, scientists began to wonder how microbes and all organisms regulate the complex machinery of gene expression. It would be a huge waste of energy to transcribe all genes all the time whether the protein product was needed or not. The transcription of genes must be "turned on" and "turned off" as their protein product is needed. So, how are genes regulated? Once again, microbes come to the rescue.

In 1961, two Pasteur Institute scientists, François Jacob and Jacques Monod, were among the first to propose a scientific mechanism for controlling gene expression. *E. coli* was their experimental organism and in 1965 they would be awarded the Nobel Prize in Physiology or Medicine for their groundbreaking work.

Jacob and Monod suggested that in *E. coli* and other bacterial species several DNA base sequences are needed: "Structural genes" provide the base codes for proteins (in the transcription process we described); an adjacent operator base sequence controls the expression of the structural genes; and a promoter sequence where the RNA polymerase binds. Today, we call such a DNA segment an **operon**. Also important, but not part of the operon, is a regulatory gene that controls the operator (FIGURE 4.13a).

The *lac* (lactose) operon model helps us understand how gene expression is controlled. When the nutritious disaccharide lactose is absent from the environment, there is obviously no need for *E. coli* to produce lactose-digesting enzymes. Accordingly, the microbe's regulatory gene encodes an mRNA to form a **repressor protein** (FIGURE 4.13b). The repressor protein binds to the operator. Although the RNA polymerase can still bind to the promoter, its movement down the DNA is blocked by the repressor protein attached to the operator. Therefore, the RNA polymerase cannot reach the structural genes and bringing about the transcription of the mRNA that would produce the enzymes to break down lactose. (The effect is somewhat like placing a log across a roadway.) The result is that *E. coli* does not produce lactose-digesting enzymes.

Then, at some later time, lactose enters the microbe's environment (FIGURE 4.13c). In this context, lactose is known as an inducer because it will induce or "turn on" transcription of the structural genes. Because the repressor protein is always made, the inducer (lactose) binds to the repressor protein and changes the protein's shape, thus making it unable to bind to the operator. With the repressor protein neutralized, the operator is free and, after binding to the promoter, the RNA polymerase can pass across the operator and transcribe the structural genes. The mRNA is then translated, the enzymes soon appear, and digestion of lactose to glucose and galactose begins. When the lactose has been totally digested, the lack of any more lactose allows the repressor protein to revert back to its original shape and bind to the operator, effectively shutting off gene expression of the *lac* operon.

The mechanisms involving the control of bacterial operons are relatively simple compared to the control systems in eukaryotic cells. Eukaryotic microbes such as protists and fungi have more genes, and more levels of gene regulation are therefore required. Nor is the regulation of gene expression confined to the time of transcription.

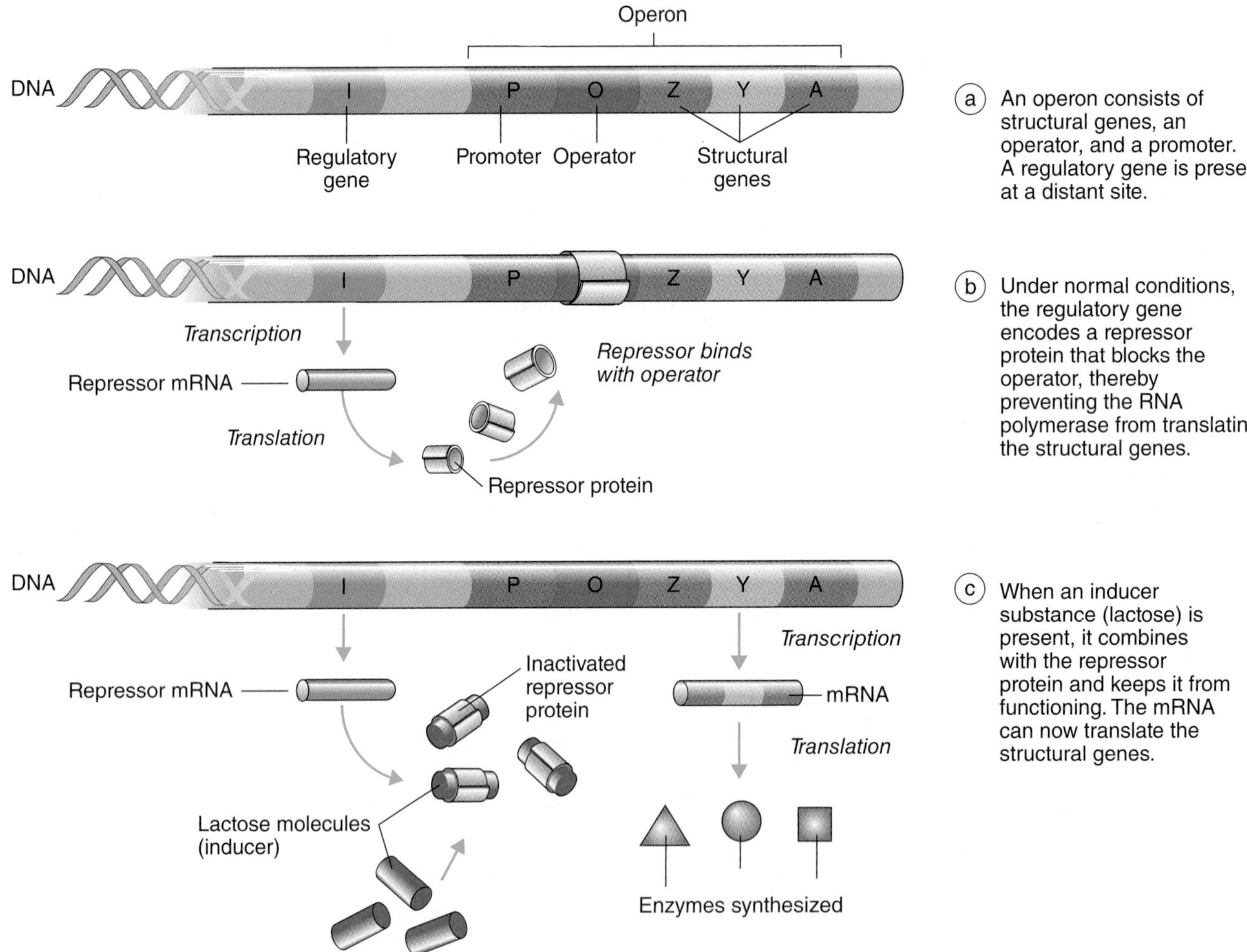
Operon
DNA
I
P
O
Z
Y
A
Regulatory gene
Promoter
Operator
Structural genes
(a) An operon consists of structural genes, an operator, and a promoter. A regulatory gene is present at a distant site.
DNA
Transcription
Repressor mRNA
Translation
Repressor binds with operator
Repressor protein
(b) Under normal conditions, the regulatory gene encodes a repressor protein that blocks the operator, thereby preventing the RNA polymerase from translating the structural genes.
DNA
Repressor mRNA
Inactivated repressor protein
Transcription
mRNA
Translation
Lactose molecules (inducer)
Enzymes synthesized
(c) When an inducer substance (lactose) is present, it combines with the repressor protein and keeps it from functioning. The mRNA can now translate the structural genes.

FIGURE 4.13 **The Bacterial Operon and Gene Regulation.** (**a**) An operon consists of a group of structural genes that are under the control of an operator. (**b**) If the repressor protein binds to the operator, it prevents RNA polymerase from being able to transcribe the structural genes. (**c**) If, on the other hand, the repressor is inactivated by an inducer like lactose, the RNA polymerase is free to travel down the DNA and transcribe the structural genes.

Regulation can take place at many levels of gene expression, such as in the processing of mRNA molecules, the transport of mRNA molecules to the cytoplasm, and the binding of mRNA molecules to the ribosome.

4.4 Genes and Genomes: Human and Microbial

In April 2003, exactly 50 years to the month after Watson and Crick announced the structure of DNA, a publicly financed, $3 billion international consortium of biologists, industrial scientists, computer experts, engineers, and ethicists completed one of the most ambitious projects to date in the history of biology. The **Human Genome Project (HGP)** had succeeded in mapping the **human genome**—that is, the 3 billion nucleobases (equivalent to 750 megabytes of computer data) and some 19,000 genes in a human cell were identified and strung together (sequenced) in the correct order. By comparison, an *E. coli* cell has only about 4.7 million bases and about 4,300 genes. However, only 2% of the human genome codes for proteins and over 99% of the genome is identical from one human to another.

Potential Benefits from the Human Genome Project

The HGP has already had profound impact on biomedical research and medicine. It is now possible to look for abnormal or defective genes associated with dozens of genetic conditions, such as fragile X syndrome, neurofibromatosis, inherited colon cancer, Alzheimer's disease, and familial breast cancer. The day is rapidly approaching when knowing one's personal genome will modify the diagnosis, monitoring, and treatment of diseases.

Sequencing the human genome also will have an impact on molecular medicine. Already doctors are talking about spending less time treating symptoms and more time looking for the most fundamental causes of disease. In fact, rapid and more specific diagnostic tests will allow for earlier and better treatment of countless maladies. The HGP also is ushering in the new field of "pharmacogenomics" where doctors, based on a patient's genetic profile, will be able to tailor novel therapeutic regimens and immunotherapy treatments using new drug products especially designed for each patient (FIGURE 4.14). Not only will these drugs and treatments be more efficient, but they also will come with fewer side effects and drug safety issues that plague many patients' therapy today. And as the ultimate disease "cure," gene therapy is becoming

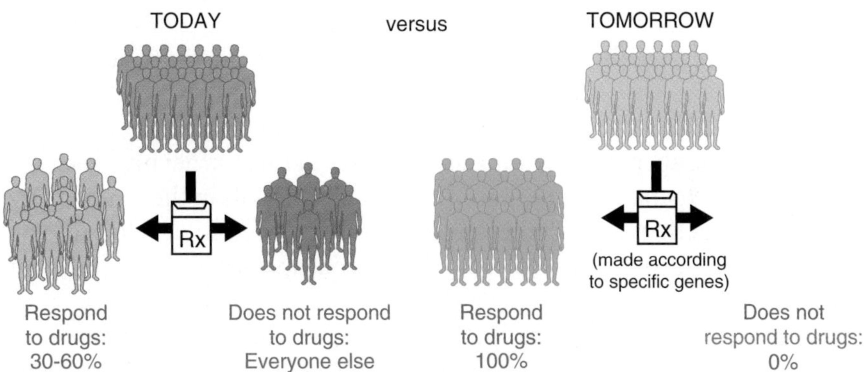

FIGURE 4.14 **Pharmacogenomics.** The field of pharmacogenomics combines the techniques of medicine, pharmacology, and genomics to develop new drug therapies and diagnostic procedures based on one's personal genome. The aim is to provide such patients with safest and most effective treatment that in the past caused varied responses.

more frequent. If one can identify a defective gene in one's genome, the techniques are getting better to augment or even replace of defective gene with a copy of the "good" gene.

On the flip side, with greater genetic knowledge comes the problem of public access to that information. Society soon will need to make some hard choices when considering who has legal rights to your medical information. For example, will companies or individuals in the nonmedical professions, such as your insurance company or employer, have the right to know your risks of developing certain conditions and to avoid hiring you or restrict the kinds of work you may do?

But let's look beyond the medical aspects of our human genome and discover some of the amazing microbial information we are gleaning from the DNA sequences of our human genome.

The Human Genome's "Microbial Ancestors"

With the sequencing of the human genome, one interesting project has been to compare the human nucleotide DNA sequences to known bacterial and viral DNA sequences—a discipline called **comparative** genomics. Are there any similarities in our genes and microbial genes?

■ genomics: The identification and study of gene sequences in an organism's DNA.

Some comparative genomic studies indicate as many as 200 of our 19,000 genes are essentially identical to those found in members of the domain Bacteria; 25% or some 6,000 of our genes are found in fungal yeasts. However, we did not acquire these genes directly from these microbial species, but rather they are genes picked up by our early ancestors. So important were these genes, they have been preserved and passed along from organism to organism and generation to generation throughout evolution; they are life's oldest genes. For example, several researchers suggest some genes coding for brain signaling chemicals and for communication between cells did not evolve gradually in human ancestors; rather, these ancestors acquired the genes directly from bacterial organisms. This provocative claim remains highly controversial though and more work is needed to better analyze this possibility.

Other startling discoveries have also been made. Most surprising is that almost 10% of our genome (and substantial parts of other animal genomes) is composed of self-replicating fragments of viral DNA called **human endogenous retroviruses (HERVs)**. These viruses were presumably able to insert their DNA directly into a human chromosome in reproductive cells thousands or perhaps millions of years ago and these viral fragments have been transmitted to future generations ever since—and these are the remnants we see in our DNA today. Most HERVs no longer have an effect in the human body to cause disease but undoubtedly have affected human evolution.

That said, a few HERV sequences may still play a critical role in our bodies today. For example, at least six of these viral genes interact in the normal functioning of the placenta. One HERV gene codes for a protein that allows cells in the outer layer of the placenta to fuse together, a needed event for a fertilized egg to develop into a fetus. Other HERV fragments may play roles in cancer, multiple sclerosis, and rheumatoid arthritis.

Further research has identified other virus "signatures," called "endogenous viral elements" or EVEs in the human genome, as well as in fungal and plant genomes. Finding EVEs in such divergent organisms suggests some viral families have ancestries that go back almost 100 million years and represent but a small subset of what existed in the past.

So, viral genomes might be a legacy of long distant infections and epidemics, and genomics analyses make it clear our genome is a hybrid of vertebrate and microbial genetic information. Certainly as more studies are undertaken, microbial genomics will have much more to tell us about our distant past and evolution as a species.

Sequencing Microbial Genomes

If the human genome was represented by a rope 2 inches in diameter, it would be 32,000 miles long, which is almost one and a half-times the circumference of the Earth. The genome of a bacterial species like *E. coli* at this scale would be only 1,600 miles long, which is about the distance from Denver to Philadelphia. So, being about 1/20 the size of the human genome, microbial genomes have been easier and much faster to sequence. In May 1995, the first complete genome of a free-living organism was sequenced: the 1.8 million bases (1.8 Mb) in the genome of the bacterial species *Haemophilus influenzae*. In a few short months, the genome for a second organism, *Mycoplasma genitalium* was reported. This reproductive tract pathogen has one of the smallest known bacterial genomes, consisting of only 580,000 bases and 485 protein-coding genes. As mentioned earlier, *E. coli* has about 4,300 protein-coding genes.

■ *Haemophilus influenzae*
hē-mä′-fil-us in-flü-en′zī

■ *Mycoplasma genitalium*
mī-kō-plaz′-mä jen′-i-tä-lē-um

As more genomes were sequenced, Craig Venter and his team then at The Institute for Genomic Research (TIGR) in Maryland wanted to know: What is the minimal number of genes an organism actually needs to grow and reproduce? They answered this question by disrupting the *M. genitalium* genes one-by-one to discover what genes were not essential. The analysis suggested *M. genitalium* needs some 382 of the 485 protein-coding genes (under laboratory conditions). So, scientists may be close to finding the genetic "essence of life," that is, the minimal genetic information needed to operate a cell—bacterial at least.

By the end of 2013, the genomes of almost 8,000 species (primarily microbial) had been completely sequenced (FIGURE 4.15). And the number keeps growing. Although 60% of these sequencing projects represent clinically important human microbial pathogens, sequences by themselves do not tell us much. So, what practical use might be derived from this information?

Microbial Genomics and the Understanding of the Microbial World

Microorganisms have existed on Earth for about 3.8 billion years, although we have known about them for little more than 300 years. Over this long period of evolution, they have become established in almost every environment on Earth, make up a significant percentage of Earth's biomass, and, although they are the smallest organisms on the planet, they influence—if not control—some of the largest events. Yet, until recently, we did not know a great deal about any of these microbes.

With the advent of **microbial genomics**, the discipline of sequencing, analyzing, and comparing microbial genomes, we are beginning to understand the workings and interactions of the microbial world. Some potential consequences from the understanding of microbial genomes are outlined below.

Safer Food Production. Because microorganisms play important roles in our foods both as contamination and spoilage agents, understanding how they get into the food product and how they produce dangerous foodborne toxins, will help produce safer foods (FIGURE 4.16). However, a major limitation with traditional food safety surveillance

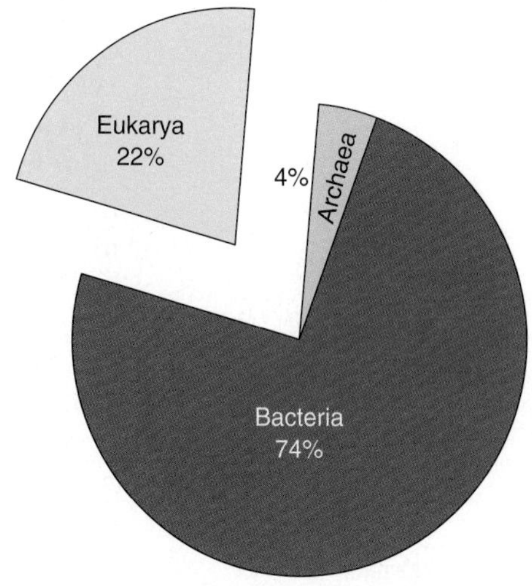

FIGURE 4.15 **Genomics Activities.** The sequencing of bacterial genomes has far outpaced those of the archaeal and eukaryotic organisms.

Source: Data from http://www.genomesonline.org

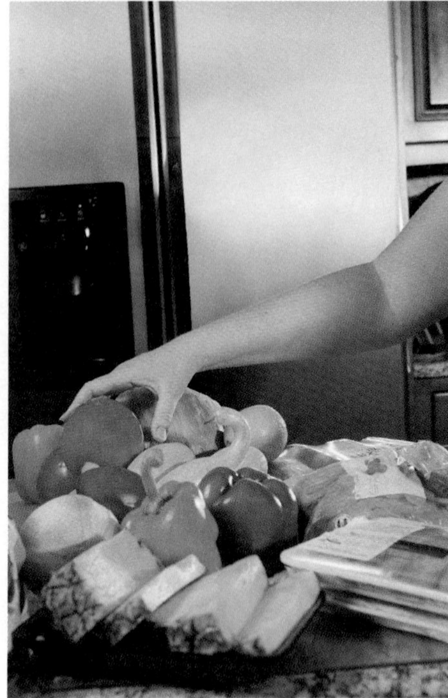

FIGURE 4.16 **Food Safety.** Maintaining clean and safe foods to eat is an increasingly difficult problem. Microbial forensics can help to rapidly detect potential pathogen contamination.

Courtesy of Amanda Mills/CDC

is that food-contaminating or spoiling microbes can only be identified after a long time-consuming approach often requiring a number of days for the organisms to grow. In addition, separate tests need to be run for each potential foodborne pathogen. This whole method of detecting foodborne pathogens can be greatly speeded up with microbial genomic technologies, enabling a quick and reliable prediction of the safety of our foods. In addition, microbial sequencing could be applied as a tool to trace potential food contamination between farm and table, a real problem today for fruits and vegetables coming from both within and outside the United States.

Microbial Forensics. The advent of the anthrax bioterrorism events of 2001 and the continued threat of bioterrorism has led many researchers to look for ways to more efficiently and more rapidly detect the presence of such "microbial bioweapons."

Many of the diseases caused by these potential biological agents cause no symptoms for at least several days after infection, and when symptoms appear, initially they are flulike. Therefore, it can be difficult to differentiate between a natural outbreak and the intentional release of a potentially deadly pathogen. Such concerns have led to a relatively new and emerging area in microbiology called **microbial forensics**. This discipline attempts to recognize patterns in a disease outbreak, identify the responsible pathogen, control the pathogen's spread, and discover the source of the pathogenic agent.

If the "outbreak" is the result of a purposeful release—a bioterrorism attack—then tracking down the source of the microbe (and perpetrator) is critical. For example, the anthrax letter attacks of 2001 generated panic among the public and showed the need to establish "attribution" (who is responsible for the crime) for fear that another such attack might occur. The 2009 swine flu pandemic at one point was proposed by some to be the result of an accidental release of the virus from a research lab doing vaccine experimentation. Microbial forensic investigations did not support this claim.

The science behind microbial forensic investigations are essentially the same as any other forensic investigation as they involve a crime scene(s) investigation, evidence collection, chain of custody for the collection, handling and preservation of evidence, interpretation of results and—unique to the scientist—court presentation. Importantly, microbial forensic data must be undeniable and must hold up to the scrutiny of judges and juries in a court of law in order to bring a perpetrator(s) to justice and to deter future attacks.

Microbial forensics also is involved in other types of medical and hygiene cases. Unique pathogen genetic sequences can be used to track infections in hospitals. For example, did a patient have the infection when admitted or was it acquired in the hospital? Cases of medical negligence, where a hospital's inadequate or improper hygiene can lead to a patient contracting a postsurgical or hospital-acquired infection (and perhaps die), could be settled through forensics analysis. Outbreaks of foodborne disease have brought lawsuits against companies alleging negligence in sanitary practices. And the potential for intentional contamination through foodborne terrorism is certainly possible.

In all these cases, tracing the infecting microbe to the company or person(s) of origin is critical. This places the fields of genetic engineering, biotechnology, and microbial genomics at the forefront of this emerging field. In fact, the ability to manipulate genomes is increasing at an ever rapid pace and this is no clearer than in the creation of nonnatural life as **A Closer Look 4.3** reports.

- *Mycoplasma mycoides* mī-kō-plaz′-mä mī-coi′-dēz
- *Mycoplasma capricolum* mī-kō-plaz′-mä ka-pri-cō′-lum
- *Mycoplasma laboratorium* mī-kō-plaz′-mä la-bôr-a-tôr′-ēum

A CLOSER LOOK 4.3

A Bacterial Imposter

Invasion of the Body Snatchers is a classic sci-fi film of the 1950s (remade in 1978 and 1994). In the original, the town's doctor discovers that many of his patients are being invaded and replaced by emotionless alien imposters and he must try to combat and quell the deadly, indestructible threat.RNA as a "gene" and acting as a ribozyme.

In 2010, a bacterial imposter was announced—but of a purposeful kind. Researchers at the J. Craig Venter Institute (JCVI) in Rockville, Maryland and San Diego, California reported they were able to carry out genome transplantation and in the process transform one bacterial species into another species. In their work, the researchers built the entire genome of *Mycoplasma mycoides* from scratch; that is, they fed the genetic sequence into a DNA synthesizer and produced fragments that were assembled into the complete functional genome.

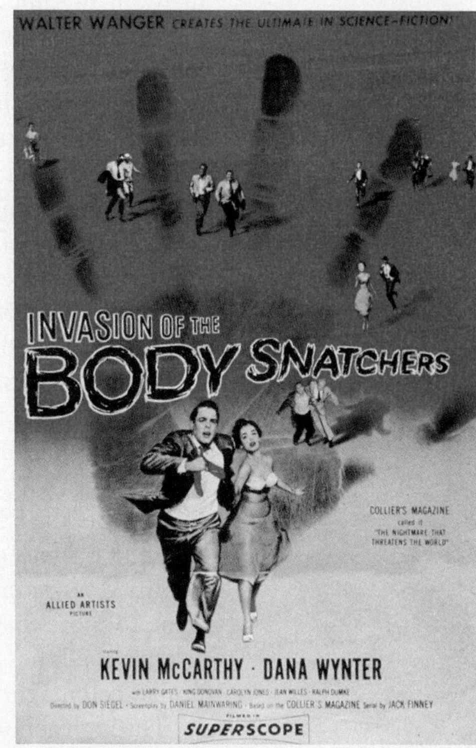

© Photos 12/Alamy

The Venter team then transferred the synthetic genome into a genome-less *Mycoplasma capricolum* cell. The result was that the transplanted genome then dictated the behavior of the recipient *M. capricolum* cell—it had the characteristics of *M. mycoides* and was able to reproduce.

JCVI refers to this as the world's first "synthetic life form." However, it is not really a "synthetic life form" because the recipient cell was a normal *M. capricolum* cell and the genome was a normal (but synthetic) genome of *M. mycoides*. So, it is like erasing a computer's old operating system and reprogramming it with a new operating system written from scratch. Perhaps it is better to say that they have made the world's first "nonnatural life form."

So, what is to be gained from this type of "genetically engineered genome transplantation?" The JCVI scientists' goal was not to create some deadly, indestructible group of alien imposters as in *Invasion of the Body Snatchers,* but to endow a synthetic genome with genes that in a recipient cell will allow those cells to carry out useful functions. In other words, one could start with a naturally occurring microbe and strip it down to its minimal genome. With such an organism, the researchers could then add foreign genes to build an organism with unique biological properties and commercial utility. For example, genes could be added to the cells to produce ethanol, hydrogen, or another form of renewable biofuels. According to Venter, such "artificial life" could also go a long way toward helping resolve the climate crisis by building organisms that could soak up carbon dioxide from the atmosphere. Certainly there are numerous other genes that could be transplanted in a similar manner—hopefully for the good of humankind.

Which brings up a more related *Invasion* scenario. Beyond the amazing breakthroughs that might come in the near future from genome transplantation, some critics are concerned that such "synthetic life forms" could be turned into biological weapons by inserting genes that make the cells the ultimate doomsday pathogen. As this area of synthetic biology continues to advance, certainly there are bioethical issues to consider.

Potential Benefits from the Human Microbiome Project

Existing within, on, and around every living organism, and in most all environments on Earth, are microorganisms. Yet, almost 99% of the bacterial and archaeal species found within us and in the environment will not grow when cultured. Consequently, most organisms found in the soil, in oceans, and even in the human body have never been seen or named—and certainly not studied. Yet these organisms represent a wealth of genetic diversity.

Advances in DNA sequencing technologies have created a new field of research for analyzing this uncultured majority. The field is called **metagenomics** (*meta* = "beyond"), and it refers to identifying the genome sequences within mixtures of organisms in a community (the **metagenome**); that is, "beyond" what can be cultured. Perhaps the most ambitious and complex endeavor to date was the **Human Microbiome Project (HMP)**. Completed in 2012, the findings have provided us with unprecedented information about the complexity of the microbial communities, called the human **microbiome**, within our body.

Within the body of a healthy adult, microbial cells are estimated to far outnumber human cells (FIGURE 4.17a); that is, there are some 100 billion microbes (primarily bacterial species) within the body, which itself is composed of about 35 billion cells. In terms of weight, an adult body carries two to three pounds of bacterial cells. Prior to the HMP, this microbial community could not be broadly studied because of the inability to grow the organisms in culture. Therefore, we had little understanding of its influence on human development, physiology, immunity, and nutrition.

To date, the HMP has generated 3.5 terabytes (3.5 trillion bytes) of data, or more than 1,000 times the amount of data produced by the HGP. And the scientists have identified more microbes than they had ever imagined, with any one person being home to more than 1,000 bacterial species. Thankfully, rather than making people sick, most all these resident microbes are living peacefully among their bacterial and human cell neighbors. Let's see a few examples.

Through genome sequencing, the skin has been found to harbor a diverse population of bacteria (FIGURE 4.17b). It is dominated by bacterial species belonging to one of four phyla and these species can be found unevenly distributed over three different skin sites: oily (e.g., the forehead and back), moist (e.g., navel, groin, and inner elbow), and dry (e.g., forearm, buttocks, and hand).

The gut, primarily the large intestine, is jam-packed with microbes such that half of one's stool is bacterial. However, the hundreds of microbial species composing the gut community actually help us stay healthy. The microbes influence how an individual reacts to various drugs and why some individuals are susceptible to certain infectious diseases while others are resistant. Most interestingly, when these gut microbial residents go awry and the community structure breaks down in an individual, the likelihood for the development of chronic diseases and conditions like irritable bowel disease, asthma, and even obesity, increase.

Where did this microbiome come from in the first place? Quite simply, as a newborn passes through the birth canal, it picks up the mother's vaginal microbiome. Then, over the next two to three years, the baby's body microbiome matures and grows while its immune system develops in concert, learning not to attack the bacteria and recognizing them as friendly. Interestingly, genomic analyses suggest a baby born by Caesarean section (C-section) starts out with a different microbiome because there is no contact with the mother's vaginal microbiome. Although it is not yet known whether the baby's microbiome remains different from that in those born vaginally, gut differences may partially explain, in part, why some C-section infants have a harder time fighting off infections, have more allergies, and are more prone to developing asthma and type 1 (juvenile) diabetes.

Much more work lies ahead to completely understand how the human microbiome affects our health. As one of the HMP microbiologists said, "We are scratching at the surface now."

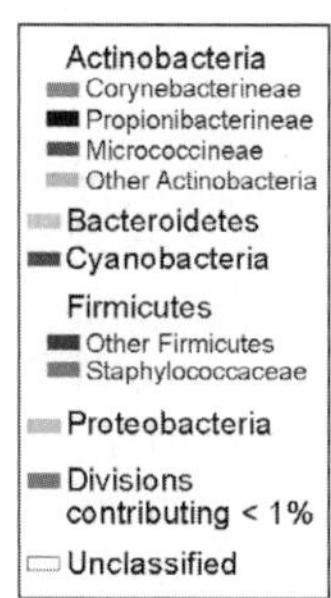

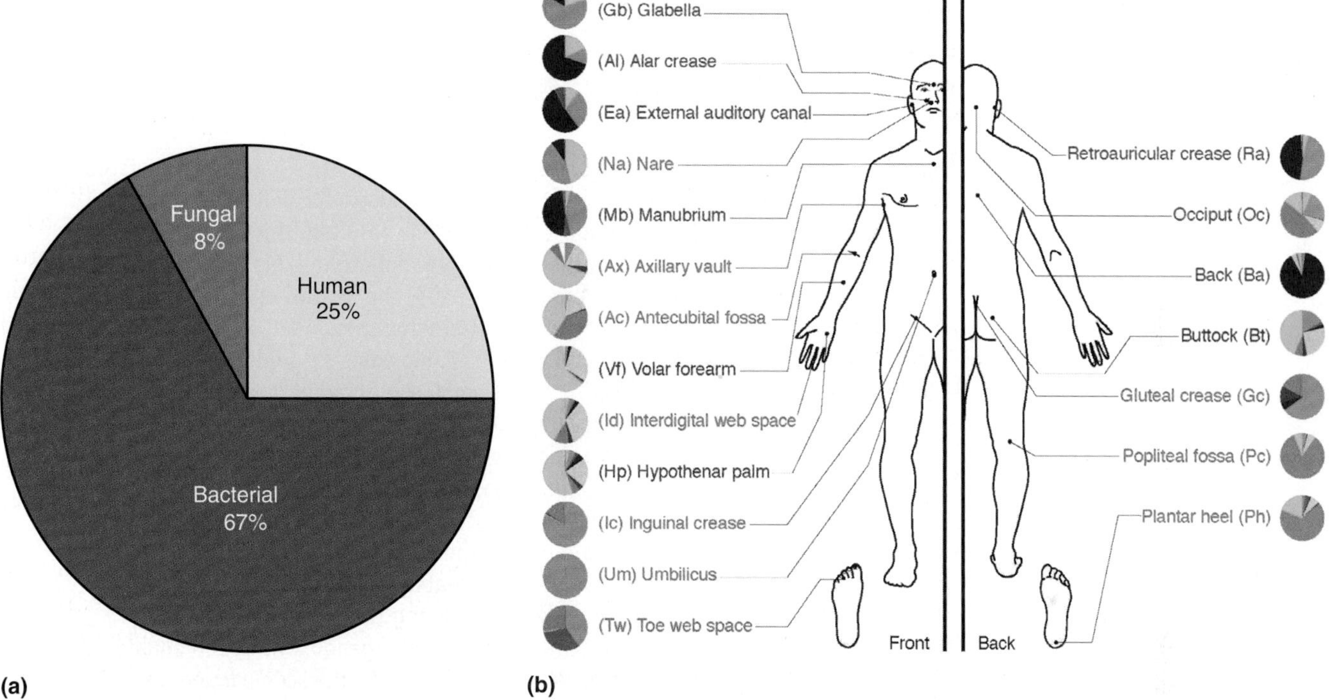

FIGURE 4.17 **Cells in the Human Body and the Topographical Distribution of Resident Skin Microbes.** (**a**) The human body is composed of large numbers of microbes, primarily bacterial and fungal. (**b**) The presence of specific bacteria is governed by the microenvironment of each of the three skin sites (oily, moist, and dry).

Courtesy of Darryl Leja, NHGRI.

A Final Thought

For many decades, scientists pondered the nature of the gene. Try as they might, however, they could not imagine how DNA could be involved. They thought it too simple ("too stupid" some said) for the sophisticated function of heredity.

But by the 1950s, compelling experiments pointing to DNA as anything but stupid. Then, when Watson and Crick announced their model of DNA's structure, a flurry of brilliant research started snowballing and scientists cracked the code that makes DNA so smart. Next, they were cloning genes and examining their structures, while researching the proteins they encode. Then, they were ready to map the genome—human and microbe. What they have accomplished has made heads spin in the scientific community.

Deciphering the human genome has been compared to landing an astronaut on the moon. The identification of the human microbiome worked out through the HMP was no less awesome. However, the milestone of stepping onto the lunar surface was the grand finale to years of space research, while the milestone of sequencing the genomes of humans and microbes is but the stepping-off point on an incredible journey that will take decades to complete. The sequencing of the human genome and human microbiome has vast and still unrecognized implications for people's lives. It is safe to assume scientists will not be turning off their sequencing machines or reading help-wanted ads for many decades to come. Today, we stand on the threshold of knowledge so powerful that we will wonder how we ever got along without it.

Questions to Consider

1. Try to put yourself in Griffith's position in 1928. Genetics is poorly understood, DNA is virtually unknown, and bacterial biochemistry has not been clearly defined. How would you explain the remarkable results of his experiments?
2. The "central dogma" is a supposedly firm principle that explains how genes function in cells. With the emergence of AIDS, the central dogma has come into question because of the ability of the AIDS virus to convert RNA into DNA. What is the central dogma and why has the principle come into question?
3. In one sense, DNA is a relatively simple molecule, having only four different subunits. And yet, DNA can specify extraordinarily complex proteins having at least twenty different amino acids in chains of many thousands or more. How does DNA accomplish this seemingly impossible task?
4. It has been said that the names of Watson and Crick are indelibly etched in the history of twentieth-century biology as the "discoverers" of DNA. Knowledgeable students of biology know, however, that Watson and Crick did not "discover" DNA. Which names should be remembered along with theirs, and why?
5. Determining the sequence of bases in the human genome has been equated to landing a human on the moon. Many would dispute that comparison and suggest that elucidating the human genome is far more important. What do you think they have in mind?
6. Tube A of DNA contains the following percentages of bases: 23.1% A, 26.9% C, 26.9% G, and 23.1% T. By contrast, tube B has 32.3% A, 32.3% C, 17.7% G, and 17.7% T. Which tube has the double-stranded DNA and which has the single-stranded DNA. Why?
7. You maintain that enzymes make proteins, but your colleague suggests that proteins make enzymes. Which of you is correct, and what is the basis for your conclusion?

Key Terms

Informative facts are necessary for the expression of every concept, and the information for a concept is founded in a set of key terms. The following terms form the basis for the concepts of this chapter. On completing the chapter, you should be able to explain and/or define each one.

anticodon
bacteriophage (phage)
central dogma
chromosome
codon
comparative genomics
deoxyribonucleic acid (DNA)
DNA polymerase
double helix
exon
gene
gene expression
genetic code
genomics
heredity
human endogenous retrovirus (HERV)
human genome
Human Genome Project (HGP)
Human Microbiome Project (HMP)
intron
messenger RNA (mRNA)
metagenome
metagenomics
microbial forensics
microbial genomics
microbiome
nucleotide
operon
promoter site
repressor protein
ribonucleic acid (RNA)
ribosomal RNA (rRNA)
ribosome
RNA polymerase
semiconservative replication
transcription
transfer RNA (tRNA)
translation

This picture shows Grand Prismatic Spring located in Wyoming's Yellowstone National Park. The bands of green, yellow, and orange surrounding the pool represent microbial mats containing extremely heat-tolerant bacterial and archaeal species growing in different temperature gradients (87°C to 63°C). The deep blue color of the water in the center of the pool is created by the scattering of blue light and contains no microbes.

5 The Prokaryotic World: The Bacteria and Archaea Domains

Looking Ahead

The prokaryotes are among the most successful organisms on Earth. Over the billions of years of their existence, they have evolved to occupy every conceivable niche on Earth; in fact, they influence almost everything we experience. In this chapter, we shall get to know the prokaryotes as we examine their structures, growth, and diversity.

On completing this chapter, you should be capable of:

- Identifying the three major shapes of prokaryotic cells.
- Explaining the reason for staining cells and describing simple and Gram staining.
- Identifying and explaining the function of bacterial surface structures and appendages, and cytoplasmic structures.
- Describing binary fission and explaining the four phases of a bacterial growth curve.
- Writing a short paragraph describing the unique properties of endospores.
- Differentiating between broth and agar culture media.
- Comparing selective and enriched media.
- Giving examples illustrating the diversity within the domain Bacteria and domain Archaea.

The Earth came into being an almost incomprehensibly distant 4.5 billion years ago. For the first several millions of years of its existence, Earth was a ball of molten rock. Then, as millennia passed, a thin crust formed over the hot core, and violent volcanic activity filled the days and nights. Now the Earth was awash with energy: There was radiation from the sun, lightning from intense electrical storms, and heat from radioactive decay and the ever-present volcanic eruptions. In the incessant rain and the tropically warm oceans, organic molecules were forming—amino acids and nucleotides—the building blocks that one day would compose the molecules of life.

For its first billion years, the Earth was barren of life. Then, about 3.8 billion years ago, microscopic cells, the first living organisms, came into being (FIGURE 5.1). Although scientists are uncertain how these cells arose, they are reasonably sure the first life forms were tiny, single-celled creatures, with little evidence of internal structure. Organisms like these would evolve into the **prokaryotes**, a name that reflects the absence of a cell nucleus (pro = "before"; karyo = "nucleus").

The prokaryotes were the only inhabitants of the Earth for almost 2 billion years, nearly half of the planet's existence. And what a group they were: Growing wildly in the oxygen-free atmosphere of Earth, the ancient prokaryotes chemically combined hydrogen with the carbon of carbon dioxide to form methane (not unlike what we find in a swamp today). Then, the photosynthetic bacteria evolved. As shown in Figure 5.1, these forms, called **cyanobacteria**, were responsible for the "great oxidation event" that occurred some 2.5 billion years ago. Cyanobacteria used their chlorophyll pigments in the process of photosynthesis to capture light energy and produce carbohydrates as energy-storage compounds. In so doing, they dramatically increased the atmospheric concentration of oxygen at that time from 1% to around 10%. Later, with the rise of photosynthetic eukaryotes (algae and plants), the oxygen level fluctuated somewhat before stabilizing at the 21% present today.

Then, about 2 billion years ago, the single-celled **eukaryotes**, typified by simple algae and other protists having cell nuclei and complex internal structures, had successfully evolved and joined the prokaryotes. Around 1.3 billion years ago, the first multicellular eukaryotes evolved followed by the emergence of animals about 500 million years later. Although the eukaryotes were larger organisms, the prokaryotes were not threatened because 3 billion years of evolution allowed

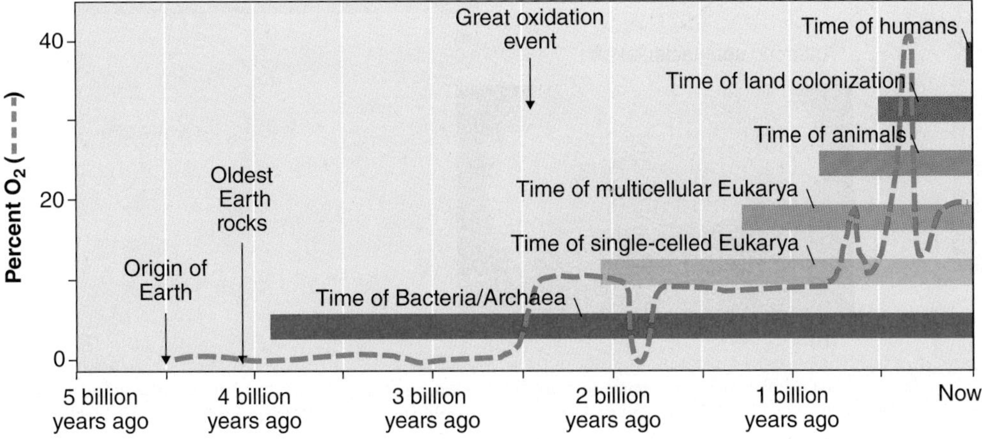

FIGURE 5.1 **The Evolution of Life on Earth.** This timeline shows the relative amount of time that various organisms have existed on Earth. The Bacteria and Archaea have been in existence for a notably longer period of time than any other group. The change in atmospheric oxygen (dashed line) began with the great oxidation event and the expansion of the cyanobacteria.

them to remain well-established, dominant organisms. And the results have been spectacular, for present-day bacterial and archaeal species have come to occupy every conceivable niche on Earth. Every crack, every crevice, every cranny—whether at the bottom of the 6-mile deep Mariana trench in the Pacific, in subglacial lakes of Antarctica, or in the blazing Atacama desert of South America—contains some form of prokaryotic life (FIGURE 5.2).

It should not be surprising then that prokaryotes occupy a critical place in the web of life. Many species make major contributions to the mineral balance of the world by metabolizing the nutrients in freshwater, marine, and terrestrial environments. They reside in soil and influence Earth's ecology by breaking down the remains of dead plants and animals and recycling the elements. And certain bacterial species do what few other species of organisms on Earth can do—they trap nitrogen from the air and convert it to substances used by plants to make protein, protein that ultimately winds up on our plates as grains, meat, and dairy products.

In industrial and biochemical laboratories, bacterial species are both workhorses and "lab rats." Growing in enormous numbers in mammoth fermentation tanks, they carry on their day-to-day chemical routines and yield products of substantial value. For example, they manufacture organic compounds, produce fermented foods, synthesize antibiotics and vitamins, and serve as biological factories for genetic engineering.

Although some bacterial species do pose a threat to humans by causing infectious disease, as our constant companions, prokaryotes have impacted our lives in many ways that are beneficial and in some ways that have changed society and history. These remarkable and fascinating microbes are the subject matter of this chapter. As a starter, read **A Closer Look 5.1**.

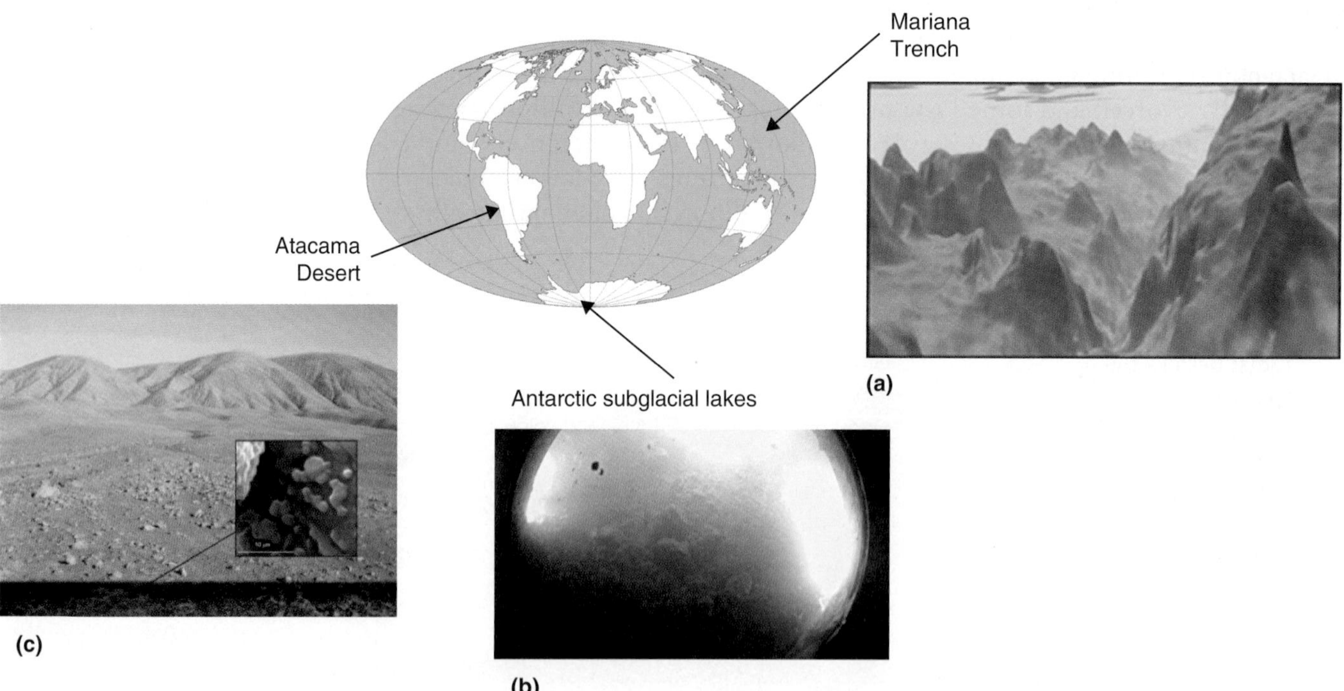

FIGURE 5.2 **Extremes of Prokaryotic Colonization.** The prokaryotes (Bacteria and Archaea) can be found all around the world, including in some fairly remote places such as (a) at the bottom of the Mariana Trench, (b) in lakes sealed under meters of Antarctic ice, and (c) in the dry Atacama desert in Chile.

(a) Courtesy of NGDC/NOAA; (b) Courtesy of Alberto Behar/JPL-Caltech/NASA; (c) Courtesy of Parro et al./CAB/SINC

A CLOSER LOOK 5.1

Bacteria in Eight Easy Lessons[1]

Mélanie Hamon, an assistante de recherché at the Institut Pasteur in Paris, says that when she introduces herself as a bacteriologist, she often is asked, "Just what does that mean?" To help explain her discipline, she gives us, in eight letters, what she calls "some demystifying facts about bacteria."

Basic principles: Their average size is 1/25,000th of an inch. In other words, hundreds of thousands of bacteria fit into the period at the end of this sentence. In comparison, human cells are 10 to 100 times larger with a more complex inner structure. While human cells have copious amounts of membrane-contained subcompartments, bacteria more closely resemble pocketless sacs. Despite their simplicity, they are self-contained living beings, unlike viruses, which depend on a host cell to carry out their life cycle.

Astonishing: Bacteria are the root of the evolutionary tree of life, the source of all living organisms. Quite successful evolutionarily speaking, they are ubiquitously distributed in soil, water, and extreme environments such as ice, acidic hot springs or radioactive waste. In the human body, bacteria account for 10% of dry weight, populating mucosal surfaces of the oral cavity, gastrointestinal tract, urogenital tract and surface of the skin. In fact, bacteria are so numerous on earth that scientists estimate their biomass to far surpass that of the rest of all life combined.

Crucial: It is a little known fact that most bacteria in our bodies are harmless and even essential for our survival. Inoffensive skin settlers form a protective barrier against any troublesome invader while approximately 1,000 species of gut colonizers work for our benefit, synthesizing vitamins, breaking down complex nutrients and contributing to gut immunity. Unfortunately for babies (and parents!), we are born with a sterile gut and "colic" our way through bacterial colonization.

Tools: Besides the profitable relationship they maintain with us, bacteria have many other practical and exploitable properties, most notably, perhaps, in the production of cream, yogurt and cheese. Less widely known are their industrial applications as antibiotic factories, insecticides, sewage processors, oil spill degraders, and so forth.

Evil: Unfortunately, not all bacteria are "good," and those that cause disease give them all an often undeserved and unpleasant reputation. If we consider the multitude of mechanisms these "bad" bacteria—pathogens—use to assail their host, it is no wonder that they get a lot of bad press. Indeed, millions of years of coevolution have shaped bacteria into organisms that "know" and "predict" their hosts' responses. Therefore, not only do bacterial toxins know their target, which is never missed, but bacteria can predict their host's immune response and often avoid it.

Resistant: Even more worrisome than their effectiveness at targeting their host is their faculty to withstand antibiotic therapy. For close to 50 years, antibiotics have revolutionized public health in their ability to treat bacterial infections. Unfortunately, overuse and misuse of antibiotics have led to the alarming fact of resistance, which promises to be disastrous for the treatment of such diseases.

Ingenious: The appearance of antibiotic-resistant bacteria is a reflection of how adaptable they are. Thanks to their large populations they are able to mutate their genetic makeup, or even exchange it, to find the appropriate combination that will provide them with resistance. Additionally, bacteria are able to form "biofilms," which are cellular aggregates covered in slime that allow them to tolerate antimicrobial applications that normally eradicate free-floating individual cells.

A long tradition: Although "little animalcules" were first observed in the 17th century, it was not until the 1850s that Louis Pasteur fathered modern microbiology. From this point forward, research on bacteria has developed into the flourishing field it is today. For many years to come, researchers will continue to delve into this intricate world, trying to understand how the good ones can help and how to protect ourselves from the bad ones. It is a great honor to be part of this tradition, working in the very place where it was born.

---

[1]Republished with permission of the author, the Institut Pasteur, and the Pasteur Foundation. The original article appeared in *Pasteur Perspectives* Issue 20 (Spring 2007), the newsletter of the Pasteur Foundation, which may be found at www.pasteurfoundation.org © Pasteur Foundation.

5.1 Cell Structure: Form and Function

Prokaryotes differ structurally from the eukaryotes. In very broad terms, a bacterial cell, such as *Escherichia coli*, is much smaller and appears less complex structurally than most eukaryotic organisms, such as an amoeba or algal cell.

■ *Escherichia coli*
esh-e'r-ē'kē-ä kō'lī (or kō'lē)

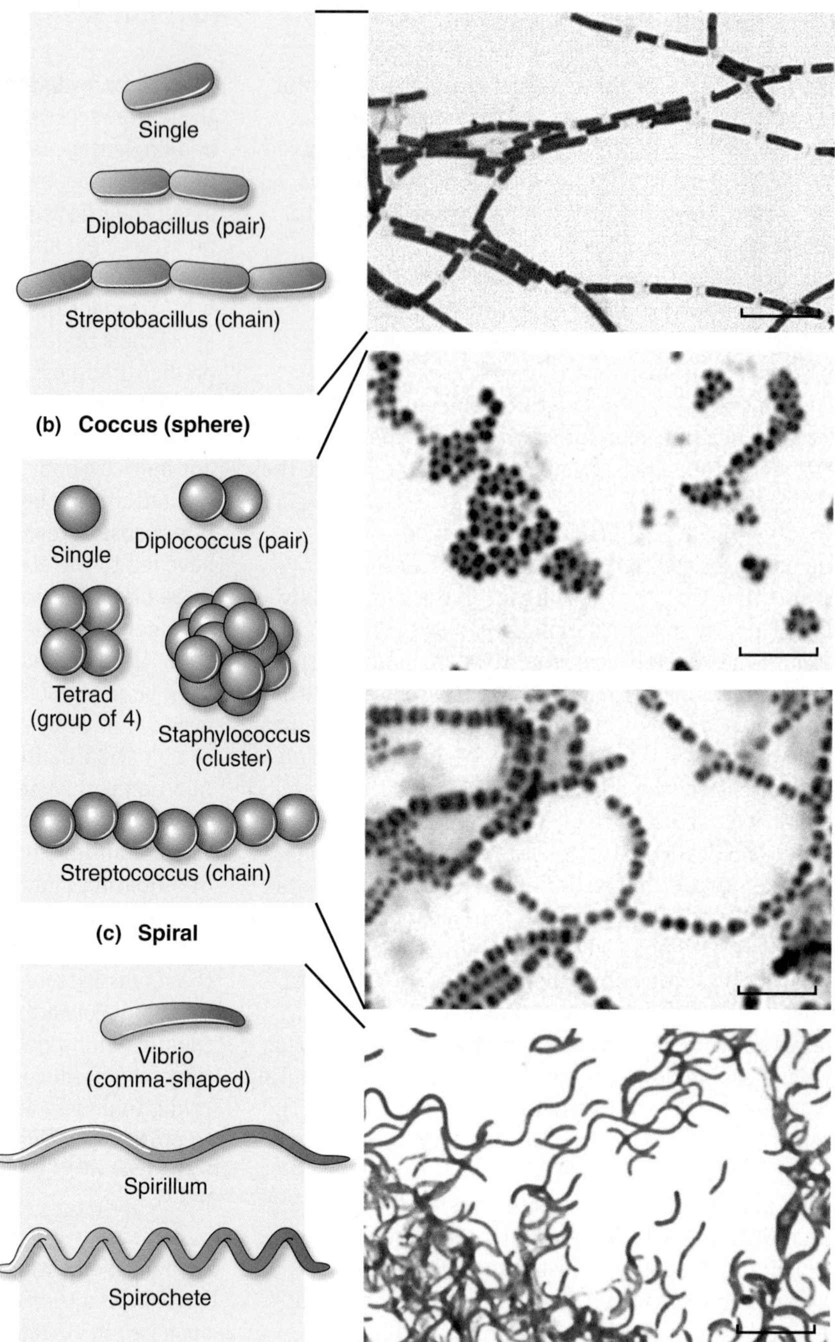

FIGURE 5.3 **Bacterial Cell Shapes and Arrangements.** Many bacterial cells have a bacillus (**a**) or coccus (**b**) shape. Most spiral shaped-cells (**c**) are not organized into a specific arrangement (All bars = 10 µm.).

Courtesy of Dr. Jeffrey Pommerville

Prokaryotic Cell Morphology

Cell **morphology** refers to the shape and structure of an organism. Bacterial cells generally occur in variations of three major forms or shapes (FIGURE 5.3): Rod-shaped bacterial cells are known as **bacilli** (sing., bacillus). They vary in size, with the shortest ones measuring about 0.5 µm in length and the longer ones about 20 µm. Other bacterial cells are spherical and are called **cocci** (sing., coccus). They measure roughly 1 μm in diameter; and still other bacterial cells have a long, spiral form that can be bent (**vibrio**), wavy and flexible (**spirillum**), or "cork-screw" shaped and rigid (**spirochete**).

Figure 5.3 also illustrates the typical cell arrangements found among the bacillus- and coccus-shaped species. Bacilli may be single cells, arranged as a pair (diplobacillus; *diplo* = "two"), or in a short or long chain (streptobacillus; *strepto* = "chain"). The cocci are most often arranged in a pair (diplococcus), a packet of four cells (tetrad; *tetra* = "four"), a cluster of cells (staphylococcus; *staphylo* = "cluster"), or in a chain of spheres (streptococcus). The arrangement can be important to the identification of certain bacterial species. For example, the bacterial species causing gonorrhea (*Neisseria gonorrhoeae*) is a diplococcus, while the species causing "strep throat" (*Streptococcus pyogenes*) is a streptococcus. Note that sometimes, as in the second example, the genus name may also describe the cell's shape and/or arrangement. In medical labs, technicians can use a microscope to distinguish these arrangements and assist physicians in their diagnoses.

■ *Neisseria gonorrhoeae*
nī-se'rē-ä go-nôr-rē' ī

■ *Streptococcus pyogenes*
strep-tō-kok'kus pī- ä j'en-ez

Other forms exist as well. For example, cyanobacteria consist of elongated microscopic cells usually occurring in filaments, which may measure a meter or more in length. Although the cells are connected at their outer walls, each cell operates independently. The cells of archaeal organisms, besides having the typical rod and sphere shapes, can also be square, triangular, or star-shaped.

Studying the structures of prokaryotic cells helps us to realize that they are much more than dots when observed with the light microscope. Especially as we survey the structural makeup of bacterial cells, watch for the various functions the bacterial structures perform. These functions give us a sense of their activities.

Staining Procedures

To study cell morphology, the cells need to be observed with the light microscope. Unfortunately, most prokaryotic cells lack color, so it can be very difficult to see the individual cells. This challenge was first solved in the early 1880s when scientists such as Robert Koch and Paul Ehrlich started using colored stains or dyes to improve their ability to see the tiny cells. These colored compounds have a positive electrical charge and are attracted to cells, which have an overall negative electrical charge. Use of a single stain characterizes the so-called **simple stain technique** (FIGURE 5.4a). However, the stain preparation kills the cells. The light microscope images of bacterial cells in Figure 5.3 are the result of simple staining.

The most common staining technique was developed in the 1880s by Christian Gram, a Danish physician. Called the **Gram stain technique**, the procedure uses two contrasting stains and not only stains most bacterial cells, but also permits them to be assigned into one of two groups: the **gram-positive** bacteria or the **gram-negative** bacteria. The bacterial cells to be studied are first stained with crystal violet and then with Gram's iodine solution (FIGURE 5.4b). At this point, all the bacterial cells are blue-purple due to the crystal violet. Now, the stained cells are washed with ethyl alcohol, which removes the color from the gram-negative cells only; the gram-positive cells remain blue-purple. Because the gram-negative cells are now colorless and would be difficult to see in the light microscope, the final step involves the application of a red dye called safranin. This stain is picked up by the gram-negative cells, causing them to appear orange-red in color; the gram-positive cells remain blue-purple. After completing the staining, microscopic observation reveals the color of the bacterial cells and tells us whether they are Gram positive or Gram negative (FIGURE 5.5).

Knowing whether a bacterium is Gram positive or Gram negative is important medically because it helps identify certain characteristics of a bacterial species. Many

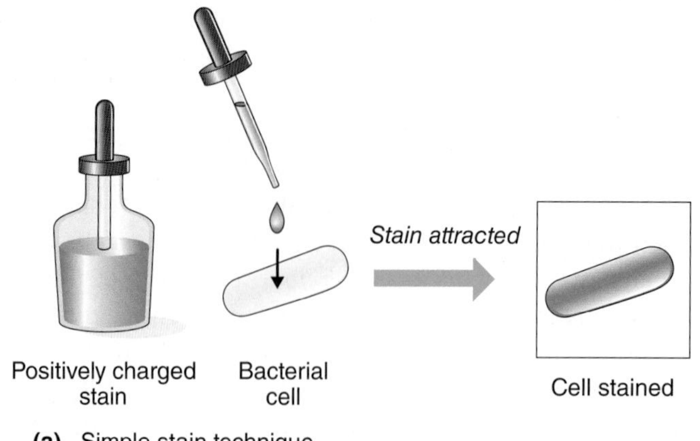

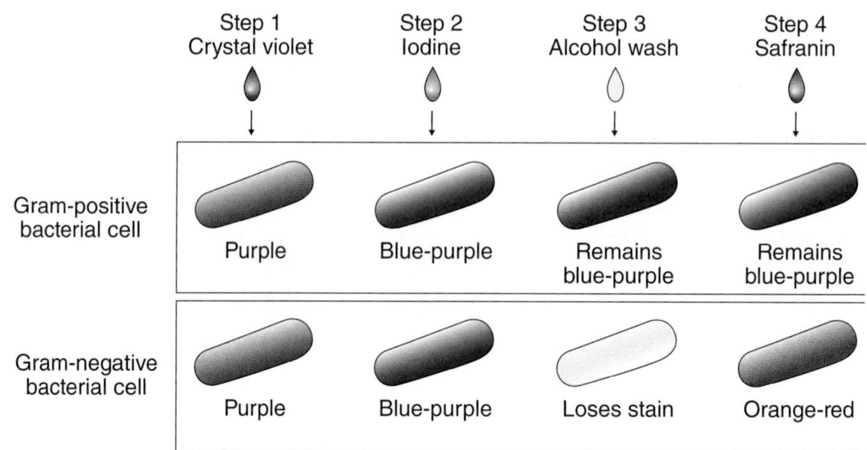

■ urethral: Referring to the tube that carries urine (and semen in males) out of the body.

FIGURE 5.4 **Stain Reactions in Microbiology.** (**a**) In the simple stain technique, mixing the positively charged dye with the negatively charged bacterial cells stains the cells and makes them visible in the light microscope. (**b**) The Gram stain technique stains gram-positive cells blue-purple and gram-negative cells orange-red.

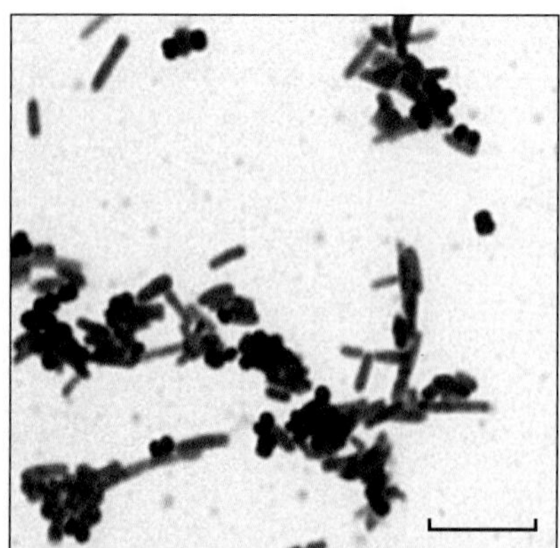

FIGURE 5.5 **Gram-Stained Bacterial Cells.** Following the Gram stain procedure, the coccus-shaped cells stain blue-purple while the gram-negative rods stain orange-red. (Bar = 6 μm.)

Courtesy of Dr. Jeffrey Pommerville

antibiotics, such as penicillin, are effective against gram-positive species but less so against gram-negative species, and certain antiseptics and disinfectants affect gram-positive bacteria but not gram-negative bacteria (or vice versa). In addition, shape together with Gram staining can be diagnostic. For example, if a urethral specimen taken from a male patient is Gram stained and found to contain gram-negative diplococci, it almost certainly is *N. gonorrhoeae* and gonorrhea is the diagnosis.

Surface Structures and Appendages

Bacterial cells have several structures on or projecting from the cell surface (FIGURE 5.6). The "cell envelope," which consists of the cell wall and the cell membrane, will be discussed first.

Cell Wall. With few exceptions, all bacterial cells are encased in a **cell wall**, which contains a tough mesh of polysaccharide and protein called **peptidoglycan**. Peptidoglycan lends rigidity and strength to the cell wall and is found in no other organism. Gram-positive cells have multiple layers of peptidoglycan (which may contribute to their ability to resist the alcohol

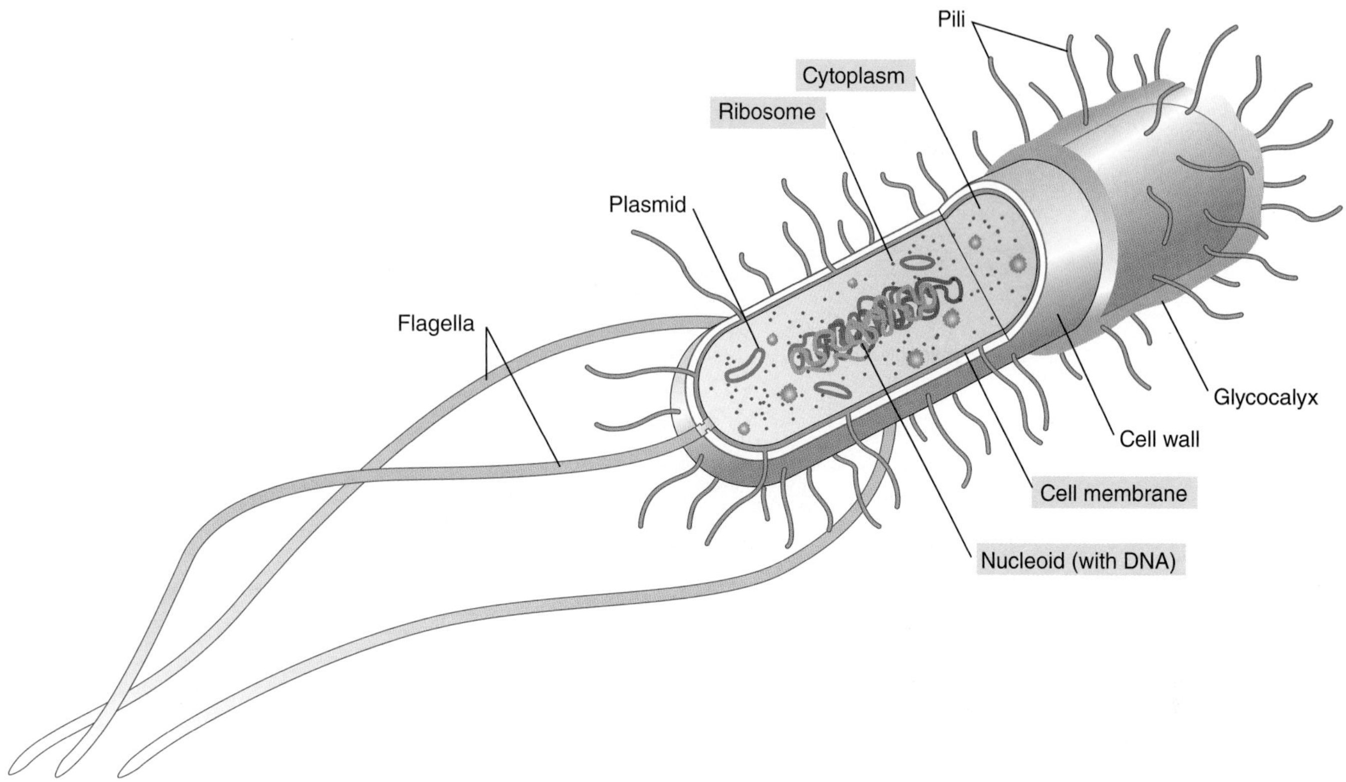

FIGURE 5.6 **Bacterial Cell Structure.** The structural features of a composite, "idealized" bacterial cell are shown. Structures highlighted in blue are found in all prokaryotes.

wash used in Gram staining), reinforced with sugar-alcohol molecules called **teichoic acid** (FIGURE 5.7a). Antibiotics, such as penicillin and its relatives (e.g., ampicillin, amoxicillin, methicillin, and numerous others) target the peptidoglycan and prevent the bacterial cells from assembling peptidoglycan chains, leaving the organism with only a cell membrane. Without a cell wall, internal water pressure soon causes the cell to swell and burst.

The walls on gram-negative cells are quite different (FIGURE 5.7b). There is but a single chain or two of peptidoglycan beneath a covering called the **outer membrane**. The outer half of this membrane forms a lipid and polysaccharide layer called the **lipopolysaccharide (LPS)**. In gram-negative bacterial cells, such as *E. coli*, penicillin and its relatives, as well as many antiseptics and disinfectants cannot cross the outer membrane, making such bacterial species more resistant to the effects of penicillin. However, other antibiotics can penetrate the wall and inhibit internal metabolic reactions. In addition, the lipid portion of the LPS, when released in the human body as the result of an infection, acts as a poison and triggers an overactive immune reaction producing fever and possibly life-threatening shock.

■ shock: A loss of blood pressure and inadequate blood circulation.

Cell Membrane. The structure of the bacterial **cell membrane** is similar to its counterpart (the plasma membrane) in eukaryotic cells. The essential feature is a double layer of phospholipids (a phospholipid bilayer) with protein molecules suspended in the phospholipids at the surface and often spanning the bilayer (see Figure 5.7). Some of these proteins function as enzymes during chemical reactions, while others transport substances across the cell membrane. Because of its ever-changing nature, the membrane is called a "**fluid mosaic**" because the proteins often change position (the mosaic) within a fluid phospholipid bilayer.

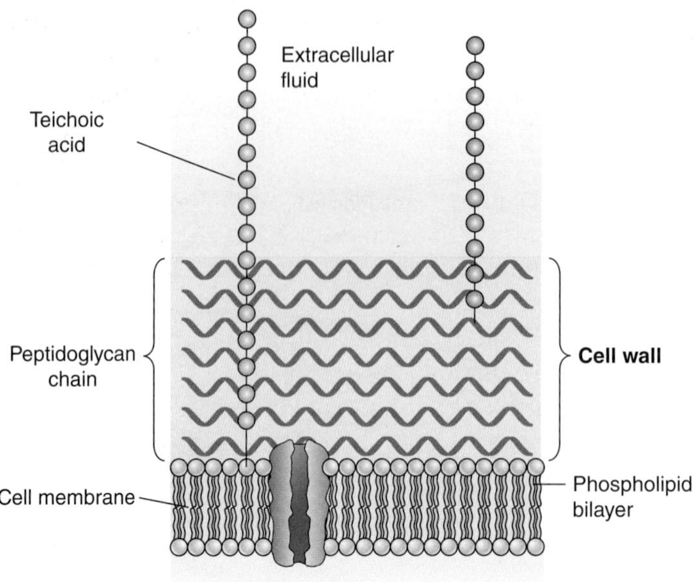

(a) A section of a gram-positive cell wall.

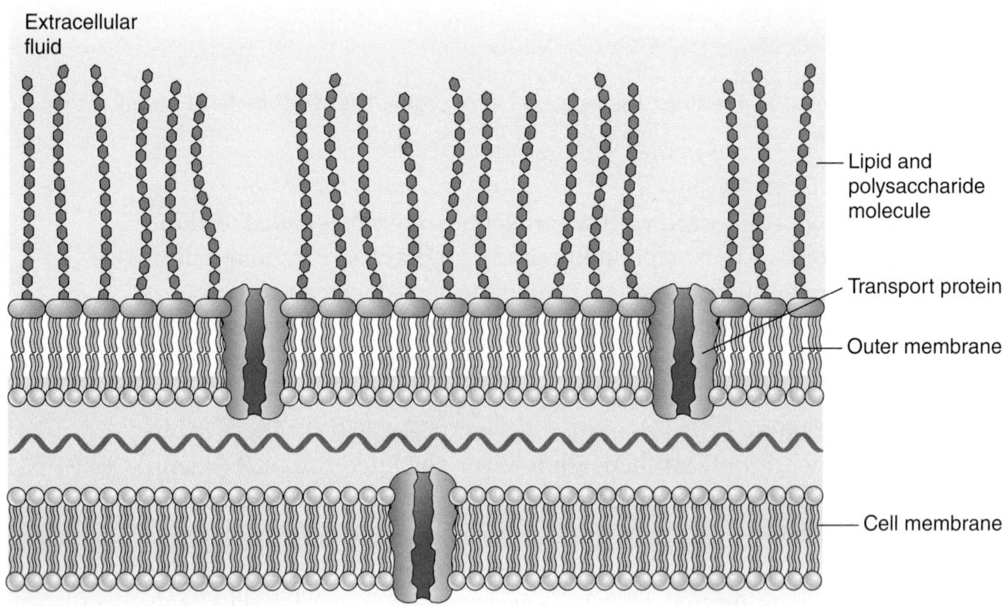

(b) A section of a gram-negative cell wall.

FIGURE 5.7 **A Comparison of the Cell Walls of Gram-Positive and Gram-Negative Bacterial Cells.** (**a**) The cell wall of a gram-positive bacterial cell is composed of multiple peptidoglycan chains intermixed with teichoic acid molecules. Below the wall is the cell membrane that consists of a phospholipid bilayer in which are embedded transport proteins. (**b**) In the gram-negative cell wall, the peptidoglycan chains are few and there are no sugar-alcohol molecules. Moreover, an outer membrane overlies the peptidoglycan layer in the periplasmic space. Note that the outer membrane contains special transport proteins and the outer half is unique in containing lipopolysaccharide (LPS) molecules.

Notice in Figure 5.7b that gram-negative cells have a gap between the outer membrane and the cell membrane. This gap, called the **periplasmic space**, is an active and important processing center where nutrients too large to pass through the cell membrane are first broken down. The chain or two of peptidoglycan also occurs here.

Glycocalyx. Some bacterial species have a **glycocalyx**, a ca coat outside of the cell envelope. Known as a "capsule" if tightly b cell, or a "slime layer" if loosely bound, the glycocalyx provides prote the cell, shielding it from drying (desiccation), chemicals, and environm stresses.

The glycocalyx also provides for attachment to surfaces. For example, the capsule of *Streptococcus mutans* provides a means for the cells to attach to tooth enamel in pockets between the teeth and gums and sets up a situation favorable for the development of dental caries. Held in place, the bacterial cells break down sugars and other carbohydrates we consume and produce large amounts of acid that gradually eats away at the tooth enamel, forming a depression, or cavity.

■ St strep

Flagella. Many bacterial cells also contain hair-like appendages anchored to the cell wall and cell membrane. One such structure, the **flagellum** (pl., flagella) is a long rigid filament of protein that rotates like a propeller and propels a bacterial cell through its liquid surroundings. This is different from the flagella of some eukaryotes, which are covered by the cell's plasma membrane and propel the cell by beating back and forth.

Various bacterial species have either a single flagellum, a tuft of flagella at one end of the cell, or flagella covering the entire cell surface (FIGURE 5.8a). Using flagella, a bacterium such as *E. coli* is capable of traveling about 2,000 times its body length in an hour. Calculations for a 5 foot 10 inch human walking at about 2.25 miles per hour would yield about the same speed. The bacterial flagella help the cells navigate toward nutrients or away from potentially toxic chemicals.

Pili. Another hair-like structure called a **pilus** (pl., pili) helps bacterial cells attach to tissues or other surfaces. Pili are short, rigid cylindrical protein rods about 1 μm in length and about 7 nm thick (FIGURE 5.8b). When causing infection, pathogens having pili can attach to tissues, facilitating an infection process. **A Closer Look 5.2** provides an "unsettling" example.

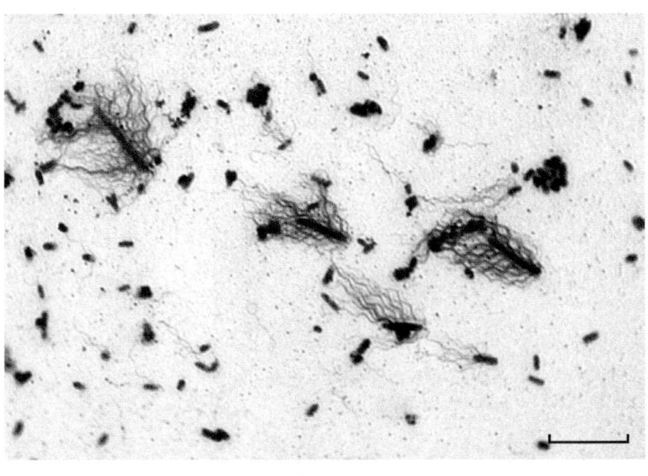

(a)

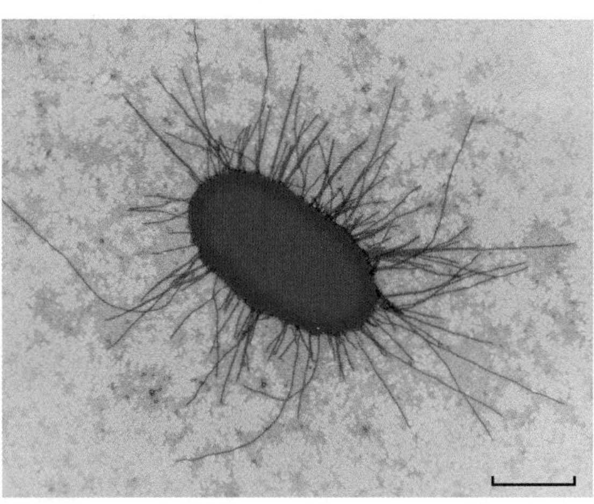

(b)

FIGURE 5.8 **Bacterial Surface Appendages.** (**a**) A light microscope image showing stained bacterial flagella that extend from the cell surface. Note that the flagella are many times the length of the bacterial cell and have a characteristic wavy arrangement. (Bar = 10 μm.) (**b**) A false-color transmission electron microscope image of an *Escherichia coli* cell with many pili. (Bar = 1 μm.)

(a) Courtesy of Dr. Jeffrey Pommerville. (b) © Dennis Kunkel Microscopy, Inc./Visuals Unlimited, Inc.

ng

ond

the

se

ept

away by ... ause

intestinal distress- ... onal

sixty were out to verify or prov...

On the fateful day, the experiment beg... ner the students nor the health professionals knew who was rinking the diarrhea cocktail and who was getting the 'free pass"; it was a double-blind experiment. Then came the waiting. Some volunteers experienced no symptoms, but others felt the bacterial onslaught and clutched at their last remaining vestiges of dignity. For some, it was three days of hell, with nausea, abdominal cramps, and numerous bathroom trips; for others, luck was on their side (investing in a lottery ticket seemed like a good idea at the time).

When it was all over, the numbers appeared to bear out the theory: The great majority of volunteers who drank the cocktail with mutated bacterial pili experienced no diarrhea, while most of those who drank the cocktail with normal bacterial pili had attacks of diarrhea, in some cases, real doozies.

In the end, all appeared to profit from the experience: The scientists had some real-life evidence that pili contribute to infection; the students made their sacrifice to science and pocketed $300 each; and the local supermarket had a surge of profits from sales of toilet paper, Pepto-Bismol®, and Imodium®.

Cytoplasmic Structures

Although the prokaryotic cell interior is structurally less complex than that in a eukaryotic cell, many small molecules and other organic substances are dissolved in the jelly-like cytoplasm. This includes amino acids and proteins, sugars, and nucleotides as well as numerous minerals and growth factors. In addition, there are a few larger subcompartments and structures (see Figure 5.6).

Nucleoid and Plasmids. The genetic information of prokaryotes is contained in a single **chromosome** existing as a closed loop of tightly compacted deoxyribonucleic acid (DNA) in the cytoplasm. The interior area where the bacterial chromosome is found is called the **nucleoid** and it lacks the surrounding membranes typical of the eukaryotic cell nucleus (FIGURE 5.9). Because bacterial cells have only one chromosome, bacterial species, such as *E. coli*, are appealing as research tools in biochemical genetics—scientists can isolate the single chromosome and study its activity without

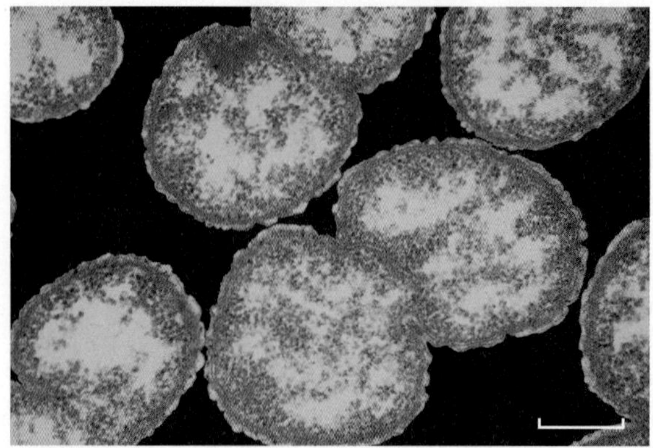

FIGURE 5.9 **The Bacterial Nucleoid.** Bacterial cells contain a nucleoid that contains the circular chromosome (DNA). In this false-color transmission electron microscope image, the nucleoid is the smooth, yellow-colored area seen near the center of each cell. The brown dots are ribosomes. (Bar = 1 µm.)

worrying about the other chromosomes, as they must do when working with eukaryotic cells. The thousands of genes, the functional units of DNA, making up the full complement of genetic information is called the bacterial **genome**.

In addition to a chromosome, many bacterial species have **plasmids** that are much smaller loops of DNA than the bacterial chromosome and replicate independently of the chromosome. Suspended in the cytoplasm, the plasmids contain genes encoding proteins that are not essential for everyday survival of the organism. Rather, plasmids often contain information for antibiotic resistance, to produce toxins, and become resistant to environmental chemicals. Plasmids also have been manipulated in modern DNA technology where they can be outfitted with new genes and then inserted into other bacterial cells, where their new genes are activated to encode the protein desired by the biochemist. This imaginative and innovative process underlies much of the current revolution in biotechnology.

Ribosomes. All prokaryotes have in their cytoplasm the necessary structures for producing proteins. These are the **ribosomes**, the tiny structures consisting of RNA and protein capable of translating the information from the genes into proteins needed for cell metabolism, growth, and reproduction.

5.2 Prokaryotic Reproduction and Survival: Binary Fission and Endospores

When we talk about microbial growth of prokaryotes, we are referring to the increase in numbers of cells in the population. Therefore, before examining microbial growth, we need to look at reproduction.

Prokaryotic Reproduction

Most prokaryotic species reproduce by a relatively straightforward method called **binary fission**. Prior to division, a cell , such as *E. coli*, increases in size and its enzymes replicate its DNA to yield two chromosomes (FIGURE 5.10). Binary fission then involves the extension of a new cell wall and cell membrane inward from the margin of the cell. As the wall and membrane come together, the adjoining walls and membranes fuse, splitting the original cell into two new daughter bacterial cells, each cell having one chromosome. (The effect is somewhat like forming two sausages from a single, longer sausage.)

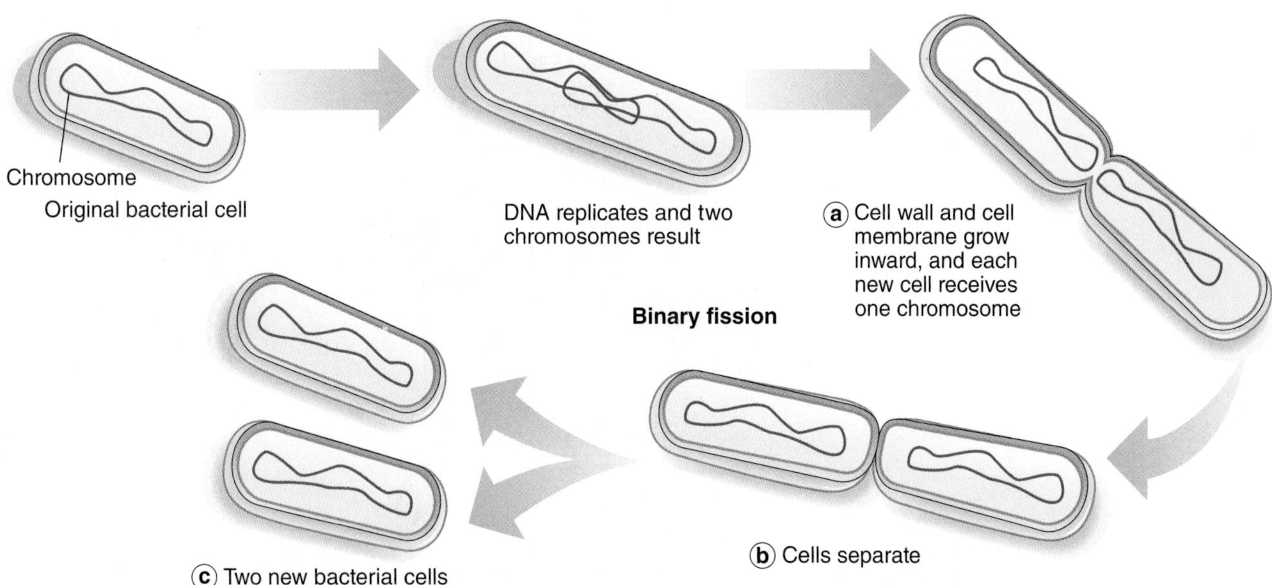

FIGURE 5.10 **Binary Fission in Bacteria.** After a bacterial cell has replicated its DNA (chromosome), the fission process (**A–C**) essentially pinches the cell in two as the cell wall and cell membrane converge.

The frequency of bacterial reproduction is quite extraordinary. For example, under ideal growth conditions, the common intestinal bacterium *E. coli* can undergo binary fission and produce two new daughter cells every 20 minutes (by comparison, a human liver cell takes about 20 hours to produce two new daughter cells). Therefore, within 1 hour, a single bacterial cell can become 8 cells; in 2 hours, there will be 64 cells; and in only 7 hours, there will be a population of more than 2 million cells (FIGURE 5.11). One enterprising mathematician has calculated that a single *E. coli* cell reproducing every 20 minutes could yield in 36 hours enough bacterial cells to cover the face of the Earth a foot thick!

If these calculations are correct, why are we not smothered in a sea of bacterial cells? The answer is that bacterial cells are susceptible to the same dynamics as any population of plants, animals, or microbes. Eventually, nutrients become scarce, waste products accumulate, water is in short supply, the environmental temperature changes, and from the bacterium's point of view, things get stressful. Fewer cells reproduce, more bacterial cells die, and the population stabilizes or might actually decline in number.

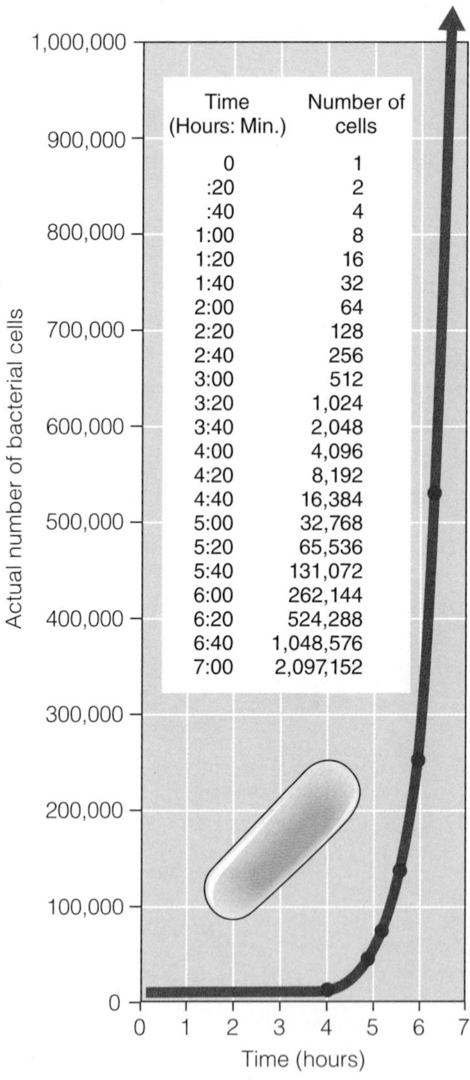

Time (Hours: Min.)	Number of cells
0	1
:20	2
:40	4
1:00	8
1:20	16
1:40	32
2:00	64
2:20	128
2:40	256
3:00	512
3:20	1,024
3:40	2,048
4:00	4,096
4:20	8,192
4:40	16,384
5:00	32,768
5:20	65,536
5:40	131,072
6:00	262,144
6:20	524,288
6:40	1,048,576
7:00	2,097,152

FIGURE 5.11 **A Skyrocketing Bacterial Population.** The number of *Escherichia coli* cells progresses from one cell to more than 2 million cells in a mere seven hours. Only the depletion of nutrients, a buildup of wastes, or some other limitation will halt the growth of the bacterial cell population.

Such a rise and potential fall of microbial cells in a population can be presented graphically as a **growth curve**. As an example, let's use a bacterial species, like *E. coli*, growing in a nutrient solution.

As described above, a population of bacterial cells will increase in number as the cells undergo binary fission. The growth in numbers is plotted with logarithmic numbers (powers of 10) because normal ordinal numbers would go off the scale very rapidly. Plotting the number of bacterial cells versus time produces a growth curve such as the one shown in FIGURE 5.12. Various phases in the curve can be identified and show the dynamics of the population as time passes. There are four phases to a growth curve.

Lag Phase. The first phase of growth is called the **lag phase**. During this "tooling up" phase, no increase in cell number is observed. In the broth tube, the *E. coli* cells are synthesizing cell parts and enzymes for digesting the nutrients in the broth. The actual length of the lag phase depends on the metabolic activity of the microbial population. They must grow in size, take up nutrients, and replicate their DNA—all in preparation for binary fission.

Log Phase. The next phase is referred to as the **logarithmic (log) phase**. Here the microbes grow and divide at the maximum possible rate. An *E. coli* population, for instance, doubles at regular intervals and the number of cells rises smoothly in an upward direction when the logarithms of the actual numbers of bacteria are plotted. In broth, the bacterial cells initially have plenty of nutrients, oxygen gas, and other factors needed for growth. In fact, the population of bacterial cells will be so vigorous that the broth becomes cloudy (turbid) as the bacterial cells multiply. If done on a nutrient agar plate, visible masses called **colonies** appear, each colony consisting of millions of bacterial cells. Because the population is at its biochemical optimum, research experiments are generally performed during the log phase. Vulnerability to antibiotics is also highest at this stage of growth because many

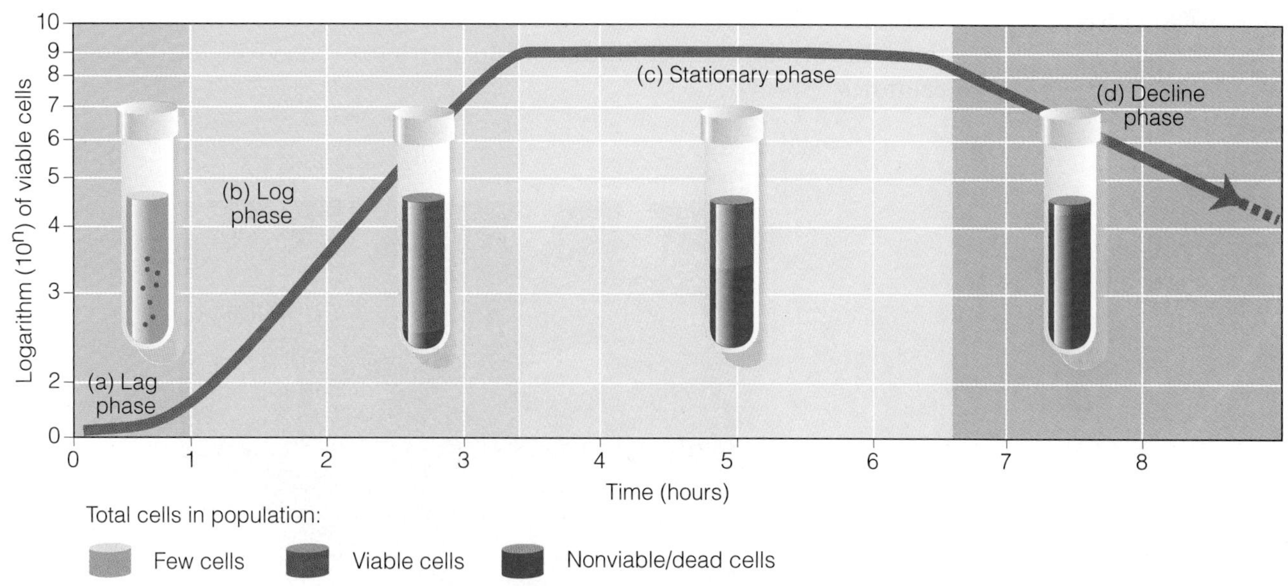

FIGURE 5.12 **The Growth Curve for a Bacterial Population.** (**a**) During the lag phase, the population numbers remain stable as bacteria prepare for division. (**b**) During the logarithmic (log) phase, the numbers double with each generation time. Environmental factors later lead to a stationary phase (**c**), which involves a stabilizing population size. (**d**) The decline (death) phase is the period during which cell death becomes substantial.

antibiotics affect metabolic processes like protein synthesis in actively dividing cells.

Stationary Phase. In the confines of a broth tube (or agar plate), nutrients needed for growth are not unlimited. Therefore, after a period of time, the microbial population enters a **stationary phase** as the nutrient supply dwindles. During this phase, many cells are dying and the number of dead cells balances the number of new ones. Besides a scarcity of nutrients, the decline in growth rate may be due to the accumulation of waste products, and for some organisms needing oxygen gas, the gas is now in short supply. Thus, the curve on the graph flattens out.

Death Phase. The final phase of the population's history is the **decline (death) phase**. Here the *E. coli* cells are dying off rapidly and the population size is decreasing significantly. For most species, like *E. coli*, the history of the population comes to an end with the death of the last cell. However, if the organism is a species of *Bacillus* or *Clostridium*, the vegetative cells form endospores, which can remain dormant but alive until nutrients return.

Endospores

Bacillus
bä -cil'lus

Clostridium
klos-tri'dē-um

A few bacterial species, particularly members of the genera *Bacillus* and *Clostridium*, have the ability to produce an extraordinarily resistant structure called the **endospore** (FIGURE 5.13). This structure is usually formed when nutrient stress is encountered in the environment. The spore is formed during an involved process during which the mother cell produces an internal spore (i.e., endospore) that contains a complete set of genetic information surrounded by a very thick peptidoglycan set of walls. Once formed and mature, the dormant spore is released with the death of the mother cell.

Endospores are probably the most resistant biological structures known. Desiccation has little effect on the spore. By containing only 10% water, endospores are heat resistant and undergo very few chemical reactions. These properties make them difficult to eliminate from contaminated medical materials and food products. For example, endospores can remain viable in boiling water (100°C) for 2 hours. When placed in 70% ethyl alcohol, endospores have survived for 20 years. Humans can barely withstand 500 rems of radiation, but endospores can survive one million rems. In this dormant condition, endospores can "survive" for decades or even centuries.

rem (Roentgen Equivalent in Man): A measure of radiation dose related to biological effect.

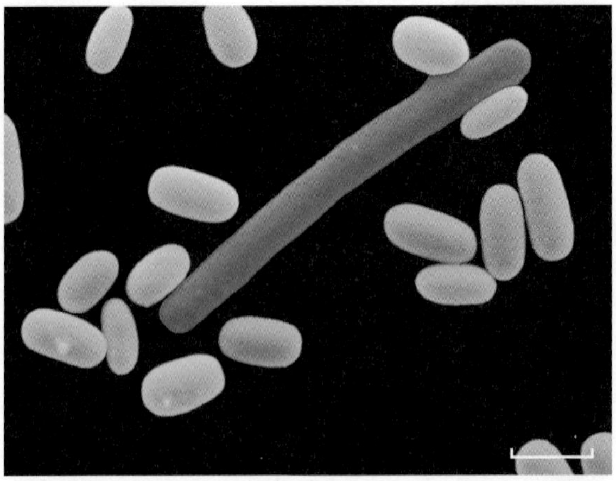

FIGURE 5.13 **Bacterial Endospores.** This false-color scanning electron microscope image shows a mother cell and several endospores. Any one mother cell produces only one endospore before disintegrating and releasing the spore. (Bar = 1 μm.)

A few serious human diseases are the result of endospores. The most newsworthy has been *Bacillus anthracis*, the agent of the 2001 anthrax bioterror attack that occurred through the mail. This potentially deadly disease develops when inhaled endospores germinate in the lungs. The resulting cells secrete two deadly toxins. Likewise, botulism and tetanus are diseases caused by different species of *Clostridium*. Clostridial endospores often are found in soil, as well as in human and animal intestines. However, the environment must be free of oxygen for the spores to germinate to growing cells. A puncture wound contaminated with endospores provides an environment where the endospores germinate and the living cells then produce tetanus toxin. In a sealed can of food inadvertently contaminated with endospores, the spores will germinate and the living cells produce botulism toxin.

Bacillus anthracis
bä -cil'lus an-thrā'sis

5.3 Prokaryotic Growth: Culturing Bacteria

With this understanding of bacterial populations, let's now look at the methods used to grow microbes in the lab. Again we will primarily examine bacterial species.

Growth Methods

The ability to grow bacterial cells in the laboratory is essential to many applications of microbiology. Bacterial organisms are cultivated on or in materials called culture media (sing., medium). A **culture medium** is a water solution containing various nutrients that promote the growth of one or more species. The medium generally contains a source of energy (for example, glucose), plus sources of carbon, nitrogen, and other essential nutrients.

Culture media may be used as liquids or gels. In liquid form, the culture medium is called a **broth**. A typical example is nutrient broth, a "beef/vegetable soup" containing extracts or digests from various animal or plant sources. Other nutrients, such as vitamins, may also be present. The gel, or semisolid, form of a culture medium is referred to somewhat imprecisely as an agar. This is because it consists of a broth solidified with **agar**, a complex carbohydrate derived from algae. Agar is not used as a nutrient by most bacterial species; rather, it is simply a solidifying substance that remains solid at temperatures as high as 80°C (human body temperature, for comparison, is 37°C). **Nutrient agar** (broth in solidified agar) is an example of an agar medium. When the cells reproduce and grow in number on the agar surface, they form colonies (FIGURE 5.14).

FIGURE 5.14 **Bacterial Colonies.** Bacterial colonies cultured on blood agar in a culture dish. Blood agar is a mixture of nutrient agar and sheep blood cells. It is often used for growing bacterial colonies.

Courtesy of Dr. J. J. Farmer/CDC

Many nutrient broth and nutrient agar culture media are general purpose because they support the growth of many different microbial species. Microbiologists have also developed a number of special media, called **selective media**, for cultivating specific types of bacterial species. Such a medium will encourage (select for) the growth of desired species while discouraging the growth of other species. In addition, some species are said to be "fastidious;" that is, they require specific nutrients not normally found in most culture media. To stimulate growth of fastidious species, microbiologists may add special nutrients to standard nutrient agar to produce a so-called **enriched medium**. A real life example is described in **A Closer Look 5.3**.

The extensive research into bacterial culture media has resulted in culture "recipes" for cultivating most human pathogens as well as many environmental species. However, most bacterial species cannot be grown in any known culture medium

A CLOSER LOOK 5.3

"Enriching" Koch's Postulates

On July 21–23, 1976, some 5,000 Legionnaires attended the Bicentennial Convention of the American Legion in Philadelphia, PA. About 600 of the Legionnaires stayed at the Bellevue Stratford Hotel. As the meeting was ending, several Legionnaires who stayed at the hotel complained of flu-like symptoms. Four days after the convention, an Air Force veteran who had stayed at the hotel died. He would be the first of 34 Legionnaires over several weeks to succumb to a lethal pneumonia, which became known as Legionnaires' disease.

As with any new disease, epidemiologists or "disease detectives" look for the source of the disease. The epidemiologists with the Centers for Disease Control and Prevention (CDC) had an easy time tracing the source back to the Bellevue Stratford Hotel. Their studies also try to identify the causative agent. Using Koch's postulates, CDC staff collected tissues from lung biopsies and sputum samples. However, no microbes could be grown on any known culture medium. By December 1976, they were no closer to identifying the infectious agent.

How can you verify Koch's postulates if you have no infectious agent? It was almost like being back in the times of Pasteur and Koch. Why was this bacterial species so difficult to culture?

After trying 17 different culture media formulations, the infectious agent was finally cultured. It turns out the unknown infectious agent was a bacterial species that had fastidious growth requirements. When the initial agar medium was enriched with 1% hemoglobin and 1% isovitalex, small, barely visible colonies were seen after five days of incubation at 37°C. CDC investigators then realized the hemoglobin was supplying iron for the bacterial cells and the isovitalex was a source of the amino acid cysteine. Using these two chemicals in pure form, along with charcoal to absorb bacterial waste, and an atmosphere of 2.5% CO_2, bacterial growth was significantly enhanced (see figure). From a microscope examination of these cultures, gram-negative rods were confirmed and the organism was appropriately named *Legionella pneumophila*.

With an enriched medium to pure culture the organism, susceptible animals (guinea pigs) could be injected as required by Koch's postulates. *L. pneumophila* then was recovered from infected guinea pigs, verifying the organism as the causative agent of Legionnaires' disease.

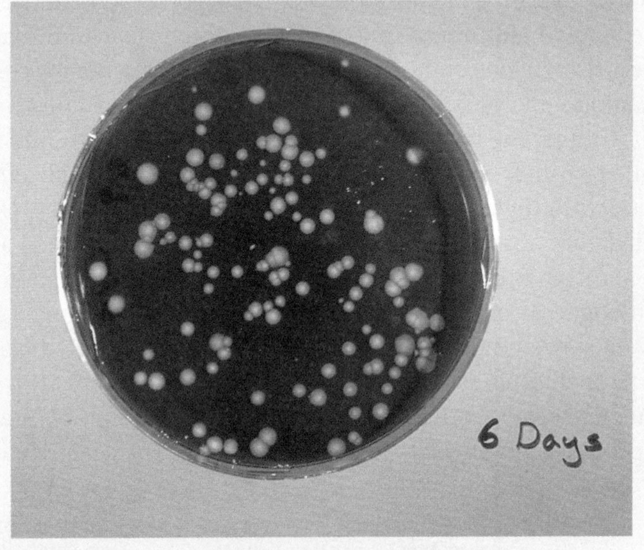

Colonies of *L. pneumophila* on an enriched medium.

Courtesy of Dr. Jim Feeley/CDC

yet devised. The cells of these species often can be observed when vie microscope, but when placed on an agar medium, no colonies develop. isms are called **viable but noncultured (VBNC)**, meaning the cells are al not reproduce in the unfamiliar culture environment. Presumably, some sp unique growth requirement is missing and no known enriched medium has ing factor(s). In fact, only about 1% of all known bacterial species can be c This means 99% of all bacterial species represent VBNCs, which again highlig tremendous diversity of bacterial and prokaryotic species that exist, which is the for the last section of this chapter.

5.4 The Spectrum of Prokaryotes: The Domains Bacteria and Archaea

You now know the vast majority of bacterial species remain unknown, with no name or available method for cultivation. In fact, a gram of ordinary garden soil contains thousands of bacterial species, only a few of which can be cultivated and studied.

Currently, there are some 7,000 known bacterial and archaeal species spread over about 30 defined phyla. In this section, we will highlight a few phyla and groups using the "tree of life," which separates all living organisms into one of three domains: the Bacteria, Archaea, and Eukarya (FIGURE 5.15). The current thinking is that an as-yet-unidentified ancestral microbe called the "last universal common ancestor" (LUCA) gave rise to both the evolutionary line ("tree branch") leading to the domain Bacteria and the line leading to the domains Archaea and Eukarya. Coming off the branches are the "twigs" representing specific groups of organisms.

■ Archaea
ar-kē'ä

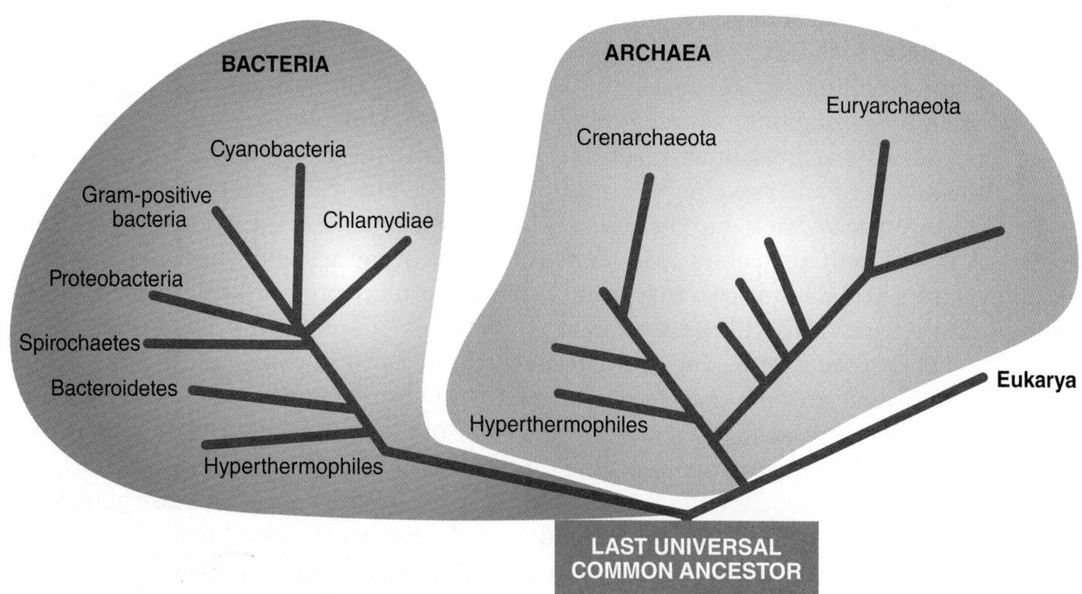

FIGURE 5.15 **The Tree of Life for the Bacteria and Archaea.** The tree shows several of the bacterial and archaeal lineages discussed in this chapter.

Domain Bacteria

The Bacteria have adapted to the diverse environments on Earth, inhabiting the air, soil, and water, and they exist in enormous numbers on the surfaces of virtually all plants and animals. They can be isolated from Arctic ice, thermal hot springs, the fringes of space, and the tissues of animals. Bacterial species, along with their archaeal relatives, have so completely colonized every part of the Earth that their mass is estimated to outweigh the mass of all plants and animals combined. Let's look briefly at some of the more common phyla and other groups, which give us a sense of the bewildering diversity that actually exists.

■ Cyanobacteria
sī′-an-ō-bak-tėr′-ē-ä

■ bloom: A sudden increase in the number of cells of an organism in an environment.

Photosynthetic Bacteria. The members of the **Cyanobacteria** thrive in freshwater ponds and in the oceans and exist as unicellular, filamentous, or colonial forms. The pigments they contain give a blue-green, yellow, or red color to the organisms and the periodic redness of the Red Sea is due to blooms of those cyanobacterial species containing large amounts of red pigment.

The phylum Cyanobacteria is unique among bacterial groups because its members carry out photosynthesis similar to unicellular algae and plants using the light-trapping pigment chlorophyll. Their evolution on Earth was responsible for the "great oxidation event" some 2.5 billion years ago, the consequences of which transformed life on the young planet (see Figure 5.1). Such photosynthetic species are known as **autotrophic** microbes because they synthesize their own food (*auto* = "self;" *troph* = "feeder"; hence, "self-feeder"). By providing organic matter at the base of the food chain, the cyanobacterial species occupy a key position in the nutritional patterns of nature. In addition, chloroplasts probably evolved from a free-living cyanobacterial ancestor that took up residence in an evolving eukaryotic cell millions of years ago.

In addition to photosynthesis, some cyanobacterial species carry out **nitrogen fixation**. In this process, specialized cells called **heterocysts** within the cyanobacterial filament take up nitrogen from the atmosphere and convert it to ammonia and other nitrogen-containing substances used within the organism as well as within marine and aquatic plants during the synthesis of organic molecules. Because cyanobacterial species carry out photosynthesis as well as nitrogen fixation, they are among the most independent organisms on Earth.

Heterotrophic Bacteria. We now turn to bacterial species representing **heterotrophic** microbes (*hetero* = "other"), meaning they obtain their organic food molecules from other sources. Many do this by playing key roles as **decomposers**, organisms that break down chemical compounds in the environment, using some for their own use and recycling the remaining carbon, nitrogen, sulfur, phosphorus, and other nutrients back into the soil, water, or atmosphere for use by other organisms. Although only a small fraction of the thousands of species of heterotrophic bacteria on the earth cause disease in humans, a few of the more prominent pathogens as well as other members will be mentioned.

■ *Shigella*
shi-gel′lä

■ *Salmonella*
säl-mōn-el′lä

■ *Yersinia*
yėr-sinē-ä

■ *Vibrio*
vib′rē-ō

■ rickettsiae
ri-ket′sē-ä

■ **Proteobacteria.** The **Proteobacteria** (*proteo* = "first") contains the largest and most diverse group of bacterial species. The phylum includes many familiar gram-negative genera, such as *Escherichia* and some of the most recognized human pathogens, including genera responsible for food poisoning (*Shigella, Salmonella*), the plague (*Yersinia*), cholera (*Vibrio*), and the sexually-transmitted disease gonorrhea (*Neisseria*). It is likely the mitochondria of the Eukarya evolved from a free-living ancestor of the Proteobacteria that took up residence in an evolving eukaryotic cell many millions of years ago.

Among the other members of the Proteobacteria are the **rickettsiae** (sing., rickettsia), which are called **obligate intracellular parasites**, meaning they can only reproduce once inside a host cell because they need the nutrients the host cell provides. These tiny bacterial cells are transmitted among humans primarily by arthropods, and are the agents for diseases such as Rocky Mountain spotted fever and typhus fever.

■ arthropod: An animal having jointed appendages and segmented body (e.g., ticks, lice, fleas, mosquitoes).

Many other species are of medical, industrial, or environmental importance. Among the more interesting are members of the genus *Pseudomonas*. One species, *P. aeruginosa*, is responsible for "hot tub rash" where the cells in contaminated water infect the skin. In the soil, other species of the genus produce a large variety of enzymes that contribute to the breakdown of pesticides and similar waste chemicals.

■ *Pseudomonas aeruginosa* sū-dō-mō'näs ä -rū-ji-nōs ä

An intriguing member of the Proteobacteria is *Bdellovibrio* that preys on other gram-negative species. A *Bdellovibrio* cell attaches to the surface of the prey, then rotates and bores a hole through the prey's outer membrane. Now it takes control of the prey and grows in the space between the outer membrane and cell membrane, killing the prey in the process.

■ *Bdellovibrio* del'ō- vib'rē-ō

Finally, it is worth mentioning one more species, *Serratia marcescens*, a rod-shaped species distinguished by the blood-red pigment it produces when it forms colonies. Its "blood" has historical significance, as **A Closer Look 5.4** details.

■ *Serratia marcescens* ser-rä'tē-ä mär-ses'sens

■ **Bacteroidetes.** The **Bacteroidetes** is another phylum of gram-negative bacterial rods that lives in oxygen-free environments. The genus *Bacteroides* lives in the stomach of cows and helps them digest the cellulose in plant cell walls, a chemical process accomplished by few other organisms. The human gut microbiome contains several

■ *Bacteroidetes* bak-te˙-roi'dētēz

A CLOSER LOOK 5.4

The Blood of History

Because of its characteristic blood-red pigment and propensity for contaminating bread, *Serratia marcescens* has had a notable place in history. For example, the dark, damp environments of medieval churches provided optimal conditions for growth on sacramental wafers used in Holy Communion. At times, the appearance of "blood" was construed to be a miracle. One such event happened in 1264 when "blood dripped" on a priest's robe. The event was later commemorated by Raphael in his fresco The Mass of Bolsena. Unfortunately, religious fanatics used such episodes to institute persecutions because they believed that heretical acts caused the "blood" to flow.

In 332 BC, Alexander the Great and his army of Macedonians laid siege to the city of Tyre in what is now Lebanon. The siege was not going well. Then one morning, blood-red spots appeared on several pieces of bread. At first it was thought to be an evil omen, but a soothsayer named Aristander indicated the "blood" was coming from within the bread. This suggested that blood would be spilled within Tyre and the city would fall. Alexander's troops were buoyed by this interpretation and with renewed confidence they charged headlong into battle and captured the city. The victory opened the Middle East to the Macedonians and their march did not stop until they reached India.

Then in 1819, Bartholemeo Bizio, an Italian pharmacist, demonstrated that the "bloody" miracles were caused by a living organism. Bizio named it *Serratia* after Serafino Serrati, a countryman whom he considered the inventor of the steamboat. The name *marcescens* came from the Latin word for "decaying," a reference to the decaying of bread that lead Alexander the Great to victory.

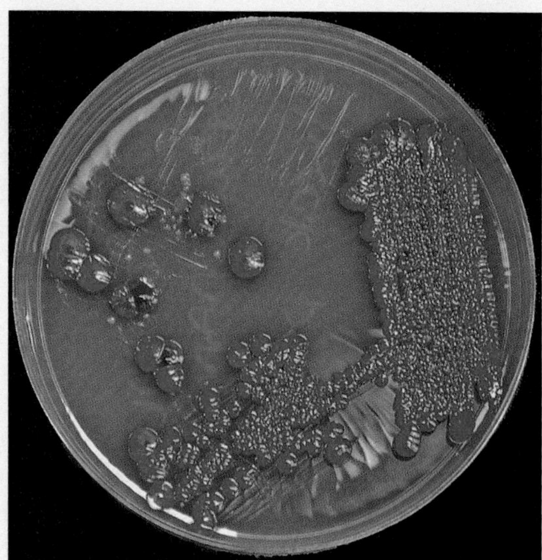

***Serratia marcescens* colonies growing on agar have the look of blood.**

Courtesy of Dr. Jeffrey Pommerville.

Bacteroides species, which are very helpful in our digestive processes. Interestingly, if an individual's population of Bacteroidetes dips too low and relatively more bacterial species from the Firmicutes increase (see below), the individual may be more predisposed to gain weight. How the Bacteroidetes could prevent obesity or if their presence is merely preferentially selected by intestinal conditions in those who are not obese, requires more investigation.

The gram-positive bacterial groups rival the Proteobacteria in diversity and are divided into two phyla.

■ **Firmicutes.** The **Firmicutes** (*firm* = "strong"; *cuti* = "skin") consists of species whose cells are Gram positive and thus have multiple chains of peptidoglycan in the cell wall (see Figure 5.7a). *Bacillus* and *Clostridium* species are endospore-formers and responsible for anthrax and tetanus, respectively. These bacterial species produce poisonous substances called **toxins** that interfere with the chemical activities of cells and tissues. For example, the tetanus toxin produces a powerful poison interfering with the relaxation of muscles after they contract. The muscles undergo severe spasms and "lock" into place ("lockjaw" is one effect). Another toxin producer is *Clostridium botulinum*, the organism that causes botulism.

■ *Clostridium botulinum*
klos-tri′dē-um bot-ū-lī-num

Species within the genera *Streptococcus* and *Staphylococcus* are responsible for several mild to life-threatening human illnesses. For example, *Staphylococcus aureus* produces an enzyme that coagulates blood plasma, forming a protective layer of clotted material around the invading bacterial cells; this layer blocks the body's attempt to attack the bacterial pathogen and destroy the invading cells.

■ *Staphylococcus aureus*
staf-i-lō-kok′kus ô′-rē-us

Also within the Firmicutes is the genus *Mycoplasma,* which lacks a cell wall but is otherwise related to the gram-positive bacterial species. Among the smallest free-living bacterial organisms, one species causes a form of pneumonia while another mycoplasmal illness represents a sexually-transmitted disease.

■ *Mycoplasma*
mī-kō-plasmä

Among the firmicutes is the genus *Lactobacillus*. *Lactobacillus* species live in the female genital tract and help guard against infection by other microbes. Other species are used in the large-scale manufacturing of cheese, sour cream, yogurt, and other fermented milk products.

■ *Lactobacillus*
lak-tō-bä-sil′lus

■ **Actinobacteria.** Another phylum consisting of gram-positive species is the **Actinobacteria**. Often called the "actinomycetes" (*actino* = "radiating"; *mycet* = "fungus"), these soil organisms form a system of branched (radiating) filaments that somewhat resemble the growth form of fungi. Actinomycetes form very resistant spores at the tips of their filaments, a feature that allows them to remain alive in difficult environments such as soil (to which they give a characteristic mustiness). However, these are not endospores and lack the extended dormancy endospores possess. The genus *Streptomyces* is the source for important antibiotics, including tetracycline, erythromycin, and neomycin.

■ *Streptomyces*
strep-tō-mīsēs

Another medically important genus is *Mycobacterium,* one species of which is responsible for tuberculosis. Because of the chemical content of their cell walls, these rod-shaped cells are very difficult to Gram stain. However, a different staining procedure, called the **acid-fast technique**, uses heat or other agents to force stain into the cytoplasm. The bacterial cells then stain red.

■ **Chlamydiae.** Roughly half the size of the rickettsiae, members of the phylum **Chlamydiae** also are obligate, intracellular parasites and are cultivated only within living cells. Most species are pathogens and one species causes the gonorrhea-like sexually transmitted disease (STD) called chlamydia. In 2013, more than 1.4 million cases of chlamydia were reported to the Centers for Disease Control and Prevention (CDC), the largest number of cases ever reported for any condition.

■ *Chlamydiae*
kla-mi-dē-ē

■ **Spirochaetes.** The species in the phylum **Spirochaetes** consist of gram-ne cells possessing a unique cell body coiled into a long helix that moves in a cork pattern (see Figure 5.3c). The ecological niches for the spirochetes are diverse, i ing: free-living species found in mud and sediments; species inhabiting the dig tracts of insects; and the pathogens found in the urogenital tracts of vertebrates. A the human pathogens are *Treponema pallidum,* the causative agent of syphilis of the most common STDs—and a species of *Borrelia,* which is transmitted b or lice and is responsible for Lyme disease.

■ **Other Phyla.** Several lineages branch off near the root of the domain. The mon link between these organisms is that they are **hyperthermophiles**; they g high temperatures (70°C–85°C) such as in hot springs and other hydrotherma Some scientists believe an ancient hyperthermophile might have been the LUC which the three domains of life arose.

Domain Archaea

We close our discussion of prokaryotic diversity by considering the other domain of prokaryotes, the Archaea which, although they often look like bacterial cells in general structure, have an unusual wall and membrane structure, and unique physiology and biochemistry.

Archaeal organisms are found throughout the biosphere. Many genera are **extremophiles**, growing best at environmental extremes, such as very high temperatures, high salt concentrations, or extremely acidic environments (see Chapter opening image). However, many more species exist in very cold environments. Although there also are archaeal genera that thrive under more modest conditions, there are no known species causing disease in any plants or animals. The archaeal genera can be placed into one of two phyla.

■ **Euryarchaeota.** The **Euryarchaeota** contain organisms with varying physiologies, many being extremophiles. Some groups, such as the **methanogens** (*methano* = "methane"; *gen* = "produce"), are killed by oxygen gas and therefore are found in marine and freshwater environments (and animal gastrointestinal tracts) devoid of oxygen gas. The production of methane (natural) gas is important in their energy metabolism. In fact, these archaeal species release more than 2 billion tons of methane gas into the atmosphere every year. About a third comes from the archaeal species living in the stomach (rumen) of cows.

■ Euryarcheota
ur-ē-ark-ē′ō-ta

Another group is the **extreme halophiles** (*halo* = "salt"; *phil* = "loving"). They are distinct from the methanogens because they require oxygen gas for energy metabolism and need high concentrations of salt (up to 30% NaCl) to grow and reproduce. They would be found in such places as Utah's Great Salt Lake.

A third group is the hyperthermophiles (*hyper* = "high;" *thermo* = "heat") that grow optimally at temperatures above 80°C (remember water boils at 100°C). They are typically found in volcanic terrestrial environments and deep-sea hydrothermal vents where the temperature can be as high as 113°C; indeed, they will not grow if the temperature drops below 90°C because it gets too cold! Many archaeal species also grow in very acidic environments at a pH of 1.0 (the acidity of fuming sulfuric acid).

■ **Crenarchaeota.** The second phylum, the **Crenarchaeota**, consist mostly of hyperthermophiles, typically growing in hot springs and marine hydrothermal vents. Other species are dispersed in open oceans, often inhabiting the cold ocean waters (–3°C) of the deep sea environments and polar seas.

■ Crenarchaeota
kren-ark-ē′ō-ta

TABLE 5.1 summarizes some of the characteristics shared or are unique among the three domains.

TABLE 5.1 **Some Major Differences Between Bacteria, Archaea, and Eukarya**

Characteristic	Bacteria	Archaea	Eukarya
Cell nucleus	No	No	Yes
Chromosome form	Single, circular	Single, circular	Multiple, linear
Chlorophyll-based photosynthesis	Yes (cyanobacteria)	No	Yes (algae)
Peptidoglycan cell wall	Yes	No	No
First amino acid in a protein	Formylmethionine	Methionine	Methionine
Ribosome sedimentation value	70S	70S	80S
Ribosome sensitivity to diphtheria toxin	No	Yes	Yes
Growth above 80°C	Yes	Yes	No
Growth above 100°C	No	Yes	No
Pathogens	Yes	No	Yes

A Final Thought

When you pick up this book for the first time, you may experience a moment anticipating your first look at microbes. Perhaps you leaf through the pages, turn back to see a photograph a second time, then pause and think: "Is that all there is? Little rods and circles?"

Our first encounter with microbes is often a disappointing (and perhaps exasperating) experience. Since early childhood, we have been taught to loathe and despise "bacteria," and we have been schooled in all the dastardly deeds they do. "Wash your hands" we are admonished; "Don't eat it if it falls on the ground." And on and on. We expect bacterial life to be filled with grotesque and fearsome monsters. We expect them to rank with death and taxes. But they turn out to be little sticks and rods, not very dangerous-looking at all.

So, perhaps we need to wipe away any preconceived notions of "bad bacteria" and start rebuilding our views—a process that will take us through the complexity of bacterial structures, give us insight into their chemistry, point up their remarkable genetics, and stress their importance in food production, industrial manufacturing, and soil ecology.

It's going to take some time to absorb the importance of the prokaryotic and eukaryotic microbes to our lives, but for now try to appreciate the prokaryotes for more than the tiny rods you see on these pages or under the microscope. The electron microscope has yielded a wealth of information about their structure (as this chapter demonstrates) and studying prokaryotic structure gives us a clue to what they do. This "what they do" part explains their importance to us and society.

Questions to Consider

1. Suppose a bacterial species had the opportunity to form a glycocalyx, a flagellum, a pilus, or an endospore. Which do you think it might choose? Why?
2. Extremophiles are of interest to industrial corporations, who see these bacterial species as important sources of enzymes that function at temperatures of 100°C and in extremely acidic conditions (the enzymes have been dubbed "extremozymes"). What practical uses can you foresee for these enzymes?
3. Several years ago, public health officials found the water in a Midwestern town was contaminated with sewage bacteria. The officials suggested homeowners boil their water for a couple of minutes before drinking it. Would this treatment remove all traces of bacterial cells from the water? Why?
4. About 100 to 250 grams (3 to 8 ounces) of feces are excreted by a human adult each day. Thirty percent of the solid matter consists of bacterial cells that were present in the colon. If that is the case, about how much bacterial mass do we "produce" in a week? In a year? How can this be possible?
5. Suppose this chapter on the structure and growth of prokaryotes had been written in 1940, before the electron microscope became available. Which parts of the chapter would probably be missing?
6. "Bacteria are all the same. Once you've seen one, you've seen 'em all!" What evidence could you present to counter this statement?
7. There are thousands of bacterial species, yet with few exceptions, all of them have variations of three shapes: the rod, the sphere, and the spiral. Do you find this strange? Why or why not?
8. A bacterial species has been isolated from a patient and identified as a gram-positive rod. Knowing it is a human pathogen, what structures would it most likely have? Explain your reasons for each choice.
9. Some evolutionary biologists argue that bacterial pathogens have acted as great "slate wipers" of history (e.g., cholera, plague, typhoid fever, and syphilis) and are the key agents of natural selection for the human species; that is, in a sense, they help improve our species by selecting the fitter individuals through the distasteful task of infectious disease. Do you agree or disagree with this statement? (Note: you might want to read Dan Brown's 2013 mystery thriller "Inferno" published by Doubleday).

Key Terms

Informative facts are necessary for the expression of every concept, and the information for a concept is founded in a set of key terms. The following terms form the basis for the concepts of this chapter. On completing the chapter, you should be able to explain and/or define each one:

acid-fast technique
Actinobacteria
agar
autotrophic
bacillus (pl., bacilli)
Bacteroidetes
binary fission
broth
cell membrane
cell wall
Chlamydiae
chromosome
coccus (pl., cocci)
colony
Crenarchaeota
culture medium
Cyanobacteria
decline (death) phase
decomposer
endospore
enriched medium
eukaryote

Euryarchaeota
extreme halophile
extremophile
Firmicutes
flagellum (pl., flagella)
fluid mosaic
genome
glycocalyx
gram-negative
gram-positive
Gram stain technique
growth curve
heterocyst
heterotrophic
hyperthermophile
lag phase
logarithmic (log) phase
lipopolysaccharide (LPS)
methanogen
morphology
nitrogen fixation
nucleoid
nutrient agar
obligate intracellular parasite
outer membrane
peptidoglycan
periplasmic space
pilus (pl., pili)
plasmid
prokaryote
Proteobacteria
ribosome
rickettsiae
selective medium
simple stain technique
spirillum
Spirochaetes
spirochete
stationary phase
teichoic acid
toxin
viable but noncultured (VBNC)
vibrio

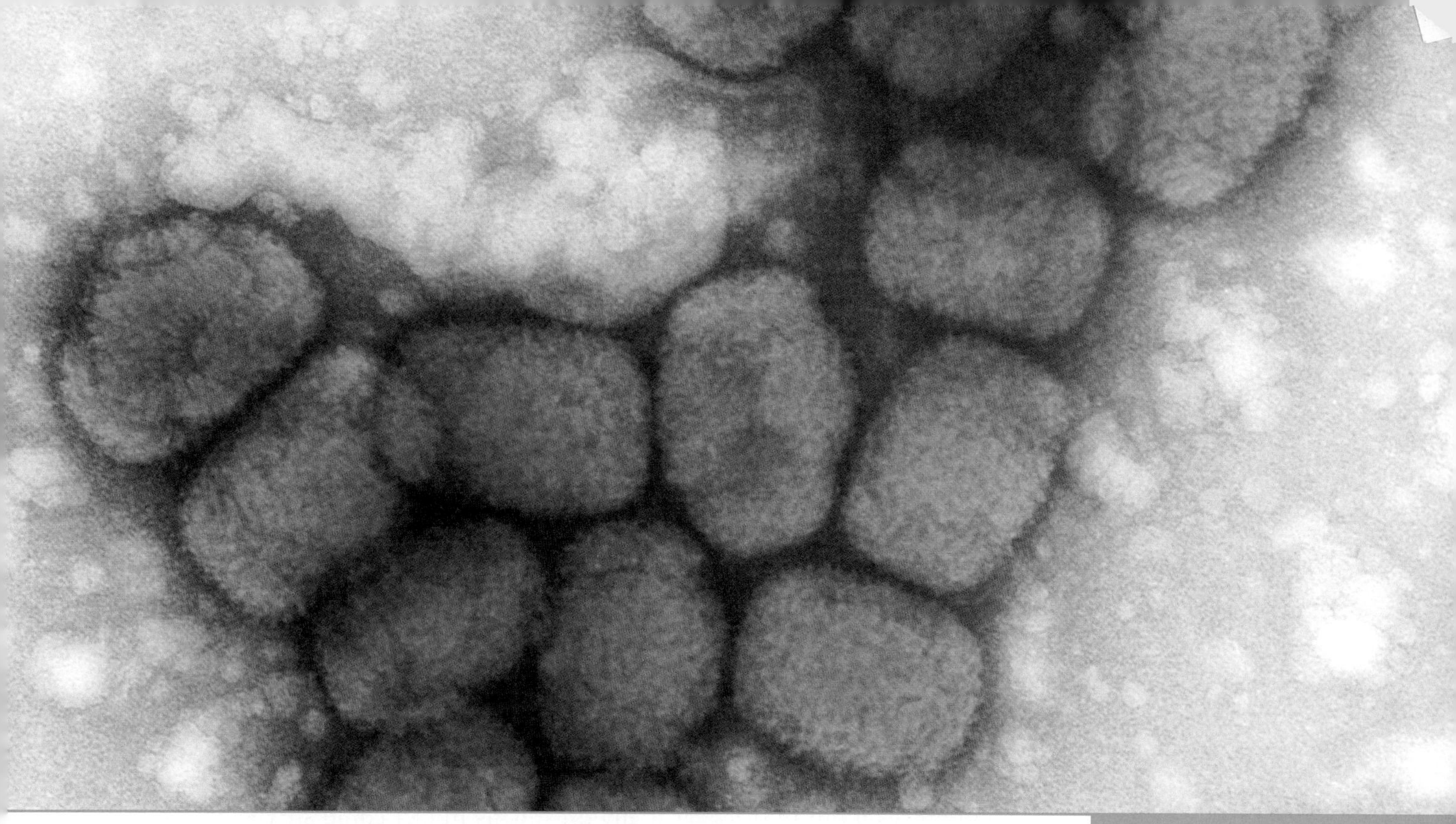

Viruses: At the Threshold of Life

6

This false-color transmission electron microscope image shows several smallpox viruses. This virus represents just one of the viruses that infect humans. To date, the smallpox virus is the only human virus that, through global vaccinations, has been eradicated.

Courtesy of Dr. Fred Murphy/CDC

Looking Ahead

In the microbial world, viruses stand out for their simplicity, extraordinarily small size, distinctive method of replication, and diversity. We highlight these characteristics in this chapter.

On completing this chapter, you should be capable of:

- Discussing the early discovery of viruses.
- Drawing the structure and labeling the components of enveloped and nonenveloped viruses.
- Differentiating between the five stages of virus replication.
- Discussing the use of antiviral drugs and vaccines.
- Explaining the two ways by which an emerging virus may arise.
- Describing how viruses can cause tumors.
- Discriminating between viroids and prions.

Under the Naica hills in northern Mexico lay some amazing caves. These gigantic caverns, some the length of a football field and two stories high, contain 50 ton crystals of gypsum that were formed 500,000 years ago (FIGURE 6.1a). Called the Cave of Crystals, the caverns are buried 300 meters below the desert surface, in sweltering temperatures of 45°C (113°F) to 50°C (122°F) with saturating humidity.

When first explored, these caves appeared devoid of life, especially microbial life. But water was present, so in 2009 Dr. Curtis Suttle and his group from the University of British

(a)

(b)

FIGURE 6.1 **The Virosphere.** Viruses can be found wherever life exists. This includes isolated environments deep in the earth such as (**a**) the Cave of Crystals and (**b**) huge, open expanses such as the world's oceans. The larger, fluorescent dots are bacterial cells and the smaller spots are bacterial viruses. (Bar = 10 µm.)

(a) © Javier Trueba/MSF/Science Source. (b) Courtesy of Rachel Parsons, Bermuda Institute of Ocean Science.

Columbia set out to discover if any organisms in fact could survive in the hellish conditions of the caves. Water samples were collected from the cave pools, taken back to the lab in British Columbia, and prepared for microscopy. What the samples contained stunned the researchers.

In the water samples from the caves, bacterial cells were identified. And more startlingly, in those samples were tiny, geometric forms; forms that had been cut off from the outside world for perhaps millions of years. The geometric forms were viruses! And there were lots of them, perhaps up to 200 million viruses per drop of water.

Almost halfway around the world, Monika Häring and other microbiologists from several European universities were studying acidic hot springs (85°C to 95°C) in Pozzuoli, Italy. When enrichment cultures were prepared from hot spring samples, several archaeal organisms were identified along with virus particles. Meanwhile, much further south in the Antarctic's Lake Limnopolar, a freshwater lake that is frozen nine months of the year, Spanish scientists found bacterial cells, protists, and—yes—viruses in the lake. Again huge numbers were seen, representing perhaps 10,000 different types, making the lake one of the most diverse virus communities in the world. Even the world's oceans are filled with viruses. For example, Rachel Parsons from the Bermuda Institute of Ocean Sciences and her collaborators have detected enormous virus populations in the western Atlantic (FIGURE 6.1b).

Around the world, the number of viruses exceeds 10^{31} (that's 1 followed by 31 zeros). Most of these infect microbes but there are plenty that infect plants and animals as well. In fact, the planet's most abundant "coterrestrials" are the viruses. This world of viruses, the so-called **virosphere**, exists, as Suttle's lab explains, "*wherever life is found; it is a major cause of* [illness and] *mortality, a driver of global geochemical cycles, and a reservoir of the greatest unexplored genetic diversity on Earth.*" And viruses are a major driving force for evolution because each virus infection has the potential to introduce new genetic information into just about any organism in the domains Bacteria, Archaea, and Eukarya, or into their own progeny viruses. The virosphere is huge, incredibly diverse, and has a tremendous impact beyond infection and disease.

Every science has its borderland where the known and visible merge with the unknown and invisible. Startling discoveries often emerge from this uncharted realm, and certain objects manage to loom large. In the borderland of microbiology, viruses exist at the threshold of life. Although many individuals speak of "live virus vaccines" and many scientists "grow viruses" in their laboratories, viruses have a level of simplicity that places them somewhere between living objects and chemical molecules. As we will see in the pages ahead, viruses truly do not fit into our traditional vision of living organisms.

6.1 Viruses: Their Discovery and Structure

In the period between 1880 and 1915, a wealth of discoveries occurred in the field of microbiology. Researchers finally pinpointed the bacterial organisms causing tuberculosis, typhoid fever, syphilis, and many other infectious diseases; they improved methods of sanitation and food preservation; and they increased our understanding of the importance of microbes in the environment. This period was the Golden Age of microbiology.

At the turn of the 20th century, there were several researchers whose work was somewhat unproductive. These investigators devoted their energies to discovering the microbes causing such diseases as measles, rabies, polio, and hepatitis. Using the methods of Pasteur and Koch bore little fruit because nothing grew in the culture plates. As it turned out years later, these diseases were not caused by bacterial pathogens, but by viruses. And in the period around 1900, neither the microscopes nor the laboratory methodologies were capable of working with viruses. We rarely know who these unfortunate researchers were, but we can be certain they suffered much frustration and disappointment because their work was largely unrewarded and the true nature of viruses remained a mystery.

The Development of Virology

One of the first scientists to study viruses was Dmitri Ivanowsky. In 1892, Ivanowsky was studying tobacco mosaic disease, a disease that causes tobacco leaves to shrivel and die (FIGURE 6.2a). Attempting to isolate the "bacterial" cause of the disease, Ivanowsky crushed diseased tobacco leaves and filtered them to separate their particulate debris from their juices. He used a filter to trap what he thought were the smallest known bacteria. To his surprise, Ivanowsky found that the clear juice coming through the filter contained the infectious agent. He had no idea what was in this juice, and he suggested a "filterable virus" caused tobacco mosaic disease, meaning that whatever caused the disease could pass through the filter.

Ivanowsky
I-van-ow′-skē

Some years later, in 1898, the Dutch investigator Martinus Beijerinck independently showed that when the juice was diluted many times, the infectious agent was still present when passed through a porcelain filter. Beijerinck called the unknown agent *contagium vivum fluidum*, or "contagious living fluid." How it caused tobacco mosaic disease remained a mystery.

Beijerinck
Bē′-yer-ink

Real progress in identifying viruses would not occur until the 1930s when scientists discovered they could form crystals of tobacco mosaic virus, suggesting viruses were some sort of chemical molecule. However, viral diseases were accompanied by fever, a response by the immune system, and other symptoms associated with microbial infection, and so scientists continued to debate the nature of viruses. While the debate continued, virologists discovered a way to cultivate viruses in fertilized eggs and other living cells. This breakthrough allowed them to obtain large quantities of viruses for study purposes.

Then, in the 1950s the electron microscope came on the scene and scientists were finally able to see viruses. Soon they had an image of the tobacco mosaic virus—it

(a)

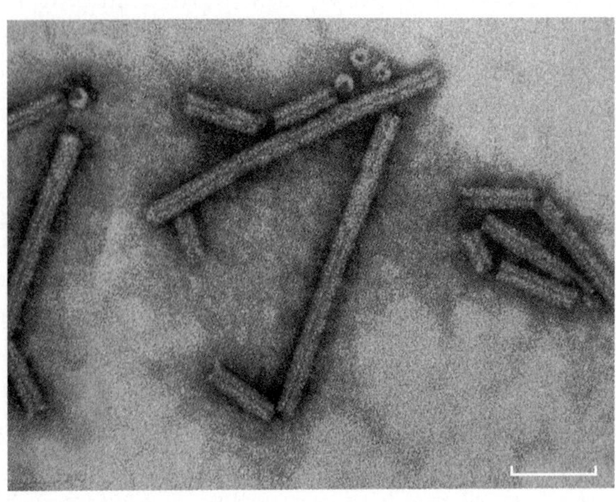

(b)

FIGURE 6.2 **Tobacco Mosaic Disease and the Virus.** (**a**) An infected tobacco leaf exhibiting the mottled or mosaic appearance caused by the disease. (**b**) This false-color transmission electron microscope image of the tobacco mosaic virus shows the rod-shaped structure of the virus particles. (Bar = 80 nm.)

(a) Courtesy of Clemson University - USDA Cooperative Extension Slide Series, Bugwood.org. (b) © Dennis Kunkel Microscopy, Inc./Phototake/Alamy.

appeared as a long thin rod (FIGURE 6.2b). Indeed, it appeared viruses were tiny bacteria. Later research, however, revealed that viruses were much different from bacterial cells. Much different, to be sure.

The Structure of Viruses

If you were to stop 100 people on the street and ask if they recognize the word virus, all 100 would probably nod their heads knowingly and say "yes." They might tell you about the flu they recently had, or mention the "computer virus" that crashed their system! Were you to ask the same individuals to describe a flu virus, they might answer it's "a tiny microbe," or "something you need a microscope to see," or "a germ." After a couple of moments, they would probably scratch their heads and admit they are familiar with the word "virus," but they don't really know what a virus is (unless, of course, they had recently read this chapter).

So what is a virus? **Viruses** are small, obligate intracellular particles; that is, they are tiny agents that must infect a cell in order to produce more viruses. This is because they lack the cellular machinery for generating energy and the components necessary for metabolism. Viruses, therefore, must find an appropriate host cell in which they can replicate—and, as a result, often cause disease.

As FIGURE 6.3 illustrates, among the smallest viruses are the polio viruses, which are 28 nanometers (nm) in diameter (28 billionths of a meter), while the larger ones, such as smallpox viruses, measure about 250 nm in diameter, about the size of the smallest bacterial cell. It is difficult for the human mind to comprehend the scale of a nanometer (a billionth of a meter), because we are used to lengths of inches, feet, and yards. Consider, however, that 500 or more viruses could fit inside a single bacterial cell.

Viruses basically have one of three shapes (FIGURE 6.4). Some viruses exist in the form of a **helix**, similar to a tightly wound coil (they are said to have "helical symmetry"). The tobacco mosaic and rabies viruses are examples of helical viruses. Numerous viruses have the shape of an icosahedron (they are said to have "icosahedral symmetry"). An **icosahedron** (*icos* = "20"; *edros* = "sided) is a geometrical figure with

20 equal-sized triangular faces. Among the icosahedral viruses are the polio viruses and the herpes simplex viruses (HSV). Other viruses have more complex shapes ("complex symmetry"). In Figure 6.4, notice the smallpox virus has the shape of a brick, with a swirling pattern of tubes on its surface. Also notice the viruses that attack bacterial cells, the **bacteriophages** (or simply **phages**), have a shape that is partially helical and partially icosahedral.

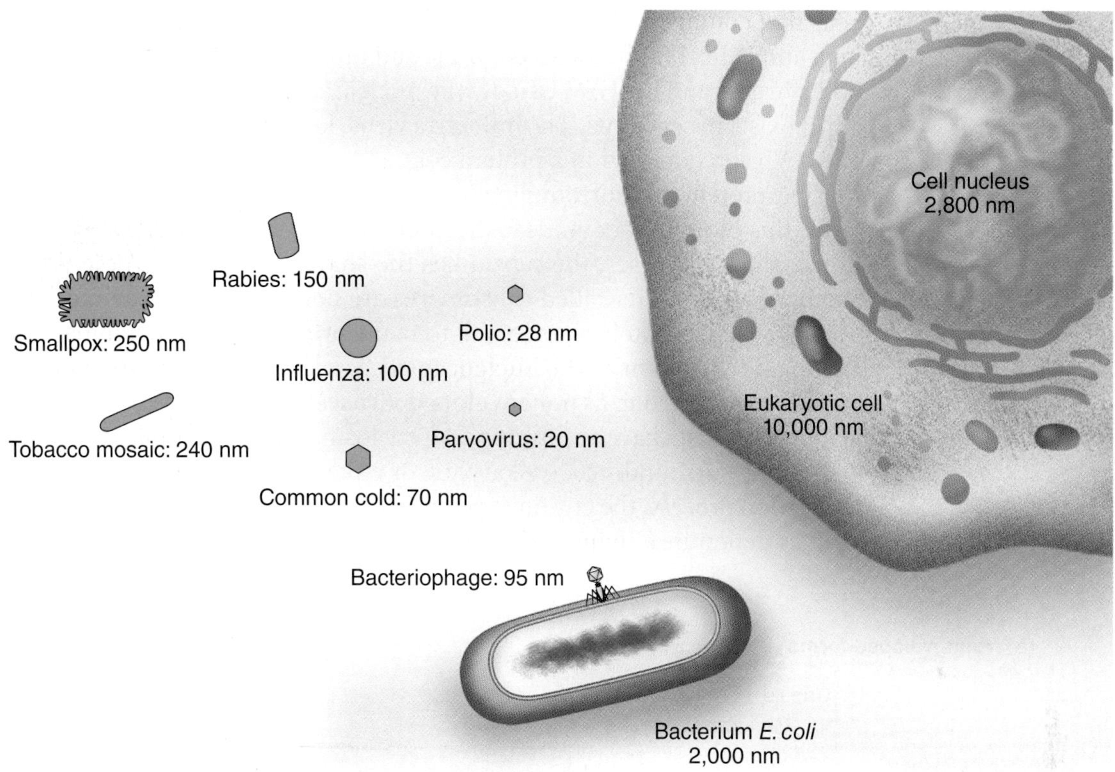

FIGURE 6.3 **Size Relationships Among Cells and Viruses.** The sizes (not drawn to scale) of various viruses relative to a eukaryotic cell, a cell nucleus, and the bacterium *Escherichia coli*.

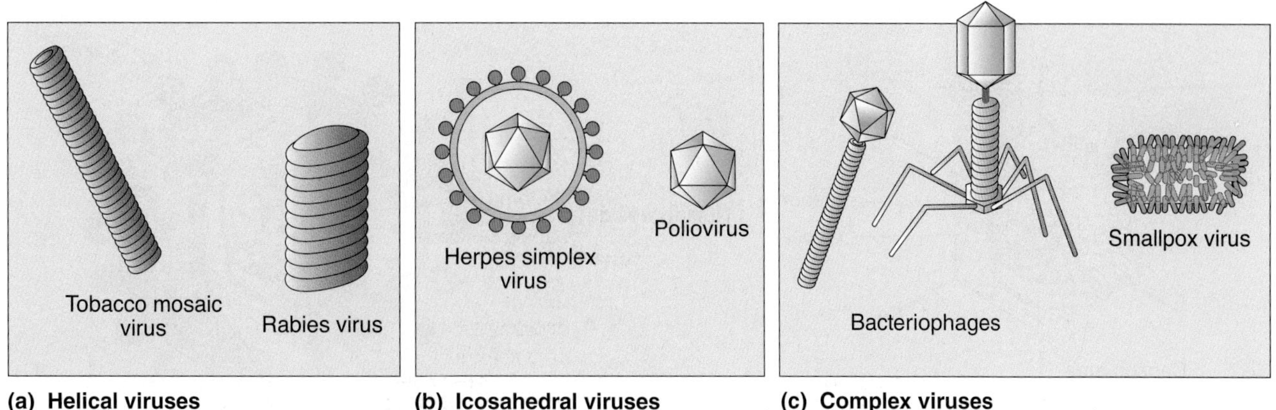

FIGURE 6.4 **Viral Shapes.** Viruses exhibit variations in form, which includes (**a**) helical symmetry, (**b**) icosahedral symmetry, and (**c**) complex symmetry.

The Components of Viruses

Plant, animal, and human cells are highly specialized, with various subcompartments (e.g., mitochondria and ribosomes) in which different cellular functions are performed. The prokaryotes lack most of these structures (exception is the ribosomes) yet still have a cytoplasm in which metabolism occurs. Viruses are extraordinarily different. All viruses consist simply of a core of nucleic acid enclosed by some type of covering (FIGURE 6.5).

The nucleic acid core of the virus, known as the viral **genome**, consists of either deoxyribonucleic acid (DNA) or ribonucleic acid (RNA), but not both unlike all organisms that contain both types of nucleic acid. Sometimes the nucleic acid molecule is long and helical (as in tobacco mosaic and rabies viruses); but often it is folded, as in polio viruses and the HSV. Usually the nucleic acid is a single molecule, but occasionally it exists in segments. The influenza virus, for example, consists of eight segments of RNA, each enclosed in a protein coat.

The protein coat surrounding the genome of the virus is called a **capsid**. In a helical virus, this protein layer binds to the nucleic acid and maintains the shape of the helix. In icosahedral viruses, the capsid has the shape that identifies the virus symmetry. Smaller protein units called **capsomeres** are bound together chemically to form the capsid, much as patches are joined to make up a quilt. The combination of genome and capsid is known as the **nucleocapsid**. Such viruses, consisting of just a nucleocapsid, are referred to as **nonenveloped viruses**. Several were depicted in Figure 6.4.

Other viruses have an **envelope**, an enclosing structure similar to the cell or plasma membrane surrounding a prokaryotic or eukaryotic cell. Acquired during the virus' replication process, the envelope protects the nucleocapsid and helps these **enveloped viruses** penetrate a living cell during the replication process, as we shall see shortly.

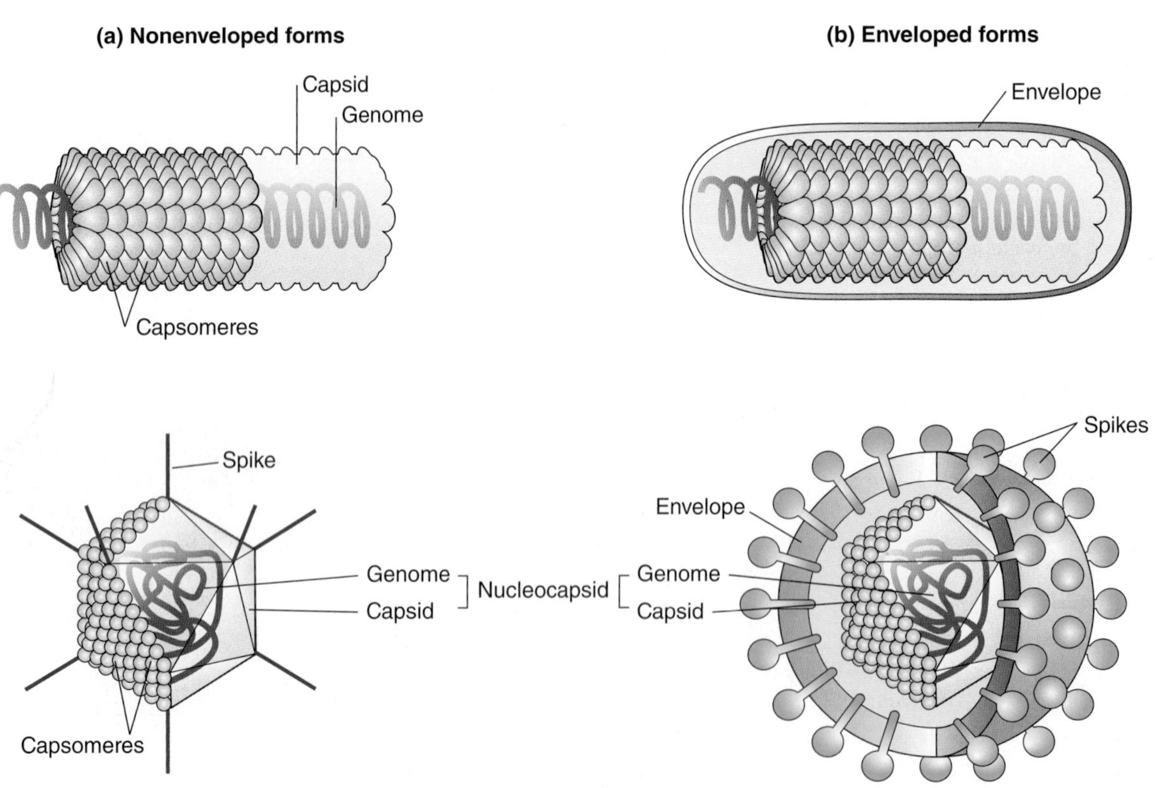

FIGURE 6.5 **The Components of Viruses.** (**a**) The structure of nonenveloped viruses having helical or icosahedral symmetry. (**b**) The structure of enveloped viruses having helical or icosahedral symmetry.

In many viruses, the capsid or envelope has protein projections called **spikes**. Sticking out like spines on a cactus, the spikes help the virus contact its host cell, and then assist the viral nucleocapsid in its infection of the host cell. In Figure 6.5b, notice the nonenveloped virus (e.g., polio virus) has spikes extending out from the capsid and the enveloped virus (e.g., HSV) contains spikes on the envelope. In both viruses, the spikes help the virus "dock" on its host cell. For viruses lacking spikes, other surface proteins play this recognition role.

In summary, a virus consists of a core of nucleic acid (the genome), a covering of protein (the capsid), and, in some cases, an enclosing envelope. You will note that the term "cytoplasm" has not been mentioned; nor has there been discussion of anything other than nucleic acid and protein. This is because there is no chemistry going on within a virus, there is no intake of nutrients, and there are no subcompartments. Viruses do not increase or decrease in size and they have no metabolism. In the environment, viruses are inert particles. However, **A Closer Look 6.1** describes some recently discovered amazing exceptions to the rule.

There is, however, one process viruses do, and they do it particularly well—they replicate. During replication, an infecting virus biochemically programs a host cell to

A CLOSER LOOK 6.1

Biological Oxymorons

An oxymoron is a pair of words that seem to refer to opposites, such as "jumbo shrimp," "passive aggressive," and, for you zombie lovers, "living dead." In this chapter, we said viruses are submicroscopic infectious agents. Always true? So, how about the oxymoron: "giant viruses?"

In 2002, researchers at the Centre National de la Recherche Scientifique (CNRS) in France were looking for bacterial cells in freshwater samples when they stumbled upon a "microbe" in a waterborne amoeba they thought was a bacterial cell. The "microbe" had double-stranded DNA and protein fibrils extending from the surface, making the "microbe" some 750 nm (0.75 μm) in diameter. However, it was discovered that this infectious microbe was actually a nonenveloped virus, making it the first virus large enough to be seen with an ordinary light microscope. It was named the Mimivirus ("microbe-mimicking" virus; see figure).

The Mimivirus genome contains more than 1,000 protein-coding genes; the influenza viruses and HIV each have around ten genes. It is even larger than the genome in some bacterial species. Like other viruses though, it cannot convert energy or replicate on its own.

The Mimivirus contains both DNA *and* RNA, something viruses by definition are not supposed to contain. It also has several genes shared between all three domains of life—Bacteria, Archaea, and Eukarya. When the virus replicates in an amoeba, it constructs a gigantic "replication factory" and uses all its own genes and proteins for the assembly of new virions. In addition, the Mimivirus can be "infected" by its own virus, called Sputnik, which uses the Mimivirus replication factory for its own replication.

Since then other giant viruses have been discovered, one called Mamavirus was discovered in an Antarctic lake. However, in 2011 the biggest virus yet discovered was found in the sea off Chile. Called Megavirus, it has more than 1,100 genes. Its host organism has not yet been identified.

Besides being giant viruses, these discoveries blur the lines between viruses and living organisms. Thus, future studies of the giant viruses may shed light on the origin of all life forms. And importantly, in science there are always exceptions, including oxymorons.genome

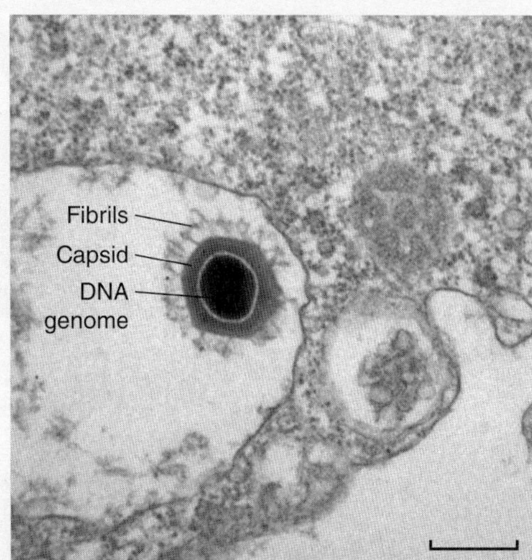

Transmission electron microscope image of the Mimivirus. (Bar = 500 nm.)

Courtesy of Didier Raoult, Rickettsia Laboratory, La Timone, Marseille, France simultaneously produce hundreds, sometimes thousands of copies of the virus. So, unlike organism reproduction where one cell gives rise to two cells, virus replication simultaneously gives rise to hundreds of new viruses. In so doing, viruses usually destroy the cell they infected. Partly for this reason, a prominent scientist has called viruses "*bad news wrapped up in protein.*" We shall see how replication takes place in the next section.

6.2 Virus Replication: A Massive Production and Assembly Line

Outside a host cell, a virus is an inert particle; when it encounters and infects an appropriate host cell, however, it transforms the host cell into a highly efficient production and assembly line. Inside the host cell, the virus uses its own genetic blueprints (genome) to take over the cell's metabolic machinery for the sole purpose of producing more viruses.

The Stages of Virus Replication

The replication process by which the viral genome produces more viruses can be broken down into five stages (FIGURE 6.6).

1. Attachment. The first event of virus replication involves **attachment** of the virus to an appropriate host cell. For example, hepatitis viruses replicate only in liver cells, and flu viruses replicate only in cells of the respiratory tract. This specificity of attachment derives partly from the presence of spikes or other proteins on the surface of the virus that recognize protein "receptors" in the plasma membrane of the host cells. In the host cells, these receptors represent the chemical "lock" to which spike proteins must fit.

2. Penetration. Once attached, the envelope of enveloped viruses, such as HSV, "blends" with the host membrane, much like bringing olive oil together with corn oil. This blending (or fusion of membranes) allows **penetration** of the nucleocapsid into the cytoplasm.

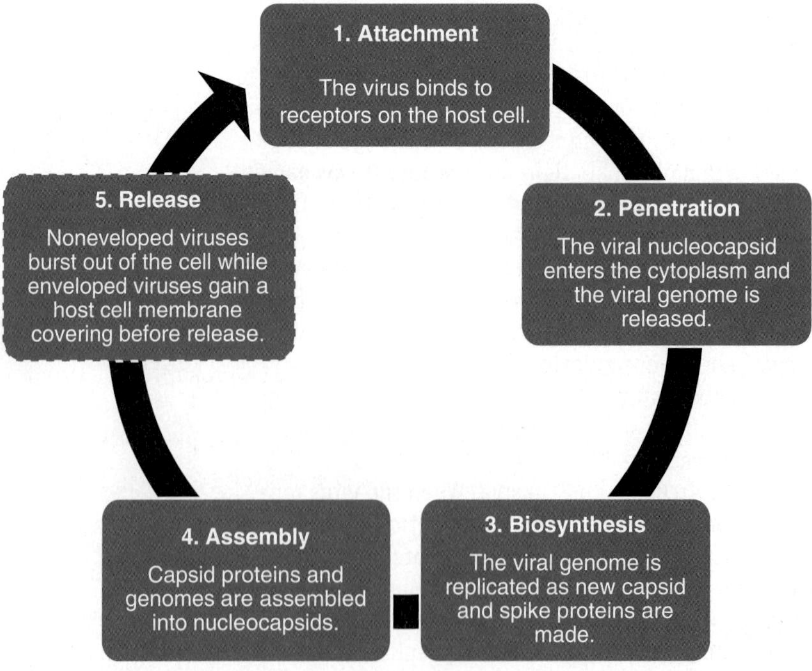

1. Attachment
The virus binds to receptors on the host cell.
2. Penetration
The viral nucleocapsid enters the cytoplasm and the viral genome is released.
3. Biosynthesis
The viral genome is replicated as new capsid and spike proteins are made.
4. Assembly
Capsid proteins and genomes are assembled into nucleocapsids.
5. Release
Noneveloped viruses burst out of the cell while enveloped viruses gain a host cell membrane covering before release.

FIGURE 6.6 **The Steps of Virus Replication.** These five steps are common to all viruses.

In contrast, many nonenveloped viruses, like the polioviruses, enter the host cell through a process called **endocytosis**. Here, the host cell engulfs the virus, which brings the genome and capsid into the cell cytoplasm. In still other cases, the capsid of the nonenveloped virus is left outside the host cell, and only the genome passes into the cell. Once within the cytoplasm, the capsid disassembles and releases the DNA or RNA genome.

3. Biosynthesis. With the blueprint to make new viruses now available, the manufacture (**biosynthesis**) of new viral parts can begin using the host cell's metabolic machinery. For DNA viruses, like HSV, messenger RNA (mRNA) molecules are transcribed and carry the genetic code for synthesizing the viral proteins (capsids and spikes). On the other hand, for RNA viruses like the polioviruses, the RNA acts as an mRNA molecule to encode the necessary proteins. In both types of viruses, through the process of translation, proteins soon appear in the cell's cytoplasm. Some of these proteins function as enzymes to hook nucleotides together and synthesize new viral genomes. Other enzymes are used to stitch together the amino acids for viral protein capsids. And some enzymes are used to break down parts of the host cells to supply the building blocks for the new viruses.

4. Assembly. Once biosynthesis of viral parts is complete, these parts are assembled into hundreds of new viral particles. Depending on the virus, **assembly** may occur in the cell nucleus or the cytoplasm.

5. Release. In the final stage of virus replication, called **release**, the nucleocapsids of some viruses push through a cellular membrane (i.e., nuclear or plasma membrane) of the host cell, forcing a portion of the membrane to surround and cover the nucleocapsid. This represents the new viral envelope. For nonenveloped viruses, often their large number is sufficient to burst open the host cell and release the nucleocapsids.

Let's look at these events visually in FIGURE 6.7 that illustrates the replication events for HSV. Replication takes place over a period of a few hours. As with all viruses, HSV does not bring along any ribosomes, the biochemical "workbenches" where amino acids are joined to make proteins; instead, the ribosomes of the host cell must be used. The virus has no energy-rich molecules to fuel the chemical reactions of biosynthesis and assembly, so it depends on the energy-rich molecules available in the host cell's cytoplasm. The virus has no amino acids it can contribute to protein synthesis, so it must use the amino acids in the host cell. The usurping of host cell resources puts a substantial drain on the host cell, and in many cases the host cell is weakened and dies. As it does, the new viruses are released and are ready to infect more cells.

For naked viruses, so many viruses are produced that the release process simply ruptures the host cells. For enveloped viruses, like HSV, the envelope is built from components of one of the cell's membranes: After the nucleocapsid forms, it moves toward the perimeter of the cell and forces its way through the membrane during the release stage. This action, which disrupts the membrane from within, occurs hundreds or even thousands of times as viruses make their exit. Normally, the cell could repair such holes, but with such a large number of viruses being produced and the cell's carefully balanced chemical processes being disrupted, virus release often represents the coup de grace for the cell.

For both naked and enveloped viruses, this viral replication cycle is termed a **productive infection** because new virus replication started immediately after infection.

Defense Against Viruses

If we stop for a moment and look back, we can develop a broad picture of how viral disease in humans takes place. An essential feature of viral disease is the destruction of body cells. As the cells are destroyed, the critical functions performed by those cells come to an end. Consider, for example, what happens to a person with a hepatitis B

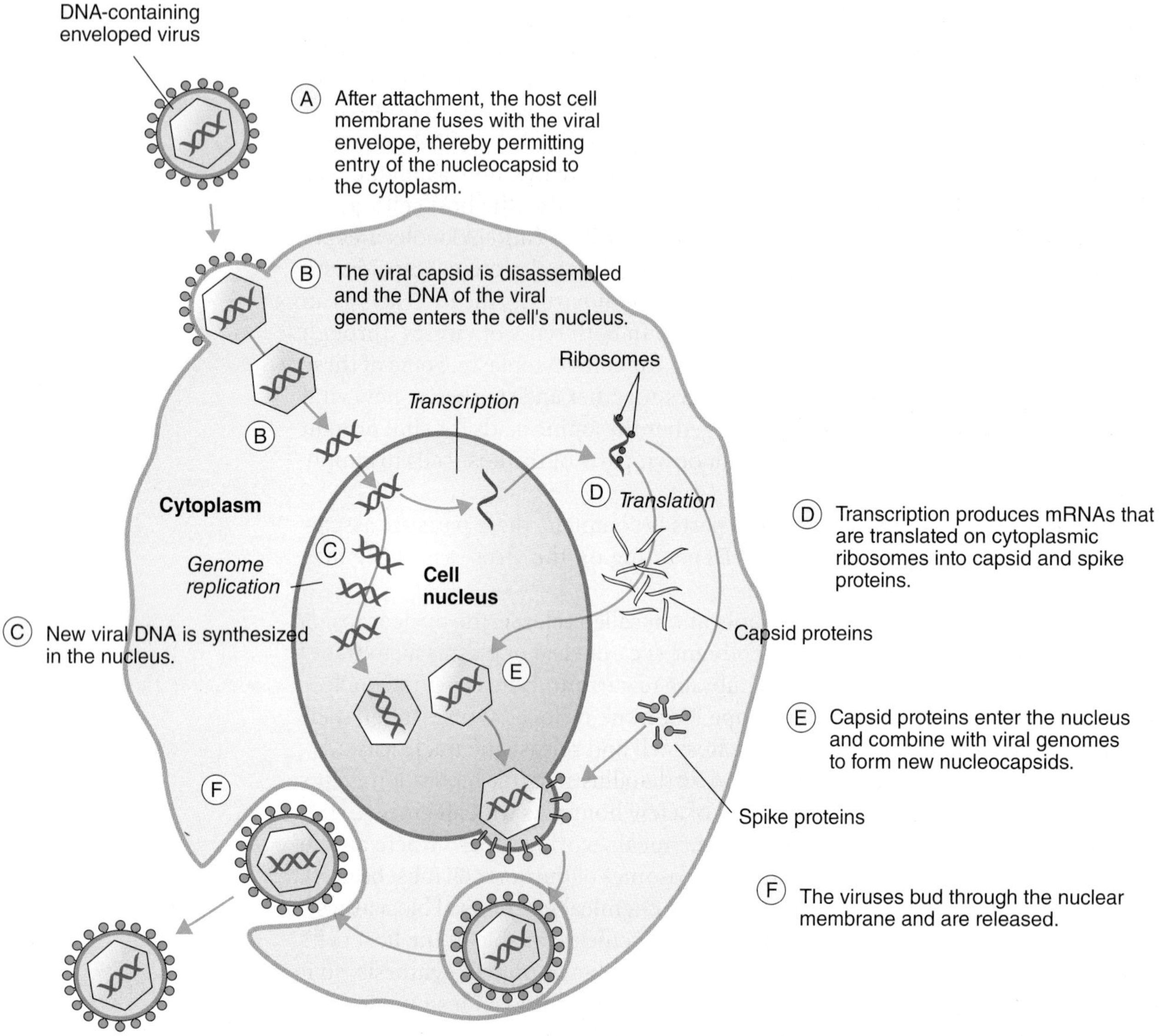
DNA-containing enveloped virus
(A) After attachment, the host cell membrane fuses with the viral envelope, thereby permitting entry of the nucleocapsid to the cytoplasm.
(B) The viral capsid is disassembled and the DNA of the viral genome enters the cell's nucleus.
Ribosomes
Transcription
(B)
Cytoplasm
(D) Translation
(D) Transcription produces mRNAs that are translated on cytoplasmic ribosomes into capsid and spike proteins.
(C)
Genome replication
Cell nucleus
(C) New viral DNA is synthesized in the nucleus.
Capsid proteins
(E)
(E) Capsid proteins enter the nucleus and combine with viral genomes to form new nucleocapsids.
(F)
Spike proteins
(F) The viruses bud through the nuclear membrane and are released.

FIGURE 6.7 **Replication of a DNA Virus.** The replication process illustrated here is for a herpes simplex virus (such as one causing cold sores or fever blisters on the lips). Genome replication and assembly occur in the cell nucleus and biosynthesis in the cytoplasm.

infection: Viruses replicate within the liver cells, and as the cells die, the liver tissue slowly collapses. The chemistry of the liver is radically altered: it cannot process amino acids for the body's use, and protein metabolism is blocked; production of bile drops off, and fat digestion is affected; liver cells are not available to process carbohydrates, and carbohydrate chemistry is disrupted; the liver is normally a storehouse for many vitamins, and its degeneration changes the body's vitamin balance; and on and on. Eventually, without treatment, liver function ceases.

Or consider acquired immunodeficiency syndrome (AIDS). In this disease, the human immunodeficiency virus (HIV) destroys immune cells critical to immune system control. Normally, these immune cells react to infection by bacterial pathogens, other viruses, infecting protists, and pathogenic fungi. They also hold in check the microbes that are harmless under normal circumstances (the so-called opportunistic microbes). However, as HIV replicates in these immune cells, the cells are slowly destroyed and the patient suffers numerous and varied infections of the lungs, brain, intestine, and blood because the immune system is incapable of fighting off all the

■ opportunistic: Referring to pathogens that only cause disease when the person's immune system is weakened.

pathogen attacks. Unless viral replication is interrupted in some way, such as through the use of antiviral drugs (see below), the patient will continue to suffer these illnesses, which can become life threatening.

These descriptions of viral disease paint a rather bleak picture of the interaction between viruses and their host cells. And yet we do not usually die of colds, chickenpox, herpes infections, mononucleosis, or numerous other viral diseases. Thus, we must be able to defend ourselves from most viral diseases, but the question is how? The answer is through immune system defenses, drug therapy, by vaccination, or some combination of these.

Immune System Defenses. One defense against viral attack is our immune system, especially through its production of antibodies. **Antibodies** are protein molecules synthesized by the immune system in response to the presence of foreign material, such as viruses, in the body. Molecular components of a virus, especially the capsid proteins and spikes, stimulate certain immune cells to produce huge numbers of highly specific antibody molecules that bind to the invading viral particles. This binding prevents the viruses from attaching to their host cells by covering the spikes or proteins on the viral capsid or envelope. Without an ability to attach to host cells, viral replication will not occur.

But what if host cells do get infected? Then, another immune defense mechanism involving white blood cells attack infected cells. These so-called cytotoxic cells then produce enzymes capable of destroying the infected cells and the viruses within them.

Another natural defense specifically against viruses is an antiviral molecule called **interferon**, which is naturally produced in the body when a virus infection occurs. These interferons alert neighboring cells of a potential impending viral attack so those cells can put up an antiviral defense. Today, one of the interferons, interferon alfa, can be produced by genetic engineering and used to treat certain viral diseases such as hepatitis B and C. Interferon is not 100% perfect because all of us still get flus, colds, and other viral infections. However, without the ability of our body to produce interferon, we would all suffer many more viral infections and more severe diseases. A more complete discussion of the immune system is in another chapter.

Drug Therapy. Beyond these natural defenses, there are several synthetic drugs that can be used against some viruses. Today, there is a whole arsenal of antiviral drugs designed to prevent HIV replication and the development of AIDS. Although the drugs allow HIV-infected individuals to lead almost normal lives, the virus is still hiding out in their bodies. Therefore, if such an individual stopped taking the drugs, viral replication would again commence.

■ synthetic drug: A man-made medicine produced by a pharmaceutical company or research lab.

Another useful antiviral drug is acyclovir (known commercially as Zovirax®). This drug inhibits DNA replication of HSV and chickenpox viruses. There are also a few antiflu drugs on the market, the most common one in the United States being Tamiflu® (oseltamivir phosphate). Tamiflu does not prevent infection, but if taken early in the infection may reduce virus release and help prevent virus spread.

Importantly, with the exception of hepatitis C, there are no antiviral drugs that will cure a viral infection. The drugs we have today only lessen the chances of getting infected or, if infected, may make the symptoms less severe and the duration of the disease shorter.

Viral Vaccines. As we have noted, one of the primary defenses against a virus infection is the production of antibodies by the body's immune system. Ideally, it makes good sense to have the antibodies available in the bloodstream and tissues even before the virus enters the body. This is the theory underlying viral vaccines. For example, suppose a person is vaccinated for this year's seasonal flu. By exposing the person to crippled flu viruses (the vaccine), the immune system is triggered to produce antibodies capable of binding to the "real" flu virus. So, later in the fall or spring, should the vaccinated individual be exposed to infectious flu viruses similar to the crippled viruses in the vaccine, the antibodies produced to the vaccine will immediately bind to the flu viruses and prevent the disease from taking hold.

Various kinds of vaccines for viral diseases are available. In one kind, scientists chemically destroy the virus particles, as we just mentioned for the seasonal flu vaccine. Such vaccines are said to be **inactivated**, meaning the viruses are crippled and cannot actively replicate in cells (some people refer to the viruses as "dead"). Another example of an inactivated vaccine is the Salk polio vaccine. Such inactivated vaccines must be given by injection to ensure a successful immune response.

Another type of antiviral vaccine contains viruses capable of infecting cells and replicating but at such an extremely low rate, disease symptoms usually do not develop. The vaccines stimulate the immune system for a long period of time, eliciting a stronger response than inactivated viruses. The viruses in these vaccines are sometimes called "live" viruses, but the preferred adjective is **attenuated**. Attenuated vaccines are routinely used to prevent chickenpox, measles, mumps, and rubella. Unfortunately, some attenuated vaccines carry a miniscule element of risk because the vaccine might contain a reactivated virus that could infect body cells. Another polio vaccine, the Sabin polio vaccine, is one example. Today, polio vaccines in much of the world consist of inactivated polio viruses although the attenuated vaccine is often used in parts of the developing world where wild polio virus still exists and raising the potential incidence of infection. Both vaccines have been instrumental in attempting to eradicate polio as described in **A Closer Look 6.2**.

The vaccines described above are called **whole-agent vaccines** because they contain the whole agent—the whole virus particle. A third kind of viral vaccine is called a **subunit vaccine** because the vaccine contains only a genetically engineered fragment (subunit) of the virus. It does not contain the whole virus. The highly successful vaccine against hepatitis B is an example of a subunit vaccine made from only the capsid proteins of the virus. It contains no genetic material, so the vaccine cannot cause the disease.

When Viruses Don't Replicate Immediately

Some viruses may not immediately replicate when they enter a host cell and do not produce productive infections. The scientific name for this phenomenon is **latency**. For example, HSV and the virus causing chickenpox (varicella zoster virus; VZV) do initially cause a productive infection leading to cold sores on the lips (HSV) or chickenpox (VZV). Then, these DNA viruses go latent and their genome hides out, but does not replicate, in neural ganglia. At a later time, the genome becomes active and initiates another crop of cold sores or, in the case of VZV, shingles later in life. Both these diseases are discussed in another chapter.

■ ganglion: A dense cluster of nerve cells.

Another particularly well studied form of latency occurs during HIV infection. HIV has RNA in its genome. After the virus has entered its host cell, the capsid is stripped away and the RNA is released (FIGURE 6.8). Next, a viral enzyme called **reverse transcriptase** uses the RNA as a template (a model) and synthesizes a complementary strand of DNA. The DNA then is transported to the cell nucleus, where it integrates into the cell's DNA. Such a fragment of viral-derived DNA incorporated into a chromosome is called a **provirus**. From its site within the nucleus, the provirus can remain latent and silent. Later, the provirus reactivates and new viruses will be encoded by the provirus. The new viruses will be released from the

A CLOSER LOOK 6.2

The Last Push to Polio Eradication?

A global effort to eradicate polio began in the 1988 as the Global Polio Eradication Initiative. Stimulated by the recent success in eradicating smallpox from the globe and polio from the Americas in 1979, the World Health Organization, Rotary International, the Centers for Disease Control and Prevention, and UNICEF spearheaded the effort to rid the world by 2000 of a virus responsible for the majority of paralysis and disability in children. Although this goal was not met, significant progress was made. By 2000, polio cases had fallen by more than 99%, to fewer than 1,000 and the disease was only endemic in four countries—Afghanistan, Pakistan, India, and Nigeria. Thus, the Initiative succeeded in reducing cases through routine immunization of infants.

Eradication has not been completed for several reasons. Areas of naturally occurring (wild-type) polio virus have been difficult to eliminate in the endemic countries. Conflicts and wars have made it hard to move freely through the countries to ensure immunization has been accomplished. In India, the problem has been trying to vaccinate an enormously overcrowded and growing population, where poor sanitation and diarrheal diseases abound.

For these reasons, and that an increasing number of polio-free areas were becoming reinfected, the stakeholders in the global eradication of polio launched the Strategic Plan 2010–2012 for eradicating wild polio virus. Unfortunately, that plan did not reach its goal so a new Polio Eradication and Endgame Strategic Plan 2013–2018 has been developed to eradicate all types of polio disease simultaneously—both due to wild poliovirus and due to vaccine-derived polioviruses. Confidence in the plan appears high as global leaders and philanthropists have pledged over 75% of the $5.5 billion needed over these 6 years to accomplish the goal. An encouraging sign was announced in early 2014 when India, considered the most challenging place to attempt polio eradication, marked its third anniversary without a reported polio case. Hopefully, the new Strategic Plan 2013–2018 will soon bring the end to another devastating human disease.

Discussion Point

According to the Polio Eradication and Endgame Strategic Plan 2013–2018, $5.5 billion is needed to finally eradicate polio by 2018. What might happen if polio is not eradicated and immunizations were to stop? What does it say about potential future immunization campaigns for eradicating other diseases if the polio initiative fails?

Global Polio Cases, 2014. Cases are limited to three endemic countries, but outbreaks in neighboring countries have occurred from importation.

Data from: Polio Global Eradication Initiative; http://www.polioeradication.org/Dataandmonitoring.aspx

cell and capable of infecting more immune cells, eventually seriously reducing the number of immune cells. As already mentioned, there soon comes a point at which the body cannot mount an effective immune response against other pathogens and the person develops AIDS.

A few latent viruses are believed to be responsible for some types of cancer: For instance, scientists believe cancer-causing viruses, called **oncoviruses**, play a role in some forms of leukemia, liver cancer, and cervical cancer (see below).

In many plant cells, latency does no apparent harm to the host cells. Indeed, the viruses are often responsible for remarkably beautiful variations in the plant's flowers. Petunias, carnations, and tulips, for example, owe their unusual patterns of pigmentation to proviruses in the cell. These plants usually grow more slowly than uninfected plants, but they show few other symptoms of viral presence. The viruses replicate only in certain cells of the plant and do not destroy these host cells.

■ leukemia: A cancer of the body's blood-forming tissues, where the bone marrow produces abnormal white blood cells, which don't function properly.

■ cervical cancer: A type of cancer occurring in the cells of the cervix—the lower part of the uterus that connects to the vagina.

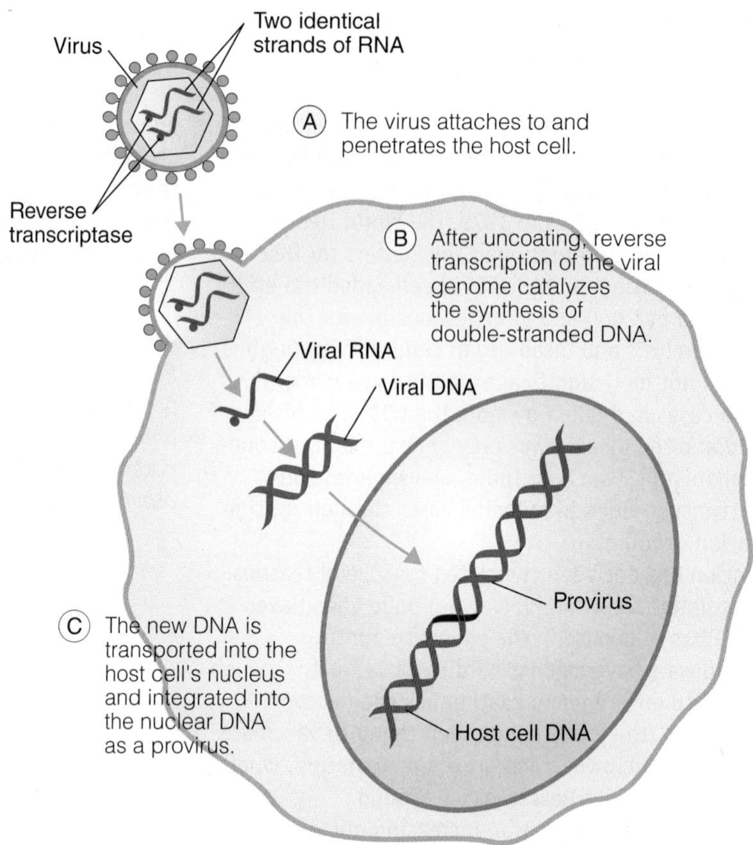
Virus
Two identical strands of RNA
Reverse transcriptase
(A) The virus attaches to and penetrates the host cell.
(B) After uncoating, reverse transcription of the viral genome catalyzes the synthesis of double-stranded DNA.
Viral RNA
Viral DNA
Provirus
(C) The new DNA is transported into the host cell's nucleus and integrated into the nuclear DNA as a provirus.
Host cell DNA

FIGURE 6.8 **The Formation of a Provirus.** Shown here is human immunodeficiency virus (HIV) infection. The single-stranded (ss)RNA genome can be reverse transcribed into double-stranded (ds)DNA by a reverse transcriptase enzyme. The DNA then enters the cell nucleus and integrates into a chromosome as a provirus.

6.3 New Viral Diseases: Where Are They Coming From?

Almost every year a newly emerging influenza virus descends upon the human population. Other viruses not even heard of a few decades ago, such as HIV, Hantavirus, and West Nile virus, are often in the news. Where are these viruses coming from?

Emerging Viruses

The United States is at greater risk than ever from **zoonotic diseases**, which are diseases transmitted from other animals to humans (TABLE 6.1). Many of these **emerging infectious diseases** are the result of viruses appearing for the first time in a population or rapidly expanding their range with a corresponding increase in detectable disease. Many are transmitted by insects such as ticks, fleas, and mosquitoes. One example is the West Nile virus, a mosquito-borne disease that marched across the United States between 1999 and 2009. It is now endemic across the continental United States. Dengue fever, another mosquito-borne disease, causes thousands of cases of illness in U.S. territories and American travelers, and 100 million infections worldwide every year.

■ endemic: Referring to a constant presence of disease or persistence of an infectious agent at a low level in a population.

The Centers for Disease Control and Prevention (CDC) is preparing for the potential emergence of chikungunya disease that is sustained by human-mosquito-human transmission of the chikungunya virus. It infected more than 265,000 people in an outbreak on the French island of Réunion in the Indian Ocean, as well as 1,400,000

TABLE 6.1 Examples of Emerging Viruses

Virus	Emergence Factor
Influenza	Mixed pig and duck agriculture, mobile population
Dengue fever	Increased population density, environments that favor breeding mosquitoes
Sin Nombre (Hantavirus)	Large deer mice population and contact with humans
Ebola/Marburg	Human contact with fruit bats
HIV	Increased host range, blood and needle contamination, sexual transmission, social factors
West Nile	Mosquito transported unknowingly to New York City
Nipah/Hendra	Human contact with flying foxes (bats)
SARS-associated	Contact with horseshoe bat
Chikungunya	Spread through new mosquito vectors and global travel

people in India in 2006. In 2007, transmission was reported for the first time in Europe, in a localized outbreak in northeastern Italy. With so many people traveling around the globe, it would be just a matter of time before the chikungunya virus reached North America. In fact, in late 2013 the virus and the disease were reported for the first time in the Caribbean. And, as expected, by mid-2014, almost 400 travel-associated cases and two locally transmitted cases (Florida) have been reported. Certainly both numbers will have increased substantially by the time you read this.

Looking ahead, scientists and microbiologists are wondering what diseases may reemerge as climate change causes temperature shifts, allowing insects like mosquitoes to survive in farther north and south latitudes than current temperatures permit.

But no matter how these viruses and zoonotic diseases are transmitted, what caused their emergence?

Jumping Species

One way "new" viruses arise is through one virus mixing genes from its genome with those from another to produce a new, unique combination of genes. Take for example influenza. Influenza viruses are notorious for swapping genes. The "swine flu" of 2009–1010 was the result of gene mixing between a strain of an avian flu virus, a human flu virus, and a swine flu virus. The new combination was unique and lead to a flu pandemic.

■ pandemic: An illness occurring over a wide geographic area of the world and affecting an exceptionally high proportion of the population.

"New" viruses also arise as a result from one of the driving forces of evolution—**mutation**, a permanent change to the genetic information in a gene such that it alters the genetic message. Although mutations often are lethal to the organism or virus, occasionally a mutation confers a benefit. In the case of HIV, beneficial mutations have resulted in new virus strains resistant to many of the antiviral drugs used to treat the disease. With fast virus replication rates, it does not take long for a beneficial mutation to spread, making drug therapy more challenging.

Even if a new virus has emerged, it must encounter an appropriate host to replicate and spread. It is believed the smallpox and measles viruses both evolved thousands of years ago from cattle viruses, while flu viruses probably originated in ducks and pigs. HIV almost certainly evolved from a monkey (simian) immunodeficiency virus.

So, most emerging viruses are not "new" in the sense of appearing from nowhere. Rather they are the result of genome mixing and mutation between already present animal viruses. The new genome now has a better ability to infect humans and cause disease. That means at some point the animal virus had to make a species jump. What could facilitate such a jump to humans?

Today, population pressure is pushing humans into new uninhabited or less inhabited areas around the world where potentially deadly animal viruses may be lurking. For example, many of the viruses responsible for hemorrhagic fevers jumped from rodents to humans as a result of increased agricultural practices that, for the first time, brought infected rodents into contact with humans.

hemorrhagic fever: A life-threatening illness caused by any of several viruses causing high fever and bleeding disorders that can lead to low blood pressure and death.

An increase in the size of the animal host population carrying a viral disease also can "explode" as an emerging viral disease. Prior to 1993, the deer mouse population in the Four Corners area of the United States (Arizona, Colorado, New Mexico, and Utah) had a low endemic infection with a virus called the Hantavirus. Then, in the spring of 1993 the American Southwest experienced a wet season, providing ample food for deer mice. The deer mouse population exploded, including those mice infected with the Hantavirus. Their activities left behind mouse feces and dried urine containing the virus that could now be spread through aerosols to unsuspecting humans. The deaths of 14 people with a mysterious respiratory illness in the Four Corners area in the spring of 1993 eventually were attributed to a Hantavirus infection, which progressed to Hantavirus pulmonary syndrome capable of causing a fatal respiratory disease.

So, emerging viruses are not really new. They are simply evolving from existing viruses and, through human changes to the environment, are given the "opportunity" to spread or to increase their range. **A Closer Look 6.3** describes the transmission of an emerging disease through the international pet trade.

6.4 Tumors and Cancer: A Role for Viruses

Cancer is indiscriminate. It affects humans and animals, young and old, male and female, rich and poor. In the United States, the American Cancer Society (ACS) projects over 580,000 Americans died of cancer in 2013, almost 1,600 people a day. Cancer remains the second most common cause of death in the U.S. (after cardiovascular disease), accounting for nearly one of every four deaths. In addition, more than 1.6 million new cancer cases were diagnosed in 2013. Worldwide, the World Cancer Fund International estimates more than 8 million people die of cancer each year and there will be more than 21 million new cancer cases diagnosed annually by 2030. Cancer is a very complex topic, so we will only summarize the basics and then describe the role of viruses in tumor and cancer development.

The Uncontrolled Growth and Spread of Abnormal Cells

Cancer starts with the uncontrolled reproduction of cells; that is, the frequency of cell division is greater for cancer cells than for normal cells. Cancer cells in some way escape controlling factors and, as they continue to multiply, form an enlarging cluster. Eventually, the cluster grows into an abnormal, large mass of cells called a **tumor**. Normally, the body will respond to a tumor by surrounding it with layers of

A CLOSER LOOK 6.3

The First American Case of Monkeypox

Monkeypox is a rare viral disease occurring mostly in central and western Africa. It is called "monkeypox" because it was first found in 1958 in laboratory primates. Blood tests of animals in Africa later found other types of animals also had monkeypox. In 1970, the first case of monkeypox was reported in humans (see figure).

The monkeypox virus is a DNA virus in the same group as the smallpox virus. People get monkeypox from an infected animal through bites or contact with the animal's blood, body fluids, or its rash. The disease also can spread from person to person through large respiratory droplets during long periods of face-to-face contact, or by touching body fluids of a sick person or objects such as bedding or clothing contaminated with the virus.

The first outbreak of monkeypox in the Western Hemisphere occurred in the Midwest in June 2003. On April 9, 2003, a Texas animal distributor received a shipment of some 800 small mammals from Accra, Ghana. This included rope squirrels, Gambian giant rats, and dormice.

Twelve days later, an Illinois distributor received Gambian rats and dormice from the Texas distributor. Unknown to the distributor, an infected Gambian rat was housed with prairie dogs that the distributor had before the prairie dogs were sold to distributors in six states. Additional prairie dogs were sold at animal swap meets.

Starting on May 15, doctor reports from the Midwest came in that people were exhibiting symptoms similar to but much milder and less contagious than smallpox. Individuals became ill with a fever, respiratory symptoms, and swollen lymph nodes. A rash developed, which progressed into raised bumps filled with fluid. The rash started on the face and spread across the body. Eventually the rash crusted over and the scabs fell off. The illness lasted for two to four weeks.

Over the next few weeks, outbreaks of monkeypox were identified in Wisconsin, Illinois, and Indiana where the Texas distributor had sent prairie dogs to pet dealers. On June 20, the last case was reported.

On June 30, 2003, a final report to the Centers for Disease Control and Prevention (CDC) identified 71 cases of monkeypox from Wisconsin, Illinois, Indiana, Missouri, Kansas, and Ohio. There were no deaths, although 18 were hospitalized.

This outbreak of monkeypox underlines the need to closely screen and protect the public from exotic animals exported to the United States. This time the infection was fairly mild as it was caused by a weak monkeypox strain; the next time, it or another disease might be contagious and lethal.

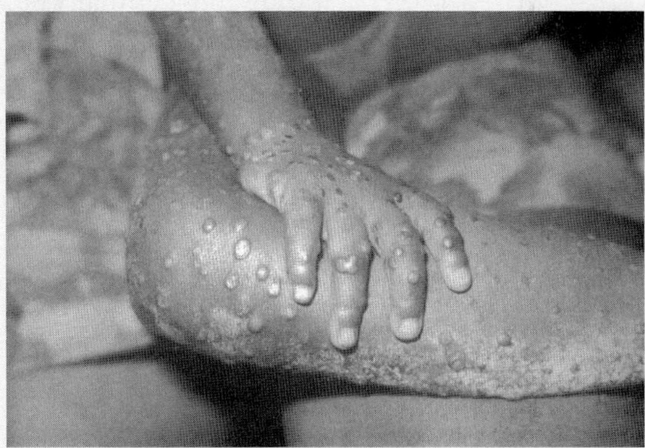

Human infection with monkeypox in Liberia, Africa in 1971.

Courtesy of CDC

connective tissue. Such an "encapsulated" tumor is designated **benign** and is usually not life threatening. If, however, the cells multiply too rapidly and break out of the capsule and spread, the tumor is described as **malignant**. The individual now has **cancer**, a reference to the radiating spread of cells resembling a crab (*cancer* = "crab"). The term **oncology** (*oncos* = "a mass") is the study of cancer.

Cancer cells differ from normal cells in three major ways: They grow and undergo cell division more frequently or for a longer period of time than do normal cells; they stick together less firmly than normal cells; and they undergo dedifferentiation. **Dedifferentiation** means the cells revert to an early stage in their development, often becoming formless cells dividing as rapidly as early embryonic cells. Moreover, cancer cells fail to exhibit contact inhibition; that is, they do not stop growing when they contact one another, as normal cells do. Rather, they overgrow one another to form a tumor. Sometimes, they **metastasize**, or spread, to form new tumors in a broad variety of body tissues. There appears to be no boundary limiting the growth of a malignant tumor.

How can a mass of cells bring disease to the body? By their sheer numbers, cancer cells invade and erode local tissues, thereby interrupting normal functions and damaging organs. For example: a tumor in the kidney may block the tubules and prevent the flow of urine during excretory function; a brain tumor might cripple this organ by compressing the nerves and interfering with nerve impulse transmission; and a tumor in the bone marrow could disrupt blood cell production. In addition, tumor cells rob the body's normal cells of vital nutrients to satisfy their own metabolic needs. Some tumors block air passageways; others interfere with immune system functions so that microbial diseases take hold. Ultimately, malignant tumors weaken the body until it fails.

The Involvement of Viruses

Some human cancers are the result of genetic abnormalities (mutations) in the body. However, the World Health Organization (WHO) estimates 60% to 90% of all human cancers are associated with **carcinogens**, chemicals and physical agents capable of producing cellular changes leading to cancer (FIGURE 6.9). Among the known carcinogens are the hydrocarbons found in cigarette smoke as well as asbestos, nickel, certain pesticides, and environmental pollutants in high amounts. Physical agents include ultraviolet (UV) radiation and X rays.

So what's this all have to do with viruses? First, realize that most viruses and viral infections do not lead to tumor formation or cancer. Of all human cancers, researchers believe up to 20% are a direct or indirect result of viruses (TABLE 6.2). When these tumor viruses are transferred to test animals or cell cultures, an observable cellular

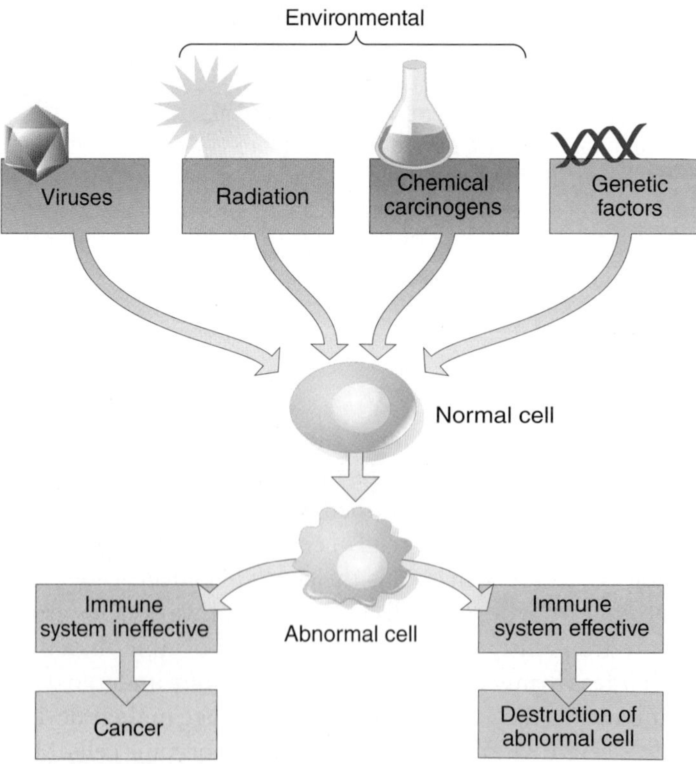

FIGURE 6.9 **The Onset of Cancer.** Viruses and carcinogens are among the factors that can induce a normal cell to become abnormal. When the immune system is effective, it destroys abnormal cells and no cancer develops. However, when the abnormal cells evade the immune system, a tumor may develop and become malignant, spreading to other tissues in the body.

transformation occurs involving morphological, biochemical, or growth patterns of the normal cells. Such a change to an abnormal cell involves a complex, multistep sequence of events, of which viruses play one part.

Of special note is cervical cancer, which along with liver cancer, are responsible for 80% of virus-associated cancers. Cervical cancer is the second most common cancer in women under age 35 and is caused by several subtypes of the human papilloma virus (HPV) that are often detected through a Pap smear. There is now a vaccine, called Gardasil®, which provides almost 100% protection against infection by the two most common HPV strains responsible for 70% of cervical cancers and two other strains responsible for 90% of genital warts. Note: The vaccine can only help prevent the infections if the vaccine is given before a person has been exposed to the tumor virus. The vaccine does not cure cervical cancer in women who already have the disease.

- **Pap smear** A test to detect cancerous or precancerous cells of the cervix, allowing for early detection of cancer.
- **genital warts** A common type of sexually transmitted infection that affects the moist tissues of the genital area.

How Viruses Transform Cells

The mechanism by which viruses transform normal cells into tumor cells remained obscure until the **oncogene theory** was developed in the 1970s (FIGURE 6.10). According to this theory, a set of cellular genes called **proto-oncogenes** that normally control cell growth and cell division can be reprogrammed to trigger tumor formation. These genes normally trigger cell division when necessary and remain silent when cell division is not required. Should one or more of these genes be permanently "turned

TABLE 6.2 Human Oncoviruses and Their Effects on Cell Growth

Oncogenic Virus	Benign Disease	Effect	Associated Cancer
DNA Tumor Viruses			
• Human papilloma virus (HPV)	Benign warts Genital warts	Encodes genes that inactivate cell growth regulatory proteins	Cervical, penile, and oropharyngeal cancer
• Merkel cell polyoma virus (MCPV)	Unknown	Being investigated	Merkel cell carcinoma
• Epstein-Barr virus (EBV)	Infectious mononucleosis	Stimulates cell growth and activates a host cell gene that prevents cell death	Burkitt lymphoma Hodgkin lymphoma Nasopharyngeal carcinoma
• Herpesvirus 8 (HV8)	Growths in lymph nodes	Forms lesions in connective tissue	Kaposi sarcoma
• Cytomegalovirus (CMV)	Mononucleosis syndrome	Stimulates genes involved with tumor signaling	Salivary gland cancers
• Hepatitis B virus (HBV)	Hepatitis B	Stimulates overproduction of a transcriptional regulator	Liver cancer
RNA Tumor Viruses			
• Hepatitis C virus (HCV)	Hepatitis C	Chromosomal aberrations, including enhanced chromosomal breaks and exchanges	Liver cancer
• Human T-cell leukemia virus (HTLV-1)	Weakness of the legs	Encodes a protein that activates growth-stimulating gene expression	Adult T-cell leukemia/ lymphoma

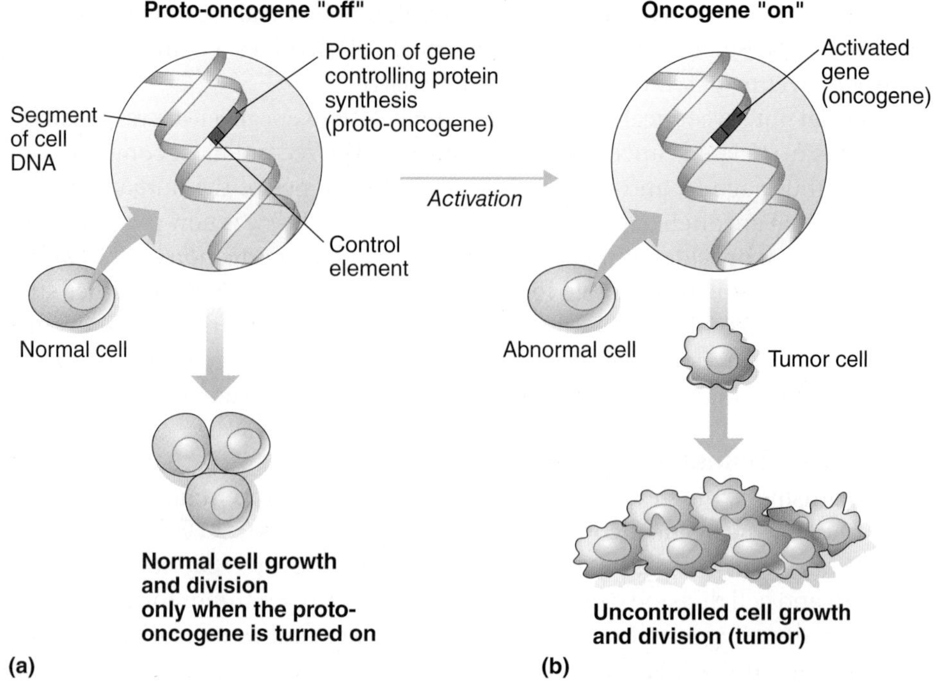

FIGURE 6.10 **The Oncogene Theory.** The oncogene theory helps explain the process of tumor development. (**a**) The normal cell grows and divides without complications. Within its DNA, it contains proto-oncogenes that are "turned on" only when cell divisions are needed. (**b**) When the genes are activated by viruses or other factors, they revert to oncogenes, which "turn on." An abnormal cancer cell may result. The oncogenes encode proteins that regulate the transformation from a normal cell to a tumor and may contribute to cancer formation.

on" through some interaction with a mutagen or virus, the genes, now referred to as **oncogenes**, are capable of starting the transformation of a normal cell into a tumorous or cancerous cell.

Viruses interact with host cells in one of three ways to trigger tumors or a cancerous condition.

First, some tumor viruses actually carry a viral oncogene (*v-onc*) in their genome. When the tumor virus infects a host cell, release of the viral genome causes cell transformation as soon as *v-onc* is expressed (FIGURE 6.11a). The oncogene's protein product drives cellular transformation, instructing the cell to continually divide and a tumor results.

In contrast, other tumor viruses carry no viral oncogene. However, after host cell penetration, the viral genome may insert near or actually within a proto-oncogene in the host genome. If so, then the provirus can regulate the activity of the proto-oncogene, converting it into an active oncogene. The result is that the oncogene expresses its protein product, which in turn again drives cellular transformation and uncontrolled cell division (FIGURE 6.11b). However, because these tumor viruses insert their genome randomly within the DNA of the host genome, the chance of insertion near a proto-oncogene is low.

The third way viruses influence tumor formation is through the loss of host gene activity. Cells normally contain several **tumor suppressor genes** whose function is to inhibit transformation and tumor formation. If a cell converts to an abnormal cell, the tumor suppressor proteins trigger a programmed cell death (FIGURE 6.12a). Some viruses carry genes whose products disrupt host DNA replication control by binding to and inactivating the proteins coded by tumor suppressor genes (FIGURE 6.12b). Loss of tumor suppressor gene activity then can result in uncontrolled cell divisions by one of the two previously described mechanisms.

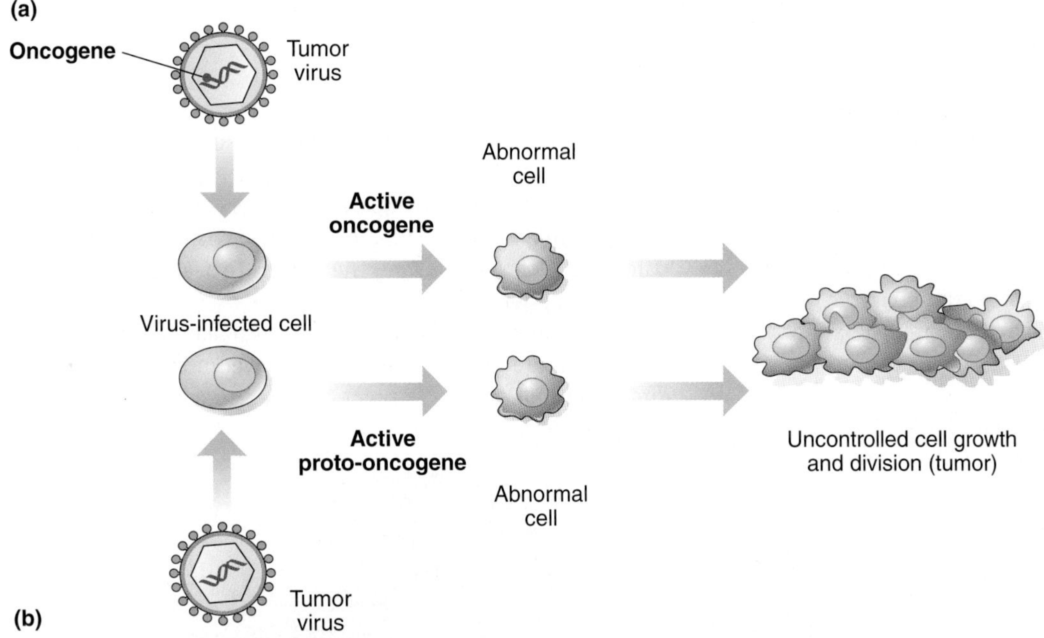

FIGURE 6.11 **Viral Oncoviruses.** Viruses causing tumors can do so by (**a**) carrying an oncogene. (**b**) Other oncoviruses not carrying an oncogene can insert next to or near a proto-oncogene as a provirus and activate the cell's proto-oncogene(s) and a transformation process.

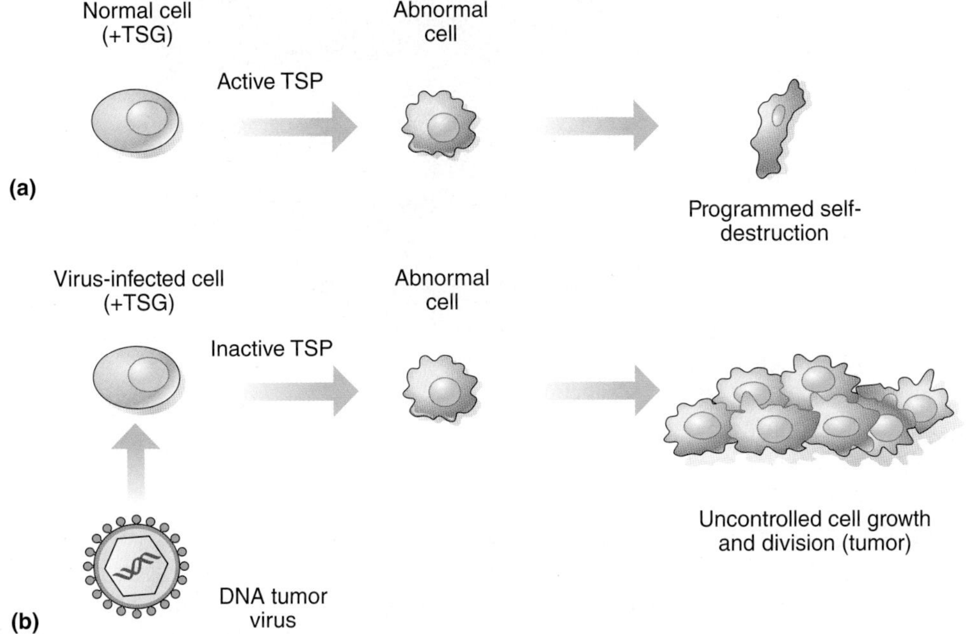

FIGURE 6.12 **Tumor Viruses and Tumor Suppressor Gene Action.** (**a**) All human cells contain tumor suppressor genes (TSG) that function to produce tumor suppressor proteins (TSP) capable of triggering programmed cell death if the cell converts to a tumor cell. (**b**) Infection by some tumor viruses can carry genes whose protein products will inactivate TSPs, permitting uncontrolled cell division and tumor formation.

In all three cases, virus infection leads to a disruption of normal growth control and a stimulation of cell division, leading to tumor formation. This natural ability of viruses to get inside cells and deliver their viral information has been manipulated by scientists to deliver genes to cure genetic illnesses. **A Closer Look** 6.4 considers the pros and cons of using viruses for this purpose.

A CLOSER LOOK 6.4

The Power of the Virus

Ever since the power of genetic engineering made it possible to transfer genes between organisms, scientists and physicians have wondered how they could use viruses to cure disease. One potential way is to use viruses against cancer. In 1997, a mutant herpesvirus was produced capable of only replicating in tumor cells. Because viruses make ideal cellular killers, perhaps these viruses would kill the infected cancer cells. Used on a terminal patient with a form of brain cancer, the virotherapy worked, as the patient has remained free of brain cancer.

Additional "oncolytic" (cancer-killing) viruses are being developed. Some cause the cancer cells to commit cell suicide, while others deliver chemicals (e.g., radioactive materials, a toxic drug, or a gene that codes for an anticancer protein) that on infection are capable of destroying the cancer cells.

Perhaps the most risky form of virotherapy is trying to correctly insert a gene into a patient to reverse a genetic disorder. In 2002, four young French boys suffering with a nonfunctional immune system were given virotherapy designed to ferry into the immune cells the gene needed to "turn on" the immune system. The therapy worked, but the virus-inserted gene disrupted other cellular functions, leading to leukemia (cancer of the white blood cells) in two of the boys. So, the idea of using virotherapy certainly works, but delivering the virus without complications and targeting the genes to the correct place can be a difficult assignment.

Another problem is the hurdle of overcoming a normal immune system response to the infecting virus. Like any infection, the immune system mounts an attack on the injected viruses, which, therefore, may be destroyed before the viruses reach their target. In 1999, an 18-year-old patient, Jesse Gelsinger, died as a result of an immune reaction that developed in response to the adenovirus used to treat a rare metabolic disorder. Importantly, 17 patients had been successfully treated using the same type of virus prior to Gelsinger. One technique researchers have devised to overcome the immune system defenses is to coat the viruses with an "extracellular envelope" invisible to the immune system yet allowing the viruses to search out and find their target.

Today, virotherapy is in high gear with many early clinical trials in progress to examine the power of the virus to directly or indirectly kill cancer cells and to deliver genes to correct genetic disorders.

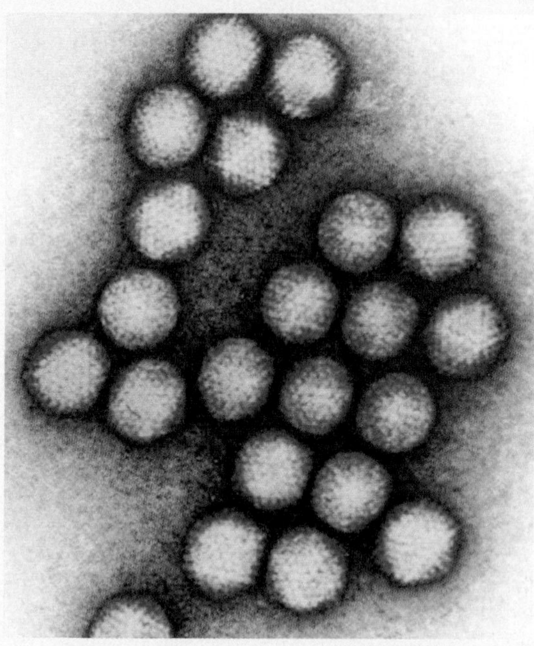

False-color transmission electron microscope image of adenoviruses.

Courtesy of Dr. G. William, Jr./CDC

6.5 Virus-Like Agents: Viroids and Prions

When viruses were discovered, scientists believed they were the ultimate infectious particles. It was difficult to conceive of anything smaller than viruses as agents of disease in plants and animals. However, this perception was revised when scientists discovered new disease agents—the subviral particles referred to as "virus-like agents;" that is, the viroids and prions.

Viroids and Prions Are Infectious Particles

Viroids are infectious agents of plants composed of RNA without any protein coat. The viroid genomes are so small they are incapable of coding for any known proteins. When the particles infect plants, they cause stunted growth and abnormal development. More than two dozen crop diseases are associated with viroids.

■ prion
prē′-on

Prions are infectious protein particles capable of causing a number of neurodegenerative diseases in mammals, including "mad cow disease" in cattle and variant Creutzfeldt-Jakob disease (vCJD) in humans. Prions are truly unique because: they can survive the heat, radiation, and chemical treatment that normally inactivate viruses; they are composed solely of protein; and they contain no known nucleic acids. So, how do they replicate and cause disease?

Infectious prions, called PrP^{Sc}, are apparently deviant versions of the normal and harmless cellular prions (PrP^{C}) found on the surface of cells in the central nervous system as well as in other mammalian tissues. If one is infected with PrP^{Sc}, these abnormal prions cause the normal PrP^{C} to fold into the abnormal PrP^{Sc} shape. A domino effect then slowly occurs where the new PrP^{Sc} proteins cause additional PrP^{C} proteins to misfold. The accumulation over time of these misshapen proteins into clumps blocks the molecular processes of protein recycling in brain tissue, which causes a buildup of faulty proteins and eventually cell death. PrP^{Sc} infected brain tissues develop a sponge-like appearance with empty areas of dead tissue (FIGURE 6.13). In cases of vCJD, the patient eventually develops neurological symptoms, including: unsteady gait and sudden jerky movements; memory loss; and persistent pain and odd sensations in the face and limbs. There is no treatment or cure for the disease and once symptoms set in, the patient usually dies within one year.

So then, how does one initially get infected with PrP^{Sc}? Almost all reported cases of vCJD occurred in the United Kingdom between 1995 and 2008 with the peak in 2000. Investigations discovered these infected individuals ate meat processed from cattle, many of which had PrP^{Sc} infections (mad cow disease). Thus, transmission of PrP^{Sc} came from slaughtered cattle whose meat products had been introduced into the human food supply. There have been 176 vCJD deaths in the United Kingdom since 1995. In the United States, there have been no reported cases of vCJD and there have been only 19 infected animals identified in Canadian and American cattle. Importantly, protection measures are in place to quickly identify and remove any mad cow cattle and to ensure products from those animals are not used for human food or animal feed.

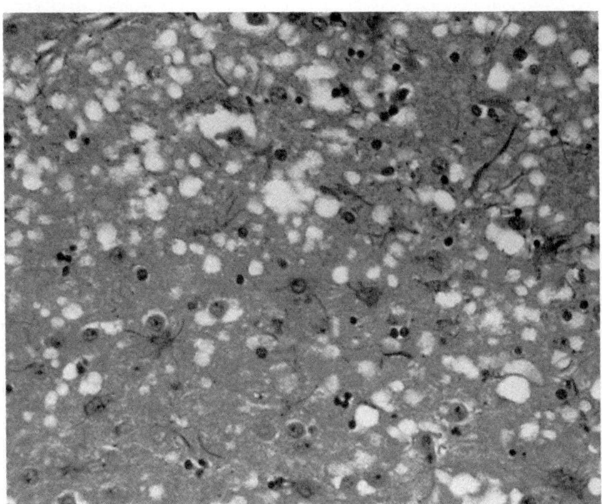

FIGURE 6.13 **Prion Infection.** A stained image showing the "spongy" degeneration of gray matter (the clear areas) in the brain characteristic of human and animal prion diseases.

Courtesy of APHIS photo by DR. Al Jenny/CDC

A Final Thought

When one of us (J.P.) was a young assistant professor of biology, the question of whether viruses are "alive" came up for discussion at an informal social gathering. As several people had had some beer to drink, the conversation became quite animated supporting one or the other viewpoint on viruses being alive. Perhaps one of the best answer was, "Who cares! We treat viral diseases the same way regardless of whether they are alive or not."

But the question has continued to pique my interest philosophically to this day. And now I put it to you. Are viruses alive? Although this textbook often refers to viruses as microbes for the sake of convenience, we have avoided references to "live" or "dead" viruses. Instead, we have used the words "active" for replicating viruses and "inactive" for viruses not replicating. Perhaps we should consider viruses to be inert chemical molecules with at least two properties of living organisms—the ability to replicate and adapt. Thus, viruses are neither totally inert nor totally alive, but somewhere on the threshold of life. Your thoughts?

Questions to Consider

1. A textbook author referring to viruses once wrote: "Certain organisms seem to exist only to reproduce, and much of their activity and behavior is directed toward the goal of successful reproduction." Would you agree with this statement? Can you think of any creatures other than viruses that fit the description?
2. Oncogenes have been described in the literature as "Jekyll and Hyde genes." What factors may have led to this label, and what does it imply? In your view, is the name justified?
3. Researchers studying the bacteria that live in the oceans have long been troubled by the question of why bacteria have not saturated the oceanic environments. Based on the material in this chapter, what might be a reason?
4. In broad terms, the public health approach to dealing with bacterial diseases is treatment and cure. Thinking of a disease like AIDS, what is the nature of the general public health approach to this and similar viral diseases? Explain your answer.
5. Bacteria can cause disease by using their toxins to interfere with important body processes, by overcoming body defenses (such as phagocytosis), by using their enzymes to digest tissue cells, or by other similar mechanisms. Most viruses, by contrast, do not encode toxins, cannot overcome body defenses, and produce no digestive enzymes. How, then, do viruses cause disease?
6. When Ebola disease broke out in Angola, Africa in 2005, the death toll was high, but the epidemic was short-lived. By comparison, when influenza breaks out at the start of winter, the toll is low, but the epidemic lasts for 6 or more months. From the standpoint of the viruses, what dynamics do you see in these two types of epidemics?
7. When discussing the multiplication of viruses, virologists prefer to call the process replication, rather than reproduction. Why do you think this is so? Would you agree with virologists that replication is the better term?
8. How have revelations from studies on viruses, viroids, and prions complicated some of the traditional views about the principles of biology?

Key Terms

Informative facts are necessary for the expression of every concept, and the information for a concept is founded in a set of key terms. The following terms form the basis for the concepts of this chapter. On completing the chapter, you should be able to explain and/or define each one.

antibody
assembly
attachment
attenuated
bacteriophage (phage)
benign
biosynthesis
cancer
capsid
capsomere
carcinogen
dedifferentiation
emerging infectious disease
endocytosis
envelope
enveloped virus
genome
helix
icosahedron
inactivated
interferon
latency
malignant
metastasize
mutation
nonenveloped virus
nucleocapsid
oncogene

oncogene theory
oncology
oncovirus
penetration
prion
productive infection
proto-oncogene
provirus
release
reverse transcriptase
spike
subunit vaccine
transformation
tumor
tumor suppressor gene
viroid
virosphere
virus
whole-agent vaccine
zoonotic disease

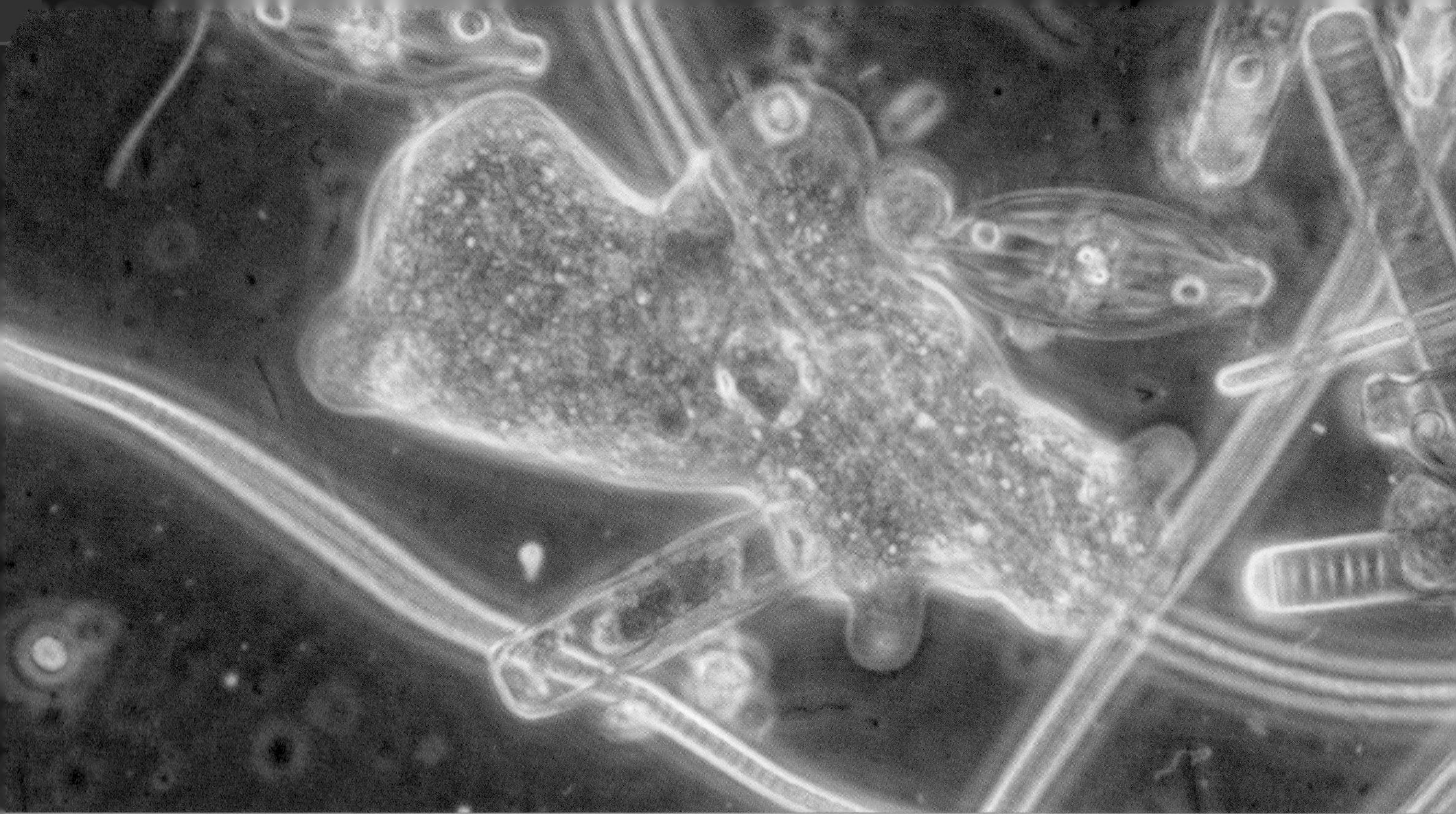

7 The Protists: A Microbial Grab Bag

In 1674, Antony van Leeuwenhoek was the first to observe microbial life when he examined a drop of pond water through his microscope. After seeing images of living animalcules, similar to this modern-day light microscope image of diverse protists, no wonder he remarked "... *no more pleasant sight has yet met my eye than this of so many thousands of living creatures in one small drop of water, all huddling and moving, but each creature having its own motion.*"

Looking Ahead

The protists are a group of eukaryotic microbes that include the protozoa, certain types of algae, and other unicellular organisms. This chapter surveys the significance of these microbes and describes some of the complex and interesting structures they possess.

On completing this chapter, you should be capable of:

- Describing the origins of the mitochondria and chloroplasts.
- Identifying the characteristics of the protists.
- Distinguishing between the animal-like, plant-like, and fungal-like protists.

Heralded at the time as "*The Eighth Wonder of the World*" and "*One of the few achievements which may properly be called epoch-making*," the construction of the Panama Canal would cut the transit time to ship goods and products quickly and cheaply between the Atlantic and Pacific coasts. By cutting the transit time from over 30 days to about two weeks, global trade and commerce would be stimulated. However, construction was being hampered by the presence of some tropical diseases. One of these was malaria, a disease that doomed France's attempt at canal construction in 1883. The United States took over construction in 1904 but malaria was still a major problem. By 1906, of the 26,000 employees working on the canal, over 21,000 had been hospitalized with malaria. Successful completion of the canal would have to control this and other mosquito-borne diseases.

In 1878, Alphonse Laveran, a military doctor in France's Service de Santé des Armées was posted in Algeria at a military hospital where malaria was then a serious problem. In the 1700s and early 1800s, people thought malaria was caused by bad air from marshlands (in Italian, *mala* = "bad" and *aria* = "air"). However, following the acceptance of Louis Pasteur's germ theory, a bacterial origin for malaria was proposed. Working from his anatomical observations and studying the lesions in organs and in blood of patients with malaria, by 1880 Laveran was convinced a protozoan parasite was the agent causing malaria, not a bacterial pathogen. After much debate, by 1890 the parasite origin of malaria was confirmed. But how was the parasite transmitted to people?

Ronald Ross was a British officer in the Indian Medical Service in Calcutta (FIGURE 7.1). In 1892 he became interested in malaria and set out to prove Laveran's hypothesis identifying mosquitoes as the agent of transmission. The breakthrough came in 1897 when Ross was dissecting the stomach tissue of a mosquito that had taken a blood meal from a malarious patient. In microscope samples from the mosquito's stomach tissue Ross observed the malaria parasite. Further studies verified his findings and proved mosquitoes transmit the parasite to humans. The problem of malaria transmission was solved.

Ross' discovery had tremendous impact on development programs in the tropics, including Panama. In 1904, Colonel W. C. Gorgas of the U. S. Army Medical Corps headed the Isthmian Canal Commission that was given the task of implementing a mosquito control program (insecticide spraying) in the Canal Zone to lessen the cases of malaria, both for the Panamanian people and for the workers on the construction project. The result of this control program was a dramatic decrease in malaria hospitalizations and deaths. Although malaria continued to be a challenge throughout the entire construction project, the Panama Canal was completed in 1913 and officially opened to ship commerce in 1914.

Once more we see the power of the microbe and its influence on human events and history. We also see again the tenacious spirit of scientific investigation and discovery that so characterizes the human spirit. If Laveran and Ross had not been intrigued about malaria and the source of its transmission, it is possible that, like the French attempt, American construction of the canal would not have been completed and global ocean commerce between the Atlantic and Pacific Oceans would have remained lengthy and economic development delayed.

DR. RONALD ROSS, C.B., THE HERO OF THE MOSQUITO THEORY OF MALARIA.

FIGURE 7.1 **Ronald Ross.** Major Ronald Ross discovered mosquitoes transmitted malaria to humans.

Courtesy of the National Library of Medicine

Malaria has been infecting humans for more than 5,000 years. During the 1700s, Europeans suffered wave after wave of malaria and few regions were left untouched. Even American pioneers settling in the Mississippi and Ohio valleys suffered great losses from the disease. Today, between 300 and 500 million of the world's population suffer from malaria, which exacts its greatest toll in Africa. The World Health Organization (WHO) estimates more than 800,000 people die from malaria every year. Children are the most susceptible because their immune system has not had time to develop partial immunity from mild malaria infections. So, in 2014 one child died every 30 seconds from malaria! No infectious disease of contemporary times can claim such a dubious distinction.

■ *Plasmodium* plaz-mō'dē-um

Today, we know malaria is caused by one of four species in the genus *Plasmodium*. It is one of the protists we shall meet as we continue to explore the world of microbes—the eukaryotic members—and continue to witness how microbes affect society.

7.1 Eukaryotic Cells: Their Origin

The word "protist" comes from Greek stems meaning "very first," and identifies the protists as the very first eukaryotes to appear on Earth about 2 billion years ago. These "newbies" would not only be the ancestors to present day protists but also to the fungi, plants, and animals. How such an ancestral eukaryote came into being is a puzzle that has received much attention in contemporary biology.

The Evolution of the Eukaryotic Cell and Endosymbiosis Theory

Today, all eukaryotic cells contain a variety of membrane-enclosed structures called **organelles**. This includes the cell nucleus, an endomembrane system, and mitochondria (plus chloroplasts in the algae and plants). The prevailing wisdom points to a theory that suggests eukaryotic cells evolved through a series of structural changes to an ancestral cell. In this theory, two major events are proposed.

The first event involved an ancestral prokaryote developing the ability to move its plasma membrane inward, thereby forming a series of inward folds (FIGURE 7.2a). These folds may have eventually surrounded the hereditary material to form a cell nucleus as well as the internal membrane system we see in all eukaryotic cells today. The resulting advantage to the ancestral cell was to divide metabolic processes into separate organelles in which complex chemistries would not be competing and interfering with one another. In addition, the additional membrane provided more surface or "workbench space" on which metabolic reactions could occur.

The second event in the evolution of an ancestral eukaryotic cell has more supporting evidence. Championed by the late Lynn Margulis and her colleagues at the University of Massachusetts, this event proposes a way ancestral cells acquired the energy organelles, the mitochondria and chloroplasts. The theory says mitochondria and chloroplasts arose through a process called endosymbiosis. A **symbiosis** is a living together of two or more organisms, so **endosymbiosis** refers to a living together where one organism (partner) lives inside (*endo*) the other.

The evolution of the mitochondria and chloroplasts is thought to have occurred in the following way (FIGURE 7.2b). A primitive eukaryotic cell engulfed a heterotrophic bacterial cell or was invaded by such a cell. In either case, rather than breaking down (digesting) the bacterial cell, the eukaryote retained it in the cytoplasm and relinquished the responsibility for energy metabolism to this microbe. Then, as the two partners became more dependent on one another for energy and survival, they evolved into a single organism with the bacterial partner eventually evolving into the present day mitochondrion.

Carrying the endosymbiosis idea further, a similar scenario can be drawn for the evolution of the chloroplast in algae and plants (see Figure 7.2b). Current thinking points to a similar uptake of a photosynthetic bacterial cell, perhaps a member of the Cyanobacteria. Again, tolerance of the other partner permitted the symbiosis and provided the "new organism" with a way to manufacture its own food (organic

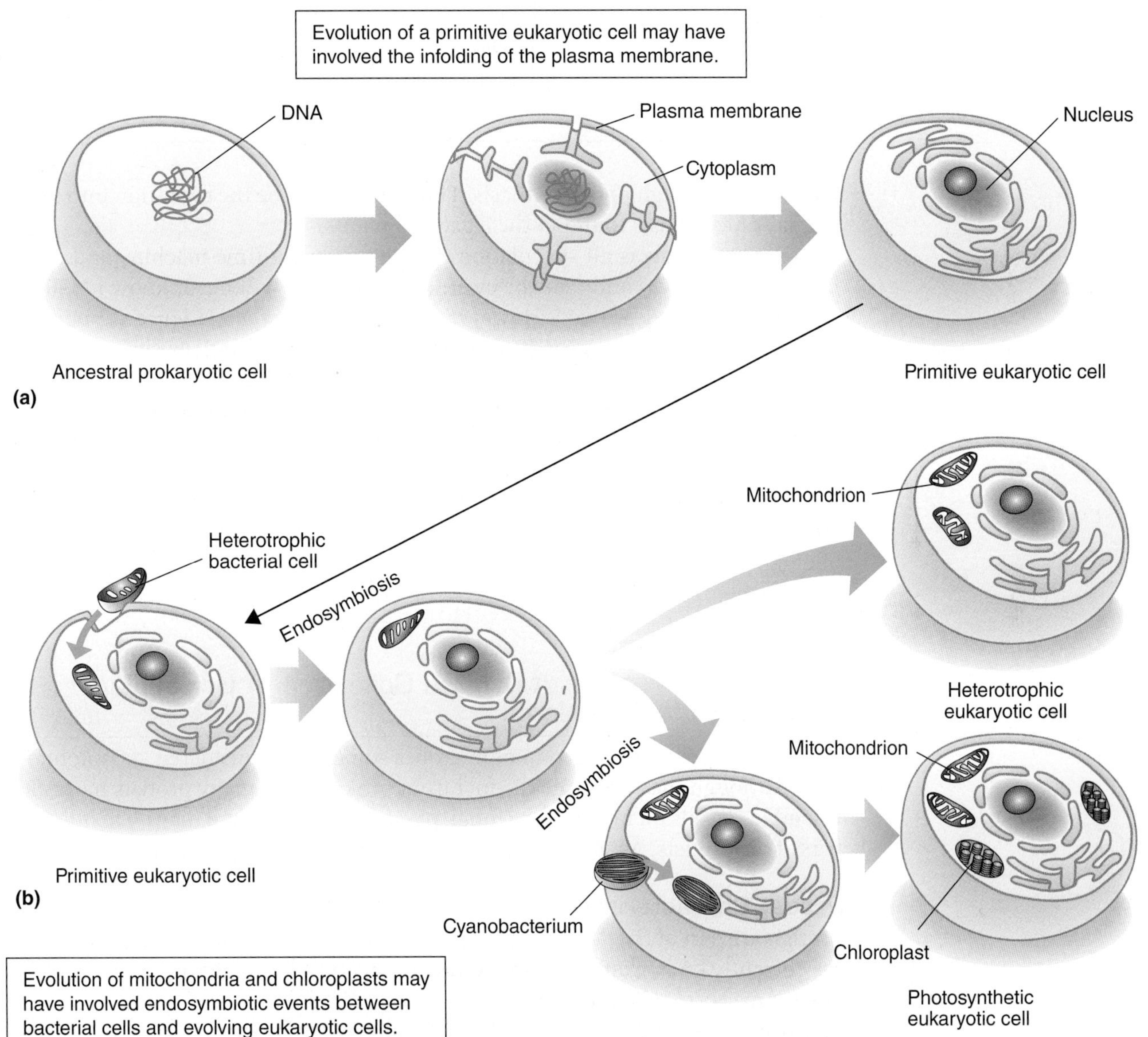

FIGURE 7.2 **Internal Membranes and the Endosymbiosis Theory.** Two major events have been proposed to explain the evolution of the first eukaryotic cell. (**a**) Internal membranes may have evolved when an ancestral prokaryote developed the ability to fold its plasma membrane inward. (**b**) The mitochondrion and chloroplast may have originated from an independent heterotrophic bacterium cell and an autotrophic cyanobacterium that were taken into primitive eukaryotic cells through endosymbiosis.

TABLE 7.1 **Similarities Between Mitochondria, Chloroplasts, and Bacteria**

Characteristic	Mitochondria	Chloroplasts	Bacteria
Average size	1–5 μm	1–5 μm	1–5 μm
Nuclear envelope present	No	No	No
DNA molecule shape	Circular	Circular	Circular
Ribosomes	Yes; bacterial-like	Yes; bacterial-like	Yes
Protein synthesis	Make some of their proteins	Make some of their proteins	Make all of their proteins
Reproduction	Binary fission	Binary fission	Binary fission

molecules) through photosynthesis. The result would be the evolution into the present day chloroplasts found in the algae and plants.

Of course, this is all speculation. Unless we had a time machine and could go back 2 billion years and "see the event or events," we will never know for sure how the eukaryotic cell evolved. However, there is substantial evidence supporting the endosymbiosis theory. As TABLE 7.1 shows, the biochemical and physiological similarities between mitochondria, chloroplasts, and bacterial cells are too striking to ignore. All contain DNA and RNA, and the ribosomes of mitochondria and chloroplasts are very similar to those in bacterial species. In addition, the organelles and bacterial cells "reproduce" themselves through binary fission.

With the evolution of an endomembrane system and energy organelles, eukaryotic cells could become more complex structurally and evolve into a diverse domain of organisms.

7.2 The Protists: A Diverse Collection of Unicellular Eukaryotic Organisms

The protists are a mixed group of microbes sometimes considered "taxonomic misfits." Although they are quite distinct from prokaryotes, many protists more closely resemble various members of the fungi, plants, or animals than they do to other members of the protists. As such, it is challenging to identify unique protist characteristics.

Protist Characteristics

Protists are part of the domain Eukarya (FIGURE 7.3). Over 200,000 species of protists are currently known and they are exceeded only by the prokaryotes in the number of environments to which they have adapted.

Although all protists are unicellular organisms, some exist as colonies of cells and are referred to as colonial organisms. While most species of protists are microscopic, some species have diameters close to a millimeter. As we shall see, protists practice all known modes of nutrition. Some are animal-like and fungal-like **heterotrophs**, capturing their food by engulfing other organisms, or by absorbing simple organic nutrients across their plasma membranes. Some of these are parasites that cause human

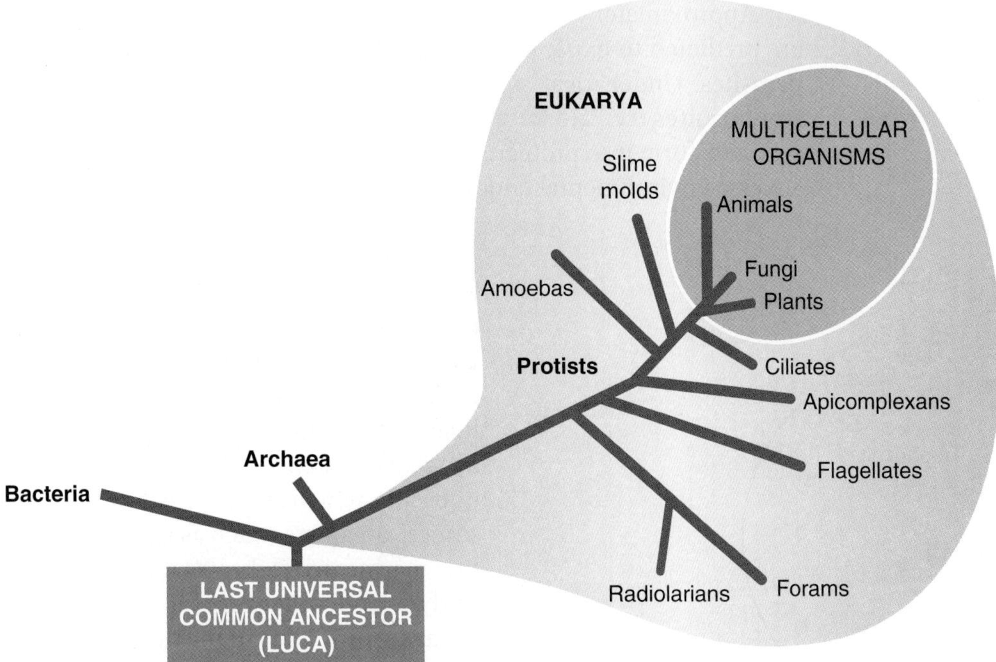

FIGURE 7.3 **The Tree of Life.** The prokaryotic domains Bacteria and Archaea share the tree of life with the eukaryotes. Note that most of the domain Eukarya consists of an enormous diversity of protists.

and animal disease. Other protists, such as the algae, are plant-like **autotrophs**, using their chloroplasts and chlorophyll pigments to trap the Sun's energy and, through photosynthesis, form carbohydrates such as glucose. There are even some protists that are **mixotrophs**, which behave as autotrophs at one time and then change their nutrition and act as heterotrophs at another time. As we said, the protists are a real "mixed bag" of organisms.

Asexual reproduction is the most common means of population growth; that is, an increase in protist numbers. But sexual reproduction also occurs—it is among the evolutionary traits first appearing in the protists. Sexual exchanges vary greatly: Some species fuse their entire cells, while other species come together to exchange only their nuclei.

Finally, it is even hard to characterize protist habitats because they too are very diverse. Most of their habitats are moist and many species reside in ponds, lakes, and oceans. Others do fine in moist soils and leaf litter. And as we said earlier, there are pathogens and parasites, such as the malarial parasite, that take up temporary or permanent residence in a host organism.

With this diversity of characteristics in mind, let's examine a few groups of protists to illustrate their diversity. For convenience, we will group these organisms by lifestyle: the protozoa, unicellular algae, slime molds, and the water molds (yes, these last two are protists, not fungi).

7.3 The Protozoa: Animal-Like Protists

The protozoa are an old lineage of heterotrophic microbes and are so named because they were once believed to be the first animals (*proto* = "very first;" *zoa* = "animal"). These protists lack cell walls, ingest food particles (e.g., bacterial cells and other protists) or absorb nutrients in the environment, and most have some form of motility.

Approximately 8,000 species of protozoa have been named and another 35,000 are predicted to exist; most are found in aquatic environments, in moist soil, or as parasites. Under ideal conditions, the protozoa exist as active feeding cells called **trophozoites** (FIGURE 7.4) As nutrients become scarce, however, the cells of some species transform into protective, dormant bodies known as **cysts**, which can withstand adverse environmental conditions.

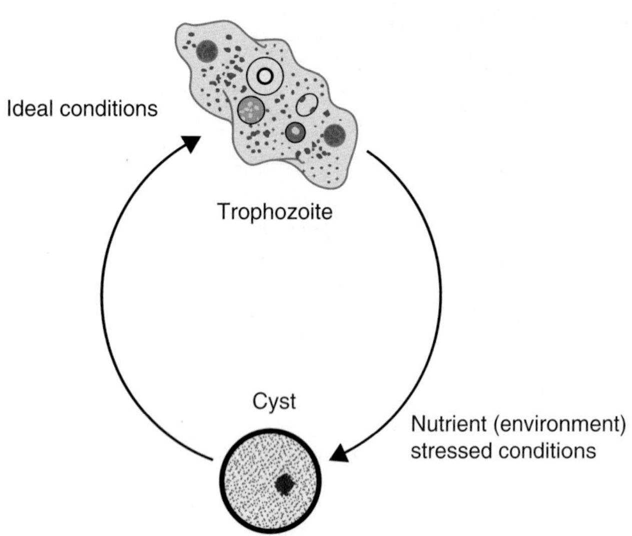

FIGURE 7.4 **A Protozoan Life Cycle.** Many species of protozoa alternate between an actively feeding trophozoite stage and a dormant cyst. When conditions are harsh (e.g., desiccation or low nutrients), the trophozoites will secrete a thick wall and enter into a dormant period characterized by low metabolic activity. The cysts will convert back into trophozoites when conditions are again favorable.

As essential participants to nutrient recycling in the soil, protozoa act with other microbes to decompose the remains of dead animals and plants and recycle the nutrients. In addition, many species of protozoa are part of the **zooplankton**, the animal-like component of aquatic food chains. Feeding on microscopic cyanobacteria and unicellular algae composing the phytoplankton (see below), the zooplankton convert the phytoplankton components into nutrients, which are absorbed and digested by other "consumer" organisms such as sponges, jellyfish, worms, and tiny marine invertebrates (animals without backbones). And on land, other protozoa perform the same nutrient-releasing functions in the digestive tracts of cattle, goats, and other so-called ruminant animals.

Because the diversity of protists has made their classification difficult and one needing more study, we shall use terms that describe types of cell movement called motility to distinguish the four groups of protozoa (FIGURE 7.5).

Protists with Pseudopodia

Amoebas have often been portrayed in popular writing as blobs of cytoplasm (some cells are larger than 2 mm) because of their flexibility: They have no definite form and are constantly changing shape by sending out **pseudopodia** (sing., pseudopodium). Pseudopodia are cellular extensions projecting outward only to then contract (FIGURE 7.6a), the movement allowing the cells to creep over their environment such as the bottom of a pond. Pseudopodia also are used to capture prey, which are then digested internally. Reproduction in the amoebas occurs primarily through asexual reproduction where one cell simply divides into two cells.

■ *Entamoeba histolytica* en-tä-mē'bä his-to-li'ti-kä

■ dysentery: A disease of the lower intestine (colon) that is characterized by severe diarrhea, inflammation, and the passage of blood and mucus.

■ gastroenteritis: An inflammation of the stomach and the intestines, causing vomiting and diarrhea.

■ *Acanthamoeba* a-kan-thä-mē'bä

None of the amoebas is photosynthetic, so all species must obtain their nutrients from preformed organic matter, the heterotrophic mode of nutrition. As such, a few pathogenic species exist. A serious human pathogen is *Entamoeba histolytica*, the cause of amoebic dysentery. The disease is more common in tropical areas where poor sanitary conditions exist. Transmitted by contaminated food and water or swallowing *E. histolytica* cysts picked up from contaminated surfaces or fingers, the parasite is responsible for intestinal ulcers and sharp appendicitis-like pain, representing a form of gastroenteritis. The disease is responsible for up to 100,000 deaths globally every year.

Another health problem is caused by a species of *Acanthamoeba*. These protozoa can cause a rare infection of the cornea of the eye (keratitis) in people who wear contact lenses. This rare disease usually results from the contact lenses becoming contaminated with the organism after improper cleaning and handling of the contacts. Other amoebas are found in some home humidifiers, where when aerated and breathed into the respiratory tract, they may cause an allergic reaction called "humidifier fever."

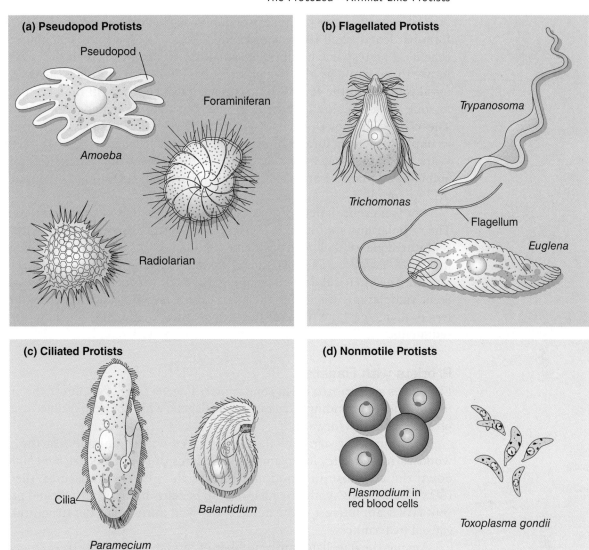

(a) Pseudopod Protists

Pseudopod
Amoeba
Foraminiferan
Radiolarian

(b) Flagellated Protists

Trichomonas
Trypanosoma
Flagellum
Euglena

(c) Ciliated Protists

Cilia
Paramecium
Balantidium

(d) Nonmotile Protists

Plasmodium in red blood cells
Toxoplasma gondii

FIGURE 7.5 **The Four Major Groups of Protozoa.** The organelles of motion (pseudopodia, flagella, or cilia) are shown for members of each of the four groups. The nonmotile protists lack such structures.

FIGURE 7.6 **The Pseudopodia-Containing Protists.** (**a**) A light microscope image of an amoeba cell showing its flexible shape characterized by the extension of pseudopodia. The cytoplasm contains the typical eukaryotic organelles, including the cell nucleus. (Bar = 100 μm.) (**b**) The White Cliffs of Dover, England are composed of the remains of foraminiferans that thrived in the oceans millions of years ago. (**c**) A light microscope image of radiolarians, showing their very delicate glassy skeletons. (Bar = 60 μm.)

Other protists also produce pseudopodia. The **foraminiferans** (called **forams**) form hardened, shell-like casings called "tests" composed of calcium carbonate (chalk). The tests have numerous small pores through which the pseudopodia move in and out as the organism feeds (*foramen* = a "passageway"). The fossil remains of the forams build up as dense deposits on the ocean floor, and in some areas, geologic upthrusting has brought these deposits to the surface. The White Cliffs of Dover in England, shown in FIGURE 7.6b, are a well-known example. Moreover, forams flourished during the Paleozoic era, about 225 million years ago, when many oil fields were forming from the remains of prehistoric animals and plants. Foram shells therefore serve as depth markers for geologists drilling for oil.

Another group of heterotrophic protists with pseudopodia are the **radiolarians**. The radiolarians are found exclusively in the sea, where they live within finely sculptured glassy skeletons of silicon dioxide reminiscent of Christmas tree ornaments (FIGURE 7.6c). The skeletons are as varied as snowflakes and have elaborate, radiating geometrical designs, making them a favorite subject of photographers. Some radiolarians are among the largest protists, with skeletons several millimeters in diameter. Like those of the forams, these skeletons contribute to oceanic sediments.

Protists with Flagella

Flagellates are protozoa having one or more **flagella** (*sing*., flagellum), the long, hairlike organelles providing movement. The flagella whip about (*flagellum* = "a whip"), propelling the cell forward.

Trichonympha
trik-ō-nimf′ä

An impressive array of flagellates lives within animals. In the gut of a wood-eating termite, for example, flagellates of the genus *Trichonympha* break down the cellulose in wood and release the glucose for use by the termite (FIGURE 7.7a). This symbiotic relationship benefits both the termite and protozoan (but is of scant interest to the homeowner who must repair the structural damage caused by termites).

Trypanosoma brucei
tri-pa′nō-sō-mä bru¯s′ē

Trypanosoma cruzi
tri-pa′nō-sō-mä kruz′ē

Among the flagellates are a number of human pathogens. *Trypanosoma brucei* and *Trypanosoma cruzi* are the causes of African and American sleeping sickness, respectively (FIGURE 7.7b). *T. brucei* is transmitted by tsetse flies while *T. cruzi* is transmitted by cockroach-like triatomid bugs. When introduced into the bloodstream of a human, the cells invade the brain tissue, where they cause a coma-like condition. American sleeping sickness, also called Chagas disease (named after the Brazilian physician Carlos Chagas who discovered the parasite in 1909), is also accompanied by a severe heart infection. Charles Darwin, renowned for his theory of evolution, is believed to have acquired this disease during his travels in South America. Today, 8 to 11 million cases of American sleeping sickness are reported each year in Mexico, Central America, and South America. There also are some 300,000 cases reported each year in the United States, primarily in people who have emigrated from Latin America.

Giardia lamblia
jē-är′dē-ä lam′lē-ä

Giardia lamblia, the bane of campers, hikers, and backpackers, is also a flagellate (FIGURE 7.7c). This species is contracted from contaminated water, especially in mountain streams and lakes. It causes nausea, gastric cramps, and a foul-smelling watery diarrhea. The disease affects wild animals, and they may be the source

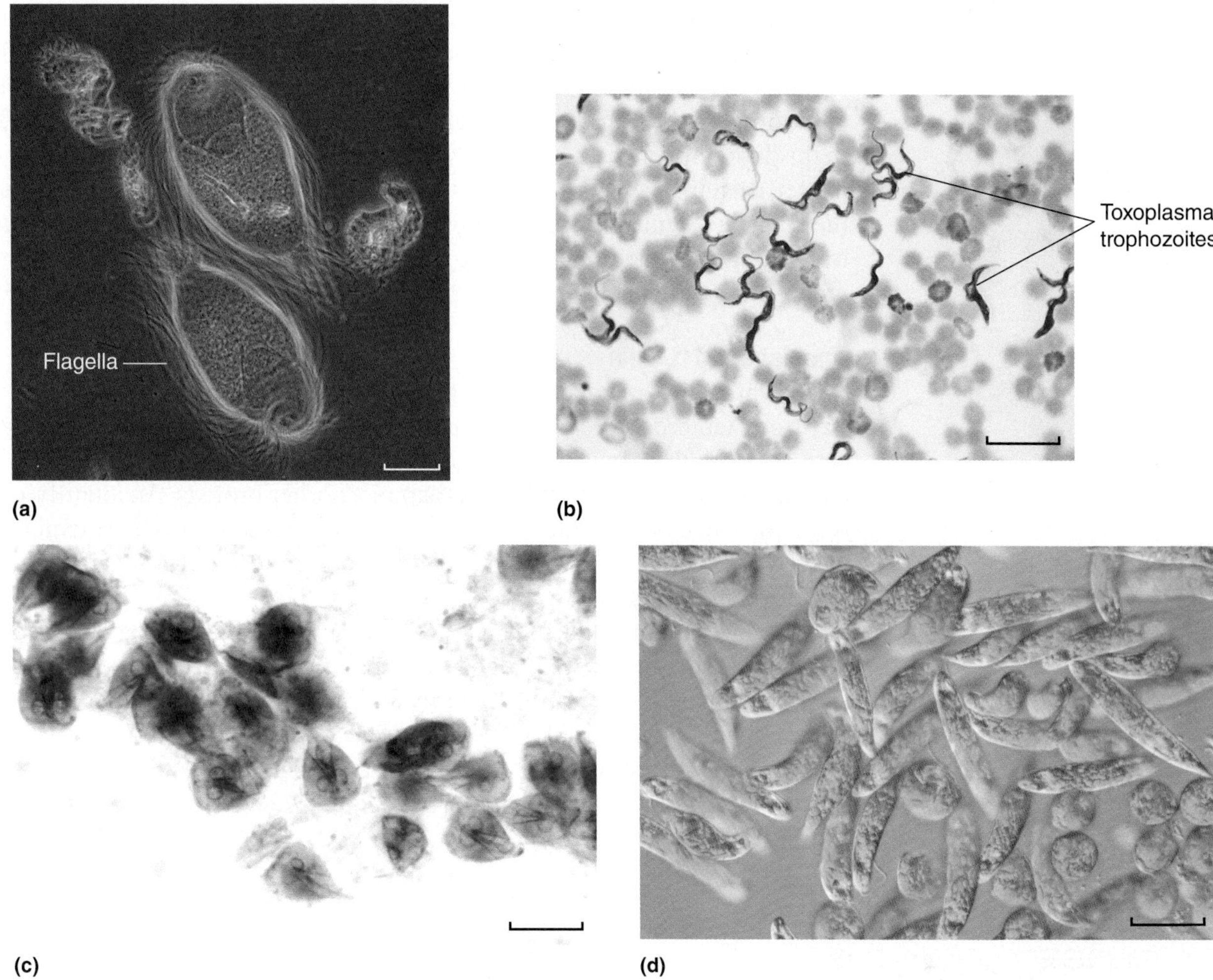

FIGURE 7.7 **Light Microscope Images of Flagellated Protozoa.** (**a**) *Trichonympha* is found in the guts of termites. Each cell has many flagella used for motility. (Bar = 30 μm.) (**b**) Stained cells of *Trypanosoma* in a blood smear. The cells appear wavy due to the undulating membrane and flagella. (Bar = 20 μm.) (**c**) Stained *Giardia* cells are pear-shaped and have flagella. (Bar = 10 μm.) (**d**) *Euglena* cells are flagellated and have chloroplasts for photosynthesis. Without sunlight, these cells act as heterotrophs. (Bar = 20 μm.)

of water contamination. Antony van Leeuwenhoek, among the first to visualize the microbial world, is believed to have seen and described *Giardia* from his own stools.

Trichomonas vaginalis is the protozoan species that causes trichomoniasis ("trich"). It is a sexually transmitted disease that affects over 2 million Americans annually. Patients suffer intense itching, burning pain, and a frothy discharge from the reproductive tract. The disease is more common in women than men, and it can be easily treated and cured with antiprotistan drugs.

■ *Trichomonas vaginalis*
trik-o-mō′n- äs vaj-in-al-is

The final heterotrophic flagellate to mention is *Leishmania*, which is transmitted by the sandfly. This parasite has caused illness in military personnel in the Middle East, as **A Closer Look** 7.1 chronicles.

■ *Leishmania*
lish-mä′-nē-ä

■ *Euglena*
ū-glē′-nä

To finish off this group of protists, a few words should be mentioned about the genus *Euglena*. This common type of freshwater protist has very flexible nutrient requirements: In sunlight, it is fully autotrophic, synthesizing its own nutrients using its chloroplasts to manufacture organic compounds by photosynthesis (FIGURE 7.7d). In the dark, the cell loses its photosynthetic pigments and feeds exclusively on organic matter, thereby displaying a heterotrophic mode of nutrition. When returned to light, the *Euglena* resynthesizes its photosynthetic pigments and once again becomes autotrophic. Such a nutritional lifestyle is referred to as mixotrophic because the organism mixes its feeding habits depending on its environment.

Protists with Cilia

The **ciliates** are an extremely diverse group of heterotrophic protozoa ranging in size from a microscopic 10 μm to a huge 3 mm (about the same relative difference as between a football and a football field). Their cells have a greater variety of specialized organelles and compartments than most other protists. An identifying character is the presence of hair-like **cilia** (*sing*., cilium) that structurally resemble very short flagella. Because ciliates are typically found in pond water, the cilia help propel the cell through the water.

The movement of the cilia is coordinated by a primitive nerve network running beneath the surface of the cell. The cilia beat in synchronized waves much as a rowing crew coordinate their paddling in a race. This ciliary movement is coordinated very precisely, allowing the cell to move forward or backward or turn. Cilia also serve a sensory function, because they appear to transmit stimuli and coordinate rapid movements in the ciliate. For example, when a ciliate encounters an unpleasant stimulus or an obstacle, it backs off rapidly.

A CLOSER LOOK 7.1

The "Baghdad Boil"

Over the years our troops have been serving in Iraq and Afghanistan, the United States Department of Defense (DoD) has identified more than 2,000 reported cases of cutaneous leishmaniasis (CL) among military personnel The disease, "affectionately" called the "Baghdad boil" (see figure), is caused by the flagellate *Leishmania major,* which is endemic in Southwest/Central Asia. It is transmitted by a sand fly when female flies bite an unsuspecting victim. Within a few weeks after being bitten, a boil-like sore appears at the bite site. Luckily, most cases of CL can be successfully treated with a drug containing antimony. Globally, more than 1 million new cases of CL occur each year.

The DoD has implemented prevention measures to decrease the risk of CL. These procedures included improving hygiene conditions, instituting a CL awareness program among military personnel, using permethrin-treated clothing and bed nets to kill or repel sand flies, and applying insect repellent containing 30% DEET to exposed skin. These measures, according to the DoD's Medical Surveillance Monthly Report, have reduced the number of reported cases.

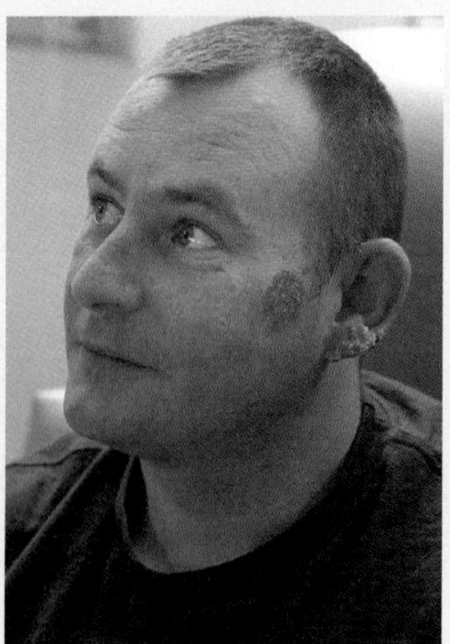

Cutaneous leishmaniasis.

As mentioned, ciliates display astonishing complexity in structure and behavior, and this is no clearer than in the genus *Paramecium* (FIGURE 7.8a) This slipper-shaped microbe is enclosed by a **pellicle**, a covering consisting of an outer membrane and an inner layer of closely packed structures including defensive organelles called **trichocysts**. Trichocysts are expelled from the pellicle as sharp, dart-like fibers (*tricho* = "hair") that are driven forward at the tip of a long expanding shaft and are used to catch prey. *Paramecium* and some other protists also have in the cytoplasm **contractile vacuoles**, membrane-enclosed compartments used to "bail out" excess water from the cell.

Paramecium
pär-ä-mē′-sē-um

Another unique characteristic of the ciliates is the presence of two types of cell nuclei: a large **macronucleus** contains hundreds of chromosome pieces, the genes of which are only used to regulate everyday cell metabolism; and one or more **micronuclei**, containing two complete sets of genes and are necessary for sexual reproduction. Both function in an elaborate form of sexual behavior called **conjugation**, a sexual process in which two cells fuse and swap micronuclei. At the conclusion of conjugation, the cells separate, each now having altered (recombined) DNA that gives new genetic characteristics to each cell. The micronuclei also give rise to a new macronucleus. Ciliates also can undergo asexual reproduction through a binary fission process.

Not all ciliates are motile. Some, like *Vorticella*, are attached to objects by a stalk and its cilia beat to produce a current in the surrounding water to bring food close (FIGURE 7.8b). The only ciliate known to be a human parasite is *Balantidium coli*. If one swallows its cysts, the cysts produce trophozoites in the intestine, causing a rare type of dysentery.

Vorticella
vôr-ti-cel′-lä

Balantidium coli
bal-an-tid-ē-um kō-lī

Nonmotile Protists

The fourth and final group of protozoa we shall consider is the **apicomplexans**, so named because at the cell's tip (the apex) there is a "complex" of organelles. In the adult stage, the parasites have no structures for motility. Virtually all apicomplexans are parasitic, and many of them cause serious diseases in humans and other animals.

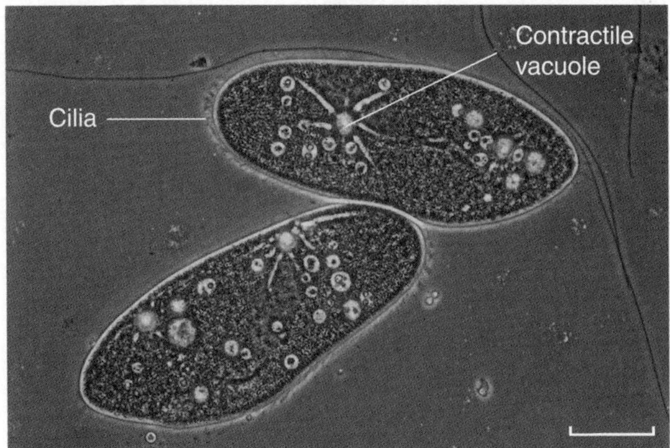

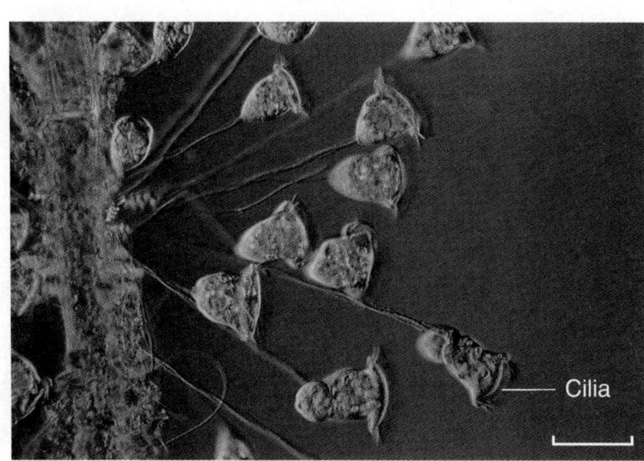

(b)

FIGURE 7.8 **Light Microscope Images of Ciliates.** (**a**) On these two *Paramecium* cells many cilia are present. In the cytoplasm are the typical eukaryotic organelles and compartments. A contractile vacuole is seen in each cell. (Bar = 50 μm.) (**b**) Sessile ciliates like *Vorticella* attached to a surface by a stalk. Cilia are present around the feeding apparatus.(Bar = 30 μm.)

Certainly the most life-threatening of the apicomplexans (and probably among all microbes) is the genus *Plasmodium* that, as the chapter opener describes, causes malaria. This blood disease accounts for over 300 million cases of illness per year worldwide and the WHO and other health organizations consider malaria the most important parasitic disease in today's world. Several species of *Plasmodium* can cause the disease, and mosquitoes are involved in their transmission. The cycle involves two hosts (FIGURE 7.9). An infected mosquito (the **definitive host**) bites (takes a blood meal from) an unsuspecting individual, transmitting the parasites to the human. In that person (the **intermediate host**), the parasites attack the liver where they reproduce asexually before moving on to infect and destroy red blood cells. The destruction of red blood cells causes a wave of intense cold followed by intense fever, the so-called "malaria attack." Extensive anemia develops, and the hemoglobin from ruptured red blood cells makes the urine dark (giving rise to the name "blackwater fever"). If an uninfected mosquito now takes a blood meal from the infected individual, some parasite cells are transferred to the mosquito where sexual reproduction occurs and the new parasites mature into a form capable of being transmitted to another unsuspecting person, starting another cycle of infection.

Death from malaria is due to a number of factors related to the anemia. Red blood cell fragments and clusters accumulate in the small vessels of the brain, kidneys, heart, liver, and other vital organs and cause clots to form. Heart attacks, cerebral hemorrhages, and kidney failure follow.

For centuries, quinine has been the mainstay treatment for malaria, but as resistance to the drug has increased, scientists have been searching for new antimalarial drugs. Another drug, artemisinin, is effective in curing malaria and mefloquine has been used as a preventative drug. Vaccines are still being developed and tested.

An interesting slant on malaria and its relation to sickle cell disease is presented in **A Closer Look** 7.2.

■ *Toxoplasma gondii*
toks-ō-plaz′-mä gon′-dē-ē

One other apicomplexan of contemporary concern is *Toxoplasma gondii*, the agent of toxoplasmosis (called "toxo"). Up to 50% of the world's population carries the

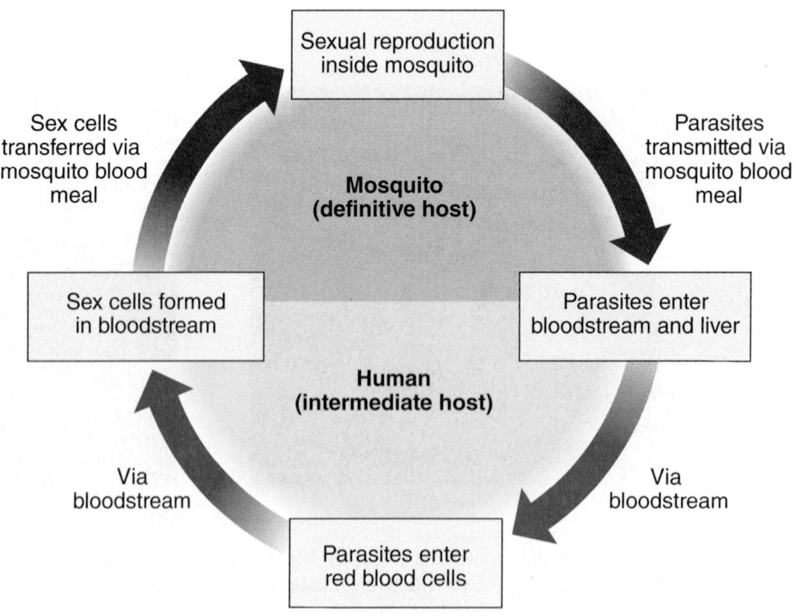

FIGURE 7.9 **The Life Cycle of *Plasmodium*.** The malarial parasite *Plasmodium* requires two different hosts (definitive and intermediate) in which specific stages of the life cycle occur.

A CLOSER LOOK 7.2

Sickle Cell Disease, Malaria, and Evolution

Some potentially deadly human genetic diseases remain rare in the population but they still exist. Why is that? If a rare genetic disease can kill, why haven't those individuals eventually died and ended the disease? The answer is that sometimes what seems like a definite disadvantage to an organism can actually be beneficial. Take for example sickle cell disease.

Normally red blood cells (RBCs) in a human are small, disk-shaped cells. However, in sickle cell disease the individual's RBCs may take on a crescent or sickle shape. The reason for this is that the protein hemoglobin in the RBCs carrying oxygen has undergone a mutation. If an individual has inherited the mutated gene from just one parent, the individual is heterozygous (Hh); that is, the person has one good gene (H) and one mutant gene (h). If the person has inherited a copy of the mutated gene from both parents, the person is said to be homozygous recessive (hh); they have two copies of the sickle cell gene.

When the oxygen content of the blood is low, such as during physical stress from exercising, the bad hemoglobin molecules in the RBCs cluster into long rods that cause the cells to take on a sickle shape. In Hh individuals, this is usually not a medical emergency because many RBCs are still normal in shape. However, in hh individuals all of the cells will take on the sickle shape, which clog blood vessels and lead to physical weakness and organ damage. Thus, it can be life threatening.

About 10% of African Americans have the heterozygous (Hh) sickle-cell trait and one out of 400 is hh. So why hasn't the evolutionary process eliminated this recessive trait? Is there an advantage of being Hh? And the answer is yes, depending on where such an individual lives.

It ends up that when the malarial parasite, *Plasmodium*, infects RBCs with a sickle shape, the oxygen content is low and the parasite is doomed because it too needs oxygen for reproduction. Therefore, if an individual, especially a child, lives in a malarious part of the world (see map), infection will lead to a lower parasite density and a more mild disease. Therefore, the sickle-cell trait would confer an advantage to heterozygous (Hh) and especially homozygous (hh) children who are geographically at a higher risk of contracting malaria than they are of suffering the effects of sickle-cell disease (even though the hh state is still very dangerous). Certainly, outside the malarial zones (such as the United States and other nonmalarious parts of the world), this adaptation is a liability.

Again, this evolutionary response shows us the power of the microbe—in this case to influence human evolution.

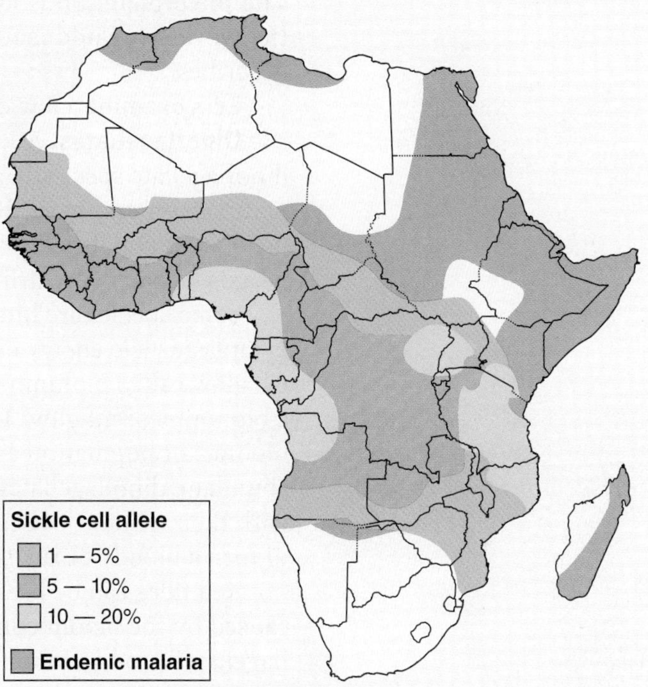

A map of Africa showing the frequency of the sickle-cell trait and the areas of endemic malaria.

parasite, including 50 million Americans, making toxo one of the most common human parasitic infections.

Toxo also is a blood disease often acquired through contact with parasite-containing cat feces, such as when changing the litter box. The feeding *T. gondii* trophozoites can also be consumed in improperly cooked infected beef or from contaminated soil while working in the garden. Healthy adults experience a mild flu-like illness, but a pregnant woman can transmit *T. gondii* cells across the placenta, and a child may be born with serious nerve damage. Infected persons with weakened immune systems, such as AIDS patients, can experience seizures and brain lesions.

7.4 Other Protists: More Diversity in the Grab Bag

To finish off our whirlwind tour of protist diversity, we will examine a few other groups that have great significance, are just interesting for study, or have changed the course of history and politics.

Unicellular Autotrophs

The importance of unicellular autotrophs to society cannot be overstated. As a major type of life in the seas, they comprise drifting or wandering communities of microbes that along with the cyanobacteria make up the **phytoplankton** (*phyto* = "plant;" *planktos* = "wandering") (FIGURE 7.10). The phytoplankton lives near the ocean surface and uses the Sun's energy during photosynthesis to generate much of the molecular oxygen available in the atmosphere. Virtually every animal on land or in the sea depends directly or indirectly on phytoplankton for food, and every animal on land as well as in the sea depends on the oxygen gas these algae produce. The phytoplankton is key to running the oceanic food chains. In fact, about half the world's organic matter is produced by phytoplankton, as **A Closer Look** 7.3 describes.

Let's examine a few of the more prominent members.

Dinoflagellates. A key member of the phytoplankton is the **dinoflagellates**. All dinoflagellate species have cells encased in rigid walls composed of cellulose coated with silicon and have the capacity to perform photosynthesis. They can move by using their two flagella; one moves the cell forward while the other whirls the cell on its axis ("dino" = "whirling").

Some species are **bioluminescent**; that is, chemical reactions occurring in their cytoplasm yield energy in the form of light, allowing the organisms to give off a greenish glow called bioluminescence. On a clear night, the sea can literally light up with a bioluminescent glow. Unfortunately, dinoflagellates sometimes undergo an "algal bloom," or population explosion. When the ocean is warm and nutrients are overly abundant, dinoflagellates experience a burst of reproductive activity and fill the water with their trillions of descendants. So many cells are present that the water appears to turn a bloody or rusty color. These are the so-called **red tides** (FIGURE 7.11a).

Red tides can be hazardous to aquatic life. The depletion of oxygen in the water caused by the bloom contributes to the death of plants and certain species of dinoflagellates produce a toxin responsible for massive fish kills. In addition, the toxin concentrates in mollusks such as mussels, clams, and scallops. When ingested by humans, the toxin may cause transient neuromuscular disturbances, such as tingling and numbing of the lips, tongue, and fingertips, followed by uncertain balance, lack

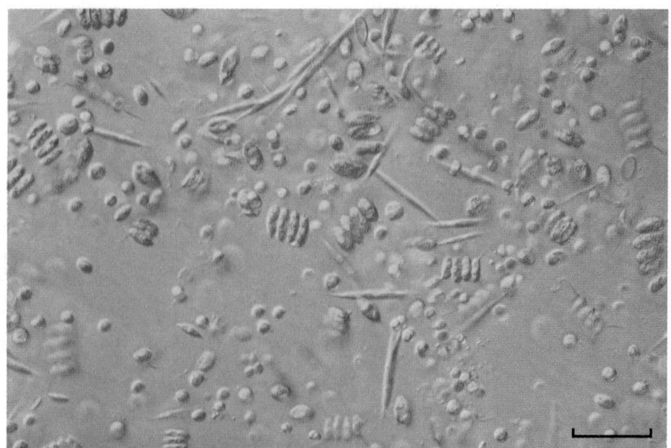

FIGURE 7.10 **Phytoplankton.** A light microscope image of green algae that compose part of the marine phytoplankton. Note the broad variety of shapes and sizes. (Bar = 10 µm.)

A CLOSER LOOK 7.3

Tuna Sandwiches

Yum! A tuna fish sandwich for lunch. Take a can of tuna, add some mayonnaise and celery, and you have what has been called "the mainstay of almost everyone's American childhood." In fact, in the United States, 52% of canned tuna is used for sandwiches. But let's examine the microbial contribution to that tuna sandwich.

"What eats what" in the world is often shown as a food web with an organism's feeding relationships linked by arrows, which always point from the organism being eaten to the one that is doing the eating (see diagram a). Organisms in food webs can also be divided into trophic levels composing a trophic pyramid where organisms are grouped by the role they play in the food web (see diagram b). Keeping in mind that trophic levels are rarely this simple because organisms often feed at more than one trophic level), we can say:

- The first level forms the base of the pyramid and is made up of producers, in this case the phytoplankton, the most abundant and widespread producers in the marine environment.
- The second level is made up of consumers. In our figure, the zooplankton in the marine environment consumes phytoplankton.
- The higher trophic levels include carnivorous consumers such as small and midsize fish, along with other consumers like squid and octopus. In simple terms, the zooplankton is eaten by smaller fish, smaller fish are eaten by midsize fish, and midsize fish are eaten by tuna, sharks, and other consumers.
- In our food web diagram, the top level consumers are humans.

On average, only 10% of the energy from a trophic level is transferred to its consumer (the rest is lost as waste and heat energy). Therefore, that tuna sandwich you ate came from: one kilogram of tuna, which had to eat 10 kg of midsized fish; each midsized fish had to eat 100 kg of small fish; each small fish had to eat 1,000 kg of zooplankton, which in turn had to eat 10,000 kg of phytoplankton. In other words, the energy you obtained from your tuna fish sandwich can be traced back to 10,000 kg of phytoplankton.

Whew! Those calculations made me hungry. I better finish my tuna fish sandwich and get more energy—and yes, thank you phytoplankton!

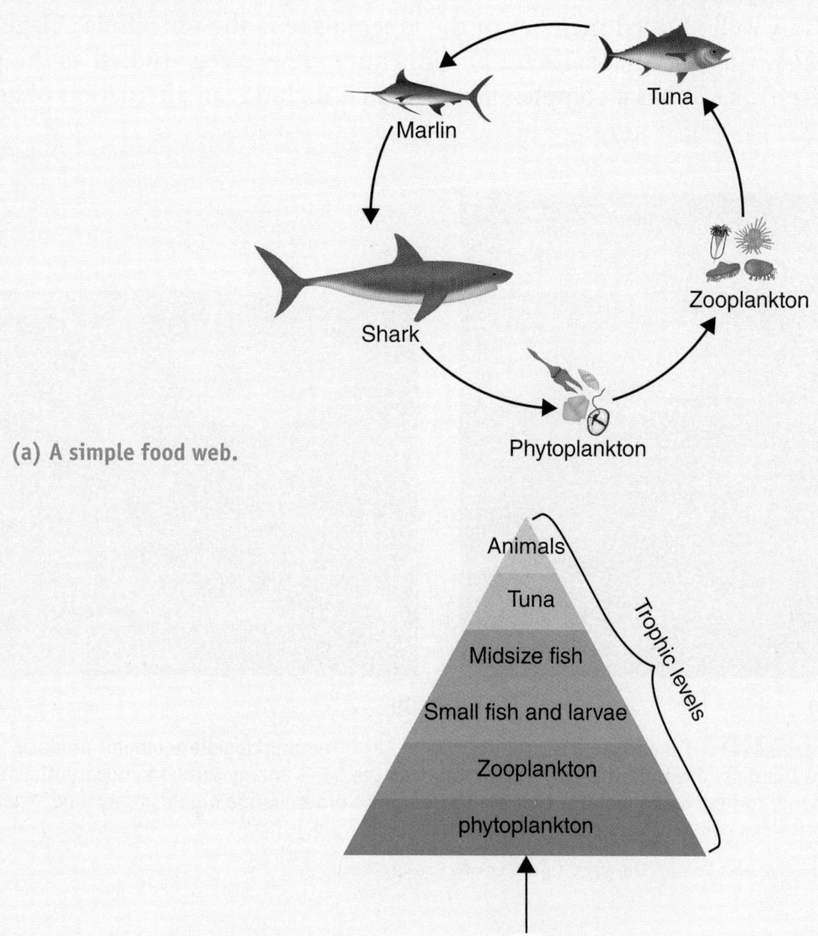

(a) A simple food web.

(b) Trophic pyramid and energy.

of muscular coordination, slurred speech, and difficulty in swallowing. There is no known antidote to the toxin.

Diatoms. Other unicellular autotrophs called **diatoms** are golden-brown and yellow-green in color (FIGURE 7.11b). Diatoms are another member of the phytoplankton and share with the dinoflagellates foundation roles in the oceanic food chain. They are distinguished by their exquisitely beautiful and intricate shells made of silicon dioxide that overlap much like the two halves of a shoe box. These delicate glasslike shells are perforated to allow contact between the cell and its external environment.

Diatoms have economic importance because they are used as filtering, polishing, and insulating materials. For example, those who have swimming pools or aquaria have probably used **diatomaceous earth**, a filtering material that removes contaminants in the water. It has also been used as, among other things, a mild abrasive in products including toothpaste, in cat litter, as a pesticide for bed bugs, as a thermal insulator for fire resistant safes, and as an anti-caking agent for animal feed. The diatomaceous earth, resulting from millions of years of accumulation of diatom shells, is mined from underwater beds or from ancient dried lake bottoms.

Green Algae. Perhaps the most recognized group of unicellular autotrophs is the **green algae**. An important difference between these protists and those autotrophs previously discussed is that the green algae have carotenoid pigments as well as unique variants of the "green grass" chlorophyll molecules called chlorophylls *a* and *b*. Most biochemists believe the multicellular algae and complex plants evolved from a unicellular green alga because both plants and green algae have both carotenoids and chlorophylls *a* and *b*, and the green algae synthesize starch and store it in their chloroplasts, a characteristic shared only with plants.

■ carotenoid: An orange or yellow pigment in chloroplasts that broadens the spectrum of colors (wavelengths) that can be used for photosynthesis.

A well-studied member of the green algae is the unicellular, flagellated genus *Chlamydomonas* (FIGURE 7.12a). This microbe is often studied as the prototypical green alga. It has a complex life cycle that includes an alternation of generations, a

■ *Chlamydomonas* klam-i-dō-mō'äs

(a)

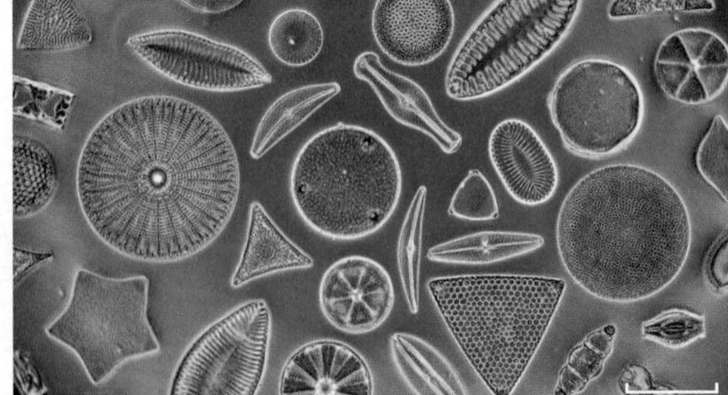

(b)

FIGURE 7.11 **Dinoflagellates and Diatoms.** (**a**) Often dinoflagellate blooms produce "red tides." (**b**) Diatoms, like the other phytoplankton, trap the Sun's energy through photosynthesis and use the energy to form carbohydrates that are passed on to other marine organisms as food. Again, note the marvelous geometric shapes of the cells. (Bar = 100 μm.)

(a) © Photoshot Holdings Ltd/Alamy. (b) © Scenics & Science/Alamy.

characteristic found in multicellular green algae and in complex plants (another bit of evidence for the ancestral relationship mentioned above). Another well-known member of the green algae is the genus *Volvox* that is organized into colonies of cells (FIGURE 7.12b). Each somatic cell forming the surface of the hollow ball resembles a *Chlamydomonas* cell. The green balls inside the sphere represent daughter colonies that will be released when the parent colony ruptures. Although not directly related to any multicellular organism, *Volvox* suggests how multicellularity might have evolved. Scientists have traced the origin of multicellular land plants to about 400 million years ago. It is conceivable that some 400 million years ago, when multicellularity evolved, one or more kinds of green algae were among the first to try this experiment in community living. From this humble beginning, all of today's land plants may have come into being.

Volvox
vōl′-voks

Slime Molds

As you have probably noted, some protists—such as protozoa—bear a striking resemblance to animals, while other protists—such as green algae—have similarities to plants. It is appropriate, therefore, to include a discussion of **slime molds**, which have animal-like, protist-like, and fungus-like characteristics. They are called "slime molds" because they are slimy (but much more attractive than the name implies!) and have a threadlike structure similar to that of fungal molds. In fact, the slime molds used to be classified with the fungi until more detailed molecular studies showed they were more allied with the protists. Indeed, the slime molds are a prime reason why the beginning of this chapter referred to the protists as a "grab bag of disparate creatures." Scientists recognize two groups of slime molds: the acellular slime molds and the cellular slime molds.

The **acellular slime molds** are also called "plasmodial slime molds" because the cells can form a huge, multinucleate "cell" (called a **plasmodium**; not to be confused with the genus name *Plasmodium* causing malaria) visible with the naked

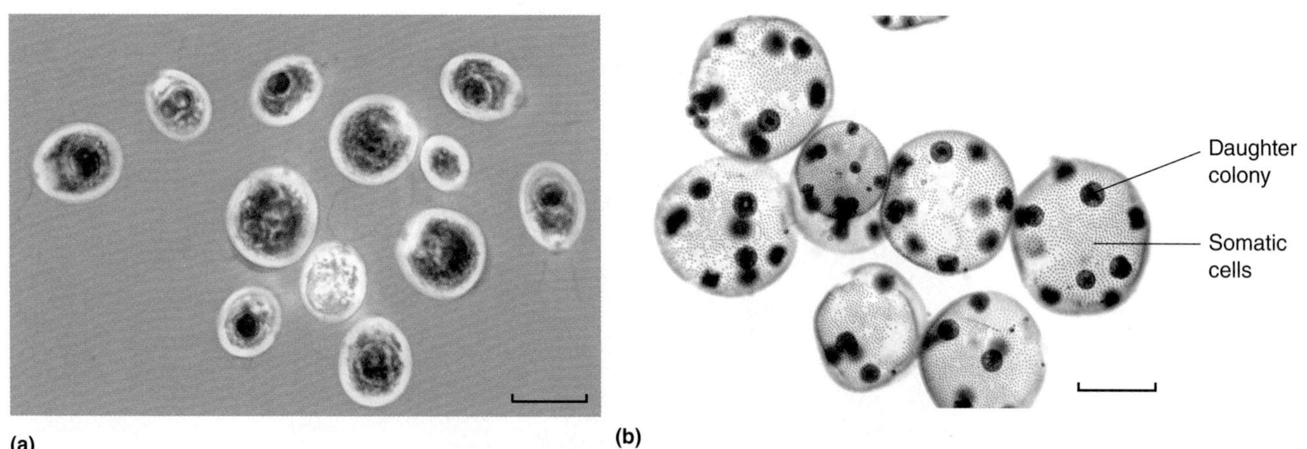

FIGURE 7.12 **Light Microscope Images of Green Algae.** (**a**) *Chlamydomonas* is an example of a flagellated, unicellular green alga. (Bar = 10 μm.) (**b**) *Volvox* is a colonial genus consisting of *Chlamydomonas*-like somatic cells and internal daughter colonies. (Bar = 300 μm.)

(a) © M I (Spike) Walker/Alamy. (b) Courtesy of Dr. Jeffrey Pommerville.

eye (FIGURE 7.13a). A single plasmodium can grow large enough to cover an entire log, although it will be only a millimeter thick. The plasmodium is the feeding stage, absorbing dead organic matter. When the habitat dries out, the plasmodium produces **fruiting bodies**, spore-producing stalks that extend upward (FIGURE 7.13b). The spores germinate into sex cells (gametes), which, through fertilization, develop into another multinucleated plasmodium.

The **cellular slime molds** differ from the acellular slime molds. When food and water are plentiful, the individual slime mold amoebae move about, engulfing bacteria, organic debris, and other available food sources (FIGURE 7.14). When food is in short supply, however, the cells come together within a cellulose sheath and congregate to form a large, many-celled stage called a "slug" (for its resemblance to a common

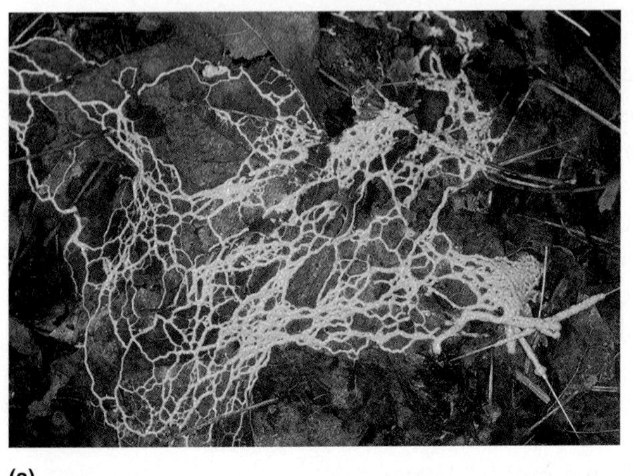

(a)

(b)

FIGURE 7.13 **Acellular Slime Molds.** (**a**) In the feeding stage, a large multinucleate plasmodium forms to absorb nutrients from soil and leaf litter. (**b**) When nutrients become scarce, spores are formed on long stalks produced from the plasmodium. The spores germinate to initiate development of another plasmodium.

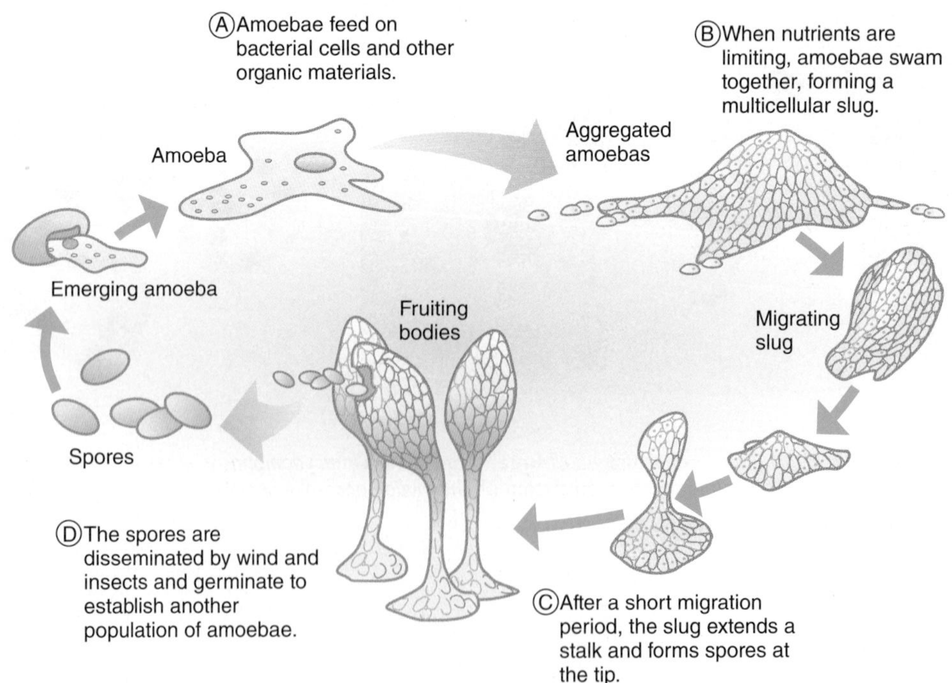

Ⓐ Amoebae feed on bacterial cells and other organic materials.
Ⓑ When nutrients are limiting, amoebae swam together, forming a multicellular slug.
Amoeba
Aggregated amoebas
Emerging amoeba
Fruiting bodies
Migrating slug
Spores
Ⓓ The spores are disseminated by wind and insects and germinate to establish another population of amoebae.
Ⓒ After a short migration period, the slug extends a stalk and forms spores at the tip.

FIGURE 7.14 **The Life Cycle of a Cellular Slime Mold.** The amoeboid form of this organism resembles a protozoan, while the fruiting body stage resembles a fungus.

animal garden slug). The entire mass (colony) of cells migrates as a single unit, a behavior that will help to disperse the species.

But the cycle is not yet finished. The slug transforms itself into an upward extending stalk, resembling a fungus. Within the tips of these stalks, spores are produced and eventually released for wide dispersal. Each spore germinates and forms another amoeba-like cell to begin the process anew.

Water Molds

Water molds are fungus-like protists living in water and damp soil and causing the unsightly and furry parasitic growths that plague fish in home and commercial aquaria. The organisms have thread-like bodies reminiscent of some fungi. Perhaps the most notable characteristics of the group is the presence of cellulose cell walls, the storage of food as starch, and presence of flagellated spores, called **zoospores**. These spores have two flagella—one directed forward, the other backward.

Several species of water molds are of significant economic importance as plant pathogens. One species causes downy mildew of grapes and another has destroyed hundreds of thousands of oak trees in California. Still another species, *Phytophthora infestans*, is responsible for late blight of potatoes, a disease that destroyed the Irish potato crop during the extremely damp years of the mid-1800s and led to mass Irish immigration. This infection, which is described in **A Closer Look** 7.4, is perhaps the best example of how microbes have impacted the political, economic, and social fabric of several nations, especially North America.

■ *Phytophthora infestans*
fī¯-tof'thô-rä in-fes'tans

A CLOSER LOOK 7.4
"Black '47"

Early Spanish explorers to the New World discovered the potato in Central and South America and brought it back to Europe. Because most of the plant is poisonous, except the tuber, it wasn't until about 1800 that Europeans grew the potato and its tuber as a food crop. Especially in Ireland, the potato grew well in the moist, cool climate. Although Ireland of the 1840s was an impoverished country of tenant farmers, the farming of potatoes brought a population explosion and the country went from about 4.5 million people in 1800 to more than 8 million by 1845, the population depending in great part on the potato season after season.

Early in the 1840s, heavy rains and dampness portended calamity. Then, on August 23, 1845, *The Gardener's Chronicle and Agricultural Gazette* reported: "*A fatal malady has broken out amongst the potato crop. On all sides we hear of the destruction. In Belgium, the fields are said to have been completely desolated.*" Beginning as black spots, the disease, called late blight, decayed the leaves and stems, and left the potatoes a rotten, mushy mass with a peculiar and offensive odor (see figure). Even the harvested potatoes rotted. The water mold *Phytophthora infestans* was running rampant through the potato fields.

The winter of 1845 to 1846 was a disaster for Ireland. Farmers left the rotten potatoes in the fields, allowing the disease to spread. For food, farmers first ate the animal feed and then the farm animals. They also devoured the seed potatoes, leaving nothing for spring planting. After 2 years, the late blight seemed to slacken, but in 1847 ("Black '47"), an unusually cool and wet year, it returned with a vengeance and once again destroyed the Irish potato crop within days.

Between 1845 and 1860, over 1 million Irish people died from starvation. At least 1.5 million Irish left the land and emigrated, mostly to the eastern United States. As great waves of Irish immigrants came to the United States, their Irish American descendants in the next century would influence American culture and politics. From Boston policemen, to workers on the transcontinental railroad, to political leaders (President John Kennedy was an Irish-American), American society would never be the same. And to think—it all resulted from the water mold *Phytophthora infestans,* which remains a difficult organism to control even today.

Potatoes covered in *Phytophthora*.

A Final Thought

In an article in *Discover* magazine several years ago, the editor gave this most interesting description for a protist. See if you can identify the organism and disease described.

> *"It races through the bloodstream, hunkers down in the liver, then rampages through red blood cells before being sucked up by its flying, buzzing host to mate, mature, and ready itself for another wild ride through a two-legged motel."*

If you identified *Plasmodium* and malaria—you are correct! Although we have been highlighting the positive aspects of microbes in society, occasionally we are jarred into the realization that microbes can also make life very difficult—and sometimes even life threatening. Microbes can cause massive kills of plants, animals, and humans, and public health officials must be continually vigilant because, despite all our efforts to understand their physiology, biochemistry, genetics, evolution, and ecology, microbes will continue to pop into our midst and leave a trail of destruction.

One more quote by a chemist makes a nice analogy:

> *"If you dip a tennis ball in the ocean, the water dripping from the tennis ball represents all that is known; the ocean represents all that is waiting to be learned."*

Questions to Consider

1. Suppose a research scientist discovered that a toxic chemical was wiping out the populations of diatoms and dinoflagellates (significant members of the phytoplankton) in the oceans of the world. What horror story could occur if this were true?
2. You and a friend, who is three months pregnant, stop at a hamburger stand for lunch. Based on your knowledge of toxoplasmosis, what helpful advice can you give your friend? On returning home, you notice that she has two cats. What additional information might you be inclined to share with her?
3. The flagellated protozoan *Giardia lamblia* is named for Alfred Giard, a French biologist of the late 1800s, and Vilem Dusan Lambl, a Bohemian physician of the same period. Unfortunately, this information does not tell us much about the organism (except who did much of the descriptive work). In contrast, the names of other protozoa in this chapter tell us much. What are some examples of more informative names?
4. Suppose you have a protist that is a protozoan. What one characteristic could you use to identify to which one of the four groups it belongs? How would that characteristic permit the identification?
5. Members of the genus *Euglena* illustrate the difficulty in assigning the designation animal-like or plant-like to protists. Why is this the case? Why could *Euglena* or a similar organism be considered the basic stock of evolution?
6. How does studying slime molds help one to understand the developmental cycle of animals? Other than their rapid division, what other characteristics of slime mold cells might provide researchers with insight into cancer cells?
7. Protists of the genus *Chlamydomonas* are among the few microbes that display an alternation of generations. What is the importance of this characteristic in the life of the microbe?

Key Terms

Informative facts are necessary for the expression of every concept, and the information for a concept is founded in a set of key terms. The following terms form the basis for the concepts of this chapter. On completing the chapter, you should be able to explain and/or define each one.

acellular slime mold
alga (pl. algae)
amoeba
apicomplexan
autotroph
bioluminescent
cellular slime mold
ciliate
cilium (pl. cilia)
conjugation
contractile vacuole
cyst
definitive host
diatomaceous earth
diatom
dinoflagellate
endosymbiosis
flagellate
flagellum (pl., flagella)
foraminiferan (foram)
fruiting body
green alga (pl., algae)
intermediate host
macronucleus
micronucleus
mixotroph
organelle
pellicle
phytoplankton
plasmodium
protozoan (pl., protozoa)
pseudopodium (pl., pseudopodia)
radiolarian
red tide
slime mold
symbiosis
trichocyst
trophozoite
unicellular alga
water mold
zooplankton
zoospore

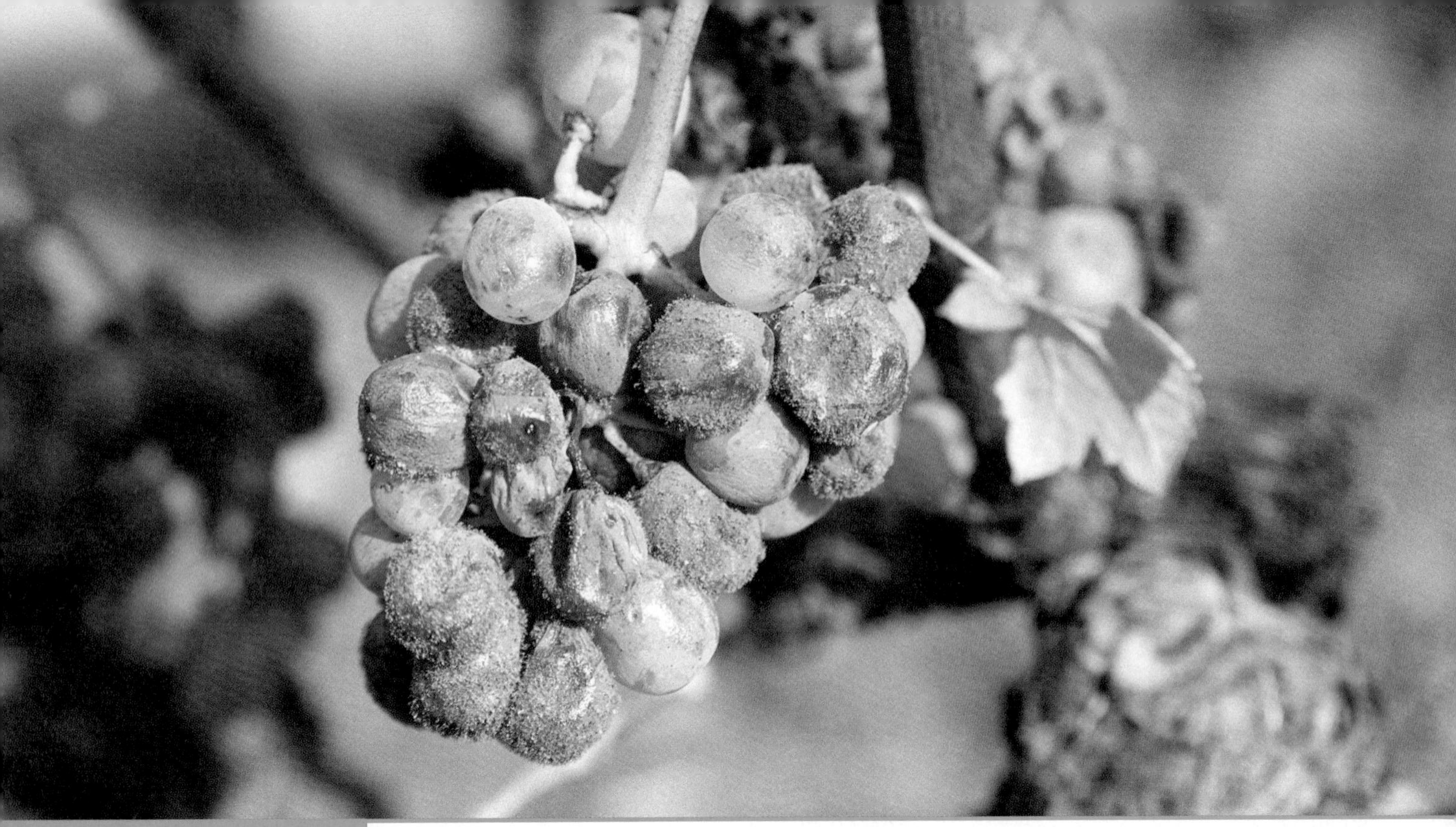

8 Fungi: Yeasts, Molds, and Mushrooms

Fungi can both harm and help us. Some, like *Botrytis cinerea*, can help us make delicious drinks! In the winemaking world, a *Botrytis* infection is known as the "Noble Rot." In cool and damp conditions, the fungus can infect Semillon grapes (as pictured here), concentrating the grape juices into an extremely sweet liquid that is turned into highly sought-after dessert wines like Sauternes from France and Tokaji from Hungary.

Looking Ahead

Before microbiology was established as a separate discipline of biology, fungi were studied in botany courses because they were thought to be plants. In contemporary biology, however, fungi are considered eukaryotic microbes, as we shall see in the pages ahead.

On completing this chapter, you should be capable of:

- Recognizing the structure of fungi and describing their mode of growth and nutrition.
- Giving examples of the five phyla of fungi.
- Summarizing the symbolic relationships within lichens, mycorrhizae, and endophytes.
- Distinguishing the different types of skin and respiratory tract infections caused by fungi.

In August 2005, Hurricane Katrina left unimaginable devastation everywhere along the south-central coast of the United States. In New Orleans, thousands of homes were flooded and left sitting in feet of water for weeks. Many health experts were concerned about outbreaks of infectious diseases like cholera and West Nile fever, and gastrointestinal illnesses. Health experts feared homes sitting in stagnant water could become breeding grounds for microorganisms.

Thankfully, most of these infectious disease scenarios did not occur. However, what did break out in many of the parishes in New Orleans was mold. There was mold on walls, ceilings, cabinets, clothes, and just about anything that provided a source of moisture and nutrients (FIGURE 8.1). Molds grew into carpets of spore-forming colonies everywhere. As one photojournalist described, "*...to see mold growing everywhere. Green, black, and blue mold growing on every surface in the houses. If the water line had been 8 feet up the wall, the mold was growing 10 feet up. There was mold on TVs, bookcases, countertops, dining tables, chairs, stove tops. Literally every surface had mold growing on it.*"

FIGURE 8.1 **A wall of mold.**

Courtesy of Andrea Booher/FEMA

If you see small spots of mold in your home, a dilute bleach solution will do a great job to kill and eliminate the problem. But what if an entire home and its contents are one giant "moldy culture dish"? More than likely, most of these homes have been demolished (or stripped to the framing) and furniture and other home contents destroyed. Even many home items, like beds, couches, or cabinets above the water level were covered in mold due to the prolonged humidity and had to be replaced. Most health officials told residents to follow the same slogan used for potentially spoiled food: "*When in doubt, throw it out.*"

At Tulane Hospital, the first floor was covered with mold, which made cleanup and reopening of many wards a very difficult task. In homes, as well as offices and hospitals, molds were discovered growing in ventilation systems and the ventilation ducts. If the ventilation fans were turned on, literally hundreds of billions of mold spores would be blown and spread to new areas to germinate and grow. In fact, 6 weeks after Katrina, the mold spore count in the air at various sites in New Orleans was as high as 102,000 spores per cubic meter—twice the number considered normal for New Orleans.

What about illnesses or disease from breathing the mold spores? By November 2005, many New Orleans residents who had returned were suffering from upper respiratory problems—the residents "affectionately" called it the "Katrina cough." In fact, residents with asthma, bronchitis, and allergies who had left New Orleans were asked not to return just yet. Although no respiratory outbreaks occurred, a 5-year study found that restoration workers had moderate increases in sinus and lung inflammations, but no significant increase in lung infections.

Although New Orleans after Katrina was an exception, fungi are usually an unheralded and often overlooked group of microbes. In fact, they are a critical link in the web of life on Earth, serving as important decomposers of organic matter. They make incalculable contributions to ecosystems by liberating nutrients from organic materials and making those nutrients available to insects, worms, bacteria, and myriad other organisms within the environment. Without fungi (and bacteria), the nutrients in organic materials would be locked up; the cycles of elements would grind to a halt; the fertility of soil would decline precipitously; and ecosystems would collapse.

The **fungi** (sing., fungus) consist of the microscopic molds and yeasts, and the macroscopic mushrooms. In this chapter, we will encounter many beneficial fungi such as those used to make antibiotics or in commercial and industrial processes. We also will identify and discuss several human diseases caused by fungi.

However, our study of the fungal world begins with a focus on the structures, growth patterns, and reproduction of fungi—something quite unique from the bacterial and protist groups of microbes.

8.1 The Fungi: Structure, Growth, and Reproduction

The fungi are a diverse group of eukaryotic microorganisms. Some 75,000 species have been named and identified, and scientists estimate another 1.5 million are waiting to be discovered. After a mold was first described by Robert Hooke (FIGURE 8.2), these microbes were collected and studied by botanists who considered them simple plants. In fact, until the mid-1900s, those wishing to study the fungi would typically have to enroll in a botany course, even though it was clear fungi do not carry out photosynthesis or have cell walls made of cellulose. Mainly for these and other reasons, fungi have since been placed in their own kingdom Fungi, within the domain Eukarya in the "tree of life." The study of fungi is called **mycology** (*myco* = "fungus"; *ology* = "the science of") and a person who studies fungi is a mycologist.

FIGURE 8.2 **The First Drawing of a Mold.** This is a drawing made by Robert Hooke in 1665 of a mold he observed growing on the sheepskin cover of a book.

© National Library of Medicine

Structure of Fungi

Being eukaryotic microbes, the fungi have the characteristic assemblage of eukaryotic organelles, including a cell nucleus, mitochondria, and ribosomes. The fungal cell, like a yeast cell, is usually several times larger than most bacterial cells (FIGURE 8.3a). Although the yeasts often exist as single cells, the molds and mushrooms are composed of threadlike filaments called **hyphae** (sing., hypha) whose walls are composed of **chitin**, a polysaccharide not found in plants or prokaryotes. As a hypha lengthens, it branches and forms an interwoven mass of hyphae called a **mycelium** (pl., mycelia), as shown in (FIGURE 8.3b). Often this mycelium escapes our attention because it usually is subterranean where it can maximize its contact with moisture and organic matter.

Fungal Growth and Nutrition

Being single cells, yeasts can grow in cell size and through asexual reproduction, called budding (see below) form a larger population of cells. For molds, the mycelium is the

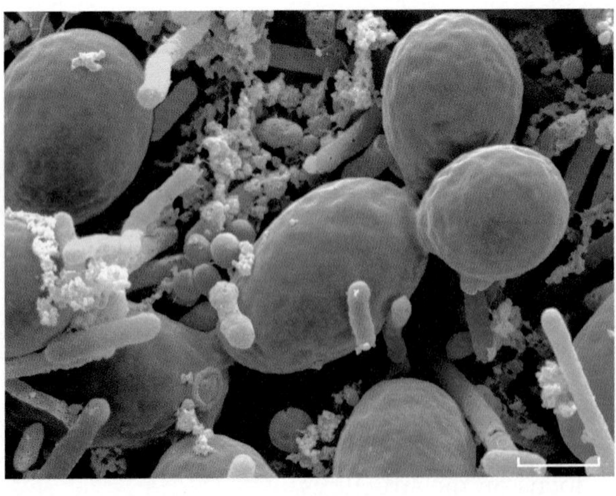

(a)

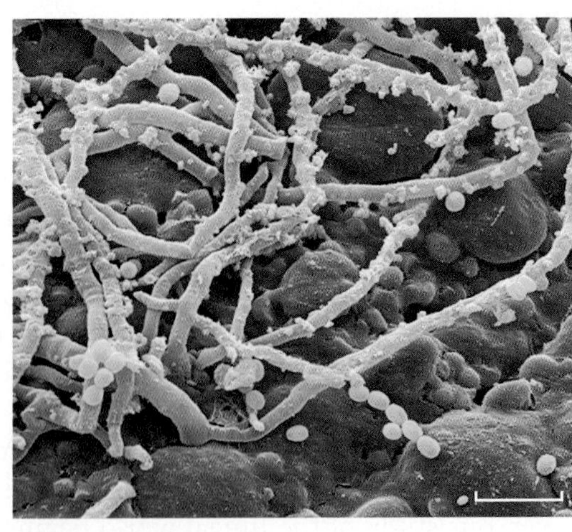

(b)

FIGURE 8.3 **The Fungal Form.** (a) A false-color scanning electron microscope image of yeast (greenish cells) and bacterial cells (various colors) demonstrating the difference in cell size. (Bar = 2 µm.) (b) False-color scanning electron microscope image of fungal hyphae (yellow filaments). (Bar = 10 µm.)

(a) © Medical-on-Line/Alamy. (b) © SPL/Science Source.

"feeding network. Hyphal growth extends out from the tip where continued growth brings the organism in contact with new food sources.

Being heterotrophic organisms, yeasts and molds digest organic matter by excreting enzymes into the environment and then absorbing the small organic products of digestion across the cell walls. For example, some fungal species can produce the enzyme **cellulase** and use it to decompose cellulose (the principal polysaccharide in wood). When cellulose is digested, it yields glucose molecules, which are extremely useful nutrients and sources of energy. Some species of fungi also have the ability to produce other enzymes capable of breaking down other plant wall fibers. That is why in a forest one typically finds molds growing on leaf litter or a rotting log (FIGURE 8.4). Many of the materials broken down by these decomposers not only produce useful nutrients but also generate vast quantities of organic matter that can be recycled. This was amazingly true some 250 million years ago as **A Closer Look** 8.1 describes.

■ decomposer: An organism that breaks down dead or decaying matter.

Many fungal species live under conditions that are acidic (pH between 5 and 6). For this reason, fungal contamination can be common in acidic foods such as sour cream and cheeses, and on damaged citrus fruits and vegetables (FIGURE 8.5a). In some cases, this "contamination" can be helpful: the flavor of blue cheese is due to the presence of the fungus *Penicillium roqueforti* (the blue streaks in the cheese are the fungus).

■ *Penicillium roqueforti* pen-i-cil'-lē-um rō-kō-for'tē

FIGURE 8.4 **A Result of Fungal Activity.** Fungi that are decomposers grow on damp timber and produce enzymes to digest the cellulose and other wall polysaccharides.

Fungal Reproduction

Most known species of fungi reproduce by both asexual and sexual methods. If you read the introduction to this chapter, all those molds in the homes in New Orleans were examples of asexual reproduction and fungal growth on a mind-boggling scale!

Reproduction in fungi involves **sporulation**, the process of spore formation. It usually occurs in structures called **fruiting bodies**, which represent the part of a fungus in which spores are formed and from which they are released. These structures may be asexual and invisible to the naked eye, or sexual structures, such as the macroscopic mushrooms.

In asexual reproduction, thousands of spores are produced and they are all genetically identical (baring a mutation). Many asexual spores develop within sacs or vessels while other fungi produce unprotected spores called **conidia** (sing., conidium) (FIGURE 8.6a). Such asexual spores are extremely light and are blown about in huge numbers by wind currents. In yet other fungi, spores may form simply by fragmentation of the hyphae, the freed fragment called an **arthrospore** (*arthro* = "joint"). The fungi causing athlete's foot multiply and spread in this manner.

Many yeasts reproduce asexually by **budding**. In this process, the cell becomes swollen at one edge and a new cell called a **blastospore** develops (buds) from the parent cell (FIGURE 8.6b). Eventually, the spore breaks free to live and grow independently, while the parent cell continues to produce additional blastospores.

Once free of the parent or fruiting body, the spores can be carried in water or blown by the wind many miles from their origin. If the spores land in an appropriate environment having

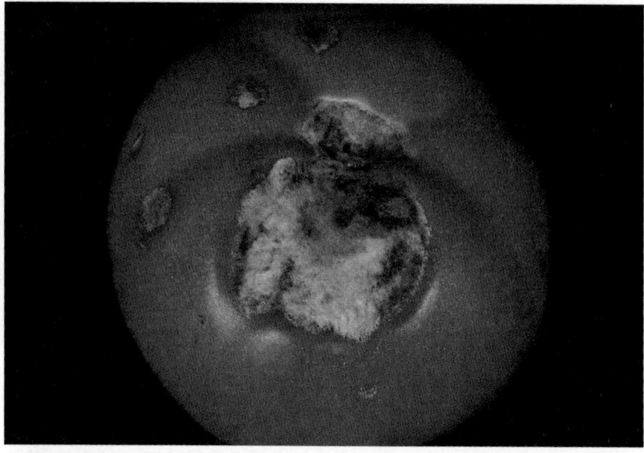

FIGURE 8.5 **Fungal Growth on Foods.** Some molds typically grow and reproduce on damaged or overripe fruits and vegetables, such as this tomato.

A CLOSER LOOK 8.1

When Fungi Ruled the Earth

About 250 million years ago, at the close of the Permian period, a catastrophe of epic proportions visited the Earth. Scientists believe over 90% of animal species in the seas vanished. The great Permian extinction, as it is called, also wreaked havoc on land animals and cleared the way for dinosaurs to inherit the planet.

But land plants managed to survive, and before the dinosaurs came, they spread and enveloped the world. At least, that is what paleobiologists traditionally believed.

Now, however, they are revising their theory and finding a significant place for the fungi. Dutch scientists from Utrecht University have presented evidence suggesting land plants were decimated by the Permian extinction and for a brief geologic span, dead wood covered the planet. During this period, they suggest the fungi emerged and wood-digesting species experienced a powerful spike in their populations (see figure). Support for their theory is offered by numerous findings of fossil fungi from the post-Permian period. The fossils are bountiful and they come from all corners of the globe. Significantly, they contain fossilized hyphae, the once active feeding forms.

And so fungi proliferated wildly and entered a period of feeding frenzy where they were the dominant form of life on Earth. It's something worth considering next time you kick over a mushroom growing on a rotting log.

A mushroom species growing on a rotting log.

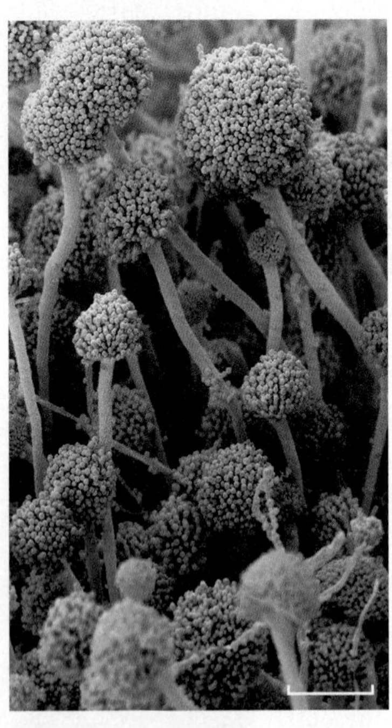

(a)

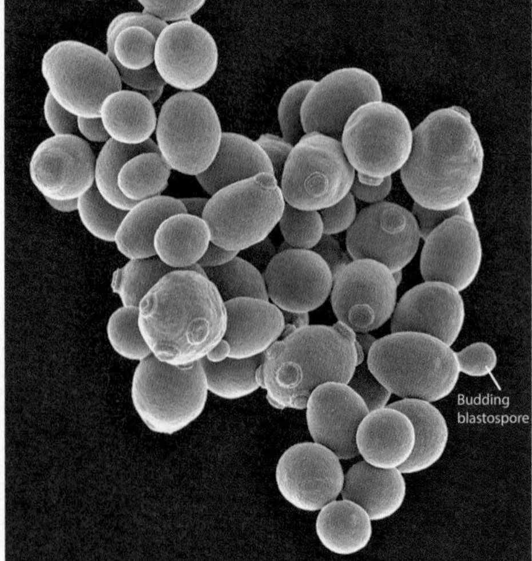

(b)

FIGURE 8.6 **Asexual Reproduction.** (**a**) False-color scanning electron microscope image of a mold fruiting body containing thousands of spores (conidia) at the tips of the filaments. (Bar = 20 μm.) (**b**) False-color scanning electron microscope image of yeast cells showing the budding process that produces blastospores. The rings (arrows) represent "bud scars" (birth marks) where a parent cell had previously produced a blastospore. (Bar = 8 μm.)

moisture and nutrients, they will germinate into new unicellular yeast cells or produce a new hypha that will elongate and form a new mycelium.

Asexual reproduction is advantageous because it provides huge numbers of spores, each of which can become a new mycelium (e.g., New Orleans and Katrina). However, the spores are genetically identical; therefore, if an environmental agent can destroy one, it will destroy all of them.

Through sexual reproduction, many fungi also produce spores, which are often contained within a visible fruiting body. Perhaps the most recognized fruiting body is the mushroom, such as the store-bought white mushroom *Agaricus bisporis* (FIGURE 8.7). Here, hyphae that represent opposite (+ and –) "mating types" (similar to opposite sexes in animals) come together and fuse. From the "mating," a mushroom will eventually arise and **basidiospores** will be produced on the gills forming the underside of the mushroom cap. These spores will be released and dispersed by water or wind. Some spores will come to rest in a nutritious environment where they germinate and divide to form new hyphae and (+ and –) mycelia.

Agaricus bisporis
ä-gär′i-kus bī-spōr′us

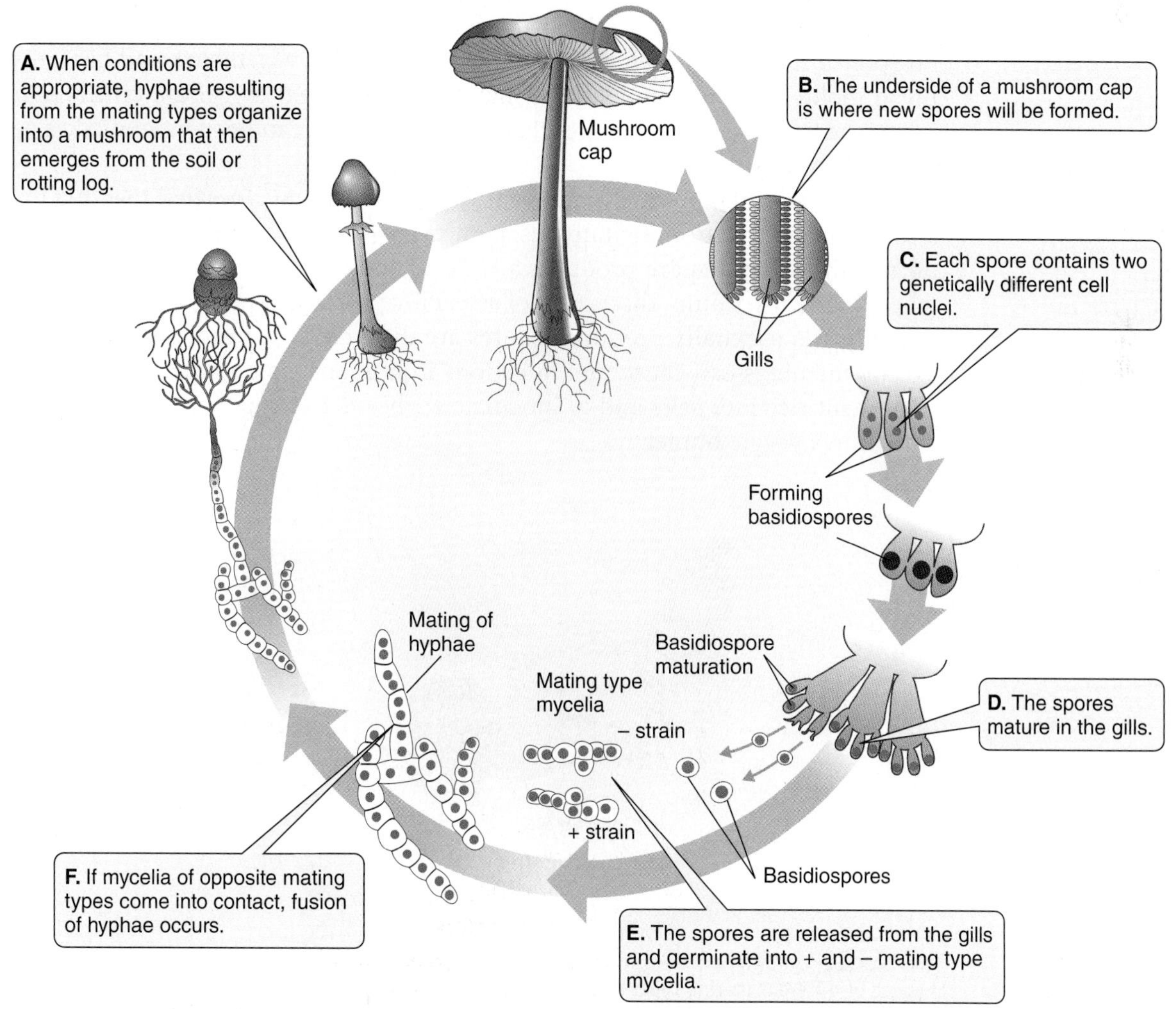

FIGURE 8.7 **The Life Cycle of a Typical Mushroom.** Basidiospores are produced from the underside of the mushroom cap.

Because the cell nuclei were genetically different in each mating type, the spores produced will have a new genetic composition. Unlike asexual spores, this genetic variability may permit a spore to germinate and survive in a formerly hostile environment.

8.2 Classification of Fungi: Cataloging the Diverse Groups

Historically, fungal classification within the kingdom Fungi in the domain Eukarya was based on either structural differences, or physiological and biochemical patterns. However, DNA analyses and genome sequencing are becoming important tools for drawing new evolutionary relationships among fungal groups or phyla. In this section, we will highlight a few of the phyla.

The Chytridiomycota

The oldest known fungi are related to certain members of the **Chytridiomycota**, commonly called **chytrids**. Chytrids are predominantly aquatic, and produce flagellated reproductive cells called **zoospores**.Until recently, their impact on the environment appeared to be of little significance. **A Closer Look 8.2** provides a different assessment.

The Zygomycota

Members of the phylum **Zygomycota** ("zygomycetes") make up about 1% of the described species of fungi. Distributed throughout the world, the zygomycetes inhabit terrestrial environments. Familiar representatives include molds typically growing on spoiled fruits with high sugar content or on acidic vegetables, and fast-growing bread molds. On these and similar materials, the fungal mycelium typically grows inside the food, secretes enzymes that digest the organic molecules, and takes up the nutrients by absorption.

During sexual reproduction, sexually opposite mating types fuse together to from a thick-walled, environmentally resistant **zygospore** (FIGURE 8.8). After a period of dormancy, the zygospore produces a fruiting body that releases spores, while elsewhere in the mycelium, thousands of asexually produced spores are produced. Both sexually and asexually produced spores are dispersed on wind currents. Although several members can cause rare infections in humans, some other species are used to ferment rice into sake and to inoculate soybeans to make tempeh, a soy product similar to a veggie burger.

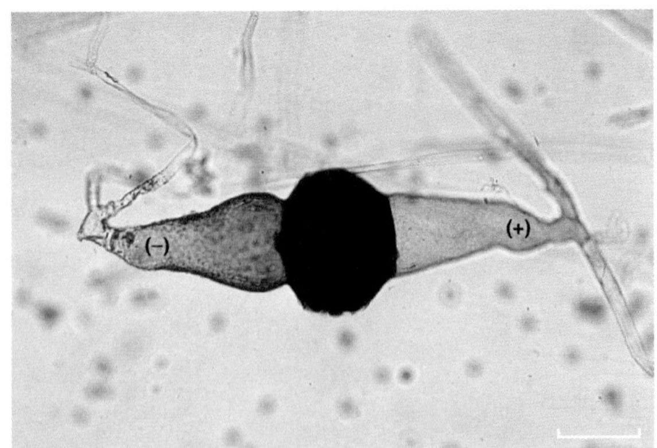

FIGURE 8.8 **Zygomycota Reproduction.** In this light microscope image, sexual reproduction between hyphae of different mating types (+ and –) has produced a darkly pigmented zygospore. (Bar = 30 μm.)

A CLOSER LOOK 8.2

The Day the Frogs Died

Amphibians are the oldest class of land-dwelling vertebrates, having survived the effects that brought the extinction of the dinosaurs. Now, the amphibians are facing their own potential extinction from a different, infectious effect. In 2010, scientists estimated that more than 100 frog species have gone extinct worldwide and many others are facing the same outcome. Why are the frogs dying?

In the early 1990s, researchers in Australia and Panama started reporting massive declines in the number of amphibians in ecologically pristine areas. As the decade progressed, massive die-offs occurred in dozens of frog species and a few species even became extinct. Once filled with frog song, the forests were quiet. "They're just gone," said one researcher.

By 1998, infectious disease was identified as one of the reasons for the decline. More than 100 amphibian species on four continents, including Central and South America and Australia, were infected with a chytrid called *Batrachochytrium dendrobatidis*. This fungus, the only one known to infect vertebrates, uses the frog's keratinized skin as a nutrient source. The skin infection alters the frog's skin permeability, which leads to a fatal osmotic imbalance and heart failure. Roughly one third of the world's 6,000 amphibian species are now considered under threat of extinction due to the disease chytridiomycosis (see figure). But, where did the chytrid come from?

Researchers believe the chytrid may have originated in African frogs exported around the world. The hypothesis is that some infected African frogs escaped (they do not die from the infection) and passed the fungus to hardier ones, like bullfrogs, which in turn infected more susceptible frog species. The global amphibian trade was the mode of transmission.

Are frog deaths from chytridiomycosis a sign of a yet unseen shift in the ecosystem, much like a canary in a coal mine? Some believe this type of "pathogen pollution" may be as serious as chemical pollution. It represents an alarming example of how an emerging pathogen could potentially wipe out a whole species and eventually a whole ecosystem.

A species of harlequin frog, one of many species being wiped out by a chytrid infection.

The Ascomycota

■ *Batrachochytrium dendrobatidis*
ba-tra-kō-kit'-rē-um
den-drō-ba-ti¯'-dis

The phylum **Ascomycota** ("ascomycetes") is the largest phylum with 65,000 described species. The name ascomycete is derived from the presence of an **ascus** (*pl.*, asci), a tiny spore-containing sac (*asco* = "sac") formed during the life cycle of the fungus. Most ascomycetes are decomposers, but many important plant parasites are found in this phylum, including the fungi causing powdery mildew, Dutch elm disease, chestnut blight disease, and peach leaf curl disease. Most species of ascomycetes are composed of hyphae that form a mycelium, but a few species such as the yeasts used in fermentation and baking (discussed below) are single-celled. Several species of ascomycetes are a great economic value, such as the morels and truffles, which are prized as edible fungi.

The sexual stages of most ascomycetes are less conspicuous than the asexual stages. First, the hyphae of different mating types fuse, a process that brings together genetically different cell nuclei. Then, in the flask-shaped or cup-shaped fruiting body the asci form and in each ascus eight **ascospores** are produced. The asci are often exposed on the upper surface (FIGURE 8.9).

Ascomycetes also include various species of *Penicillium*, the fungi that produce the important antibiotic penicillin. Moreover, some species of *Penicillium* are used to ripen and flavor blue cheese, Roquefort cheese, and Camembert cheese. Also in

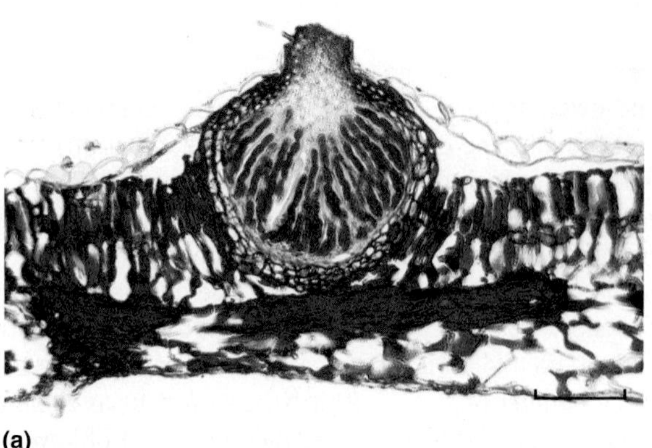

(a)

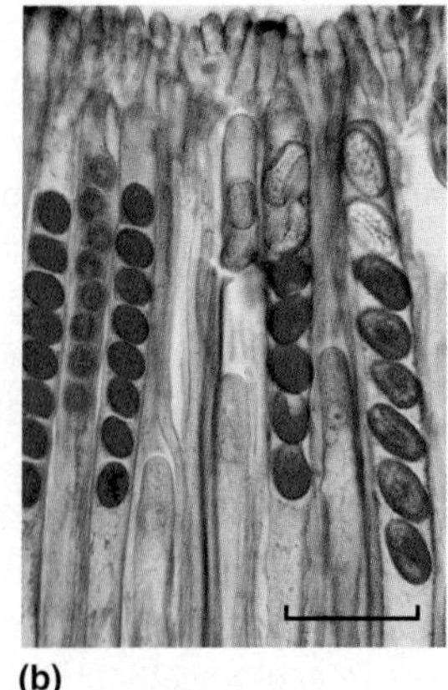

(b)

FIGURE 8.9 **Ascomycetes and Spores.** (**a**) Cross section of a stained fruiting body on an apple leaf. (Bar = 20 µm.) (**b**) A higher magnification of several asci, each containing eight ascospores. (Bar = 5 µm.)

(a) © Biodisc/Visuals Unlimited. (b) © Dr. John D. Cunningham/Visuals Unlimited.

■ *Aspergillus* a-spėr-jil′lus

■ carcinogen: A substance capable of causing cancer.

■ *Cryphonectria parasitica* krī-fō-nek-trē-ä pãr-ä-si-ti-cä

■ *Claviceps purpurea* kla′vi-seps pür-pü-rē′ä

the phylum Ascomycota are species of the common black mold *Aspergillus*. Certain *Aspergillus* species contaminate house dust and cause allergies and respiratory illness. Some *Aspergillus* species can produce dangerous toxins (called **aflatoxins**) in fungal contaminated nuts and stored grain. These toxins can poison the nervous system and are the most potent known natural carcinogen. On the positive side, *Aspergillus* species synthesize citric acid, a major component of beverages. They are also used in the production of soy sauce and vinegar.

Included in the phylum are a number of plant and human pathogens. Most notable is *Cryphonectria parasitica*, the main cause of chestnut blight, a devastating disease that has wiped out more than 4 billion chestnut trees since the early 1900s. Another ascomycete has more recently been associated with an infectious disease in bats that, as of 2013, has killed millions of hibernating bats in the eastern United States.

One particularly interesting ascomycete, *Claviceps purpurea*, has been associated with human behavioral disorders. The fungus infects rye grains, and when humans consume the contaminated grain or bread baked from it, the fungal chemical deposited in the rye causes a nervous system disorder called ergot disease. There is an intriguing body of evidence, related in **A Closer Look 8.3**, indicating ergot disease may have been a contributing factor in the Salem witchcraft trials. Ironically, *C. purpurea* is now cultivated for medical purposes. In small quantities, its chemical products can be used to treat migraine headaches or to induce childbirth.

The Basidiomycota

The **Basidiomycota** ("basidiomycetes") consist of some 32,000 described species. This includes puffballs, shelf fungi, earthstars, stinkhorns, and jelly fungi. The best-known members of the phylum are the gill fungi or mushrooms. **Mushroom** is the common name for the spore-producing fruiting body, which is composed of densely packed hyphae. As detailed earlier in Figure 8.7, the mycelium forms underground and the mushrooms arise from the mycelium. Importantly, some mushrooms are highly poisonous as **A Closer Look 8.4** points out.

The basidiomycetes also include certain plant parasites that cause rust and smut diseases. Rust diseases are so named because of the distinct orange-red color on the infected plant, a color derived from pigments in the parasitic fungi (FIGURE 8.10a).

A CLOSER LOOK 8.3

"The Work of the Devil"

As an undergraduate, Linda Caporael was missing a critical history course for graduation. Little did she know that through this class she was about to provide a possible answer for one of the biggest mysteries of early American history: the cause of the Salem Witch Trials. These trials in 1692 led to the execution of 20 people who had been accused of being witches in Salem, Massachusetts (see figure).

Linda Caporael, now a behavioral psychologist at New York's Rensselaer Polytechnic Institute, in preparation of a paper for her history course had read a book where the author could not explain the hallucinations among the people in Salem during the witchcraft trials. Caporael made a connection between the "Salem witches" and a French story of ergot poisoning in 1951. In Pont-Saint-Esprit, a small village in the south of France, more than 50 villagers went insane for a month after eating ergotized rye flour. Some people had fits of insomnia, others had hallucinogenic visions, and still others were reported to have fits of hysterical laughing or crying. In the end, three people died.

Caporael noticed a link between these bizarre symptoms, those of Salem witches, and the hallucinogenic effects of drugs like LSD, which is a derivative of ergot. Could ergot possibly have been the perpetrator in Salem too?

During the Dark Ages, Europe's poor lived almost entirely on rye bread. Between 1250 and 1750, ergotism, then called "St. Anthony's fire," led to miscarriages, chronic illnesses in people who survived, and mental illness. Importantly, hallucinations were considered "the work of the devil."

Toxicologists know eating ergotized food can cause violent muscle spasms, delusions, hallucinations, crawling sensations on the skin, and a host of other symptoms—all of which, Linda Caporael found in the records of the Salem witchcraft trials. Ergot thrives in warm, damp, rainy springs and summers, which were the exact conditions Caporael says existed in Salem in 1691. In addition, parts of Salem village consisted of swampy meadows that would be the perfect environment for fungal growth. And because rye was the staple grain of Salem, it is not a stretch to suggest that the rye crop consumed in the winter of 1691–1692 could have been contaminated by large quantities of ergot.

Caporael concedes that ergot poisoning can't explain all of the events at Salem. Some of the behaviors exhibited by the villagers probably represent instances of mass hysteria. Still, as people reexamine events of history, it seems just maybe ergot poisoning did play some role—and, hey, not bad for an undergraduate history paper!

A witch trial in Massachusetts, 1692.

© North Wind Picture Archives/Alamy

(a)

(b)

FIGURE 8.10 **Rusts and Smuts.** (**a**) Close-up of a bright orange patch of cedar apple rust growing on an apple tree leaf. (**b**) Corn Smut growing on an ear of corn in a corn field.

(a) © Picture Hooked/Nigel Downer/Alamy. (b) © Marvin Dembinsky Photo Associates/Alamy.

A CLOSER LOOK 8.4

A Word of Caution

In ancient Rome, mushrooms were the food of the gods, and only the emperors were permitted to partake of their pleasures. Today, exotic mushrooms enjoy an equally high reputation among the world's gourmets. Some experts know how to spot them in the wild, but for amateurs, the key word is "caution"—in mushroom hunting, ignorance is disaster. In fact, every year in the United States such hunting results in some 9,000 cases of mushroom poisoning being reported; children under 10 years of age account for the majority of cases.

Mushrooms come in a huge variety of colors and forms. Psilocybin mushrooms, also known as psychedelic (magic) mushrooms, contain psychoactive chemicals. They have been used since prehistoric times by many cultures in religious rites. In today's society, they are used recreationally for their psychedelic effects. Other interesting wild mushrooms are the jack-o-lantern fungus, known for its luminous gills; the beefsteak fungus, whose cap resembles a piece of raw beef; and the bird's nest fungus, in which the fruiting body and its spores look like a bird's nest with eggs. On the toxic side, about 100 of the 2000 known mushroom species can cause fatal mushroom poisoning and death. High on the list of dangerous mushrooms with portentous names: *Amanita verna*, the destroying angel and *Amanita phalloides*, the death cap (see figure).

For those who insist on eating wild mushrooms, mycologists recommend joining a local mycological society, reading extensively, and treading lightly into this hobby. According to the Minnesota Mycological Society:

"There are old mushroom hunters,
And there are bold mushroom hunters,
But there are no old, bold mushroom hunters."

***Amanita phalloides*, the "death cap" mushroom.**

Amanita verna
am-an-ī′tä ver-nä

Amanita phalloides
am-an-ī′tä fal-loi′dez

Wheat, oat, and rye plants, as well as lumber trees such as white pines, are susceptible. Smut diseases get their name from dark-pigmented fungi that give a black sooty appearance to infected plants such as corn, blackberry, and various grains (FIGURE 8.10b). These diseases bring about untold millions of dollars of crop damage and loss each year.

The Imperfect Fungi

Some fungi either lack a sexual cycle or such a cycle has never been observed. Such fungi are labeled as **imperfect fungi** because the lack of a known sexual cycle is said to be "imperfect." For this reason, these fungi do not fit into any of the other phyla, although they probably are related to the ascomycetes or basidiomycetes.

Many fungi pathogenic to humans are found in this group. These fungi usually reproduce by forming conidia or fragments of hyphae, which cling to surfaces. For example, fragments of the athlete's foot fungus are picked up from towels used by infected individuals or from shower room floors, as we will discuss below.

8.3 Symbiotic Relationships: A Win-Win Association

Many fungal species live in a close relationship called a **symbiosis**. Often this symbiotic relationship with other species in nature is mutually beneficial, which is called "mutualism." Three such relationships are noted below.

Lichens

Lichens provide an outstanding example of mutualism. **Lichens** are organisms that represent an association between a fungus and a photosynthetic partner, usually a cyanobacterium or a unicellular green alga. Of the approximately 15,000 species of lichens, most of the fungal partners belong to the phylum Ascomycota. About 90% of a lichen's mass is fungal.

In lichens, specialized fungal hyphae either penetrate or encase the photosynthetic cells (FIGURE 8.11a and b). The photosynthetic partner is generally not exposed to the light directly, but enough light penetrates tshe layers of fungal hyphae to make photosynthesis possible. In addition, certain lichens include cyanobacteria, which trap atmospheric nitrogen gas and incorporate it into organic molecules used for protein production. To establish a new population, a lichen breaks into fragments, called **soredia** (sing. soredium), which are blown or carried away where they give rise to new populations.

Lichens can survive with only 2% water by weight (compared to about 90% by weight for other organisms), which means they are extremely resistant to harsh environments. They grow in such diverse settings as arid desert regions and Arctic zones, as well as on bare soil and tree trunks. Lichens are often the first organisms to occur in rocky areas (FIGURE 8.11c), and their biochemical activities begin the process of rock breakdown and soil formation, a process that yields an environment in which mosses and other plants can gain a foothold.

Mycorrhizae and Endophytes

There are also two interesting and extremely important symbiotic roles that fungi play with plants.

Mycorrhizae. **Mycorrhizae** are mutualistic associations between fungi and vascular plants (e.g., trees, flowers, vegetables). The term mycorrhizae literally means "fungus-roots" And, in fact, over 90% of all vascular plants are believed to have a fungal symbiont. Through the mutualistic relationship, the fungus attaches to the

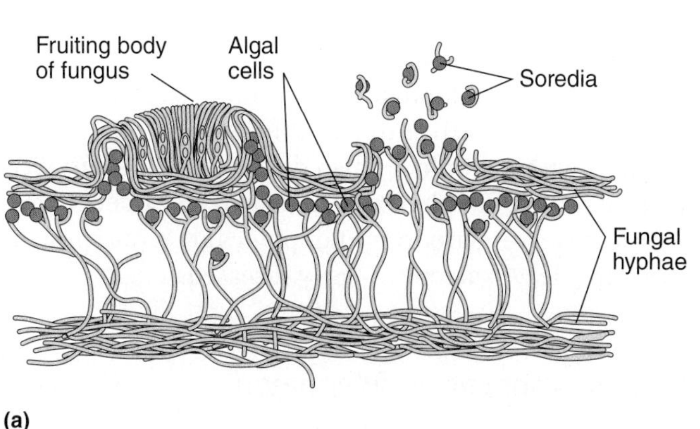

(a)

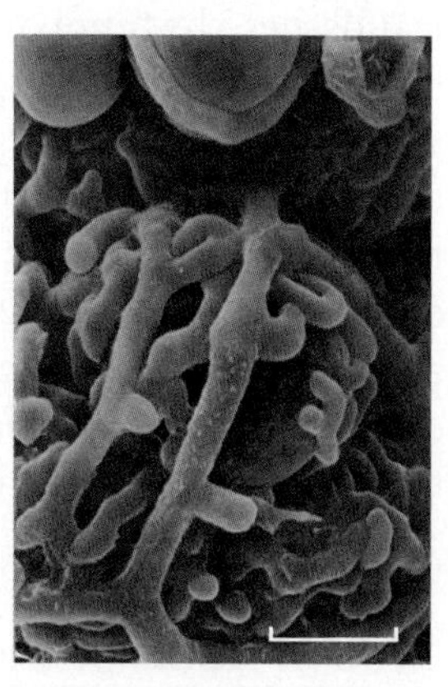

(b)

(c)

FIGURE 8.11 **Lichens.** (**a**) A cross section of a lichen, showing the upper and lower surfaces where tightly coiled fungal hyphae enclose photosynthetic algal cells. On the upper surface, a fruiting body has formed. Airborne clumps of algae and fungus called soredia are dispersed from the ascocarp to propagate the lichen. Loosely woven fungi at the center of the lichen permit the passage of nutrients, fluids, and gases. (**b**) A false-color scanning electron microscope image of the close, intimate contact between fungal hyphae (orange) and an alga cell (green). (Bar = 2 μm.) (**c**) A typical lichen growing on the surface of a rock. Lichens are rugged organisms that can tolerate environments where there are few nutrients and extreme conditions. The brown fruiting bodies can be seen.

(b) © V. Ahmadjian/Visuals Unlimited. (c) Courtesy of Jeffrey Pommerville.

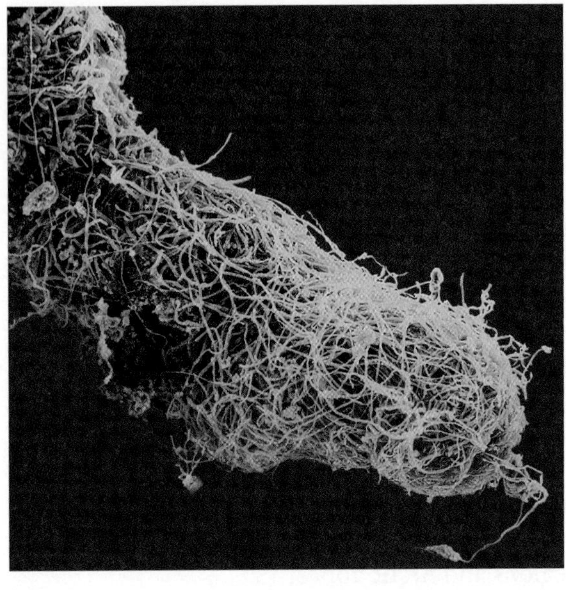

(a)

(b)

FIGURE 8.12 **Mycorrhizae and Their Effect on Plant Growth.** (**a**) Mycorrhizae surround the root of a *Eucalyptus* tree in this false-color scanning electron microscope image. These fungi are involved symbiotically with their plant host, such as aiding in mineral metabolism. (**b**) An experiment analyzing the mycorrhiza effects on plant growth. Which plant or plants (CK, GM, GE) do you believe has a mycorrhiza association?

surface or penetrates the roots of the plant, supplying the plant with more nutrients (especially phosphorus) than it could absorb through its roots alone; meanwhile, the plant supplies the fungus with products of photosynthesis that provide the raw materials for its metabolism (FIGURE 8.12a). The fungus also absorbs water and passes it to the plant, a great advantage to the plant when the soil is dry and sandy. In fact, plants with a mycorrhiza association tend to grow larger and more vigorously than plants lacking a fungus, especially in soils of poor nutrient quality (FIGURE 8.12b).

Endophytes. Besides the mycorrhizae, most plants also contain fungal **endophytes**, which are fungi living and growing symbiotically within plants, especially leaf tissue. The fungi do not cause disease but, rather like mycorrhizae, they provide better or new growth opportunities for the plant. Here are two examples.

When endophyte spores are placed on tomato, watermelon, or wheat seedlings, the tomato and watermelon seedlings with endophytes survived the stresses of high temperatures (50°C) or drought conditions whereas the seedling without endophytes died. Although the wheat seedlings with endophytes also died, the plants survived about a week longer than those without endophytes. In Panama, researchers have discovered that chocolate-tree (cacao) leaves harboring endophytes are more resistant to pathogen attack, while fungus-free leaves are more easily diseased.

8.4 Yeasts: Industrial Fermentation Machines

In 1897, two German chemists, Eduard and Hans Buchner, were preparing yeast as a nutritional supplement for medicinal purposes. They ground yeast cells and collected the cell-free "juice." To preserve the juice, they added a large quantity of sugar (as was commonly done at that time) and set the mixture aside. Several days later Eduard noticed an unusual alcoholic aroma coming from the mixture. Excitedly, he called to his brother, "Hans, du wirst das nicht glauben!" ("Hans, you won't believe this!") One taste confirmed their suspicion: The sugar had fermented to alcohol. The discovery by the Buchner brothers was momentous because it demonstrated a chemical substance inside yeast cells brings about fermentation.

■ fermentation: A metabolic process that converts sugar to acid, gases, and/or alcohol.

There are many different types of **yeast**, which usually refers to any fungus growing as single cells. Such yeasts are found in both the ascomycete and basidiomycete phyla.

Saccharomyces

The "true yeasts" with which most of us are familiar belong to the genus *Saccharomyces* (*saccharo* = "sugar"), a fungus in the Ascomycota that has the ability to ferment sugars. The most commonly used species are *S. cerevisiae* and *S. ellipsoideus*, the former used for bread baking (Baker's yeast) and alcohol production, the latter for alcohol production.

The *Saccharomyces* yeasts are single-celled, oval microbes about 8 μm in length and 5 μm in diameter. They usually reproduce by the asexual process of budding (see Figure 8.6). The cytoplasm of *Saccharomyces* is rich in B vitamins, a factor making yeast tablets valuable nutritional supplements (ironized yeast) for people with iron-poor blood. The baking industry relies heavily upon *S. cerevisiae* to supply the texture in breads. During the dough's rising period, yeast fermentation breaks down glucose and other carbohydrates, producing carbon dioxide gas. The carbon dioxide expands the dough, causing it to rise. In addition, the protein-digesting enzymes in yeast partially digest the gluten of the flour to give bread its spongy texture.

■ *Saccharomyces cerevisiae* sak-ä-rō-mī′-sēs se-ri-vis′-ē-ī

■ *Saccharomyces ellipsoideus* sak-ä-rō-mī′-sēs ē-lip-soi-′dē-us

■ gluten: A type of protein found in many grains, cereals, and breads.

Yeasts are plentiful where there are orchards or fruits (the haze on an apple is a layer of yeasts). In natural alcohol fermentations, wild yeasts of various *Saccharomyces* species are crushed with the fruit; in controlled fermentations, *S. ellipsoideus* is added to the prepared fruit juice. The fruit juice bubbles profusely as carbon dioxide evolves and as the oxygen is depleted, the yeast metabolism shifts to fermentation and the production of consumable ethyl alcohol. The huge share of the American economy taken up by the wine and spirits industries is testament to the significance of the fermentation yeasts.

S. cerevisiae probably is the most understood eukaryotic organism at the molecular and cellular levels. Its complete genome was sequenced in 1997, the first eukaryote to be completely sequenced. *S. cerevisiae* contains about 6,000 genes. It might appear initially that *S. cerevisiae* would have little in common with human beings. However, both are eukaryotic organisms with a cell nucleus, chromosomes, and a similar mechanism for cell division. *S. cerevisiae,* therefore, has been used to better understand cell function in animals. In addition, potential drugs useful in disease treatment also can be screened using yeast cells. A yeast mutant, for example, with the equivalent of a human disease-causing gene, can be treated with potential therapeutic drugs to identify a compound able to restore normal function to the yeast cell gene. And yeast cells have been key to the development of some human vaccines, such as the hepatitis B vaccine.

So, *Saccharomyces* has a long and useful role both in products intimate to society and to the improvement of human health.

8.5 Fungi and Human Diseases: From Skin to Lungs

To finish off this chapter, we will examine a few fungal diseases, called **mycoses** (sing. mycosis), and the fungal pathogens responsible.

Skin (Surface) Infections

Fungal infections of the skin and body surfaces are very common and include athlete's foot, jock itch, ringworm, and yeast infections.

Tinea Infections. One of the more frequent fungal skin ailments is commonly referred to as athlete's foot, jock itch, and numerous other names, including ringworm. The name depends on where on the body the infection occurs. In premodern times, people mistakenly thought the infections were caused by worms—hence, "ringworm" (FIGURE 8.13a). Thus, today these afflictions often are still referred to as "tinea infections" (*tinea* = "worm").

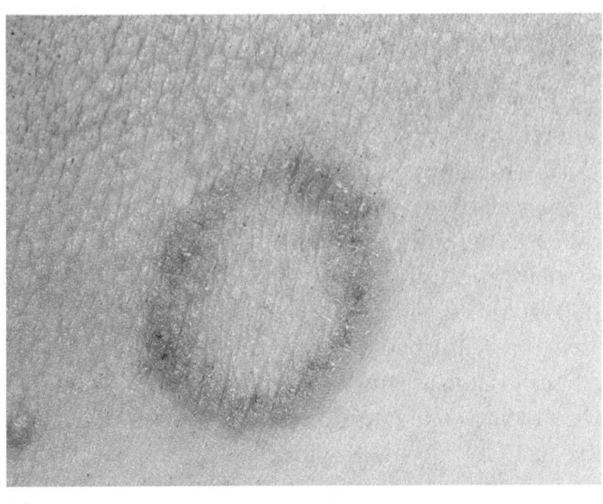

(a)

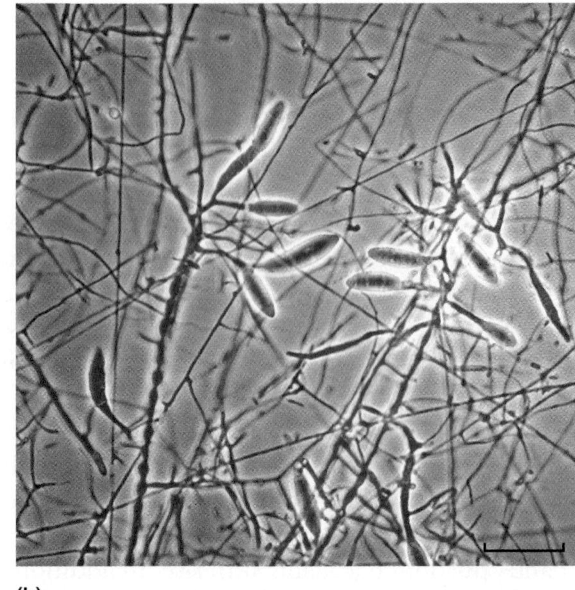

(b)

FIGURE 8.13 **Ringworm.** (**a**) Ringworm is caused by a fungal skin infection in which the mycelium spreads out under the skin surface producing the red inflammation. It is not caused by parasitic worms. (**b**) Light microscope image of *Microsporum* mycelium and spores. (Bar = 15 μm.)

Epidermophyton ep-ē-der-mō-fi'ton

Trichophyton trik-ō-fī'ton

Microsporum mī-krō-spô'rum

Tinea infections are surface mycoses caused by one of three ascomycete genera (*Epidermophyton*, *Trichophyton*, and *Microsporum*) (FIGURE 8.13b). The spores are transmitted by direct contact with infected persons or even household pets. They also are spread by indirect contact as the organisms can survive for long periods of time on shower room floors or mats as well as on combs, hats, towels, and numerous other objects. Typical signs and symptoms depend on the infection site but often involve blisters on the skin, along the surface of the nails, or in the webs between the fingers or toes.

Over-the-counter antifungal powders and ointments, including miconazole (Desenex®, Micatin®), and tolnaftate (Tinactin®) are available to treat various forms of the infection.

Candida albicans kan'-did-ä al-'bi-kanz

Candidiasis. The disease called **candidiasis** is caused by ascomycete yeasts in the genus *Candida*, the most common species being *Candida albicans*. These yeasts are oval-shaped cells when they infect body tissues, but grow as filamentous hyphae called "pseudohyphae" in a vaginal infection (FIGURE 8.14). Small amounts of *C. albicans* are normally found on the skin and mucous membranes of the mouth, intestine, and vagina and cause no infection; however, overgrowth of these organisms can lead to illness.

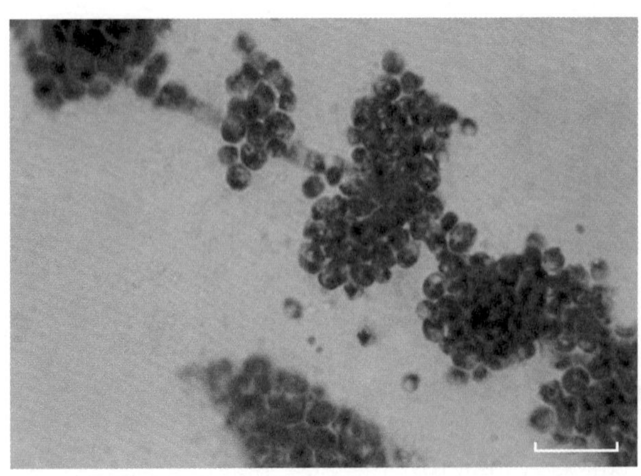

FIGURE 8.14 *Candida albicans.* A light microscope image of stained *C. albicans* cells from a vaginal swab. (Bar = 60 μm.)

Candidiasis in the vagina (vulvovaginal candidiasis) is commonly referred to as a "yeast infection." Normally, the yeasts are held in check by the lactobacilli and other resident bacteria normally found in the surface of the vagina. Many of these bacterial species produce lactic acid and other acids that create an inhospitable environment for *C. albicans*. However, when a woman takes antibiotics for some other condition, or there is trauma to the vagina, the lactobacilli may be reduced. Acid production soon disappears and the *C. albicans* cells flourish. Other predisposing factors are corticosteroid treatment, pregnancy, diabetes, and tight-fitting garments (which increase the local temperature and humidity). Transmission of the yeast can occur during sexual intercourse.

The symptoms of yeast infection include itching sensations, burning internal pain, and a white "cheesy" discharge. Treatment is usually successful when women use antifungal antibiotics.

Oral candidiasis that develops in lining of the mouth or throat is called "oral thrush." Again, if there is an overgrowth of *C. albicans*, the organism proliferates and forms small, white flecks that grow together to become soft, crumbly, white patches on the tongue and mucous membranes of the oral cavity. Although oral thrush can affect anyone, it most often occurs in babies and in individuals who use inhaled corticosteroids or have weakened immune systems. It may also occur as the result of using antibiotics, which can disrupt the normal microbiota in the body. Oral antifungal medications are prescribed for treatment.

Respiratory Tract Infections

Additional mycoses affect other body parts in humans, with a primary infection in the lungs often spreading to other body areas. In many of these fungal diseases, a weakened immune system contributes substantially to the occurrence of the infection.

Cryptococcosis. One of the most dangerous fungal diseases of the human lungs is **cryptococcosis.** It is caused by *Cryptococcus neoformans*, a basidiomycete found in soil and in pigeon droppings. Gusts of wind bring the fungal spores to the human respiratory tract, and from there the fungus may pass into the bloodstream and enter the meninges and lead to meningitis. Piercing headaches, neck stiffness, and paralysis usually follow. Resistance to the disease depends on a healthy immune system. Indeed, in AIDS patients, whose immune systems have been destroyed by the viral infection, the fungus can cause serious illness and death.

■ *Cryptococcus neoformans* krip'tō-kok-kus nē-ō-fôr'manz

■ meninges: The three membranous coverings of the brain and spinal cord.

Valley Fever. Travelers to the dry regions of the southwestern United States and parts of Central and South America should be on the alert for respiratory symptoms associated with infection by the genus *Coccidioides*. This fungus naturally exists in the soil and the spores can be carried with air into the lungs (FIGURE 8.15), where they cause an influenza-like illness that in rare instances may spread through the body and progress to meningitis. Residents of the San Joaquin Valley of California and dry regions of the southwestern United States have been cautioned to watch for "**valley fever**" symptoms, especially after dust storms or in places where there is major soil disturbance. During most of the 1980s, about 450 annual cases of valley fever were reported to the Centers for Disease Control and Prevention (CDC). In 1991, the number jumped to over 1,200 cases, and in 2013, the number of reports was more than 10,000, the majority occurring in California and Arizona.

■ *Coccidioides* kok-sid-ē-oi-dēz

***Pneumocystis* pneumonia.** Currently the most common cause of nonbacterial pneumonia in Americans with weak immune systems is ***Pneumocystis* pneumonia (PCP)**. The causative organism, *Pneumocystis jirovecii,* remained in relative obscurity until the 1980s, when it was recognized as the cause of death in over 50% of AIDS patients.

This ascomycete fungus is transmitted person to person by droplets from the respiratory tract, although transmission from the environment also can occur. A wide cross section of individuals harbors the fungus without symptoms, mainly because of the control imposed by the immune system. However, when the immune system is suppressed, as in AIDS patients, *Pneumocystis* cells fill the alveoli and occupy the air spaces. A nonproductive cough develops, with fever and difficult breathing. Progressive deterioration leads to consolidation of the lungs and respiratory failure. The current treatment for severe PCP is with antifungal drugs.

■ *Pneumocystis jirovecii* nü-mō-sis-'tis yer-ō-vet-zē- ē

■ alveolus: A tiny thin-walled air sac found in large numbers in each lung, through which oxygen enters and carbon dioxide leaves the blood.

■ consolidation: Formation of a firm dense mass in the alveoli.

Other Lung Infections. Other fungal lung diseases are caused by two ascomycetes, *Histoplasma capsulatum* and *Blastomyces dermatitidis*. The spores of both pathogens are associated with bird droppings, particularly near barns and sheds,

■ *Histoplasma capsulatum* his-tō-plaz'mä kap-sulä'tum

■ *Blastomyces dermatitidis* blas-tō-mı˘'sez de˘r-mä-tit'i-dis

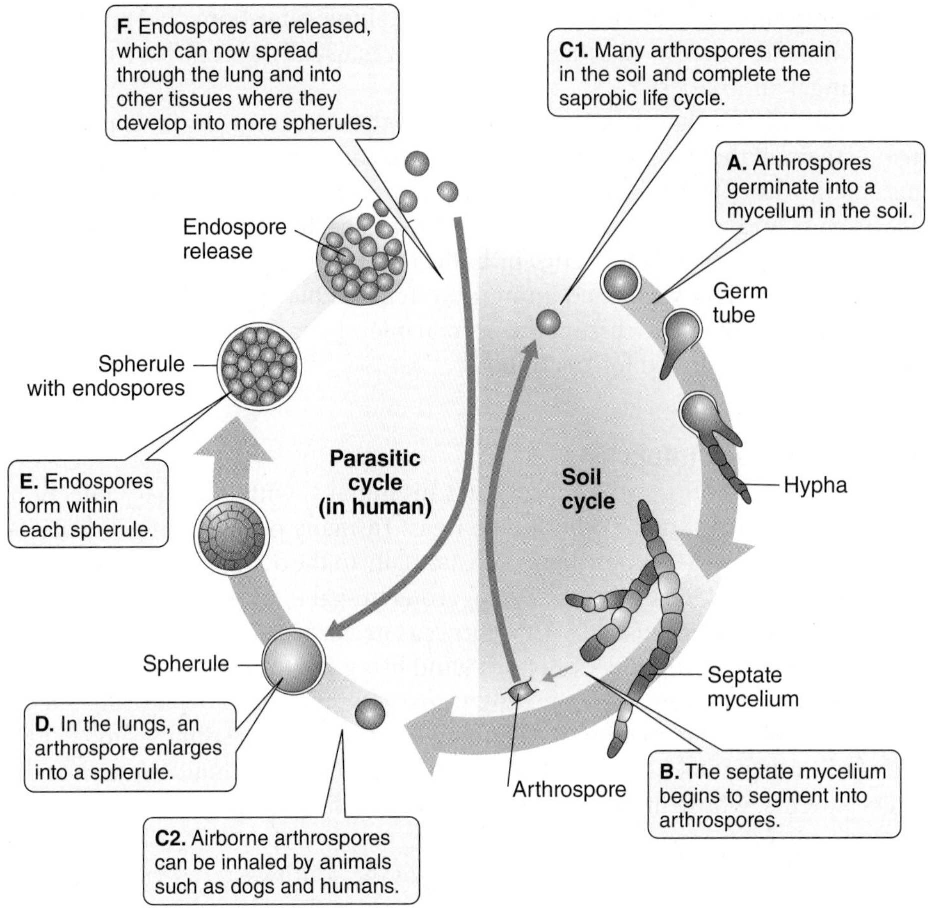

F. Endospores are released, which can now spread through the lung and into other tissues where they develop into more spherules.
C1. Many arthrospores remain in the soil and complete the saprobic life cycle.
A. Arthrospores germinate into a mycellum in the soil.
Endospore release
Germ tube
Spherule with endospores
Parasitic cycle (in human)
Soil cycle
Hypha
E. Endospores form within each spherule.
Spherule
Septate mycelium
D. In the lungs, an arthrospore enlarges into a spherule.
Arthrospore
B. The septate mycelium begins to segment into arthrospores.
C2. Airborne arthrospores can be inhaled by animals such as dogs and humans.

FIGURE 8.15 **The Life Cycle of *Coccidioides*.** Outside the body, the fungus goes through cycles of reproduction in the soil. However, the arthrospores entering the human respiratory tract go through a parasitic cycle, producing endospores capable of forming more spherules and infecting other tissues such as the skin, bone, and central nervous system.

and their spores are transmitted in breezes and wind gusts. They cause mild respiratory illnesses, unless the immune system has been weakened; in that case, the illness can involve multiple body organs and become quite serious and even fatal.

A Final Thought

In this chapter we have learned about fungi and the fungal diseases affecting humans. In the global picture, fungi infect billions of people every year, yet their significance to worldwide infectious disease remains unrecognized. Although there is no accurate record keeping for the diseases caused by the fungal pathogens of humans, taken as a whole, medical experts believe these pathogens kill as many people every year as tuberculosis or malaria. Part of the increasing incidence of fungal diseases is due to immunosuppressive diseases, like AIDS, which weaken the immune system's response to opportunistic infections such as cryptococcosis and other respiratory diseases we discussed in this chapter. And this increased incidence of fungal diseases is not something limited to humans.

In the last 20 years, an extraordinary number of fungal diseases have caused severe die-offs and extinctions. As described in this chapter, frog populations, especially in Central America, are being decimated by a fungal disease and the bat population in North America could be wiped out by a deadly fungus. In addition, collapse of honeybee colonies in 2010 was linked to a fungus and insect virus coinfection that may have been responsible for 40% to 60% of disappearing beehives since 2005. So, other animals, as well as humans, are at risk for life-threatening fungal infections.

Questions to Consider

1. You decide to make bread. You let the dough rise overnight in a warm corner of the room. The next morning you notice a distinct beer-like aroma in the air. What are you smelling, and where did the aroma come from?
2. Fungi are extremely prevalent in the soil, yet we rarely contract fungal disease by consuming fruits and vegetables. Why do you think this is so?
3. A student of microbiology proposes a scheme to develop a strain of bacteria that could be used as a fungicide. Her idea is to collect the chitin-containing shells of lobsters and shrimp, grind them up, and add them to the soil. This, she suggests, will build up the level of chitin-digesting bacteria. The bacteria would then be isolated and used to kill fungi by digesting the chitin in fungal cell walls. Do you think her scheme will work? Why?
4. Mr. A and Mr. B live in an area of town where the soil is acidic. Oak trees are common, and azaleas and rhododendrons thrive in the soil. In the spring, Mr. A spreads lime on his lawn, but Mr. B prefers to save the money. Both use fertilizer, and both have magnificent lawns. Come June, however, Mr. B notices mushrooms are popping up in his lawn and brown spots are beginning to appear. By July, his lawn has virtually disappeared. What is happening in Mr. B's lawn, and what can Mr. B learn from Mr. A?
5. A woman is complaining of scaly ring of blisters occurring on the legs, especially in the area around the shin. Questioning her reveals that she has five very affectionate cats at home. What disease does she have and what would be your suggestion to her?
6. In a suburban community, a group of residents obtained a court order preventing another resident from feeding the flocks of pigeons regularly visited the area. Microbiologically, was this action justified? Why?
7. On January 17, 1994, a serious earthquake struck the Northridge section of Los Angeles County in California. Over the next two months 170 cases of valley fever were identified in adjacent Ventura County. This number was almost four times the previous year's number of cases. What is the connection between the two events?

Key Terms

Informative facts are necessary for the expression of every concept, and the information for a concept is founded in a set of key terms. The following terms form the basis for the concepts of this chapter. On completing the chapter, you should be able to explain and/or define each one.

aflatoxin
arthrospore
Ascomycota (ascomycete)
ascospore
ascus (pl., asci)
Basidiomycota (basidiomycete)
basidiospore
blastospore
budding
candidiasis
cellulase
chitin
Chytridiomycota (chytrids)
conidium (pl., conidia)
cryptococcosis
endophyte
fruiting body
fungus (pl., fungi)
hypha (pl., hyphae)
imperfect fungi
lichen
mushroom
mycelium (pl. mycelia)
mycology
mycorrhiza (pl., mycorrhizae)
mycosis (pl., mycoses)
Pneumocystis pneumonia (PCP)
soredium (pl., soredia)
sporulation
symbiosis
valley fever
yeast
zoospore
Zygomycota (zygomycete)
zygospore

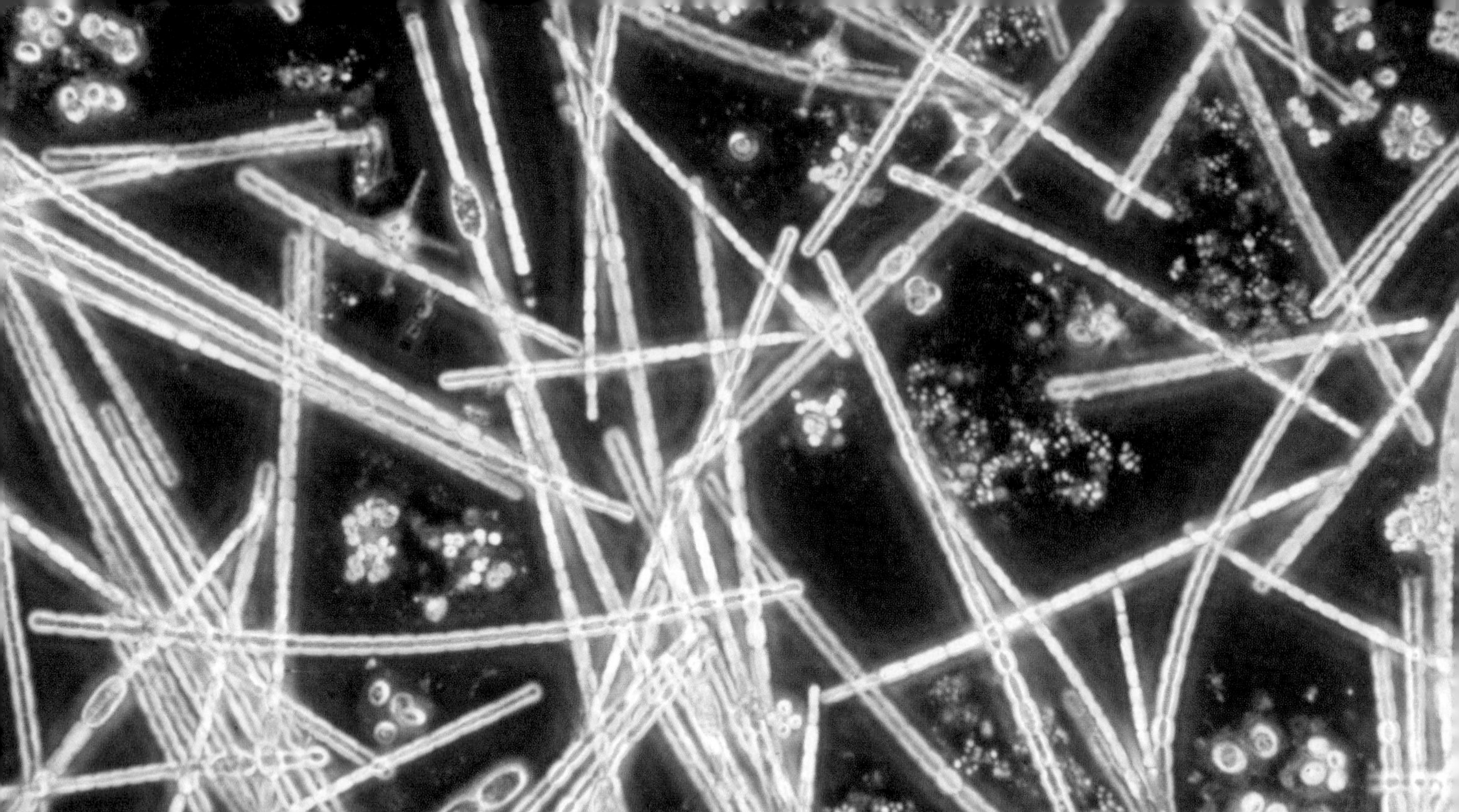

9 Growth and Metabolism: Running the Microbial Machine

All food and energy for living organisms originates from photosynthesis. The sun fuels photosynthesis by land plants and, as shown in this image, equally by the phytoplankton (cyanobacteria and unicellular algae) in the oceans. If one can find magic in life, it is in the regulation of energy flow, a process called metabolism that starts with the energy producers such as the phytoplankton.

Looking Ahead

Microbes share with humans many aspects of growth and metabolism. However, there are numerous aspects that set microbes apart and lend them uniqueness, as we will see in this chapter.

On completing this chapter, you should be capable of:

- Contrasting microbial groups based on (a) temperature, (b) oxygen, (c) pH, and (d) salt requirements.
- Differentiating between anabolism and catabolism.
- Identifying the characteristics of enzymes.
- Justifying the need for ATP energy by cells.
- Distinguishing between the stages of aerobic cellular respiration.
- Describing anaerobic respiration.
- Defining and explaining the importance of fermentation.
- Summarizing the two major reactions of photosynthesis.

Books have been written about it; movies have been made; even a radio play (the War of the Worlds) in 1938 about it frightened thousands of Americans. What is it? Martian life. In 1877 the Italian astronomer, Giovanni Schiaparelli, saw lines on Mars, which he and others assumed were canals built by intelligent beings. It wasn't until well into the twentieth century that this notion was disproved. Still, when we gaze at the Red Planet, we wonder: Did life ever exist there?

We are not the only ones wondering. Astronomers, geologists, and many other scientists have asked the same question. Today, microbiologists have joined their other science colleagues, wondering if microbial life once existed on the Red Planet or, for that matter, elsewhere in our Solar System.

Could microbes, as we know them here on Earth, survive on Mars where the temperatures are far below 0°C, the atmosphere contains little oxygen gas, and the surface is bombarded with ultraviolet radiation? Researchers here on Earth have placed microbes, known to survive in extremely cold earthly environments, in a device simulating the Martian environment. Their results suggested members of the domain Archaea, specifically the methanogens, could grow in the cold, low oxygen atmosphere, especially if they were buried just under the soil surface.

There are in fact many microbes here on Earth capable of surviving and growing in very extreme environments (TABLE 9.1). In fact, many of these so-called **extremophiles** must live in an extreme environment or they will die—some environments not so different from Mars (FIGURE 9.1). If life (as we know it) did or does exist on Mars, it almost certainly was or is composed of bacterial and/or archaeal organisms.

In 2004, NASA sent two spacecraft to Mars to look for indirect signs of past life. Scientists here on Earth monitored instruments on the Mars rovers, *Spirit* and *Opportunity*, designed to search for signs suggesting water once existed on the planet. Some findings suggest there are areas where salty seas once washed over the plains of Mars, creating a life-friendly environment. *Opportunity* even found evidence for ancient shores on what once was a sea.

In 2008, another spacecraft, the *Phoenix Mars Lander* arrived on Mars and soon detected water ice near the Martian soil surface. This was tantalizing news because life, as we know it, does require water for metabolism and growth. Then, in July 2013 NASA reported that *Opportunity* discovered clay minerals in an ancient rock, which

TABLE 9.1 Some Microbial Record Holders

Some Microbial Record Holders
Hottest environment (Juan de Fuca ridge)—121°C: Strain 121 (Archaea)
Coldest environment (Antarctica)—15°C: Cryptoendoliths (Bacteria and lichens)
Highest radiation survival—5MRad, or 5000x what kills humans: *deinococcus radiodurans* (Bacteria)
Deepest—3.2 km underground: Many bacterial and archaeal species
Most acid environment (Iron Mountain, CA)—pH 0.0 (most life is at least a factor of 100,000 less acidic): *Ferroplasma acidarmanus* (Archaea)
Most alkaline environment (Lake Calumet, IL)—pH 12.8 (most life is at least a factor of 1000 less basic): Proteobacteria (Bacteria)
Longest in space (NASA satellite)-6 years: *Bacillus subtilis* (Bacteria)
High pressure environment (Mariana Trench)—1200 times atmospheric pressure: *Moritella, Shewanella,* and others (Bacteria)
Saltiest environment (Eastern Mediterranean basin)—47% salt (15 times human blood saltiness): Several bacterial and archaeal species

FIGURE 9.1 **The Martian Surface?** This barren-looking landscape is not Mars but the Atacama Desert in Chile. It looks similar to photos taken by the Mars rovers *Spirit, Opportunity*, and *Curiosity*.

suggests water once flowed through the area. And if there was water, there might have been microbial life.

Now another exploration of the Red Planet is underway, as the Mars rover *Curiosity* analyzes dozens of Martian rock and soil samples trying to determine what happened to the Martian atmosphere that at one time may have been hospitable to life. Only more studies and perhaps human travel to Mars will answer the question of Martian microbial life.

Whether microorganisms are here on Earth in a moderate or extreme environment, or on Mars, there are certain physical and chemical requirements they must possess to survive and grow. In this chapter, we explore the process of microbial growth, examine some of the varied physical conditions for growth, and then study some of the universal metabolic pathways microorganisms, and all life, possess and use to stay alive. Because much of the chapter emphasizes the role of carbohydrates in metabolism, it might be worthwhile to review the material on the organic molecules (especially carbohydrates).

9.1 Microbial Growth: Physical Factors

Microbes are metabolic machines with the potential for explosive growth and reproduction. Sometimes we suffer the consequences of microbial growth, such as when microbes grow in our bodies, but at other times scientists are able to take advantage of microbial growth by harvesting the valuable products of their **metabolism**, the sum of all biochemical processes occurring in a living cell. For example, sewage plant operators stimulate bacterial cells, algae, and other microbes to grow and break down sewage and return the by-products of their metabolism to the soil. And industrial microbiologists grow microbes in building-sized vats to obtain vitamins, amino acids, and other products valuable for human nutrition and for society.

A major objective of microbiologists is to manipulate microbial growth so they can better use microbes both to enhance the quality of human life and to minimize the microbes' harmful effects. In addition, the study of microbial growth gives researchers a glimpse into the characteristics of microbes and helps them understand microbial life processes. In doing so, they gain a better appreciation for all life processes.

The growth of microbial populations was described in a previous chapter. Here we need to examine the factors that govern that growth.

Water and Temperature

The growth of a microbial population, whether in a broth tube, agar plate, or the natural environment, can be significantly influenced by aspects of the physical environment. Because the microbial cytoplasm is water-based, a liquid environment is absolutely required if life is to continue (FIGURE 9.2). (Notable exceptions are bacterial and fungal spores, which survive arid environments).

Another significant aspect is temperature: It influences the rate of enzyme activity, and enzymes are protein molecules responsible for all chemical reactions taking place in microbes (as we discuss below). If the temperature is too low, the rate of the chemical reactions is reduced; if the temperature is too high, enzymes may be broken down by the heat and the reactions will cease.

As mentioned in the chapter introduction, microbes have adapted to most environments on Earth and that means they have adapted to different temperatures (FIGURE 9.3). Those microbes growing best at temperatures between 0°C/32°F and 20°C/68°F are said to be **psychrophiles** (*psychro* = "cold;" *phil* = "loving"); these "psychrophilic" microbes are found in Arctic and Antarctic environments as well as deep below the ocean surface. Not surprising, considering 70% of the Earth is covered by oceans having deep water temperatures below 5°C/41°F.

Another group of "cold-loving" microbes are the **psychrotrophs** (*troph* = "nourish"), which have a slightly higher temperature range than the psychrophiles. Psychrotrophs

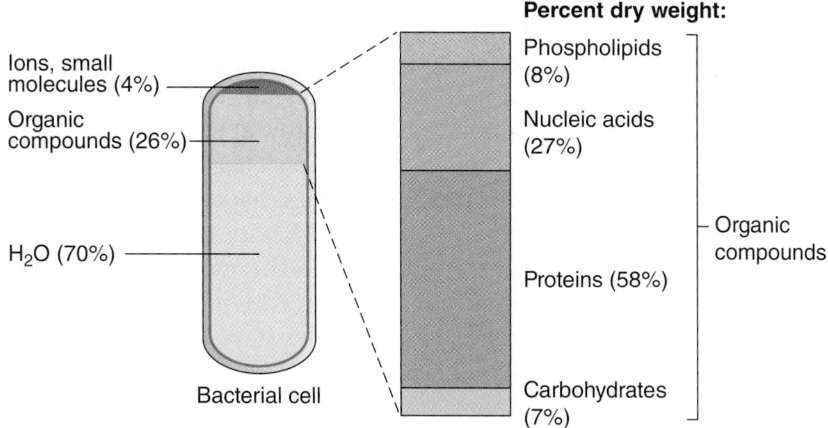

FIGURE 9.2 **Water and Organic Compounds in Bacterial Cells.** Bacterial cells are about 70% water. The other 30% are ions, small molecules, and organic compounds. Dry weight refers to the weight of materials after all the water has been removed.

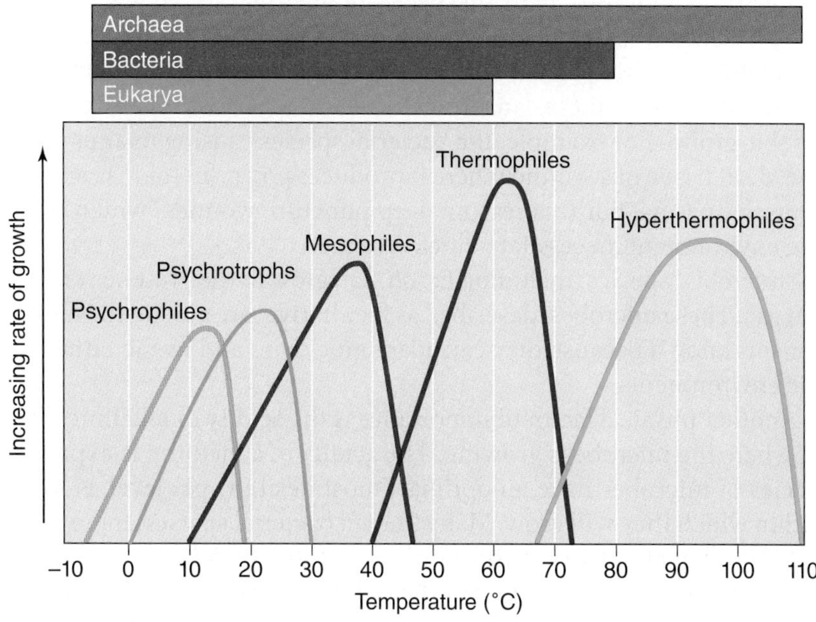

FIGURE 9.3 **Growth Rates for Different Microorganisms in Response to Temperature.** Temperature optima and ranges define the growth rates for Bacteria, Archaea, and Eukarya. The growth rates, however, decline quite rapidly to either side of the optimal growth temperature.

grow well in the refrigerator (at 5°C) and can spoil foods such as milk. For example, streptococci grow in milk and deposit acid, causing it to become sour. Although these microbes do not represent a threat to health, their presence makes the milk undesirable to the eye, nose, and taste buds. On the other hand, some dangerous "psychrotrophic" species also grow in refrigerated foods and deposit their diarrhea-inducing toxins in the contaminated food. Consumption of the contaminated food without proper heating usually leads to food poisoning. One example is the bacterial species *Campylobacter jejuni*, the most frequently identified cause of infective diarrhea in humans.

■ *Campylobacter jejuni*
kam′-pi-lō-bak-tėr jē-jū′-nē

Another group of microorganisms is described as **mesophiles** (meso = "middle"), growing at temperatures between 10°C/50°F and 45°C/113°F. Because the body temperature of warm-blooded animals, including humans, is about 37°C/98.6°F, mesophiles grow well in the body—most human pathogens are mesophiles. "Mesophilic" species are also found in aquatic and soil environments in temperate and tropical regions of the world. *Escherichia coli* is a typical mesophile.

■ *Escherichia*
esh-ėr-ē′kē-ä kōlī (or kō′lē)

Microorganisms growing at high temperatures are said to be **thermophiles** (*thermo* = "heat"). These "thermophilic" microbes grow at temperatures of 45°C and higher, thriving in such varied environments as compost heaps, hot springs, and thermal vents in the oceans. Then, there are those species that can grow at temperatures at or above the boiling point of water (100°C/212°F), with some actually growing at an astounding 121°C/250°F; examples of these **hyperthermophiles** are species in the domain Archaea. These "hellish" environments often are hot-water vents found along rifts on the floor of the Pacific Ocean. Here the water stays in a liquid form because of the high water pressure at these depths.

Oxygen and Acidity

Besides temperature, there are additional physical factors governing microbial growth.

Oxygen. The growth of many microbes depends on a plentiful supply of oxygen gas (O_2). Microbes requiring the gas are said to be **aerobes**; humans also are "aerobic" organisms. In addition, there are species of microbes called **anaerobes** that live only in the absence of oxygen and will die when the gas is present. Such environments as landfills (tightly packed with garbage) and dense, muddy swamps provide an ideal environment for these "anaerobic" species. Some pathogens are included in the anaerobic group. For example, the bacterial species causing tetanus grows in the anaerobic, dead tissue of a wound where it produces powerful toxins causing uncontrolled muscle spasms. For this reason, deep puncture wounds, which represent an anaerobic environment, necessitate quick attention.

Some microbial species, including *E. coli*, can grow in the presence or the absence of oxygen gas. These microbes, described as **facultative**, are among the most interesting organisms known because they can adapt quickly to and live in either aerobic or anaerobic environments.

pH. Another physical factor of importance is the acidity or alkalinity of the environment where the microbe is growing. The acidity of a medium is expressed as pH. Most species of microbes have an optimal (most desired) pH level as well as a pH range within which they will grow. Many known bacterial species, for example, grow best at a pH level of about 7.0, with a range as low as 5.0 and as high as 8.0.

■ pH: A measure of the hydrogen ion (H^+) concentration of an aqueous solution. Solutions with a pH less than 7 are said to be acidic and solutions with a pH greater than 7 are alkaline (basic). Pure water has a pH of 7.

Other bacterial species grow best at the very acidic pH level of 3.0—the bacterial species that ferment cabbage and convert the cabbage to sauerkraut are examples. Such species tolerating or thriving in acid conditions are called **acidophiles**. Other "acidophilic" bacterial genera are of value in the dairy industry. Certain species of

Lactobacillus and *Streptococcus* produce the acid that converts milk to buttermilk and cream to sour cream. Importantly, these species pose no threat to good health even when consumed in large amounts. The "active cultures" in a cup of yogurt are actually acidophilic bacterial species.

■ *Lactobacillus* lak-tō-bä-sil'lus

■ *Streptococcus* strep-tō-kok'kus

Most known species of fungi are acidophilic, tolerating pH levels of about 5.0—this is why fungi contaminate acidic fruits such as oranges, lemons, and limes as well as acidic vegetables such as tomatoes. Fungi are also commonly found in cheeses, which tend to have an acidic pH.

Another group of acidophiles, called the **extreme acidophiles**, prefer pHs of 1 to 2. Close to home, the bacterial species *Helicobacter pylori* inhabits the human stomach lining where the gastric juices have a pH of 2. In a small number of these individuals, *H. pylori* is known to cause gastric ulcers and, in an even smaller number of people, can be involved in stomach cancer. Even more extreme are the bacterial and archaeal species living in the drainage water from the Iron Mountain Mine in Northern California. The water draining from the mine is the most acidic water naturally found on Earth; the pH is 0.8. Yet a few prokaryotic species form a pink film several millimeters thick that floats on the water surface. At the other extreme, there are bacterial species that "love" extremely alkaline environments, **A Closer Look 9.1** provides an example.

■ *Helicobacter pylori* hē'-lik-ō- bak-tėr pī'-lō-rē

A CLOSER LOOK 9.1

Just South of Chicago

All you need is a map, some pH paper, and a few collection vials. When in Chicago, use your map to find the Lake Calumet region just southeast of Chicago. When you arrive, pull out your pH paper and sample some of the groundwater in the region near the Calumet River. You will be shocked to discover the pH is greater than 12—almost as alkaline as oven cleaner! In fact, this might be one of the most extreme pH environments on Earth.

How did the water get this alkaline and could anything possibly live in the groundwater?

The groundwater in the area near Lake Calumet became strongly alkaline as a result of the steel slag dumped into the area for more than 100 years. Used to fill the wetlands and lakes, water and air chemically react with the slag to produce lime (calcium hydroxide). It is estimated that 10 trillion cubic feet of slag and the resulting lime has pushed the pH to such a high value.

Now use your collection vials to collect some samples of the water. Back in the lab you will be surprised to find bacterial communities present in the water. Hydrogeologists who have collected such samples have discovered some bacterial species that until then had only been found in Greenland and deep gold mines of South Africa. Other identified species appear to use the hydrogen resulting from the corrosion of the iron for energy.

How did these bacterial organisms get there? Perhaps the bacterial species have always been there and have simply adapted to the environment over the last 100 years as the slag accumulated. Otherwise, the microbes must have been imported in some unknown way.

So, once again, provide a specific environment and they will come (or evolve)—the microbes, that is.

Other Factors

In addition to temperature, oxygen gas, and pH, other physical factors such as salt and pressure may influence the growth of a microbial population. Microbes normally inhabiting high-salt environments are called **halophiles** (*halo* = "salt"). Examples of such "halophilic" species include the Halobacteria, a class of archaeal organisms that require at least 9% salt for growth (as compared to 1% for many bacterial species)—and for some, an incredibly high salt concentration of 27% is most desirable. Although many people are inclined to believe that halophiles are

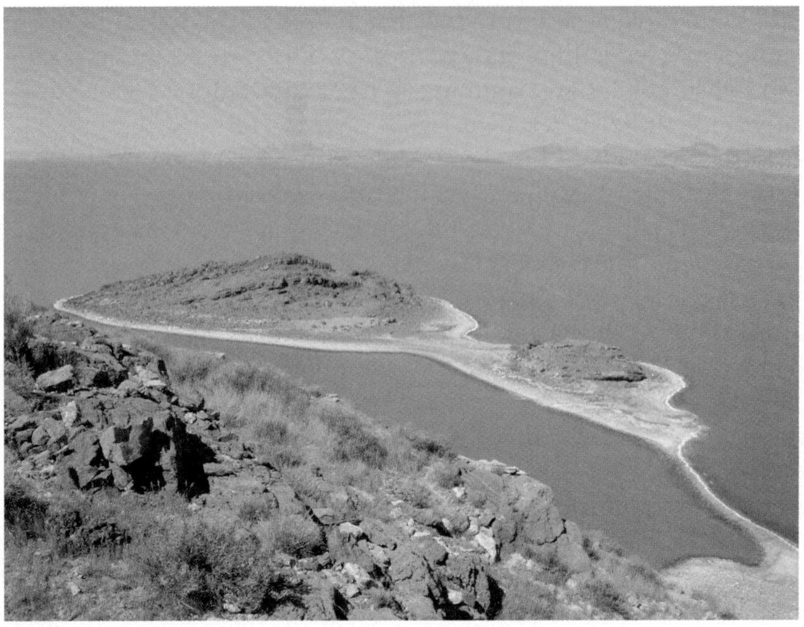

FIGURE 9.4 **Home to Halophilic Bacteria.** The Great Salt Lake in Utah provides the high-salt environment favored by many halophilic bacteria. The pinkish color of the water is due to high numbers of Halobacteria.

inhabitants of the oceans, the salt concentration of the ocean is only about 3.5%. Halophiles are more commonly found in places like Utah's Great Salt Lake where the lake's salinity has ranged from a little less than 5% to nearly 27% (beyond which water cannot hold more salt) (FIGURE 9.4).

Microbes tolerating high-pressure environments are referred to as **barophiles** (*baro* = "pressure"). "Barophilic" microbes are found at the bottoms of the oceans, some species living an astounding six miles beneath the surface where the pressure is 16,000 pounds per square inch (psi); sea level pressure is 14.7psi. Cultivating these microbes in the laboratory presents a considerable challenge because the pressure, oxygen, pH, and nutrient levels of the ocean depths must be matched.

Deinococcus radiodurans
dī-nō-kok′-kus rā-dē-ō-du′r′-anz

One last physical factor is radiation, including X rays, gamma rays, and ultraviolet (UV) radiation. UV light is a component in sunlight and is often deleterious to microbes inhabiting the upper layers of the soil. However, there are microbes that can resist the effects of radiation. One notable species is *Deinococcus radiodurans*, a bacterial species that can survive extremely high levels of radiation. Scientists hope to harness the metabolic powers of this microbe to help dispose of radioactive waste materials, which are a continuing concern as pollutants of the environment. If the scientists are successful, they will once again show how microbes can be a considerable benefit to society.

9.2 Microbial Metabolism: Enzymes and Energy

As mentioned earlier, metabolism refers to all the chemical changes occurring in a microbe during its growth. These chemical changes maintain the stability of the microbial cell, while providing a dynamic pool of chemical building blocks for the synthesis of new cellular materials, such as proteins and nucleic acids like DNA.

Metabolism also refers to the chemical changes going on in the complex cells of plants and animals (including humans). Thus, many of the concepts in this section apply equally well to more complex living things. Indeed, much of the metabolism of human cells was first discovered and studied in microbial cells.

The Forms of Cellular Metabolism

Although there are thousands of different chemical reactions going on in cells at any one time, many of these metabolic reactions fall into either of two general categories: biosynthesis reactions (also known as **anabolism**), and digestive reactions (also referred to as **catabolism**) (FIGURE 9.5).

"Biosynthesis" is a broad term that applies to any chemical process resulting in the formation of cellular structures and molecules. It is essentially a building process in which larger molecules are formed from smaller ones, a process that generally requires an input of energy. "Digestion," by comparison, is the general chemical process in which large molecules (such as food molecules) are broken down into smaller ones. In this process, chemical linkages are usually broken and energy is often liberated, much in the form of unusable heat.

The chemical reactions of metabolism are the fundamental underpinnings for such activities as movement, growth, synthesis, and use of foodstuffs. The reactions are highly organized and responsive to cellular controls. For example, a cell will produce the chemical substances necessary for the breakdown of certain carbohydrates only when those carbohydrates appear in its local environment. At other times, production of the substances is suppressed. This form of cellular economy ensures the reactions of metabolism are used efficiently. The molecules controlling and catalyzing the chemical changes of metabolism are enzymes, the topic of the next section.

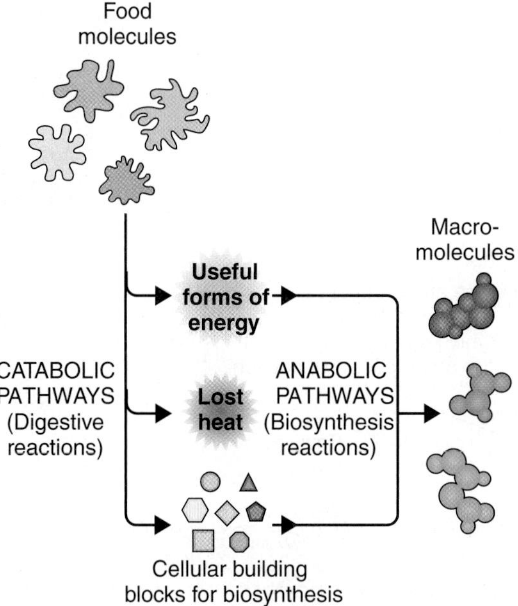

FIGURE 9.5 **Anabolism and Catabolism.** Metabolism can be broken down into anabolism, which includes those reactions that build large molecules (biosynthesis reactions) from building blocks and catabolism, which involves those reactions that tear apart larger molecules (digestion reactions) and release energy.

Enzymes

All the reactions of metabolism are brought about and controlled by a special class of molecules known as enzymes. **Enzymes** are proteins that increase the rate of a metabolic reaction while themselves remaining unchanged. They accomplish in fractions of a second what otherwise might take hours, days, or longer to happen spontaneously under normal biological conditions. For example, even though organic molecules like amino acids can interact, it is highly unlikely they would randomly bump into one another in the precise way needed for a chemical reaction to occur and for a new protein to be made. Thus, the reaction rate would be very slow were it not for the activity of enzymes.

Enzymes have several common characteristics that can be understood by looking at FIGURE 9.6. In this example, the enzyme is involved in a catabolic (digestion) reaction. Be aware there are many enzymes involved anabolic (biosynthesis) reactions as well.

1. **Enzymes are highly specific**. An enzyme that functions in one chemical reaction usually will not participate in another type of reaction. Consequently, there must be thousands of different enzymes to catalyze the thousands of different chemical reactions of metabolism occurring in a microbial cell. The explanation for this specificity is that each enzyme has a special pocket or cleft called an **active site**, which has a unique three-dimensional shape complementary to a reactant molecule (called a **substrate**). When the correct fit occurs between active site and

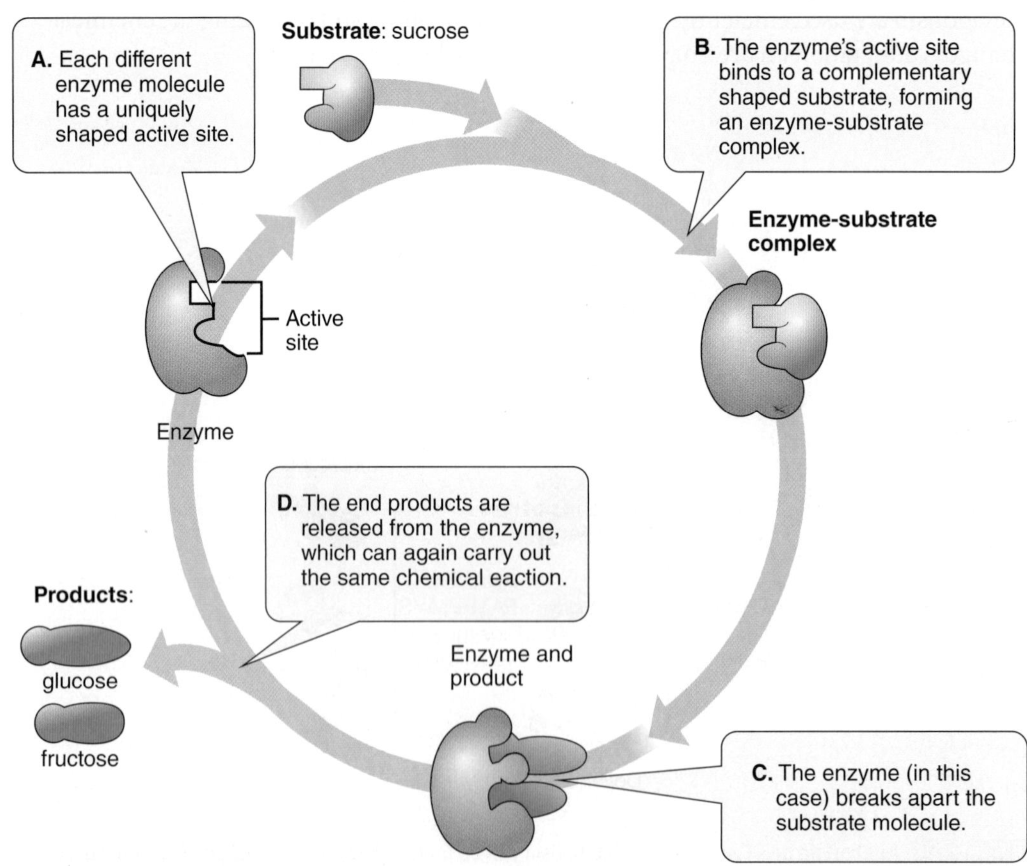

A. Each different enzyme molecule has a uniquely shaped active site.
Substrate: sucrose
B. The enzyme's active site binds to a complementary shaped substrate, forming an enzyme-substrate complex.
Enzyme-substrate complex
Active site
Enzyme
D. The end products are released from the enzyme, which can again carry out the same chemical eaction.
Products:
glucose
fructose
Enzyme and product
C. The enzyme (in this case) breaks apart the substrate molecule.

FIGURE 9.6 **The Mechanism of Enzyme Action.** Although this example shows an enzyme doing catabolism (digesting the substrate sucrose), enzymes also catalyze anabolic (biosynthesis) reactions.

substrate, the substrate is altered in some way, resulting in the formation of one or more **products**. Certain enzymes are composed of protein alone, while others are composed of protein plus a nonprotein portion called a **coenzyme**.

2. **Enzymes are reusable**. Once a chemical reaction has occurred, the enzyme is released to participate in another identical reaction. In fact, the same enzyme can catalyze the same type of reaction 100 to 1 million times each second.
3. **Enzymes are required in small amounts**. Because an enzyme can be used thousands of times to catalyze the same reaction, only tiny amounts of a particular enzyme are needed to ensure a fast and efficient metabolic reaction occurs.

Many enzymes can be identified by their names, which often end in "-ase." For example, "sucrase" is the enzyme that breaks down the sugar sucrose and "protease" digests protein. However, other enzymes do not have such descriptive names. Trypsin is also an enzyme that degrades protein.

Energy and ATP

One of the major outcomes of metabolic reactions is to establish a readily accessible supply of a substance to store energy and release the energy when needed. A steady supply of energy is necessary not only to perform the processes of life but also to maintain order. For example, consider what might take place if the librarians at a community library were to quit or retire all at once. The library might remain open, but as the days passed, more and more books would be strewn about tables or placed on the wrong shelves. Eventually, the disorder would be so great that the library could not function as a library. Without librarians, the "energy" of the library would be lost, and the functional "life" of the library would come to an end.

Physicists tell us the universe has a fixed amount of energy and this energy cannot be created or destroyed. However, energy can be converted from one form to another, such as when logs are burned in the fireplace and the energy in their molecules is released as heat and light. In the web of life, certain species of microbes and plants trap the Sun's energy and use this energy to synthesize energy-rich molecules such as glucose and other larger carbohydrates. The process by which this is accomplished is called photosynthesis (we shall consider this process near the end of this chapter). Nonphotosynthetic organisms acquire the energy-rich molecules from the photosynthetic organisms. Then, they convert the chemical energy from the carbohydrates into a usable cellular energy form known as **adenosine triphosphate** (**ATP**).

ATP molecules are somewhat like portable batteries: They can supply energy to any part of the cell where energy is needed. ATP can fuel the transport of nutrients into the cell or the elimination of waste products taking place at the cell membrane; it can be used in the synthesis of protein; it assists the movements of bacterial and protist flagella and protozoal cilia; and it energizes the reproduction process, regardless of whether it is taking place in microbial, plant, or animal cells. However, unlike a portable battery, ATP energy needs to be used immediately. It cannot be stored, so if it is not used straightaway, it will spontaneously break down and the "battery" will be useless.

ATP has been referred to as the "universal energy currency." As shown in FIGURE 9.7a, the molecule consists of adenine and the sugar ribose (called adenosine when chemically linked); the adenosine is linked to three phosphate groups (phosphorus bonded to oxygen atoms). Much of ATP's energy is released when an enzyme breaks the high-energy bond connecting the terminal phosphate group to the remainder of the molecule (FIGURE 9.7b). Most people are unaware of the importance of ATP in their lives—but without ATP, life as we know it could not continue.

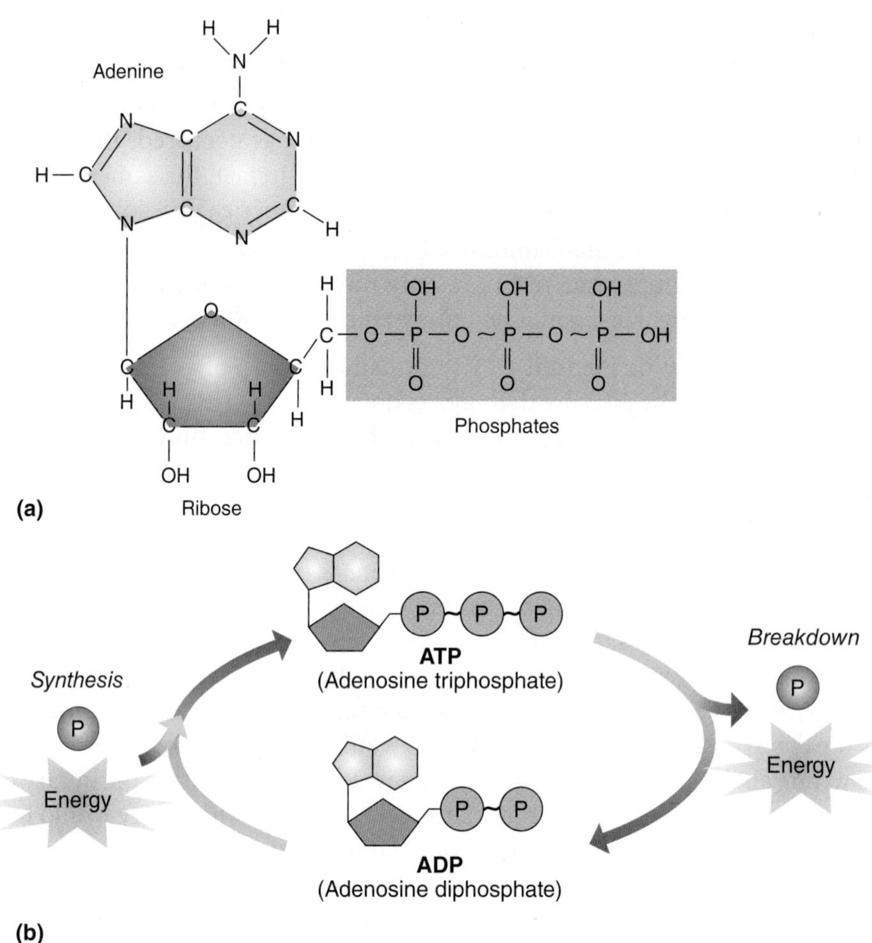

FIGURE 9.7 **Adenosine Triphosphate and the ATP/ADP Cycle.** Adenosine triphosphate (ATP) is a key energy source for microbes and all life. (**a**) The ATP molecule is composed of adenine and ribose bonded to one another and to three phosphate groups. (**b**) When the ATP molecule breaks down (right), it releases a phosphate group and much energy to do the cell's work; it then becomes adenosine diphosphate (ADP). For the synthesis of ATP (left), energy and a phosphate group must be supplied to an ADP molecule.

The step-by-step process by which carbohydrates are broken down and their energy is released to generate ATP is known as **cellular respiration**. Cell respiration is basic and essential to the metabolism taking place in all organisms. It is the key aspect of the metabolic activity we discuss in this next section.

9.3 Cellular Respiration: Providing ATP

Among the most important molecules organisms obtain from the environment are the energy-rich carbohydrates. Formed in photosynthesis by photosynthetic organisms, carbohydrates are the major source of chemical energy. Among the carbohydrates, glucose is probably the one most widely metabolized, and so we shall focus our attention on this six-carbon carbohydrate. As you will see, many of the descriptions to follow are somewhat involved, but rest assured they have been simplified as much as possible without skipping key elements of the processes.

The cell respiration process occurs in three stages: glycolysis, the citric acid cycle, and electron transport.

Glycolysis

The breakdown of glucose occurs by the process of **glycolysis**.(*glycol* = "sweet," as in a sugar; *lysis* = break). During glycolysis, each six-carbon glucose molecule is converted into smaller molecules and eventually into two three-carbon molecules called pyruvate (FIGURE 9.8). The pathway takes place in the cytoplasm of microbes and does not require oxygen gas. Each of the multiple chemical reactions in the pathway is catalyzed by a different enzyme.

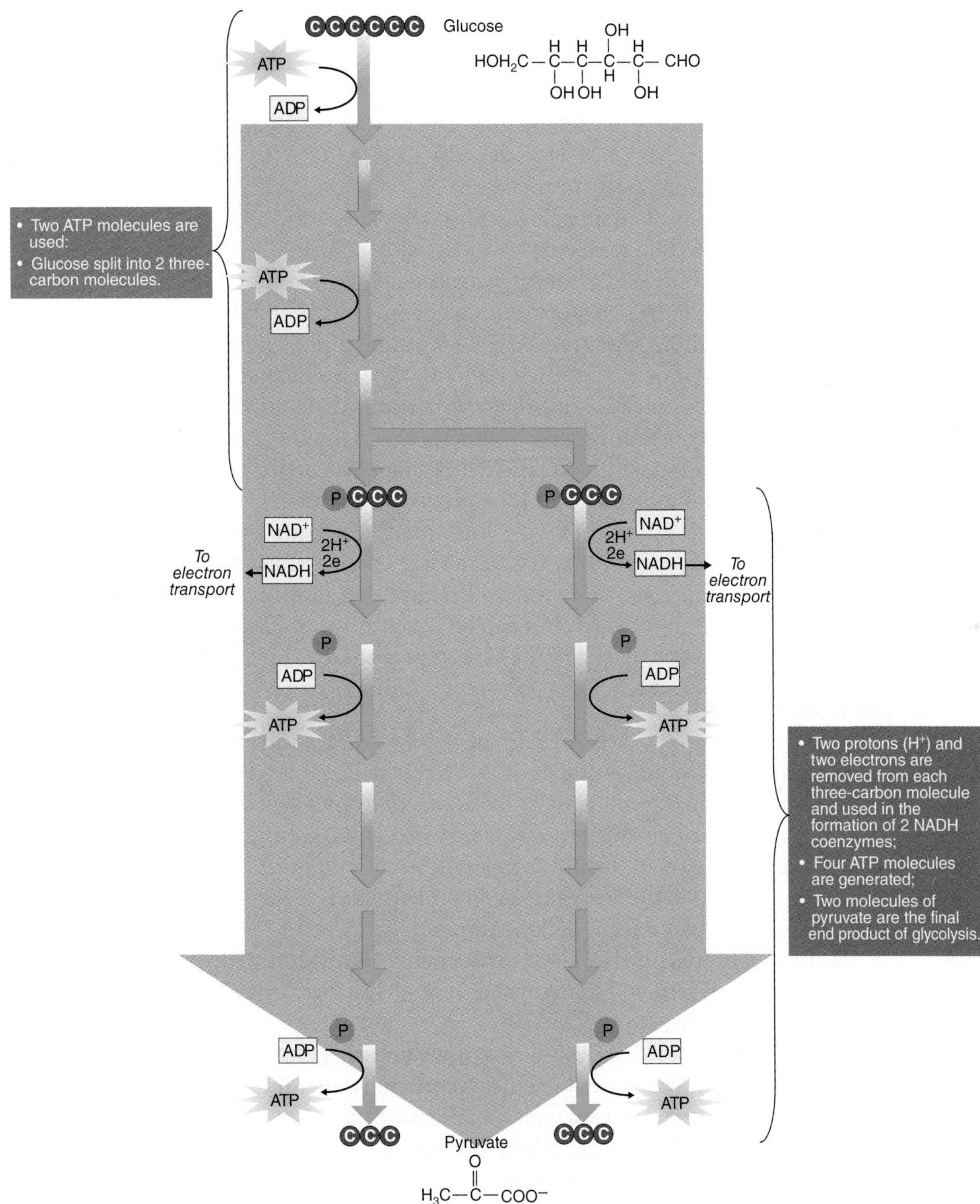

FIGURE 9.8 **The Reactions of Glycolysis.** Glycolysis is a metabolic process that converts glucose, a six-carbon sugar, into two three-carbon pyruvate products. In the process, two NADH coenzymes and a net gain of two ATP molecules occur. Carbon atoms are represented by circles. Each enzyme-catalyzed reaction is represented by an arrow.

Glycolysis is important because a small amount of ATP is generated and some of the energy in the sugar molecules is transferred to a coenzyme called NAD (nicotinamide adenine dinucleotide). In the next paragraphs, we shall go through the key steps, so use Figure 9.8 as your metabolic roadmap.

The pathway of glycolysis begins with a molecule of glucose. (We should pause and note, however, that many other carbohydrates as well as various amino acids and numerous lipids can enter the pathway and be metabolized to acquire some ATP.) In the initial reactions, enzymes bring about a number of changes in the glucose molecule, resulting in the splitting of glucose into two three-carbon molecules. Along the way, two ATP molecules are "invested" in order to "energize" the glucose molecules. In the second half of the glycolysis process, several important events occur. An enzyme strips two electrons from each three-carbon molecule and deposits them in the coenzyme NAD. The NAD molecules each acquire one proton (or hydrogen nucleus) as well and become NADH. The NADH molecules are used later in the electron transport (note "To electron transport" in the figure).

In the next series of chemical reactions, enzymes bring about additional conversions of the three-carbon molecules (we shall skip many of the details), and pyruvate eventually forms at the end of the glycolysis pathway. Two notable events have happened along the way: In two steps, enough energy has been released to generate the formation of ATP molecules from ADP (adenosine diphosphate) and phosphate. In all, four ATP molecules have formed (two on the left side and two on the right side of the pathway) from each original glucose consumed. Thus, the two ATP molecules invested in the early reactions have been returned to the cell, and the cell has profited on its investment—the two invested ATP molecules have returned four ATP molecules, for a net gain of two molecules. But the best is yet to come.

We should stop at this juncture and note that we are discussing some rather involved and detailed biochemistry. A general appreciation of this biochemistry is relevant to the undergraduate student because this (and what is to follow) is the chemistry by which most microbes and all other forms of life get their energy in the form of ATP. And, as we have emphasized, where there is no energy, there is no life.

Citric Acid Cycle

Following glycolysis, the second stage of cellular respiration occurs through a cyclic series of chemical reactions called the **citric acid cycle**. It is termed a "cycle" because the substance formed at the end of the series of events serves as the starting point for another round of chemical reactions. In protists and fungi, the chemistry of the citric acid cycle occurs in the membranes of the mitochondria because the necessary enzymes are located in these organelles. In prokaryotic cells, which lack mitochondria, the reactions take place at the cell membrane.

At the end of glycolysis, there is still energy trapped in the pyruvate molecules and it is this energy that will be extracted through the reactions of the citric acid cycle. The process is presented in FIGURE 9.9.

The end product of glycolysis—two pyruvate molecules—is considered the starting point. However, the pyruvate molecules do not yet enter the cycle. Instead, they undergo a transformation in which an enzyme removes a carbon atom from each molecule (and releases each as carbon dioxide gas), then combines the remainder of each molecule with a coenzyme called "coenzyme A." This transformation results in two acetyl-coenzyme A (acetyl-CoA) molecules and is the transition step linking glycolysis to the citric acid cycle.

The importance of the transition step is that during the formation of each acetyl-CoA, a NAD molecule accepts two electrons and a proton to become NADH. The NADH molecules, along with those from glycolysis, are later used for electron transport.

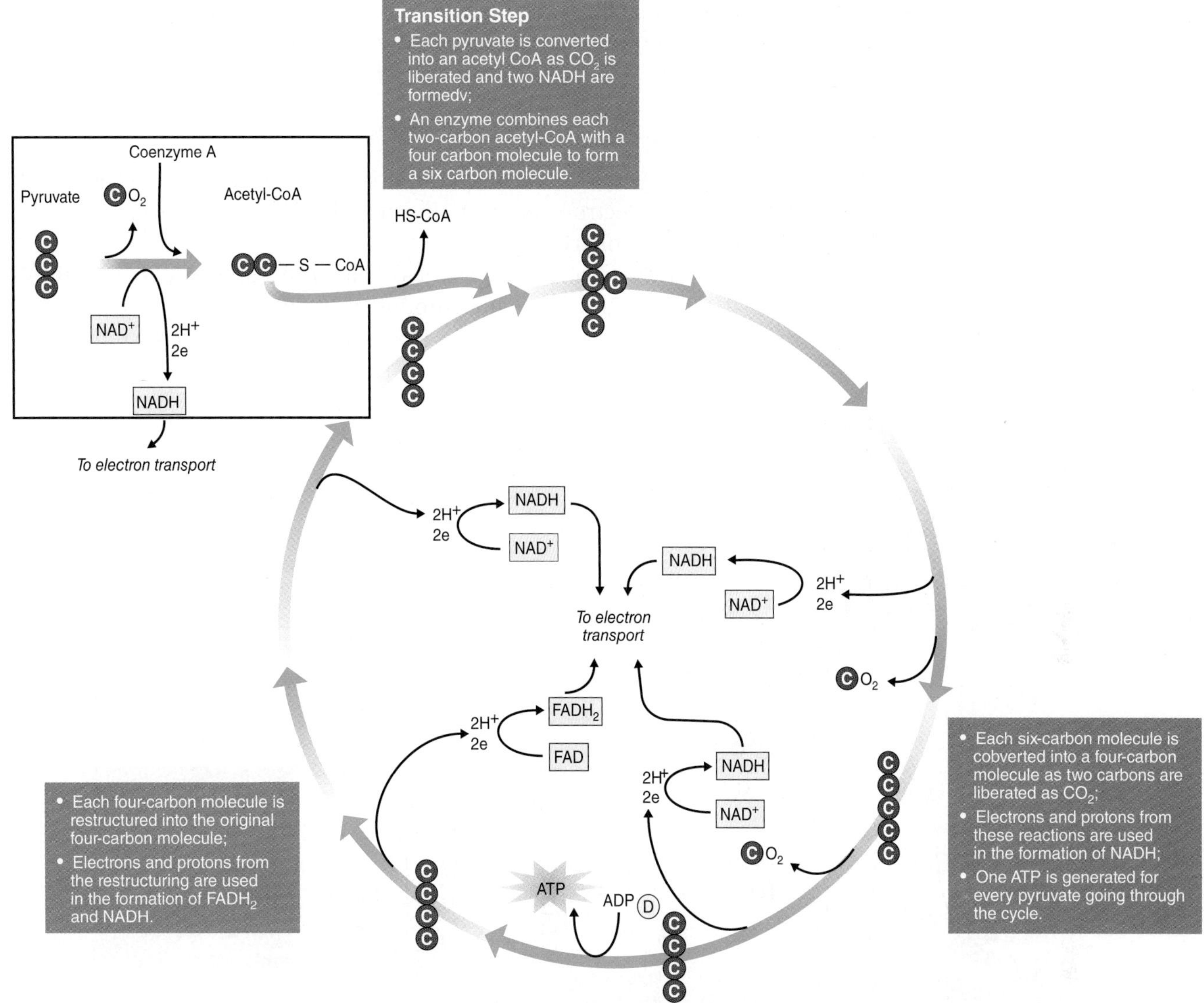

Transition Step
• Each pyruvate is converted into an acetyl CoA as CO_2 is liberated and two NADH are formedv;
• An enzyme combines each two-carbon acetyl-CoA with a four carbon molecule to form a six carbon molecule.
Coenzyme A
Pyruvate
CO_2
Acetyl-CoA
— S — CoA
HS-CoA
NAD^+
$2H^+$
2e
NADH
To electron transport
NADH
NAD^+
To electron transport
FADH2
FAD
ATP
ADP
• Each six-carbon molecule is cobverted into a four-carbon molecule as two carbons are liberated as CO_2;
• Electrons and protons from these reactions are used in the formation of NADH;
• One ATP is generated for every pyruvate going through the cycle.
• Each four-carbon molecule is restructured into the original four-carbon molecule;
• Electrons and protons from the restructuring are used in the formation of $FADH_2$ and NADH.

FIGURE 9.9 **The Reactions of the Citric Acid Cycle.** Pyruvate from glycolysis combines with coenzyme A to form acetyl-CoA (transition step). This molecule then joins with a four-carbon molecule to form a six-carbon molecule. Each turn of the cycle releases CO_2, produces ATP, and forms NADH and $FADH_2$ coenzymes. Each enzyme-catalyzed reaction is represented by an arrow.

Each acetyl-CoA molecule now enters into the citric acid cycle reactions as shown in Figure 9.9. First, an enzyme combines the acetyl-CoA molecule with a four-carbon molecule to form a six-carbon acid called citric acid (thus the naming of the cycle). In the next series of steps, enzymes convert the six-carbon molecule into a five- and then a four-carbon molecule. As you follow around the cycle, you will see that the four-carbon molecule undergoes a series of chemical rearrangements eventually reforming the four-carbon molecule that started the cycle.

Along the way, notice two carbon atoms were lost as carbon dioxide gas in the six-carbon to four-carbon conversion. Also two ATP molecules are generated. But, like glycolysis, no oxygen gas was used.

Most noteworthy, in several steps of the cycle more NADH coenzymes are produced. Besides the NADH coenzymes, in one reaction an electron pair and a proton are taken up by another coenzyme FAD (flavin adenine dinucleotide). The coenzyme thus becomes $FADH_2$. Like the NADH we discussed previously, the $FADH_2$ coenzymes will be used in the electron transport system.

Now, the last stage of cellular respiration is to use all the NADH and $FADH_2$ coenzymes generated from glycolysis and the citric acid cycle. And a powerful last stage this will be.

Electron Transport

The third and last stage of cellular respiration is called **electron transport** and the system that transports electrons occurs in the mitochondria (in eukaryotes) or along the cell membrane (in prokaryotes).

In this electron transport system, a series of electron transfers occur, beginning with the NADH and $FADH_2$ molecules produced in glycolysis and the citric acid cycle (FIGURE 9.10). The participants in the electron transport system include

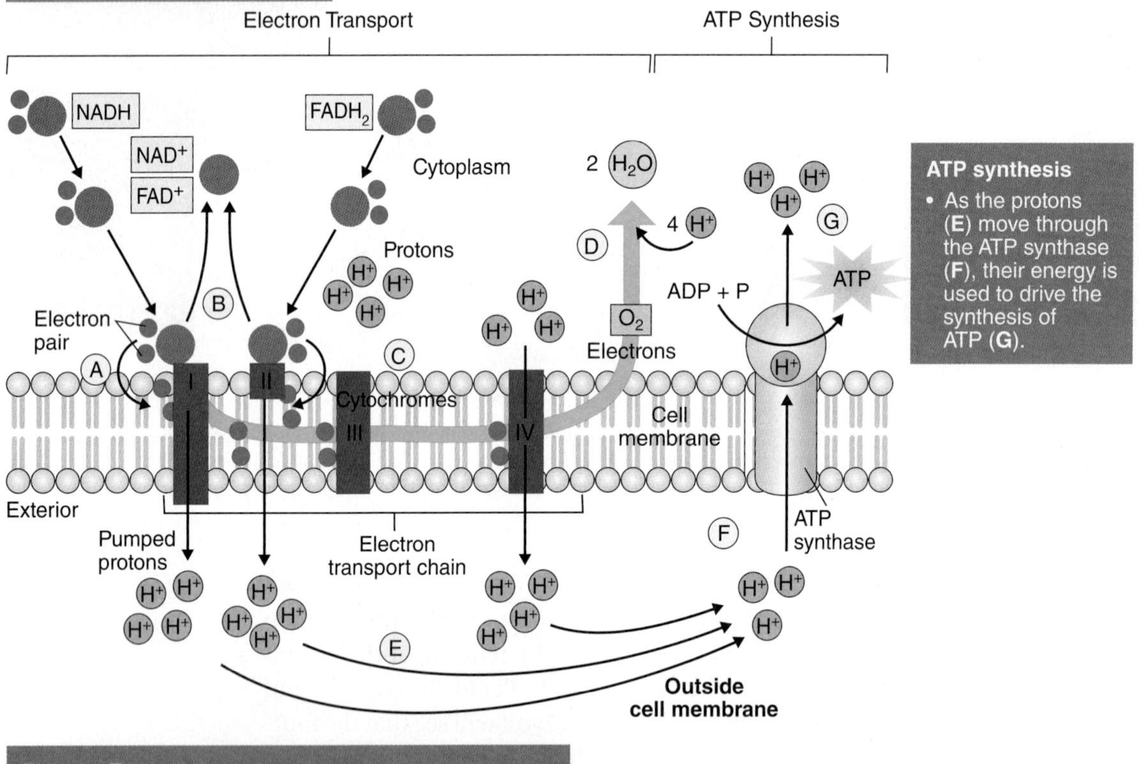

FIGURE 9.10 **Electron Transport and ATP Synthesis in Bacteria.** Originating in glycolysis and the citric acid cycle, coenzymes NADH and $FADH_2$ transport electron pairs to the electron transport chain in the cell membrane, which fuels the transport of protons (H^+) across the cell membrane. Protons then reenter the cell through a protein channel in the ATP synthase enzyme. ADP molecules join with phosphates as protons move through the channel, producing ATP.

A CLOSER LOOK 9.2

"It's Not Toxic to Us!"

It's hard to think of oxygen as a poisonous gas, but several billion years ago, oxygen was as toxic as cyanide. One whiff by a cellular organism, and a cascade of chemical reactions was set into motion in its cytoplasm. Death followed quickly.

Difficult to believe? Not if you realize ancient organisms relied on fermentation and anaerobic chemistry for their energy needs. They absorbed organic materials from the environment and digested them to release the available energy. The atmosphere was full of methane, hydrogen, ammonia, carbon dioxide, and other gases. But no oxygen. And it was that way for hundreds of millions of years.

Then the cyanobacteria came on the scene and brought with them the ability to perform photosynthesis. Chlorophyll pigments evolved, and the cyanobacteria could trap energy from the sun and convert it to chemical energy in carbohydrates. But there was a downside—oxygen was a necessary waste product of the process; and the oxygen was deadly.

But not deadly to those organisms able to adapt. As millions of species of organisms died off in the now-toxic oceans and atmosphere, a few species evolved to live on because they had the enzymes to tuck oxygen atoms safely away in nontoxic chemical compounds. And then, surprise, some organisms even evolved the ability to use oxygen in an election transport system and tap carbohydrates for large amounts of energy. And so the citric acid cycle and the electron transport system came into existence.

Also coming into existence were millions of new species, some merely surviving, others thriving in the newly expanding oxygen-rich environment. The face of planet Earth was changing rapidly as anaerobic and fermenting species declined and aerobic species evolved. A couple of billion years would pass, and then, finally, a particularly well-known species of oxygen-breathing creature evolved—humans.

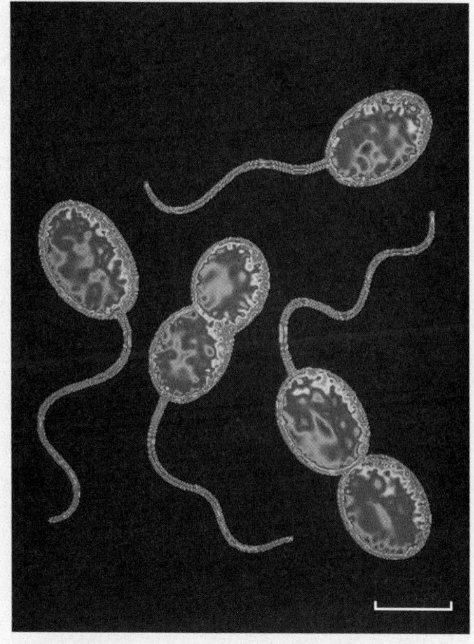

A false-color transmission electron microscope image of anaerobic bacterial cells. (Bar = 2 μm.)

various molecules: Among these are a coenzyme and a series of complex compounds known as cytochromes. **Cytochromes** are iron-containing cell proteins that receive and give up electron pairs much like a biochemical bucket brigade. The coenzymes and cytochromes are the vehicles through which the energy of electron pairs is released.

Looking at Figure 9.10 (we have lettered the steps A–G for easier tracking). First, the NADH and $FADH_2$ molecules from glycolysis and the citric acid cycle pass their electron pairs to the first cytochrome (A). (In this way, the NAD and FAD molecules are regenerated and return to glycolysis and the citric acid cycle to be used again [B].) The biochemical "bucket brigade" continues by passing electron pairs from one cytochrome to the next, each passing releasing some energy (C). The last cytochrome then passes the electron pairs to oxygen atoms. Each oxygen atom now takes on two protons from the surrounding environment and becomes a molecule of water (D). Thus, the electron pair, once in NADH or FADH, has passed through the electron transport system, ultimately winding up in a water molecule.

Let's pause for a moment and note the importance of oxygen in this process. In aerobic microbes, the oxygen atom is the only substance capable of accepting electrons at the end of the transport chain. If oxygen atoms were absent, the cytochromes could not give up their electrons, and the series of electron transfers would soon grind to a halt. Without electron transport, the synthesis of ATP would also cease (as we will now see), and death would ensue. Therefore, when oxygen gas is used for cellular respiration, the process as we have described is called **aerobic respiration**. It is interesting to note that oxygen was not always used for metabolism on Earth, as **A Closer Look** 9.2 explains.

But what is the importance of moving all these electrons? As the electrons pass among the various components of the electron transport system, the energy released by the electrons "pumps" protons (hydrogen ions, or H^+) across the membrane (E). This proton pumping results in a high concentration of protons outside the membrane and a low concentration in the cytoplasm. The unequal distribution results in protons flowing back through the membrane into the cytoplasm (F). Importantly, the protons do not just randomly pass through the membrane but rather flow through special pores in the membrane containing the enzyme **ATP synthase**. During their rush, the protons supply enough energy to allow the ATP synthases to hook together ADP molecules and phosphate groups to form ATP molecules (G).

In summary, aerobic respiration uses glucose and oxygen gas to produce ATP.

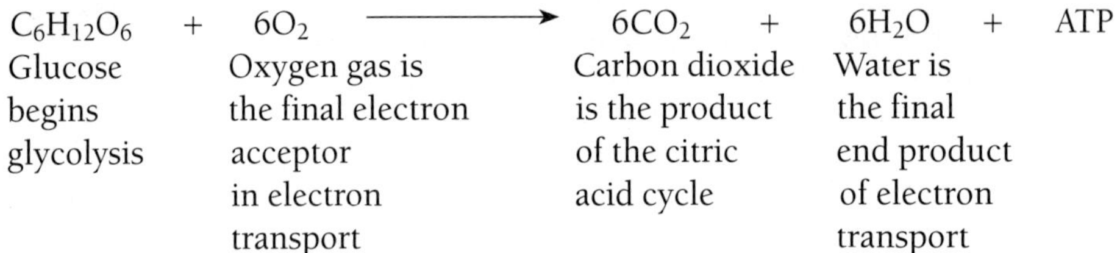
$C_6H_{12}O_6$ + $6O_2$ → $6CO_2$ + $6H_2O$ + ATP
Glucose begins glycolysis
Oxygen gas is the final electron acceptor in electron transport
Carbon dioxide is the product of the citric acid cycle
Water is the final end product of electron transport

The yield of life-sustaining ATP molecules from electron transport is considerable (FIGURE 9.11). For each NADH molecule that delivers its electrons to the transport chain, enough energy is released to synthesize three ATP molecules. Each FADH has enough energy to generate two ATP molecules in electron transport. So, totaling up the NADH and $FADH_2$ coenzymes produced in glycolysis and citric acid cycle (including the transition step): there was a total of 10 NADH and 2 $FADH_2$ molecules resulting from one glucose molecule being broken down. Therefore, 34 molecules of ATP can be generated. Add to this the net gained two ATP molecules made directly in glycolysis and the two ATP molecules in the citric acid cycle, and the ATP total becomes considerable—38 molecules of ATP for cell use for every glucose molecule consumed. Not bad for an initial investment of two ATP molecules!

Although we have concentrated on glucose in this chapter (it is believed to be the richest source of chemical energy on Earth), numerous other carbohydrates—such as lactose, sucrose, maltose, and starch—are used in microbial metabolism as sources of energy. These carbohydrates do not have separate metabolic pathways. Rather, they

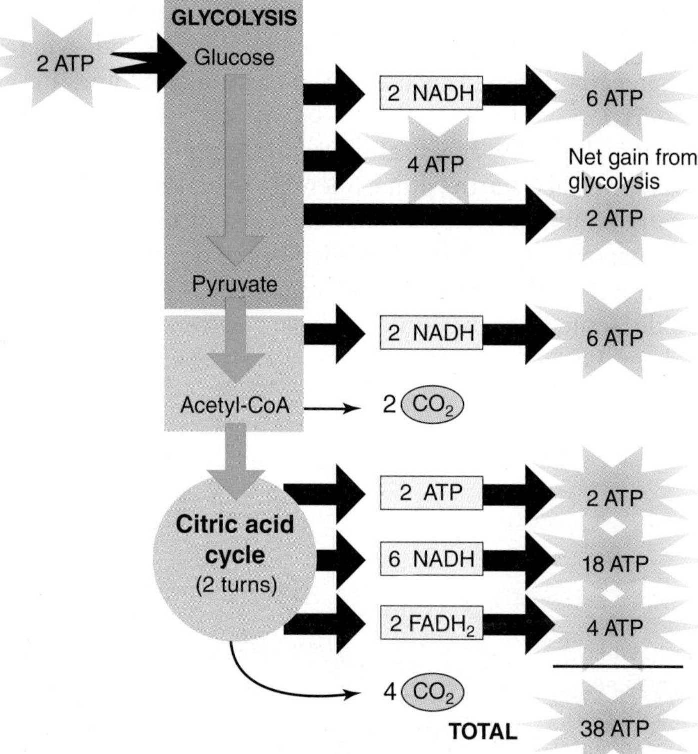

FIGURE 9.11 **The ATP Yield from Aerobic Respiration.** In a microbial cell, 38 molecules of ATP can result from the metabolism of a molecule of glucose. Each NADH molecule accounts for the formation of three molecules of ATP; each molecule of $FADH_2$ accounts for two ATP molecules.

use the basic pathways we have explored and enter them at various points, thus lending efficiency and economy to the metabolism.

The biochemistry of cell respiration is how microbes and other living things obtain their energy for life. A single glucose molecule yields over three dozen ATP molecules that the cell can use as portable batteries. This energy runs the microbial machine—indeed, the chemical machinery of all living things.

Anaerobic Respiration

As we discovered earlier in this chapter, many species of microbes can live under anaerobic conditions; that is, in environments where there is little or no oxygen gas. These microbes still need to make ATP, so how do they do it? Through the process of anaerobic respiration.

Anaerobic respiration uses alternatives to oxygen as a final electron acceptor in the electron transport chain. For example, when it lives anaerobically, *E. coli*, uses a nitrate ion (NO_3^-) as an electron acceptor in place of oxygen. After acquiring electrons, the nitrate ion releases an oxygen atom to become the nitrite ion (NO_2^-). This chemistry is an essential feature of the nitrogen cycle occurring in soils.

Other electron acceptors used by microbes include sulfate ions (SO_4^-) and carbon dioxide molecules (CO_2). When microbes use sulfate ions as electron acceptors, they convert the ions to molecules of hydrogen sulfide (H_2S) gas. Hydrogen sulfide has the odor of rotten eggs, and it gives a horrid smell to foods or soils where the microbes are living. When carbon dioxide is used as an electron acceptor, it is converted into methane gas (CH_4).

Anaerobic metabolism was a useful way of obtaining energy in the billions of years before oxygen filled Earth's atmosphere. Indeed, the anaerobes practicing this type of chemistry remind us that life today can exist in an oxygen-free environment. Thousands of anaerobic species are classified in the domains Bacteria and Archaea, and they continue to exist deep in the soil, in marshes and swamps, and in landfills. On the plus side, sometimes the stink can have an advantage, especially if you are a microbe. **A Closer Look** 9.3 explains.

A CLOSER LOOK 9.3

Microbes "Raise a Stink"

We are all familiar with body odor and bad breath. When one does not maintain a level of cleanliness or hygiene, bacterial species on the skin surface or in the mouth can grow out of proportion and, as they metabolize compounds like proteins, they produce noticeably unpleasant, smelly odors.

On a more environmental level, what about the smells often coming from rotting materials, such as in a landfill? In fact, it is the microbes decomposing the garbage that produce the noxious aromas. This got some marine biologists to thinking: when competing in the environment with other animal scavengers for food, do bacteria produce the odors to repel or deter animal species from consuming important food resources, especially with respect to marine ecosystems? Their hypothesis: Decaying food resources become repugnant to larger animal species like crabs or fish.

To test their hypothesis, the research team baited crab traps near Savannah, Georgia, with menhaden, a typical bait-fish for crabs. Some traps contained microbe-laden menhaden that had been allowed to rot for one or two days, while other traps contained freshly thawed samples having relatively few microbes. When the traps were inspected, those with fresh samples had more than twice the number of animals per trap than did the traps with microbe-laden menhaden. Lab studies with stone crabs showed they also avoided the microbe-laden, rotting food, but readily consumed the freshly thawed menhaden.

To examine the role of bacterial organisms in the avoidance behavior by stone crabs (see figure), some menhaden was allowed to rot in water without the antibiotic chloramphenicol while other samples contained the antibiotic in the water to prevent or inhibit microbial growth. Again, in this study the crabs readily ate the antibiotic-incubated menhaden but avoided the menhaden without antibiotic; the bacterial organisms were in some way responsible for the aversion.

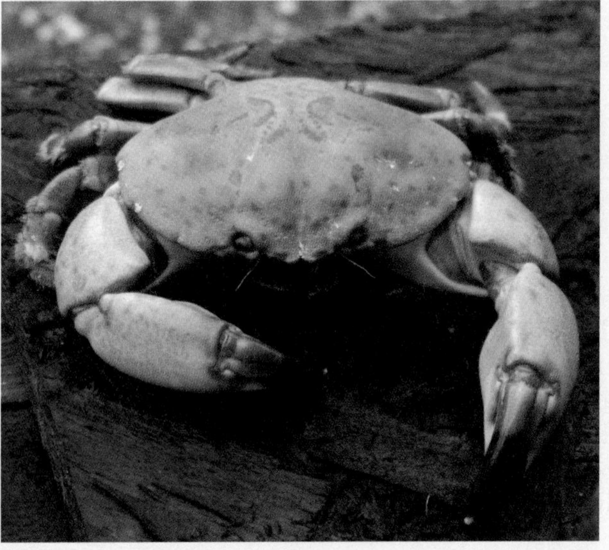

A stone crab.

Courtesy of Catherine Billick, www.flickr.com/photos/catbcorner/

Finally, the researchers used chemical extracts prepared from the microbe-laden menhaden and mixed these chemical substances with freshly thawed menhaden. Again, the crabs were repelled. Exactly what chemical compounds were responsible for the behavior were not evident from the study. But it appears bacterial species not only act as decomposers and pathogens in the environment, but they also can compete—very successfully—with relatively large animal consumers for mutually attractive food sources.

9.4 Fermentation: A Metabolic "Safety Net"

Suppose you are a microbe and you live in an environment without oxygen gas and without all the other electron acceptors needed for cellular respiration. Without any of these, electron transport cannot function and NADH and FADH coenzymes cannot be recycled to NAD and FAD. Without these coenzymes, ATP synthesis will quickly come to a halt and because ATP cannot be stored, the end result would be a quick death—unless your microbial talents happen to include some other way to generate ATP. And, in fact, some microbes have this ability; they can carry out a process called fermentation.

Fermentation is an anaerobic process in which the pyruvate formed in glycolysis is transformed not into acetyl-CoA and sent through the citric acid cycle, but rather into other final end products such as alcohol, acids, and carbon dioxide gas. Fermentation thus provides a "safety net" by allowing some microbes, such as fungal yeast cells, to survive in the absence of oxygen and other cellular respiration electron acceptors until the needed electron acceptors are again present. Here is how fermentation works using yeast cells as an example.

When yeast cells (*Saccharomyces* species) metabolize glucose by glycolysis, they produce pyruvate (FIGURE 9.12a). Two ATP molecules is the net gain and two NAD molecules are converted to NADH. To keep glycolysis going and producing two ATP molecules per glucose, the NADH coenzymes have to be recycled to NAD. Normally, this would occur through the electron transport chain, but that process is not functional now. Therefore, pyruvate must be converted into some other end product to reform NAD. For yeast cells, pyruvate is converted into ethyl alcohol. In so doing, enzymes in the yeast cells remove the two electrons and the proton from the NADH molecule, resulting in the regeneration of NAD for reuse in glycolysis as shown in Figure 9.12a.

Saccharomyces sak-ä -rō-mī′-sēs

Fermentation by yeast cells has a special significance for humans because the end product is alcohol. If grapes are the starting material, grape alcohol (wine) is the result; if barley is used, barley alcohol (beer) results; and if potatoes are used, the end product is potato wine, which is then distilled to produce vodka. The products of fermentation are as varied as the mind can imagine. In addition, carbon dioxide is liberated in the conversion of pyruvate to alcohol, and this carbon dioxide accounts for the bubbles appearing in champagne and beer and the gas making dough rise when bread is made.

But yeasts are not the only microbes capable of performing fermentation; numerous bacterial species can produce other fermentation end products (FIGURE 9.12b). For example, species of *Streptococcus* and *Lactobacillus* do not have the enzymes to produce alcohol, but they do have the enzymes to convert pyruvate into lactic acid. In the dairy industry, this acid converts condensed milk into yogurt. Cheeses obtain their taste from the mixture of acids produced by microbial fermentation during the ripening process. And such products as acetone, butyl alcohol, vitamins, many antibiotics, and numerous amino acids are also produced by fermentation brought about by microbes. Another chapter details the processes behind these products and their uses.

In summary, all these microbes carrying out fermentation are just trying to keep glycolysis running and make two ATP molecules per glucose until "better days" return and the electron acceptors for cell respiration are again present. Therefore, even though the end product, whether it is alcohol, lactic acid, or acetone, may be important to us, it is of no significance to the microbe. Rather, by performing fermentation metabolism, the microbe has a steady supply of NAD to run glycolysis and crank out ATP.

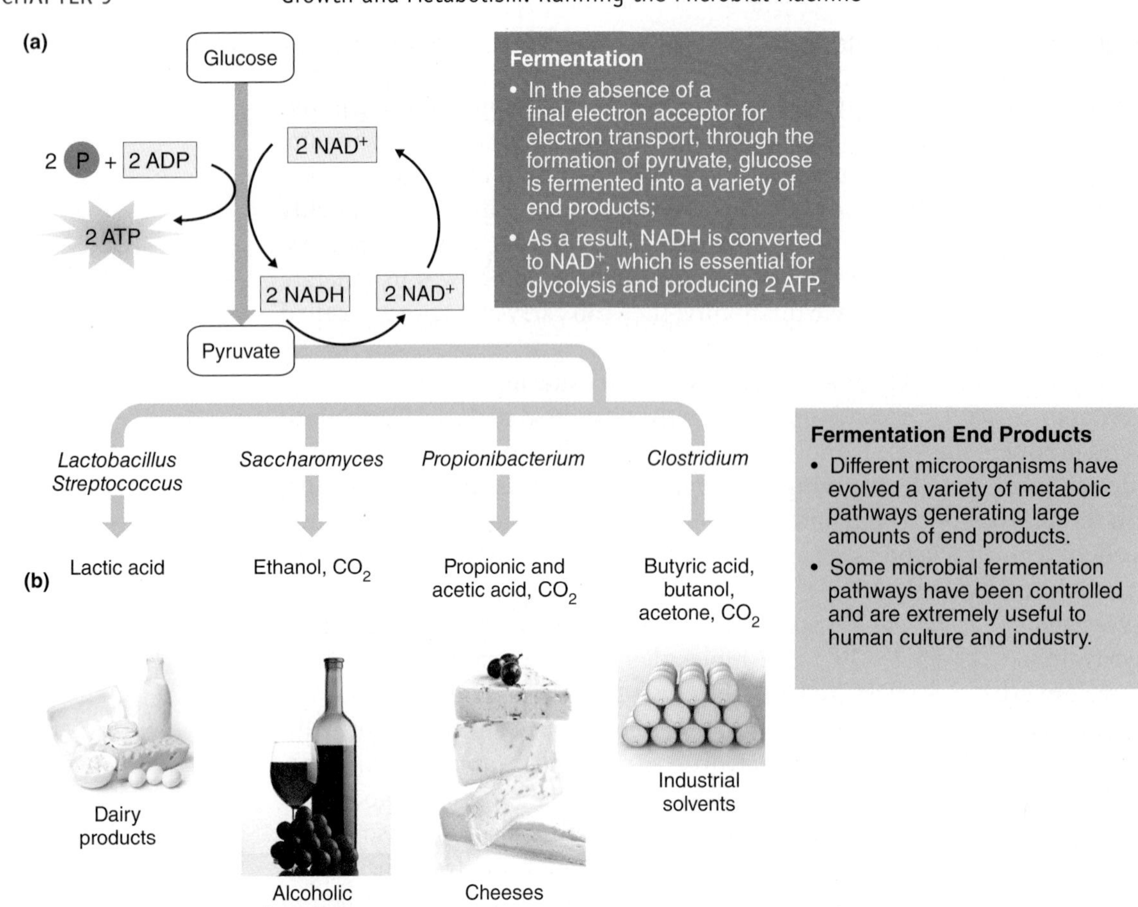

FIGURE 9.12 **Microbial Fermentation.** (**a**) Fermentation is an anaerobic process that converts NADH to NAD by converting pyruvate into an end product (**b**). The presence of NAD allows glycolysis to continue and generate a net gain of two ATPs for each glucose consumed.

(inset a) © niderlander/ShutterStock, Inc. (inset b) © drKaczmar/ShutterStock, Inc. (inset c) © Elena Elisseeva/ShutterStock, Inc. (inset d) © Esteban De Armas/ShutterStock, Inc.

9.5 Photosynthesis: An Anabolic Process

Cellular respiration and fermentation provide all living cells with the ATP energy they need to carry out the numerous life-giving chemical reactions that occur as part of metabolism. But as we alluded to earlier in this chapter, much of the chemical energy (carbohydrates) to perform ATP synthesis comes from photosynthesis. So, let's wrap up this chapter by examining some of the anabolic (biosynthesis) reactions characteristic of photosynthesis.

Photosynthesis is the biosynthetic process in which light energy is captured and used to construct molecules of carbohydrates. It takes place on and near the cell membrane in the cyanobacteria and in the chloroplast in the autotrophic protists. These microbes have light-absorbing pigments to trap the energy in sunlight (light energy) and eventually convert it to chemical energy.

Photosynthesis is conveniently divided into the energy-trapping reactions and the carbon-trapping reactions.

In the energy-trapping reactions of photosynthesis, the energy in sunlight is harvested by chlorophyll pigments that absorb light (FIGURE 9.13a). The light "energizes" electron pairs, and they "jump out" of the chlorophyll molecules. The electrons move through an electron transport system (much as they do in cell respiration), jumping from one electron carrier to another. ATP is manufactured in the process and involves a flow of protons across a membrane (as in cell respiration). In addition, once they have passed through a series of cytochrome molecules, the electrons are taken up by another coenzyme called NADP (nicotinamide adenine dinucleotide phosphate), rather than NAD. The NADP molecule then takes on a proton and is converted to NADPH. Thus, the two main products of the energy-trapping reactions are ATP and NADPH.

Before we leave the energy-trapping reactions, there is one extremely important reaction to mention. These reactions started with the loss of two electrons from a chlorophyll molecule. If the energy-trapping reactions are to continue, these electrons must be replaced. Notice in Figure 9.13a that the cyanobacteria and unicellular algae obtain replacement electrons by breaking down water molecules. Once the electrons have been incorporated, the remaining oxygen atoms from the water molecules are released to the atmosphere. Over billions of years of history, this oxygen has accumulated, and it now accounts for 20% of the gaseous content of the atmosphere. It is the oxygen we breathe to keep our metabolism going. To be sure, green plants contribute mightily to the atmosphere's oxygen content, but about 50% of the contribution comes from the microbes.

To finish the story, the ATP and NADPH generated in the energy-trapping reactions are used in the carbon-trapping reactions (FIGURE 9.13b). In these reactions, an enzyme binds a carbon dioxide molecule to a five-carbon molecule, resulting in a six-carbon unstable molecule. The latter immediately splits into two three-carbon molecules, which are converted into other three-carbon molecules. This latter conversion uses the energy in ATP and the NADPH from the energy-trapping reactions. Now, the three-carbon molecules undergo a series of enzyme-catalyzed conversions and some join to form a molecule of glucose. The cycle is complete once the other three-carbon molecules are formed back into the starting five-carbon molecule.

In summary, in the process of photosynthesis the energy of light has been trapped and used to form glucose molecules.

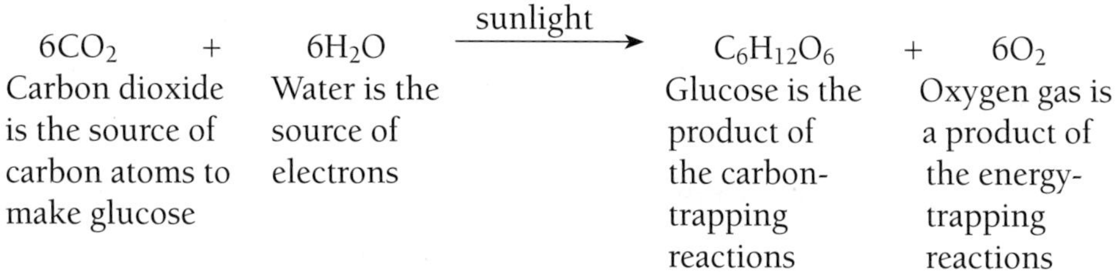
$6CO_2 + 6H_2O \xrightarrow{\text{sunlight}} C_6H_{12}O_6 + 6O_2$
Carbon dioxide is the source of carbon atoms to make glucose
Water is the source of electrons
Glucose is the product of the carbon-trapping reactions
Oxygen gas is a product of the energy-trapping reactions

These glucose molecules can be stored as starch and used later for glycolysis reactions and ATP production.

And so we come to the end of our biochemical journey (whew!). It has been long and at times complicated, but if you have stuck it out, you have a sense of the chemical machinery of microbial life. It is not too difficult to know microbiology, but it is rather difficult to know what makes microbes tick. And that's what this chapter has been all about—the chemical machinery hidden from the images we can see under a microscope or in a photograph.

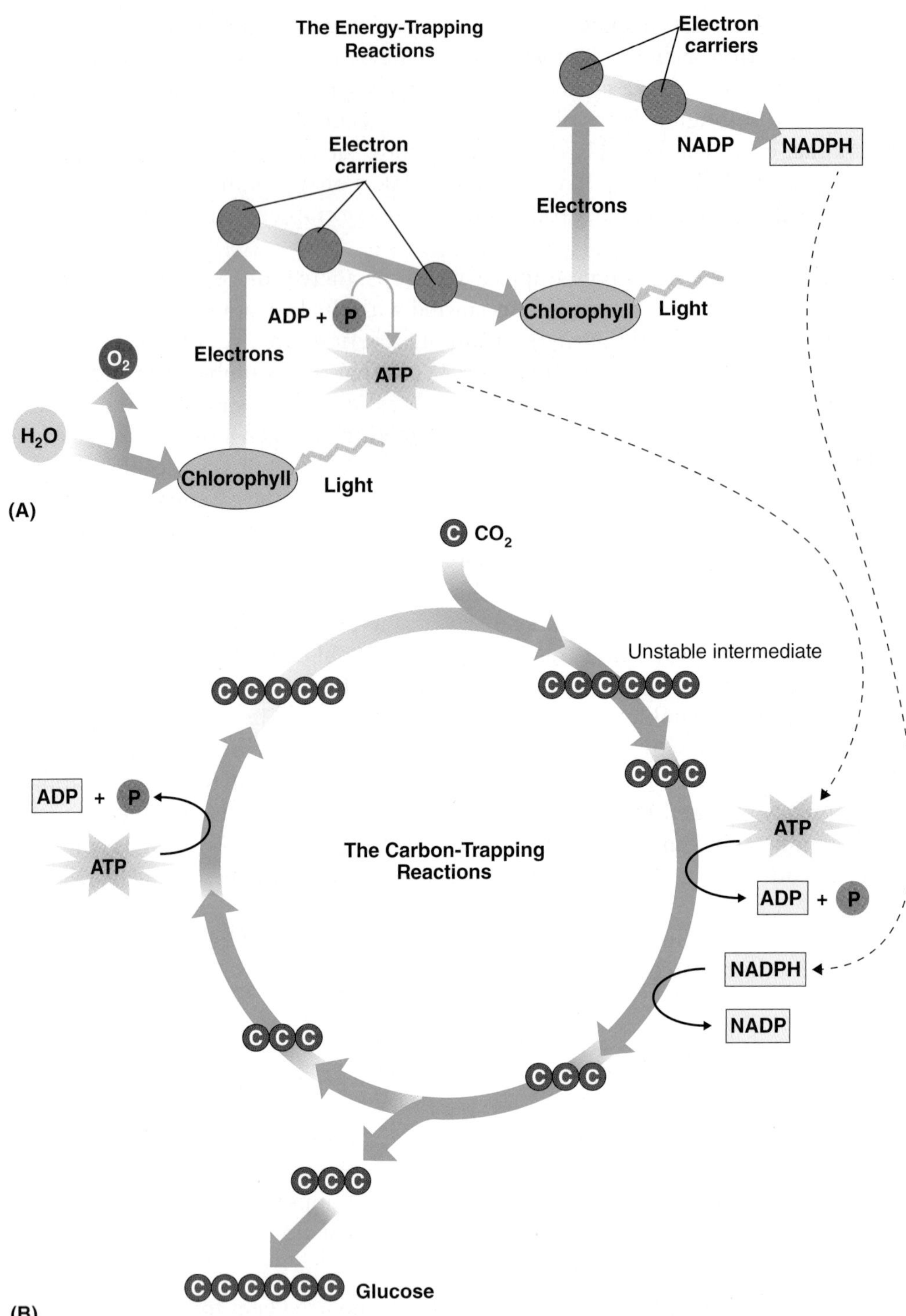

FIGURE 9.13 **Photosynthesis in Microbes.** (**a**) The energy-trapping reactions generate ATP and NADPH. (**b**) The carbon-trapping reactions use carbon dioxide gas (CO_2) along with the ATP and NADPH from the light-trapping reactions to yield glucose and reform the starting five-carbon molecule.

A Final Thought

This chapter contains some of the most complicated concepts you will encounter in microbiology. We have tried to simplify them as best we could. The concepts are also among the most fundamental because this energy metabolism applies to all life—bacterial, plant, or human. We hope you can step back and see the forest as well as the trees. Virtually all living organisms use enzymes for their chemistry and ATP for their energy; many species obtain their energy from glucose metabolism and most of them depend ultimately on the carbohydrates produced in photosynthesis.

Along with these observations comes the realization that certain mechanisms and molecules have been preserved in microbes, plants, and animals virtually intact through the billions years of evolution. This was one of the great revelations of the twentieth century; it revealed the kinship of all living things. Indeed, one of the corollary benefits of studying the metabolism of microbes is that you come away with a better understanding of the biochemistry of all life, including humans. Nowhere is this more apparent than in this chapter. There are certainly variations, but there is also an underlying kinship among all forms of life. To understand one form is to understand them all.

Questions to Consider

1. An organism is described as a facultative, heterotrophic, mesophile. How might you translate this complex microbiological jargon into a description of the organism?
2. During the filming in 1997 of the movie *Titanic,* researchers discovered at least 20 different bacterial and archaeal species literally consuming the ship, especially a rather large piece of the midsection. What type of environmental conditions are these bacterial and archaeal species subjected to at the wreck's depth of 12,600 feet?
3. To prevent decay by bacterial species and to display the mummified remains of ancient Egyptian pharaohs, museum officials often place the mummies in sealed glass cases where oxygen has been replaced with nitrogen gas. Why do you think nitrogen is used?
4. When you (and all humans) exhale, you give off carbon dioxide gas. What reactions in the body are the sources of this gas?
5. One of the most important steps in the evolution of life on Earth was the appearance of certain organisms in which photosynthesis takes place. Why was this critical?
6. A student goes on a college field trip and misses the microbiology exam covering microbial metabolism. Having made prior arrangements with the instructor for a make-up exam, he finds one question on the exam: "Discuss the interrelationships between anabolism and catabolism." How might you have answered this question?
7. ATP is an important energy source in all organisms, and yet it is never added to a microbial growth medium or consumed in vitamin pills or other growth supplements. Why do you think this is so?

Key Terms

Informative facts are necessary for the expression of every concept, and the information for a concept is founded in a set of key terms. The following terms form the basis for the concepts of this chapter. On completing the chapter, you should be able to explain and/or define each one.

acidophile
active site
adenosine triphosphate (ATP)
aerobe
aerobic respiration
anabolism
anaerobe
anaerobic respiration
ATP synthase
barophile
catabolism
cellular respiration
citric acid cycle
coenzyme
cytochrome
electron transport
enzyme
extreme acidophile
extremophile
facultative
fermentation
glycolysis
halophile
hyperthermophile
mesophile
metabolism
photosynthesis
product
psychrophile
psychrotroph
substrate
thermophile

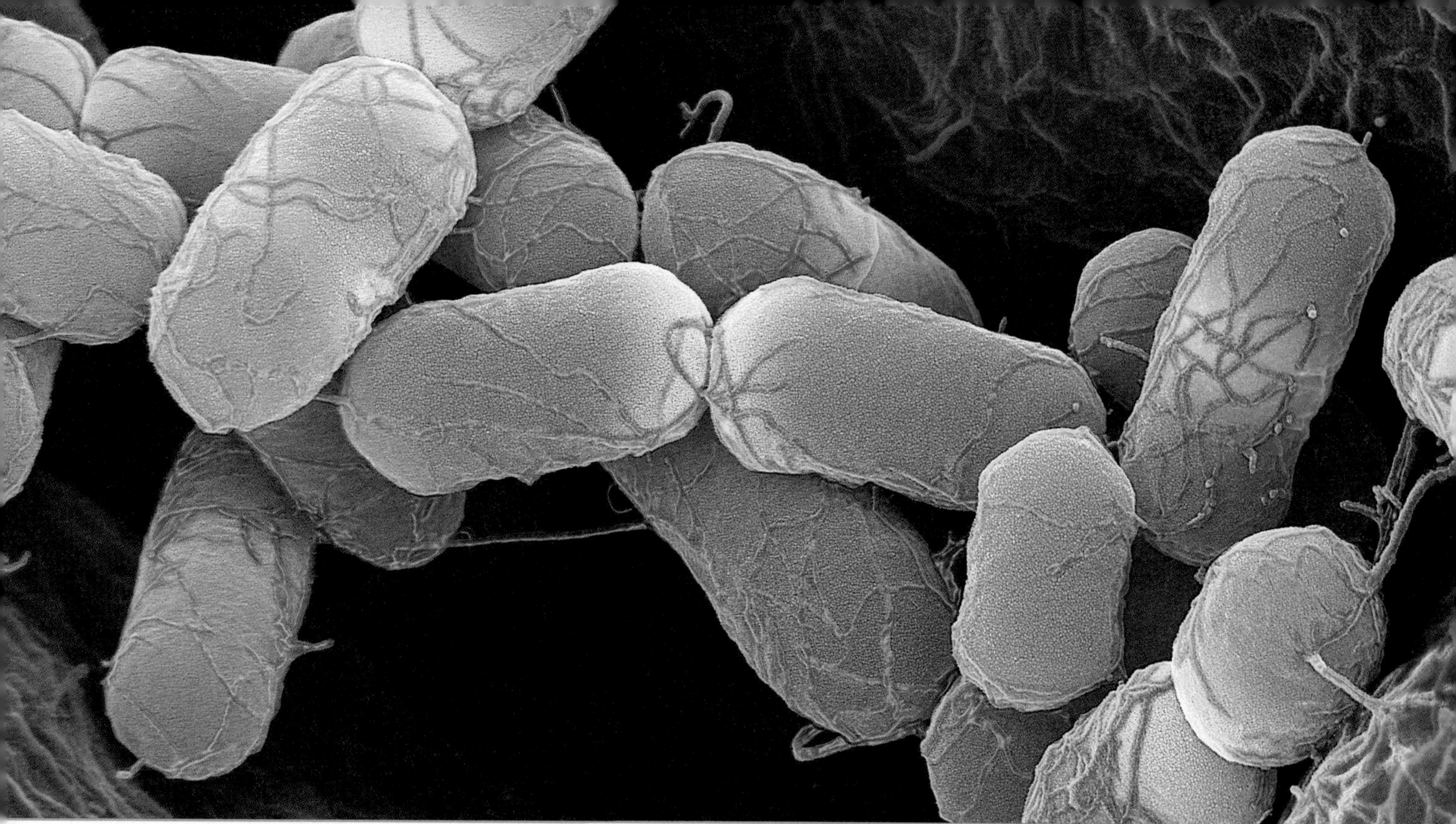

Microbial Genetics: From Genes to Genetic Engineering

10

Looking Ahead

Among the unique characteristics of bacterial organisms is their ability to acquire genes from other organisms. This characteristic is extraordinarily rare in the world of living organisms. Together with genetic changes resulting from mutation, the acquisition of new genes can have profound practical consequences, as we shall discuss in this chapter.

On completing this chapter, you should be capable of:

- Identifying the two types of DNA found in bacterial cells and describing their replication.
- Defining mutations, and distinguishing the various causes of mutations and the role of mutagens.
- Comparing and contrasting the three forms of gene recombination through horizontal gene transfer.
- Recalling the beginnings of genetic engineering and discussing the implications for recombinant DNA molecules in biotechnology.

To understand better the inner workings of genetics, model systems that are genetically simple and grow fast often have been used. That is certainly true for *Escherichia coli* shown in this false-color scanning electron microscope image. Although we often associate this bacterial species with food poisoning, most strains are harmless. In fact, millions of *E. coli* cells live in our gut and help us stay healthy. And scientists use the organism as a model system for all kinds of research, especially in the field of genetics.

In 1969, an extremely serious form of bacterial dysentery broke out in the Central American country of Guatemala. Bacterial dysentery affects the human intestinal tract, resulting in watery diarrhea containing blood or mucus, together with waves of intense abdominal pain and stomach cramps. The disease is caused by various species of *Shigella*, a gram-negative rod; the disease is often called shigellosis. In Guatemala, the outbreak spread

Shigella
shi-gel′lä

quickly, with high infection rates in all age groups but with mortality rates highest in young children.

■ electrolyte A mineral or salt in blood and other body fluids that regulates nerve and muscle function.

During the *Shigella* outbreak, patients were given salt tablets or oral salt solutions to help replace lost water and electrolytes. In addition, various patients also received one of four different antibiotics to kill the bacterial pathogen. As the months passed, however, physicians became increasingly frustrated as they attempted to treat the disease. They discovered that none of the four antibiotics to treat dysentery was effective—neither tetracycline, nor chloramphenicol, nor sulfanilamide, nor streptomycin could eliminate the *Shigella* infection. The pathogen had changed genetically and, without useful antibiotics, physicians lost a major weapon in their fight to stop the epidemic. By the third year of the outbreak, more than 112,000 people had been affected and at least 12,000 died.

The outbreak of shigellosis illustrates what can happen when antibiotic-resistant bacterial pathogens emerge in a population or community. At the time of the Guatemala outbreak, it was unusual to find infectious agents with resistance to one drug, much less four. Now you may say, "Well that was 1968. Things must have changed in the last 46 years in terms of global antibiotic resistance." In fact, they have. The problem of antibiotic resistance is much worse today and represents one of the most pressing issues facing medicine and society on a global scale. Over all these years, antibiotic use has certainly been beneficial and has saved tens of thousands of lives. Their value in patient care has been enormous. However, antibiotics have been used so widely and indiscriminately that the infectious organisms have devised ways to become resistant to the very drugs designed to kill the pathogens. As a result, the Centers for Disease Control and Prevention (CDC) says, "*People infected with antimicrobial-resistant organisms are more likely to have longer, more expensive hospital stays, and may be more likely to die as a result of the infection.*"

In this chapter, we shall learn how bacterial organisms like *Shigella* might have acquired this resistance, as we explore the concept of microbial genetics. We shall begin our study of microbial genetics by exploring the nature of the bacterial chromosome, a DNA molecule that can be altered by a change in its genes or through the acquisition of new genes. Either of these processes could lead to bacterial drug resistance. We shall also study the genetics of viruses because viruses are intimately associated with at least one type of bacterial alteration.

■ *Escherichia coli* esh-ėr-ē-'kē-ä kō'lī (or kō'lē)

Included under the umbrella of microbial genetics is the topic of genetic engineering, one of the most extraordinary technological advances of all time. In fact, the editors of *Time* magazine said, "*Genetic engineering is the most powerful and awesome skill acquired by man since the splitting of the atom.*" The fruits of this technology are remarkable, and we have only begun to see what is possible. In fact, the contributions of microbial genetics as a whole have been numerous, diverse, and far reaching. Model microbial systems, such as *Escherichia coli*, have helped establish the principles of molecular biology and provided the biochemical tools routinely used in genetics and genetic engineering. And yet, despite this profound legacy, the field of microbial genetics continues to provide new insights and tools to better understand genes and heredity.

10.1 Bacterial DNA: Chromosomes and Plasmids

Modern-day bacterial species enjoy the fruits of all the genetic changes their ancestors have undergone over the 3.8 billion years of their existence. Because of their diverse genes, there are bacterial species able to thrive in the snows of the Arctic or the boiling hot vents at the bottom of the oceans. No other organism can compete with bacterial organisms for sheer numbers—a pinch of rich soil has more bacterial cells than all the people living in the United States today. And considering the fantastic multiplication rate of these cells (a new generation in some cases as fast as every half hour), one can easily see how a useful genetic change (such as drug resistance) can be propagated quickly through a bacterial population.

Later in this chapter, we shall study "mutations," one of the processes responsible for the genetic changes that have resulted in the myriad bacterial forms we observe on Earth today. So, let's begin our study by first considering the bacterial chromosome. A quick review on nucleic acids and their components would be helpful to understanding the chemistry of DNA as we progress through this chapter.

The Bacterial Chromosome

Most of the genetic information (**genome**) in a bacterial cell is located within a single chromosome. The chromosome consists of DNA existing as a double helix in a closed loop. This loop structure is unlike the linear (shoestring) form of chromosomes of eukaryotic organisms. Moreover, the bacterial chromosome exists in a region of the cytoplasm called the **nucleoid** that is without any surrounding membrane. Recall that eukaryotic cells have a nuclear envelope surrounding the chromosomes. The bacterial chromosome occupies about half of the total volume of the cell and, extended its full length, measures about 1.5 mm long, about 1,000 times the length of the bacterial cell containing it.

How a 1.5-mm-long chromosome could fit into a 2.0-µm *E. coli* cell was poorly understood until scientists discovered the chromosome takes on a structure having a number of tightly packed loops. To do this, DNA loops, each consisting of about 10,000 bases, attach to one another at anchorage points and form a "flower" structure, as shown in FIGURE 10.1a. The very tight packing of the DNA accounts for its explosive release when the cell membrane and wall are broken (FIGURE 10.1b).

The chromosome of *E. coli* is one of the most intensely studied. Distributed along the chromosome are sites to which genetic activity can be traced. Each site, called a "locus" (*pl.*, loci), consists of one or more genes. The chromosome overall has some 4,300 genes (viruses, by contrast, have as few as 7 genes and a human cell has some 19,000 genes scattered on 23 pairs of chromosomes).

Prior to cell division, the bacterial chromosome must replicate to form another complete molecule so each daughter cell resulting from division will have a full set of genetic instructions. Because it is a closed loop, a unique type of DNA replication occurs (FIGURE 10.2). First, the DNA is anchored to a fixed point on the cell membrane

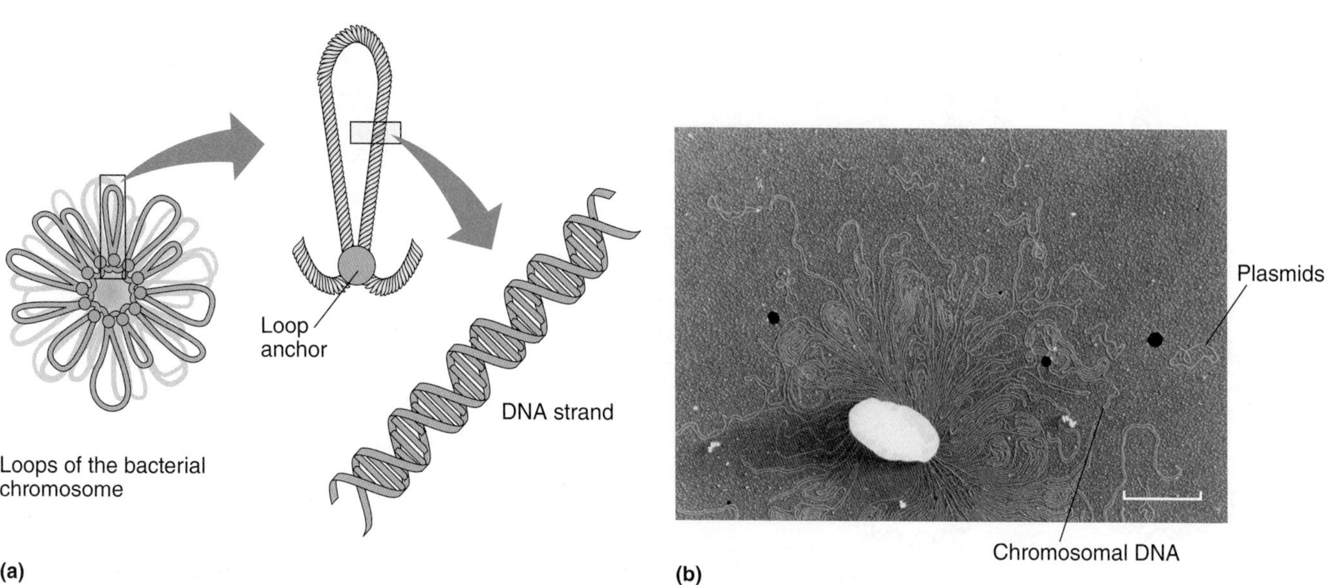

FIGURE 10.1 **Bacterial DNA.** (**a**) The loops in the structure chromosome, viewed head-on. The loops in the DNA help account for the compacting of a large amount of DNA in a relatively small bacterial cell. (**b**) A false-color transmission electron microscope image of an *E. coli* cell immediately after disruption. The tangled mass is the organism's DNA. (Bar = 2 µm.)

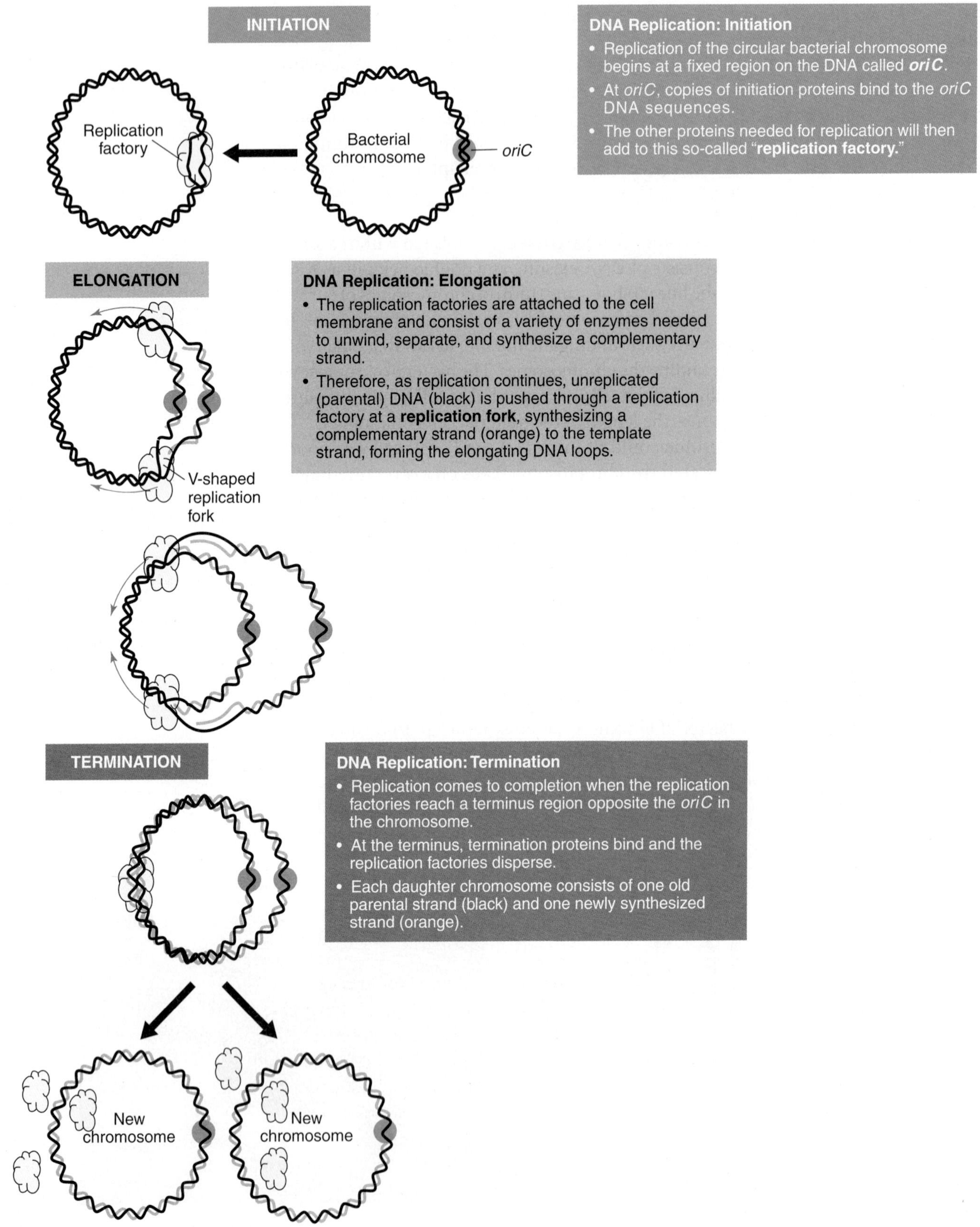

INITIATION

Replication factory

Bacterial chromosome

oriC

DNA Replication: Initiation

- Replication of the circular bacterial chromosome begins at a fixed region on the DNA called ***oriC***.
- At *oriC*, copies of initiation proteins bind to the *oriC* DNA sequences.
- The other proteins needed for replication will then add to this so-called "**replication factory.**"

ELONGATION

V-shaped replication fork

DNA Replication: Elongation

- The replication factories are attached to the cell membrane and consist of a variety of enzymes needed to unwind, separate, and synthesize a complementary strand.
- Therefore, as replication continues, unreplicated (parental) DNA (black) is pushed through a replication factory at a **replication fork**, synthesizing a complementary strand (orange) to the template strand, forming the elongating DNA loops.

TERMINATION

DNA Replication: Termination

- Replication comes to completion when the replication factories reach a terminus region opposite the *oriC* in the chromosome.
- At the terminus, termination proteins bind and the replication factories disperse.
- Each daughter chromosome consists of one old parental strand (black) and one newly synthesized strand (orange).

New chromosome

New chromosome

FIGURE 10.2 **Replication of the *E. coli* Chromosome.** DNA replication involves the addition of complementary bases to the parental (template) strand within the replication forks.

and it is from this point, called the "origin of replication," that DNA replication starts. Enzymes now open and start to unwind and then separate, or "unzip," the two DNA strands. This produces two V-shaped **replication forks**, one running clockwise and the other counterclockwise around the DNA loop.

Next, enzymes belonging to a complex called **DNA polymerase** synthesize new strands of DNA using each old parental strand as a template. Thus, during DNA synthesis, nucleotides complementary to those on the original (template) strand (i.e., adenine is complementary to thymine, and guanine to cytosine) are paired as the replication forks advance.

Finally, when the replication forks reach the opposite side of the loop from where they started, replication ends and the two chromosomes separate from one another. Each new DNA molecule now becomes fully twisted to form a new double helix (i.e., a chromosome).

This method of replication is known as **semiconservative replication** because one strand of parent DNA is conserved in the new DNA, while the other strand is newly synthesized.

Plasmids

In the cytoplasm of many bacterial cells is a number of closed loops of DNA called **plasmids**. Plasmids exist as independent DNA molecules apart from the bacterial chromosome (FIGURE 10.3). They contain about 2% of the total genetic information of some bacterial cells, and they multiply independently of the chromosome. Numerous bacterial species, especially gram-negative ones like *E. coli*, are known to have one or more plasmids, but other than some fungal yeasts, eukaryotic forms of life appear to lack plasmids.

Plasmids are not essential for the everyday existence of a bacterial cell. However, they may confer a selective advantage to the microbe. For example, many plasmids carry genes facilitating resistance to antibiotics. These and other plasmids permit bacterial cells to transfer their genetic material to another cell, as we shall see. By moving from cell to cell, these "traveling plasmids" may have allowed the *Shigella* species to accumulate multidrug resistance.

Plasmids also can contain genes encoding the production of toxic proteins. The proteins, called **bacteriocins**, have a deleterious effect on other bacterial species and give the toxin producer an advantage in the struggle for survival. Bacterial toxins that affect human cells may also be encoded by genes in a plasmid.

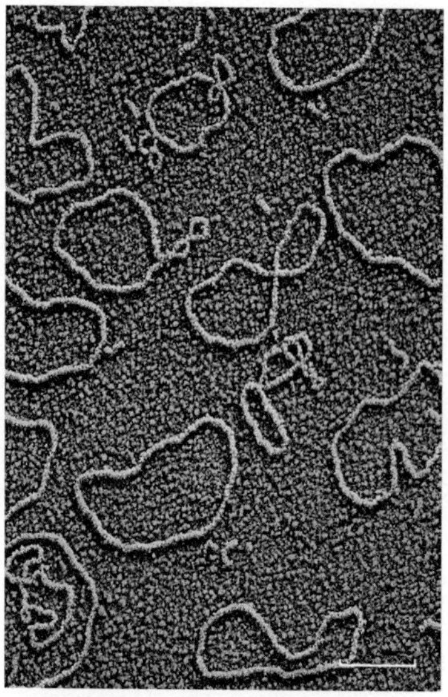

FIGURE 10.3 **Bacterial Plasmids.** A false-color transmission electron microscope image of bacterial plasmids. The plasmids are closed loops of double-stranded DNA. (Bar = 20 nm.)

10.2 Gene Mutations: Subject to "Change Without Notice"

With rare exceptions, humans go through their lives with virtually the same genes; what is inherited from one's parents is essentially what each individual has throughout life. This is not necessarily the case in the prokaryotic world because the genetic information in the bacterial plasmids and chromosome is subject to change. Such changes can occur by two major methods: mutation and recombination. In both cases, the changes can have substantial impact not only on the bacterial organism but on human society as well.

A **mutation** is a permanent change in an organism's DNA. Such a change can involve disruption of the nucleotide sequence in a gene or loss of significant parts of the gene. This action can result in the placement of the wrong amino acids in a protein molecule being formed. Such mutations are most easily noticed when the mutation causes a change in the observable characteristics or **phenotype** of the organism.

A microbe with a normal, nonmutated characteristic is called the "wild type," while an organism with the mutation is called the "mutant." For example, this could be a change in the morphology of the microbe or its nutritional requirements, a bacterium's resistance to an antibiotic or a virus' resistance to an antiviral drug.

Causes of Mutations

In the environment, mutations occur regularly from spontaneous changes to the DNA. Such **spontaneous mutations** are often the result of random DNA replication errors. It has been estimated, for example, that at least one mutation occurs for every billion replications of a bacterial population. Because a population of bacterial cells the size of a pinpoint has well over a billion cells, probability tells us at least one mutant cell should exist in this population. The surviving mutant may then multiply, and a new population of bacterial cells will emerge, all of which have the new characteristic. When we consider how fast some bacterial species can replicate, we begin to understand how an antibiotic-resistant population can emerge in a relatively brief period.

Induced mutations are the result of exposure to **mutagens**, physical agents or chemical substances capable of inducing mutations in cells. Ultraviolet (UV) light, a component of sunlight, is a known physical mutagen. When soil-dwelling bacterial cells absorb UV light, thymine and cytosine nucleobases next to one another in the same strand of DNA bind together (FIGURE 10.4). With these bases joined in unnatural linkages, DNA replication and/or transcription is blocked. Incidentally, UV light is often used in a germicidal lamp for disinfection purposes because it quickly kills bacterial cells by altering their DNA.

Many chemicals in the environment may also act as mutagens. Aflatoxin, a chemical produced by a mold growing on peanuts and grain, and benzopyrene, a component of industrial smoke, cause the deletion (loss) of nucleotides from, or

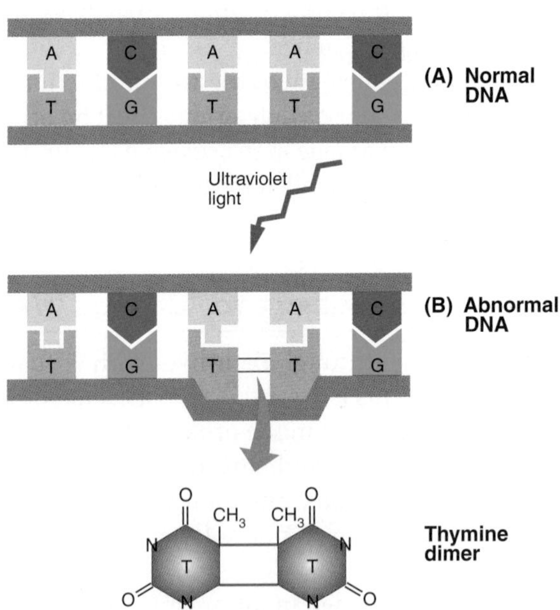

FIGURE 10.4 **Ultraviolet Light and DNA.** (**a**) When cells are irradiated with ultraviolet (UV) light either naturally or through experiment, the radiation may affect the cell's DNA. (**b**) UV light can cause adjacent thymine molecules to pair within the DNA strand to form a thymine dimer.

insertion (addition) of extra nucleotides into, the DNA during replication. With missing or additional nucleotides, the entire gene sequence changes, producing a different genetic message. Because the genetic message encodes the amino acid sequence for a protein, the protein, if produced, will be flawed in some way. Nitrous acid is another chemical mutagen capable of causing an incorrect pairing of bases in the DNA (FIGURE 10.5).

As mentioned, mutations can be the result of a gain or loss of nucleotides. Much smaller mutations, where one nucleotide is substituted for another, are called **point mutations**. Often these mutations are said to be "neutral mutations" because the change in no way affects the structure or function of the protein made. A "missense mutation" occurs if the substitution creates a faulty or less functional protein or a protein with a new function. Point mutations can also result in a "nonsense mutation" where the change has actually stopped the production of the protein.

You might ask, "Could a bacterial cell, such as in the *Shigella* case in the chapter opener, become resistant to an antibiotic through a mutation?" Great question. And the answer is yes. Many antibiotics kill bacterial cells by disrupting a critical cell function, often by binding to a protein so the protein cannot function properly. However, if the bacterial cell has a "beneficial" mutation, such that the protein is still functional but the antibiotic cannot now bind to the altered protein, the mutant bacterial cell survives and its reproduction generates a whole population of resistant cells. This is one scenario for the *Shigella* example. The general process of antibiotic resistance is discussed in another chapter.

Another possible result of mutation is discussed in **A Closer Look 10.1**.

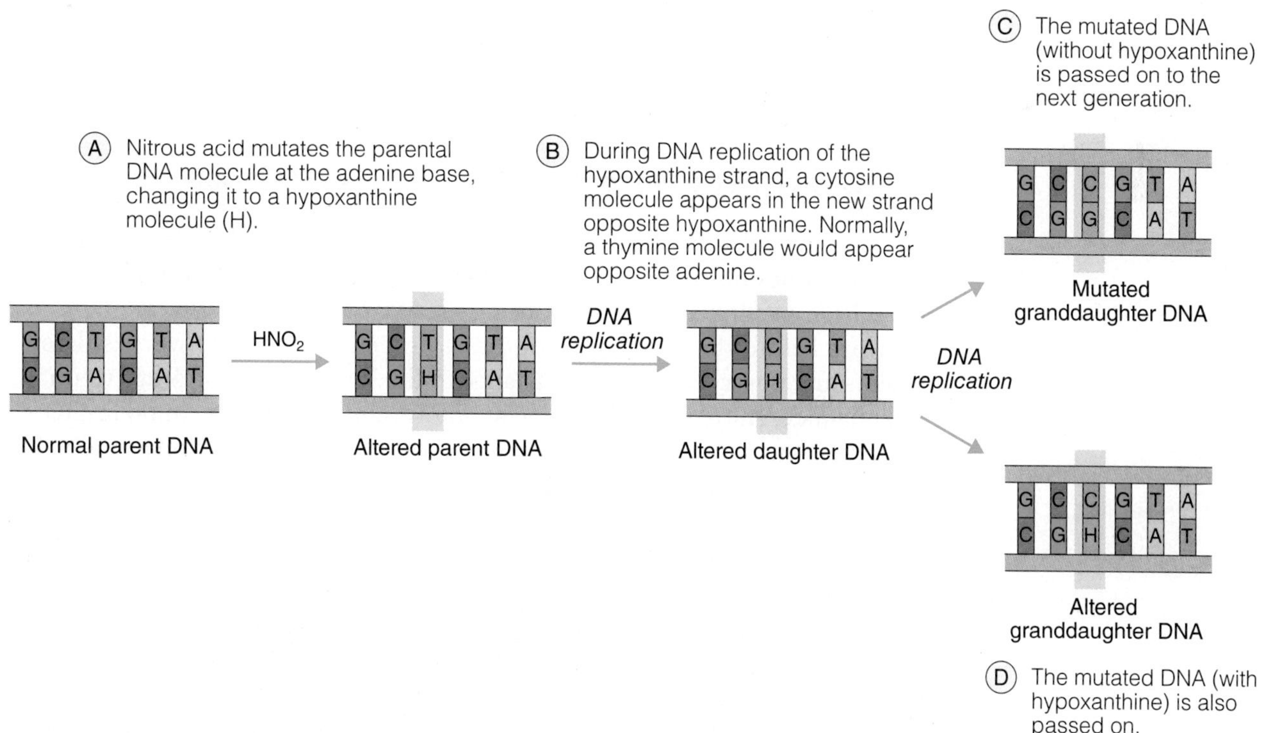

FIGURE 10.5 **How Nitrous Acid Causes Bacterial Mutations.** Nitrous acid induces an adenine to hypoxanthine change. After replication of the hypoxanthine-containing strand, the daughter DNA has a mutated C–G base pair.

A CLOSER LOOK 10.1

Three Genes

Could the Black Death of the 1300s have resulted from three defective genes? Could 25 million Europeans have succumbed to plague because of those three genes? Could the entire course of Western civilization have changed based on three genes?

Possibly so, maintain researchers from the federally-funded Rocky Mountain laboratory in Montana. A research group led by Joseph Hinnebush has reported that three genes in the plague bacillus are absent in a harmless form of the organism. And it is possible that the entire story of plague's pathogenicity revolves around these three genes.

The scenario goes like this: Bubonic, septicemic, and pneumonic forms of the plague are caused by *Yersinia pestis*, a rod-shaped bacterium transmitted by the rat flea (see figure). When the flea is infected, the bacterial cells amass in its foregut and obstruct its gastrointestinal tract. Soon the flea is starving, and it uncontrollably starts biting humans and rodents, feeding on their blood. During the bite, the flea regurgitates the mass of bacterial cells into the victim's bloodstream, which spreads the bacterial cells through the body.

The three genes enter the picture at the very beginning. The harmless plague bacilli have genes encouraging the microbes to remain in the midgut of the flea. Pathogenic plague bacilli, by contrast, do not have the genes, allowing the bacterial cells to migrate to the foregut of the flea and form a plug of packed bacilli; these are the organisms that are passed on to the next plague victim.

Sometimes it is dangerous to oversimplify matters, and this may be one of those times. Still, scientists are inclined to reduce concepts to their least common denominators. And if the tragic Black Death reduces to three genes, then so be it.

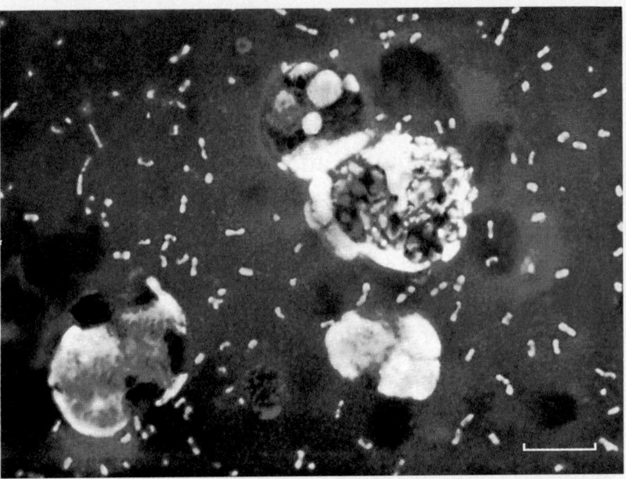

Stained cells (yellow) of *Yersinia pestis* cells in the blood. (Bar = 10 um.)

■ *Yersinia pestis*
yėr-sin′-ē-ä pes′-tis

10.3 Genetic Recombination: Remixing Genes

One of the horrific aspects of the Guatemala epidemic was *Shigella's* resistance to all four antibiotics normally used to treat bacterial dysentery. How a bacterial species might have accumulated these numerous resistances is as follows.

Years before the Guatemala outbreak, species of *Shigella* were isolated from patients and observed to have resistance to the same quartet of drugs (tetracycline, sulfanilamide, chloramphenicol, and streptomycin) as would be found in Guatemala. The patients with the drug-resistant *Shigella* also had in their colon a strain of *E. coli* with resistance to the same four drugs. Because identical mutations resulting in the same four resistances in two different bacterial species is about as likely as winning the lottery twice in the same week, the researchers believed resistance to the four drugs must have been transferred from *E. coli* to *Shigella*.

Investigators discovered that when they mixed together liquid suspensions of drug-resistant *E. coli* and drug-sensitive *Shigella*, within a few days the *Shigella* cells, once sensitive to the four antibiotics, were now resistant to the same four drugs as the *E. coli* cells. The transfer of drug resistance from *E. coli* to *Shigella* had taken place.

In the following decades, investigators realized that gene transfers can occur between bacterial cells of the same species as well as between different bacterial species. In fact, today we know this production of new combinations of genes, called **genetic recombination**, is rampant. For example, biochemists have calculated that *E. coli* has acquired nearly 20% of its genome from other bacterial species. A nonpathogenic species, *Thermotoga maritima*, has acquired about 25% of its genome from other

■ *Thermotoga maritima*
thėr′-mō-tō-gä mar-i-tē′-mä

prokaryotes in both the Bacteria and Archaea domains. The surprisingly large extent of gene swapping and gene acquisition accounts in large measure for the problem of antibiotic resistance, as we shall see in the following paragraphs.

Unlike the normal vertical transfer of genetic information through binary fission, genetic recombination in bacteria occurs through a mechanism called **horizontal gene transfer**, the direct intercellular transfer of DNA from a donor cell to a recipient cell. Bacterial species have three ways of accomplishing this type of genetic recombination: conjugation, transduction, and transformation (FIGURE 10.6). Let's take a look at each.

Conjugation

In the recombination process called **conjugation**, two live bacterial cells come together and the donor cell transfers some of its genetic material to a recipient cell (FIGURE 10.7). Specifically, the donor is known as an F^+ cell because it has an **F factor** (plasmid) containing about 20 genes mostly associated with conjugation. The recipient is known as an F^- cell because it lacks an F factor.

In gram-negative species, like *E. coli*, the process of conjugation begins with the F^+ cell forming a **conjugation pilus** that contacts the recipient cell. The pilus shortens to bring the two cells close together and a conjugation bridge connects the two cytoplasms.

Once connected, an enzyme nicks one strand of the DNA of the F factor and the single DNA strand of the plasmid passes through the bridge to the recipient cell. When it arrives in the recipient cell, enzymes synthesize a complementary strand of DNA and a double helix forms. The double helix bends to form a loop and becomes a plasmid, thereby completing the conversion of the recipient cell to an F^+ cell. Meanwhile, back in the donor cell, a complementary new strand of DNA forms and unites with the leftover strand of the original F factor.

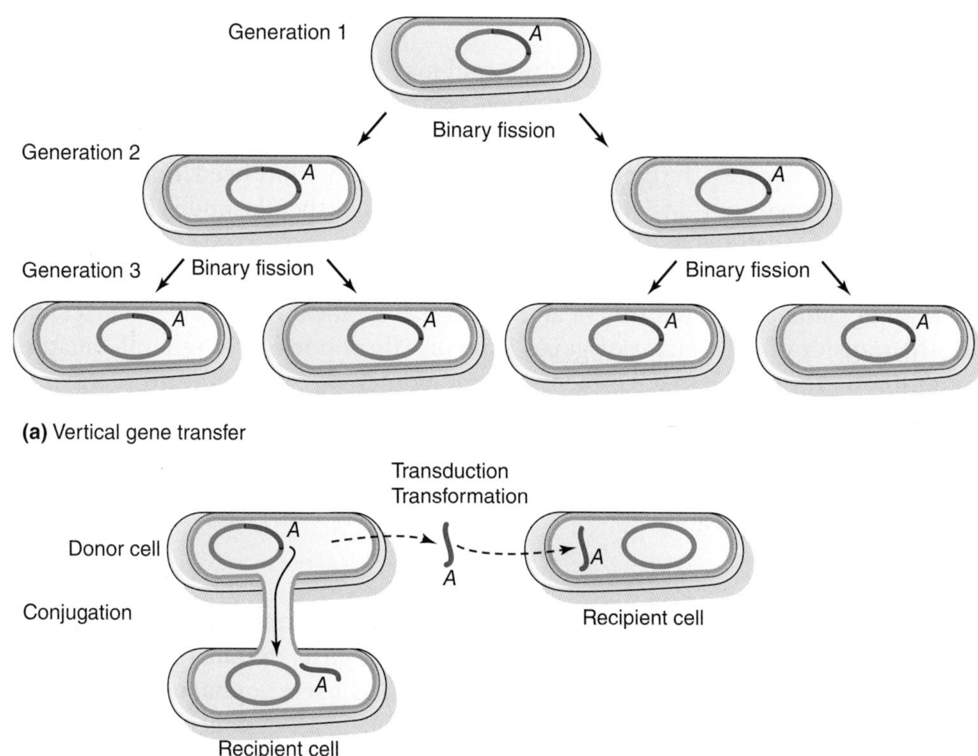

FIGURE 10.6 **Genetic Recombination.** Unlike vertical gene transfer (**a**), genetic recombination (**b**) involves horizontal gene transfer through conjugation, transduction, or transformation.

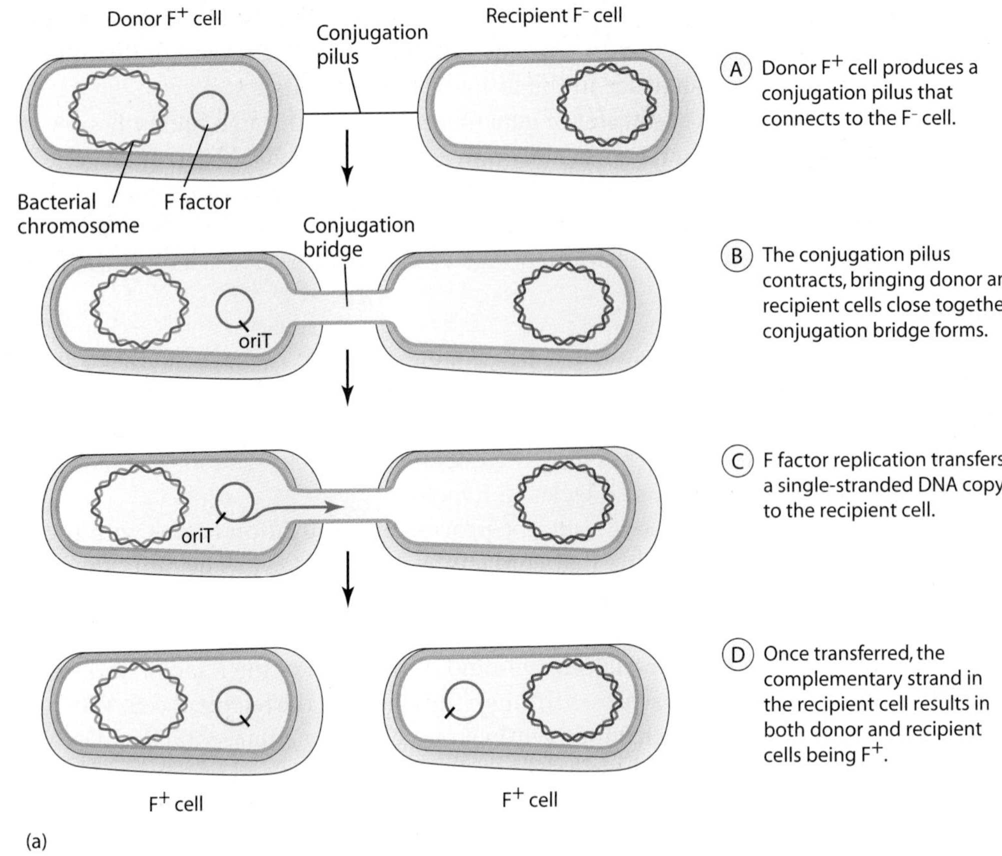

Donor F+ cell
Conjugation pilus
Recipient F- cell
A Donor F+ cell produces a conjugation pilus that connects to the F- cell.
Bacterial chromosome
F factor
Conjugation bridge
oriT
B The conjugation pilus contracts, bringing donor and recipient cells close together; conjugation bridge forms.
oriT
C F factor replication transfers a single-stranded DNA copy to the recipient cell.
D Once transferred, the complementary strand in the recipient cell results in both donor and recipient cells being F+.
F+ cell
F+ cell
(a)

FIGURE 10.7 Bacterial Conjugation. When the F factor (plasmid) is transferred from a donor (F+) cell to a recipient (F-) cell, the F- cell becomes an F+ cell as a result of acquiring a copy of the F factor.

In another form of conjugation, a plasmid may become integrated into the bacterial chromosome of the donor cell. Such "high-frequency recombination" (Hfr) donors during conjugation with an F- cell begin to transfer some chromosomal genes to the recipient. Because the conjugation bridge is usually short lived, not all the host chromosome genes are transferred and, in fact, some of the plasmid genes may not be transferred as well.

■ *Salmonella* säl-mōn-el′-lä

Conjugation has been observed in numerous genera of bacteria, including *Shigella*, *Escherichia*, *Salmonella*, and others and is of great medical importance. For example, the transfer of an F factor or genes from an Hfr donor containing information for antibiotic resistance to a recipient cell can confer that resistance to the recipient. In fact, resistance to multiple antibiotics can be gained in this way. Add to this that other types of plasmids or Hfr donor genes may confer new pathogenic factors (e.g., toxins, enzymes) to recipients, conjugation can readily increase the pathogenicity of a bacterial population.

Transduction

A second recombination process involving horizontal gene transfer is **transduction**. This mechanism involves the assistance of a bacterial virus to transfer the bacterial DNA to a recipient cell. Bacterial viruses are known as **bacteriophages** (**phages**, for short). In the replication cycle of many phages, the phage DNA penetrates the bacterial host cell and immediately starts the replication of new phages. When this happens, the bacterial cell serves as a biochemical factory for the production of hundreds or thousands of new phages, and during the virus replication process the bacterial chromosome is cut into many small fragments. Once the viruses are formed, they are released from the cell, which is destroyed in the process.

Sometimes during the virus replication process, some newly forming phages may mistakenly contain a bacterial gene along with an incomplete set of viral genes or, as shown in (FIGURE 10.8), the virus contains only a few fragments of chromosomal DNA from the bacterial cytoplasm rather than viral DNA. In either case, these are called

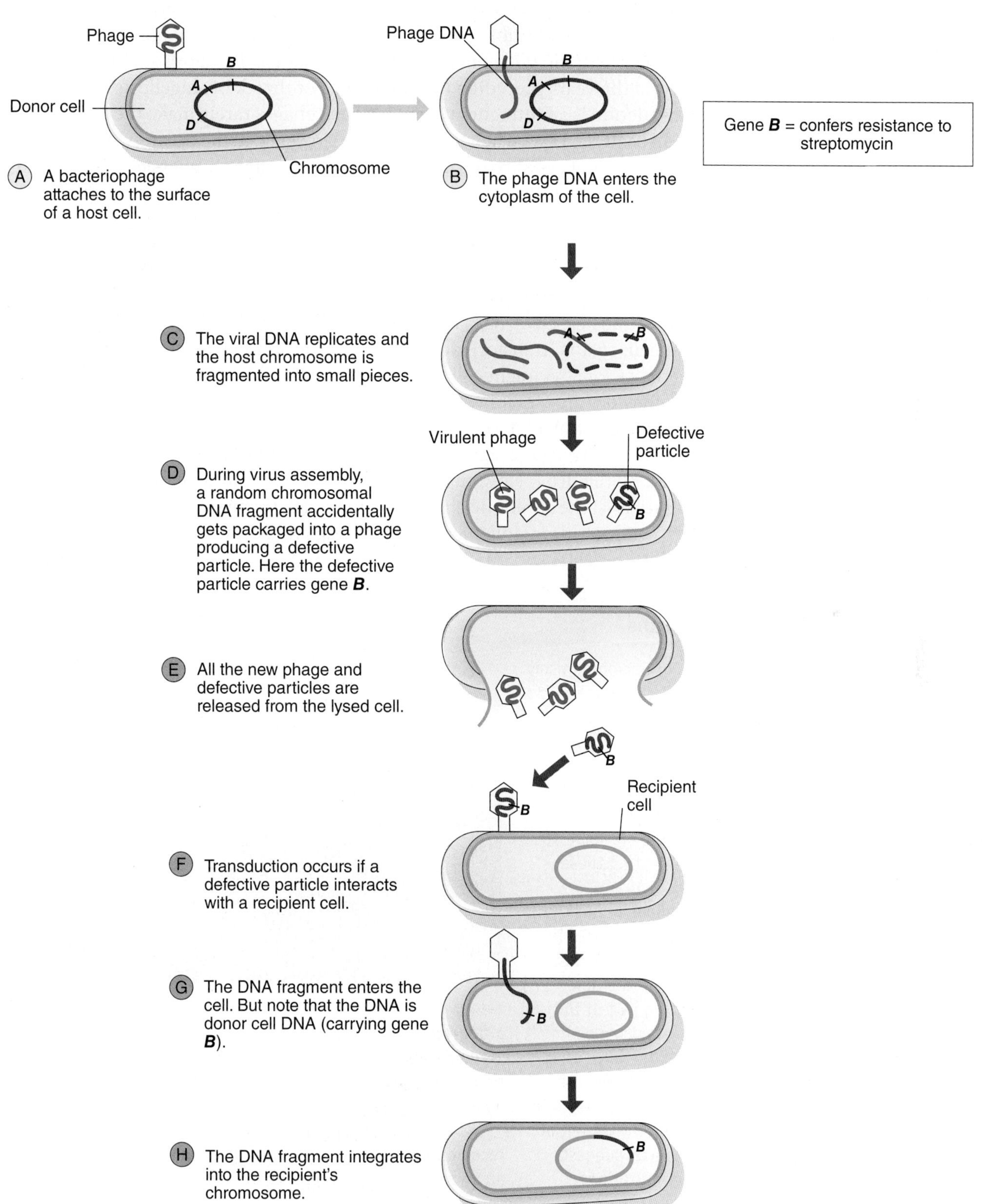

Phage
Donor cell
B
A
D
Chromosome
A A bacteriophage attaches to the surface of a host cell.
Phage DNA
B The phage DNA enters the cytoplasm of the cell.
Gene B = confers resistance to streptomycin
C The viral DNA replicates and the host chromosome is fragmented into small pieces.
Virulent phage
Defective particle
D During virus assembly, a random chromosomal DNA fragment accidentally gets packaged into a phage producing a defective particle. Here the defective particle carries gene B.
E All the new phage and defective particles are released from the lysed cell.
Recipient cell
F Transduction occurs if a defective particle interacts with a recipient cell.
G The DNA fragment enters the cell. But note that the DNA is donor cell DNA (carrying gene B).
H The DNA fragment integrates into the recipient's chromosome.
Transduced cell

FIGURE 10.8 **Transduction.** Defective phages (particles) can transfer a few bacterial genes to a recipient cell.

defective particles because they do not carry a complete set of viral genes although they are still capable of infecting another bacterial cell. After release of the viruses, if one of the defective particles enters another bacterial cell, the bacterial DNA (and any viral DNA) may integrate into the recipient cell's chromosome.

Looking again at Figure 10.8, let's suppose the defective particle acquired a gene for resistance to the antibiotic streptomycin. When the DNA is inserted into the recipient's chromosome, the antibiotic resistance gene will now be a permanent part of the recipient's genome and will confer antibiotic resistance. Perhaps something like this happened in the years preceding the Guatemala outbreak of shigellosis.

Transduction is an infrequent event because phages rarely make the "mistake" of picking up bacterial DNA when they replicate. However, in the oceans where there are unimaginable numbers of bacterial cells and even greater numbers of phages, transduction can be a significant form of genetic recombination, as **A Closer Look 10.2** points out.

Transformation

The third form of genetic recombination is **transformation**. During this form of horizontal gene transfer, DNA released from a dead bacterial cell is acquired by a living recipient cell; that is, transformation is another form of acquisition of new genetic characteristics for a recipient cell.

Let's consider an example of how transformation can occur. A patient has been treated for an infection of the colon caused by a pathogenic strain of *E. coli* (e.g., *E. coli* O157:H7). The physician used an antibiotic called gentamicin and the bacterial cells were killed by the drug. Now suppose this strain of *E. coli* had a gene enabling the microbes to resist the antibiotic chloramphenicol (the gene would have protected the microbes against chloramphenicol if the physician had prescribed that antibiotic).

A CLOSER LOOK 10.2

Gene Swapping in the World's Oceans

Many of us are familiar with the accounts of microorganisms in and around us, but we are less familiar with the massive numbers of microbes in the world's oceans. For example, microbial ecologists estimate there are an estimated 10^{29} bacterial and archaeal cells in the world's oceans. At the Axial Seamount, a Pacific deep-sea volcano, researchers have discovered some 3,000 archaeal species and more than 37,000 different bacterial species. Also, there are some 10^{30} bacteriophages in the oceans capable of infecting these oceanic microbes.

As described in the narrative, during phage replication, sometimes phages by mistake carry pieces of the bacterial chromosome (rather than viral DNA) from the infected cell to another recipient cell. In the recipient cell, the new DNA fragment can be swapped for an existing part of the recipient's chromosome. It is a fairly rare event happening only once in every 100 million (10^8) virus infections. That doesn't seem very significant until you now consider the number of phages and susceptible bacterial cells existing in the oceans. Working with these numbers and the potential number of virus infections, microbial ecologists suggest that if only one in every 100 million infections brings a fragment of DNA to a recipient cell, there are about 10 million billion (that's 10,000,000,000,000,000 or 10^{16}) such gene transfers per second in the world's oceans or about 10^{21} infections per day!

We do not understand what all this genetic recombination means. What we can conclude is there's an awful lot of gene swapping going on!

Courtesy of Dr. Jeffrey Pommerville.

However, the physician chose gentamicin and the bacterial cells were killed. Cell killing though resulted in fragments of the *E. coli* DNA being strewn about in the intestinal tract of the patient.

Now, through drinking contaminated water, a few *Shigella* cells enter the patient's intestine. The *Shigella* cells have the ability to acquire fragments of the *E. coli* DNA present in the nearby environment. So, the *Shigella* cells pick up the gene for resistance to chloramphenicol. They then incorporate the new DNA into their own DNA and become transformed and are now resistant to chloramphenicol, as (FIGURE 10.9) illustrates. Next, they pass out of the intestine in the feces and accumulate in soil or water, where they reproduce to a sizable bacterial population, all of which have resistance to chloramphenicol.

Over the course of years, decades, and generations, cells of this *Shigella* species could conceivably acquire resistance to numerous antibiotics and become a horrific sleeping giant in the environment. Perhaps, during that fateful period of 1968, such a sleeping giant found its way to thousands of patients in Guatemala, where it caused misery and death.

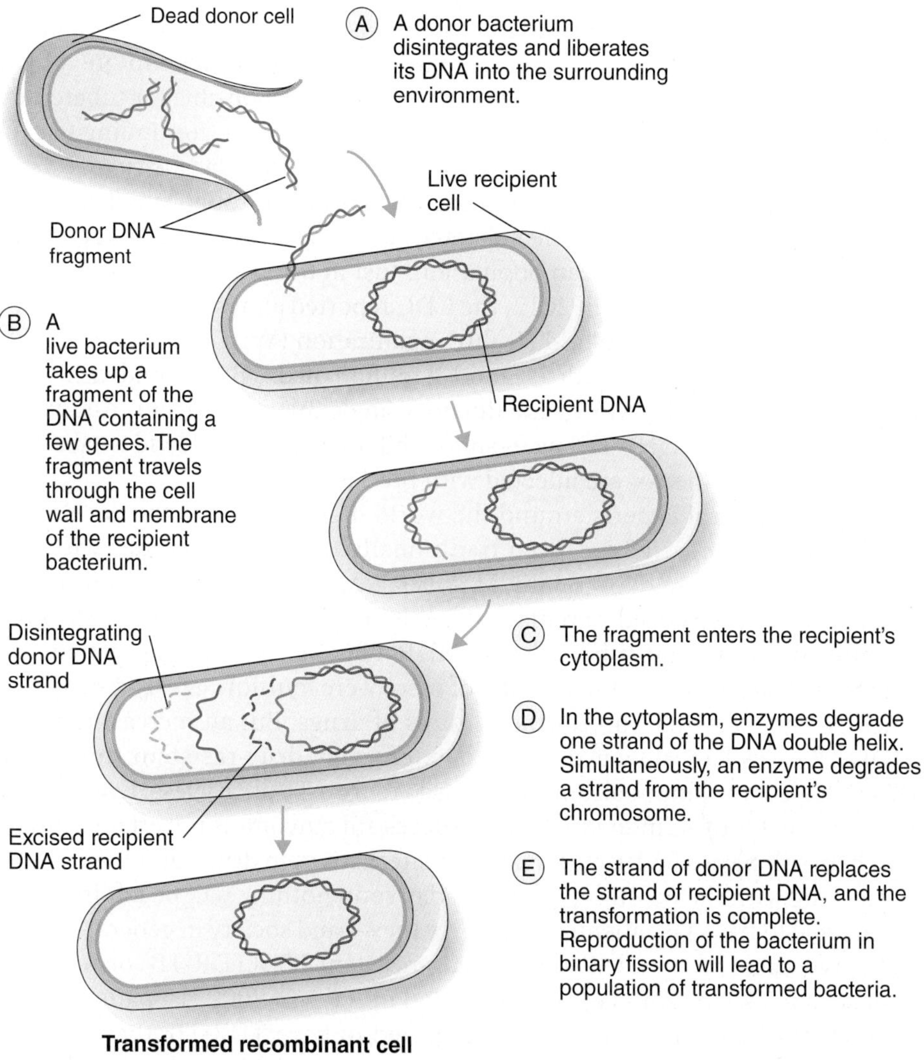

FIGURE 10.9 **Bacterial Transformation.** Transformation is the process in which DNA from the environment binds to a live recipient cell, passes into the recipient, and incorporates into the recipient's chromosome.

Transposons

Another possible form of genetic transfer involves movable genetic elements called **transposons**. Transposons are small segments of DNA that have the ability to move from one position to another in the bacterial chromosome and have been termed "jumping genes." The segments simply jump from one chromosomal site to another, from a chromosome to a plasmid, or even from a plasmid to a chromosome. Thus, transposons may have an adverse effect by interrupting the genetic coding sequence of a gene, causing the DNA to encode an incorrect protein or, in some cases, no protein at all.

Because transposons can be found in plasmids, they also can be transferred to a recipient cell through something like conjugation. Therefore, if a transposon in a plasmid carries information for antibiotic resistance, plasmid transfer will carry the transposon to the recipient and confer to that cell the antibiotic resistance. There, the transposon may then jump to the recipient's chromosome. So, there is yet another example of genetic recombination and the mixing of genes.

In Today's World

Now we fast-forward to today's world. Antibiotic resistance has become one of the most serious problems confronting microbiologists and the medical community. An alarming number of bacterial species have evolved resistance to one or more antibiotics. High on the list of concerns is the microbe responsible for tuberculosis.

■ *Mycobacterium tuberculosis* mī-kō-bak-ti′rē-um tü-bėr-kū-lō′-sis

Tuberculosis (TB) has been with us for thousands of years, continuing to evolve and resist our best drugs. The disease is caused by *Mycobacterium tuberculosis*, a small, aerobic, nonmotile rod that spreads from person to person through air. (The specifics of the disease are discussed in another chapter.) Over the centuries, TB has continued to be a "slate wiper" in the human population. Just in the last two centuries it killed an estimated 1 billion people. In 2012, the CDC reported almost 10,000 new cases in the United States, while the World Health Organization (WHO) reported that globally 8.7 million people fell ill with TB and 1.4 million died. TB is second only to HIV/AIDS as the greatest killer worldwide due to a single infectious agent. And perhaps most frighteningly, the WHO estimates some 2 billion people worldwide—about 30% of the world's population— are infected with the TB bacillus.

What has medical experts around the world most worried is the development of antibiotic resistance. TB has been traditionally treated for decades with drugs able to cure the disease. Now, many *M. tuberculosis* strains are resistant to these drugs and other drugs used to replace the original ones. Such patients are said to have **multidrug-resistant tuberculosis (MDR-TB)**. In 2010, among patients who had previously been treated for TB, 20% of cases were multidrug resistant. MDR-TB has necessitated a switch to yet other groups of drugs. But an increasing number of MDR-TB cases have now evolved into **extensively drug-resistant tuberculosis (XDR-TB)**, meaning almost all antibiotics used to treat TB are useless. Few treatment options remain for these individuals and a successful outcome is uncertain. Although researchers still have a few useful treatment alternatives in drug combinations, they are grappling with the possibility that one day soon nothing will be left in the antibiotic arsenal to treat TB patients. And worry they—and society in general—should. In 2007, two cases of so-called "totally drug-resistant TB" (TDR-TB) or "extremely drug-resistant TB" (XXDR-TB) were reported in Italy. In 2009, 15 patients in Iran and in 2012, 12 patients in India were reported to have TDR-TB; these strains of the pathogen were completely resistant to all known available antibiotics. This (and all antibiotic resistance) is due to mutations and genetic recombination through horizontal gene transfer.

10.4 Genetic Engineering: Deliberate Transformation

If you have read the previous sections, you know microbes, through the processes of mutation and genetic recombination, can modify their genetic material. The problem of multidrug resistant species like *Shigella* and *M. tuberculosis*, is but one of the many effects of genetic alterations.

However, these negative aspects of genetic recombination are far outweighed by the positive practical aspects it offers for medicine and society. Knowledge of microbial genetics helps scientists understand antibiotic resistance, but it also makes available a technology few could have imagined a generation ago. By applying the principles of microbial genetics in the laboratory, scientists found they could control mutation and genetic recombination almost at will. During the 1970s, researchers discovered they could alter bacterial DNA and mimic the processes of nature, especially transformation. Soon, they were cutting and splicing DNA, removing and inserting genes, and opening new vistas of pure and applied research. In this concluding section, we will explore how genetic engineering emerged from studies in microbial genetics. Other chapters describe the fabulous fruits of this technology.

The Beginnings of Genetic Engineering

Genetic engineering changes the genetic makeup of a cell or an organism by removing genes or by introducing new DNA prepared in the laboratory. This ability to shuffle genes artificially was born in the early 1970s.

Until the early 1970s, scientists lacked the biochemical tools to transfer genes and DNA between cells. It was at this time they discovered and isolated a group of bacterial enzymes called **restriction endonucleases** capable of cutting the DNA molecule at specific (restricted) points into small pieces or fragments (FIGURE 10.10a). Importantly, scientists could use these "biochemical scissors" to cut a bacterial chromosome at specific points into small fragments. Today, genetic engineers have hundreds of such endonucleases, each able to cut at different DNA sequences (FIGURE 10.10b).

Scientists could now use these enzymes to cut DNA molecules on demand, regardless of the source of the DNA. For example, enzyme X would cleave a DNA molecule at sequence X, regardless of whether the DNA was from a plant, animal, bacterium, or virus. Knowing that bacterial cells could thrive even though they had foreign DNA, scientists were now ready to use restriction endonucleases to alter DNA in a test tube.

The First Recombinant DNA Molecule

In 1971, scientists used an endonuclease to cut open the DNA molecule from a virus called SV40. Then, they spliced the viral DNA into an *E. coli* chromosome that also had been cut open with the same endonuclease (FIGURE 10.11). The cutting of the DNA molecules produced "**sticky ends**," which represent DNA bases able to complementary pair when brought together. In doing so, they constructed the first **recombinant DNA molecule**; that is, a hybrid DNA molecule containing DNA from two different sources.

Using Recombinant DNA Molecules

Scientists soon discovered how to artificially transfer plasmids into bacterial cells, mimicking the natural process of transformation. Now, they could move foreign genes into new organisms.

The first successful transfer used plasmid-derived recombinant DNA molecules using *E. coli* plasmids. After the double-stranded molecules were cut open at one point with an endonuclease, a gene from a species of African toad was inserted. It encoded a protein found in the ribosomes. The scientists then inserted these recombinant DNA molecules (plasmids) into *E. coli* cells growing

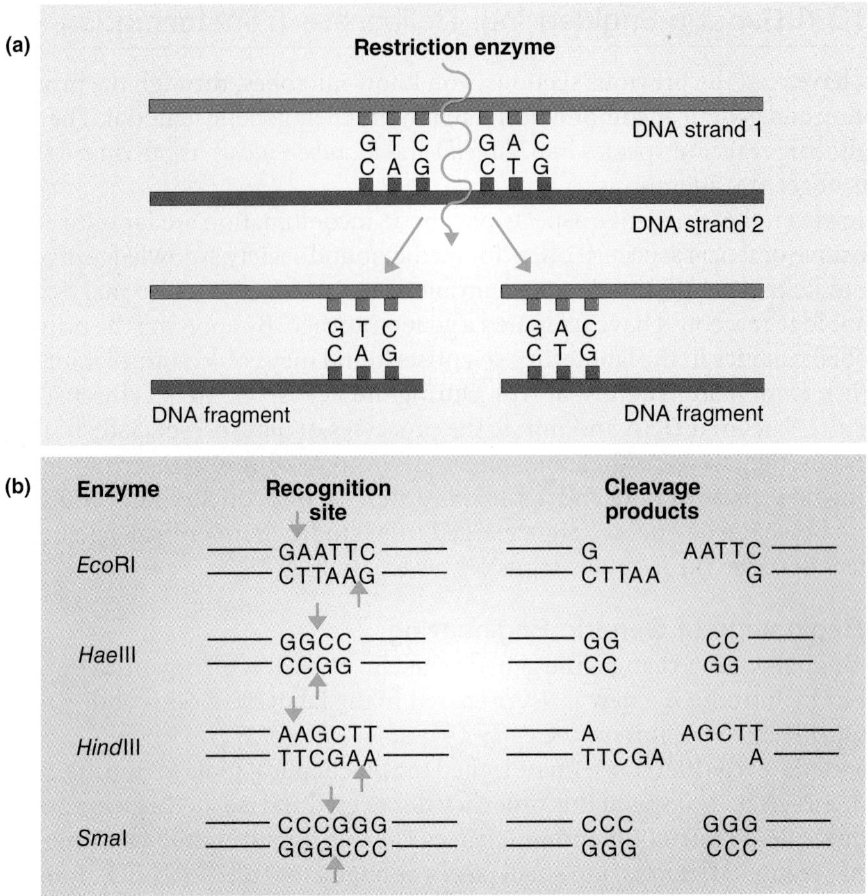

FIGURE 10.10 **Restriction Endonucleases.** (**a**) A restriction endonuclease is an enzyme that cuts through two strands of a DNA molecule at a specific site to produce two fragments. (**b**) The recognition sites of several such endonucleases. Note that *Eco*RI and *Hind*III leave dangling ends in the cleavage products. Dangling ends ("sticky ends") are desirable because such fragments can attach to other fragments more easily.

in broth. The plasmids started replicating inside the bacterial cells within minutes, indicating the bacterial cells could act as tiny factories presumably producing the toad protein. When they did a biochemical analysis of the proteins in the broth, they found a new protein was present; it was the ribosome protein of the toad. A successful transfer of vertebrate genes to a bacterial species (using a plasmid) had been accomplished—a procedure mimicking natural genetic recombination (horizontal gene transfer)—specifically, the transformation process. It was the beginning of the era of genetic engineering.

The Implications

Once these pioneering experiments were reported, molecular biologists were quick to see the implications of genetic engineering. If they could find the appropriate human gene, they could integrate it into a bacterial plasmid, and insert the plasmids into bacterial cells. The cells would then produce and secrete the desired protein. Depending on the gene selected, the genetic engineering process could be used to produce valuable commercial products and treat various medical conditions. The first successful commercial/medical product for human use was insulin, the protein responsible for regulating the amount of glucose in the blood. A lack of insulin can lead to diabetes. **A Closer Look 10.3** describes how the engineering achievement was accomplished.

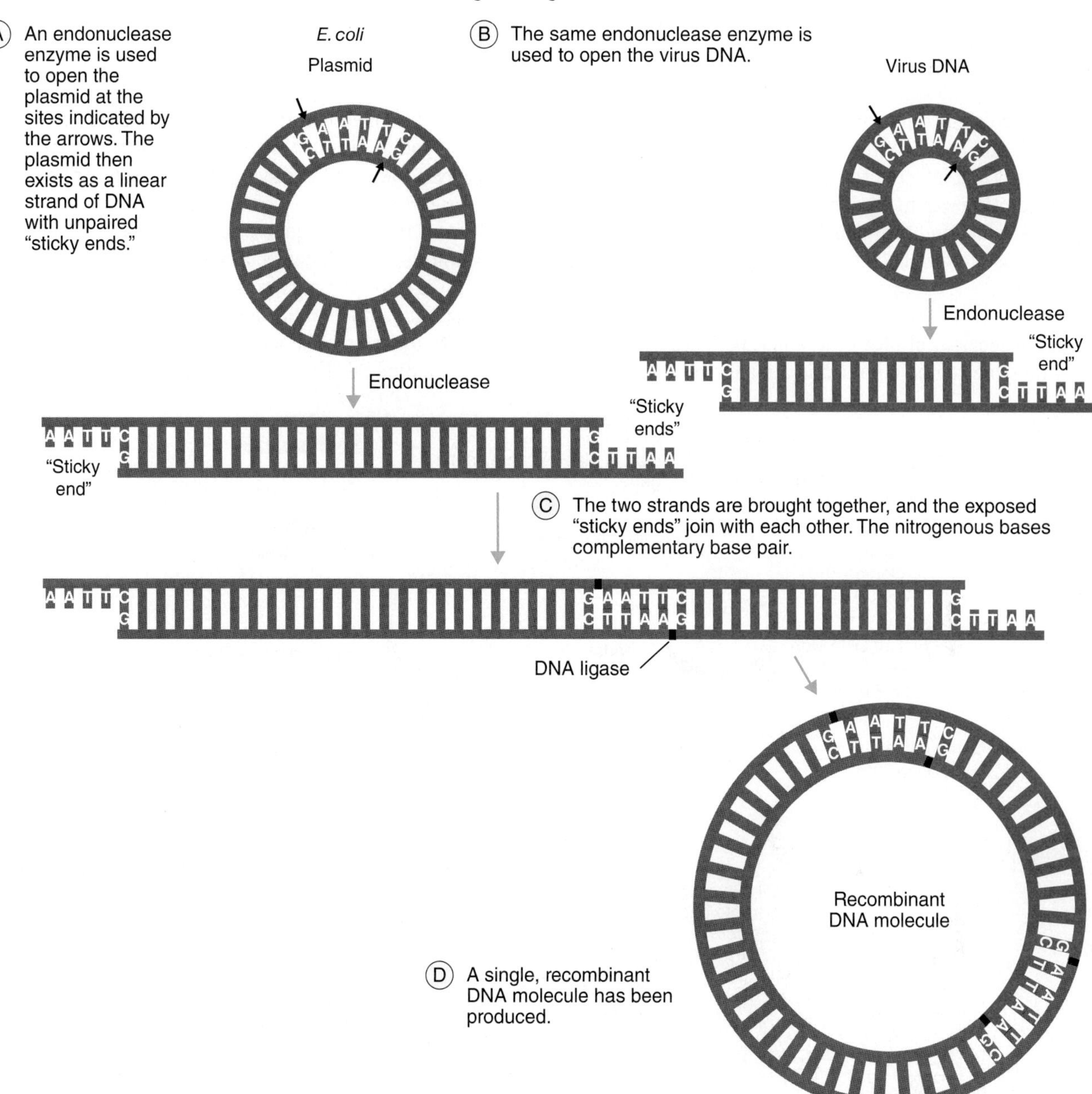

(A) An endonuclease enzyme is used to open the plasmid at the sites indicated by the arrows. The plasmid then exists as a linear strand of DNA with unpaired "sticky ends."
E. coli
Plasmid
(B) The same endonuclease enzyme is used to open the virus DNA.
Virus DNA
Endonuclease
"Sticky end"
"Sticky ends"
Endonuclease
"Sticky end"
(C) The two strands are brought together, and the exposed "sticky ends" join with each other. The nitrogenous bases complementary base pair.
DNA ligase
Recombinant DNA molecule
(D) A single, recombinant DNA molecule has been produced.

FIGURE 10.11 **Construction of a Recombinant DNA Molecule.** In this construction, two unrelated plasmids (loops of DNA) are united to form a single recombinant DNA molecule.

A CLOSER LOOK 10.3

Sweet News for Diabetics

Today, over 18 million people in the United States have diabetes, a group of diseases resulting from abnormally high blood glucose levels. In individuals with type I diabetes (formerly called insulin-dependent or juvenile diabetes), the disease is the result of the immune system destroying the pancreas' ability to make insulin. Without insulin, there are increased levels of glucose in the blood and urine, which can cause increased fatigue and long-term damage to organs.

Diabetics must receive regular injections of insulin to control their blood glucose level. Before 1982, diabetics received purified insulin extracted from the pancreas of cattle and pigs, or even cadavers. However, these sources often posed a problem because animal insulin is not human

(*continued*)

A CLOSER LOOK 10.3
Sweet News for Diabetics (continued)

insulin and could trigger allergic reactions. In addition, animal insulin could contain unknown disease-causing viruses that had infected the animal previously.

The solution was to produce human insulin using genetic engineering techniques by cloning the human gene for insulin into bacterial cells. The basic steps were (see figure):

A. The segment of DNA containing the human insulin gene can be obtained in a number of ways (not described here); obtain the bacterial plasmid and cut it open with an endonuclease that leaves the same "sticky ends" as present on the insulin gene.
B. Insert the insulin gene into a bacterial plasmid to form the recombinant plasmid.
C. Insert through transformation the recombinant plasmids into *Escherichia coli* cells.
D. Grow the cells in large, sterile stainless steel vessels containing all the nutrients and optimal conditions

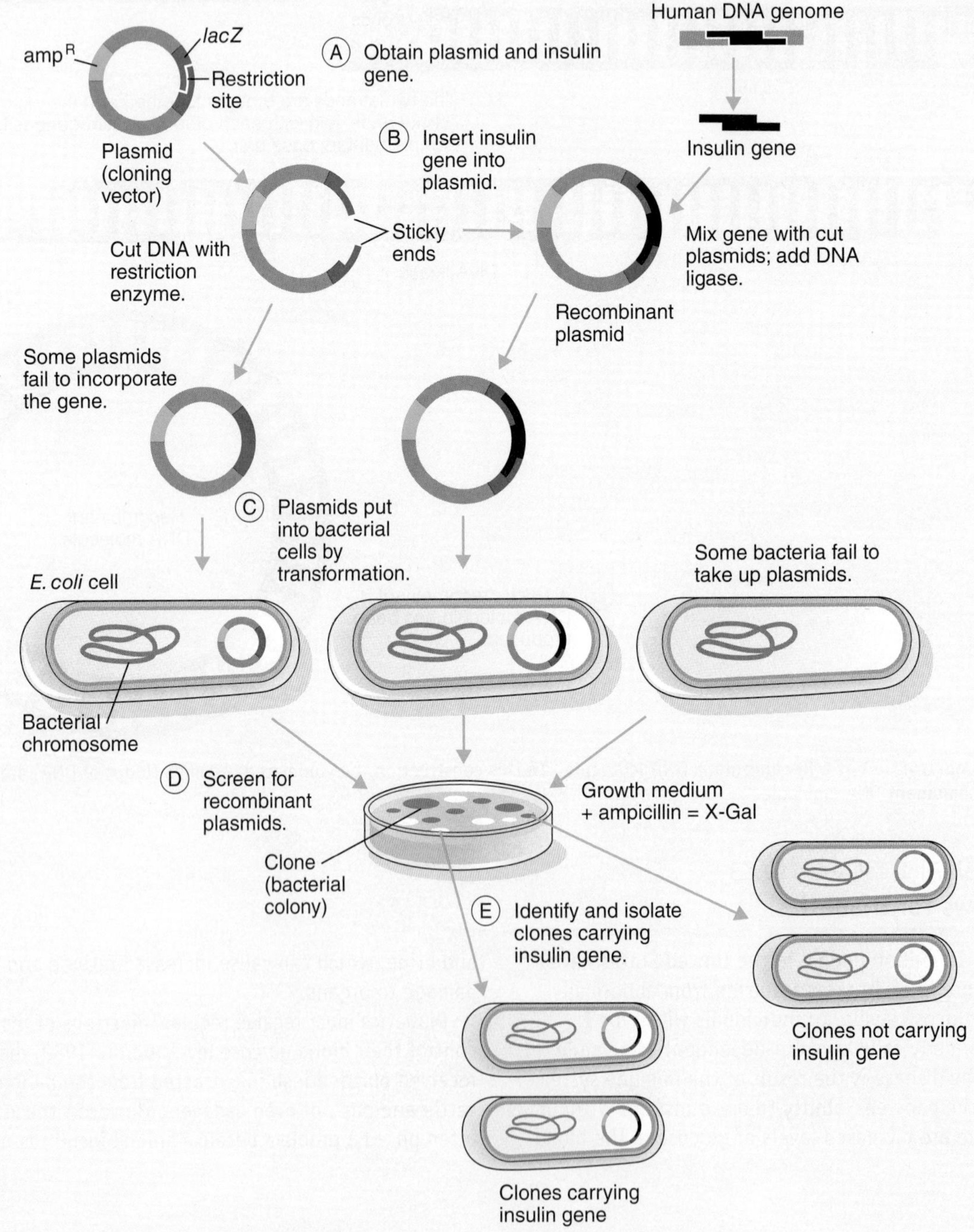

An outline of the steps to engineer the insulin gene into *Escherichia coli* cells.

A CLOSER LOOK 10.3

Sweet News for Diabetics (continued)

for bacterial growth; the cells divide and produce a large population of genetically identical cells all containing the plasmid.

E. When the insulin has been produced, harvest the insulin from the growth medium. Then, purify and package for distribution.

In 1982, Eli Lilly marketed the first synthetic human insulin, called Humulin. Importantly, genetic engineered human insulin is identical to the insulin naturally made inside the human body. Today, most people with type 1 diabetes manage their blood glucose levels with multiple daily insulin injections or by using an insulin pump.

Today, hundreds of companies worldwide are working on the commercial and practical applications of genetic engineering (FIGURE 10.12). Many of the genetically engineered products are either proteins expressed by the recombinant DNA in a host organism or in the target organism (FIGURE 10.13). The protein products are numerous and diverse.

These are just a sampling of the extraordinary products coming from the techniques of genetic engineering. It is envisioned that the commercial application of genetic engineering using living organisms, what is referred to as **biotechnology**, will offer solutions to problems of food production, synthesis of new medicines, pollution abatement, and other human and societal concerns. Scientists envision new enzymes to dissolve oil spills, new vaccines against emerging diseases, and new drugs and medicines to treat illnesses. The only limits to what could be accomplished lay in the scope of the human imagination. Other chapters will present detailed explorations of the breathtaking advances in agriculture, medical diagnostics, forensic science, and pharmaceutical research stemming from studies in microbial genetics.

Today, we find ourselves at the midst of a new world of biotechnology, a world derived from the fundamental research in microbial genetics. Perhaps Louis Pasteur in 1871 saw the future lying 100 years ahead by saying: "*There does not exist a category of science to which one can give the name applied science. There are sciences and the applications of science, bound together as the fruit of the tree which bears it.*"

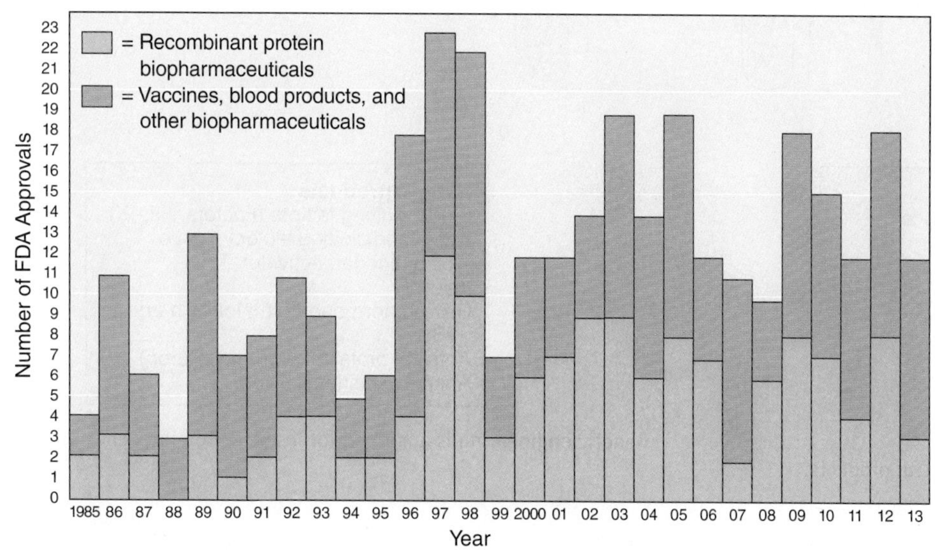

FIGURE 10.12 **FDA Approvals of New Pharmaceutical Products, 1985–2013.** This graph shows the number of approvals of new biopharmaceutical products since 1985.

Data from: Biotechnology Information Institute (http://www.biopharma.com/approvals_2011.html).

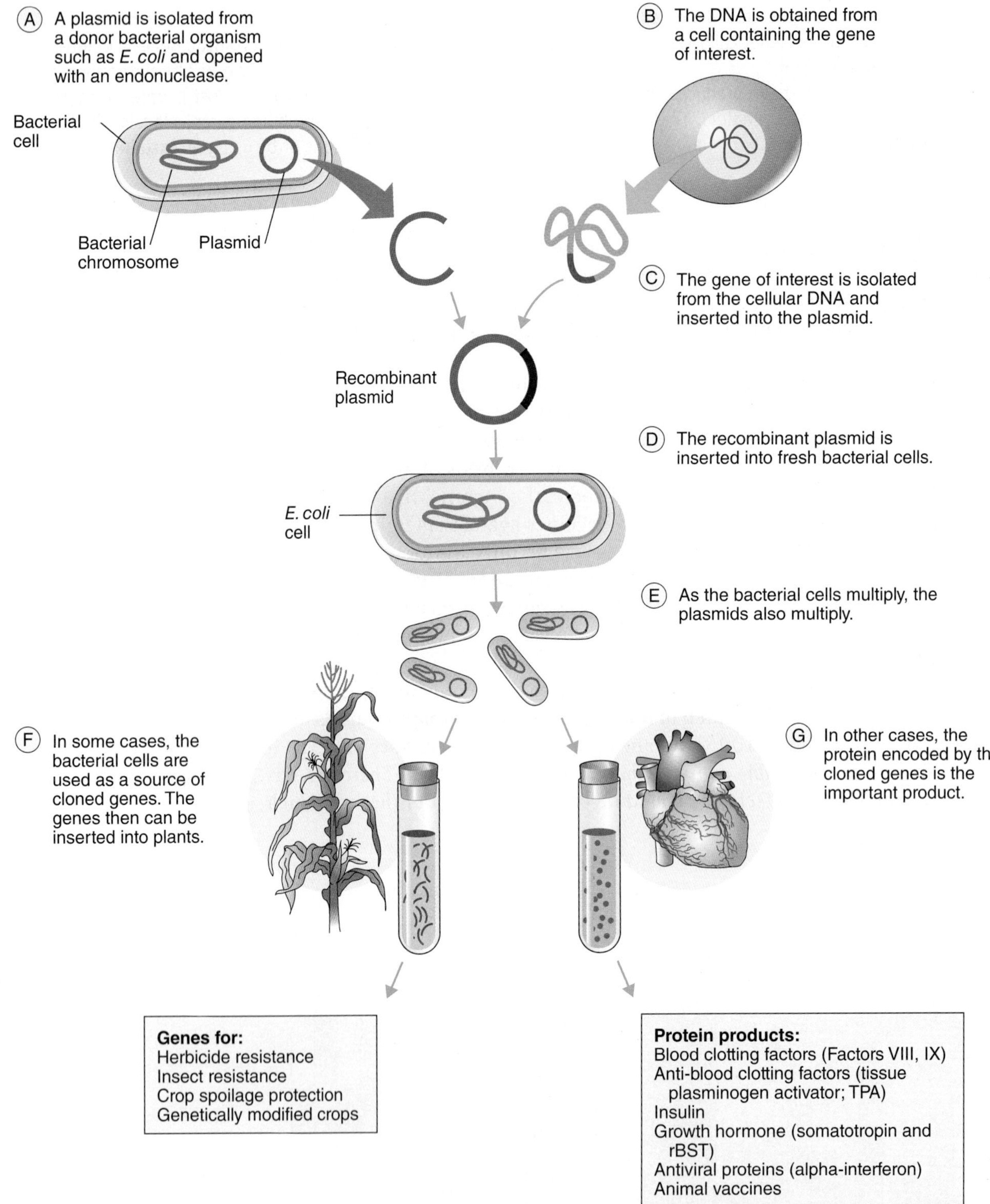
A A plasmid is isolated from a donor bacterial organism such as *E. coli* and opened with an endonuclease.
B The DNA is obtained from a cell containing the gene of interest.
Bacterial cell
Bacterial chromosome
Plasmid
C The gene of interest is isolated from the cellular DNA and inserted into the plasmid.
Recombinant plasmid
D The recombinant plasmid is inserted into fresh bacterial cells.
E. coli cell
E As the bacterial cells multiply, the plasmids also multiply.
F In some cases, the bacterial cells are used as a source of cloned genes. The genes then can be inserted into plants.
G In other cases, the protein encoded by the cloned genes is the important product.
Genes for:
Herbicide resistance
Insect resistance
Crop spoilage protection
Genetically modified crops
Protein products:
Blood clotting factors (Factors VIII, IX)
Anti-blood clotting factors (tissue plasminogen activator; TPA)
Insulin
Growth hormone (somatotropin and rBST)
Antiviral proteins (alpha-interferon)
Animal vaccines

FIGURE 10.13 **Developing New Products Using Genetic Engineering.** Genetic engineering is a method for inserting foreign genes into bacterial cells and obtaining chemically useful products.

A Final Thought

So you decide you'd rather be 6 foot 2 than 5 foot 1. You decide to go out and get some "tall" genes—maybe you pick them up at a used-gene shop, or order them from a catalogue, or buy them at a local Wal-Mart. "That's crazy," you say. "People can't get new genes or change their genetic makeup."

Maybe you can't shop for such genes (yet), but with the advances being made in biotechnology today, it is possible to change one's phenotype, such as one's height. Medically, one can receive human growth hormone (HGH) that, like insulin, has been produced in bacterial cells. With genetically engineered HGH, growth hormone therapy is available for children with idiopathic short stature; in fact, such therapy seems to be effective in reducing the deficit in height as they grow to adults. But what constitutes short stature and who should be treated medically? Should growth hormone be made available to parents who want their "normal-height" child to grow very tall so he/she can be a basketball player? These are the types of questions society will have to confront as the genetic engineering and biotechnology revolution continues.

idiopathic Referring to a disease or disorder that has no known cause.

It is fairly common to read in biology books about the "higher" and "lower" forms of life. Typically, humans and all animals and plants are cast as "higher" forms, while microbes, like the prokaryotes, are considered "lower" forms. Really? Today, with our increased understanding of the prokaryotic world, how can bacterial organisms be "lower" than any other forms of life? And with the advances in biotechnology, we are ever so more dependent on them. Further, unlike us and most "higher" forms of life, bacterial cells are not confined by what they inherit. They collect foreign genes; they exchange genes; they store genes; and when they need them, they use their new genes. And that's just one of the extraordinary features of the exceptional bacteria.

Questions to Consider

1. In hospitals, it was common practice to clear the air bubble from a syringe by expelling a small amount of the syringe contents into the air. One microbiologist estimated that this practice resulted in the release of up to 30 liters of antibiotic into a typical hospital's environment annually. How might this have brought about the appearance of antibiotic-resistant mutants in hospitals?
2. The author of a general biology textbook writes in reference to the development of antibiotic resistance: "The speed at which bacteria reproduce ensures that sooner or later a mutant bacterium will appear capable of resisting the antibiotic." How might this mutant bacterial cell have come into existence? Do you agree with the statement? Does this bode ill for the future use of antibiotics?
3. Which of the recombination processes (transformation, conjugation, or transduction) would be most likely to occur in the natural environment? What factors would encourage or discourage your choice from taking place?
4. In 1976, an outbreak of pulmonary infections among participants at an American Legion convention in Philadelphia led to the identification of a new disease, Legionnaires' disease. The bacterium responsible for the disease had never before been known to be pathogenic. From your knowledge of bacterial genetics, can you postulate how it might have acquired the ability to cause disease?
5. Some scientists believe mutations are the single most important event in evolution. Do you agree? Why or why not?
6. In 2011, researchers discovered bacterial cells possessing human genes. They found that more than 10% of a *Neisseria gonorrhoeae* population, the species

responsible for gonorrhea, contained a human DNA fragment. This fragment was not found in other species of *Neisseria*. How might a bacterial cell have acquired human DNA?

7. It is not uncommon for students of microbiology to confuse the terms reproduction and recombination. How do the terms differ?
8. In your opinion, are the prokaryotes, as exemplified by the bacteria, "lower" forms of life? Give examples to support your answer

Key Terms

Informative facts are necessary for the expression of every concept, and the information for a concept is founded in a set of key terms. The following terms form the basis for the concepts of this chapter. On completing the chapter, you should be able to explain and/or define each one.

bacteriocin
bacteriophage (phage)
biotechnology
conjugation
conjugation pilus
defective particle
DNA polymerase
extensively drug-resistant tuberculosis (XDR-TB)
F factor (plasmid)
genetic engineering
genetic recombination
genome
horizontal gene transfer
induced mutation
multidrug resistant tuberculosis (MDR-TB)
mutagen
mutation
nucleoid
phenotype
plasmid
point mutation
recombinant DNA molecule
replication fork
restriction endonuclease
semiconservative replication
spontaneous mutation
"sticky end"
transduction
transformation
transposon

11 Controlling Microbes: From Outside and Within the Body

Household and industrial cleaning products are diverse and formulated for every cleaning need. Many are especially formulated to disinfect surroundings, such as the kitchen counter or the bathroom toilet bowl in our homes, the food-processing area in a meat factory, or the floors in a hospital ward. The aim with all these products is to maintain sanitary conditions. In fact, today there are numerous physical methods or chemical agents available to control or eliminate microbes in the environment and drugs to use within us.

Looking Ahead

Microbes have the ability to grow and multiply at extremely high rates. Often, this is desirable, as, for example, in industrial situations; on occasion, however, it can be a problem, as when pathogenic microbes grow in our bodies. In this case, microbial growth represents a hazard to good health, and it must be controlled and preferably eliminated. Several methods of control from both outside and cure from inside the body are discussed in this chapter.

On completing this chapter, you should be capable of:

- Comparing the five ways physical methods used to control microbial growth.
- Identifying the general principles used for disinfection.
- Describing the different chemical agents and gases used to control microbial growth.
- Recalling the early history of antibiotic discovery.
- Explaining the mode of action of the penicillins and the development of penicillin resistance.
- Identifying the mode of action for other antibiotics.
- Assessing the challenge of antibiotic resistance in the medical field, and identifying the abuses and misuses of antibiotics today.

For personal hygiene, washing our hands, taking regular showers or baths, brushing our teeth with fluoride toothpaste, and using an underarm deodorant are common practices we use to control microorganisms on our bodies. In our homes, we try to keep microbes in check by cooking and refrigerating foods, and cleaning our kitchen counters and bathrooms with disinfectant chemicals, all in the hopes of preventing infection and disease. In the community, we expect sanitary public bathrooms, clean streets, and safe drinking water. Should we still get sick and develop an infectious disease, we often depend on antibiotics or other antimicrobial drugs to pull us through.

However, until the early twentieth century, the living conditions in most of the world, including hospitals, was far from clean and healthy. Disease outbreaks and epidemics were common, as many believed disease was due to vapors in the air and garbage in the streets. Such perceptions in the early 1800s led an English lawyer and social organizer named Edwin Chadwick to pioneer efforts to safeguard public health. What developed was the Great Sanitary Movement, which attempted to control filth, odor, and contagion in cities by building sewers, removing garbage, and providing clean water. Although it was more a social movement than medical intervention, the movement represents the inception of modern public health. In fact, a British medical journal in 2007 polled its readers asking: "What has been the most important medical milestone since 1840?" Sanitation was the top vote getter.

When Louis Pasteur put forward and Robert Koch validated the germ theory of disease in the late 1800s, insights about microorganisms added considerably to the understanding of disease and disease transmission. However, it did little to help the infected patient. Then, in the 1940s, thanks to earlier efforts by Paul Ehrlich and Alexander Fleming, antimicrobial agents, including antibiotics like penicillin, burst on the scene, and a revolution in medicine began. Doctors could now cure patients with bacterial infections. As a matter of fact, discovery of antibiotics was a close second to sanitation in the British journal poll mentioned above.

In this chapter, we will first examine various sanitation methods available for controlling populations of microbes. The discussion will focus on two general types of control: physical and chemical methods. Then, we will turn our attention to antimicrobial prophylaxis and the use of antimicrobial drugs as the mainstays of our healthcare delivery system for treating infectious diseases within our bodies. Applied on a broad scale, sanitation and antimicrobial therapy together constitute a major deterrent to infection and disease.

■ prophylaxis: A measure taken to maintain health and prevent the spread of disease.

11.1 Physical Methods of Control: From Hot to Cold

The Citadel, a novel by A. J. Cronin, follows the life of a young British physician named Andrew Manson. In the 1920s, Manson arrives at a small coal-mining town in Wales and almost immediately encounters an epidemic of typhoid fever. When his first patient dies of the disease, Manson becomes distraught. For a while, he believes there is nothing he can do to alter the course of the epidemic. But then, in a moment of insight, he realizes microbes succumb to the effects of intense heat, and before long, he has built a huge bonfire. Into the fire go all his patients' bed sheets, clothing, and personal effects. To his delight, the epidemic subsides shortly thereafter.

Heat has long been known as a fast, inexpensive, and reliable way of controlling microbes. Heat causes biochemical changes in microbes' organic molecules (such as enzymes and structural proteins), and it drives off water, creating a lethally arid environment.

Heat is among the physical methods we shall study in this section. In most cases, the physical methods are designed to achieve **sterilization**, a term that implies the destruction or removal of all forms of life. Sterilization is an absolute term without exceptions: An object cannot be "partially sterilized"; either it is sterilized or it remains "contaminated" with some form of microbial life.

A notable consideration when using physical control methods is the cellular structures that lend resistance to these methods. Certain bacterial species in the genera *Bacillus* and *Clostridium* produce endospores whose multiple wall and protein layers prevent heat and other physical agents from affecting a relatively dry interior. The fungi, too, produce thick-walled spores in their reproductive cycle and several protist species have the ability to form cysts; however, these spores and cysts don't have the extreme heat resistance of endospores.

Moist and Dry Heat Methods

Most microbes live within a specific range of temperatures, and although they survive at the limits of these ranges, they cannot live beyond those points. At high temperatures, for instance, microbial proteins are inactivated as their three-dimensional shape changes to two-dimensional forms, a process called **denaturation**. In the denatured form, proteins such as enzymes are inoperative, and the microbes die as their chemical reactions grind to a halt.

Moist Heat. Moist heat is among the most widely used methods for controlling microbes. For instance, the intense heat of the steam generated by boiling water at 100°C kills most microbes in a few seconds, the notable exception being bacterial endospores that require two hours or more (FIGURE 11.1). Moist heat denatures proteins, as noted above, and it disrupts the integrity of cell membranes by affecting their proteins or causing chemical changes in their lipids.

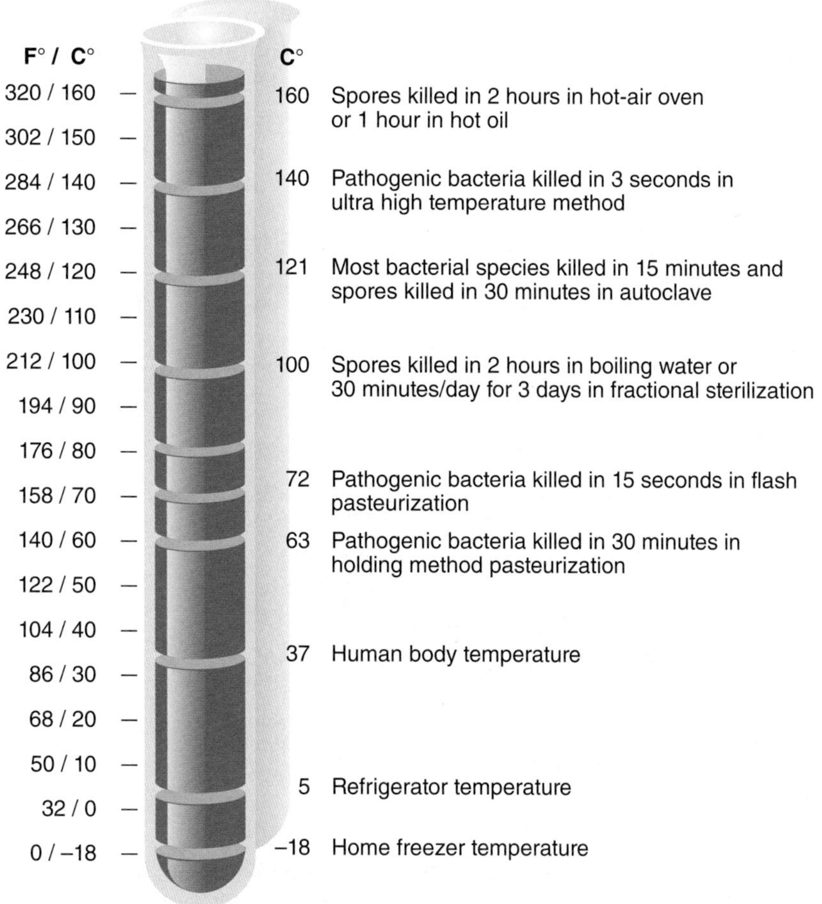

FIGURE 11.1 **Temperature and the Physical Control of Microorganisms.** Notice that materials containing bacterial endospores require longer exposure times and higher temperatures for killing.

Pressure is used to efficiently raise the temperature of steam above 100°C. As the pressure increases, the temperature rises and the destruction of microbes increases proportionally. An apparatus called the **autoclave** (FIGURE 11.2) increases steam pressure to 15 pounds per square inch (psi) above atmospheric pressure. Under these conditions, the temperature of the steam rises from 100°C to 121°C, sterilizing most materials in about 15 minutes. These materials include metal instruments, glassware, microbial media, hospital and laboratory equipment, and virtually anything else able to withstand the high temperature and pressure. Important exceptions are plastics that would melt and certain chemical substances (e.g., vaccines, antibiotics) that would be inactivated.

The final example of moist heat involves the process of **pasteurization**, which reduces a bacterial population in liquids, such as milk, and destroys organisms capable of causing spoilage and human disease. Spores are not affected by pasteurization.

One method for milk pasteurization, called the **holding** (or **batch**) **method**, involves heating the liquid at 63°C for 30 minutes. A similar heating procedure employed in industry is the **flash pasteurization method** that keeps the liquid at 72°C for 15 seconds. For decades, such pasteurization methods have been aimed at destroying *Mycobacterium tuberculosis,* long considered the most heat-resistant bacterial species. More recently, attention has shifted to destruction of *Coxiella burnetii,* the agent of Q fever, because this bacterial pathogen has a higher resistance to heat. Both pathogens can infect cattle and thus make their way into the dairy industry. Because both organisms are eliminated by pasteurization, dairy microbiologists assume other pathogenic bacterial species also are destroyed. The holding and flash pasteurization methods do not completely eliminate all relatively benign, nonpathogenic microorganisms capable of souring improperly stored liquids like

■ *Mycobacterium tuberculosis*
mī-kō-bak-ti'rē-um tü-bėr-kū-lō'sis

■ *Coxiella burnetii*
käks'ē-el-ä bėr-ne'tē-ē

(a)

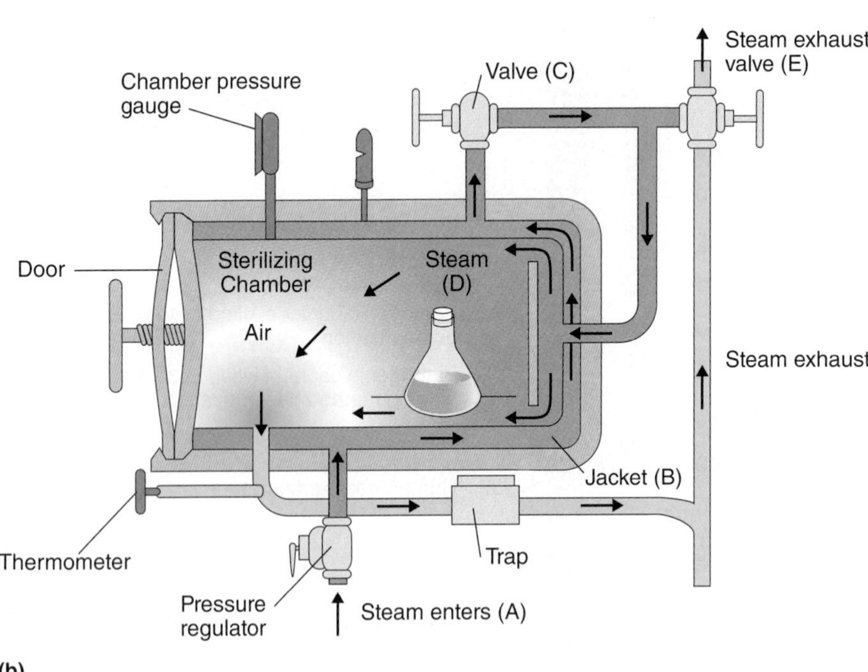

(b)

FIGURE 11.2 **Operation of an Autoclave.** (**a**) An autoclave can be used to sterilize liquid materials. (**b**) Steam enters through the port (a) and passes into the jacket (b). After the air has been exhausted through the vent, a valve (**c**) opens to admit pressurized steam (**d**) that circulates among and through the materials, thereby sterilizing them. At the conclusion of the cycle, steam is exhausted through the steam exhaust valve (**e**).

(a) © Huntstock, Inc/Alamy.

milk. These pasteurization methods also are used to prevent pathogen contamination in fruit juices, cream, yogurt, and alcoholic beverages.

Many liquid products today have a 6 to 9 month shelf life because they have been "ultra-pasteurized" or subjected to what is called the **ultra-high temperature** (UHT) **method**. These products are heated to 140°C for 3 seconds and packaged in sterilized containers. The drawback is that some people say UHT products, especially milk, can have a burned taste. So, why is "organic milk" often subjected to UHT? **A Closer Look 11.1** investigates.

Dry Heat. The other method of heat employs dry heat. Dry heat is used in an oven. The dry heat from the hot air penetrates less rapidly than moist heat, and the sterilizing temperatures tend to be more extreme and the times longer. To achieve sterilization, a temperature of 160°C to 170°C must be applied to a population of microbes for a period of two or more hours. In this case, the means of destruction involves changes in the chemical nature of molecules, and the process usually requires more substantial energy input than for denaturation. Nevertheless, for sterilizing such things as powders, oily materials, and dry instruments, an oven's dry heat is efficient.

Using a direct flame as dry heat can incinerate (burn to ashes) microbes very rapidly as the flame can generate temperatures greater than 1,000°C (1,832°F). In the past, the bodies of disease victims were burned to prevent spread of the

A CLOSER LOOK 11.1

Does Milk Stay Fresher Longer If It's Organic?

If you ever go shopping for milk at the local market, especially one specializing in "natural and organic foods," you might notice that the "Best by" date expires much sooner on a carton of regular milk than on a carton of organic milk. In fact, the date on the organic milk may be some 3 weeks longer than on the regular milk, which typically is 5 to 7 days from the store delivery date. So, being organic, does that ensure a longer shelf life?

The fact that the milk is organic has nothing to do with its longer shelf life. Labeling the milk as "organic" only means the cows on the dairy farm were not given antibiotics or hormones like bovine growth hormone (BGH), which stimulates a cow's milk production. The reason it has a longer shelf life is due to the pasteurization process. Organic milk is subjected to the ultra-high temperature (UHT) process (ultra-pasteurized) where the milk is heated to 140°C for 3 seconds (see figure). This kills all the microorganisms that may be in the liquid—it is sterile. Most regular milk today is subjected to the flash pasteurization method where the milk is at about 72°C for 15 seconds. This "high temperature, short duration" process does not kill all microbes that may be in the milk; only the pathogens have been eliminated. Because there are psychrotolerant bacterial species, the milk can spoil if left on the refrigerated shelf too long.

Regular milk also could be subjected to UHT. However, it usually is not because it has to travel only a short distance to market; organic products, on the other hand, are not often produced throughout the country, so they have further to travel to reach the consumer. Therefore, UHT preserves the product longer. Although appearing more frequently in the United States, room-temperature Parmalat milk is a product of UHT and can be found commonly in Europe, Mexico, and other parts of the world. The verdict? "Organic" is not defined as "longer shelf life." If shelf life is important to you, simply look for products treated by UHT.

A carton of organic milk.

Courtesy of Dr. Jeffrey Pommerville.

plague. Similar to Cronin's *The Citadel*, today disposable hospital gowns and certain plastic apparatus may be incinerated. It still is common practice to incinerate the carcasses of cattle that have died of anthrax and to put the contaminated field to the torch because anthrax spores cannot be destroyed adequately by other means. The 2001 outbreak of foot-and-mouth disease in British cattle required the mass incineration of thousands of cattle as a means to stop the spread of the disease (FIGURE 11.3).

Although heat is a valuable physical agent for controlling or eliminating microorganisms, sometimes it is impractical to use. For example, no one would suggest removing the microbial population from a tabletop by using a blowtorch! In addition, heat-sensitive solutions cannot be sterilized in an autoclave. In instances such as these, and numerous others, a heat-free method must be used.

Radiation

Various kinds of radiation exert destructive effects on microbes by disrupting the nucleic acid components of their cytoplasm (FIGURE 11.4). **Ultraviolet radiation (UV light)**, for example, interacts with the DNA, linking together adjacent thymine bases. This disrupts DNA replication and transcription. The damaged organisms can no longer produce critical proteins or reproduce, and eventually die. UV light also excites the electrons of other molecules in microbes and brings about biochemical changes leading to death. It is a useful sterilizing agent for dry surfaces such as a tabletop or flat instrument, and it can be used in a closed environment such as an operating room to lower the microbial population of the air.

UV light effectively reduces the microbial population where direct exposure takes place. It is used to limit airborne or surface contamination in a hospital room, morgue, pharmacy, toilet facility, or food service operation. It is noteworthy that UV light from the sun may be an important factor in controlling microorganisms in the air and at the soil surface, but it may not be effective against all bacterial spores. Ultraviolet light does not penetrate liquids or solids, and it can lead to human skin cancer.

Other kinds of radiation, including X rays and gamma rays, are also useful for sterilization. X rays and gamma rays are about 10,000 times more energetic than UV light, and they induce electrons and protons to jump out of the molecules they strike. This process creates ions, which are atoms or molecules lacking one or more electrons

FIGURE 11.3 **Incineration Is an Extreme Form of Dry Heat.** Nearly four million hoofed animals infected with or exposed to foot-and-mouth disease in England in 2001 were incinerated and buried.

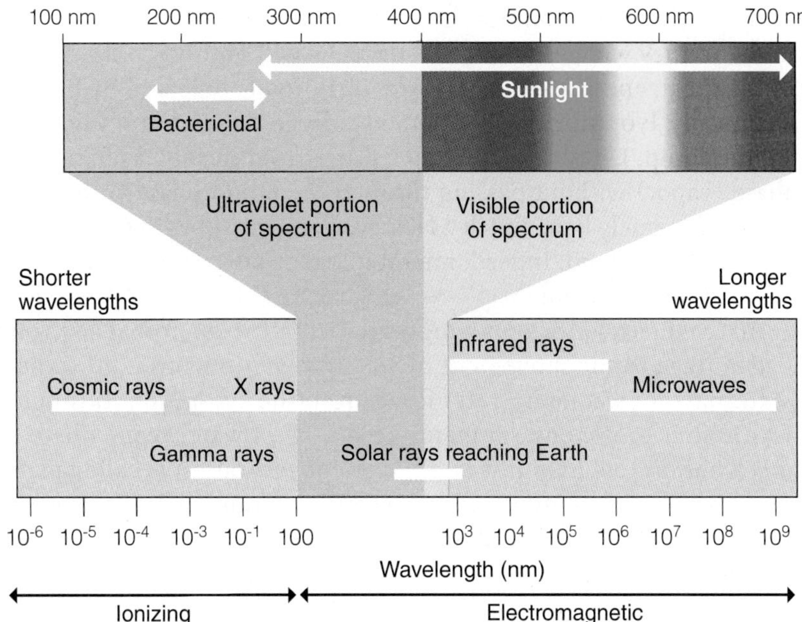

FIGURE 11.4 **The Ionizing and Electromagnetic Spectrum of Energies.** The complete spectrum is presented at the bottom of the chart, and the ultraviolet and visible sections are expanded at the top.

(the high-energy radiation is therefore called **ionizing radiation**). Water molecules are particularly susceptible: Ionizing radiation causes them to ionize to hydroxyl ions (OH^-) and hydronium ions (H_3O^+). These ions react with other chemical compounds and bring about chemical changes in the microbial cytoplasm, particularly in the bonds holding the bases to one another in DNA.

Ionizing radiations also have been approved for controlling microorganisms, and for preserving foods. Today, irradiation is used as a food preservation method in more than 40 countries for over 100 food items, including potatoes, onions, cereals, flour, fresh fruit, and poultry. The U.S. Food and Drug Administration (FDA) has approved the irradiating of poultry and red meats such as beef, lamb, and pork. In 2008, the FDA approved the irradiation of fresh and bagged spinach, and iceberg lettuce, to reduce potential foodborne illness. Irradiation also has been used to prepare many meals for the U.S. military and the American astronauts (FIGURE 11.5). The irradiation level used on American products is called a **pasteurizing dose** because the level is not intended to eliminate all microbes in the food, but, like pasteurization, will eliminate the pathogens. And, importantly, the irradiated food is not radioactive and therefore of no danger to the consumer.

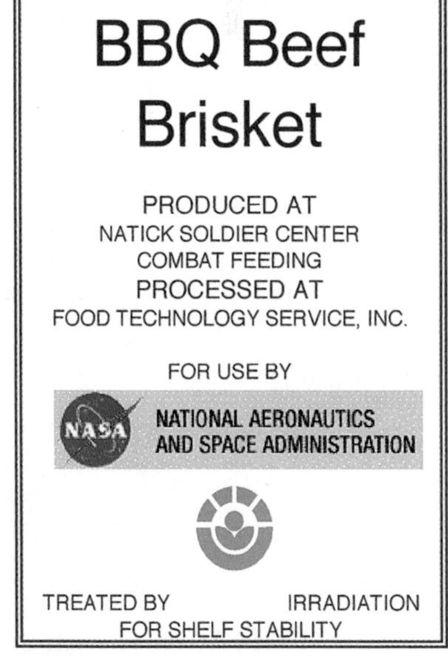

FIGURE 11.5 **Food Irradiation.** Many otherwise perishable foods eaten by NASA astronauts were prepared by irradiation.

Courtesy of US Army Natick Soldier Center.

Drying

Another nonheat method for controlling microbial populations is drying (also called **desiccation**). Humans have used drying to control microbial growth well before heat or radiation methods were developed. For example, tradition has it that the Peruvian Incas of the Andes Mountains preserved potatoes and other foodstuffs by placing them for several weeks on high mountainsides, where they dried in the open air.

Drying is still used today to prepare foods such as cereals, grains, and numerous other products for storage in the home pantry. Because water is required for most chemical reactions in microbes, it follows that microbes cannot grow where water is very limited or absent. However, many types of microbes, especially bacterial spores, remain alive under these conditions, and if water is introduced to the environment, they will germinate and the cells will multiply.

Hikers, campers, and backpackers often carry foods prepared by freeze-drying, or more technically, **lyophilization**. The food is frozen; then the water is drawn off with a vacuum pump. Lyophilization takes water from the solid phase (ice) to the gaseous phase (vapor) without passing through the liquid phase (water). A freeze-dried product is extremely light and dry. However, any microbes in the product will be preserved as well as the food. Indeed, microbiologists often preserve their microbial cultures by freeze-drying them.

An alternative method for removing water from the microbial environment is by taking advantage of the movement of water. When microbes are exposed to a high-salt external environment, water flows from the cytoplasm, through the cell membrane, and out to the environment. This flow of water from where it is in a higher concentration to where it is in a lower concentration is called **osmosis**. In the salty environment, the microbes shrivel and die quickly. Food processors use salt for preserving meats, fish, and numerous other types of foods. Sugar can be used as alternatives to salt.

Filtration

Microbes can be removed from a liquid solution by the process of **filtration**. During filtration, liquid passes through a porous material that traps microbes in its submicroscopic pores, as shown in FIGURE 11.6. For heat-sensitive solutions such as the vaccines and antibiotics mentioned earlier, filtration may be a useful alternative.

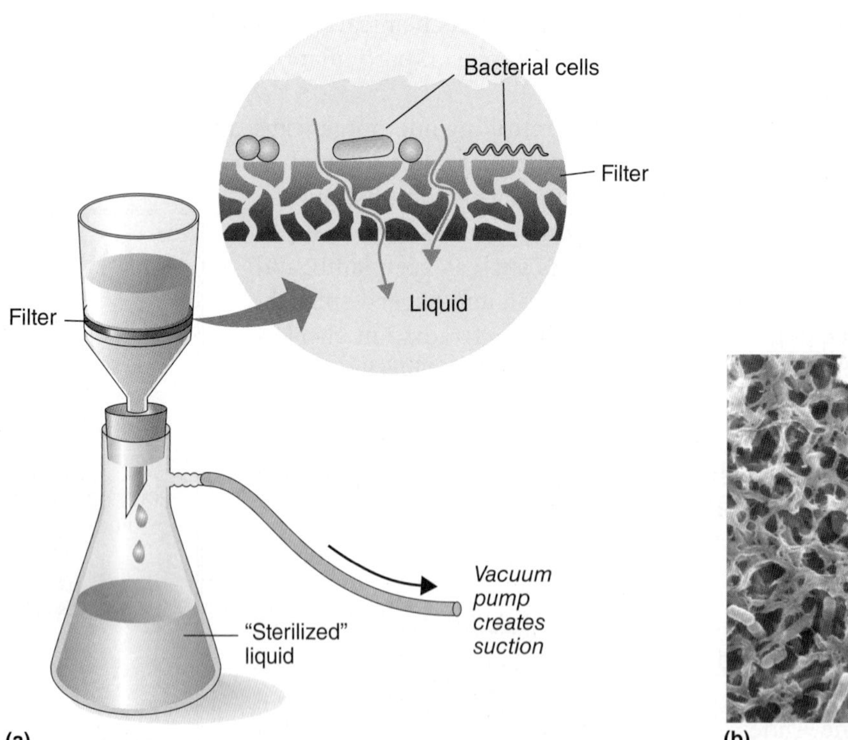

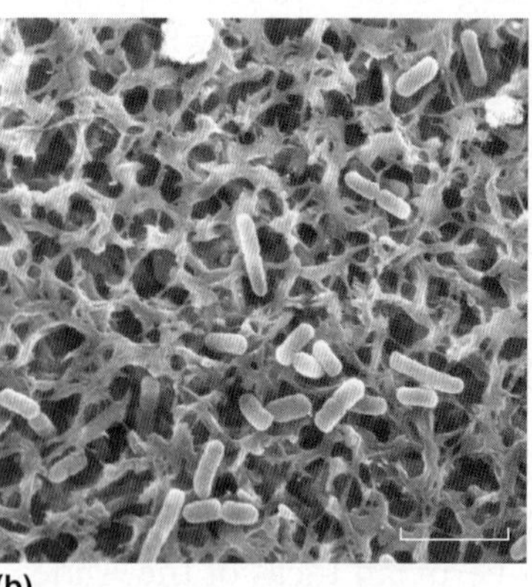

FIGURE 11.6 **The Principle of Filtration.** Filtration is used to remove microorganisms from a liquid. The effectiveness of the filter is proportional to the size of its pores. (**a**) Bacteria-laden liquid is poured into a filter, and a vacuum pump helps pull the liquid through and into the flask below. The bacterial cells are larger than the pores of the filter, and they become trapped. (**b**) A scanning electron micrograph of *Escherichia coli* cells trapped in the pores of a 0.45-µm membrane filter. (Bar = 5 um.)

(b) Courtesy of Pall Corporation.

Because filters function by mechanical entrapment, they are manufactured in a range of sizes to fit the microbes to be removed. For example, if bacterial cells are to be trapped, then a filter with a pore size of about 0.25 μm might be used (because the smallest bacteria are above this size range). However, the filtered liquid would not be sterilized because it could contain viruses, which can pass through such pores. To trap viruses, a much smaller pore size is needed.

Air also can be purified to remove microorganisms. Most of us have some type of air filter in our home air conditioning/heating system to trap dust and microscopic particles, like smoke, mold spores, and pollen. Hospitals and facilities that need to maintain a high standard of air quality use **high efficiency particulate air (HEPA) filters**, which consist of a mat of randomly arranged fibers able to trap particles, microorganisms, and spores. HEPA filters are critical in preventing of the spread of airborne bacterial and viral organisms and, therefore, infection. Typically, medical-use HEPA filtration systems also incorporate UV light units to kill off the live bacteria and viruses trapped by the filter.

Low Temperatures

Though not a sterilization method, low temperatures can be employed as a physical agent to control microbial populations. At low temperatures, such as in a refrigerator or freezer, enzyme activity diminishes, and microbial reproduction slows considerably (see Figure 11.1). The mobility of molecules through biological membranes is also reduced, and the rate of chemical reactions in the cytoplasm is lowered. Low temperatures, though, only slow microbial growth; they do not kill microbes, except if ice crystals manage to tear cells apart. Therefore, freezing or refrigerating foods preserves the foods but does not completely eliminate microbial populations.

11.2 Chemical Methods of Control: Antiseptics and Disinfectants

The practices of sanitation and chemical control of microbes are not new. The Bible often refers to cleanliness and prescribes certain dietary laws to prevent consumption of possibly contaminated foods. The Egyptians used resins and aromatic chemicals for embalming even before they had a written language, and other ancient peoples burned sulfur for deodorizing and sanitary purposes.

Medicinal chemicals came into widespread use in the 1800s. As early as 1830, for example, the U.S. Pharmacopeia listed tincture of iodine as a valuable antiseptic, and soldiers used it in plentiful amounts during the Civil War. Copper was valuable for preventing fungal disease in plants, and mercury was sometimes used for treating syphilis, as Arabian physicians had suggested centuries before. Moviegoers have probably noticed that American cowboys practiced the art of disinfection by pouring whiskey into wounds—between drinks, that is.

■ pharmacopeia: A book describing drugs, chemicals, and medicinal preparations.

The chemical control of microbes received a considerable boost in the 1860s with the work of Joseph Lister, a Professor of Surgery at Glasgow Royal Infirmary in Scotland. Lister was puzzled as to why more than half his amputation patients died—not from the surgery—but rather from postoperative infections. Hearing of Pasteur's germ theory, Lister hypothesized that the surgical infections resulted from germs in the air. Knowing that carbolic acid had been effective on sewage control, in 1865 he used a carbolic acid spray in surgery and on surgical wounds. The result was spectacular—the wounds healed without infection. His technique would not only revolutionize medicine and the practice of surgery, but also lead to the practice of **antisepsis**, the use of chemical methods for disinfection of external living surfaces, such as the skin.

General Principles

To interrupt the spread of microbes, chemical controls are applied in such diverse locations as the hospital environment, the food-processing plant, and the typical household. While physical methods are used to achieve sterilization, **disinfection** represents a method to lower microbial populations and kill most, but not necessarily all, pathogens. The chemicals used are called disinfectants or antiseptics. **Disinfectants** are chemical compounds formulated for use on inanimate (lifeless) objects, while **antiseptics** are meant for use on the surface of the body (FIGURE 11.7). Although certain chemical agents are strong enough to sterilize, exposure to the air following treatment reintroduces contamination. Therefore, for both disinfectants and antiseptics, sterilization is not a realistic objective.

No single chemical agent is ideal for controlling all microbes under all conditions. However, if an ideal chemical agent were to exist, it would possess an elaborate array of characteristics as summarized in TABLE 11.1. With such stringent requirements, it is not surprising that an ideal disinfectant or antiseptic does not exist.

The chemical agents currently in use for controlling microorganisms range from very simple liquid substances, such as household bleach, to very dangerous gaseous

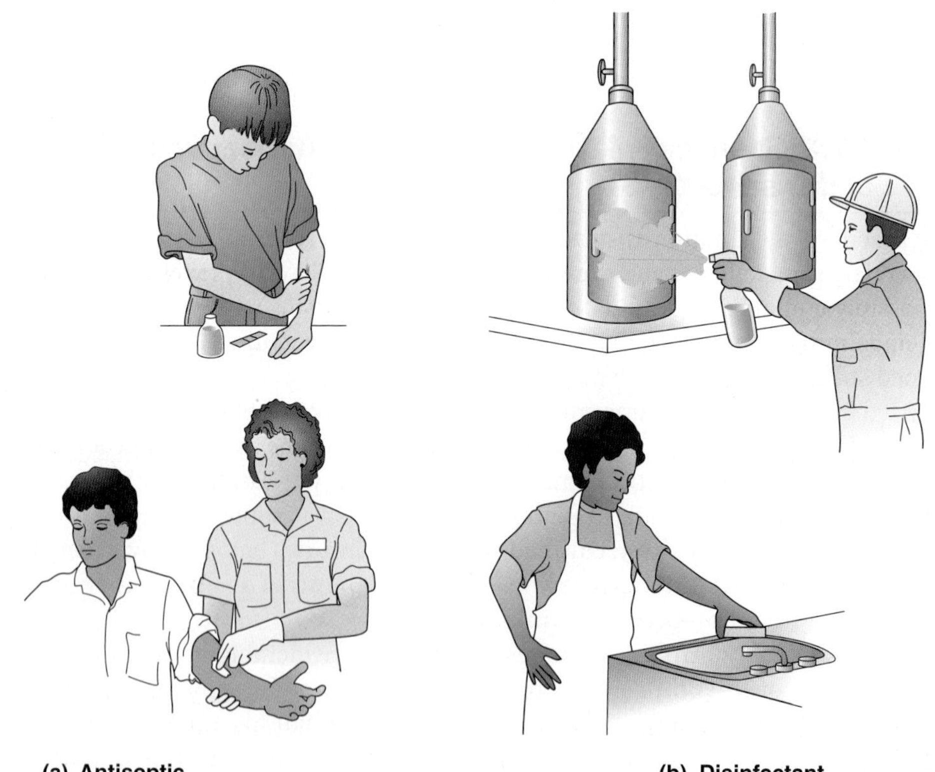

FIGURE 11.7 **Sample Uses of Antiseptics and Disinfectants.** (**a**) Antiseptics are used on body tissues, such as on a wound or before piercing the skin to take blood. (**b**) A disinfectant is used on inanimate objects, such as equipment used in an industrial process or tabletops and sinks.

TABLE 11.1 Characteristics of an Ideal Chemical Agent

Characteristics of an Ideal Chemical Agent
Kills all microbes
Dissolves in water
Remains stable on standing for extended period of time
Is nontoxic to humans and animals
Does not combine with organic matter other than in microbes
Has strongest action on microbes at room or body temperature
Penetrates surfaces efficiently
Does not corrode or rust metals or damage or stain fabrics
Is available in useful quantities and at reasonable prices

substances, typified by ethylene oxide. Many of these agents have been used for generations, while others represent the latest modern chemical products. In this section, we will survey several groups of chemical agents and indicate how they are best applied in the control of microorganisms. **A Closer Look 11.2** looks in your pantry to identify some common but surprising antiseptics.

Alcohols and Peroxides

Among the common chemical agents for microbial control are the **alcohols**. The most widely used alcohol is ethyl alcohol (ethanol), usually in a 70% solution. Ethyl alcohol denatures proteins and dissolves lipids in the cell membranes of microbes. It is the active ingredient in many popular hand sanitizers and is used as an antiseptic to treat the skin before a venipuncture or injection where the alcohol mechanically removes bacterial cells. Ethyl alcohol can be used as a disinfectant by immersing instruments in it for a minimum of 10 minutes. Overall, the chemical is effective against multiplying bacterial cells, but it has no effect on spores. Isopropyl (rubbing) alcohol is equally useful.

venipuncture: The piercing of a vein to take blood, to feed somebody intravenously, or to administer a drug.

Peroxides are chemical agents containing oxygen-oxygen single bonds. **Hydrogen peroxide** (H_2O_2; H–O–O–H), a common household antiseptic, has been used as a rinse for wounds, scrapes, and abrasions. However, when applied to such areas it foams and bubbles, as the enzyme catalase in the damaged tissue breaks down hydrogen peroxide to oxygen (the bubbles) and water. Therefore, it is not recommended as an antiseptic for open wounds. However, the furious bubbling loosens dirt, debris, and dead tissue, and the oxygen gas is effective against anaerobic bacterial species. Hydrogen peroxide breakdown also results in a reactive form of oxygen—the superoxide radical (O_2^-)—which is highly toxic to microorganisms and viruses.

New forms of H_2O_2 are more stable than traditional forms, do not decompose spontaneously, and therefore can be used topically. Such inanimate materials as soft contact lenses, utensils, heat-sensitive plastics, and food-processing equipment can be disinfected within 30 minutes. Benzoyl peroxide is an active ingredient in teeth whitening products and at low concentrations (2.5%) is used to treat acne.

A CLOSER LOOK 11.2

Antiseptics in Your Pantry?

Today we live in an age when alternative and herbal medicine claims are always in the news, and these reports have generated a whole industry of health products that often make unbelievable claims. With regard to "natural products," are there some that have genuine medicinal and antiseptic properties?

Honey

For nearly three decades, Professor Peter Molan, associate professor of biochemistry and director of the Waikato Honey Research Unit at the University of Waikato, New Zealand, has been studying the medicinal properties of and uses for honey. Its acidity, between 3.2 and 4.5, is low enough to inhibit many pathogens. Its low water content (15% to 21% by weight) means that it osmotically ties up free water and "drains water" from wounds, depriving pathogens of an ideal environment. In addition, in 2009–2010 other researchers discovered two proteins in honey, one interfering with bacterial cell wall synthesis and the other interacting with the bacterial cell membrane.

Unfortunately, not all honey is alike. The antibacterial properties of honey depend on the kind of nectar, or plant pollen, that bees use to make honey. At least manuka honey from New Zealand and honeydew from central Europe are thought to contain useful levels of antiseptic potency. Professor Molan is convinced that "honey belongs in the medicine cabinet as well as the pantry." In 2011, the United States Food and Drug Administration (FDA) approved wound dressings containing manuka honey.

Licorice Root

Dried licorice root has been used in traditional Chinese medicine for centuries. In 2012, scientists reported they had identified two substances in licorice root capable of killing the two most common bacterial species causing tooth decay and one species responsible for gum disease. But don't run out and start eating lots of licorice candy because the licorice root extract originally in candies has been replaced by anise oil, which has a similar flavor but no antimicrobial properties.

Wasabi

The green, pungent, Japanese horseradish called wasabi may be more than a spicy condiment for sushi. Professor Hedeki Masuda, director of the Material Research and Development Laboratories at Ogawa & Co. Ltd., in Tokyo, Japan, and his colleagues have found that natural chemicals in wasabi, called isothiocyanates, inhibit the growth of those bacterial species involved with tooth decay. At this point, these are only test-tube laboratory studies and the results will need to be proven in human clinical trials.

Cinnamon

Professor Daniel Y. C. Fung, Professor of Food Science and Food Microbiology at Kansas State University, believes cinnamon (although not an antiseptic) might be a chemical agent that can control pathogens, at least in fruit beverages. Fung's group added cinnamon to commercially pasteurized apple juice. They then added a sample of typical foodborne bacterial and viral pathogens. After one week of monitoring the juice at refrigerated and room temperatures, the investigators discovered the pathogens were killed more readily in the cinnamon blend than in the cinnamon-free juice. In addition, more bacterial organisms and viruses were killed in the juice at room temperature than when refrigerated.

The verdict? There appear to be products having genuine antimicrobial properties—and there are many more than can be described here.

Honey and cinnamon are two substances with potential antimicrobial properties.

Courtesy of Dr. Jeffrey Pommerville

Halogens and Heavy Metals

Halogens are extremely reactive elements. Two such halogens, iodine and chlorine, are used as chemical control agents.

■ tincture: A low concentration of a chemical dissolved in alcohol.

Iodine (I_2) can be used as a tincture of iodine (2% iodine in ethyl alcohol) or as iodine-detergent compounds known as iodophors. In iodophors, the detergent loosens microbes from the skin surface and the iodine kills them.

Chlorine (Cl_2) is used as chlorine gas to reduce the microbial content of water or as an organic compound called a chloramine. Like iodine, chlorine reacts with proteins, is effective against all types of microbes, and may be used in antiseptics or disinfectants depending on the formulation. Chlorine is also used in the form of sodium hypochlorite in a 5% concentration in household bleach. To disinfect clear water, the Centers for Disease Control and Prevention (CDC) recommends a half-teaspoon of bleach in 2 gallons of water, with 30 minutes of contact time before consumption. Other uses for chlorine are shown in FIGURE 11.8.

Three **heavy metals** are useful as chemical antiseptics and disinfectants: Silver (Ag) in the form of silver nitrite ($AgNO_3$) is employed as a general antiseptic. Mercury (Hg) is still used in some antiseptics for treating skin wounds, such as Mercurochrome and Merthiolate, but it is toxic. Indeed, a mercury compound named thimerosal was previously used as a preservative in vaccine preparations. As a topical antiseptic, the mercury compounds have been replaced by other agents, such as the iodophors.

Copper (Cu) in the form of copper sulfate ($CuSO_4$) is used to control cyanobacteria in swimming pools and to restore the clarity of the water. Copper sulfate is also mixed with lime to form the so-called "Bordeaux mixture" for controlling fungal growth. Copper ions also are very toxic to bacterial cells. In recent studies, hospital fixtures containing copper were found to contain 70% to 90% fewer microbes than did non-copper fixtures and using such fixtures reduced the risk of acquiring an infection by 40%.

Soaps and Detergents

Soaps and detergents are strong wetting agents and surface tension reducers; that is, they work their way between microbes and a surface and "lift" the microbes

FIGURE 11.8 **Some Practical Applications of Disinfection with Chlorine Compounds.** Different chlorine compounds have been used as disinfectants in city water supplies, swimming pools, on countertops, and in floor cleaners. At lower concentrations, chlorine can be used as an antiseptic for wound treatment.

so they can be removed with the wash water. They also dissolve the microbial cell membrane by reacting with its lipids, thereby causing leakage through the membrane and cell death. When used on cutting boards, detergents can reduce the possibility of cross contamination of foods and utensils. Soaps also remove skin oils, further reducing the surface tension and increasing the cleaning action. But are antibacterial soaps that are so common today worth the money? **A Closer Look 11.3** presents one view.

The most useful detergents to control microorganisms and many viruses are derivatives of ammonium chloride. Called quaternary ammonium compounds or, simply quats, they are used as sanitizing agents for industrial equipment and food utensils, and as disinfectants on hospital walls and floors. They also are used as disinfectants in contact lens cleaners and some mouthwashes. If you use a mouthwash, look on label and if you cannot pronounce the chemical names of the ingredients, they are probably quats. So, just shake the bottle—if it foams, it contains quats.

Phenols

Phenol (carbolic acid) and phenolics were among the first chemical agents used for microbial destruction, having been employed by Lister in his landmark experiments described earlier in this chapter. Phenol derivatives include Lysol® and hexachlorophene (the active ingredient in pHisoHex handwash). Another important phenol

A CLOSER LOOK 11.3

Are Antibacterial Soaps Worth the Money?

All of us want to be as clean as possible. In fact, hand washing is one of the best ways to protect oneself and prevent the spread of disease-causing microbes. To that end, numerous consumer product companies have provided us with many different types of antimicrobial cleaning and hygiene items. Perhaps the most pervasive are the antibacterial soaps, which usually contain about 0.2% triclosan.

It is estimated that 75% of liquid and 30% of bar soaps on the market today are of the antibacterial type. The question though is: Are these products any better than regular soaps? The short answer is—no.

Numerous studies have shown these antibacterial soaps do little against bacterial foodborne pathogens. In addition, they do nothing to reduce the chances of picking up and harboring infectious microbes.

A 2005 study gathered together over 200 families with children. Each family was given cleaning and hygiene supplies—soaps, detergents, and household cleaners—to use for one year. Half of the families (controls) received regular products without added antibacterial chemicals, while the other half used products with the antibacterial chemicals.

When the families were surveyed after one year, those using the antibacterial products were just as likely to get sick, as identified by such symptoms as coughs, fevers, sore throats, vomiting, and diarrhea.

You may say that this is not surprising, as many of these symptoms are the result of a viral infection—and the antibacterial products are not effective on viruses. However, further analysis of the families indicated there were just as many bacterial infections in the antibacterial group as there were in the control group.

A 2007 study that reviewed 27 reports on the effectiveness of triclosan-containing antibacterial soaps found antibacterial soaps were no more effective than regular soap and water.

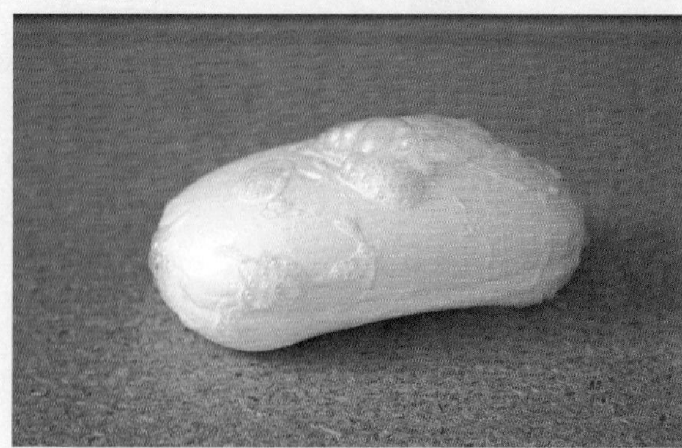

derivative is chlorhexidine, used commercially in Hibiclens, a common handwash found in hospital and clinical facilities.

A very commercialized phenol derivative in widespread use today is **triclosan**, which destroys bacterial cell membranes by blocking lipid synthesis. As a fairly mild and nontoxic chemical, it is moderately effective against bacterial pathogens (but less so against viruses and fungi). The chemical is incorporated in "antibacterial "soaps (see A Closer Look 11.3), lotions, mouthwashes, and toothpastes and in plastic and synthetic fibers used to make toys, food trays, underwear, kitchen sponges, utensils, and cutting boards. The negative side to overuse is that bacterial organisms may develop resistance to the chemical, just as they have developed resistance to antibiotics.

Aldehydes and Ethylene Oxide

The chemical agents we discussed in previous sections are used for disinfection. However, a few other chemical agents can sometimes be used as chemical sterilants.

Aldehydes are highly active agents inactivating proteins and nucleic acids. Two aldehydes, formaldehyde and glutaraldehyde, are useful as microbial control agents, but in closed environments (because the chemicals are toxic). Formaldehyde is a gas at high temperatures and a solid at room temperature. As a 37% solution it is called formalin. For over a century, formalin was used in embalming fluid for anatomical specimens (though rarely used anymore) and by morticians for disinfecting purposes. In microbiology, formalin has been used for inactivating viruses and toxins in certain vaccines.

Formalin can be used to disinfect surgical instruments, isolation rooms, and dialysis equipment. However, it leaves a residue, and instruments must be rinsed before use. Many allergic individuals develop a dermatitis reaction to this compound.

■ dermatitis: An inflammation of the skin.

Glutaraldehyde as a 2% solution destroys bacterial and fungal cells within 10 minutes and spores in 10 hours. It does not damage delicate objects, so it can be used to disinfect or sterilize optical equipment, such as the fiber-optic endoscopes used for arthroscopic surgery. Glutaraldehyde gives off irritating fumes, however, and instruments must be rinsed thoroughly in sterile water.

All chemical control agents mentioned thus far are usually used in a liquid form. One agent, though, is used in a gaseous form.

The development of plastics for use in science labs required a suitable method for sterilizing these heat-sensitive materials. In the 1950s, research scientists discovered the antimicrobial properties of **ethylene oxide**, making the plastic Petri (culture) dish and plastic (disposable) syringe possible. Ethylene oxide is a small molecule with excellent penetration capacity and it has the ability to kill endospores. However, it is toxic and explosive, so it must be used in a tightly sealed chamber flushed with inert gas for 8 to 12 hours to ensure that all traces of ethylene oxide are removed; otherwise, the chemical will cause "cold burns" on contact with the skin.

Manufacturers use ethylene oxide to sterilize paper, leather, wood, metal, and rubber products as well as plastics. In hospitals, it is used to sterilize catheters, artificial heart valves, heart-lung machine components, and optical equipment. The National Aeronautics and Space Administration (NASA) has used the gas for sterilizing interplanetary spacecraft. For sterilization purposes, ethylene oxide chambers have become the chemical counterparts of autoclaves.

11.3 Antimicrobial Drugs: Antibiotics and Other Agents

For many centuries, physicians believed heroic measures were necessary to save patients from the ravages of infectious disease. They prescribed frightening courses of purges and bloodlettings, enormous doses of strange chemical concoctions, ice water baths, starvation, and other drastic remedies. These treatments probably complicated an already bad situation by reducing the natural body defenses to the point of exhaustion. In fact, the death of George Washington in 1799 is believed to have been due to a streptococcal infection of the throat, perhaps exacerbated by the bloodletting treatment used to remove almost two liters of his blood within a 24-hour period.

But a transformation in medicine took place during the 1940s, when the antibiotics burst on the scene. Doctors were astonished to discover they could kill bacteria in the body without doing substantial harm to the body itself. The practice of medicine experienced a period of powerful, decisive therapy for infectious disease, and physicians found they could successfully alter the course of disease. The antibiotics effected a radical change in medicine, charting a new course that has been followed to the present day.

The First Antibacterials

When the germ theory of disease emerged in the late 1800s, the newly discovered information about microbes added considerably to the understanding of infectious disease. Furthermore, it increased the storehouse of knowledge available to physicians. However, it did not change the treatment of infected patients. Tuberculosis continued to kill one out of every seven people; streptococcal pneumonia was a fatal experience; and meningococcal meningitis exacted a heavy toll in human life.

Against this backdrop, scientists dreamed of chemical substances capable of killing microbes in the body without damaging the body. One such investigator was a German chemist named Paul Ehrlich. In the early 1900s, Ehrlich envisioned antimicrobial chemical substances as "magic bullets" that would seek out and destroy microbes in the body without damaging the tissues. To this end, he attempted to develop a chemical that would kill the bacterial agent of syphilis (then, as now, a serious disease). With his assistant Sahachiro Hata, Ehrlich synthesized hundreds of arsenic-based compounds and eventually identified a compound named arsphenamine. Although the compound showed promise, some physicians used it indiscriminately, resulting in overdoses for some patients and adverse reactions in the liver and kidneys. Moreover, World War I was beginning, and support for drug development was very limited. Soon, Ehrlich's magic bullet was forgotten.

Significant advances in drug therapy would not be made for another 20 years. Then, in 1932, a new chemical called prontosil showed promise against gram-positive bacterial species such as the staphylococci and streptococci. Prontosil's discoverer was a German investigator named Gerhard Domagk. In February 1935, Domagk's daughter Hildegarde became gravely ill with a blood infection after pricking her finger with a needle. Domagk decided to gamble with his new drug, and he gave her an injection of prontosil. When her health improved dramatically, the efficacy of the drug was established. Four years later, Domagk was awarded the 1939 Nobel Prize in Physiology or Medicine.

In the years thereafter, French investigators isolated sulfanilamide, the active substance in prontosil. At another period in history, scientists might have spent years researching the therapeutic value of sulfanilamide, but in the late 1930s, World War II was in progress in Europe, and sulfanilamide was a godsend to soldiers who might otherwise have died from wound-related infections. The success of sulfanilamide did not go unnoticed by medical researchers, who began an intense search for new antimicrobial substances, hoping to find something as good as or better than sulfanilamide. The word "antibiotic" was coined by Selman Waksman who discovered streptomycin, the first effective drug against tuberculosis. An **antibiotic** refers to those antimicrobial substances naturally produced by a few molds and bacterial species that inhibit growth or kill other bacterial species.

The Development of Penicillin

In 1939, René Dubos, a French-born American microbiologist, discovered soil bacteria capable of producing numerous antibacterial substances. At the same time, a group of English scientists at Oxford University were investigating an antimicrobial substance discovered many years before by a fellow Englishman. The Englishman was Alexander Fleming, a physician at St. Mary's Hospital in London.

In 1928, Fleming was performing research on staphylococci FIGURE 11.9a. Before going on vacation, he spread staphylococci on plates of nutrient agar and set them aside to incubate. On his return, he found one plate was contaminated by a green mold, which he identified as *Penicillium*. Fleming's attention was drawn to the clear area around the mold, an area where the staphylococci were unable to grow FIGURE 11.9b. Unsure what was happening, he cultivated the *Penicillium* in broth and added a small drop of the broth to a culture of staphylococci. As he looked on with astonishment, the staphylococci disintegrated before his eyes. Other gram-positive bacteria were equally susceptible to the mold broth, which he appropriately called **penicillin**. Unfortunately, Fleming could not isolate the active substance in the broth and eventually gave up work on the substance after publishing a paper describing penicillin.

■ *Penicillium*
pen-i-sil'lē-um

We now fast-forward to England in 1939 and the research group at Oxford University. Spurred by the success of sulfanilamide, the group led by Howard Florey and Ernst Chain searched the literature for antimicrobial substances and came upon

(a)

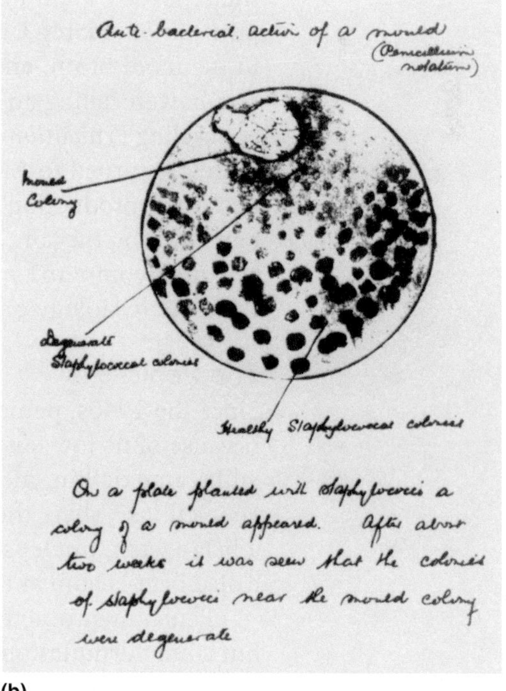

(b)

FIGURE 11.9 **Fleming and Penicillin.** (**a**) A painting by Dean Fausett of Fleming in his laboratory. (**b**) Fleming's notes on the inhibition of bacterial growth by the fungus *Penicillium*.

A CLOSER LOOK 11.4

Hiding a Treasure

Their timing could not have been worse. Howard Florey, Ernst Chain, Norman Heatley, and others of the team had rediscovered penicillin, purified it, and proved it useful in infected patients. But it was 1939, and German bombs were falling on London. This was a dangerous time to be doing research into new drugs and medicines. What would they do if there was a German invasion of England? If the enemy were to learn the secret of penicillin, the team would have to destroy all their work. So, how could they preserve the vital fungus yet keep it from falling into enemy hands?

Heatley made a suggestion. Each team member would rub the mold on the inside lining of his coat. The *Penicillium* mold spores would cling to the rough coat surface where the spores could survive for years (if necessary) in a dormant form. If an invasion did occur, hopefully at least one team member would make it to safety along with his "moldy coat." Then, in a safe country, the spores would be used to start new cultures and the research could continue. Of course, a German invasion of England did not occur, but the plan was an ingenious way to hide the treasured organism.

The whole penicillin story is well told in *The Mold in Dr. Flory's Coat* by Eric Lax (Henry Holt Publishers, 2004).

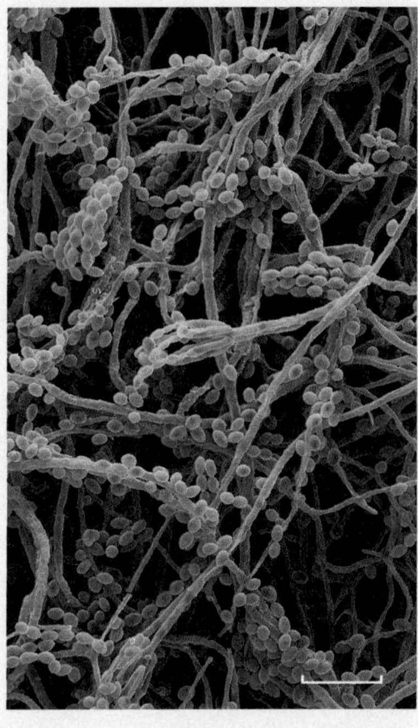

False-color scanning electron microscope image of *Penicillium* spores and hyphae. (Bar = 10 um.)

the report by Fleming. Using the newest biochemical methods for separation and purification, Florey, Chain, and their associates isolated penicillin, the active principle in the mold broth, and they began human trials. But England was at war; German bombs were falling on London, and the researchers feared for their lives, which meant concealing *Penicillium* from a possible German invasion (**A Closer Look 11.4**). They therefore turned to American companies to produce penicillin in industrial quantities. Soon production facilities in the Midwestern United States were churning out penicillin by the ton, and the Age of Antibiotics was underway. The international scientific community recognized Fleming, Florey, and Chain with the 1945 Nobel Prize in Physiology or Medicine.

The Penicillins

Since the 1940s, penicillin has remained the most widely used antibiotic, primarily because of its low cost and numerous derivatives (for example, penicillin G, ampicillin, amoxicillin, methicillin, and others). All of these relatives, referred to as the "penicillins," share the same fundamental structure, with a chemical complex called a beta-lactam nucleus at the core FIGURE 11.10. For this reason, they are sometimes called beta-lactam antibiotics.

Penicillins are active primarily against a variety of gram-positive bacterial species, but some formulations also affect gram-negative species, especially the diplococci that cause gonorrhea and meningitis. Penicillin and its derivatives function during the synthesis of the bacterial cell wall—they block the formation of the protein cross linking within the wall, thus resulting in a weak wall that gives way to internal pressures, causing the microbe to swell and burst. Penicillin is therefore most useful when bacterial cells are multiplying rapidly, as they do during an infection.

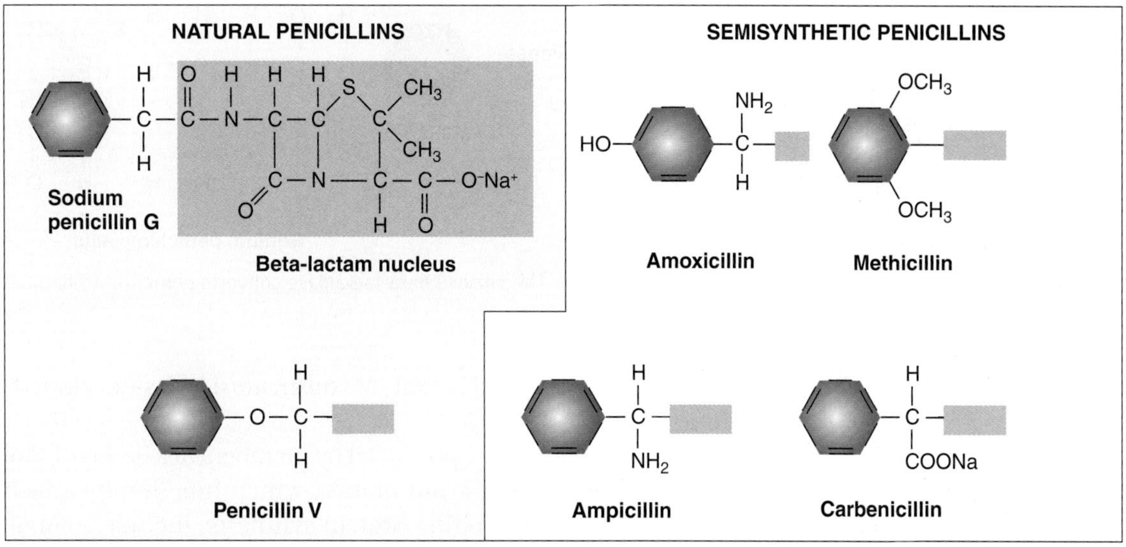

FIGURE 11.10 **Some Members of the Penicillin Group of Antibiotics.** The beta-lactam nucleus is common to all the penicillins. Different penicillins are formed by varying the side group on the molecule.

Over the decades of exposure, many bacterial species have developed resistance to penicillin, as we soon will discuss. These organisms can produce an enzyme called **penicillinase**, which converts penicillin into a harmless substance called penicilloic acid, shown in FIGURE 11.11. This ability has probably always existed in some bacterial species, but exposure to penicillin creates an environment in which penicillinase-producing forms are encouraged to emerge. As penicillin-susceptible cells die off, a form of natural selection takes place, and the rapid multiplication of penicillinase-producing bacterial cells results in populations resistant to the antibiotic.

Various penicillin derivatives, called **semisynthetic** penicillins, are produced by altering the side groups attached to the beta-lactam nucleus (see Figure 11.10). One derivative, ampicillin, is absorbed more readily than penicillin. Amoxicillin, another derivative, is more stable in stomach acid and does not bind to food, as many antibiotics are inclined to do. Other drugs in this family include methicillin and carbenicillin, both of which are used against organisms capable of developing resistance to penicillin itself. All the penicillins may induce allergic reactions in a patient, and some have been implicated in disturbance of the intestinal tract.

Cephalosporins and Aminoglycosides

A valuable alternative to the penicillin group of antibiotics is the **cephalosporin** group. The first cephalosporins were used against gram-positive cocci. They included cephalexin (Keflex) and cephalothin (Keflin). The newer members of this family are also used against gram-negative bacterial species. They include such drugs as cefotaxime (Claforan), ceftriaxone (Rocephin), and ceftazidime (Fortaz). All the cephalosporins resemble the penicillins except for a slight change to the beta-lactam nucleus. For this reason, they resist enzymes that destroy penicillins. Like penicillins, they are produced by a mold (*Cephalosporium* species), and they interfere with cell wall synthesis.

Cephalosporium
sef-ä-lō-spô'rē-um

Aminoglycoside antibiotics have traditionally been used against gram-negative bacteria. One of the first aminoglycoside antibiotics to be discovered was streptomycin. In 1943, this antibiotic caused a sensation because it helped cure patients of tuberculosis. More recently, other drugs have replaced streptomycin for

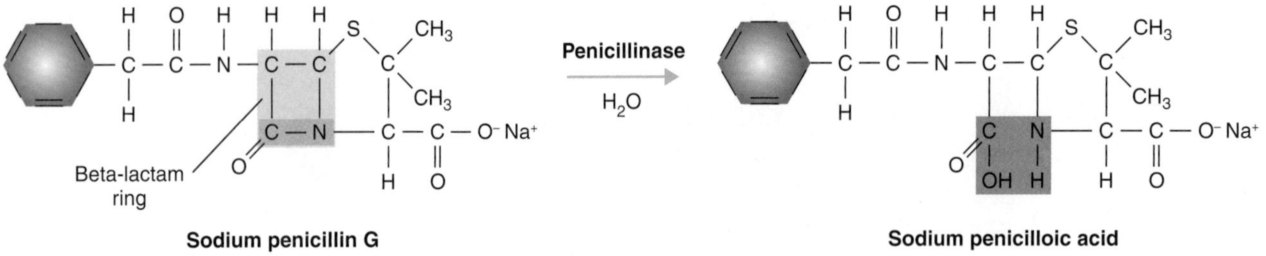

Penicillinase
H_2O
Beta-lactam ring
Sodium penicillin G
Sodium penicilloic acid

FIGURE 11.11 The Action of Beta-Lactamase on Sodium Penicillin G. The enzyme beta-lactamase converts penicillin to harmless penicilloic acid by opening the beta-lactam ring.

TB therapy because the causative agent, *M. tuberculosis*, has developed resistance to the antibiotic.

Streptomyces
strep-tō-mī′sēs

Aminoglycoside antibiotics are produced by members of a genus of mold-like soil bacteria called *Streptomyces*. These antibiotics, which function by attaching to the bacterial cell ribosomes and inhibiting protein synthesis, include: gentamicin, used against urinary tract infections; neomycin, for intestinal tract and eye infections; and many other "mycins" that are useful against gram-negative bacteria.

Broad-Spectrum Antibiotics

In 1947, scientists discovered the first **broad-spectrum antibiotic**; that is, an antibiotic capable of killing numerous types of microbes, both Gram positive and Gram negative. Isolated from a species of *Streptomyces*, this antibiotic was called chloramphenicol. Investigators discovered it was inhibitory to gram-positive as well as gram-negative bacterial species by attaching to a different site than streptomycin in the bacterial cell ribosome but still inhibiting protein synthesis. Although chloramphenicol is still used to treat certain diseases such as typhoid fever, it is very toxic and is considered a last-resort antibiotic. The drug causes bone marrow disturbances, resulting in red blood cells that lack hemoglobin.

Also in the 1940s, investigators found a broad-spectrum antibiotic named tetracycline. Tetracycline and its derivatives (minocycline, doxycycline, and others) remain the drugs of choice for many diseases caused by gram-negative bacteria. Because tetracycline has few side effects, many physicians prescribe this valuable antibiotic in trivial situations (such as for acne), and the overexposure encourages antibiotic-resistant bacteria to emerge. In the past, the antibiotic was often used in excessive quantities, which destroyed the normal bacterial microbiota in the large intestine, thus allowing yeasts to flourish (a similar phenomenon happens in the vaginal tract when too much tetracycline is used). Overuse of tetracycline may also result in a yellow-gray-brown discoloration of the teeth or stunted bones in children. Like many other antibiotics, it interferes with protein synthesis in bacteria by binding to bacterial ribosomes.

Other Antibiotics

Another antibiotic with clinical importance is erythromycin. This product of another *Streptomyces* species inhibits protein synthesis in many gram-positive bacterial species. It is particularly useful as an alternative in patients allergic to penicillin. Two other antibiotics in the same family as erythromycin are clarithromycin (Biaxin) and azithromycin (Zithromax), which are used against gram-negative bacterial pathogens.

Vancomycin has contemporary significance as a last-resort antibiotic for use against gram-positive bacteria having multidrug resistance. It is administered during serious cases of streptococcal, staphylococcal, and other gram-positive infections. Because of its damaging side effects in the ears and kidneys, vancomycin is not routinely used for trivial situations. Unfortunately, resistance to this antibiotic has also been observed, and substitute drugs have been sought.

Rifampin is a synthetic antibiotic prescribed for tuberculosis patients. It is also administered to individuals who harbor bacterial pathogens causing meningitis and for protective purposes to individuals who have been exposed to these bacterial species. Rifampin acts by interfering with RNA synthesis (transcription) in the bacterial cytoplasm. It may give an orange-red color to urine, feces, tears, and other body secretions and may cause liver damage.

Also acting on nucleic acids is the groups of antibiotics called the quinolones. These pharmaceutically designed and produced drugs block DNA replication. One of the most prescribed antibiotics in the United States is the fluoroquinolones such as ciprofloxacin (cipro) for urinary and intestinal tract infections, and gonorrhea.

Both bacitracin and polymyxin B are polypeptide antibiotics produced by *Bacillus* species. These antibiotics are generally restricted to use on the skin because, if taken internally, they are poorly absorbed from the intestine and may cause kidney damage. Bacitracin is available in pharmaceutical skin ointments and is effective against gram-positive bacteria such as staphylococci. Polymyxin B is useful against gram-negative bacilli, particularly those that cause superficial infections in wounds, abrasions, and burns. Bacitracin inhibits cell wall synthesis, while polymyxin B injures bacterial membranes.

FIGURE 11.12 summarizes the sites at which various antibiotics affect a bacterial cell.

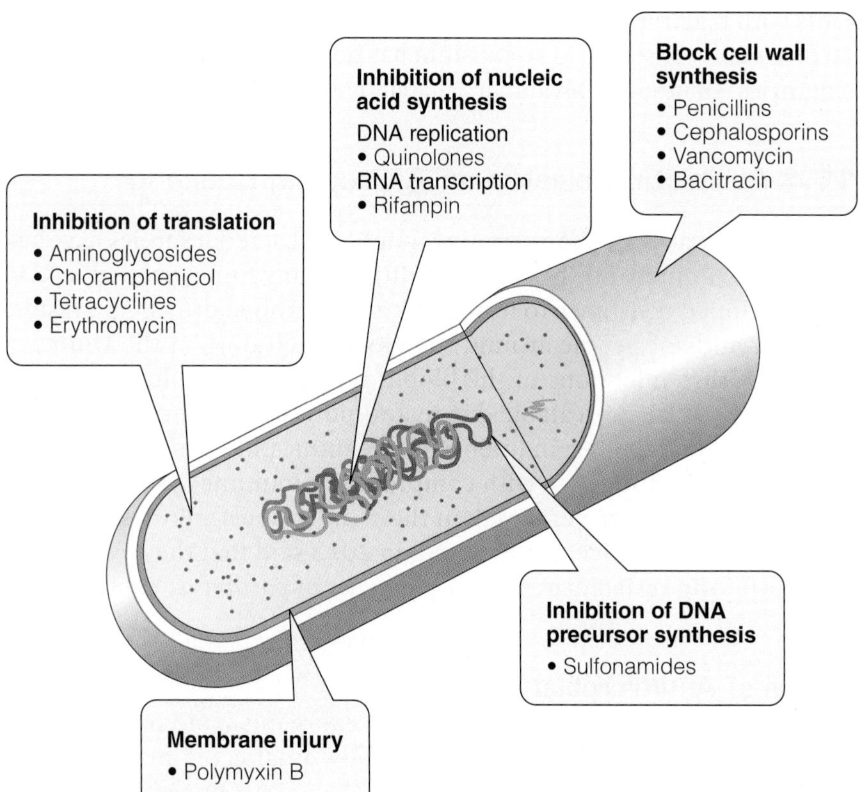

FIGURE 11.12 **The Targets for Antibacterial Agents.** There are six major targets for antibacterial agents: the cell wall, cell membrane, ribosomes (translation), nucleic acid synthesis (DNA replication and transcription), and metabolic reactions.

Antiviral, Antifungal, and Antiprotistan Drugs

Over the decades, scientists have developed other types of antimicrobial drugs to fight virus and fungal infections. To control viruses, the viral penetration, genome replication, and viral maturation processes can be interrupted with a variety of drugs, the majority having been designed and developed against AIDS and herpes virus infections, such as fever blisters and shingles. Drugs capable of blocking release of viruses from host cells also have been developed against the flu viruses.

Importantly, unlike antibiotics that can cure a patient of a bacterial infection, there are no antiviral drugs capable of curing a patient of a viral infection. The drugs may lessen the symptoms or make the disease of shorter duration, but they will not prevent or cure a disease.

When dealing with fungal disease, the medical mycologist has relatively few drugs available. One example is nystatin, which is effective against yeast infections due to *Candida albicans*. It acts by reacting with sterol compounds and thereby changing the structure of the cell membrane of the yeast. Other useful antifungal antibiotics include griseofulvin for ringworm and athlete's foot and amphotericin B for serious fungal diseases of the internal organs. Another class of antifungal drugs is the imidazoles, which also interfere with fungal cell membrane structure. Many of these compounds are used against *C. albicans*. They include clotrimazole (Lotrimin) and miconazole (Monistat).

■ *Candida albicans*
kan'did-ä al'bi-kanz

Finally, the same scenario applies to the drugs developed to fight some protistan parasitic diseases, primarily malaria, which is caused by several species of *Plasmodium*. In 2010, the WHO estimated there were about 220 million new malaria cases worldwide and over 660,000 people died, mostly (91%) in Africa. The current therapeutic drug, artemisinin, can swiftly reduce the number of *Plasmodium* parasites in the blood of patients with malaria. However, the parasite has become resistant to several antimalarial drugs and resistance to artemisinin has now been detected in four Southeast Asian countries: Cambodia, Myanmar, Thailand, and Viet Nam.

■ *Plasmodium*
plaz-mō'dē-um

11.4 Antibiotic Resistance: A Growing Challenge

During the last half a century, an alarming number of bacterial species have acquired resistance to one or more antibiotics. In addition, many fungi, protists, and viruses also have developed resistance to numerous antimicrobial drugs. As a result, hundreds of thousands of people around the globe, including in the United States, die every year from infections of the blood, intestinal tract, lungs, and urinary tract that have become untreatable due to antibiotic-resistant microbes. They are especially dangerous to those in intensive care units and burn wards as well as to children, the elderly, and those with compromised immune systems (because the body's natural defenses are very weak in these individuals). Professor Dame Sally Davies, Chief Medical Officer for England, in 2013 said that the development and spread of antibiotic resistance "*...is arguably as important as climate change for the world.*"

The Origins of Antimicrobial Resistance

Antimicrobial resistance (AMR) is when a microorganism or virus that was originally vulnerable to an antimicrobial drug is now unaffected by the drug. These resistant organisms, the so-called **superbugs** (bacterial, fungal, and protist species), and viruses have evolved ways to withstand the attack by the antimicrobials.

This means standard medical treatments become ineffective and infections persist, which is not only dangerous for the affected patient but also for society because now the superbug has an increasing chance of spreading to others in the community, nation, or even globally.

The evolution or origin of the superbugs is a natural phenomenon occurring when microorganisms are exposed to antimicrobial drugs. Microbes can evolve by chance beneficial mutations that provide the resistance mechanism (FIGURE 11.13). If a bacterial cell undergoes a mutation that changes the structure of a ribosome, and that bacterial cell now becomes exposed to an antibiotic that targets the ribosome, the organism may now be resistant to some of those antibiotics.

Perhaps more prevalent in today's world is the ability of bacterial species to undergo genetic recombination that generates AMR. With recombination, genes are randomly swapped between species through the process of horizontal gene transfer, as described in another chapter. In the process, a bacterial cell with an antibiotic resistance gene can transfer the gene or a copy of the gene to another bacterial cell previously susceptible to the antibiotic.

How might a bacterial cell become exposed to such antibiotics? Some bacterial species naturally found in Nature produce antibiotics to, among other things, limit the growth and spread of their competitors. Such antibiotic producers can typically be found in the soil, but probably just about anywhere there is competition between microbial species, some number produce antibiotics. Again, probably the biggest forcing ground for AMR is the excessive and often unnecessary prescribing by medical professionals of, and demanding by the public for, antibiotics for human illnesses. As the World Health Organization (WHO)

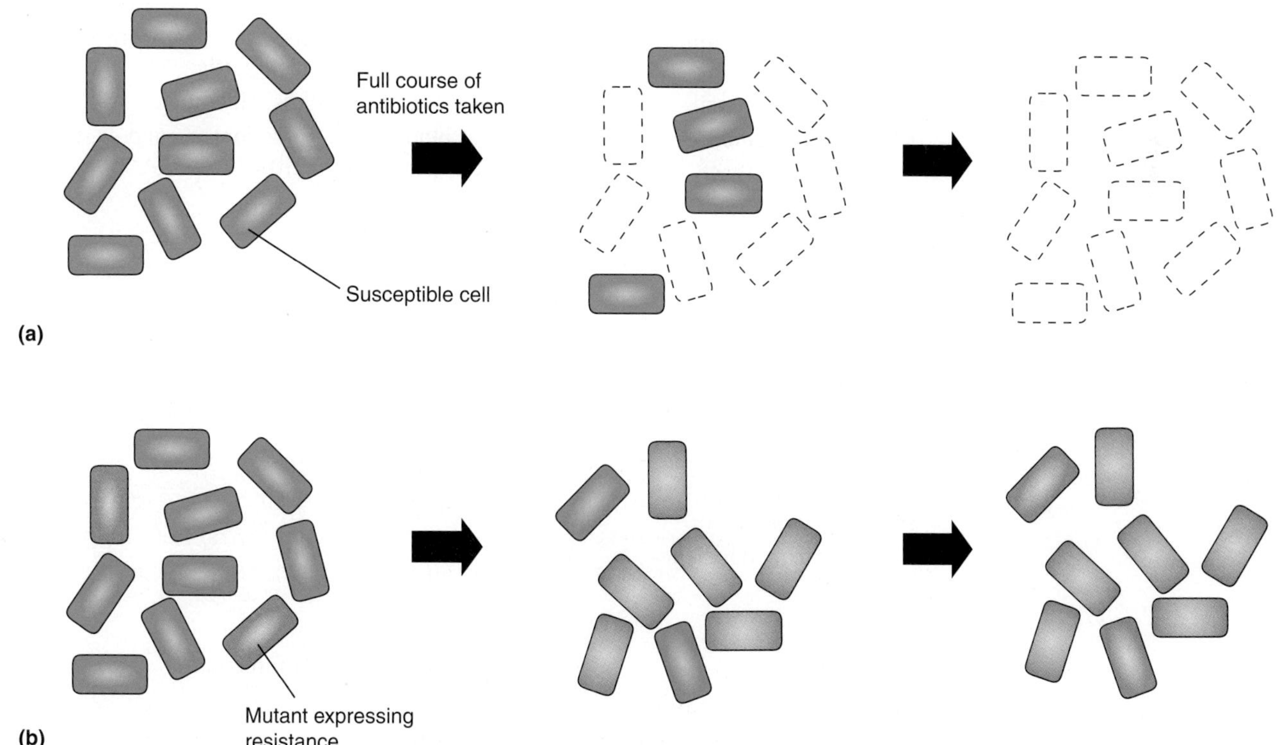

FIGURE 11.13 **The Possible Outcomes of Antibiotic Treatment.** (**a**) Ideally, with a complete course of antibiotics, all pathogens will be destroyed. (**b**) If there are some resistant cells in the infecting population, they will survive and grow without any competition.

has clearly stated, antimicrobial resistance today is a global concern—and here are some reasons why.

1. AMR Kills. Infections caused by resistant microorganisms often fail to respond to the standard treatment, resulting in prolonged illness and greater risk of death. The death rate for patients with serious infections treated in hospitals is about twice the rate in patients with infections caused by nonresistant bacterial pathogens. In addition, AMR reduces the effectiveness of treatment, meaning patients remain infectious for a longer time, which increases the risk of spreading resistant microorganisms to others.

As one example: Among the world's 12 million cases of tuberculosis (TB) in 2010, 650,000 cases involved multidrug-resistant TB (MDR-TB) strains of *M. tuberculosis*. Treatment of MDR-TB is extremely complicated, typically requiring up to 2 years of medication with toxic and expensive medicines, some of which are in constant short supply. Even with the best of care, only some 50% of these patients will be cured. And many other pathogens besides *M. tuberculosis* are developing resistance to one or more drugs at nearly the same rate. Many of these are contracted in the hospital (TABLE 11.2).

TABLE 11.2 Pathogens and Diseases Associated with AMR[1]

Pathogen	Diseases
Bacteria	
Acinetobacter baumannii[2]	Pneumonia, skin and wound infections, meningitis
Bacillus anthracis	Skin, intestinal, and respiratory diseases
Clostridium difficile	Diarrhea
Enterococcus faecium	Urinary tract infections, blood and heart infections, intestinal infections, and meningitis
Group B streptococci	Blood infections of newborns, elderly
Klebsiella pneumoniae	Pneumonia, bloodstream infections, wound or surgical site infections, and meningitis
Mycobacterium tuberculosis	Tuberculosis
Neisseria gonorrhoeae	Gonorrhea
Neisseria meningitidis	Childhood meningitis
Salmonella Typhi	Typhoid fever
Shigella dysenteriae	Diarrhea
Staphylococcus aureus	Wide variety of diseases
Streptococcus pneumoniae	Pneumonia, blood infections, sinus and middle ear infections
Viruses	
Influenza viruses	Influenza
Human immunodeficiency virus	AIDS
Fungi	
Candida albicans	Oral thrush, vaginitis, body infections
Protists	
Plasmodium species	Malaria

[1]Data are from the CDC.

[2]**Bold** = Currently cause the majority of US hospital infections.

2. AMR Threatens a Future that Will Be a Post-Antibiotic Era. If AMR continues to spread, it is very possible and probable that the infectious diseases we had been able to control and cure with antimicrobials will become untreatable and uncontrollable. In fact, a few like some tuberculosis strains essentially are untreatable today. If we run out of effective drugs, we are looking toward a post-antibiotic era. In fact, in 2012, Margaret Chan, Director-General of the WHO, said of antibiotics "*...the pipeline is virtually dry...the cupboard is nearly bare.*"

3. The Medical Community and an Uninformed Society Are Much to Blame. Very simply, AMR to a large part has been the result of abuse and misuse of antimicrobials by the medical community and by the public. Although mutations and genetic recombination are natural phenomena, certain human actions have greatly accelerated the emergence and spread of AMR.

The rise in antibiotic resistance is partly the result of improper use of antibiotics. For example, drug companies promote antibiotics heavily, patients pressure doctors for quick cures, and physicians sometimes misdiagnose infections or write prescriptions to avoid ordering costly tests to pinpoint a patient's illness.

Hospitals are another source for the emergence AMR. In many cases, physicians use unnecessarily large doses of antibiotics to prevent infection during and following surgery. This increases the possibility that other resistant strains will replace the susceptible normal microbiota destroyed by the antibiotic, causing a **superinfection**. As bacterial species have evolved and acquired more resistance to antibiotics, more and more species are becoming multidrug resistant (MDR). For example: many hospital and community strains of *S. aureus* (MRSA) are multidrug resistant; there are multidrug resistant and extensively drug-resistant strains of *M. tuberculosis* (MDR-TB and XDR-TB); and multidrug resistant strains of *Clostridium difficile* and *Acinetobacter baumannii* have increased in hospitals (see Table 11.2).

Clostridium difficile
klôs-tri'dē-um dif'fi-sil

Acinetobacter baumannii
a-si-ne'tō-bak-tėr bou-mä'nē-ē

Antibiotics also are abused in developing countries where they often are available without prescription. Some countries permit the over-the-counter sale of potent antibiotics, and large doses encourage resistance to develop.

Controlling the AMR Problem

It will never be possible to completely eliminate AMR. However, it can be controlled through a two-pronged approach.

First, as already described, we need to preserve the antibiotics we still have by using them prudently. Preventing abuse and misuse of antimicrobial drugs is something we and society as a whole can do to stem the spread of resistance. Control of AMR means taking and sustaining the following four steps.

1. **Do not ask for antibiotics.** For illnesses not treatable with antibiotics (e.g., colds, flu, most fevers, any viral infection), do not ask your doctor for an antibiotic. In fact, the physician should not prescribe antibiotics for such illnesses.
2. **Do not stop taking a prescribed antibiotic.** If an antibiotic is prescribed correctly for a bacterial illness, take the full course of therapy. Do not stop taking the antibiotic as soon as you start "feeling better." There still may be some pathogens "hanging around" in the body, so you need to make sure they all are eliminated.
3. **Do not save antibiotics for a future illness.** If for some reason you have "extra" antibiotic pills, do not save them for a future illness. At some time in the future, you might have similar symptoms of an illness. Do not assume it is the same illness you had previously and take the left over antibiotics. Also, if you have a very different illness, do not take the left over antibiotics. They probably will not work.

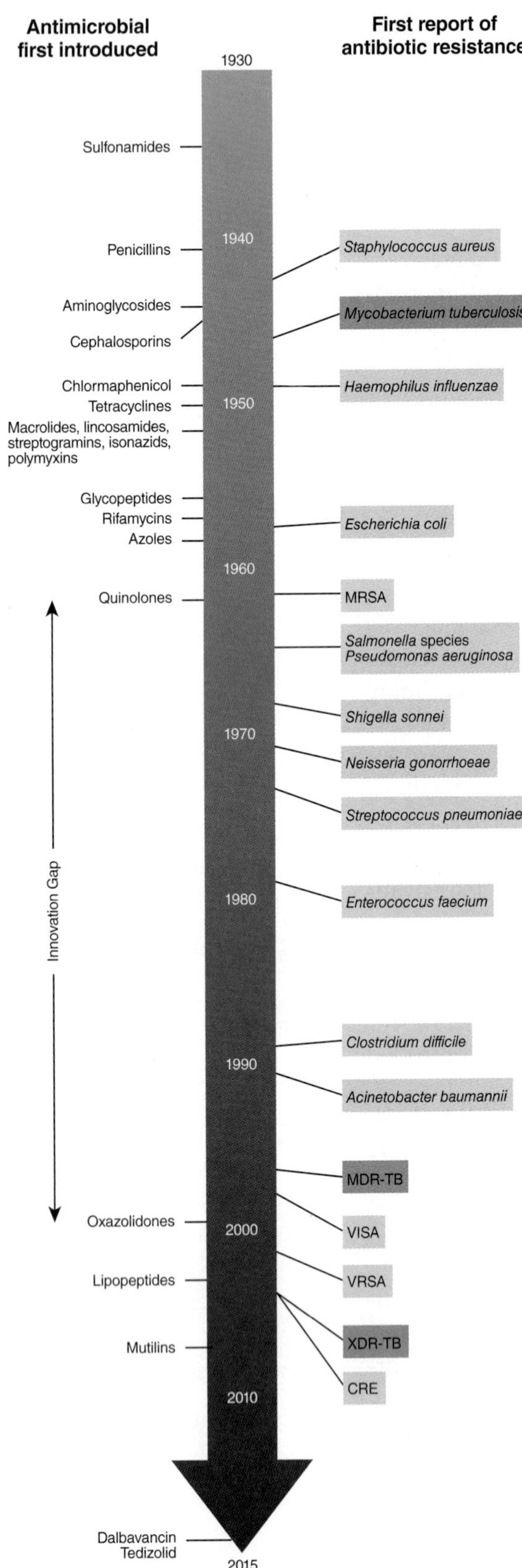

FIGURE 11.14 Timeline for Antibiotic Introduction and Appearance of Antibiotic Resistance. This timeline shows the time when major antibiotics were introduced (left side) and the approximate dates when some notable bacterial pathogens (pink = gram-negative species; purple = gram-positive species; red = acid-fast species) were identified as being antibiotic resistant (right side). MRSA (VISA, VRSA) = methicillin-resistant (vancomycin intermediary resistant; vancomycin-resistant) *Staphylococcus aureus;* MDR (XDR)-TB = multidrug-resistant (extensively drug-resistant) tuberculosis.

4. **Do not share your antibiotics with another person.** If you know someone who appears to have similar symptoms of an illness you had, don't give that person your antibiotics. You don't know if it actually is the same infection and you don't know if the person might be allergic to the antibiotic.

Appropriate use of antibiotics and completion of antibiotic treatment will limit antibiotic resistance and reduce its spread. As the CDC campaign has declared: "*Get Smart. Know when antibiotics work.*"

The second part of the approach is a bit more involved and challenging. You might ask, "*If we have fewer effective antibiotics, let's discover and develop new antibiotics to which bacterial and other microbial species are vulnerable.*" Excellent idea but it hasn't really happened as of late. Between the late 1930s and the 1960s, most all major groups of antibiotics were discovered, developed, and used. However, the following decades represent an "innovation gap" where no new antibiotic classes were introduced by pharmaceutical companies FIGURE 11.14. Importantly, during this gap, an increasing number of bacterial species became resistant to many of these antibiotics.

There are three reasons for the lack of development of new antimicrobials. First, due to the perception in the late 1960s that antibiotics would cure all infectious diseases, pharmaceutical companies and research organizations stopped or severely slowed the development of new antibiotics even though it was obvious many microbes and pathogens were becoming resistant to them. Second, the cost to develop and bring an antibiotic to market is staggering, requiring close to $1 billion over 10 years. Pharmaceutical companies say that the relatively large research cost and time for development is lost if the medication does not make it to market. And third, pharmaceutical companies find there is a much higher financial reward in developing and marketing medications for the treatment of chronic illnesses (e.g., depression, hypertension, diabetes, cancer, cholesterol, arthritis) because, unlike antibiotics, which are usually given for a short 5 to 14 day period and then discontinued, medications for chronic illnesses may be taken for a lifetime. So, the development of new antimicrobial agents is declining at a time when there is a pressing need for new agents. As Margaret Chan said in her 2012 report:

"A post-antibiotic era means, in effect, an end to modern medicine as we know it. Things as common as strep throat or a child's

scratched knee could once again kill. Some sophisticated interventions, like hip replacements, organ transplants, cancer chemotherapy, and care of preterm infants, would become far more difficult or even too dangerous to undertake. At a time of multiple calamities in the world, we cannot allow the loss of essential antimicrobials, essential cures for many millions of people, to become the next global crisis."

It is time to accelerate the development and supply of new antimicrobial drugs. Although antimicrobial drugs are not the perfect magic bullets once perceived by Ehrlich, new drug discovery may provide us with many new antimicrobial agents to fight pathogens and infectious disease.

A Final Thought

In the last 100 years, there were two periods in which the incidence of disease declined sharply. The first was in the early 1900s, when a new understanding of the disease process led to numerous social measures, such as water purification, care in food production, control of insects, milk pasteurization, and patient isolation. Sanitary practices such as these made it possible to prevent virulent microbes from reaching their human targets.

The second period began in the 1940s with the development of antibiotics and blossomed in the years thereafter, when physicians found they could treat established cases of disease. Major health gains were made as serious illnesses came under control. An outgrowth of these successes has been the belief by many people that science can cure any infectious disease. A shot of this, a tablet for that, and then perfect health. Right? Unfortunately not.

Scientists may show us how to avoid infectious microbes and doctors can control certain diseases with antimicrobial drugs, but the ultimate body defense relies on the immune system and other natural measures of resistance. Used correctly, antimicrobials provide an additional chemical defense to help our natural immune defenses overcome pathogenic microbes. The antimicrobials supplement natural defenses; they do not replace them.

The great advances in antimicrobial research should be viewed with caution. Antibiotics have undoubtedly relieved much misery and suffering, but they are not the cure-all some people perceive them to be. In the end, it is well to remember that good health comes from within, not from without.

Questions to Consider

1. Instead of saying "food has been irradiated," processors indicate it has been "cold pasteurized" Why do you believe they use this term? Do you think it is appropriate? What will it take for the food-processing industry to use the correct term? Can you think of any other euphemism like this used about foods?
2. Of all the sterilization methods reviewed in this chapter, why do you think none has been widely adapted to the sterilization of milk? Which, in your opinion, holds the most promise?
3. While on a camping trip, you find a luxury hotel has been built near the stream where you once swam and from which you drank freely. Fearing contamination of the stream, you decide it would be wise to use some form of disinfection before

drinking the water. The nearest town has only a grocery store, pharmacy, and post office. What might you purchase? Why?

4. With over 11 million children currently attending day-care centers in the United States, the possibilities for disease transmission among children have mounted considerably. Under what circumstances may antiseptics and disinfectants be used to preclude the spread of microorganisms?
5. A brochure called Operation Clean Hands lists several times when individuals should wash their hands thoroughly. One list entitled "Before you" includes "prepare food" and "insert contact lenses." A second list entitled "After you" includes "change a diaper" and "play with an animal." How many items can you add to each list?
6. One of the novel approaches to treating gum disease is to impregnate tiny vinyl bands with antibiotic, stretch them across the teeth, and push them beneath the gumline. Presumably, the antibiotic would kill bacteria that form pockets of infection in the gums. What might be the advantages and disadvantages of this therapeutic device?
7. The antibiotic issue can be argued from two perspectives. Some people contend that because of side effects and microbial resistance, medical use of antibiotics will eventually be abandoned. Others see the future development of a superantibiotic, a type of "miracle drug." What arguments can you offer for either view? Which do you see as more likely to be true?

Key Terms

Informative facts are necessary for the expression of every concept, and the information for a concept is founded in a set of key terms. The following terms form the basis for the concepts of this chapter. On completing the chapter, you should be able to explain and/or define each one.

alcohol
aldehyde
aminoglycoside
antibiotic
antimicrobial resistance (AMR)
antisepsis
antiseptic
autoclave
biological safety cabinet
broad-spectrum antibiotic
cephalosporin
denaturation
desiccation
disinfectant
disinfection
ethylene oxide
filtration
flash pasteurization method
halogen
heavy metal
high efficiency particle air (HEPA) filter
holding (batch) method
hydrogen peroxide
ionizing radiation
lyophilization
osmosis
pasteurization
pasteurizing dose
penicillin
penicillinase
peroxide
phenol
semisynthetic
sterilization
superbug
superinfection
triclosan
ultra-high temperature (UHT) method
ultraviolet radiation (UV light)

Microbes and Human Affairs

In 1944, a doctor traveling in Europe noticed that the kitchens of many homes had a loaf of moldy bread hanging in one corner. He inquired about the bread and was told that when a family member sustained a wound or abrasion, a sliver of the bread was mixed with water to form a paste, and the paste was applied to the skin. A wound so treated was less likely to become infected than one left untreated. Modern scientists speculate the moldy bread probably contained an unidentified chemical, most likely an antibiotic being produced by the mold.

Was this the first time microbes had unwittingly helped out humans with an antibiotic? Apparently not. In 1980, a graduate student from Detroit was examining slides of bone tissue when she noticed a peculiar yellow-green glow coming from the tissue. Her colleagues identified the source of the emission as the antibiotic tetracycline. What made the finding remarkable was that the bone tissue was from a 1,600-year-old Nubian mummy excavated from the floodplain of the Nile River. Scientists knew the ancient Nubians were unusually free of disease, and now they had a possible answer why—it was from the tetracycline they consumed. But how did they consume the tetracycline? The most likely source of tetracycline was in the beer the ancient Nubians drank. Scientists believe the grain used to ferment the beer contained one of the *Streptomyces* species that naturally produces tetracycline. And based on the levels of the antibiotic in their bones, this probably was no accidental contamination. The Nubians were consciously producing the drug in the beer fermentation process, which would make this one of the earliest examples of biotechnology and 1,600 years before the "modern discovery" of tetracycline!

Microbes have been helping us out for a very long time and certainly long before any of us suspected. How microbes have woven their magic in our lives will be the major theme of Part II of *Microbes and Society*. We will see evidence of their beneficial role in Chapter 12 as we visit a restaurant and experience a menu from a microbial point of view. Next, Chapter 13 reveals how microbes contaminate foods and how scientists limit the contamination. In Chapter 14, we explore how microbes make possible a stunning variety of products ranging from oral contraceptives to vitamins. Chapter 15 describes the chores microbes perform on the farm, showing how essential they are to producing our meat and dairy products, and how modern technologists use them to hold off the frost, make insecticides, and create veggie vaccines. Our look at the bright side concludes in Chapter 16, which outlines how microbes continue to protect our environment while making possible life itself.

As we all know, there is a darker side to the microbes, and we spend time discussing their disease-causing aspects in the final three chapters of this book: Chapter 17 describes the disease process and the means by which our bodies develop resistance; Chapters 18 and 19 present a "conversational" look at familiar and not-so-familiar viral and bacterial diseases.

12 Microbes and Food: A Menu of Microbial Delights

Many food and beverage products we consume and enjoy every day are either created by certain species of microbes or they are part of the production process in making the commercial product.

Looking Ahead

Although we often think of microbes as food contaminants, many species actually play vital roles in producing the foods we enjoy, as we will discover in this chapter.

On completing this chapter, you should be capable of:

- Describing the process of wine fermentation.
- Comparing the roles of microbes in the fermentation processes for olives and cheese making.
- Explaining the role of yeast in bread making.
- Outlining the fermentation process for sausages and sauerkraut.
- Discussing the process of brewing beer.

Over the decades, many food-related microbes have received bad press: They have been linked to numerous disease outbreaks associated with foods, and sometimes they even received the blame although no trace of a microbe could be found. One possible reason for this negative reputation is because people often are afraid of things they cannot see, and microbes certainly meet this criterion. To be sure, a small minority of microbial species is associated with disease in foods, but these few species hardly provide sufficient rationale for condemning the thousands of harmless species.

This chapter emphasizes the beneficial roles for the food-related microbes and their necessity for producing and processing many of the items we eat and drink. Indeed, we will encounter numerous examples of the substantial contributions microbes make to the quality of our gastronomic lives.

We will use a fine restaurant as our venue to explore the relationships between microbes and foods, for it is here that the end products of microbes come to please our palates and heighten our senses. Many of the foods we will order are culturally unique. For example, in the United States, Germany, and several other countries, sauerkraut is consumed liberally with frankfurters, sausages, and other meats; or it is eaten separately as a vegetable. By contrast, in other parts of the world, people see sauerkraut as spoiled cabbage and are quick to discard it. Ironically, what is "spoiled" food in one culture is relished in another.

A recurring theme in our exploration of microbes and foods is fermentation. **Fermentation** refers not only to the process that produces alcoholic beverages, but to any partial breakdown of carbohydrates taking place in the absence of oxygen gas. Microbes are the tools of fermentation because they produce the enzymes necessary for the chemistry to take place. A variety of organic substances such as alcohols and acids result from fermentation. These products add distinctive aromas and tastes to many foods, including those pictured in FIGURE 12.1. Moreover, they act as preservatives, making the foods safe to eat by holding in check any dangerous microbes that may be present.

Fermentation also brings nutritional benefits because the process converts complex substances to simpler nutrients our body needs. For example, we will encounter

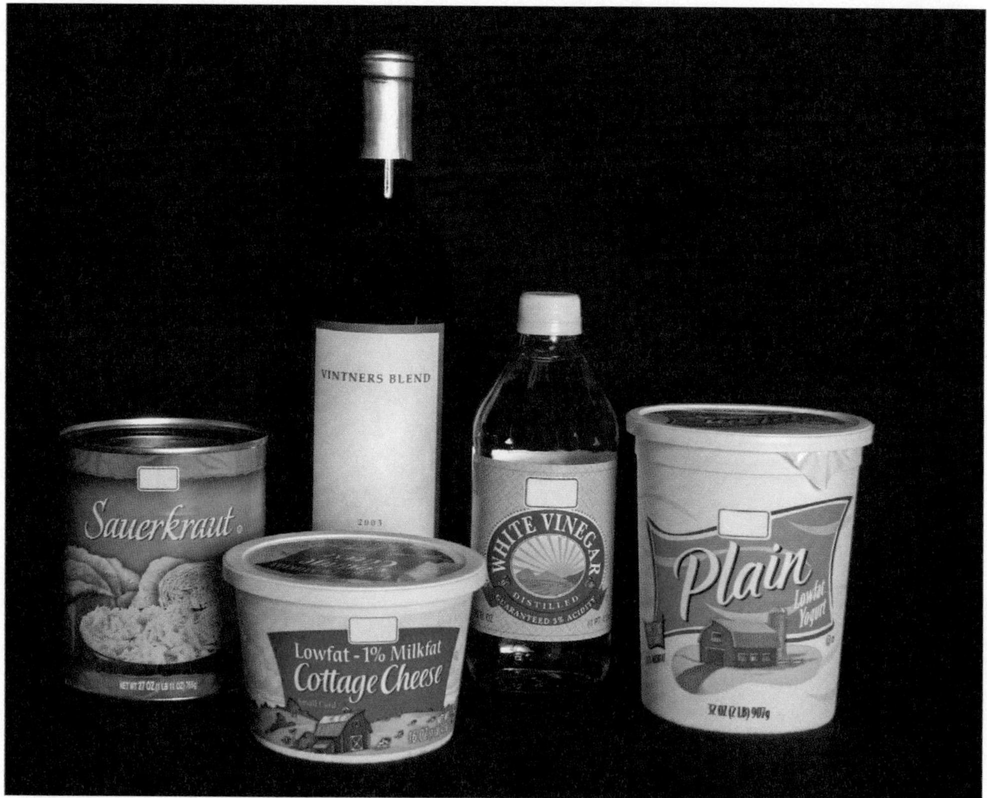

FIGURE 12.1 **An Array of Foods and Beverages Produced by Microorganisms.** Many foods and beverages are the result of microbial fermentations.

many acids produced from complex carbohydrates, such as starches. By converting the carbohydrate to an acid, the microbe preserves much of the energy originally present in the molecule. This energy can be released when the body metabolizes the acid. As another example, ethyl alcohol (in alcoholic beverages) is usually regarded as toxic, but it contains 90% of the caloric value present in glucose, the starting point of the alcohol fermentation process.

12.1 Beginning Our Meal: Bon Appétit!

With this introduction and background, we are now ready to consider how microbes help us enjoy a meal. To begin, we enter the restaurant and are seated at our table by our charming hostess, who extends the manager's greetings. Then, our attentive waiter approaches and inquires if we would like to start with a glass of wine.

A Glass of Wine

In ancient times, humans must have been awestruck by **wine**, for it was powerful enough to turn the mind, even though it began as mere grape juice. It was called *aqua vitae,* the "water of life" for much of history, because people marveled at the mystical abilities of this beverage and wondered how it came about. We may wonder, too, but for the moment, we will set aside the technicalities of fermentation. We must first decide if we prefer white wine or red wine. For making red wine, black grapes are used, while white wine can be fermented from black grapes or white grapes. Once we have made our selection, the waiter leaves to fill our order. Now we can consider how alcoholic fermentation works.

■ aerobic respiration: A set of the metabolic reactions and processes that take place in the cells of organisms to convert biochemical energy from nutrients into cellular energy in the form of ATP.

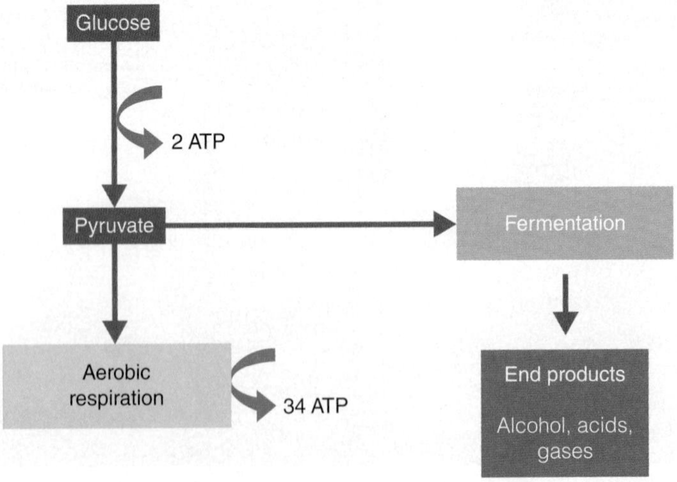

As long as oxygen is present in their environment, many microbes—including yeasts—will live contentedly. If the oxygen is removed, however, yeast cells shift their metabolism to a form of oxygen-free chemistry known as fermentation (FIGURE 12.2). Rather than shuttling pyruvate into aerobic respiration, yeasts use the acid as an acceptor of electrons and convert it to ethyl alcohol. This somewhat complex

FIGURE 12.2 **A Metabolic Map of Aerobic and Fermentation Pathways for ATP Production.** The production of ATP by microorganisms can be achieved through an aerobic respiration pathway or, in the absence of oxygen gas, some microbes can make a small amount of ATP through a fermentation pathway. The end products of that pathway are of human importance.

chemistry, which is the key to fermentation, was explored in depth in another chapter.

Before we go much further, we should note that few organisms other than yeasts can convert pyruvate to ethyl alcohol. Furthermore, this conversion is not particularly desirable for the yeast cells because they can gain more energy for their life processes by putting the pyruvate into aerobic respiration than by using it for fermentation. But the yeast cells have no choice when there is no oxygen in their environment; without using fermentation at those times, they would quickly die. For yeasts, fermentation therefore brings an evolutionary advantage in the struggle to survive—they can subsist by making a little bit of ATP; for vintners, fermentation is economically advantageous as the basis for the wine industry—a $34.6 billion industry in the United States alone in 2012. In 2011, the United States ranked 36th in per capita wine consumption (10.5 liters). At the top were several European nations where the per capita consumption topped 50 liters.

FIGURE 12.3 **Wild Yeasts.** Although cultured yeasts are usually used for wine making, when growing on the vine, grape skins are covered with wild yeast species as seen in the "coating" covering the grapes.

© Ellen Isaacs/Alamy

Wild yeasts are plentiful wherever there are vineyards, which is not unexpected because the fruit is the source of carbohydrates for energy. The haze on grapes is a layer of yeasts (FIGURE 12.3), and if those yeasts penetrate the skin and enter the soft flesh of the grape, they will ferment its carbohydrates and produce alcoholic grape juice. The same process occurs with other fruits or grains: Yeasts growing in peach juice will produce peach-flavored ethyl alcohol; yeasts in potato will produce unflavored ethyl alcohol; yeasts in orange juice will produce orange-flavored ethyl alcohol; and yeasts in barley grain will produce barley-flavored ethyl alcohol, the alcohol in beer.

The most common fermentation is performed by inoculating the grape juice with cultured yeasts, usually *Saccharomyces cerevisiae* (FIGURE 12.4). Because wild yeasts can produce undesirable qualities in a wine, they are killed by adding sulfur dioxide ("sulfites") during the production process.

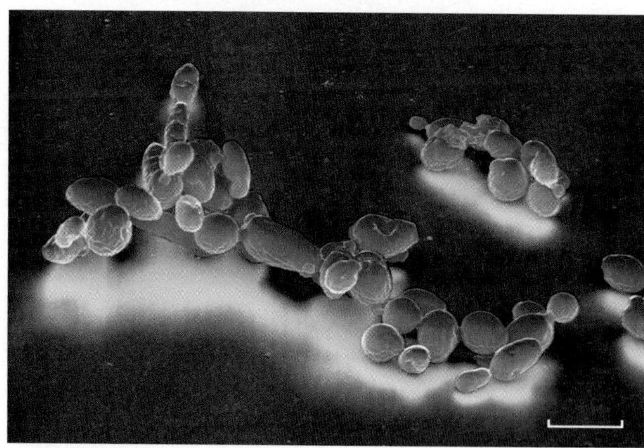

FIGURE 12.4 *Saccharomyces cerevisiae*. A false-color scanning electron microscope image of the yeast *S. cerevisiae*, the most common cultured yeast used to make wine. (Bar = 10 um.)

© Eye of Science/Science Source

First, the grape juice—known as **must**—bubbles intensely and froths from the carbon dioxide produced during aerobic respiration, as oxygen gas is still present during this step. The carbon dioxide fills all the air spaces in the juice, and, so long as the juice remains still, the environment becomes oxygen-free, or **anaerobic**. When the oxygen is depleted, the yeast cells shift their metabolism from aerobic respiration to fermentation and start producing ethyl alcohol. But there is a limit to this process—when the percentage of alcohol in the wine reaches about 14% to 18%, the population of yeast cells starts to die due to the toxic effect of the high alcohol content. Thus, no natural wines have an alcoholic content higher than about 15%. The alcohol content of individual wines varies even among the same harvest due to the subtle differences in the fermentation process. Some healthful benefits of wine are explored in **A Closer Look 12.1**.

■ *Saccharomyces cerevisiae*
sak-ä-rōmī'sēs se-ri-vis'ē-ī

It is difficult to locate the roots of winemaking, but archeological evidence suggests the techniques were well developed at least 6,000 years ago. Apparently the Egyptians taught the art of winemaking to the Greeks, who established the first notable vineyards in regions bordering the Mediterranean Sea. The Romans organized winemaking into a highly efficient operation to supply the needs of their troops, and their later conquests spread the art of winemaking to Northern Europe. Wine eventually became a preferred

A CLOSER LOOK 12.1

The Benefits of Red Wine

In 1992, scientists from the Bordeaux region of France did a population and epidemiological study and noted that while French and other Mediterranean peoples ate large amounts of fatty foods, they suffered a relatively low incidence of coronary artery disease. A 1996 report by the television news magazine program *60 Minutes* also pointed out that fatty meats, creams, butters, and sauces had little apparent effect on French hearts. This so-called "French Paradox" was due, in part, to the apparent medicinal properties of wine, especially drinking red wine. So, is red wine good for your health?

In 1996, researchers identified phenol-based compounds in red wine that inhibited the oxidation of low-density lipoproteins (LDLs) in the blood, and by doing so, prevented the buildup of cholesterol and blood platelets in the arteries. The scientists pointed out that red wine contains more phenolics than white wine, and far more than beer. The phenolics also were present in other foods (e.g., raisins and onions), but not in the quantity found in the skins of grapes used for red wine, which concentrate even more after fermentation has taken place.

Courtesy of Dr. Jeffrey Pommerville

A compound known as resveratrol, which belongs to the family of phenols, naturally exists in grapes and red wine. This compound was shown to extend the lifespan of yeast cells by up to 80% and might help explain why moderate consumption of red wine has been linked to a lower incidence of heart disease in humans and cancer in experimental mice. In fact, the effects of resveratrol remain effective at fighting cancer even after the body's metabolism has converted it into other compounds. On entering cells, the metabolized compound is reconverted to resveratrol in a more concentrated form. And in 2013, more good health news was reported about red wine. Researchers in Canada and Australia discovered 23 additional molecules in red wine that may hold more potential health benefits for wine drinkers.

Red wine and the resveratrol it contains also have an inhibitory effect on the foodborne pathogens *Escherichia coli*, *Salmonella enterica*, and *Listeria monocytogenes*. *Helicobacter pylori*, which is the main cause of stomach ulcers, is especially susceptible.

However, red wine is not for everyone. For those wishing to stay away from alcohol, alcohol-free red wines are appearing in the marketplace. They offer the opportunity to take advantage of a natural health ingredient while enjoying a glass of nature's bounty. Perhaps Louis Pasteur was correct when he said, *"Wine is the most healthful and hygienic of beverages."*

alternative to water, which was usually contaminated with microbes and often caused horrible intestinal diseases. The Romans thought so highly of wine that they adopted the Greek god of wine (Dionysus) but gave him their own name (Bacchus).

■ *Vitis vinifera*
vi'tis vi-ni'fėr-ä

Among grapes, the species *Vitis vinifera* is recognized as producing the highest-quality fruit for winemaking. Characteristics of the grapes are determined by soil and climate conditions, such as temperature, humidity, and amount of water. Thus, wine varies, and there are "vintage years" and "poor years." Although the production of alcohol by fermentation requires only a few days, the aging process may go on for weeks or months. Wooden casks have traditionally been used for aging because the wine develops its unique flavor, aroma, and bouquet with help from organic molecules in the wood (FIGURE 12.5). Today, more and more wines are being aged in giant stainless-steel tanks and the desired "flavors" are imparted by adding chips, chunks, or even whole planks of wine-barrel wood suspended inside the tank.

The basic fermentation process is varied to obtain a broad variety of wine types. To produce a dry (still) wine, for instance, most all the sugar in the grape juice is allowed to break down. Should a sweet wine be desired, some sugar is left unfermented.

(a)

(b)

FIGURE 12.5 **The Large-Scale Production of Wine.** (**a**) After separating the stems and seeds, the juice (often with the skins) goes into large, temperature controlled, stainless steel fermentation tanks for primary fermentation before barrel aging. (**b**) Many white wines are aged in stainless steel fermentation tanks.

(a) © Tips Images/Tips Italia Srl a socio unico/Alamy; (b) Courtesy of Dr. Jeffrey Pommerville.

For a sparkling wine, such as champagne, a second fermentation takes place inside the bottle: Sugar cubes are added to the wine after the first fermentation, and the yeast is encouraged to continue fermenting the sugar within the bottle. Carbon dioxide builds up and adds the sparkling bubbles to the wine. A thick bottle is needed for champagne because the gas pressure would cause ordinary glass to break, and a wire cage is used to prevent the cork from popping out. By the way, if you purchase a bottle of "cheap champagne," it does not necessarily mean that quality is lacking. Some champagnes are "cheap" because they are mass produced in large vats rather than handled as individual bottles.

Most table wines average about 10% to 12% alcohol. Exceptions are the fortified wines, such as port, sherry, and Madeira. These wines have an alcohol content approaching 22%. They are produced by adding brandy or other spirits to the wine following fermentation. Because fortified wines are generally considered dessert wines, we will pass on them for now.

12.2 First Course: The Appetizers

As we anticipate our meal, our palates will first be stimulated by fermented olives and cheese brought to our table by our waiter.

Olives

Olives have traditionally represented the abundance of life, and in many cultures the olive branch is a symbol of peace. Unfortunately, the natural taste of olives is quite bitter.

Tradition and microbial fermentation resolved the bitterness problem well before the chemistry was understood. In regions of Western Europe, unripe olives were soaked in lye (sodium hydroxide) to neutralize their bitter taste, then washed and covered with brine. Next, the olives were sealed in casks, and bacteria normally present on their skins would ferment the carbohydrates. Some weeks later, when the fermentation was complete, the tasty Spanish, or "green," olive resulted (FIGURE 12.6).

In Greece, olives were eaten without the benefit of fermentation, and so they had to be preserved for later use. This was accomplished by allowing the olives to ripen on the tree, then picking them and exposing them to the air for weeks. During this interval, chemical conversions of the tannin compounds in the olive skins created black deposits, yielding Greek, or "black," olives. The Italians modified the process by placing the black olives in salt and encouraging the naturally occurring microbes in the olive skins to carry out fermentation. And certainly lots of olives are processed into olive oil.

FIGURE 12.6 **Green and Black Olives.** Unripe green olives are fermented for several weeks while black olives are allowed to ripen before fermentation.

Courtesy of Dr. Jeffrey Pommerville

Cheese

Our Spanish, Greek, and Italian olives are accompanied by crackers and assorted cheeses. Cheese results when microbes interact with casein, the major protein in milk. In the dairy plant, the microbes produce enzymes that join with added enzymes to curdle the casein. These so-called **curds** are then separated as "unripened cheese" or the familiar cottage cheese. The remaining fluid is called **whey**.

To prepare different kinds of ripened cheese, the curds are washed, and salt is added to flavor the curds and prevent spoilage. Then, cultures of microbes are added to the curds. For example, if a fine Swiss cheese is to be made, two different bacterial species are mixed in. During the aging process, these microbes bring about the unique chemical changes in the available proteins and carbohydrates: Species of *Lactobacillus* produce acids to lend sourness, and species of *Propionibacterium* produce a variety of organic molecules as well as carbon dioxide gas. The organic compounds give Swiss cheese its distinctively nutty flavor and the carbon dioxide accumulates as the holes, or eyes, in the cheese. FIGURE 12.7 displays the complete process used to produce such cheese.

■ *Lactobacillus*
lak-tō-bä-sil'lus

■ *Propionibacterium*
prō-pē-on'ē-bak-ti-rē-um

(a) (b) (c) (d) (e) (f)

FIGURE 12.7 **A General Home Cheesemaking Process.** (**a**) The milk is mixed with the enzyme rennet and heated. (**b**) After the curd has formed, the cheese maker cuts the curd with a knife. (**c**) The curds are gently poured into and collected on a large piece of cheesecloth. (**d**) Salt is added before (**e**) being pressed in a cheese press (for hard and semi-hard cheeses). The cheese may be coated in wax before (**f**) being aged (in this case in a mini-refrigerator).

Courtesy of Rick Robinson

FIGURE 12.8 **Soft Cheeses.** Roquefort, a blue-green veined cheese contains the blue-green fungus *Penicillium roqueforti* that grows within the cheese. With Camembert, the fungus *P. camemberti* grows on the outside of the cheese. (Bars = 10 um.)

However, we may instead select a mold-ripened cheese for our appetizer. Among our choices in this category are Camembert and Roquefort cheeses (FIGURE 12.8). Camembert, a soft cheese, is made by dipping milk curds into the fungus *Penicillium camemberti*. As the fungus grows on the outside of the curds, it digests the proteins and softens the cheese. To make Roquefort, a blue-green veined cheese, the curds are rolled with spores of the blue-green fungus *Penicillium roqueforti*. The fungus penetrates cracks in the curds and grows within them, thereby creating the distinctive blue-green veins in the cheese.

Penicillium camembert
pen-i-sil′lē-um kam-am-bė r′tē

Penicillium roqueforti
pen-i-sil′lē-um rō-kō-fôr′tē

These are just a sampling of the numerous cheeses that owe their existence to microbes. We consider other cheeses as well as many other dairy products in another chapter.

12.3 The Salad Course: Of Salad and Bread

For our salad course, we select an assortment of mixed greens and add some other healthful and nutritious vegetables, such as tomatoes, red onions, and cucumbers. On our salad, we order an oil- and-vinegar dressing (vinaigrette). Although microbes cannot claim to have produced the greens or the oil, they are essential for producing the vinegar.

Vinegar

Vinegar has traditionally been made by the souring of wine (the word "vinegar" is derived from the French word *vinaigre*, which means "sour wine"). Apple cider vinegar, for instance, was once apple cider wine fermented from apple cider. And clear white vinegar began as potato starch, which was fermented to potato wine, then converted to clear vinegar. This vinegar has no taste other than the natural sour taste of acetic acid. By contrast, balsamic vinegar acquires its sweet flavor from the wood barrels in which it is aged; the barrels are made of balsam fir (FIGURE 12.9). And the flavor of wine vinegar is determined by compounds present in the original wine as well as products of bacterial growth.

FIGURE 12.9 **Balsamic Vinegar.** Aging in wooden barrels speeds the acidification process and the wood imparts characteristics flavors to the vinegar.

Acetobacter aceti a-sē′tōbak-te′r a-set′ē

The Germans were among the first industrial producers of vinegar. As early as the 1800s, people in the German countryside practiced the art of converting fruit juice to vinegar. Then, as now, they fermented the juice and sprayed the wine into a tank called a vinegar generator. Traditionally and in the present, the tanks are filled with wood shavings and gravel containing naturally occurring cultures of the bacterium *Acetobacter aceti*. The bacterium grows and multiplies on the wood shavings and gravel, and as alcohol percolates through, the bacterial enzymes convert the alcohol into acetic acid. The fresh vinegar recirculates several times through the tank before collecting at the bottom; it has an acetic acid content of about 3% to 5%.

Bread

gluten: A substance in cereal grains consisting of two proteins that add elastic texture to dough.

Of course, we have not forgotten the bread basket. For this part of our meal, we once again turn to the yeasts, especially *S. cerevisiae* (baker's yeast). This microbe is added to flour and water, the other two basic ingredients for making all the types of breads. In bread making, the yeast plays three roles (FIGURE 12.10). When yeast is added to flour and water (plus sugar, salt, and other ingredients at the whim of the bread maker) the fungus subsists on carbohydrates in the dough and produces substantial amounts of carbon dioxide gas. The carbon dioxide expands the dough and it rises seemingly by magic.

FIGURE 12.10 **Yeast and Dough Rising.** The yeast *Saccharomyces cerevisiae* causes dough to rise while developing gluten in, and adding flavors to, the dough.

Enzymes also produced by the yeast help to strengthen and develop gluten in the dough and help give bread its spongy texture (the bread is kneaded to redistribute the gluten and rid the dough of any large gas pockets). In addition, yeast cells manufacture some ethyl alcohol in the dough during fermentation, but the high temperature of baking causes the alcohol to vaporize, and it is driven off.

Lastly, yeasts are not necessarily the only microbes growing in the dough. Other microbes may be present and their metabolites can contribute to the incredible flavors in many types of bread, as described for sourdough in **A Closer Look 12.2**.

A CLOSER LOOK 12.2

Sourdough Bread—Don't Leave San Francisco Without It

Most of us are familiar with the unique taste of sourdough bread. It has a crusty surface and a chewy, but sharply acidic taste. This most distinctive bread, which certainly tastes nothing like typical white sandwich bread, supposedly has its origins with San Francisco. Local stories say the recipe was brought to the city by Basque immigrants from the Pyrenees during the heyday of the California gold rush. So, what gives the bread its distinctive flavor and taste? It is, of course, the resident microbes!

To make sourdough you need a starter culture, which is usually a piece of unused dough from a previous batch. In fact, the San Francisco bakeries say they have kept sourdough cultures going for over a century.

Besides the starter, making sourdough only requires unbleached flour, water, and salt. Most sandwich breads have at least 25 ingredients and additives—and no starter culture, meaning yeast is the only microbe present in the rising process for the dough. The result will be relatively bland tasting bread.

Therefore, the most important "ingredients"—and key to sourdough bread—are the microorganisms living in the starter. The sourdough starter is a mixed microbial population of wild yeasts and bacterial species. A teaspoon of sourdough starter contains some 50 million yeasts and 5 billion lactobacilli.

Sourdough is sour because of the acids produced by the lactobacilli. Initially, there may be other bacterial species in the dough, but they are quickly competed out as the pH reaches about 3.8. The bacterial cells also add flavorand produce the characteristic smell of sourdough bread.

Sourdough starter represents a mutualistic symbiosis between the principal bacterium *Lactobacillus sanfranciscensis* and the principal yeast, *Candida milleri*, which can tolerate the acidic conditions. So, the next time you pass through the San Francisco airport, be sure to take some *Lactobacillus sanfranciscensis*-laced bread back home—or buy a starter culture and make your own bread at home.

A sourdough baguette.

Yeast and flour can come together in an almost infinite range of variations, and the baker can lend some ingenuity to the process. Vienna bread, for example, is baked in a high-humidity oven and develops a flaky crust. Semolina flour is used to make semolina bread while potato flour is used for potato breads and rolls. Bagels are boiled in water before baking; pizza dough is modified to give it high elasticity; and pumpernickel bread is produced from rye flour and yeast-fermented molasses.

12.4 The Main Course: Beef, Sausages, and Sides

Assuming we have not overindulged during the previous courses, we are now ready for our main course. Microbes have contributed mightily to our dining experience thus far and will continue their influence as our culinary adventure unfolds.

Beef and Sausages

Our microbial menu contains two choices for the main course: teriyaki steaks and a European dish.

The teriyaki steak choice consists of flank steak marinated and then broiled or grilled in a teriyaki sauce. So what's microbial here?

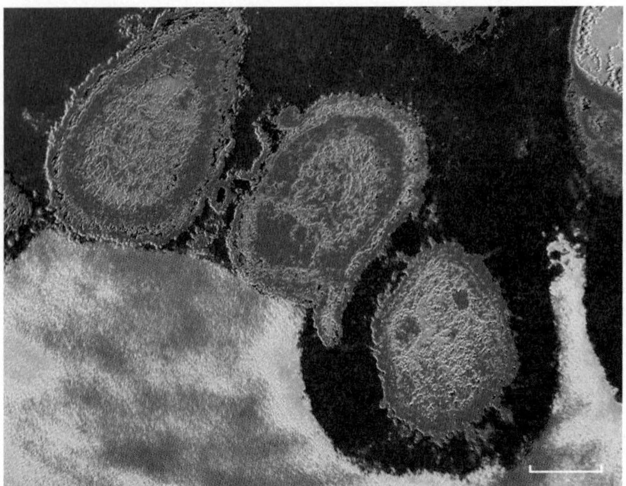

FIGURE 12.11 **A Cow's Rumen.** The rumen of a cow is home to a large variety of anaerobic microbes (false-color transmission electron microscope image). (Bar = 1 um.)

Scientifically speaking, the beef is a product of microbial growth (although an indirect product). Cattle would have little ability to manufacture protein without the intervention of microbes. Indeed, the process of converting carbohydrates in plants to proteins (muscle and meat) in beef cattle is intimately dependent on microbes. Microbes live in enormous numbers in every cow's stomach (called the rumen), where they break down plants and synthesize simple nutrients that are then used by the microbes and by the animal (FIGURE 12.11). Without these microbes, beef cattle would be unable to make more of themselves.

For marinating and cooking the steaks, our chef is using teriyaki sauce, which consists of soy sauce, mirin (a type of rice vinegar), and sugar. Mirin, being a type of vinegar, is the product of rice fermentation and is somewhat similar to the process to make soy sauce, which we will describe here. To make **soy sauce**, manufacturers begin with a starter mixture of soybeans and wheat bran that is inoculated with **koji**, the common name for the fungus *Aspergillus oryzae*. The mix is then incubated at slightly over 30°C (85°F), which allows the fungus to break down complex proteins and carbohydrates into smaller molecules (FIGURE 12.12).

FIGURE 12.12 **The Production of Soy Sauce.** Soybeans and wheat bran are inoculated with *Aspergillus oryzae* (koji) and eventually then mixed with salt containing other microorganisms.

The mixture is blended with salt brine or coarse salt. Over time, the fungus continues to grow in the mixture. Over the course of another year, the mixture continues to age as bacteria (*Lactobacillus* and *Bacillus* species) produce some lactic acid and *S. cerevisiae* yeasts add a small amount of alcohol. The liquid pressed from the mixture is soy sauce. It has a distinct meaty flavor called *umami*, which in Japanese means "pleasant savory taste." Because the fungus has contributed the predominant flavor and aroma to the soy sauce, our teriyaki steak—with its microbe-derived beef and microbe-derived soy and mirin sauces—has totally microbial roots.

The European choice for the main course consists of a variety of sausages accompanied by appropriate vegetables. Sausages generally consist of dry or semi-dry fermented meats. They include pepperoni from Italy, Thuringer from Germany, and polsa from Sweden. To produce sausages, curing and seasoning agents are added to ground meat. The meat mixture is then stuffed into casings and incubated at warm temperatures. Microbes multiply and produce a mixture of acids from the meat's carbohydrates, thereby giving the sausage its unique taste.

Pickled Sides

Having ordered the German sausage, we order some vegetable sides to go along with the sausage. One is "sour cabbage," or **sauerkraut**, as it is better known. Sauerkraut is a well-preserved and tasty form of cabbage that is an excellent source of vitamin C. Indeed, in the days of exploration, the British often took sauerkraut on their ocean voyages to help prevent scurvy (as an alternative to the more expensive citrus fruits). For example, on Captain James Cook's first world voyage (1768), among the provisions were three tons of sauerkraut!

Modern researchers have further verified the healthful benefits of sauerkraut: Scientists have noted that Polish women who had immigrated to the United States were less likely to develop breast cancer than non-Polish immigrant women. The reason?

- *Aspergillus oryzae* a-spėr-jil'lus ô'ri-zē
- scurvy: A disease resulting from insufficient vitamin C, the symptoms of which include spongy gums, loosening of the teeth, and bleeding into the skin and mucous membranes.

Polish foods, such as sauerkraut and fermented products of other cabbage family members (e.g., broccoli, cauliflower, and Brussels sprouts), contain compounds capable of blocking the activity of estrogen, which scientists say can be a possible stimulator of breast cancer. While the research is not complete, it is extremely provocative.

Species of *Leuconostoc* and *Lactobacillus* are essential to the production of sauerkraut. These are both gram-positive bacterial genera naturally found in the leaves and tissues of a head of cabbage. Sauerkraut preparation begins by shredding the cabbage and adding about 3% salt. The salt ruptures the walls of the cabbage plant cells and releases their juices, while adding flavor. Then, the shredded and salted cabbage is packed tightly into a closed container to eliminate oxygen and stimulate fermentation. About a day later, the *Leuconostoc* species begins multiplying rapidly. The cells ferment the carbohydrates and produce lactic and acetic acids. After several days, the pH content of the cabbage is an acidic 3.5. Now, species of acid-tolerant *Lactobacillus* take over. They ferment the carbohydrates further and produce additional lactic acid to reduce the pH to about 2.0 and give the sauerkraut its tangy sourness.

FIGURE 12.13 **Pickled Vegetables.** Many vegetables can be fermented and canned, including pickles (pickled cucumbers) and beets.

Courtesy of Dr. Jeffrey Pommerville

■ *Leuconostoc* lü-kō-nos'tok

Having a taste for sour foods, we have also ordered a variety of pickled cucumbers and beets to accompany the sausages. For all types of dill, sour, and sweet pickles, the fermentation is essentially similar: Manufacturers begin by placing cucumbers of any type or size in a high-salt solution, where the cucumbers change color from bright green to a duller olive-green. Then, the fermentation begins in an aging tank. The first bacterial genus to emerge is *Enterobacter*. This gram-negative rod produces large amounts of carbon dioxide gas, which takes up the air space in the tank and establishes fermentation conditions. The next bacterial genera to proliferate are *Leuconostoc* and *Lactobacillus*. They produce large amounts of acid to soften and sour the cucumbers. Yeasts also grow in the aging tank and establish many of the flavors associated with pickled cucumbers. Various herbs and spices are added to finish the process. Virtually any vegetable, including beets, can be substituted for cucumbers for an equally tasty result (FIGURE 12.13).

■ *Enterobacter* en-te-rō-bak'tėr

12.5 Washing it Down: A Refreshing Beer

The distinctive flavors of the sausage-and-sauerkraut entree are wonderfully complemented by the sparkle and tang of a glass of beer. And, not surprisingly, microbes once more play a part in our meal.

Much of the chemistry of wine fermentation applies equally well to the production of **beer**. Beer making is thousands of years old. Historians tell us that as early as 3400 BC, Egyptians were placing a tax on beer produced in the ancient city of Memphis, the capital of the Old Kingdom. In fact, companies and individuals have tried to mimic the ancient Egyptian brew, as **A Closer Look 12.3** examines.

In later generations, the Greeks brought the art of brewing beer to Western Europe, and the Romans refined the process further. Writers of the period point out that the main drink of Caesar's legions was beer because the water was so polluted that stomach illnesses would result.

During the Middle Ages, the monasteries were the centers of beer making. By the year 1200, taverns and breweries were commonplace throughout the towns of Great Britain. Centuries would pass, however, before beer would make its appearance

A CLOSER LOOK 12.3

This hqt's for You!

Beer, called hqt by the ancient Egyptians, was a very important drink. They often used beer in religious ceremonies and, because water could be a source of illness, they served beer at mealtimes to both adults and children. In fact, workmen at the great pyramids had five types of beer that they drank three times a day. It was the staple drink of both the poor (wages often were paid in beer) and the pharaohs—and was offered to the gods.

Many Egyptologists have studied beer residue from ancient Egyptian vessels in hopes of remaking the ancient brew. In 1996, Delwen Samuel, an archaeobotanist at the University of Cambridge, suggested that the ancient Egyptians used barley to make malt and a type of wheat, called emmer, instead of hops. Samuel and her colleagues tried brewing the beer using the recipe derived by the analysis. They brewed it at a modern brewery and found the beer to be fruity and sweet because it lacked the bitterness of hops. The beer was reported to have an alcoholic content of between 5% and 6%. Samuel gave her recipe to Scottish and Newcastle Breweries in London that then bottled a limited edition, 1,000-bottle batch of "King Tut Ale." It was sold at Harrods department store for $100 per bottle, the proceeds going toward further research into Egyptian beer making.

In 2002, Kirin Brewery Company Ltd., Japan's second-largest beer maker, recreated a 4,000-year-old Egyptian beer by following a recipe recorded on ancient tomb paintings. They brewed 30 liters of beer that was dark brown, contained no froth, had a strong sour taste, and an alcohol content of about 10%.

Note to be outdone, in 2012 a brew called "Pharaoh Ale" was produced based on ancient Egyptian archaeological sources. The brewers, Horst Dornbusch, Peter Egelston, and Tod Mottstated, all master brewers, reported that the Pharaoh Ale had a "minimal but fresh and appetizing bouquet. Visually, it is slightly turbid, and the color is deep reddish amber to almost light mahogany."

The latter two beers were not sold commercially—they were developed for research purposes only!

King Tut Ale.
Courtesy of Mark Nesbitt

in cans. That auspicious event took place in the United States in 1935 in the city of Newton, New Jersey. The six-pack was the logical successor.

The word "beer" is derived from the Anglo-Saxon *baere*, which means "barley." This terminology developed because beer is traditionally a product of barley fermentation. The process begins by predigesting barley grains, a process called malting (FIGURE 12.14). During malting, the barley grains are steeped in water, and naturally occurring enzymes in the barley digest the starch into smaller carbohydrates, among them maltose (also known as malt sugar).

Humulus lupulus
hū'mū-lus lū'pū-lus

Next, the malt is ground with water in the process known as mashing, and the liquid portion, or **wort**, is removed. At this point, the brewmaster adds dried flowers (hops) of the hop vine *Humulus lupulus*. Hops give the wort its characteristic beer flavor, while adding color and stability. Then, the fluid is filtered, and a species of *Saccharomyces* is added. Initially, there is intense frothing as carbon dioxide is produced during aerobic respiration. Then, the frothing subsides, and the yeast shifts its metabolism to fermentation. Alcohol production begins. The final alcoholic content of beer is approximately 4% to 5%.

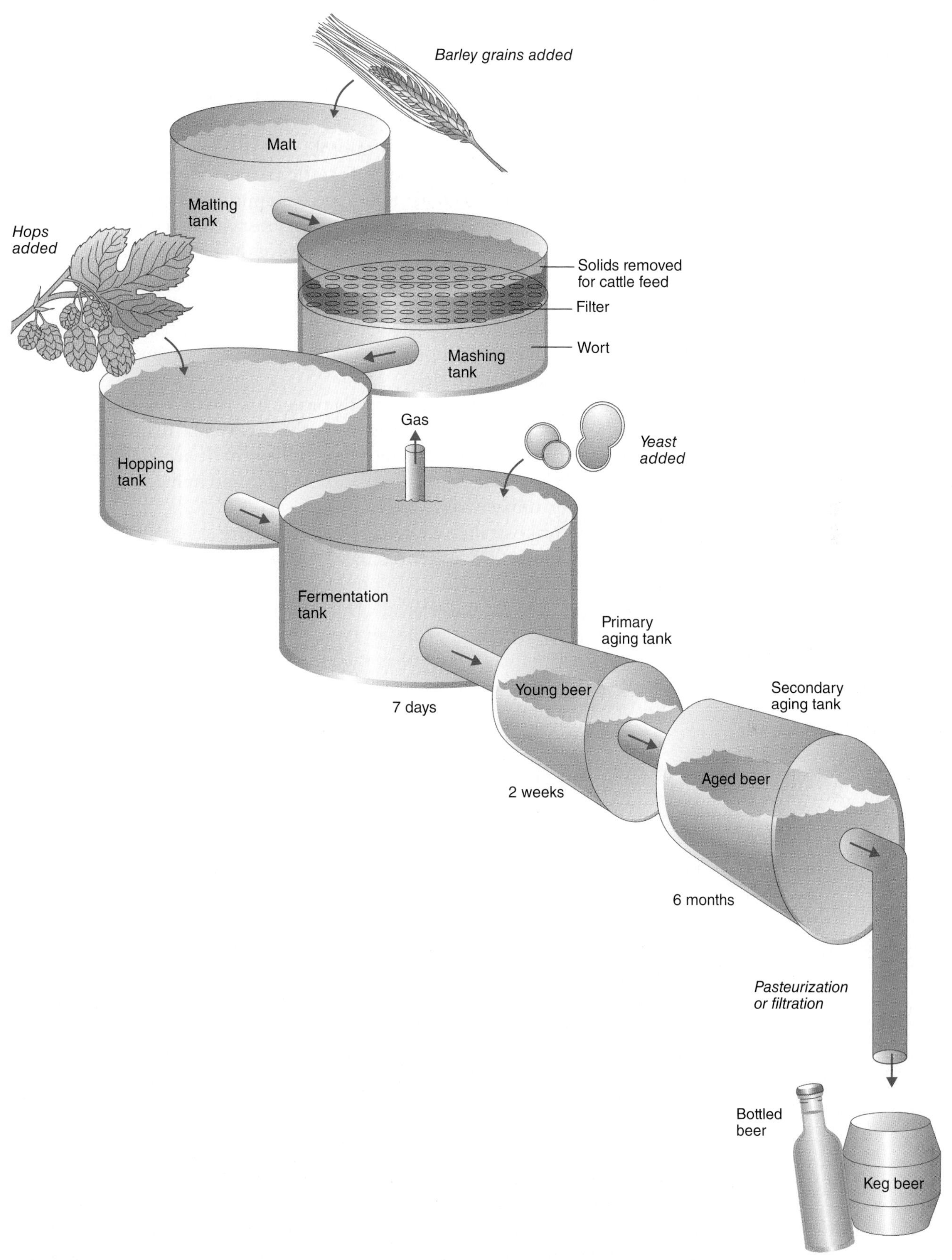

FIGURE 12.14 **A Generalized Process for Producing Beer.** Barley grains are held in malting tanks while the seeds germinate to yield fermentable sugars. The digested grain, or malt, is then mashed in a mashing tank and the fluid portion, the wort, is removed. Hops are added to the wort in the next step, followed by yeast growth and alcohol production during fermentation. The young beer is aged in primary and secondary aging tanks. When it is ready for consumption, it is transferred to kegs, bottles, or cans.

Various species of yeasts are used to produce different types of beer. For example, *S. cerevisiae* gives a dark cloudiness to beer; it is called a "top yeast" because the yeast cells are carried to the top of the vat by the extensive carbon dioxide foam. The beer it produces is an English-type ale or stout. A different species, *S. carlsbergensis*, causes a slower fermentation and produces a lighter, clearer beer with less alcohol. This microbe is called a "bottom yeast" because there is less frothing, and the yeast cells settle to the bottom. Fermentation with this yeast is carried out at a cooler temperature (approximately 15°C) than is used for ale production (approximately 20°C). The beer produced by bottom yeast is Pilsner, also called lager. Almost three-quarters of the beer produced in the world is lager beer.

Saccharomyces carlsbergensis
sak-ä-rō-mī'sēs kä-rls-bėr-gen'sis

After about seven days of fermentation, the yeast has produced a "young beer." This young beer is transferred to wooden vats for secondary aging, also called **lagering**. This process may take an additional six months, during which time the beer develops its characteristic flavor and taste.

If the beer is to be canned, it is usually pasteurized at 60°C (140°F) for a period of 55 minutes to kill the yeasts. Alternatively, the beer can be filtered before canning to remove the yeasts; in this case, the beer is called "draft beer." If the beer is to be delivered directly to an alehouse, it is placed in casks and immediately chilled, making pasteurization unnecessary.

With the teriyaki steak, we might be inclined to try sake. Though many people consider sake a wine, it is more correctly a type of beer because it is fermented from rice (a grain rather than a fruit). To produce sake, steamed rice is mixed with the fungus *Aspergillus oryzae* and set aside. During the incubation period, enzymes from the fungus break down the starches in the rice to simpler sugars. Fermentation by yeasts follows. The final product has the alcoholic concentration of a wine, about 13%, and is commonly drunk at room temperature or warmer.

But we cannot linger too long, for it is time for dessert.

12.6 The Dessert Course: Chocolates and Cherries

Try as we might, we cannot really credit microbes for the luscious assortments of cakes, pastries, and cookies on the restaurant's dessert cart. However, some of the fruit fillings and toppings owe their flavors to slight degrees of fermentation, but this would be stretching the point a bit.

However, we can draw an association between microbes and the flavors of cocoa and coffee. To prepare cocoa, the cocoa beans must first be separated from the pulp covering them in the pod. Enter the microbes. Manufacturers add a controlled culture of bacteria and yeasts to the beans, and the microbes ferment the pulp and soften it so that it is easily shed. The fermentation also lends a bit of taste to the resulting cocoa. In fact, part of the color, flavor, and aroma of chocolate is due to the action of microbes on cocoa. Chocolate lovers should recall the role microbes play when they sit down to a luscious chocolate dessert.

Coffee, as we all know, is made from coffee beans. Like the cocoa bean, the coffee bean is surrounded by a fleshy pulp. Microbes are used to ferment this pulp and assist in its removal. In addition, microbes are needed to help remove the outer skin of the coffee bean by digesting a protein called pectin that comprises a large portion of the bean skin. This protein is broken down by pectin-digesting enzymes produced by various fungi and by bacteria, including *Erwinia dissolvens*. Bacteria producing lactic acid assist the pectin removal but do not appear to add to the final flavor of the coffee.

Erwinia dissolvens
ė r-wi'-nē-ä dis-sol-venz

To express appreciation for our visit, the restaurant manager has sent over a tray of chocolate-covered cherries to enjoy with our coffee. Unknowingly, he has introduced us to yet another microbial product. The soft cherries owe their production to invertase, an enzyme obtained from *S. cerevisiae* and a species of *Bacillus*. Sweet cherries are pitted, then mixed with the enzyme and dipped into a vat of chocolate for their coating. Over the next few days, the enzyme softens the cherries and produces the tasty liquor surrounding them inside the chocolate coating. Delightful!

And in Conclusion

To conclude our meal, we have requested a snifter of brandy. **Brandy** is one of a broad variety of distilled spirits containing considerably more alcohol than beer or wine. Each type of spirit has a proof number, which is twice the percentage of alcohol in the product. A 90 proof brandy, for example, is 45% alcohol.

To produce brandy or other distilled spirits, wine fermentation must first take place. For example, a raw product such as cherries is fermented by yeasts to produce cherry wine. The wine is aged and matured in casks. Once it is ready, manufacturers heat the wine at approximately 80°C (the word "brandy" was originally derived from the Dutch word *brantewijn*, which means "burnt wine"). The heat drives off the alcohol, which is trapped in a condenser and converted back to a more concentrated liquid, which is cherry brandy (FIGURE 12.15). Wooden casks are then used to mature the brandy and introduce unique flavors from organic compounds in the wood. Brandy is the ultimate expression of fermentation and among the most expensive fermented products available.

A sip of brandy pleases the palate yet again, but it also tells us that we must be leaving soon. We have arrived at the end of our gastronomic adventure, which has been marked by good company, fine food, and intriguing insights into the microbial world.

FIGURE 12.15 **Brandy Distiller.** After the grapes or other fruits are fermented into wine, the wine is put in a distiller where the alcohol and water, and other volatile components rise. As they cool, they become a liquid again. The product is then collected and aged.

A Final Thought

When we were children, most of us were taught that Marco Polo traveled to China during the 1200s to obtain spices and explore new trade routes. "Why spices?" we may have thought. The fact is spices were not a luxury at the time. They were essential for improving the smell and taste of leftover food, which ranged from awful to sickening. Refrigeration was unknown, and canning had yet to be invented. Therefore, what couldn't be eaten often spoiled very quickly. Spices had some antimicrobial qualities, to be sure, but they also made spoiled food palatable.

An alternative to spices was fermentation. By fermenting food, people allowed it to spoil naturally. The food might taste bad, but they gradually came to accept the bad taste and found fermentation to be a worthwhile way of preventing harmful spoilage and preserving food for later use.

Although microbes can spoil food, if we use fermentation to control that spoilage, we can make a bad thing good. Indeed, as this chapter shows, some of those "spoiled" foods are really quite good!

Questions to Consider

1. One day, the students in a microbiology class presented the instructor with a basket of "microbial cheer" in recognition of her efforts on their behalf. From your knowledge of this and other chapters, can you guess some of the things that the basket contained?
2. Yeasts are sometimes known as the "schizophrenic microbes." What do you think that means?
3. Sometimes even the most careful food preparation can lead to tragedy. For many generations, humans have made sausages, but on occasion the sausages have become contaminated with *Clostridium botulinum*, and individuals have developed a deadly disease known as botulism (*botulus* = "sausage"). How do you think sausage becomes contaminated, and what characteristics of *C. botulinum* make it attracted to sausage?
4. This chapter has surveyed many foods of microbial origin, but the survey is incomplete. From your reading of other chapters and your general knowledge, what foods of microbial origin can you add to the list discussed in this chapter?
5. Wine is part of many cultures worldwide, but one of the great mysteries of the human experience is how wine was discovered. Write a scenario depicting the circumstances under which wine might have been first experienced by a human.
6. Suppose you decide to enter the pickling business. You intend to pickle tomatoes, peppers, and a host of other foods. How would you proceed with the science end of your new business?

Key Terms

The key terms in this chapter are the various foods that owe their existence to microbes. In each case, you should name the food and explain how one or more microbes contribute to its production.

aerobic respiration
anaerobic
beer
brandy
fermentation
koji
lagering
must
sauerkraut
soy sauce
vinegar
whey
wine
wort

As seen in this market in Barcelona, Spain, dried fruits and nuts are popular food items. Dried fruits contain natural antimicrobial compounds and little water, so they, along with nuts, have been considered stable and examples of nonperishable foods.

Food Preservation and Safety: The Competition

13

Looking Ahead

Although microbes are valuable allies to humans, they can also be formidable competitors because they multiply in many of the foods consumed by humans. Unfortunately, this competition can lead to rotting and decay, resulting in food spoilage and sometimes illness, as we will see in this chapter.

On completing this chapter, you should be capable of:

- Identifying the intrinsic and extrinsic factors involved with food spoilage.
- Naming some bacterial and fungal species involved with food spoilage.
- Assessing the effects on foods of food spoilage (e.g. meats, fish, eggs, dairy, fruits and vegetables, and grains).
- Identifying seven factors important to food preservation.
- Describing the challenges to the prevention of foodborne disease.

It was the annual company picnic, and the softball game was finally over. Now the serious eating could begin. The picnic table was overflowing with goodies—salads, stuffed eggs, barbecued chickens, and lots of desserts. Unfortunately, there was also an unwelcome guest at the picnic—a bacterial organism named *Salmonella* (FIGURE 13.1a).

Salmonella
säl-mōn-el'lä

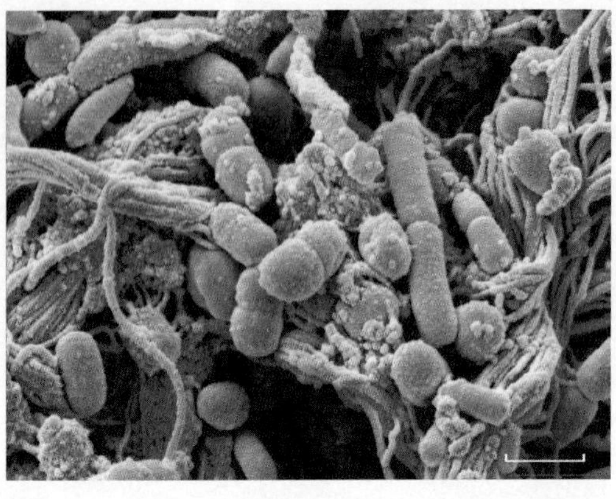

(a)

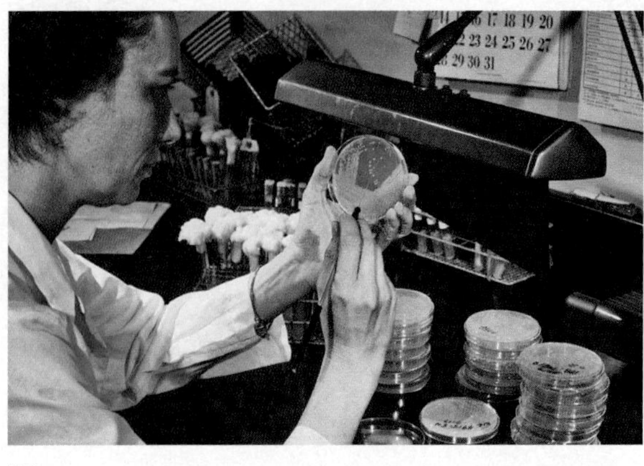

(b)

FIGURE 13.1 **Salmonella and its Identification.** (**a**) A false-color scanning electron microscope image of *Salmonella* bacterial cells contaminating chicken. (Bar = 3 um.) (**b**) A lab technician isolating *Salmonella* from a fecal specimen on an agar medium.

(a) © Scimat/Science Source; (b) Courtesy of Dr. Kokko/CDC.

During the next three days, it became apparent that the picnic would be remembered, but for all the wrong reasons: Over half the picnic attendees suffered abdominal cramps, diarrhea, headaches, and fever. When public health inspectors questioned the company employees, they found a common thread: All the sick attendees had eaten the barbecued chicken.

The inspectors then questioned the woman in charge of the chickens. She led them to a local supermarket, where barbecued chickens were sold "ready to eat." Inspectors learned that store employees had cooked the chickens—and then put them back in their original trays. It didn't take long for microbiologists to find evidence of *Salmonella* in the trays. (FIGURE 13.1b). Further questioning revealed that the woman had bought the chickens in the morning and stored them for the next 7 hours in the trunk of her car, believing they would remain cool. They did not. In fact, they quickly reached ideal temperatures for bacterial growth, and by dinnertime, at 7:00 pm, they were teeming with the pathogen.

The lessons of this incident are clear: Keep foods cold for storage; store them in clean, fresh trays after cooking; and eat them quickly after preparing them.

Incidents like this one highlight how most foods, even cooked foods, can provide favorable conditions for microbial growth. The organic matter in food is plentiful, the water content is usually sufficient, and the pH is either neutral or only slightly acidic. To a food manufacturer or restaurant owner, contaminating microbes may spell economic loss or a bad reputation. To the consumer, it often means illness or, in extreme cases, death.

The primary focus of this chapter is to examine the types of microbes able to contaminate foods and to point out how foods get contaminated. We then will discover methods to prevent spoilage.

13.1 Food Spoilage: Microbes in Action

Food spoilage has been a continuing problem ever since humans first discovered they could produce more food than they could eat in a single meal. We all know of Marco Polo's travels to China in the thirteenth century to obtain spices and explore new trade routes (FIGURE 13.2). What often is not mentioned is that the spices were more than just a luxury of the time. Spices were essential for improving the smell and taste of spoiled food because refrigeration was virtually unknown and canning was yet to be invented.

Food spoilage is a metabolic process causing foods to be undesirable or unacceptable for human consumption due to changes in sensory characteristics such as color, texture, and/or flavor. Although food spoilage is a broad topic that involves more than microbial contamination, we will emphasize spoilage caused by microorganisms and will consider spoilage of several common foods people purchase and consume.

FIGURE 13.2 **Spices.** A typical Asian spice stand in a larger spice market of today is similar to that in Marco Polo's time.

General Principles

Food spoilage is a complex process and excessive amounts of foods are lost every year. The Natural Resources Defense Council in New York reported that a family of four in the United States throws out some 25% of the food they buy (primarily grains, seafood, and fruits and vegetables) (FIGURE 13.3). Up to 75% of that waste is due to microbial food spoilage; the rest is the result of preparing, cooking, and serving too much food.

Because the greatest percentage of this spoilage is due to microbial contamination, microbiologists occupy an important place in the food manufacturing process. Their work helps to ensure the food we eat is safe and will not transmit disease. Moreover, by reducing microbial contamination, the wholesomeness of food and its shelf life are extended. Preservation methods are high on the list of tasks addressed by the

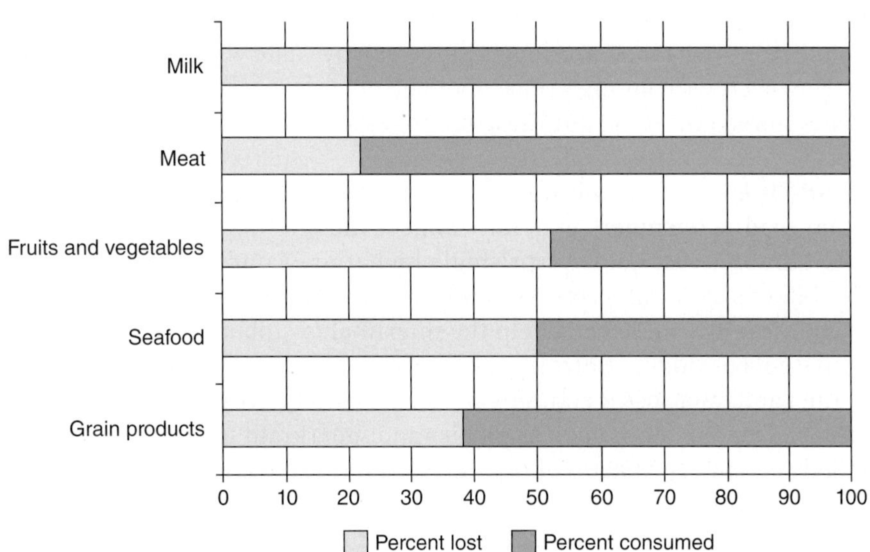

FIGURE 13.3 **Food Lost and Consumed.** Percentages of food lost and consumed for the United States, Canada, Australia, and New Zealand combined. Graph adapted from Natural Resources Defense Council Issue Paper, August, 2012.

Data from Natural Resources Defense Council Issue Paper, August, 2012.
Source: Food and Agriculture Organization 2011.

food microbiologists. In cases where food preservation has been unsuccessful, the microbiologists are responsible for detecting the offending microbes and working toward instituting better food preservation methods.

The microbial content of foods usually has qualitative as well as quantitative aspects. The qualitative aspects refer to which microorganisms are present, while the quantitative aspects refer to how many of them are present. In some cases, the quantitative requirements for a food product must be zero—a can of vegetables, for example, is expected to have no microbes in it; it must be **sterile**. By contrast, raw hamburger meat is usually considered acceptable if it contains up to 1,000 staphylococci per gram of meat because cooking kills the organisms. So, a raw hamburger might have more than 200,000 staphylococci—before cooking. And because staphylococci do multiply in contaminated meat and produce diarrhea-inducing toxins, meat processors must be vigilant to keep the amount of staphylococci as low as possible.

Several factors are important when considering microbial spoiling.

Shelf Life. The **shelf life**, the length of time a particular food may be stored, depends largely on the quantity of microbes present in it and the conditions at which it is held. For instance, because of their dry condition, foods like potato chips and dried pasta products normally have low numbers of microbes and long shelf lives. Breads and citrus fruits contain more moisture, so higher numbers of microbes will develop in them with time. Fish and eggs present the most favorable conditions for microbial growth and can contain the highest number of microbes. Therefore, these foods have the shortest shelf lives.

For packaged products, a **manufacturer code** is present on the product so if there is a contamination problem or a food outbreak, investigators can do a "source trace-back" to locate the origin of the problem or contamination.

Label Date. Packaged foods usually have an **expiration date** (label date), indicated by "Best by...," "Use by...," or "Sell by..." These designations represent the manufacturer's suggestion as to how long the product is at "peak quality" in terms of taste, smell, and/or freshness. It should not be interpreted as a safety date and, except for infant formula, the label date is not federally regulated. Manufacturers point out that eating a food product past its "sell by" date will not necessarily cause illness. However, with enough time, spoilage microbes will multiply even if a food product is stored correctly, but these microbes are not usually the ones that make us sick.

Microbial Load. The microbial content of a particular food is known as its **microbial load**. Consumers generally expect the microbial load to be low in all foods, but in some cases, it is surprisingly high. For example, a teaspoon of yogurt contains billions of harmless bacterial cells that have converted condensed milk into yogurt. In the body, these cells help the intestinal microbiome work well and help to keep pathogens under control. Such foods are often sold as **probiotics**, meaning the product and microbes it contains confer a health benefit to the consumer. Also, the microbial load in foods such as pickles and sauerkraut tends to be high because these foods contain the bacterial organisms responsible for the fermentation used to produce the product.

■ microbiome The population of microorganisms normally living in and on an environment, such as the human body.

Also, products like milk, though pasteurized to remove all pathogenic microorganisms, can still have a substantial microbial load as high as 1,000 bacterial cells per milliliter (a one gallon milk carton could have over 3.7 million bacterial cells) and it is still safe to drink.

Ready-to-eat foods often pose a health threat to consumers because any pathogens in the food will be directly consumed as compared to foods that are cooked at home before eating, which should kill any pathogens present.

Microbial Contamination. Microbial contamination of foods is usually impossible to avoid, and the general trend is to control or minimize the contamination by good food management processes, acceptable sanitary practices, rapid movement of the foods through processing plants, and well-tested preservation procedures. Contaminating microbes enter foods from various sources:

- Airborne organisms fall onto fruits and vegetables, and then penetrate to the softer inner tissues when the skin or rind is broken.
- Crops bring bacterial cells from the soil to the processing plant.
- Shellfish, such as clams and oysters, concentrate microbes in their tissues by catching them in their filtering apparatus as they strain their food from contaminated water.
- Rodents and insects transport microbes on their feet and body parts as they move about among garbage, foods, and other possibly unsanitary locations.

Human handling is also a source for microbial contamination. For instance, when meat is handled carelessly by a butcher, bacterial organisms from the slaughtered animal's intestines can mix with the meat. Moreover, if the water used to clean meat, poultry, and fish products has not been treated to eliminate microbes, they will enter the food. Of even more concern are raw vegetables purchased at supermarkets because they have been exposed to the contaminated hands of numerous shoppers. And more and more of our foods are coming from abroad, as **A Closer Look 13.1** describes, providing another potential source for foodborne illnesses.

A CLOSER LOOK 13.1

Free Market Economy and Foodborne Diseases

Today, you can go into just about any supermarket and find fresh fruit that seems out of season. For example, you can find fruits such as peaches and grapes in the middle of winter. Thanks to advances in agriculture and transportation, there has been a globalization of the food market. Peaches from Chile or raspberries from Mexico can be heading to U.S. markets within hours after being picked. It's great to have fresh fruits and vegetables, but it brings along the risk of a foodborne illness.

About 70% of many fresh fruits and vegetables we find in our local supermarket come from other countries. Sometimes these countries do not have the regulated and safe food practices found in the United States and other developed countries. Although outbreaks of foodborne illnesses are rare, they are occurring with greater regularity every year. However, actual data can be hard to collect because many people never visit a doctor for a foodborne illness.

The Centers for Diseases Control and Prevention (CDC) states that although foodborne infections have decreased by nearly 25% since 1996, "continued investments are essential to detect, investigate, and stop outbreaks promptly in order to protect our food supply."

Detecting foodborne outbreaks is not easy because fresh produce is distributed across the nation and not to just one local community. In fact, physicians and local health departments may not see a local outbreak as part of a national event. Most local health departments are accustomed to responding to a foodborne outbreak at a social function or commercial establishment. Globalization now means that local, state, and national food safety programs need to coordinate their surveillance methods and establish clear communication channels to properly recognize and quickly respond in a timely fashion to a foodborne outbreak.

Importantly, should you shy away from imported produce? Certainly not. In most cases, proper washing of fresh fruits and vegetables, or cooking produce, will eliminate most chances of contracting a foodborne illness.

Courtesy of Dr. Jeffrey Pommerville

The Conditions for Spoilage

Because food can be looked at as a culture medium for microorganisms, the food's chemical and physical properties have a significant effect on what types of microbes it will harbor (FIGURE 13.4). Conditions naturally present in foods that influence microbial growth are called **intrinsic factors**. These include:

Water Content and pH. One of the prerequisites for life is water. Therefore, to support microbial growth, food must be moist and have a minimum water content of 18% to 20%. Microbes do not grow in foods such as potato chips, dried pasta, nuts, rice, and flour because of their low water content.

Another important factor affecting microbial growth is pH. Most foods fall into the slightly acidic to neutral range on the pH scale, and numerous bacterial and mold species find these conditions ideal for growth. In foods with a pH of 5.0 or below, acid-loving molds are the predominant microbes. For example, citrus fruits generally escape bacterial spoilage but may yield to mold growth and spoilage if damaged.

Physical Structure and Chemical Composition. Another property contributing to a food's tendency to spoil is its physical structure. Whole meat is not likely to spoil quickly because microbes cannot penetrate the surface of the meat easily. However,

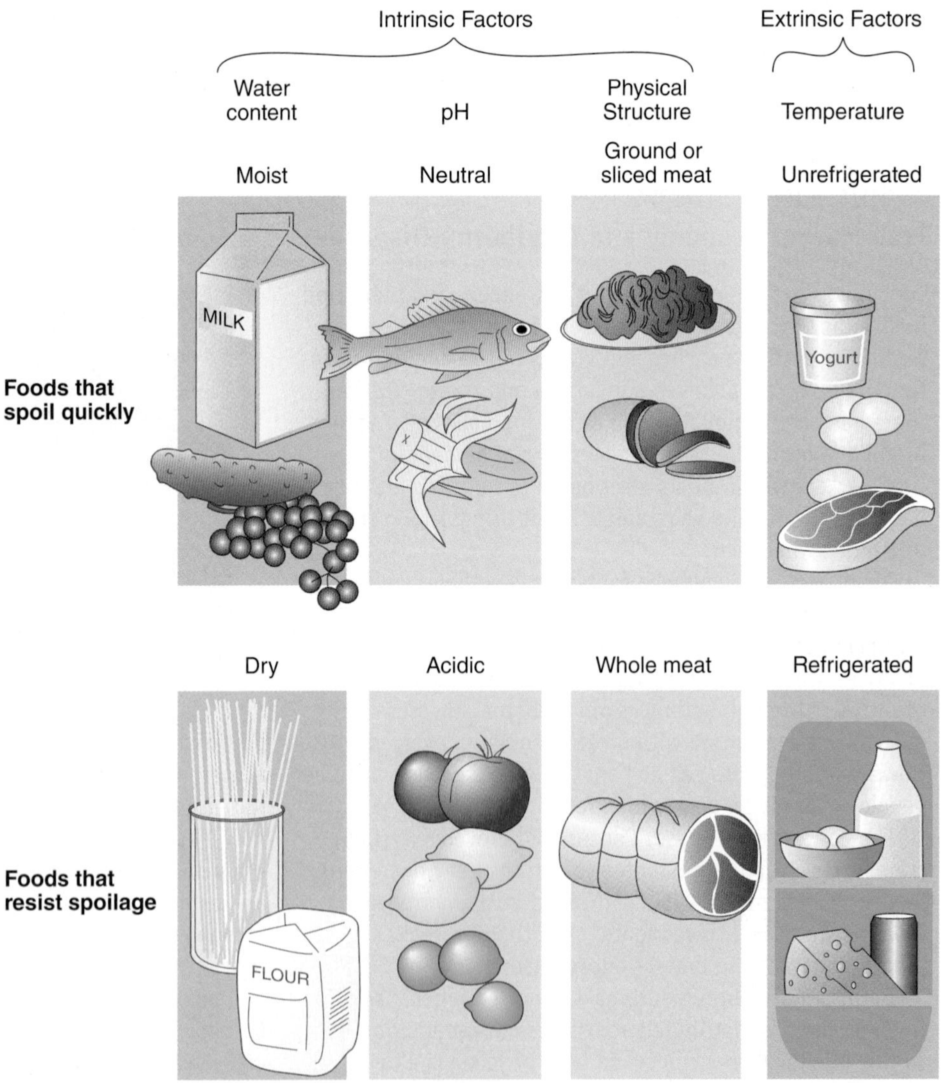

FIGURE 13.4 **Food Spoilage.** Several intrinsic and extrinsic factors determine whether foods are likely to spoil quickly or resist spoilage.

raw, ground or sliced meat can deteriorate rapidly because microbes grow within the loosely-packed structure of the moist meat as well as on the surface.

A food's chemical composition (nutrients) is another intrinsic factor determining the type of spoilage possible. Fruits, for example, can support organisms metabolizing carbohydrates, whereas meats support protein decomposers. Starch-digesting bacterial cells and molds may be found on raw potatoes and corn products.

Environmental conditions surrounding the food (food storage and packaging) make up the so-called **extrinsic factors** influencing microbial growth. These include:

Oxygen. Properly vacuum-sealed cans of food are without oxygen gas and thus do not support the growth of aerobic bacterial pathogens such as *Pseudomonas*. However, an improperly prepared food contaminated with the anaerobe *Clostridium botulinum* provides a suitable environment for growth and toxin production, and one may contract botulism if the food product is consumed. Likewise, vegetables and baked goods do not support the growth of anaerobes.

Pseudomonas sū-dō-mō'näs

Clostridium botulinum klôs-tri'dē-um bot-ū-lī'num

Temperature. Refrigerator temperature is usually too cold for the growth of most human pathogens and freezer temperature halts the growth of microbes (see Figure 13.4). However, the warm hold of a ship or a humid, hot warehouse storeroom is an environment conducive to the growth of many human pathogens. It is common knowledge that contamination is more likely in cooked food at a warm temperature than cooked food kept at refrigerator temperature.

In summary, the food industry recognizes three groups of foods, loosely defined on the basis of their growth conditions (FIGURE 13.5). **Highly perishable** foods are those that spoil rapidly. They include poultry, eggs, meats, most vegetables and fruits, and dairy products. Foods such as nutmeats, potatoes, and some apples are considered **semiperishable**, meaning they spoil less quickly. **Nonperishable** foods are often stored in the kitchen pantry. Included in this group are cereals, rice, dried beans, macaroni and dried pasta products, flour, and sugar.

13.2 Microbes Causing Spoilage: Some Prime Suspects

As described above, food spoilage can be caused by a variety of microorganisms, including both bacterial and fungal (yeast and mold) species. Some microbes can be found in many types of spoiled foods while others are more specific and limited to only certain foods. Within a specific food group, spoiling often involves a succession of different microbial populations that rise and fall as nutrients change during the deterioration process.

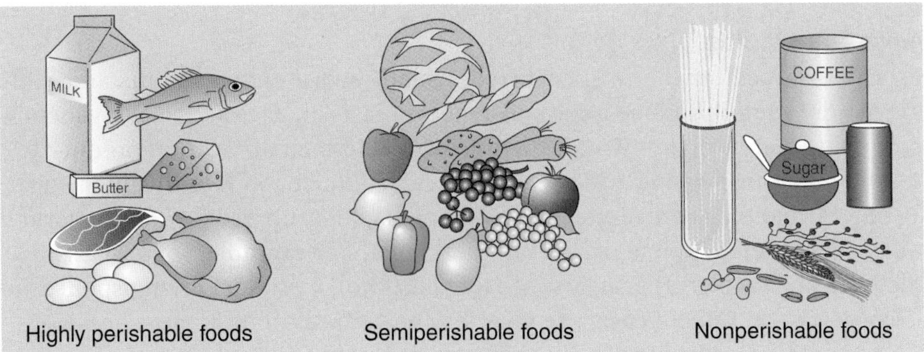

FIGURE 13.5 **The Perishability of Foods.** Examples of highly perishable, semiperishable, and nonperishable foods. The physical and chemical properties of these foods are reliable indicators of their rate of perishability.

Spoilage microbes typically come from the soil, water, or intestinal tracts of animals. They may be transmitted through the air and water, and carried by insects. Let's look at some of the "prime suspects" causing food spoilage before examining the actual food groups affected.

Bacteria

Several bacterial genera and specific species are commonly associated with foodborne illnesses.

***Bacillus* and *Clostridium*.** Some forms of food deterioration are caused by spore-formers associated with the spoilage of heat-treated foods because their spores can survive the high temperatures used for food processing. These gram-positive bacterial genera include: *Bacillus*, a spoilage agent of high or low pH canned foods, vegetables, and chilled foods and milk; and *Clostridium*, a spoilage agent associated with canned vegetables and fruits, chilled meats, and brine-cured hams.

■ *Bacillus* bä-sil'lus

***Lactobacillus* and *Leuconostoc*.** The non-sporeforming bacterial species referred to as **lactic acid bacteria (LAB)** may be associated with a variety of spoiled foods. Although some species of *Lactobacillus* and *Leuconostoc* are useful in the production of fermented foods (e.g., yogurt and sauerkraut), under low oxygen, low temperature, and acidic conditions, these microbes can cause greening of meat, and gas formation in cheeses as well as canned or packaged meat and vegetables. LAB may also produce large amounts polysaccharide that causes slime on meats and ropy spoilage in some beverages.

■ *Lactobacillus* lak-tō-bä-sil'lus

■ *Leuconostoc* lü-kō-nos'tok

***Salmonella*, *Escherichia*, and *Shigella*.** Several species or strains of *Salmonella*, *Escherichia*, and *Shigella* are human pathogens as well as being associated with spoilage. These gram-negative organisms grow with or without oxygen (facultative) and are present in many foods because they are widespread in soil, on plant surfaces, and in digestive tracts of animals. Temperature, salt concentration, and pH are the most important factors determining which, if any, of these microbes spoil foods.

■ *Escherichia* esh-ėr-ē'kē-ä

■ *Shigella* shi-gel'lä

Other bacterial genera (*Proteus*, *Serratia*, and *Enterobacter*) convert food proteins in meat and fish into amino acids and then break down the amino acids into foul-smelling end products.

■ *Proteus* prō'tē-us

■ *Serratia* ser-rä'tē-ä

■ *Enterobacter* en-te-rō-bak'tėr

Other Bacterial Genera. A few other notable bacterial genera are associated with spoilage of chilled, high protein foods such as meat, fish, and dairy products. They may not be the predominant spoilage organisms but contribute to the breakdown of food components and may produce off-odors. *Acinetobacter* is a genus often found on poultry carcasses on the processing line and has been isolated from a variety of spoiled meat and fish. *Alcaligenes* is a potential contaminant of dairy products and meat, and has been isolated from rancid butter and milk with an off-odor. These genera are found naturally in the digestive tract of some animals and also in soil and water.

■ *Acinetobacter* a-si-ne'tō-bak-tėr

■ *Alcaligenes* al'kä-li-gen-ēs

Yeasts and Molds

Yeasts are adapted for life in specialized, usually moist environments and, unlike some molds, do not produce toxic substances. Yeasts are facultative and genera like *Saccharomyces* are well known for their beneficial fermentations that produce bread and alcoholic drinks (wine and beer). They can colonize foods with a high sugar or salt content, and they contribute to spoilage of maple syrup, pickles, and sauerkraut. Fruits and juices with a low pH are another target. For example, some yeasts live in apple juice and convert the sugars into ethyl alcohol, a product giving spoiled juice an alcoholic taste. Other yeasts can grow on the surfaces of meat and cheese.

■ *Saccharomyces* sak-ä-rō-mī'sēs

Molds are decomposers and, in the environment, they are very important for recycling dead plant and animal remains. If present in foods, they are involved in the decomposition of those materials as well. They are well adapted for growth on and through solid substrates, generally produce airborne spores, and require oxygen for their metabolic processes. Most molds grow at a pH range of 3 to 8 and the spores can tolerate harsh environmental conditions but most are sensitive to heat treatment. Different mold species have different optimal growth temperatures, with some able to grow at refrigerator temperatures. Molds can produce a number of toxic substances.

Penicillium and related genera can be useful to humans in producing antibiotics and some soft cheeses. However, many species are also important spoilage organisms of fruits, vegetables, and cereal grains. Some species can grow at refrigerator temperatures and are found in jams and margarine. *Aspergillus* and related molds generally grow faster and are more resistant to high temperatures and low water concentrations than *Penicillium*. Some molds spoil a wide variety of food and non-food items (paper, leather, etc.) but are probably best known for spoilage of grains, dried beans, peanuts, tree nuts, and some spices.

■ *Penicillium* pen-i-sil'lē-um

■ *Aspergillus* a-spėr-jil'lus

13.3 The Foods We Eat: Factors Affecting Food Spoilage and Shelf Life

Most of the foods in the human diet are rich in carbohydrates, fats, and proteins. Unfortunately, these are the same nutrients microbes find nutritious. As described above, many different bacterial and fungal microbes can potentially use the nutrients in a food but some species have a competitive advantage under certain conditions. Let's examine several food categories and uncover the challenges food processors and manufacturers face to inhibit the contamination and growth of spoilage microbes.

Fresh and Processed Meats

Spoilage in meat and meat products is almost expected because of the nature of the food; meats are high in protein and fats. Muscles of healthy animals are sterile and do not contain any microbes. However, when the animals are slaughtered, the extraordinarily high nutrient content in the fresh meat can support a huge variety of microbes—if contaminated. Thus, good sanitation practices are essential.

■ sanitation The process of reducing the number of microbes to a safe level.

Importantly, the many steps in meat processing provide opportunities for microbial contamination. For example, chopping and grinding of meats can increase the microbial load as more surface area is exposed and the moisture content increases. In the processing of fresh meat, contamination may also be traced to dirty conveyor belts used to transport the meat, improper temperature control while holding the meat, and failure to distribute meat and meat products quickly. In addition, bacterial cells can enter from the animal's intestinal contents via knives and cutting blocks. So, it is not surprising fresh meat is often related to episodes of disease. One example was the 1993 outbreak of hemolytic diarrhea traced to raw, fresh hamburger meat contaminated by *Escherichia coli* O157:H7. Over 600 patrons of a fast-food restaurant were involved and four children died. Since then, almost every year there are recalls of raw ground beef due to contamination by *E. coli* or another pathogen like *Salmonella*.

An early indication of meat spoilage in the meat counter or butcher shop is the loss of red color and the appearance of a brown color with surface slime. Species of LAB, such as *Lactobacillus* and *Leuconostoc*, are often responsible for these changes. Although they are not dangerous to health, these microbes can give an abnormal taste and smell to the meat even after it has been cooked.

Most of the ground meat recalls involve spoilage and microbial contamination in vacuum-packed and sealed fresh ground meats (FIGURE 13.6a). Being a facultative organism, *E. coli* is a major cause, as the recent meat recalls attest. Contamination with LAB or *Clostridium* gives rise to off-odors, slime formation, and a greening of meats that have been vacuum packed and sealed.

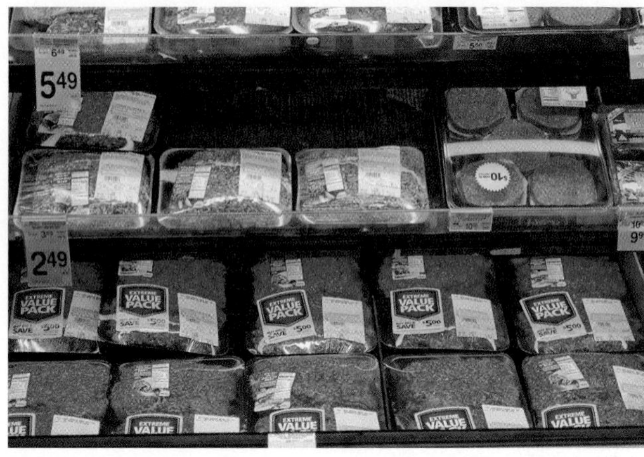

(a)

(b)

FIGURE 13.6 **Packaged Meats.** (**a**) Ground meat that has been vacuum-wrapped/sealed has been subject to most recalls. (**b**) Processed meats contain preservatives and are subject to a different group of spoilage microbes than are wrapped, ground meats.

(a) and (b) Courtesy of Dr. Jeffrey Pommerville

Processed meats, such as luncheon meats, sausages, and frankfurters (FIGURE 13.6b) often contain added salt, nitrites and/or nitrates, and preservatives, which change the microbial population involved in spoilage. Thus, they pose special hazards because the foods are handled often and contain a variety of meat products. Also, natural sausage and salami casings made from animal intestines may contain residual bacteria or spores of *Clostridium botulinum*, the causative agent of **botulism**. In the food, the spores germinate and the resulting bacterial cells produce a nerve toxin that affects the central nervous system and causes progressive muscular paralysis. However, fermented meats, such as sausages, are rarely contaminated because they contain a variety of organic chemicals toxic to spoilage microbes. Nevertheless, faulty cooking and/or cooling procedures may allow growth, and spores can be introduced with the spices added to flavor the sausages.

Cured meats such as ham, bacon, and corned beef are often treated with large doses of salt to draw water out of microbes and thereby kill them. However, species of LAB tolerate the salt and ferment carbohydrates in the cured meat to lactic acid, which sours the meat and causes off-odors and taste. In addition, the bacterial cells produce slime and gas that causes the package to puff and form hydrogen peroxide, which changes the red meat pigments to green ones and gives the meat a spoiled look.

Fish

Fish is a highly perishable food high in protein and free amino acids. Because microbes usually contaminating fish are naturally adapted to the cold environment in which fish live, refrigeration is not as effective for long periods as it is for meat; freezing, salting, or drying fish is preferred. Shellfish, such as clams, oysters, and mussels, are of particular concern because they obtain their food by filtering particles from the water. If the water is contaminated, shellfish concentrate the pathogens in their tissues. Hepatitis A, typhoid fever, cholera, and other intestinal illnesses may result from eating the raw shellfish.

Microbiologists can often trace spoilage in fish to the water from which it was taken or in which it was held. Fish taken from polluted waters usually have a high microbial load, and it is not surprising that fish caught close to land are more contaminated than fish from the deep sea. Very often, spoilage organisms concentrate in the gills of fish because these structures trap microbes as water is strained. Another source of contamination is the boxes used for storing and transporting fish. Cracks, pits, and splinters in the wood can trap microbes and spread them among the fish.

The characteristic odor associated with spoiled fish is generally due to the breakdown of the free amino acids by marine microbes. For example, when a fish is taken out of its natural habitat, it dies quickly, and the natural process of decomposition begins. Marine bacterial species degrade a compound in fish muscle tissue called trimethylamine oxide to trimethylamine, which gives rotting fish its dreadful smell.

Poultry and Eggs

Contamination in poultry and eggs may be due to human contact, but it often stems from bacterial cells that have infected the bird. Members of the genus *Salmonella*, which includes over 2,400 pathogenic strains, commonly cause diseases in chickens and turkeys, which if passed on to consumers via poultry can lead to **salmonellosis**, which is characterized by diarrhea, fever, vomiting, and abdominal cramps. Processed foods, such as chicken pot pies, whole egg custard, mayonnaise, eggnog, and egg salad also may be sources of salmonellosis.

Eggs are normally sterile when laid, but the outer waxy membrane, as well as the shell and inner shell membrane, can be penetrated by some bacterial organisms. *Proteus* species cause black rot in eggs when they break down the amino acid cysteine and produce hydrogen sulfide gas, which leads to black deposits and gives rotten eggs their horrid smell. Spoilage in eggs also occurs as the yolk develops a blood-red appearance from growth of red pigment-producing *Serratia*. In fact, the primary location of egg contamination is in the yolk rather than the white because the yolk is more nutritious and the white has an inhospitable pH of approximately 9.0. Also, lysozyme in egg white is inhibitory to gram-positive bacterial species.

Part of the problem of infected chicken and poultry products traces to the use by high-tech chicken farms of machines to remove the eggs as soon as they are laid. In so doing, manufacturers prevent normal bacterial organisms from entering the young chicks and keeping pathogenic microbes in check. One solution is to use so-called "pasteurized" eggs, which have been treated with ozone and reactive oxygen gases under high and low pressures before replacement with inert nitrogen gas. Pasteurized eggs may be sold as whole eggs in the shell or as a liquid egg product.

Dairy Products

Milk is an extremely nutritious food. It is a solution of proteins, fats, and carbohydrates, with numerous vitamins and minerals. Milk has a pH of about 7.0 and is an excellent nutrient for humans and animals, and a good growth medium for microbes. About 87% of milk is water. Another 2.5% is a protein called casein, a mixture of three different long chains of amino acids. A second protein in milk, lactalbumin, is a whey protein; that is, it remains in the clear fluid (the whey) after the casein is removed during cheese production.

Carbohydrates make up about 5% of milk. The major carbohydrate is lactose, sometimes referred to as milk sugar. Rarely found elsewhere, lactose is a disaccharide that is digested by relatively few bacterial species, and these are usually harmless. The last major component of milk is butterfat, a mixture of fats often churned into butter. When bacterial enzymes digest these fats into fatty acids, the milk or butter becomes rancid and develops a sour taste.

Milk spoilage can occur in the kitchen refrigerator or the supermarket dairy case. The bacterial organisms causing spoilage could come from the dairy farm's environment, milking equipment, processing plant equipment, employees, or the air. Milk is normally sterile in the udder of the cow, but contamination occurs as it enters the ducts leading from the udder. Species of LAB (*Lactobacillus* and *Streptococcus*) are acquired there, together with various pathogenic microbes from dust, manure, and polluted water. If not properly cooled (and pasteurized), LAB can grow to such large quantities that lactic and acetic acids accumulate. In addition, off-flavors develop and bacterial enzymes can produce chemicals giving the milk rancid odors.

■ *Streptococcus* strep-tō-kok'kus

The introduction of yeasts and molds can spoil cultured milks (e.g., yogurt, sour cream, buttermilk) because the higher acidity in these products inhibits many bacterial species. Cream, on the other hand, may become rancid if bacterial species like *Pseudomonas* and *Enterobacter* contaminate the product and proliferate.

If a wedge or block of cheese is left in the refrigerator too long, a mold might start growing and eventually spoil the cheese. Hard and semi-hard cheeses have a low moisture content (<50%) and a pH near 5.0, which limits or delays the start of growth of most molds. Soft cheeses, on the other hand, are less acidic (5.5–6.5) and have a higher moisture content (50%–80%). These cheeses may be spoiled by bacterial species in the genera *Pseudomonas*, *Alcaligenes,* and *Flavobacterium*. Spoilage problems in cheese can arise from using a low quality milk, unhygienic conditions in the processing plant, or simply mold spores landing on the cheese after the package has been opened.

■ *Flavobacterium* flā-vō-bak- tėr'-ē-um

Fruits and Vegetables

Undamaged fruits normally have microbes on their surfaces. However, once ripe, the cell walls in the fruits weaken and physical damage during harvesting may lead to breaks in outer protective layers that the surface microbes can now exploit. We are all familiar with molds growing on citrus fruits, like lemons and oranges (FIGURE 13.7). *Penicillium* and *Rhizopus* are frequently the spoilage organisms. Yeasts and some bacteria, including *Erwinia* and *Xanthomonas*, can also spoil some fruits and these may particularly be a problem for fresh cut packaged fruits.

■ *Rhizopus* rī'zōpus

■ *Erwinia* ėr-wi'-nē-ä

■ *Xanthomonas* zan'-thō-mō-nas

Fruit juices are relatively high in sugar and low in pH, which will be favorable conditions for yeast, mold, and some acid-tolerant bacterial species to grow. However, the lack of oxygen gas in most juices limits mold growth. LAB can spoil orange and tomato juices. These bacterial organisms are not as heat tolerant so they may appear as growths after the pasteurization process.

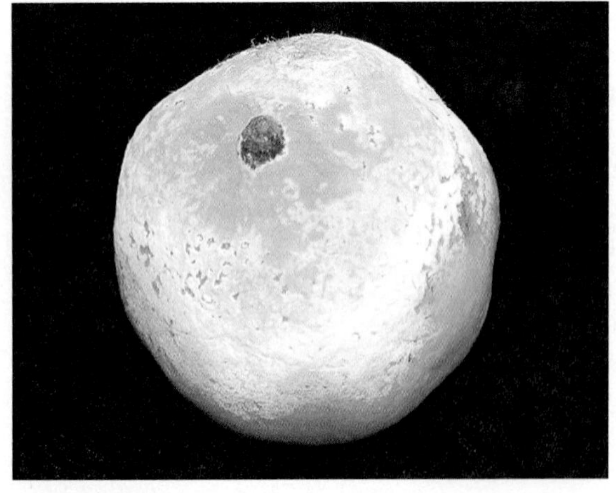

FIGURE 13.7 **Moldy Fruit.** A *Penicillium* species growing on an orange.

Courtesy of Dr. Jeffrey Pommerville

Vegetables with their neutral pH and high water content are a source of nutrients for spoilage microbes. Vegetables are certainly exposed to many soil microbes while growing, but many spoilage organisms, such as the LAB are not common in the soil. Therefore, most microbial spoilage occurs after there has been mechanical and chilling damage to plant surfaces. Species of *Erwinia* cause most cases of spoilage and can be found associated with many types of vegetables.

Bacterial spoilage first causes softening of tissues in the skin. As it is degraded, the whole vegetable may eventually disintegrate into a slimy mass. Starches and sugars are metabolized next and unpleasant odors and flavors develop along with lactic acid and ethanol. Besides *Erwinia*, several *Pseudomonas* species and LAB are important spoilage microbes.

Molds belonging to several genera, including *Rhizopus, Alternaria*, and *Botrytis*, cause a number of vegetable rots that are identified by the color, texture, or acidic products of the rot.

Alternaria äl-tėr-ner′-ēä

Botrytis bō-trī′-tis

Grains and Bakery Products

Cereal grains (e.g., corn, oats, rye, wheat, rice, barley) are naturally exposed to a variety of microbes during growth, harvesting, drying, and storage. Molds are the most common contaminants because of the low moisture levels in grains. Two examples of grain spoilage are described here.

The first contamination example is caused by the mold *Aspergillus flavus*. This fungus produces **aflatoxin**, a poisonous substance that accumulates in stored cereal grains (such as wheat) as well as peanuts and soybeans. The toxins can be consumed with grain products and meat from animals feed the contaminated grain. Scientists have implicated aflatoxins in liver and colon cancers in humans.

Aspergillus flavus a-spėr-jil′lus flā′vus

The second example of mold grain spoilage involves *Claviceps purpurea*, the cause of ergot poisoning (**ergotism**), which manifests itself in neurological symptoms. Rye plants are particularly susceptible to this type of spoilage, but wheat and barley grains may also be affected. The toxins deposited by *C. purpurea* may trigger convulsions and hallucinations when consumed (the psychedelic drug lysergic acid diethylamide or LSD was originally derived from the toxin).

Claviceps purpurea kla′vi-seps pür-pü-rē′ä

In the production of bakery products, ingredients such as flour, eggs, and sugar are generally the sources of spoilage microbes. Although most contaminants are killed during baking, *Bacillus* spores can survive in the interior of bread loaves baked at high oven temperatures. On cooling the bread, the spores can germinate and the cells start growing. Some strains of *Bacillus* are notorious in causing a bread condition called "ropiness." This is where the bread has a soft, cheesy texture with long, stringy threads. This is due to the bacterial cells degrading starch and producing a slimy polysaccharide. Yeasts can cause spoilage of breads and fruitcakes, resulting in a chalky appearance on surfaces and producing off odors.

Cream fillings and toppings in bakery products provide excellent chemical and physical conditions for microbial growth. For instance, custards made with whole eggs may be contaminated with *Salmonella* species, and whipped cream may contain microbes such as species of *Lactobacillus* and *Streptococcus*. The acid produced by the bacterial cells results in a sour taste.

13.4 Food Preservation: Keeping Microbes Out

Centuries ago, humans battled the elements to keep a steady supply of food at hand. Sometimes, there was a short growing season; at other times, locusts descended on their crops; at still other times, they underestimated their needs and had to cope with scarcity. However, experience taught humans that they could prevail through difficult times by preserving foods. Among the earliest methods was drying vegetables and strips of meat and fish in the sun. Foods could also be preserved by salting, smoking, and fermenting. One benefit of the new methods was that individuals could trek far from their native lands, and soon they took to the sea and moved overland to explore new places.

The next great advance did not come until the mid-1700s. In 1767, Lazzaro Spallanzani attempted to disprove spontaneous generation by showing that beef broth would remain unspoiled after being subjected to heat. In 1795, Nicolas Appert, a French winemaker, took note and applied this principle to the preservation of a variety of foods. He placed the foods in glass jars, sealed them with corks and sealing wax, and placed them in boiling water; not all that much different from home canning today. But neither Appert nor his contemporaries was quite sure why the food was being preserved. The significance of microbes as agents of spoilage awaited Pasteur's classic experiments with wine six decades later.

Through the centuries, preservation methods have had a common objective: to reduce the microbial population and maintain it at a low level until the food can be eaten. Modern preservation methods still have that objective. Though today's methods are sophisticated and technologically dynamic, advances have been counterbalanced by the increased complexity of food products and the great volumes of food to be preserved. Thus, the food preservation problems faced by early humans do not differ fundamentally from those confronting modern food technologists. As we will see in this section, many preservation methods are old standbys.

Heat

Heat kills microorganisms by changing the physical and chemical properties of their proteins. In a moist heat environment, proteins are denatured and lose their specific three-dimensional structure and biological activity. As their structural proteins and enzymes undergo this change, the microbes die.

The most useful application of heat in food technology is in **canning** (FIGURE 13.8). Shortly after Appert used bottles in establishing the value of heat in preservation, an English engineer named Bryan Donkin substituted iron cans coated with tin. Soon he was supplying canned meat to the British navy. In the United States, the tin can was virtually ignored until the Civil War period, but in the years thereafter, mass production of canned food began. Soon the tin can was the symbol of prepackaged convenience.

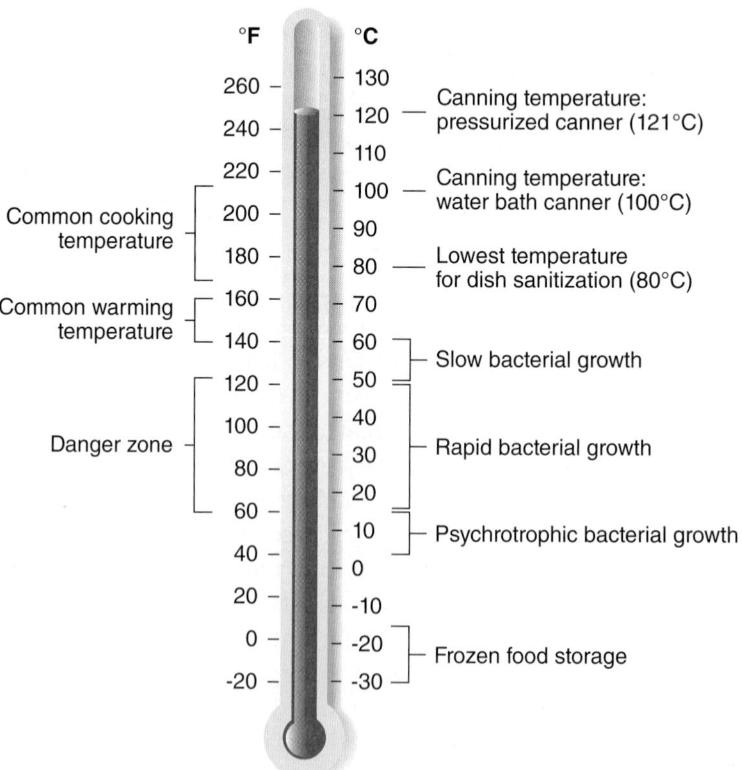

FIGURE 13.8 **Important Temperature Considerations in Food Microbiology.** Shown are several temperatures at which canning or sanitization procedures are carried out.

FIGURE 13.9 **The Industrial Canning Process.** An inspector watches over the canning process to ensure sanitary conditions are maintained.

Modern canning processes are complex. Machines wash, sort, and grade the food, then subject it to steam heat for 3 to 5 minutes (FIGURE 13.9). This last steam step, called **blanching**, destroys many enzymes in the food and prevents any further cellular metabolism. Next, the food is processed (peeled and cored, for example) and then put into the can. The air is evacuated from the can, which is then placed in a pressurized steam sterilizer similar to an autoclave at a temperature of 121°C or lower, depending on the food's pH, density, and heat penetration rate.

The "sterilizing process" used in the canning industry, called **commercial sterilization**, is designed to eliminate the most resistant bacterial spores, especially the endospores of the genera *Bacillus* and *Clostridium* as well as other microbes capable of producing spoilage under conditions of normal, nonrefrigerated storage. It is not as rigorous as the true sterilization used in a microbiology lab or hospital (FIGURE 13.10). Moreover, if a machine error leads to improper heating temperatures, a small hole allows airborne bacterial cells to enter, or a proper seal does not form, contamination may result. The contamination is usually obvious because many spoiled canned foods have a putrid odor. Most contamination of canned food is due to facultative or anaerobic bacteria (such as *Clostridium*) that produce gas and cause the ends of the can to bulge. (Can swelling can also be caused by a reaction between food acids and

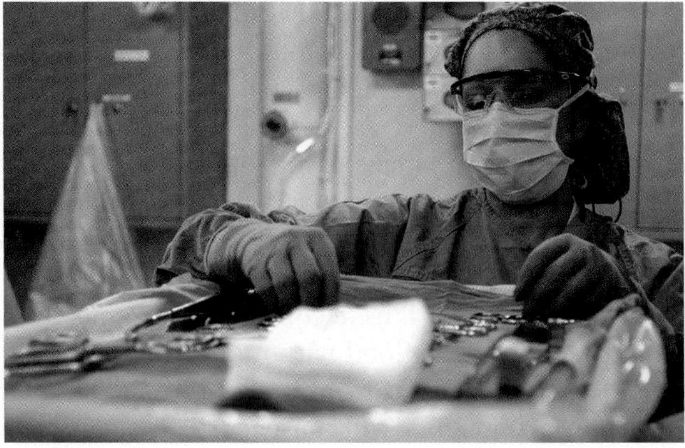

FIGURE 13.10 **Sterile Surgical Instruments.** For surgery, it is critical that all surgical instruments be sterile.

the can's metal, a reaction that leads to accumulation of hydrogen gas.) Common contaminants also include coliform bacteria, which are a group of gram-negative non-sporeforming rods (such as *Escherichia*, *Serratia*, and *Enterobacter*) that ferment lactose to acid and gas.

Growth of acid-producing microbes presents a different problem because spoilage cannot be discerned from the can's shape. Food has a flat-sour taste from the acid and has probably been contaminated by a *Bacillus* species, a species of coliform, or another acid-producing bacterial species that survived the heating. This may happen if the bacterial cell population was extremely high at the outset and the heat was unable to destroy all the microbes.

Pasteurization

The process of pasteurization was developed by Louis Pasteur in the 1850s to eliminate bacterial cells in wines. His method was first applied to milk in Denmark about 1870 and was widely employed by 1895. Although the primary object of **pasteurization** is to eliminate bacterial pathogens from milk, the process also lowers the total number of bacterial cells and thereby reduces the chance of spoilage.

The more traditional method of pasteurization, the **holding (batch) method**, involves heating the milk in a large bulk tank at 63°C for 30 minutes. Concentrated products, such as cream, are heated at the higher temperature of 68°C to ensure successful pasteurization. The more modern method of pasteurization is called the **flash method**. In this process, machines pass the milk through a hot cylinder at 72°C (161°F) for a period of 15 seconds and then cooled rapidly.

Microbes, such as the streptococci, capable of surviving pasteurization are described as **thermoduric** (heat-enduring). Such species, such as *Streptococcus lactis*, can grow slowly in refrigerated milk. When their numbers reach about 20 million per milliliter, enough lactic acid has been produced to make the milk sour. The shelf life of the milk indicated on the carton is an estimate of when souring is likely to occur.

Although most milk in the United States is pasteurized by the flash method, milk can be sterilized by exposure to pressurized steam at 140°C for 3 seconds. This **ultra-high temperature (UHT) method** results in milk with an indefinite shelf life, so long as the container remains sealed. Small containers of coffee cream are often prepared this way.

Low Temperatures

In a refrigerator or freezer, the lower temperature reduces the rate of microbial enzyme activity and thus slows their rate of growth and reproduction, while extending the shelf life of food (see Figure 13.8). Although the microbes are not killed, their numbers are kept low, and spoilage is minimized. Well before the invention of refrigerators, the Greeks and Romans had partially solved the problem of keeping things cold. They dug snow cellars in the basements of their homes, lined the cellars with logs, insulated them with heavy layers of straw, and packed them densely with snow delivered from far-off mountaintops. The compressed snow turned to a block of ice, and foods left in this makeshift refrigerator would remain unspoiled for long periods.

Modern refrigerators at 5°C (41°F) accomplish the same goal by providing a suitable environment for preserving food without destroying its appearance, taste, or cellular integrity. However, **psychrotrophic** (cold-tolerant) microbes survive in the cold, and given enough time, they will cause meat surfaces to turn green, eggs to rot, fruits to become moldy, and milk to sour. *Staphylococcus aureus* can be a problem because as the cells multiply in foods, they deposit their toxins. If the toxins are not

■ *Staphylococcus aureus*
staf-i-lō-kok'us ôrē-us

destroyed by heating, they will cause staphylococcal food poisoning, and the affected individual can expect diarrhea, nausea, and cramps.

When food is placed in a freezer at –18°C (0°F), ice crystals form, which tear and shred microbes, killing substantial numbers of cells. However, many survive, and because the ice crystals are equally destructive to food cells, the microbes multiply quickly when the food thaws. Rapid thawing and cooking are therefore recommended for frozen foods. Moreover, food should not be refrozen; during the time it takes to thaw and refreeze, microbes can deposit sufficient toxins to cause food poisoning the next time the food is thawed. Microwave cooking, which requires minimal thawing, may eliminate some of these problems.

Deep freezing at –60°C results in smaller ice crystals, and although the physical damage to microorganisms is less severe, their biochemical activity is reduced considerably. Some food producers blanch their product (apply moist heat briefly) before deep freezing, a process that reduces the number of microbes even further. A major drawback of freezing is freezer burn, which develops as food dries out from moisture evaporation. Another disadvantage is the considerable energy cost. Nevertheless, freezing has been a mainstay of preservation since Clarence Birdseye first offered frozen foods for sale in the 1920s. Approximately a third of all preserved food in the United States is frozen.

Drying

Another method of food preservation involves drying. In past centuries, people used the sun for drying, but modern technologists have developed sophisticated machinery for this purpose. For example, the spray dryer expels a fine mist of liquid such as coffee into a barrel cylinder containing hot air. The water evaporates quickly, and the coffee powder falls to the bottom of the cylinder.

Another machine for drying is the heated drum. Machines pour onto the drum's surface liquids such as soup, and the water evaporates rapidly, leaving dried soup to be scraped off. A third machine uses a belt heater that exposes liquids such as milk to a stream of hot air. The air evaporates any water and leaves dried milk solids. Unfortunately, spore-forming and capsule-producing bacterial cells resist drying.

During the past few decades, freeze-drying, or **lyophilization**, has emerged as a valuable preservation method. In this process, food is deep frozen, and then a vacuum pump draws off the water. (Water passes from its solid phase [ice] to its gaseous phase [water vapor] without passing through its liquid phase [water].) The dry product is sealed in foil and easily reconstituted with water. Hikers and campers find considerable advantages in freeze-dried foods because of their light weight and durability; however, the ancient Incas of Peru were freezing drying foods centuries earlier as told in **A Closer Look 13.2**.

Osmotic Pressure

When living cells are immersed in large quantities of a compound such as salt or sugar, water diffuses out of cells through their cell membranes and into the surrounding environment. The flow of water is called **osmosis**, and the force that drives the water is termed **osmotic pressure**.

Osmotic pressure can be used to preserve foods because water flows out of microbes as well as food cells. For example, in highly salted or sugared foods, microbes dehydrate, shrink, and die. Jams, jellies, fruits, maple syrups, honey, and similar products are preserved this way. Foods preserved by salting include ham, cod, bacon, and beef,

A CLOSER LOOK 13.2

It Started with "Stomped Potatoes"

Freeze-drying, or lyophilization, involves the removal (sublimation) of water from a substance without going through a liquid state. Such freeze-dried foods last longer than other preserved foods and are very light, which makes them perfect for everyone from backpackers to astronauts. However, the origins of freeze-dried foods go back to the ancient Incas of Peru.

The basic process of freeze-drying food was known to the pre-Columbian Incas who lived in the high altitudes of the Andes Mountains of South America. The Incas stored some of their food crops, including potatoes, on the mountain heights. First, most of the moisture was crushed out of the potatoes when the Incas stomped on them with their feet. Then, the cold mountain temperatures froze the stomped potatoes and the remaining water inside slowly vaporized under the low air pressure of the high altitudes in the Andes.

Today, a similar procedure is still used in Peru for a dried potato product called chuno. It is a light powder that retains most of the nutritional properties of the original potatoes but, being freeze-dried, chuno can be stored for up to 4 years.

On the commercial market, the Nestlé Company was asked by Brazil in 1938 to help find a solution to save their coffee bean surpluses. Nestlé developed a process to convert the coffee surpluses into a freeze-dried powder that could be stored for long periods. The result was Nescafe® coffee, which was first introduced in Switzerland. However, during World War II the process of freeze-drying developed into an industrial process. Blood plasma and penicillin were needed for the armed forces and freeze drying was found to be the best way to preserve and transport these materials.

In 1964, Nestlé developed an improved method of producing instant coffee by freeze-drying. Today, several other coffee producers use a similar process of preservation. In the late 1970s, freeze-drying was used for taxidermy, food preservation, museum conservation, and pharmaceutical production.

Today, more than 400 commercial food products are preserved by freeze-drying. In the medical field, freeze-drying is the method of choice for extending and preserving the shelf life of enzymes, antibodies, vaccines, pharmaceuticals, blood fractions, and diagnostics. The benefit of all lyophilized products is that they rehydrate easily and quickly because of the porous structure created by the freeze-drying process.

And to think—it all started with the Incas stomping on their potatoes.

A family of Qetchua Indians, descendants of the Inca, dig potatoes on a small farm high in the Andes Mountains.

as well as certain vegetables such as sauerkraut, which has the added benefit of large quantities of acid.

Chemical Preservatives

For a chemical preservative to be useful in foods, it must inhibit microbial growth while being easily broken down and eliminated by the human body without side effects. These requirements are enforced by the U.S. Food and Drug Administration (FDA). The criteria have limited the number of chemicals used as food preservatives to a select few.

A key group of chemical preservatives are **organic acids**, including sorbic acid, benzoic acid, and propionic acid. These chemicals damage microbial membranes and interfere with the uptake of certain essential nutrients such as amino acids. Sorbic acid, which came into use in 1955, is added to syrups, salad dressings, jellies, and certain cakes. Benzoic acid, the first chemical to be approved by the FDA for use in foods, protects beverages, catsup, margarine, and apple cider. Propionic acid is incorporated in wrappings for butter and cheese, and it is added to breads and bakery products, where it prevents the growth of fungi and inhibits the ropiness commonly due to *Bacillus* species. Other natural acids in foods add flavor while serving as preservatives. Examples are lactic acid in sauerkraut and yogurt, and acetic acid in vinegar.

The process of smoking with hickory or other woods accomplishes the dual purposes of drying food and depositing chemical preservatives. By-products of smoke, such as aldehydes, acids, and certain phenol compounds, effectively inhibit microbial growth for long periods of time. Smoked fish and meats have been staples of the human diet for many centuries.

Sulfur dioxide is used as a preservative for dried fruits, molasses, and juice concentrates. Used in either gas or liquid form, the chemical retards color changes on the fruit surface, while reducing microbial spoilage. Another sulfur compound, sulfurous acid, is used to prevent growth of the lactobacilli able to sour wine. The two sulfur compounds are commonly known as sulfites. Unfortunately, the FDA estimates that over 1% of the U.S. population is sensitive to sulfites, including over a million individuals who suffer from asthma.

Radiation

Though some of the public is apprehensive about irradiated foods, various forms of radiation have received FDA approval for preserving foods (TABLE 13.1). For instance, gamma rays are used to extend the shelf life of fruits, vegetables, fish, and poultry from several days to several weeks. This form of radiation also increases the distance fresh food can be transported and significantly extends the storage time for food in the home. Gamma rays are a high-frequency type of electromagnetic energy emitted by the radioactive isotopes cobalt-60 and cesium-137. Health officials are quick to

TABLE 13.1 Foods Approved for Irradiation[a,b]

Year	Food
1964	Wheat flour and potatoes
1983	Dried spices
1985	Additional dried spices, herb and vegetable seasonings, dry and hydrated enzymes, pork
1986	Fruit and vegetables
1990	Poultry
1997	Beef, lamb, pork, and byproducts
2000	Fresh shell eggs, sprouting seeds, fruit and vegetable juices
2002	Imported fruit and vegetables (for pest treatment)
2005	Ready-to-eat foods (deli meats, frozen foods)
2008	Packaged salads (for pathogen treatment), molluscan shellfish (oysters, clams, mussels)
Pending	Crustacean shellfish (shrimp, crabs, lobster), unrefrigerated beef, lamb, pork, horsemeat, and byproducts

[a]Approvals by the U.S. Food and Drug Administration and Department of Agriculture.

[b]Data from: Food and Water Watch, Washington, DC.

note that such irradiation of food does not cause food to become radioactive. The radiation kills microbes by reacting with and destroying microbial DNA and other key organic compounds.

Food processors using isotopes store the pellets by encapsulating them in double-layered stainless steel pencil-like tubes arranged in racks under water. When food is to be irradiated, the racks are withdrawn from the water, and the food is passed through the radiation field. Importantly, the gamma rays penetrate the food and cause microbial death. There is no radiation in the food and the nutritional losses are similar to those occurring with cooking and/or freezing.

13.5 Preventing Foodborne Disease: The Challenges

As the story opening this chapter illustrates, food can be a method for transferring microbes as well as a culture medium for their growth. Unsuspecting consumers are affected by either the microbes or the toxins they have produced in the food. When the microbes themselves are transferred to consumers, a **food infection** is established; when their toxins are consumed, a **food poisoning** (or food intoxication) occurs.

Food infections are typified by typhoid fever, cholera, and several types of **gastroenteritis** (inflammation of the stomach and intestines), including salmonellosis and shigellosis, all of which are of bacterial origin. Amoebiasis and giardiasis represent foodborne diseases (dysentery and gastroenteritis) caused by protists. Viral infections caused by the hepatitis A virus (hepatitis A) and the noroviruses (gastroenteritis) also occur. Food poisonings include staphylococcal food poisoning, botulism, and clostridial food poisoning. Full discussions of these diseases are presented in other chapters, so we will not examine them here. Foodborne diseases can be nasty experiences, so the next section presents some suggestions on how to avoid them.

Helpful Suggestions

The CDC estimates more than 48 million Americans get sick (1 person in every 6), over 128,000 are hospitalized, and 3,000 people die each year from foodborne illnesses. However, efforts by the meat industry are attempting to reduce the number of foodborne illnesses (FIGURE 13.11). For example, *Escherichia coli* O157:H7 infections declined 25% between 2010 and 2011, the largest decline ever. Since 1999, *E. coli* O157:H7 infections have declined over 65%, while others, such as those caused by *Salmonella* species, have remained relatively stable.

In many of these cases, the foodborne disease might have been avoided if some basic precautions had been taken. For example, unrefrigerated foods are a prime source of staphylococci and *Salmonella* species (FIGURE 13.12), so perishable groceries such as meats and dairy products should not be allowed to warm up while other errands are performed. Also, a thermometer should be used to ensure the refrigerator temperature remains below 5°C (41°F) at all times.

Another way to avoid foodborne disease is to be aware that skin infections are often caused by some of the same staphylococci causing human disease (e.g., typhoid fever); it is therefore prudent to cover any skin infections while working with foods. Moreover, the hands should always be washed thoroughly before and after handling raw vegetables (or salad fixings) to avoid cross-contamination of other foods. It is

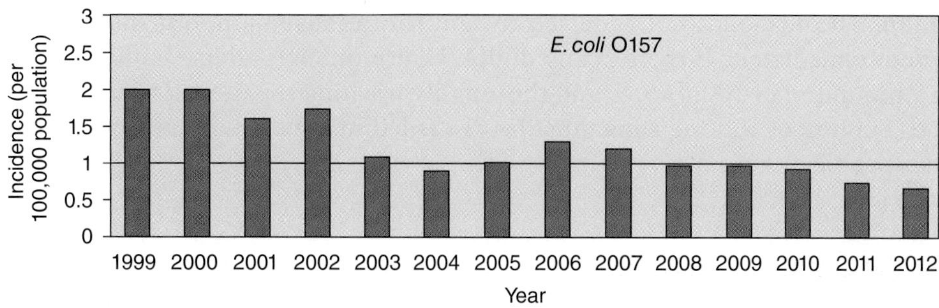

(a)

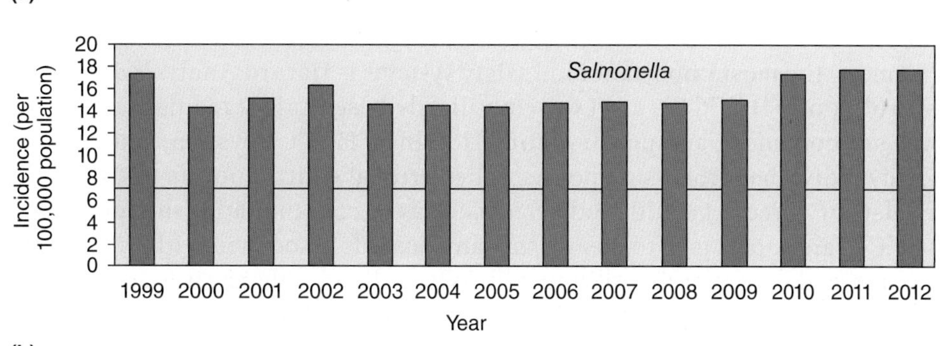

(b)

FIGURE 13.11 Incidence of Selected Foodborne Illnesses—United States, from 1999–2012. The incidence of foodborne illness caused by (**a**) *E. coli* 0157:H7 and (**b**) *Salmonella*. The national health objective is to reach a level where the incidence is no more than 1 per 100,000 people for *E. coli* and 6.8 per 100,000 for *Salmonella*.

Source: FoodNet/CDC.

wise to cook meat from a frozen or partly frozen state; if this is not possible, the meat should be thawed in the refrigerator. Cutting boards should be cleaned with hot, soapy water after use, and old cutting boards with cracks and pits should be discarded.

Studies indicate that leftovers are implicated in most outbreaks of foodborne disease. It is therefore important to refrigerate leftovers promptly and keep them no more than a few days. Thorough reheating of leftovers, preferably to boiling, also reduces the possibility of illness.

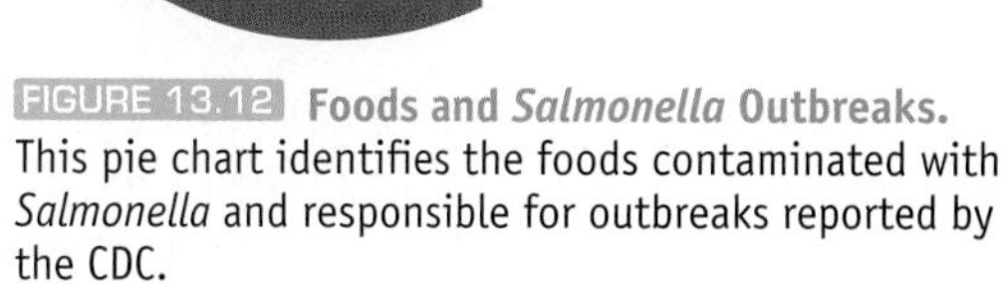

FIGURE 13.12 Foods and *Salmonella* Outbreaks. This pie chart identifies the foods contaminated with *Salmonella* and responsible for outbreaks reported by the CDC.

Reproduced from CDC Vital Signs

Many instances of foodborne disease occur during the summer months, when foods are taken on picnics, where they cannot be refrigerated. As a general rule, foods containing eggs, such as custards, cream pies, pastries, and deli salads, should be left off the picnic menu. For outdoor barbecues, one dish should be used for carrying chicken, hamburgers, or steaks to the grill and another dish for serving them. Many of these principles apply equally well to fall and winter tailgate parties.

Over 90% of botulism outbreaks reported to the CDC are traced to home-canned food. To prevent this sometimes fatal foodborne disease, health officials urge homemakers to use the pressure method to can foods. Reliable canning instructions should be obtained and followed stringently. Canned foods suspected of contamination should not be tasted to confirm the suspicion, but should be discarded immediately. If there is any doubt, health officials advise boiling the food for a minimum of 10 minutes and thoroughly washing the utensil used to stir the food. Bulging or leaking cans must be discarded in a way that will not endanger other people or animals.

HACCP Systems

Fueled by consumer awareness, the entire food industry has been placed under a food-safety spotlight overseen by the FDA and the U.S. Department of Agriculture (USDA).

Among the most important food safety systems is **Hazard Analysis and Critical Control Point (HACCP)**, a set of scientifically based safety regulations enforced in the seafood, meat, and poultry industries. In an HACCP system, manufacturers identify individual processing points, called **critical control points (CCPs)**, where the safety of a product could be affected, such as by contaminating microorganisms. The CCPs are supervised to ensure that any hazards associated with the operation are contained or, preferably, eliminated. When all possible hazards are controlled at the CCPs, the safety of the product can be assumed without further testing or inspection.

The standard regulations require food processors to monitor and control eight key sanitation areas, including the condition and cleanliness of utensils, gloves, outer garments, and other food contact surfaces; the prevention of cross-contamination from raw products and unsanitary objects to foods; and the control of employee health conditions that could result in food contamination. If the system and all the CCPs are kept up to standards, then the competition between humans and microbes will tip in our favor.

So, how are your sanitary practices at home? **A Closer Look 13.3** provides the test.

A CLOSER LOOK 13.3
Keeping Microorganisms Under Control

How knowledgeable are you concerning food safety in your home or apartment?

If you are, do you actually practice what you know? Take the following quiz (honestly) and then we will see where microbes lurk, and where you need to eliminate or keep them under control.

1. Your refrigerator should be kept at what temperature?
   a. 0°C
   b. 5°C
   c. 10°C
2. Frozen fish, meat, and poultry products should be defrosted by _____.
   a. setting them on the counter for several hours
   b. microwaving
   c. placing them in the refrigerator
3. After cutting raw fish, meat, or poultry on a cutting board, you can safely _____.
   a. reuse the board as is
   b. wipe the board with a damp cloth and reuse it
   c. wash the board with soapy hot water, sanitize it with a mild bleach, and reuse it
4. After handling raw fish, meat, or poultry, how do you clean your hands?
   a. Wipe them on a towel
   b. Rinse them under hot, cold, or warm tap water.
   c. Wash them with soap and warm water.
5. When was the last time you sanitized your kitchen sink drain and garbage disposal?
   a. Last night
   b. A few weeks ago
   c. Cannot remember

A CLOSER LOOK 13.3

Keeping Microorganisms Under Control (continued)

6. Leftover cooked food should be _____.
   a. cooled to room temperature before being put in the refrigerator
   b. put in the refrigerator immediately after the food is served
   c. Left at room temperature overnight or longer
7. How do you clean your kitchen counter surfaces that are exposed to raw foods?
   a. Use hot water and soap, then a bleach solution.
   b. Use hot water and soap.
   c. Use warm water.
8. How was the last hamburger you ate cooked?
   a. Rare
   b. Medium
   c. Well-done
9. How often do you clean or replace your kitchen sponge or dishcloth?
   a. Daily
   b. Every few weeks
   c. Every few months
10. Normally dishes in your home are cleaned _____.
   a. by an automatic dishwasher and air-dried
   b. after several hours soaking and then washed with soap in the same water
   c. right away with hot water and soap in the sink and then air-dried

So, how did you do?

A Final Thought

Today, the FDA is responsible for ensuring the safety of about 80% of the food we eat with the rest (meat, poultry, and some eggs) being under the jurisdiction of the USDA. Although the HACCP system is in place to inspect domestically produced food, the FDA only manages to inspect 1% to 2% of all imported food. This is of concern because with the growing globalized food supply, Americans now get 15% of their food (66% of the fresh fruits and vegetables) from another country. And these numbers continue to grow each year, meaning that if the imported food is not thoroughly inspected, microbial contamination and unsafe chemicals could be associated with the imported food.

To address this gap in food inspections and food safety, in July 2013 the FDA proposed new rules requiring all American importers to inspect the food they are importing and to make sure it is safe for human consumption. In the first significant update to the FDA safety authority in 70 years, the proposed law would require importers to subject all imported food (except seafood and fruit juices, which have a different set of rules) to the same safety standards as those used to ensure the safety of domestically produced food.

The next time you shop at the supermarket, stop and take note of the broad variety of imported foods, which we consider are safe to eat without making us ill. To maintain this confidence, better and more stringent inspection and import practices are needed. Hopefully, by the time you read this, the proposal will be law.

Questions to Consider

1. Chicken and salad are two items on the dinner menu at home, and you are put in charge of preparing both. You have a cutting board and knife for slicing up the salad items and cutting the chicken into pieces. Which task should you perform first? Why? What other precautions might you take to ensure that dinner is not remembered for the wrong reason?
2. Having read this chapter, you decide to outfit your refrigerator with UV lights. What are you trying to accomplish?

3. It is a hot Saturday morning in July. You get into your car at 9:00 am with the following list of chores: Pick up the custard eclairs for tonight's dinner party, drop off clothes at the cleaners, buy the ground beef for tomorrow's barbecue, deliver the kids to the Little League baseball game, pick up a broiler at the poultry farm. Microbiologically speaking, what sequence should you follow in doing these chores?
4. It is 5:30 pm and you arrive on campus for your evening college class. You stop off at the cafeteria for a bite to eat. At this time of the day, which foods might you be inclined to avoid purchasing?
5. To avoid *Salmonella* infection when preparing eggs for breakfast, the operative phrase is "scramble or gamble." How many foods can you name that use uncooked or undercooked eggs and that can represent a health hazard?
6. A standard set of recommendations and regulations exists for individuals who work in restaurants and cook food for customers (i.e., food handlers). However, very few regulations exist for individuals who pick fruits or vegetables in the fields. What regulations would you recommend for such workers?
7. Foods from tropical nations tend to be very spicy, with lots of hot peppers, spices, garlic, and lemon juice. By contrast, foods from cooler countries tend to be much less spicy. Why do you think this pattern has evolved over the ages?
8. On Saturday, a man buys a steak and a pound of liver and places them in the refrigerator. On Monday, he must decide which to cook for dinner. Microbiologically, which is the better choice? Why?
9. Suppose you had the choice of purchasing "yogurt made with pasteurized milk" or "pasteurized yogurt." Which would you choose? Why? What are the "active cultures" in a cup of yogurt?

Key Terms

Informative facts are necessary for the expression of every concept, and the information for a concept is founded in a set of key terms. The following terms form the basis for the concepts of this chapter. On completing the chapter, you should be able to explain and/or define each one:

aflatoxin
blanching
botulism
canning
coliform bacteria
commercial sterilization
critical control point (CCP)
ergotism
expiration date
extrinsic factor
flash method
food infection
food poisoning
food spoilage
gastroenteritis
HACCP
highly perishable
holding (batch) method
intrinsic factor
lactic acid bacteria (LAB)
lyophilization (freeze drying)
manufacturer code
microbial load
microbiome
mold
nonperishable
organic acid
osmosis
osmotic pressure
pasteurization
probiotics
psychrotrophic
salmonellosis
semiperishable
shelf life
sterile
thermoduric
ultra-high temperature (UHT) method
yeast

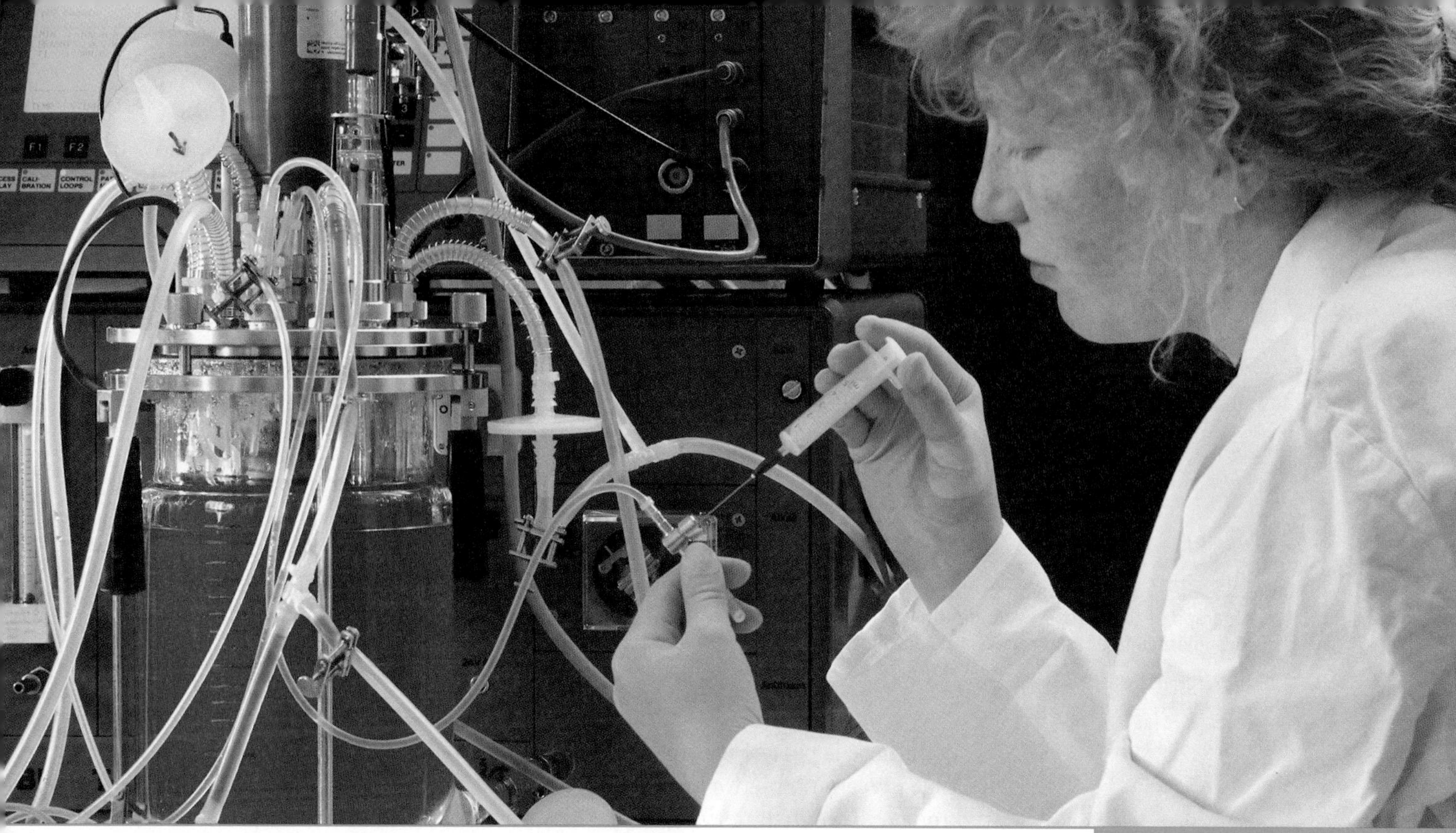

Biotechnology and Industry: Microbes at Work

14

Today, microbes can be genetically engineered and harnessed to produce a large variety of substances for humans and society. Here, a technician is sampling a small bioreactor used to grow genetically engineered microbes. Microbes in larger bioreactors can be used to produce medicines, vaccines, antibiotics, and nutrients that are added to foods, as well as to many other common products we consume or use every day.

Looking Ahead

The use of microbes has brought many breakthroughs in microbiology and DNA research and has resulted in a wealth of innovative industrial and biotechnological applications. But this is not surprising because microbes have been used in industry for many decades to synthesize a broad variety of valuable products. We will combine the old and the new in this chapter as we survey some of the goods and services provided by microbes to enhance the quality of our lives and society in general.

On completing this chapter, you should be capable of:

- Distinguishing between primary metabolites and secondary metabolites.
- Discussing the industrial role of microbes in the production of antibiotics, vitamins, enzymes, and biofuels.
- Describing how bacterial cells are genetically engineered to produce a specific product.
- Comparing and contrasting traditional vaccines, and differentiating them from genetically engineered vaccines.
- Describing the uses of biotechnology in the diagnostic lab.
- Discussing the uses of metagenomics.

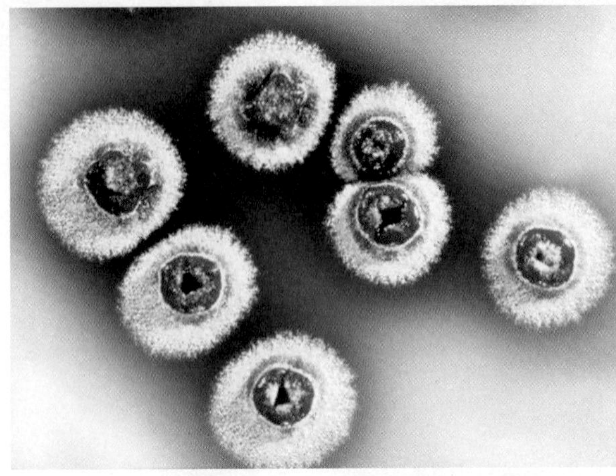

FIGURE 14.1 ***Streptomyces coelicolor* Colonies.** In this photograph of *S. coelicolor,* colonies growing on agar are secreting droplets of liquid containing antibiotics.

Courtesy of John Innes Centre www.jic.ac.uk

Streptomyces coelicolor
strep-tō-mī'sēs kō'lē-ku-lėr

Medicinal microbe's genome sequenced. How often have we read in the newspaper or heard from the news media about an organism's genes (genome) being sequenced or its DNA being mapped? It must be significant, right? After all, it made the news! But what is the underlying significance of such sequencing? Here is a good example.

In late spring of 2003, a group of British scientists announced they had mapped the genome of a very important bacterial species, *Streptomyces coelicolor.* This is a gram-positive organism commonly found in the soil (FIGURE 14.1). The mapping project began in 1997 and took 6 years to complete (this could be done much faster today) partly because the organism's genome is one of the largest ever sequenced. It has 8.6 million base pairs and some 7,825 genes. On completion of the sequencing, one of the scientists on the project said, "*It is a fabulous resource for scientists.*" Why?

Well, here is part of the answer. *S. coelicolor* belongs to a group of bacterial species that is responsible for producing over 65% of the naturally known antibiotics used today. By analyzing the genome of *S. coelicolor* and other *Streptomyces* species, additional metabolic pathways may be discovered for the production of other, yet unidentified and perhaps novel antibiotics. In fact, since the genome was sequenced, researchers have identified 18 gene clusters they suspect are involved with the production of different antibiotics. If correct, it may be possible for genetic engineers and biotechnologists to transform the organism into an "antibiotic factory" capable of churning out new drugs to replace the dwindling supply of usable antibiotics to which pathogens are not yet resistant. Using genetic engineering techniques, they could rearrange gene clusters and perhaps produce even more useful and potent antibiotics than naturally possible.

But there is more. *S. coelicolor* is a close relative of the pathogens causing tuberculosis, leprosy, and diphtheria. By comparing genomes, scientists hope to learn why *S. coelicolor* is not pathogenic, while the other three are pathogens. Differences in the genome of *S. coelicolor* might be important in understanding the infectious nature of its relatives and perhaps providing information to design new antibiotics through genetic engineering to attack these pathogens.

Today, our dependency on industrial products and biotechnological substances derived from microbial metabolism has never been greater. **Industrial microbiology** depends on using microbes that naturally carry out the production of a desired end product, such as antibiotics. The key here is using microbes to which genetic alterations or metabolic improvements have been made that substantially increase the amount of product produced to a commercially profitable level. **Biotechnology** can be defined as using the techniques of genetic engineering to alter the genome of microbes so they can produce substances they otherwise would not naturally produce, such as human hormones. The purpose of both fields is to design, develop, and produce commercial products for the benefit of humans, other organisms, and society as a whole.

In this chapter, we will examine some of the products of industrial microbiology and biotechnology in the context of the microbes that are their essential underpinnings. Some of the principles underlying the new technologies are complex, to be sure, but we must be able to comprehend them if we are to understand the advances being reported almost daily in the public media.

14.1 Microbes and Industry: Working Together

Long before biotechnology came into existence, the use of microbes for human welfare was commonplace. In fact, microbes were being used well before people realized they were taking advantage of these tiny organisms. For example, wine and beer making have been key activities of human culture and society for millennia. As these processes are covered in another chapter, let's examine some of the modern-day uses of microbes and see their relationships to industrial microbiology.

The Industrial Microbes

There are numerous industrial and commercial products derived from microbes. These include antibiotics, vitamins, amino acids, food additives, and, as we already mentioned, alcoholic beverages. Such a large variety of substances means a large group of microbes is involved.

Many microbes, both prokaryotic and fungal, can be described as "metabolic specialists" and industrial microbiologists take advantage of these specialists to produce desired end products or byproducts (called **metabolites**) of metabolism. In identifying these specialists, certain attributes of the microbe are critical.

- The microbe must produce the required metabolite in high quantities to be commercially profitable. Because the microbe may not always produce the metabolite in high enough yield, the organism must be amenable to genetic mutation or be genetically "tweaked" to produce high yield metabolite-producing strains.
- For industrial purposes, the microbe must grow rapidly in large-scale cultures and produce the metabolite in a short time.
- It is helpful if the industrial microbe is a spore-former to make inoculation of the culture relatively straightforward.
- The microbe must be capable of fast growth.
- Importantly, the microbe should not be a human or animal pathogen.

Producing Metabolites and Growing Microbes in Mass

Microbial growth, which is discussed in another chapter, typically goes through several phases and their ability to produce certain metabolites depends on what phase of growth they are in. Microorganisms may produce one of two types of metabolites that are important to industrial microbiology (FIGURE 14.2). **Primary metabolites** are directly involved in the normal growth, development, and reproduction of the microbe. They are formed during the log phase of growth. For example, pyruvate and the end products of the microbial fermentation pathways are examples of primary metabolites.

Secondary metabolites form the bulk of products of industrial interest. These metabolites are often species-specific but are not essential for growth and reproduction. Looking again at Figure 14.2, notice that secondary metabolites usually are produced near or at the end of the microbe's log phase or in stationary phase.

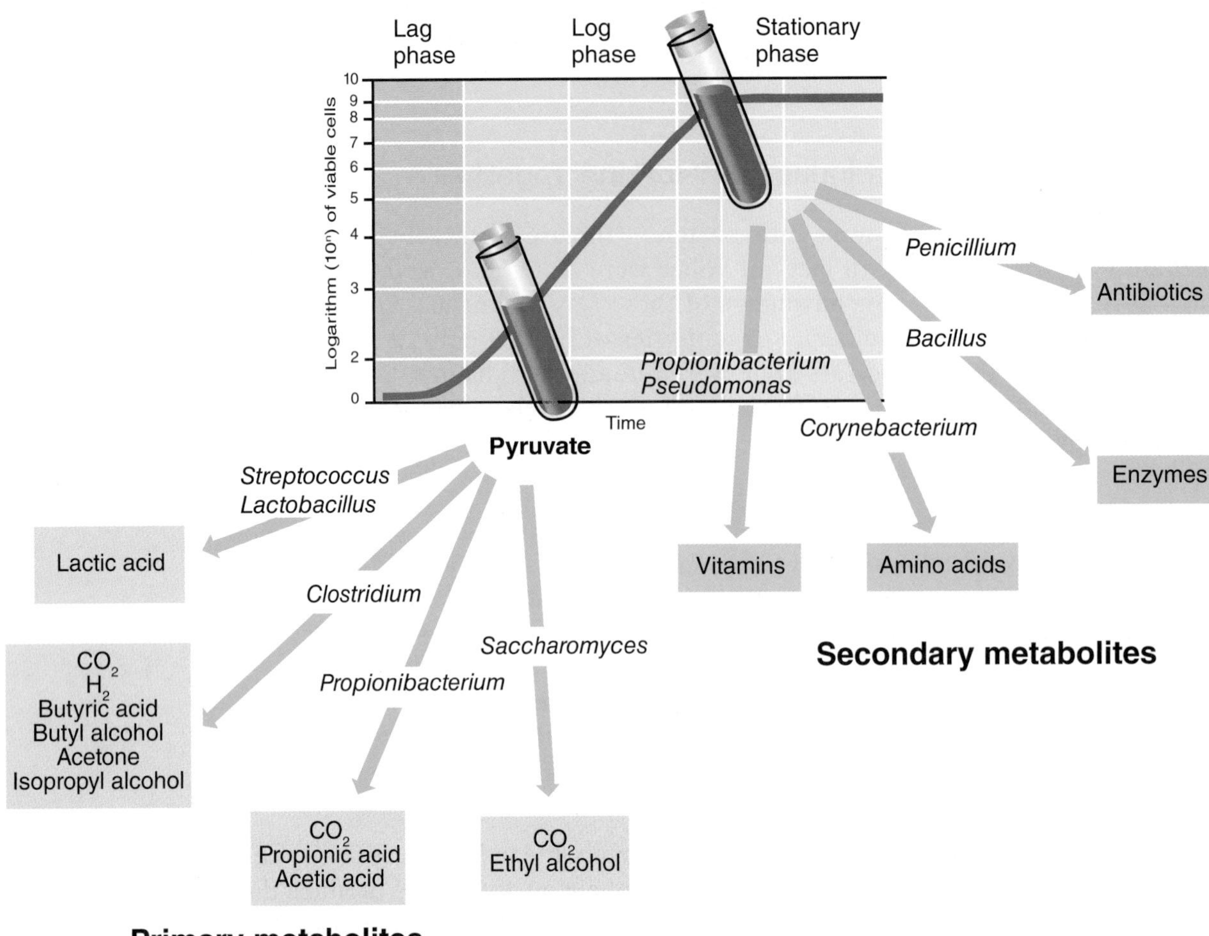

FIGURE 14.2 **Metabolite Synthesis.** In culture, primary metabolites are mainly produced during the active log phase of growth. Secondary metabolites are produced near the end of the log phase and or during stationary phase.

Most all antibiotics, as well as some vitamins, amino acids, and enzymes are examples of secondary metabolites.

To produce a significantly useful amount of metabolite from microbes at the industrial scale, tremendously large numbers of cells are needed. On this large scale, industry often uses the term **industrial fermentation**, which refers to any procedure used to grow large masses of microbes, be it aerobic or anaerobic.

Needing huge microbial populations means scaling up microbial metabolic processes to a level significant enough to produce the valuable metabolite. Such metabolic processes are eventually carried out in large, stainless steel tanks called **fermentors** or **bioreactors** (FIGURE 14.3). Technicians add the desired microbes that had been growing in a flask or smaller fermentor to the sterile growth medium in the industrial fermentor. Environmental conditions in the fermentor are continually computer monitored to maintain the desired growth conditions and to ensure the microbes reach a state in which maximal metabolite is produced. Temperature, pH, oxygen, and nutrients—all of the typical physical and chemical conditions that apply to growing microbes in a culture tube—apply here, just on a much grander scale.

Once the maximal amount of metabolite has been produced, the liquid in the fermentor is filtered, the product is extracted, and the purity determined. It then can be packaged and sent to the marketplace.

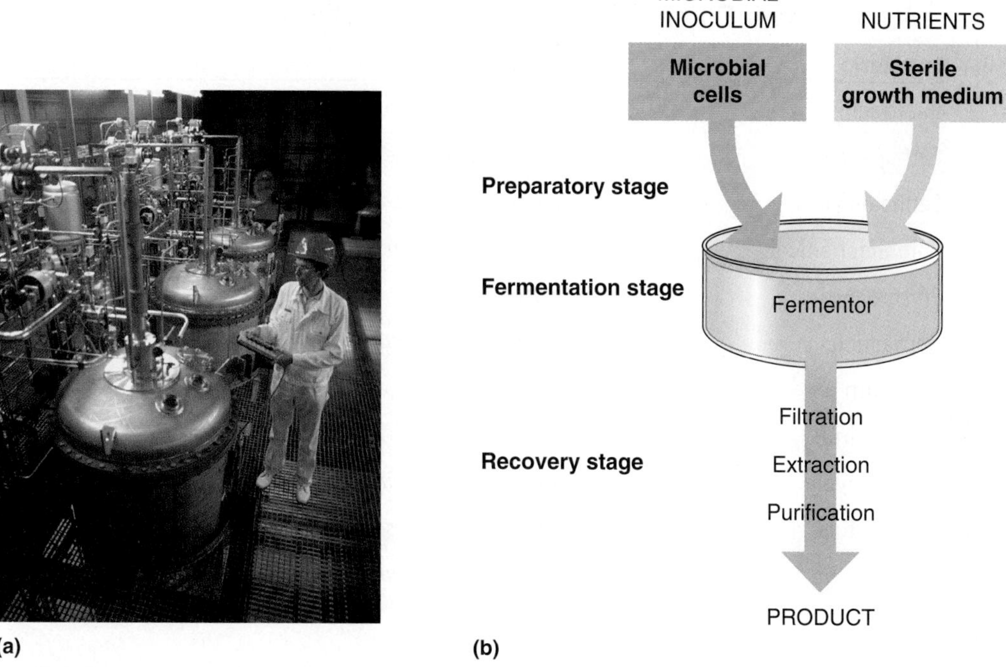

FIGURE 14.3 **The Industrial Fermentation Process.** (**a**) A pharmaceutical technician monitors a series of fermentors to ensure a maximum yield of metabolite. (**b**) A general industrial process scheme for metabolite production and recovery.

14.2 Industrial Microbiology: The Products

Having seen how microbes are grown on the industrial scale, let's now survey some of the products of industrial microbiology.

Antibiotics

Antibiotics are chemical products (or their synthetic derivatives) of microbes used to kill other microbes. Well over 5,000 antibiotic substances have been described and approximately 100 antibiotics are available to medical practitioners. Although important and useful antibiotics are supplied by *Penicillium* and *Bacillus* species, many antibiotics are produced by species of *Streptomyces*, as mentioned in the opener to this chapter – and listed in TABLE 14.1.

■ *Penicillium* pen-i-sil′lē-um

■ *Bacillus* bä-sil′lus

Antibiotic production, like other secondary metabolites, is carried out in huge fermentors similar to those used in brewing. A typical tank may hold 40,000 to 200,000 liters (10,500 to 52,000 gallons) of medium. Rather than inoculating the medium with bacterial spores, some newer technologies use small fragments of bacterial filaments or fungal hyphae as the inoculum. As the carbon source in the medium becomes exhausted at stationary phase, the microbes start producing the antibiotic. After several weeks, the cells are filtered out and the antibiotic is extracted from the medium for further purification.

Of course, industrial production of antibiotics assumes you have a microbe capable of producing a new antibiotic. The traditional screening method for discovering new antibiotics was rather straightforward: Microbes were isolated (usually from the soil) and cultivated in the laboratory at various temperatures and in a variety of growth media. Then, biochemists would attempt to harvest, purify, and identify the organic products of the microbes' metabolism and see if any of those byproducts had any antibiotic activity. By these methods, novel compounds were discovered, but only rarely. Few proved to be medically useful and able to make it to commercial production. In fact, pharmaceutical

TABLE 14.1 Some Commercially Produced Antibiotics

Antibiotic	Source (microbe)
Bacitracin	*Bacillus licheniformis* (bacterium)
Carbapenem	*Streptomyces cattley* (bacterium)
Cephalosporin	*Cephalosporium acremonium* (fungus)
Chloramphenicol	*Streptomyces venezuelae* (bacterium)
Erythromycin	*Streptomyces erythreus* (bacterium)
Gentamicin	*Micromonospora purpurea* (bacterium)
Penicillin	*Penicillium chrysogenum* (fungus)
Polymyxin B	*Bacillus polymyxa* (bacterium)
Streptomycin	*Streptomyces griseus* (bacterium)
Tetracycline	*Streptomyces rimosus* (bacterium)
Vancomycin	*Amycolatopsis orientalis* (bacterium)

companies have been known to have company employees traveling to distant countries take along cellophane bags with instructions for bringing home soil samples. Then, even if a drug does look promising, time and costs in developing the new antibiotic can be staggering. It can take more than 10 years and up to $1 billion (not a misprint!) to get a truly useful drug approved by the United States Food and Drug Administration (FDA) and to market.

■ *Penicillium chrysogenum* pen-i-sil'lē-um krī-so'gen-um

The first commercial production of an antibiotic used *Penicillium chrysogenum* to produce penicillin. In the early 1940s, Howard Florey and Ernst Chain developed methods for the mass production of penicillin based on Alexander Fleming's initial discovery of penicillin in 1928. However, they quickly realized they needed to improve the quantity of penicillin that *P. chrysogenum* naturally produces, as it was taking tens of liters of culture to produce just enough penicillin to treat one patient for one day. Eventually, through the use of mutagenic agents, they were able to obtain mutant strains of *P. chrysogenum* that were producing much higher levels of the antibiotic. In the end, their work saved the lives of countless soldiers during World War II. Today, penicillin can be produced in massive amounts in fermentors such as the one shown in FIGURE 14.4.

FIGURE 14.4 **Penicillin Production.** Penicillin production today in an aseptic production hall.

© RGB Ventures LLC dba SuperStock/Alamy

Some antibiotics produced in industry come in different forms. We will use penicillin as our example. First, there are the natural penicillins, such as penicillin G that is normally produced by the fungus. As years passed and more and more penicillin was used to treat infectious diseases, scientists discovered that pathogens were developing resistance to these penicillins; antibiotic resistance was on the rise. Therefore, biochemists started changing the structure of penicillin by chemically or enzymatically altering its structure. These **semisynthetic** penicillins, like ampicillin and methicillin, had clinical advantages because they were more useful on a broader group of pathogens and the pathogens were susceptible to these new forms. Unfortunately, the pathogens also quickly developed resistance to these penicillins as well. Still, semisynthetic penicillins, and those of other antibiotics, became the predominant forms produced industrially.

Vitamins and Amino Acids

Vitamins and amino acids are nutrients used in the food industry and as nutritional supplements (**nutraceuticals**) in the human diet and animal feeds.

Vitamins. The industrial production of vitamins ranks second only to antibiotics in total yearly commercial sales. Although most vitamins today are made by chemical processes, a few are easier to produce using microbes. Two of these are riboflavin (vitamin B_2) and cobalamin (vitamin B_{12}).

A deficiency of vitamin B_2 in the human diet can lead to cracking and reddening of the lips, inflammation of the mouth and mouth ulcers, sore throat, and iron-deficiency anemia. Although it is found naturally in many foods, such as milk, cheese, leaf vegetables, and mushrooms, riboflavin is added to white flour, breads, and breakfast cereals as a dietary supply (FIGURE 14.5). World production of riboflavin, a product of the fungal mold *Ashbya gossypii*, is around 1 million kilograms (1,000 tons) each year.

Vitamin B_{12} plays a key role in brain and nervous system function. Like riboflavin, it cannot be made by the human body and must be taken in from foods or supplements (see Figure 14.5). If a person, most commonly an older individual or vegetarian (plants do not produce B_{12}), does not consume sufficient B_{12}, a severe deficiency can lead to fatigue, depression, memory loss, and loss of taste and smell. The deficiency also leads to an inability to produce sufficient numbers of red blood cells, resulting in anemia. Therefore, it is important that these people eat breads, cereals, and other foods fortified with B_{12}. The vitamin is produced by high yield species of *Propionibacterium*, *Pseudomonas*, or *Streptomyces*, and the world production of vitamin B_{12} is on the order of 10 million kilograms (10,000 tons) per year.

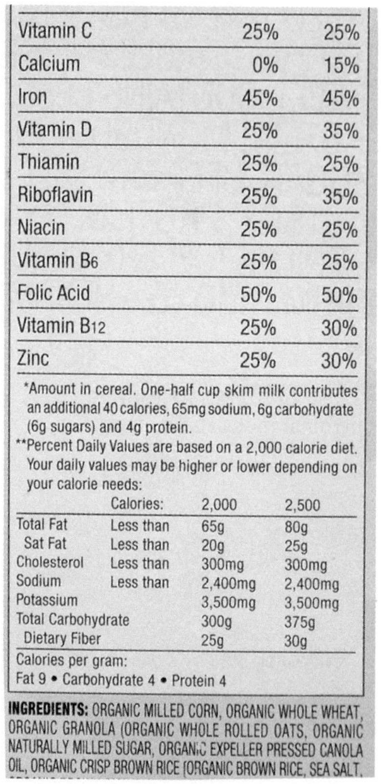

Amino acids. Amino acids are essential to the building of proteins in all organisms. Commercially, they are important microbial products used extensively as additives in the food and animal husbandry industries, and as nutritional supplements in the nutraceutical industry.

The most important commercial amino acid is glutamic acid that is a product of an industrial fermentation process using *Corynebacterium glutamicum*. Annually, hundreds of millions of kilograms of glutamic acid are produced industrially using this bacterial species, and sales of the amino acid are in the billions of dollars. The most familiar use of glutamic acid is as a flavoring agent in many cuisines in the form of monosodium glutamate (MSG) (FIGURE 14.6a).

Another widely used amino acid that is produced by microbes is lysine. Lysine, also a product of *C. glutamicum*, is used as a food additive, particularly in bread. Because this amino acid is not synthesized by the human body, it must be obtained in the foods we eat (it is an essential amino acid). This nutritional requirement creates a strong consumer demand for lysine. Indeed, lysine is a key U.S. export product.

Two other microbially produced amino acids, aspartic acid and phenylalanine, are particularly noteworthy in the beverage industry. They are used to synthesize the artificial sweetener aspartame, which is a nonnutritive and nonsugar sweetener of diet soft drinks, chewing gum, and hundreds other foods sold as low-calorie or sugar-free products (FIGURE 14.6b). Some people and groups have associated aspartame and MSG with neurological disorders and other health problems. Numerous scientific studies have failed to find a link.

Industrial Enzymes

Enzymes are well known as the biological catalysts that bring about metabolic reactions, while themselves remaining unchanged. Over the decades, microbiologists have identified and extracted specific enzymes from microbes and put those enzymes to practical use. In 2013, the market value of all industrially produced enzymes was estimated to be well over $8 billion.

Vitamin C	25%	25%
Calcium	0%	15%
Iron	45%	45%
Vitamin D	25%	35%
Thiamin	25%	25%
Riboflavin	25%	35%
Niacin	25%	25%
Vitamin B6	25%	25%
Folic Acid	50%	50%
Vitamin B12	25%	30%
Zinc	25%	30%

*Amount in cereal. One-half cup skim milk contributes an additional 40 calories, 65mg sodium, 6g carbohydrate (6g sugars) and 4g protein.
**Percent Daily Values are based on a 2,000 calorie diet. Your daily values may be higher or lower depending on your calorie needs:

	Calories:	2,000	2,500
Total Fat	Less than	65g	80g
Sat Fat	Less than	20g	25g
Cholesterol	Less than	300mg	300mg
Sodium	Less than	2,400mg	2,400mg
Potassium		3,500mg	3,500mg
Total Carbohydrate		300g	375g
Dietary Fiber		25g	30g

Calories per gram:
Fat 9 • Carbohydrate 4 • Protein 4

INGREDIENTS: ORGANIC MILLED CORN, ORGANIC WHOLE WHEAT, ORGANIC GRANOLA (ORGANIC WHOLE ROLLED OATS, ORGANIC NATURALLY MILLED SUGAR, ORGANIC EXPELLER PRESSED CANOLA OIL, ORGANIC CRISP BROWN RICE (ORGANIC BROWN RICE, SEA SALT,

FIGURE 14.5 Vitamin Additives. Some vitamins, such as riboflavin (vitamin B_2) and vitamin B_{12} that are added to many foods we eat, are produced through industrial fermentation using microbes.

Courtesy of Dr. Jeffrey Pommerville

■ *Ashbya gossypii*
ash'bē-ä gos-sip'ē-ē

■ *Propionibacterium*
prō-pē-on'ē-bak-ti-rē-um

■ *Pseudomonas*
sū-dō-mō'näs

■ *Corynebacterium glutamicum*
kôr'ē-nē-bak-ti-rē-um
glü-tam'-i-kum

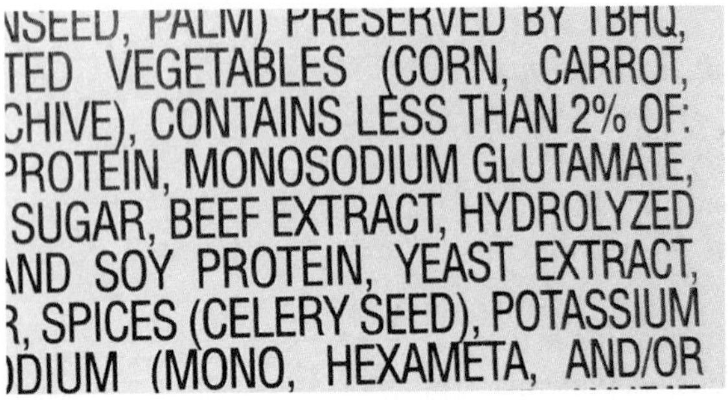

(a)

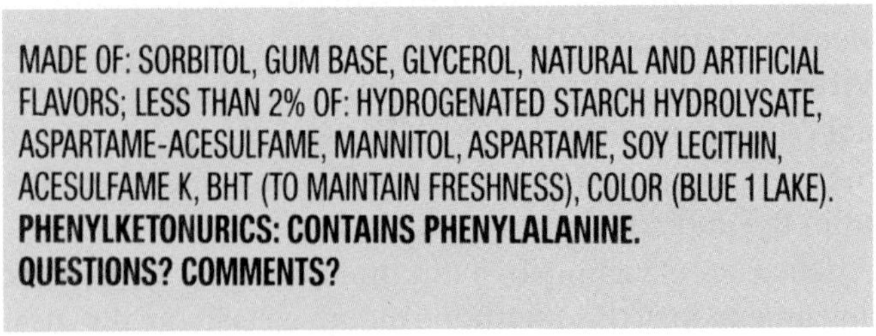

(b)

FIGURE 14.6 **Chemical Additives.** (**a**) The flavoring agent monosodium glutamate and (**b**) the artificial sweetener aspartame are products of microbial industrial fermentation.

(a) © Jones & Bartlett Learning; (b) Courtesy of Dr. Jeffrey Pommerville

■ *Trichoderma*
trik'ō-dėr-ma

■ *Bacillus subtilis*
bä-sil'lus su'til-us

■ *Aspergillus niger*
a-spėr-jil'lus nī'jėr

Microorganisms produce many different enzymes, mostly in small amounts for the cell's own internal use. However, some microbial enzymes are excreted into the environment. These so-called **exoenzymes** are produced in much larger amounts to digest large organic molecules (e.g., cellulose, proteins, and starches) into smaller ones that can be transported into the cell. These enzymes have great value in the food and health industries and in the laundry and textile industries. For example, an enzyme from *Trichoderma* breaks down the cellulose in blue denim jeans, giving the jeans a "stone-washed" appearance.

Among the most useful industrial enzymes is **amylase**. This enzyme breaks down starch molecules into simple sugars, including the disaccharide maltose. Several species of microbes, including *Bacillus subtilis* and *Aspergillus niger*, produce large amounts of amylase during their normal cycles of growth. Bakers add amylase to dough to promote the breakdown of starch to sugar, after which the sugar is used by yeasts. Amylase is also important to the digestion of starch at the beginning of the process of beer production. In household products, the enzyme is used in many spot removers to breakdown the starches in plant material that soils clothing.

Amylases are also important in the beverage industry. When amylase breaks down starch to glucose, the glucose can then be converted by another enzyme to fructose, which is a much sweeter sugar than glucose. If the starting starch comes from corn, the final product is high-fructose corn syrup, which is used as a sweetener in numerous juices and soft drinks (FIGURE 14.7). Worldwide production is over 10 billion kilograms (11 million tons) per year.

Protein-digesting enzymes of industrial significance are best represented by the **proteases**. Proteases digest proteins into smaller fragments (peptides) and amino acids. In the baking industry, proteases encourage the breakdown of gluten proteins in flour, thereby increasing the nutritional value of the bread. Proteases are also in the tenderizers sprinkled on meat before cooking. Here they break down the protein fibers and release the juices. In laundry products, proteases are used as spot removers in dry cleaning for anything that contains protein, such as egg, blood, and milk. And proteases help remove the organic matter from leather during the bating process.

Rennin is an enzyme used to curdle the proteins in milk, an early step in the manufacture of cheese. Historically obtained from the stomach lining of a calf, rennin is now

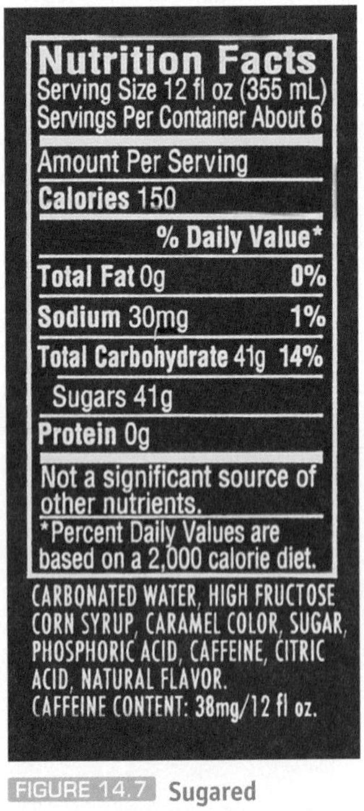

FIGURE 14.7 **Sugared Beverages.** Many juices and soft drinks have added high fructose corn syrup, another microbial product of industrial microbiology.

Courtesy of Dr. Jeffrey Pommerville

produced by fungi grown in industrial plants. And for breaking down compounds in aromatic ("strong-smelling") vegetables, the fungus *A. niger* produces galactosidase. This enzyme is the major ingredient in Beano®, a commercial product used to relieve bloating and gas.

Organic Acids

Citric acid is one of the most widely encountered acids found in consumable items such as soft drinks (see Figure 14.7). Hundreds of thousands of tons of citric acid are produced annually in the United States and this organic molecule is used in such diverse products as soft drinks, candies, frozen fruits, and wines. Citric acid is also used to tan leather, to electroplate metals, and to activate slow-flowing oil wells (these wells are often clogged with iron deposits, which can be broken down by pumping citric acid into the hole). Most citric acid is produced by the normal metabolism of the fungus *A. niger*. The citric acid is naturally excreted making the fungus unwittingly a partner in an industrial process.

Still another microbial product is lactic acid. Several *Lactobacillus* species produce this acid from the whey portion of milk derived from cheese production. Lactic acid is used as a flavoring and preservative agent in many foods. It is also employed to finish fabrics, prepare hides for leather, and dissolve lacquers.

■ *Lactobacillus*
lak-tō-bä-sil′lus

Other Microbial Examples

Many industrial catalysts occur most efficiently at higher temperatures. Until recently, those industry processes were limited in scope because most enzymes are destroyed (denatured) at those higher temperatures. Then came the discovery of archaeal organisms and specifically the "hyperthermophiles" that prefer to grow at higher temperatures, typically around 80°C to 90°C. These organisms contain heat-stable enzymes, called **extremozymes**, to perform the metabolic needs of these organisms living in extreme environments. Other extremozymes in other prokaryotes are resistant to acids, bases, and high salt. A common industrial use of these enzymes is in laundry products. In research, techniques for amplifying specific DNA sequences depends on the use of extremozymes tolerant to the higher temperature (72°C) used in the sequencing process.

Biofuels

Biofuels are renewable biological materials, such as plant carbohydrates, vegetable oils, and treated municipal and industrial wastes that contain potential energy in forms that can be released for useful purposes. Gasoline and diesel are examples of ancient biofuels made from decomposed plants and animals that have been buried in the ground for millions of years. Today's biofuels are similar, except they are made from living or recently living materials specifically grown as a potential biofuel.

The idea of using biofuels to meet our energy needs has been with us since cars were first invented, as Henry Ford had planned on using ethanol to fuel his Model Ts. Later, with diesel engines, peanut oil was considered to be a potential biofuel. Unfortunately, with the discovery of petroleum deposits, these ancient biofuels of gasoline and diesel became much cheaper and easier to refine than plant-originated alcohols and oil. Alternative biofuels were largely forgotten.

Today, with current oil prices and a finite time before oil reserves run dry, along with growing concerns about climate change, alternate biofuels are back "on the burner." The most common biofuel today is ethyl alcohol (ethanol). As a major industrial process, over 60 billion liters (15 billion gallons) of ethanol are produced yearly worldwide from the fermentation of various raw materials. Most of these fermentations depend on microbes. For example, in the United States, most ethanol is produced through yeast (*Saccharomyces*) fermentation of glucose obtained from cornstarch. For decades, Brazil has been turning sugarcane into ethanol, and some cars there can run on pure ethanol rather than as additive to fossil fuels. Other major biofuels being tested and used include biodiesel, made from vegetable oils, and algal fuels, including alcohols and oils produced from green algae.

■ *Saccharomyces* sak-ä-rō-mī'se-s

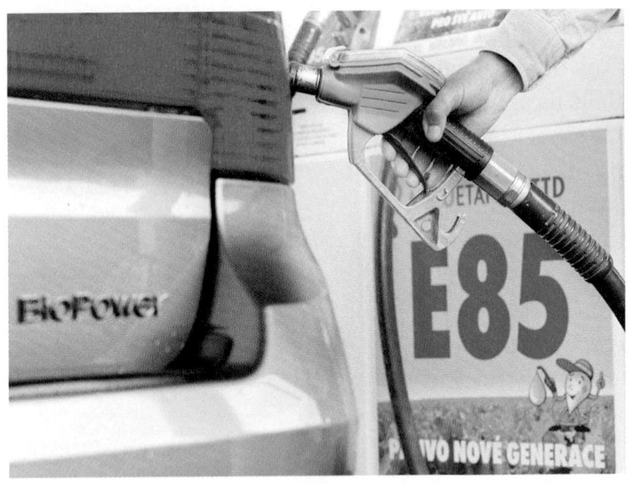

FIGURE 14.8 **Gasohol.** Ethanol-rich fuels (gasohols) such as E-85 (85% ethanol and 15% gasoline) are cleaner-burning fuels than pure gasoline.

© CTK/Alamy

Much of the gasoline in the United States is blended with ethanol to produce gasohol. Combustion of gasohol is a cleaner-burning fuel that produces lower amounts of carbon monoxide and nitrogen oxides than pure gasoline. Ethanol-rich fuels such as E-85 (85% ethanol and 15% gasoline) reduces emissions of nitrogen oxides by nearly 90% and is thus a means for reducing important pollutants in the atmosphere and reducing dependence on conventional sources of oil (FIGURE 14.8). Many major cities concerned about air pollution are retrofitting their public transportation systems, especially buses, to burn E-85.

Our energy concerns today have also spurred research on "algaculture." For example, the green alga *Botryococcus braunii* excretes hydrocarbons having the consistency of crude oil (FIGURE 14.9). However, the major problem with algal petroleum is scale: the logistics of growing oil-producing algae capable of contributing significantly to global oil demand are formidable.

■ *Botryococcus braunii* bo-trē-ō-kok'kus brawn'-ē-ē

For the future, many believe a better way of making biofuels will be from grasses, such as switchgrass, which contain more cellulose (FIGURE 14.10). Cellulose is the

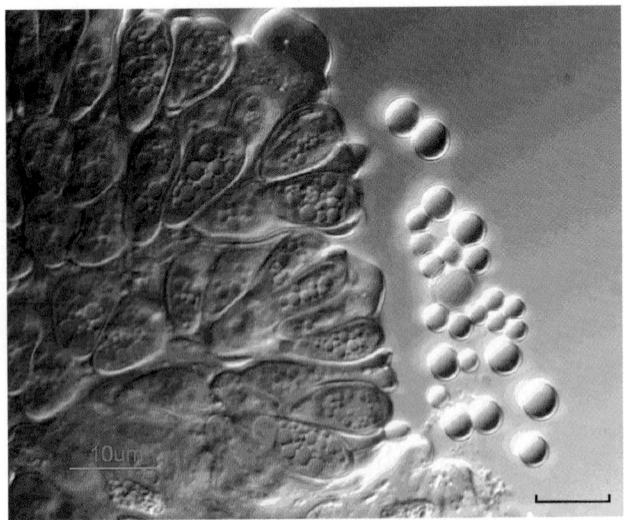

FIGURE 14.9 **Algae as Source of Biofules.** This alga, *Botryococcus braunii*, produces petroleum, as seen by the secreted oil droplets along the edge of the colony. (Bar = 8 μm.)

Courtesy of Tim Devarenne

FIGURE 14.10 **Switchgrass.** Switchgrass contains a high percentage of cellulose, a potential feedstock for bioethanol production by microbes.

© Jim Parkin/Alamy

tough material that makes up plants' cell walls, and most of the weight of a switchgrass plant is cellulose. If cellulose can be turned into biofuel, it could be more efficient than current biofuels, and emit less carbon dioxide. The role of microbes in these industrial processes can be made even more efficient when microbes form "tag teams" as **A Closer Look 14.1** points out.

A CLOSER LOOK 14.1

Tag Team Microbes

In tag team wrestling, each team usually consists of two wrestlers who at other times wrestle individually. The term "tag team" is now used in common society to describe two or more individuals who cooperate in an activity. Well, if humans can tag team, so can microbes.

Since the 1990s, scientists have searched for or have tried to engineer microbes that would break down plant materials into useful biofuels. Certainly, *Saccharomyces* has been doing this naturally since the yeast evolved. The fungus can break down glucose and, through fermentation, produce ethanol. But ethanol is not the best alternate biofuel as it tends to absorb water and, in a car engine, would damage the engine. Algae are possible sources for biofuel production, but scaling up the production process is a challenging problem today.

So scientists have been looking for other potential biofuels to replace gasoline. One possibility is isobutanol, which is a slightly larger molecule than ethanol. However, as a biofuel it may be better than ethanol because isobutanol does not mix with water and its fuel characteristics are more closely allied to those of gasoline.

To make isobutanol, scientists turned to the microbial workhorse for so many aspects of microbiology, *Escherichia coli*. The bacterial cell does not naturally produce isobutanol, but it has been genetically engineered to do so. If the new engineered strain is fed glucose, it has the biochemical ability to produce very high yields of isobutanol. But what is the best way to "feed" the bacterial species the huge amounts of glucose needed to generate substantial isobutanol? Ideally one wants to use plant materials (inedible biomass), especially cellulose because this plant cell wall polysaccharide is composed of nothing but tremendously long chains of glucose molecules. So, what is the best way to break down cellulose? Use a microbe, of course.

It turns out that the fungus *Trichoderma reesei* can efficiently break down cellulose. The fungus naturally secretes a high yield of the enzyme cellulase to digest cellulose. So, why not tag team *T. reesei* and *E. coli*?

And that is exactly what a team of researchers at the University of Michigan did in 2013.Under the guidance of research leader Xiaoxia Lin, a doctoral student, Jeremy Minty, combined *T. reesei* and *E. coli* together in a fermentor with dried corn husks, which are almost pure cellulose.

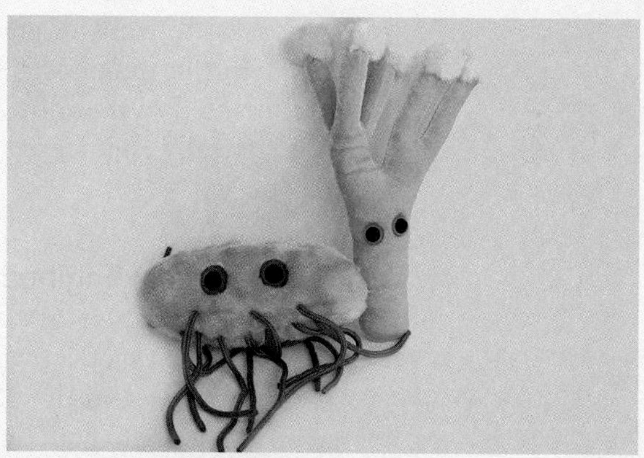

Two plush toy Giant Microbes representing *Escherichia coli* (left) and *Trichoderma reesei* (right).

Courtesy of Dr. Jeffrey Pommerville

The result: the two organisms behaved nicely; the fungus degraded the cellulose into glucose and then *E. coli* finished the job by converting the glucose to isobutanol. The tag team microbes formed a good team as neither outgrew the other in the fermentor and they maintained a state of balance.

The researchers believe that being able to do the whole process in one "pot" (a fermentor) makes the capital investment operating costs much lower than many of the alternative biofuel processes being studied today. The next step is to improve the efficiency of the conversion and increase the tag team's tolerance to isobutanol, which can be toxic. So, just like two wrestlers working together in the ring, two microbes working together in a fermentor can bring winning results.

So, we have seen that microbes play a critical role in the food and energy production through industrial microbiology. Now let's turn to the other side of the coin and examine some of the products of the biotechnology industry.

14.3 Microbes and Gene Technology: Seeing the Promise

Every so often in science, a window opens—and, suddenly, the theoretical becomes possible, then inevitable. Discoveries emerge in rapid succession, and powerful new insights drive researchers to unimagined heights. With the advent of genetic engineering and biotechnology, such a window opened in microbiology, and scientists began to see impossible dreams become reality.

Today, close to 50% of the soybean crop in the United States is planted with genetically engineered seeds. Thousands of individuals are now receiving medicines, such as anti-cancer drugs, vaccines, immunosuppressants, and other drug products that are the fruits of biotechnology. And much of this technology has been based on using microbes as production factories or borrowing the skills microbes possess to transfer genes between organisms.

In the remaining sections of this chapter, we will see how foreign genes are expressed in microbes and then consider some of the products coming from these genetically engineered microorganisms.

Genetically Engineering Bacterial Cells

Another chapter covers the historical and technical aspects of genetic engineering. In this section we are concerned with the expression of those foreign genes in microbes and the useful products produced. Consequently, let's briefly review how scientists genetically engineer a population of bacterial cells and how those genes are expressed.

anterior pituitary: The gland, situated at the base of the brain, that secretes several hormones.

One of the essential hormones produced in the body is **human growth hormone (hGH)**, which is produced by the anterior pituitary. Increased height during childhood is the most widely known effect of hGH. Should a child fail to secrete a sufficient amount of hGH, long bone growth is retarded, which can lead to growth failure, short stature, and dwarfism. Although individuals with this disorder are generally properly proportioned, they usually reach a height of only about 1.2 meters (4 feet). If diagnosed before puberty, the condition can be treated successfully with genetically engineered hGH (recombinant hGH; rhGH), which today is a fairly routine procedure for affected children.

To produce rhGH, the hGH gene is isolated from human cells and placed into a bacterial plasmid (FIGURE 14.11). The plasmid is then transferred into *E. coli* cells where, following procedures similar to those for industrial fermentation, the bacterial cells in a fermentor transcribe and translate the hGH gene into the growth hormone protein. Once the maximal yield of rhGH has been reached, it is purified and packaged for medical use. Treatments with rhGH (known commercially as Protropin®) encourage growth spurts that permit children to reach a normal height for their age. In addition, rhGH has been abused by athletes and weightlifters who believe (but unproved) that rhGH can increase physical performance. The International Olympic Committee and the National Collegiate Athletic Association (NCAA) have banned the use of rhGH. The FDA has approved rhGH only for medical purposes.

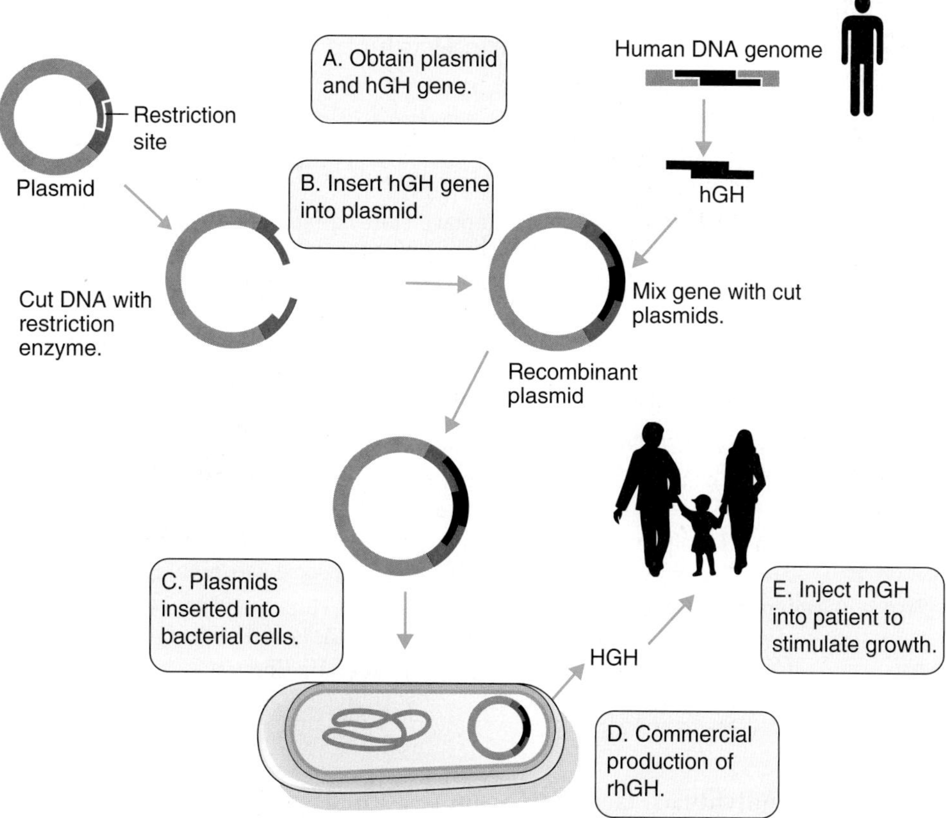

FIGURE 14.11 **Genetic Engineering and Expression of Recombinant Human Growth Hormone (rhGH).** After obtaining the gene for growth hormone, the gene is inserted into a bacterial plasmid, which is then introduced into *Escherichia coli* cells. The cells produce the rhGH, which can then be isolated, purified, and marketed for treatment of short stature (dwarfism).

14.4 The Products of Biotechnology: Proteins and Vaccines

Other therapeutic products of biotechnology, such as hormones and other proteins, follow similar schemes for protein production. All involve inserting the human gene of interest into bacterial cells (usually *E. coli*) and allowing the gene to be expressed as the protein product.

Pharmaceutical Proteins

One of the most economically profitable areas of biotechnology and the one with the highest pharmaceutical value is the production of human hormones, blood products, and enzymes to alleviate human medical conditions and deficiencies (TABLE 14.2). Here, we briefly discuss a few of these.

The historic first step in using biotechnology to relieve human disease occurred in 1982 when the FDA approved the use of the first genetically engineered (synthetic) human protein. The protein was human **insulin** (Humulin®) and with its approval individuals suffering from insulin-dependent diabetes could now better control their blood sugar levels.

Hemophilia is a rare bleeding disorder in which a person's blood doesn't clot normally because the individual is unable to synthesize an essential blood clotting protein. One of these proteins, clotting factor VIII, is required in all humans to stimulate normal blood clotting and this form of hemophilia is the most common inherited blood disorder in the United States (about 1 in 10,000 males is affected).

TABLE
14.2 Some of the Therapeutic Products of Biotechnology and Their Functions

Product	Function
Replacement Proteins and Hormones	
Factors VII, VIII, IX	Replace clotting factors missing in hemophiliacs
Growth hormone (rhGH)	Replaces missing hormone in people with short stature
Insulin	Treatment of insulin-dependent diabetes
Therapeutic Proteins, Hormones, and Enzymes	
Epidermal growth factor (hEGF)	Promotes wound healing
Granulocyte colony stimulating factor (hG-CSF)	Used to stimulate white blood cell production in cancer and AIDS patients
Interferon-alpha (IFN-alpha)	Used with other antiviral agent to fight viral infections and some cancers
Tissue plasminogen activator (TPA)	Dissolves blood clots; prevents blood clotting after heart attacks and strokes
DNase I	Treatment of cystic fibrosis

To help these individuals, factor VIII, and the other two factors, can now be produced synthetically.

Among other therapeutic products of biotechnology is **tissue plasminogen activator** (**TPA**), a protein-digesting enzyme that stimulates other body enzymes to break down a blood clot. The synthetic protein, called Activase®, is often prescribed for individuals with poor circulation or displaying signs of heart attack or stroke. Many individuals owe their continued good health to this microbe-derived protein.

Interferon (**IFN**), another therapeutic product of biotechnology, is a group of immune proteins produced by various body cells after stimulation by a viral infection. Interferon-alpha (INF-alpha) signals uninfected cells to protect themselves against viral infection by producing proteins able to prevent viral genome replication. The successful production of INF-alpha from genetically engineered bacterial cells means that individuals suffering from some viral diseases, such as hepatitis B, hepatitis C, and some forms of cancer, now have a way to fight the illnesses.

Some of the therapeutic products of biotechnology are enzymes (see Table 14.2). One, called human **DNase I**, is used to treat the buildup of DNA-containing mucus in patients with cystic fibrosis, a genetic abnormality in which thick, sticky mucus clogs the respiratory passageways of the affected individual. Pathogens, such as *Pseudomonas aeruginosa*, find the sticky mucus to be a perfect home. As some bacterial cells die, they lyse and release their DNA into the mucus, making the material more viscous. As such, the mucus contributes to respiratory tract obstructions and protects the bacterial cells from host immune defenses. Human DNase I (Pulmozyme®) therapy digests the DNA and greatly decreases the viscosity of the mucus. Now the *P. aeruginosa* cells are more susceptible to immune attack and antibiotic treatment may be more successful. Sales of this life-saving enzyme exceed $100 million annually.

■ *Pseudomonas aeruginosa*
sū-dō-mō′näs ā-rü ji-nō′sä

Today, we are witnessing the continuous rise in the number of approved protein therapeutics derived from biotechnology, and biopharmaceuticals have the potential to become first-line medicines of the future.

Old Technology Vaccines

Vaccines are substances that prepare the immune system to recognize and respond to a pathogen, resulting in protection (immunity) to that pathogen. Vaccines are composed of weakened or killed microorganisms, chemically altered toxins, or molecular parts of microorganisms. However, because these pathogen look-alikes have been altered in some way (see below), the vaccine usually does not trigger the disease and the person vaccinated does not become ill from the vaccine. Importantly, the vaccine components stimulate the immune system to produce antibodies that protect the body against a future attack by the harmful microbes. The vaccine also generates memory cells, so if the vaccinated person is exposed to the same pathogen at some later time, the memory cells act swiftly to produce antibodies, stopping the infection before it can make the individual sick. Vaccines then are simply strengthening and "educating" the body's natural infection fighting system.

Let's examine the different types of vaccines (both old technology and new biotechnology) and find out how they are made. The viral and bacterial vaccines currently in use in the United States are summarized in TABLE 14.3. We will discuss the old technologies first that use whole bacterial cells, viruses, or toxins as the immune-triggering substance.

Attenuated Vaccines. Some pathogens can be weakened (**attenuated**) such that they should not cause disease. Such attenuated vaccines contain pathogens capable of multiplying or replicating but at an extremely slow pace. Such vaccines are the closest to the natural pathogens and, therefore, they generate a strong immune response.

The downside of attenuated vaccines is that they require refrigeration to retain their effectiveness, which can present a problem in many developing nations lacking adequate refrigeration facilities. In addition, on very rare occasions, the agent in a few attenuated vaccines may revert back to their virulent form and may cause the very disease the vaccine was meant to protect against. The oral (Sabin) polio vaccine is an example of one such vaccine that is no longer used for polio immunizations in most countries. Although a healthy person with a fully functioning immune system (**immunocompetent**) clears these rare infections without serious consequence, individuals with a compromised immune system, such as AIDS patients or pregnant women, should not be given attenuated vaccines, if possible. In fact, there are no attenuated bacterial vaccines routinely used in the United States today.

Inactivated Vaccines. Another old technology strategy for preparing vaccines is to use "killed" bacterial cells or viruses. These vaccines are relatively easy to produce because the pathogens are killed by exposing them to certain chemicals or heat. However, the inactivation process alters the microbe's structure and shape so it produces a weaker immune response. The injected (Salk) polio vaccine and the hepatitis A vaccine typify such preparations of inactivated whole viruses.

Compared to attenuated vaccines, these **inactivated** vaccines are safer because the bacterial cells or viruses cannot multiply or replicate, and therefore cannot cause the disease in a vaccinated individual. The vaccines can be stored in a freeze-dried form at room temperature, making them a vaccine of choice in developing nations.

TABLE

14.3 The Principal Bacterial and Viral Vaccines Currently in Use

Disease	Route of Administration	Recommended Vaccine Usage/Comments
Contain Inactivated Whole Bacteria		
Cholera	Subcutaneous (SQ) injection	For travelers; short-term protection
Typhoid fever	SQ and intramuscular (IM)	For travelers only; variable protection
Plague	SQ	For exposed individuals and animal workers; variable protection
Contain Live, Attenuated Bacteria		
Tuberculosis (BCG)	Intradermal (ID) injection	For high-risk occupations only; protection variable
Subunit Bacterial Vaccines (Capsular Polysaccharides)		
*Meningitis (meningococcal)	SQ	Part of childhood immunization schedule
Meningitis (*H. influenzae*)	IM	Part of childhood immunization schedule; may be administered with DTaP
*Pneumococcal pneumonia	IM or SQ	Part of childhood immunization schedule; moderate protection
Pertussis	IM	Part of childhood immunization schedule
Subunit Bacterial Vaccine (Protective Antigen)		
Anthrax	SQ	For lab workers and military personnel
Toxoids (Formaldehyde-Inactivated Bacterial Exotoxins)		
Diphtheria	IM	Part of childhood immunization schedule
Tetanus	IM	Part of childhood immunization schedule; highly effective
Botulism	IM	For high-risk individuals, such as laboratory workers
Contain Inactivated Whole Viruses		
Polio (Salk)	IM	Part of childhood immunization schedule; highly effective
Rabies	IM	For individuals sustaining animal bites or otherwise exposed
Influenza	IM	Part of childhood immunization schedule
*Hepatitis A	IM	Part of childhood immunization schedule; protection for travelers and anyone at risk

TABLE
14.3 The Principal Bacterial and Viral Vaccines Currently in Use (*continued*)

Disease	Route of Administration	Recommended Vaccine Usage/Comments
Contain Attenuated Viruses		
Adenovirus infection	Oral	For immunizing military recruits
Measles (rubeola)	SQ	Part of childhood immunization schedule; highly effective
Mumps (parotitis)	SQ	Part of childhood immunization schedule; highly effective
Polio (Sabin)	Oral	Possible vaccine-induced polio
Smallpox (vaccinia)	Pierce outer layers of skin	For lab workers, military personnel, healthcare workers
Rubella	SQ	Part of childhood immunization schedule; highly effective
Chickenpox (varicella)	SQ	Part of childhood immunization schedule; immunity can diminish over time
*Shingles (zoster)	SQ	Prevention in individuals 60 years and older
Yellow fever	SQ	For travelers, military personnel in endemic areas
Influenza	IM	In a nasal spray form; only for healthy, nonpregnant individuals 2–50 years old
*Rotavirus infection	Oral	Part of childhood immunization schedule
Recombinant Viral Vaccine		
Hepatitis B	IM	Part of childhood immunization schedule; medical, dental, laboratory personnel; highly effective
*Genital warts/cervical cancer	IM	Preventative vaccine for individuals 9–26 years old

*Vaccines licensed since 2000.

Toxoid Vaccines. For some bacterial diseases, such as diphtheria and tetanus, a bacterial exotoxin is the main cause of illness. Such toxins can be inactivated chemically, with the resulting inactivated toxin, called a **toxoid**, being the vaccine. The diphtheria-pertussis-tetanus (DPT) is an example. Immunity induced by a **toxoid vaccine** allows the body to generate protective immunity and to recognize the natural toxin should the individual come in contact with the actual diphtheria or tetanus

■ exotoxin: A bacterial protein poison secreted into the environment or body by living bacterial cells.

toxin. Because toxoid vaccines are inactivated products, booster shots are necessary to re-establish strong protective immunity.

Genetically Engineered Vaccines

As mentioned, on rare occasions, a few attenuated vaccines, although very effective, might cause the disease they are meant to protect against. Inactivated and toxoid vaccines are not as effective and may require booster shots. Therefore, biotechnologists have developed new ways of producing some vaccines. The key to these genetically engineered vaccines is that they contain only a subunit or fragment of the bacterial cell or virus. With only a small piece of the pathogen, a vaccinated individual cannot possibly contract the disease.

Subunit Vaccines. Unlike the whole agent attenuated or inactivated vaccines, the strategy for a **subunit vaccine** is to have the vaccine contain only those parts (subunits) of the pathogen that stimulate a strong immune response. One successful product of the technology is the vaccine for hepatitis B. The key element of this vaccine is a protein in the surface capsid (coat) of the virus. Biotechnologists have identified the gene that encodes the protein and have inserted the gene into plasmids of the harmless baker's yeast *Saccharomyces cerevisiae*. The yeast cells then synthesize the coat protein and this material is collected and purified to make the vaccine.

■ *Saccharomyces cerevisiae* sak-ä-rō-mī'sēs se-ri-vis'ē-ī

Currently marketed as Recombivax® by one company and Engerix-B® by another, these hepatitis B vaccines are a far cry from traditional vaccines, as the new subunit vaccines contain no whole microbes of any sort. Today, they are used to immunize healthcare workers, police officers, firefighters, emergency services personnel, medical laboratory workers, and anyone else who might normally come into contact with blood or body secretions that could harbor the hepatitis B virus. The Centers for Disease Control and Prevention (CDC) and many pediatricians recommend the vaccine for newborns.

Subunit vaccines against many other viruses and pathogens are currently being developed. That said, if genetically engineered vaccines have such great promise, how come we do not have an AIDS vaccine after all these years? **A Closer Look 14.2** delves into the problem.

Before we leave the topic of genetically engineered vaccines, we should mention a new biotechnology approach to vaccines using pathogen DNA.

DNA Vaccines

Part of the renaissance in vaccine development is a strategy to use a pathogen's genome as a vaccine. These investigational **DNA vaccines** consist of plasmids genetically engineered to contain one or more protein-encoding genes from a viral or bacterial pathogen. When injected intramuscularly in a saline solution, cells in the body should take up the injected foreign DNA, commence to make proteins encoded by the foreign DNA, and display these antigens on their surface, much like infected cells do during an actual infection. Such a display should stimulate a strong immune response to protect the individual if at some later time the person comes in contact with the actual pathogen.

DNA vaccines are being studied for a variety of infectious diseases in humans, including AIDS, influenza, and hepatitis C, and for several cancers. Unfortunately, few experimental vaccine trials have produced an immune response equivalent to the other types of vaccines. For some reason, the DNA vaccines do not generate the level of protective immunity needed for one to be properly immunized.

A CLOSER LOOK 14.2

An AIDS Vaccine—Why Isn't There One after All These Years?

Producing an AIDS vaccine might appear rather straightforward: Cultivate a huge batch of human immunodeficiency virus (HIV), inactivate it with chemicals, purify it, and prepare it for marketing. In fact, this was the mentality and approach in the mid-1980s. Then along came the biotech industry and its attempts to produce a safe and effective AIDS vaccine. Researchers have identified and isolated the genes encoding many different viral proteins. These have been inserted into *E. coli* cells or other microbes, and these microbial factories have responded by producing the viral proteins. Unfortunately, for AIDS, vaccine development has involved more than people had realized. The problem here is not the biotech strategy to produce a vaccine but rather other intrinsic and extrinsic factors. Here are some major reasons why a vaccine remains elusive after all these years.

The effects of a bad batch of weakened or inactivated vaccine would be catastrophic, and people generally are reluctant to be immunized with whole HIV particles, no matter how reassuring the scientists may be. A genetically engineered vaccine would get around this hurdle as it only has some protein antigens. However, in both cases, even if the vaccine is safe, how the human body actually protects itself from pathogens such as HIV is not yet completely understood. Without that understanding, an effective vaccine cannot be developed.

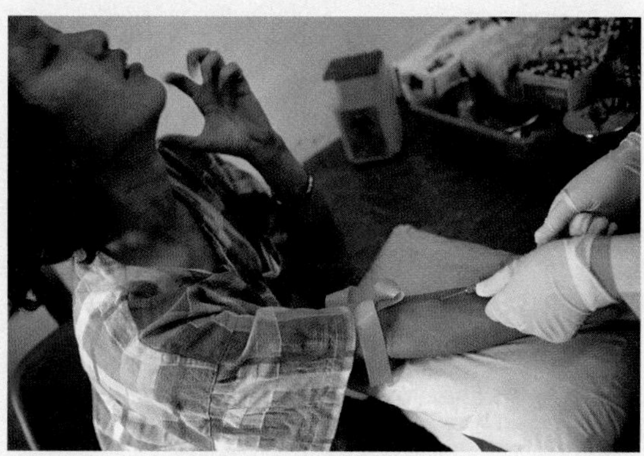

Taking a blood sample for AIDS vaccine testing.

Vaccine development also has been slow because of the high mutation rate of the virus. Some HIV strains around the world vary by as much as 35% in the capsid and envelope proteins they possess. Thus, HIV is more mutable than the influenza viruses, and no vaccine has been developed yet to make one completely immune to influenza. It is hard to make a vaccine targeted at one strain when the virus keeps changing its coat.

Another hurdle to consider is that an AIDS vaccine could actually make someone more at risk if they actually were to contract the disease. In 2002, an HIV patient whose body was holding the virus in check became infected with another strain of HIV through unprotected sex. This patient subsequently became "superinfected," meaning his immune system could keep the original strain in check, but was powerless to control the new strain. So, could a vaccine produce the same result if the individual was infected with another strain of HIV?

As of 2014, no vaccine trials have been shown to stimulate the immune system to a level necessary to destroy HIV. The number of immune cells stimulated by the vaccines is simply not up to the job. In March 2002, Anthony Fauci, Director of the National Institute for Allergy and Infectious Diseases, reported to the Presidential Advisory Council on HIV and AIDS that a "broadly effective AIDS vaccine could be a decade or more away." By 2014, minimal progress has been reported.

Researchers are working hard to discover what improvements need to be made to produce an effective DNA vaccine. That being said, in 2005, a veterinary DNA vaccine to protect horses from West Nile virus became the world's first licensed DNA vaccine.

Thus far, we have examined how biotechnology works and identified some of the products of the technology (FIGURE 14.12). We have seen how genetically modified microbes are designed and how they produce their products. We then looked at some of the uses for these protein products. To finish off this chapter, let's explore a few of the newer diagnostic procedures, detection methods, and novel techniques coming from biotechnology.

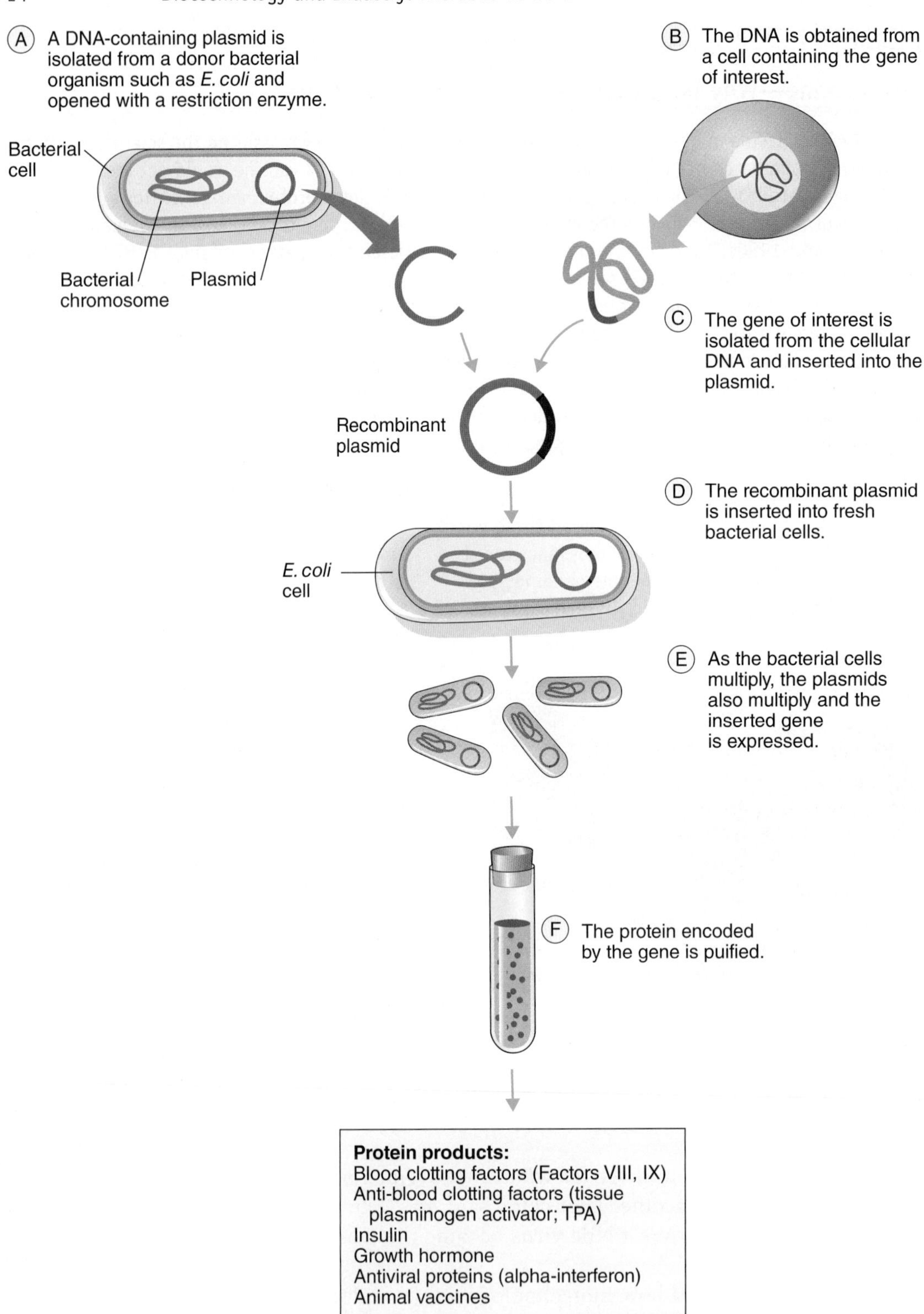

A A DNA-containing plasmid is isolated from a donor bacterial organism such as *E. coli* and opened with a restriction enzyme.
B The DNA is obtained from a cell containing the gene of interest.
Bacterial cell
Bacterial chromosome
Plasmid
C The gene of interest is isolated from the cellular DNA and inserted into the plasmid.
Recombinant plasmid
D The recombinant plasmid is inserted into fresh bacterial cells.
E. coli cell
E As the bacterial cells multiply, the plasmids also multiply and the inserted gene is expressed.
F The protein encoded by the gene is puified.
Protein products:
Blood clotting factors (Factors VIII, IX)
Anti-blood clotting factors (tissue plasminogen activator; TPA)
Insulin
Growth hormone
Antiviral proteins (alpha-interferon)
Animal vaccines

FIGURE 14.12 **Developing New Products Through Biotechnology.** Genetic engineering is a method for inserting foreign genes into bacterial cells and obtaining chemically useful products.

14.5 Microbial Detection: Probes and Diagnostics

One of the more remarkable applications of biotechnology occurs in the diagnostic laboratory when DNA samples are analyzed to determine whether target genes and/or DNA segments are present. These applications make it possible for scientists to identify microbes directly (instead of spending long hours cultivating them or looking for their telltale proteins). Furthermore, these methods allow the verification of numerous inherited diseases whose presence could previously be predicted only by guesswork.

DNA Probes

Two essential elements in DNA analyses are the DNA probe and the polymerase chain reaction. The **DNA probe** (also called the gene probe) is a synthetic fragment of single-stranded DNA that "hunts down" a complementary DNA fragment in a mass of cellular material and signals when it has located that fragment. The DNA probe reacts specifically with a target DNA fragment having complementary base pairs, much like a left hand matches up with its complementary right hand. As millions of the two kinds of DNA molecules bind, a fluorescent or radioactive substance tagged to the probe accumulates, providing a visual signal that binding has taken place.

In order to use DNA probes effectively, billions of these DNA molecules must be available. Making them is the job of the **polymerase chain reaction (PCR)**. In a highly sophisticated apparatus (often described as a molecular equivalent of a photocopier), the PCR takes a strand of DNA and multiplies it billions of times.

To begin the process, a sample of DNA is collected from a pathogen or whatever target DNA is required. In the PCR machine, the double helix is unwound at a high temperature to yield single strands of DNA (FIGURE 14.13). Then a special, heat-resistant DNA polymerase enzyme called **Taq polymerase** is added. (The enzyme is obtained from the thermophilic bacterium *Thermus aquaticus* and represents

■ *Thermus aquaticus*
thėr'-mus ä-kwä'-ti-kus

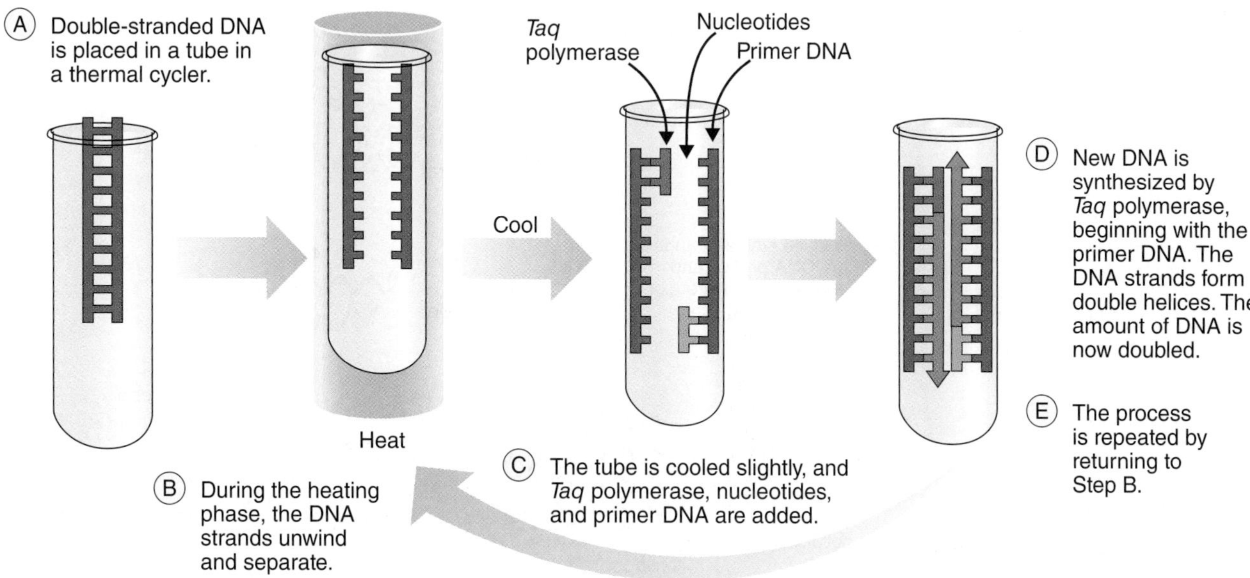

(A) Double-stranded DNA is placed in a tube in a thermal cycler.
Heat
(B) During the heating phase, the DNA strands unwind and separate.
Cool
Taq polymerase
Nucleotides
Primer DNA
(C) The tube is cooled slightly, and *Taq* polymerase, nucleotides, and primer DNA are added.
(D) New DNA is synthesized by *Taq* polymerase, beginning with the primer DNA. The DNA strands form double helices. The amount of DNA is now doubled.
(E) The process is repeated by returning to Step B.

FIGURE 14.13 **The Polymerase Chain Reaction (PCR).** PCR is a laboratory technique for quickly amplifying the amount of available DNA to enhance detection methods.

one of the industrial enzymes we discussed in the first part of this chapter.) Now the machine mixes in nucleotide molecules containing the four bases (adenine, thymine, guanine, and cytosine) together with a strand of primer DNA—the primer recognizes a complementary section of the sample DNA, binds to it, and serves as a starting point for the copying process. The Taq polymerase brings together the nucleotides to form a strand of DNA that complements the single-stranded DNA, adding the nucleotides onto the primer DNA one at a time. Next, the mixture is cooled, whereupon the new and old strands of DNA twist together to form double-stranded DNA molecules. At this point, the original number of DNA molecules has doubled. Additional rounds of heating and cooling quickly produce billions of identical DNA molecules.

Diagnostic Tests

The gene probe technology has bred a host of new diagnostic tests. One example is a diagnostic test to detect the provirus of HIV in infected cells (FIGURE 14.14). The HIV probe is produced as described above. Additionally, DNA from a patient's tissues is isolated, fragmented into single-stranded pieces, and attached to a solid surface. To perform the test, the DNA probe (whose base sequence is complementary to the DNA

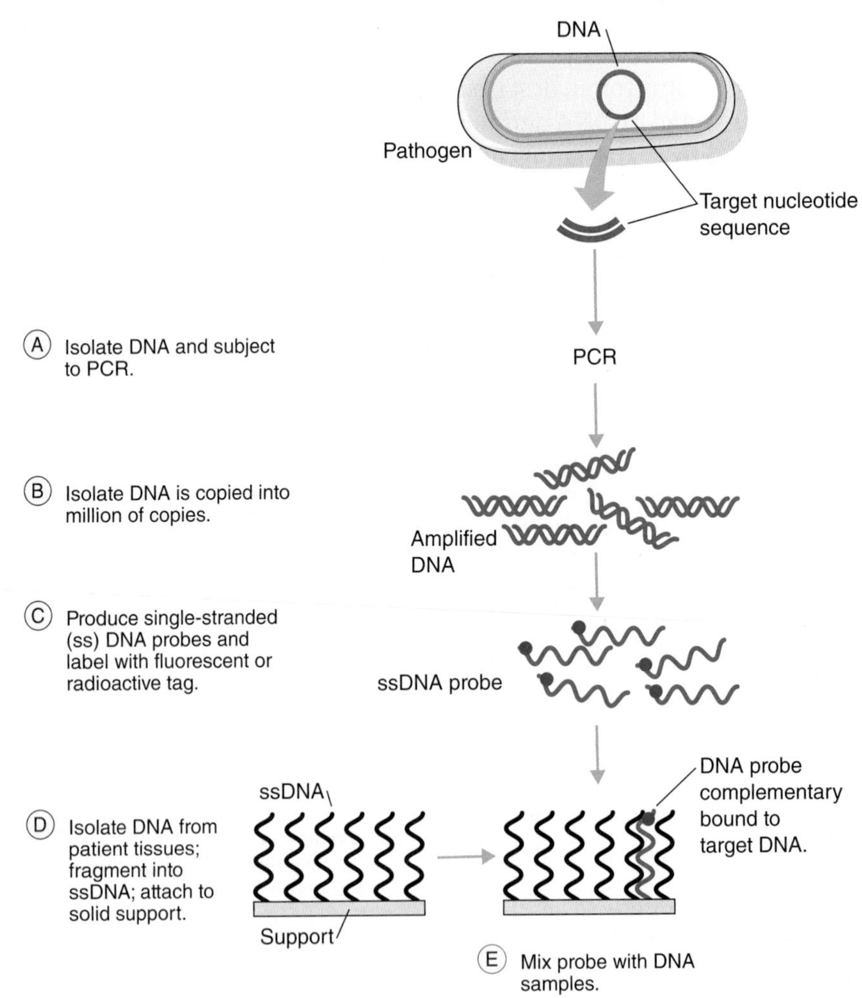

DNA
Pathogen
Target nucleotide sequence
PCR
Amplified DNA
ssDNA probe
ssDNA
Support
DNA probe complementary bound to target DNA.
(A) Isolate DNA and subject to PCR.
(B) Isolate DNA is copied into million of copies.
(C) Produce single-stranded (ss) DNA probes and label with fluorescent or radioactive tag.
(D) Isolate DNA from patient tissues; fragment into ssDNA; attach to solid support.
(E) Mix probe with DNA samples.

FIGURE 14.14 **DNA Probes.** Construction of a DNA probe and its use in disease detection and diagnosis.

of the AIDS provirus) is mixed with the patient's DNA sample. If the provirus DNA is present, the probe locates it amidst the cellular debris and binds to it as detected by the presence of the fluorescent or radioactive tag. If there is no provirus present, there will be do binding and no signal will be detected. The test has been a major boon in detecting HIV early and following the course of AIDS.

DNA diagnostic tests are also available for detecting other infectious diseases. For example, Lyme disease, a blood disease due to a bacterial infection transmitted by ticks, can be extremely difficult to cultivate. But, thanks to biotechnology, it is possible to secure blood from a patient, amplify the bacterial pathogen DNA (if present in the blood), and then use a Lyme disease DNA probe to detect it.

These new technologies make this type of pathogen detection possible and reduce the guesswork often associated with the traditional methods. In addition, the biotech diagnostic tests are rapid and allow for quicker treatment of infected patients and hopefully shortening the misery caused by many of these diseases.

A slightly different DNA diagnostic procedure can be used to identify the cause of an epidemic even before the causative agent has been isolated or cultivated from infected patients. In 1994, a severe respiratory disease occurred in the Four-Corners area of the southwestern United States (Arizona, Colorado, New Mexico, and Utah). The disease spread among Native Americans, who suffered hemorrhaging, lung infection, and kidney problems. Believing the disease was caused by a virus, scientists painstakingly mixed patient antibodies with numerous types of laboratory viruses, one at a time, until a rare virus known as the hantavirus was pinpointed. Then, to develop a diagnostic test, they generated large quantities of hantavirus DNA in a PCR machine to produce a DNA probe. Finally, to identify the disease in a patient, they took diseased tissue and used the probe to search for matching DNA. When a match was identified, they began preventive treatment, while continuing to chart the course of the epidemic. Soon afterward, a DNA probe was used to solve another microbial mystery, as **A Closer Look 14.3** explores.

A CLOSER LOOK 14.3
"Not Guilty"

Poor Christopher Columbus! Some historians have accused him of inadvertently bringing smallpox to the New World—and measles, whooping cough, and tuberculosis (TB)—and almost every other infectious disease. It makes it sound like the Santa Maria was a plague ship and the New World was a sterile place, ripe for infection!

Well, rest easy, Chris, for scientists have cleared you of bringing at least one disease—TB. While studying the remains of a mummified woman from Peru, Arthur Aufderheide of the University of Minnesota noticed in the mummies' lung tissues several lumps reminiscent of TB. He enlisted the help of a molecular biologist, who extracted DNA from the lumps and amplified it so that there was enough DNA to identify. Using a DNA probe for TB, the mummy's DNA turned out to have sequences identical to those of *Mycobacterium tuberculosis*, the causative agent of TB.

Why was that important? It ends up that the mummy was a thousand years old—that's right, one thousand years. Apparently both the woman and the TB pathogen were already here hundreds of years before Columbus arrived. However, for the other diseases, it may be a different story.

A Peruvian mummy.

14.6 Gene Technology: The Unseen Microbial World and Altering Life

With the ever-expanding biotechnology toolkit, it is now possible to "see" the microbial world that cannot be observed with a microscope. In addition, with the ability of scientists to manipulate genes and whole organisms, the time has arrived when they can change the metabolic behavior of microbial cells and perhaps even design new life.

Analyzing Communities of Microbes

Existing within, on, and around every living organism, and in most all environments on Earth, are microorganisms. Yet, some 99% of the bacterial and archaeal species found within us and in the environment will not grow on traditional growth media; we refer to these organisms as **viable, but noncultured** (**VBNC**). That means most organisms found in the soil, in oceans, and even in the human body have never been seen or named—and certainly not studied. Yet these organisms represent a wealth of genetic variety because the vast majority of genetic information on Earth is found in these noncultured microorganisms and in viruses.

There is now a genetic process for accessing the information in this uncultured majority. The process is called **metagenomics** (*meta* = "beyond") and it refers to identifying the genome sequences within mixtures of organisms in a community (the **metagenome**); that is, "beyond" what has been cultured. Metagenomics has the potential to stimulate the development of advances in fields as diverse as medicine, agriculture, and energy production. Here is how the metagenomics process is carried out (FIGURE 14.15).

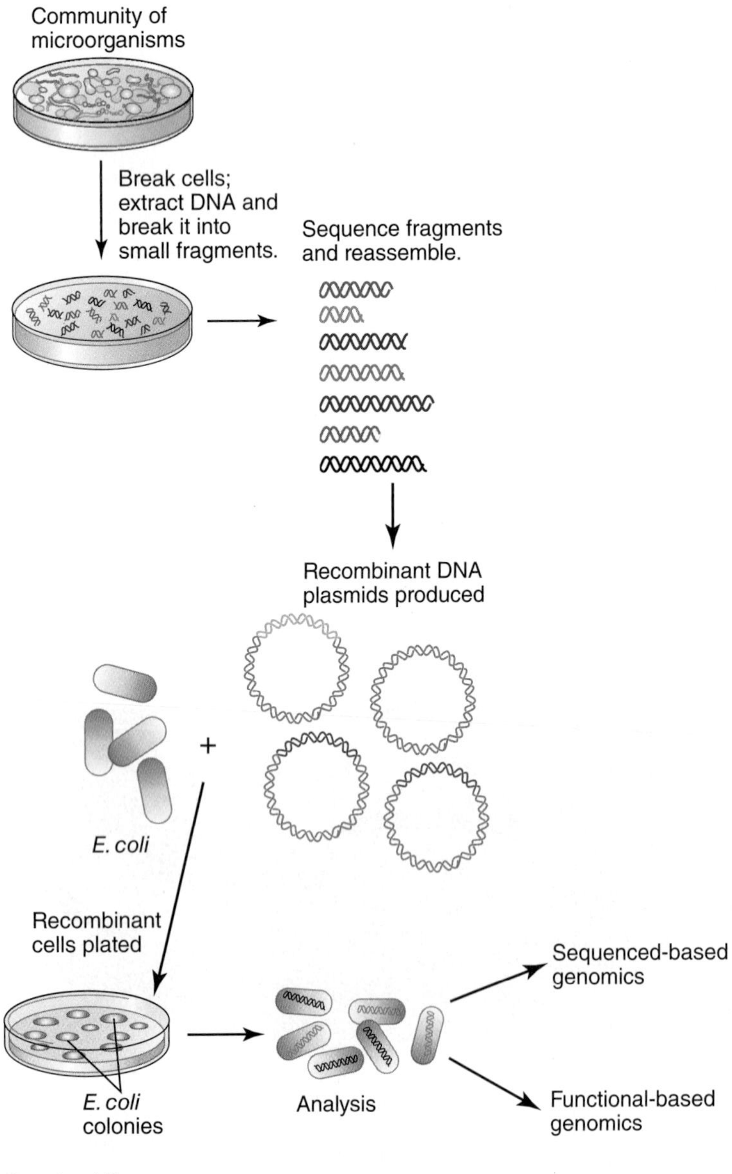

Samples of the desired community of microorganisms are collected. These samples could be from soil, water, or even the human digestive tract. The DNA is then extracted from the sample, which produces DNA fragments from all the microbes in the samples. The fragments can be amplified through PCR and introduced into plasmids, which are transferred into bacterial cells, such as *E. coli*. The metagenomic "library" produced represents the entire community DNA from the microbes sampled.

Analysis of the DNA fragments can be used to do sequence analysis and functional analysis. In "sequence-based metagenomics," after amplifying the random fragments, the fragments can be used to compare genetic relationships between organisms in the sample. More valuable to society is "functional-based metagenomics," where the gene products from the DNA fragments can be used to compare metabolic relationships within the community or to search for new enzymes, vitamins, antibiotics, or other potential chemicals of therapeutic or industrial use.

FIGURE 14.15 **The Process of Metagenomics.** Metagenomics allows the simultaneous sequencing of a whole community of microorganisms without growing each species in culture. Fragments can be either sequenced or subjected to functional analysis.

In fact, metagenomics already has opened our eyes to the diversity of microbes in ocean environments and in our body. The process also makes it possible to harness the power of microbial communities to help solve some of the most complex medical, environmental, agricultural, and economic challenges in today's world. For example:

- **Medicine.** Understanding how the microbial communities inhabiting our bodies affect human health could lead to new strategies for diagnosing, treating, and preventing diseases. Research is even suggesting these communities of microbes affect our behavior and immune system function.
- **Ecology and the Environment.** Exploring how microbial communities in soil and the oceans affect the atmosphere and environmental conditions could help us understand, predict, and address climate change. Adding to the arsenal of microorganism-based environmental tools can help in monitoring environmental damage and cleaning up oil spills, groundwater, sewage, nuclear waste, and other hazards.
- **Energy.** Harnessing the power of microbial communities might result in sustainable and ecofriendly bioenergy sources as was discussed earlier with industrial microbiology.
- **Food and Health.** Taking advantage of the functions of microbial communities might lead to the development of safer food and health products.
- **Agriculture.** Assessing the roles of beneficial microorganisms living in, on, and around domesticated plants and animals could enable detection of diseases in crops and livestock, and aid in the development of improved farming practices.
- **Biodefense.** The addition of metagenomics tools to the identification of infectious diseases will help to monitor pathogens, create more effective vaccines and therapeutics against bioterror agents, and reconstruct the "crime scene" of a bioterrorism attack.

Engineering New Skills

With the expanding tools available in biotechnology today, the possibility of altering life processes is becoming more of a reality. A blossoming area in microbial biotechnology today is **pathway engineering**, which is a field that attempts to alter and/or improve the metabolic capabilities inherent in organisms.

In today's biotech world, bacterial metabolic pathways can be altered through genetic engineering to produce a useful end product—in the example here, something as mundane as the dye for blue jeans. Blue jeans or denim is a $53 billion a year business with an estimated 450 million pairs of jeans made every year just in the United States. Blue jeans are made from blue denim and the blue in the denim comes from indigo dye (FIGURE 14.16). Originally extracted from sea snails and then plants, most indigo dye today is made from coal or oil. However, the process produces potentially toxic by-products. To be greener, scientists have considered genetically altering *E. coli* to churn out the indigo pigment.

Biotech indigo can be made through the following reconstructed pathway. *E. coli* produces the essential amino acid tryptophan, which it can then degrade with an enzyme it naturally has into a number of other end products, including indole and pyruvate (FIGURE 14.17a). Scientists also know that indole can be treated with another enzyme whose product, dihydroxy-indole (DHI) in the presence of oxygen, spontaneously yields indigo. So, the pathway engineering step was to get the gene coding this enzyme and engineer it using a plasmid into *E. coli* cells. And such a gene exists in *Pseudomonas*. When the gene was engineered into *E. coli* cells and the cells cultured on agar, they soon turned blue due to the production of indigo (FIGURE 14.17b).

FIGURE 14.16 **Indigo Dyes.** Indigo dyes, such as these for sale in Morocco, are used to dye denim blue (blue jeans).

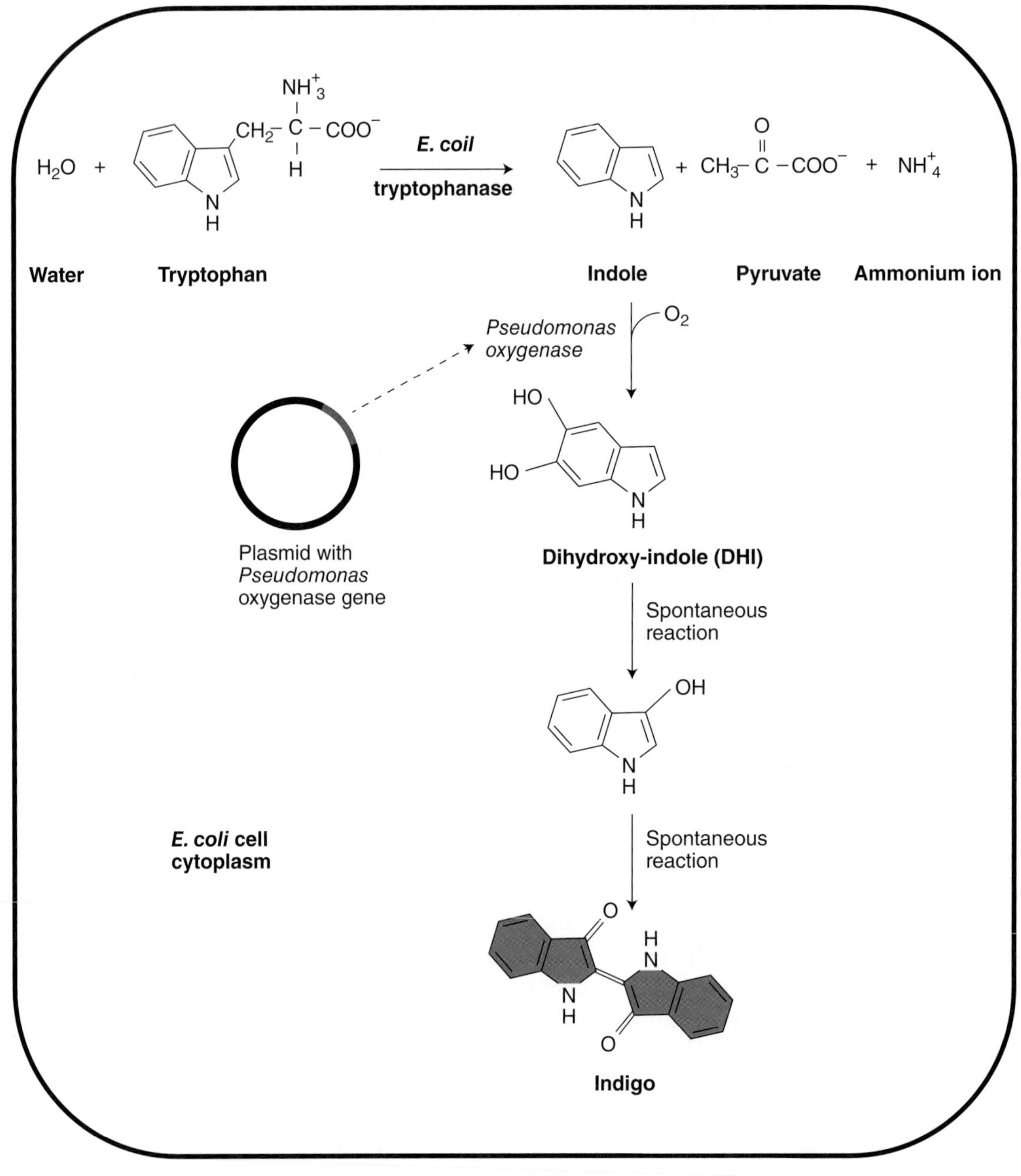

(a)

(b)

FIGURE 14.17 **Pathway Engineering in Microbes.** (**a**) Microbes, such as *Escherichia coli*, can have a metabolic pathway altered (engineered) to produce a nonmicrobial product, such as indigo. (**b**) Such cells will now appear blue as they produce indigo.

Today, approximately 16 million kilograms (18,000 tons) of indigo dye are made annually and most of it is used to stain denim. The indigo dye made through biotechnology is indistinguishable from the deep blue of the chemically made dye. So, if the cost of pathway engineering can be made more competitive with the chemical process, "biotech indigo" may be in your future pair of genes—jeans that is!

A Final Thought

For many decades, humans stood by and watched the "game of life." They marveled at the wonders of nature; they searched out and cataloged the plants, animals, and microbes of the world; and they spent exhaustive hours trying to understand how these organisms fit into the scheme of things.

Then came the age of DNA science and biotechnology. Now many of the observers became manipulators as they learned how to change the character of an organism at its most fundamental level. They isolated its DNA, changed its metabolism, and inserted the new DNA into recipient cells to see what would happen. Bacterial and yeast cells began producing human hormones, therapeutic proteins, and enzymes. In the offing were new vaccines and diagnostic tests undreamed of a generation before. Microbes were at the center of biotechnology, and their widespread use added to the positive press they enjoyed for their industrial contributions.

We stand at the brink of an adventure that will carry us through the twenty-first century and beyond. The implications of biotechnology are so colossal that the human mind has yet to imagine all of them. Biotechnology will continue to impact human lives for many centuries to come, and we can proudly tell our grandchildren we were there at the beginning. It's a wonderful time to be studying microbiology.

Questions to Consider

1. Certain bacterial cells produce many thousands of times more of specific vitamins than they require. Some biologists suggest that this makes little sense because the excess is wasted. Can you suggest a reason for this apparent overproduction in nature?
2. During the past decade, publications such as *Discover* magazine and *Scientific American* have carried articles about a renaissance in vaccines and the increased reliance future generations will have on vaccines to preserve health. What evidence of such a renaissance do you see in this chapter, and why do you suppose we will rely on vaccines even more in the future?
3. A biotechnologist suggests that one day it may be possible to engineer certain harmless bacterial cells to produce antibiotics and then to feed the cells to people who are ill with infectious disease. The bacterial cells would then serve as antibiotic producers within the body. Would you favor research of this type?
4. While studying for the exam that covers the material in this chapter, a friend asks you: "What is industrial microbiology?" How would you answer her question?
5. Textbook writers tend to oversimplify complex issues in attempting to make them understandable to students. This chapter has material on the methods of biotechnology to show the complexity of this discipline, but space limitations

have required that a host of issues be omitted. (For example, how do scientists "store" genes and how do they isolate genes from DNA?) From your knowledge of microbiology and of science in general, what societal issues must biotechnologists confront, and how might these problems be circumvented?

6. How many times in the last 24 hours have you had the opportunity to use or consume the industrial product of a microbe?
7. Although the products of biotechnology have been of great benefit in numerous arenas, they have also been abused. One example is the use of erythropoietin by athletes to increase their red blood cell counts and give them an unfair advantage at competitive events. What other abuses of products of biotechnology can you think of?

Key Terms

Informative facts are necessary for the expression of every concept, and the information for a concept is founded in a set of key terms. The following terms form the basis for the concepts of this chapter. On completing the chapter, you should be able to explain and/or define each one.

amylase
antibiotic
attenuated
biofuel
biotechnology
DNA probe
DNase I
DNA vaccine
exoenzyme
extremozyme
fermentor (bioreactor)
human growth hormone (hGH)
immunocompetent
inactivated
industrial fermentation
industrial microbiology
insulin
interferon (IFN)
metabolite
metagenome
metagenomics
neutraceutical
pathway engineering
polymerase chain reaction (PCR)
primary metabolite
protease
rennin
secondary metabolite
semisynthetic
subunit vaccine
Taq polymerase
tissue plasminogen activator (TPA)
toxoid
toxoid vaccine
vaccine
VBNC
virulent

Microbes and Agriculture: No Microbes, No Hamburgers

15

The formation of frost on a crop plant, like these strawberry plants, can seriously damage the plant and cause farmers to lose money. A bacterial species, *Pseudomonas syringae*, which is often found on the surface of plants, contains a protein on its cell wall that actually assists in the formation of ice crystals. Scientists have used biotechnology to generate an altered form of *P. syringae* without this protein. If this genetically engineered species can outcompete the native species normally on the surface of crop plants, less frost will form when the temperature dips to the freezing point, and less crop damage will occur.

Looking Ahead

Microbes play a key role in agriculture—they are essential to the production of meat, dairy products, and numerous other foods. Moreover, they help farmers fight crop pests, and they are instrumental in the cycling of elements in the soil. These are but a few of the countless places where microbes exert a powerful influence on agriculture, as the pages ahead will show.

On completing this chapter, you should be capable of:

- Explaining the importance of nitrogen and the central role of microbes in fixing nitrogen gas.
- Discussing the role of microbes to ruminant digestion.
- Illustrating the role of microbes to milk and cheese production.
- Describing how DNA is inserted into plants.
- Assessing the importance of bacterial insecticides.
- Illustrating what is meant by "pharm animals."
- Explaining what genetically modified (GM) foods are and assembling a list of the perceived pros and cons of GM foods.
- Contrasting "veggie vaccines" and edible vaccines.

The man from Delft stood in front of the assembled farmers and made an outrageous proposal: "*Don't plant your crops in the same field as last year*," he said. "*Leave the field alone for the next two years; let it lie fallow*." The year was 1887; the man was Martinus Willem Beijerinck;

the country was The Netherlands. Because agricultural land was scarce in his small country, the proposal was outrageous.

Beijerinck was a local bacteriologist. While his colleagues in France and Germany were investigating the germ theory of disease and its implications, Beijerinck was out in the agriculture fields. He noticed the land was very productive when it was freshly cleared of brush and trees and newly planted, but less productive after several years of use. Moreover, the fields yielded bountiful crops when the farmer let the field lie fallow for a couple of years. And now he thought he had the answer to these mysteries: great populations of bacterial organisms in the soil.

Beijerinck was an agricultural expert, to be sure, but he also had a solid background in chemistry, something other botanists lacked. He believed atmospheric nitrogen gas (N_2) is essential for plant growth, but he could not figure out how the nitrogen gas became part of the organic molecules in plant cells—what bridged the gap between atmosphere and biomolecules, such as proteins? Then it dawned on him—bacterial cells must be the link, and the little lumps and bumps on plant roots were the answer. Time and again he had observed great hordes of bacterial cells in the lumps and bumps that were called root nodules. He did not often see the nodules on the tended agricultural crops, but they always seemed to be in abundance on wild plants like clover that grew in untended fields (FIGURE 15.1a).

Beijerinck performed a number of laboratory experiments to test his ideas: He isolated bacterial cells from the root nodules and injected them into seedlings of different kinds of plants. Invariably, the nitrogen content of the soil rose dramatically. So, his advice that day in 1887 was direct and straightforward: Leave the field alone for a spell; plant elsewhere; let wild plants and clover grow in the field and allow their bacterial populations to enrich the nitrogen content of the soil; and later, when the field is finally planted, the crop yield will be worth the wait.

Indeed, he was right. Modern farmers know that every so often, it is important to let a field lie fallow and "refresh" itself. To the farmer, the nodules on the roots of wild plants are the important parts. To the microbiologist, it's what's inside that counts; for inside the nodules, bacterial cells perform the essential chemistry of the nitrogen cycle and bring this key element into plant proteins and other cellular molecules (FIGURE 15.1b). As we will see in this chapter, this chemical transition would not occur without bacterial organisms, and our supply of protein would quickly diminish. Indeed, without the cycling of nitrogen, life as we know it would probably come to an end.

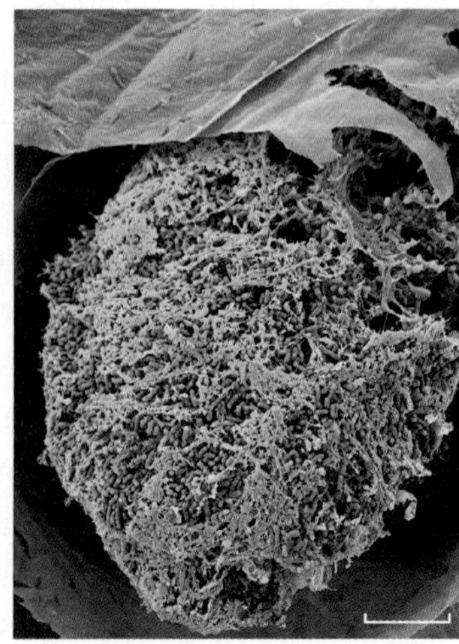

FIGURE 15.1 ***Rhizobium* Root Nodules.** (**a**) Root nodules on clover roots. (**b**) A false-color electron microscope image of a torn open root nodule. The mass of *Rhizobium* cells is evident. (Bar = 30 μm.)

(a) © Hugh Spencer/Science Source. (b) © Steve Gschmeissner/Science Source.

15.1 Microbes on the Farm: If the Environment Is Suitable, They Will Come

Next time you have a chance, go online (try YouTube) and listen to Beethoven's Sixth Symphony, known as the *Pastorale*. Sit back and enjoy this most beautiful piece of music. Envision a tree-shaded meadow, wild flowers and clover everywhere, and a herd of cows gathered in the midday sun under a sprawling oak. The idyllic scene seems too good to be true. But what's that have to do with microbiology?

This is where music and microbiology mix because in the nodules of the clover, microbes are busy at work bringing usable nitrogen to plants and to animals when they eat the plants; and in the stomachs of ruminants, such as cows, microbes are taking plant materials the cows eat and changing them so cows can make meat and dairy products. In both cases, microbes are continuing their unceasing efforts on our behalf. They are doing what they do best and indispensable for us. These processes are worthy of a fine piece of musical harmony. Let's take a closer look.

Connecting the Nitrogen Dots

In soil and water, elements such as carbon, oxygen, and hydrogen are cycled and recycled by a broad variety of plants, animals, and microbes (as discussed in another chapter); but the cycle of nitrogen is largely the province of microbes alone. Scientists know the productivity of many ecosystems is limited by the supply of nitrogen available to organisms in the soil, and one part of the **nitrogen cycle** (i.e., the conversion of nitrogen gas (N_2) to ammonia (NH_3) and organic nitrogen compounds) is performed almost exclusively by a limited group of microbial species. Without these species, the entire cycle would grind to a halt—life on this planet would eventually disappear.

But why is nitrogen metabolism so important? Nitrogen is an essential element of many organic molecules, including all the amino acids (and, consequently, all proteins) as well as all the nucleobases of nucleic acids (DNA and RNA). In all, between 9% and 15% of the dry weight of a typical cell is nitrogen.

Ironically, even though nitrogen is the most plentiful gas of Earth's atmosphere (about 80% of the air is nitrogen gas), neither plants nor animals can use this gas to synthesize the molecules they must have to survive. Instead, animals and plants depend on microbes to bring nitrogen into the cycle of life. To fill cellular needs, microbes make available approximately 200 million tons of nitrogen molecules each year.

The trapping of nitrogen gas from the atmosphere and its incorporation in organic molecules is known as **nitrogen fixation**. Two types of microbes accomplish nitrogen fixation: free-living, independent microbes and **symbiotic** microbes, which live in close relationships with other organisms. Both types of microbes produce **nitrogenase**, an enzyme that uses nitrogen to synthesize ammonia. Through a complex series of reactions, nitrogenase takes a nitrogen atom and adds three hydrogen atoms to yield ammonia (NH_3). Nitrogenase is very sensitive to oxygen and is inactivated by this gas. Therefore, nitrogen fixation takes place where conditions are **anaerobic**; that is, without oxygen gas.

The free-living nitrogen-fixers include species of *Azotobacter*, *Beijerinckia* (named for Beijerinck), and some species of *Clostridium* as well as several genera of cyanobacteria such as *Nostoc* and *Anabaena*. In soil and water, these microbes manufacture ammonia as well as a variety of amino acids, as FIGURE 15.2 illustrates. In the Arctic region, cyanobacteria are the most important nitrogen-fixers in the ecosystem. These microbes synthesize ammonia and then convert it to amino acids and proteins. Later, when the cyanobacteria die, their cells are decomposed by other bacterial species, and the nitrogen compounds are converted back to ammonia, which is made available to plants.

- *Azotobacter* ä-zo'tō-bak-tėr
- *Beijerinckia* bī-yė-rink'ē-ä
- *Clostridium* klôs-tri'dē-um
- *Nostoc* nos'tok
- *Anabaena* an-ä-bē-nä

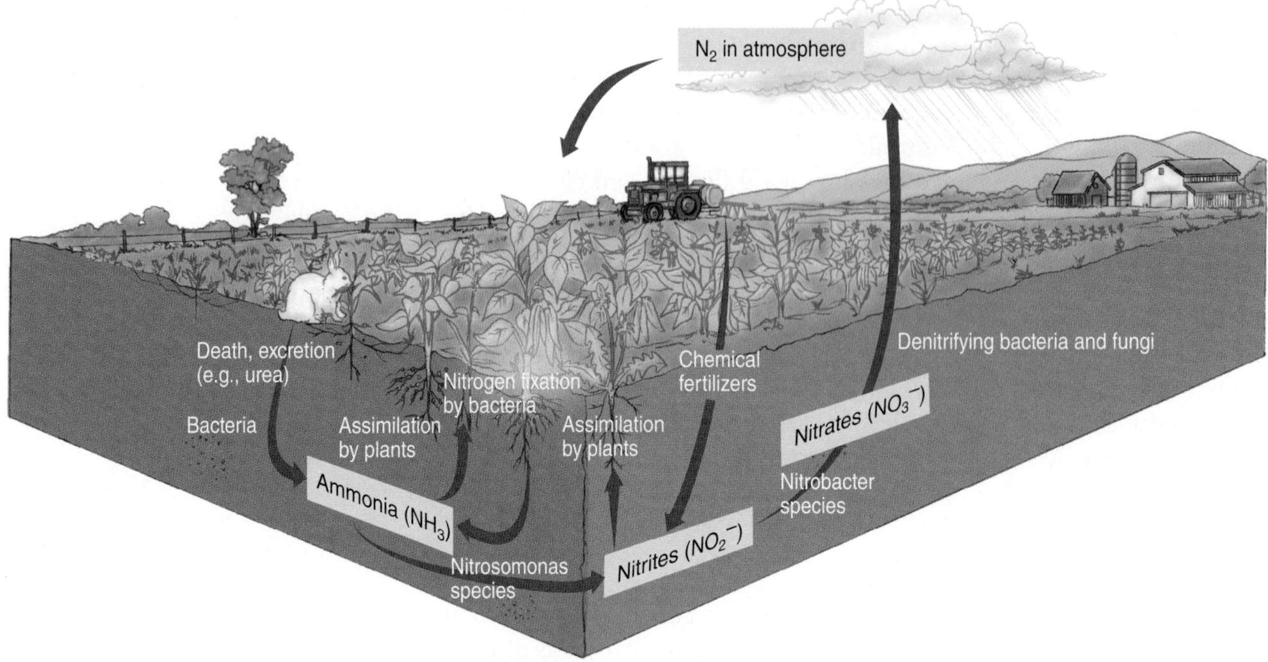

FIGURE 15.2 **Simplified Nitrogen Cycle.** Plant and animal protein and metabolic wastes are decomposed by bacteria into ammonia. The ammonia may be used by plants, or it may be converted by *Nitrosomonas* and *Nitrobacter* species to nitrate, which also is used by plants. Some nitrate is broken down to atmospheric nitrogen. This nitrogen is returned to the leguminous plants as ammonia by nitrogen-fixing microorganisms. Animals consume the plants and get their nitrogen from the proteins plants contain.

■ *Rhizobium*
rī-zō'bē-um

■ *Bradyrhizobium*
brad-ē-rī-zō'bē-um

■ legume: A plant that bears its seeds in pods.

The symbiotic nitrogen-fixers are typified by species of *Rhizobium* and *Bradyrhizobium*. These are the bacterial genera that live in root nodules of legume plants such as soybeans, alfalfa, peas, beans, and clover. The symbiotic relationship between the microbe and plant is illustrated in FIGURE 15.3. It involves a chemical attraction between the bacterial cells and the cells of the plant root, a binding to the plant root, the curling and branching of rootlets, and the entry of bacterial cells into root hairs. In quick succession, an infection thread develops; the nearby plant cells transform into a tumor-like growth, which is the nodule; and the bacterial cells assume distorted forms known as **bacteroids**.

Bacteroids cannot live independently once they have entered the symbiotic relationship with the plant. Indeed, they derive life-sustaining nutrients from the legume plant while using their nitrogenase to fix atmospheric nitrogen for the plant's benefit. Since Beijerinck's time, farmers have found that soil fertility can be enhanced dramatically when legumes are cultivated in a field. So much organic nitrogen is captured, in fact, that the net amount of nitrogen (and the net worth of the soil) increases considerably after a crop of legumes is harvested. Moreover, no artificial nitrogen-containing fertilizer is necessary when legumes have been cultivated in the field or when a crop such as clover or alfalfa is plowed under. In addition, nutritionists know full well the value of legumes in the diet, and humans are indebted to microbes for the high protein content of various kinds of peas and beans. Because cattle and other animals feed on clover and other legume plants, humans also depend on microbes for many indirect protein products of nitrogen fixation, everything from milk to hamburgers.

■ *Nitrosomonas*
nī-trō-sō-mō'näs

■ *Nitrobacter*
nī-trō-bak'tėr

But nitrogen fixation is only one aspect of microbial involvement in the nitrogen cycle. Looking back at Figure 15.2, notice that much of the ammonia deposited in the soil is converted to nitrate (NO_3) through the process of **nitrification**. In the first step of nitrification, bacterial species of *Nitrosomonas* convert the ammonia molecules to nitrite (NO_2^-). Then, the nitrite is converted by species of *Nitrobacter* to nitrate, which is used by plants for making their organic materials, including proteins. Nitrification occurs under aerobic conditions and is used by microbes as an energy-yielding process. In fact, so much energy is obtained that the bacterial cells can grow without other energy sources.

From the plant and animal point of view, one of the less beneficial aspects of the nitrogen cycle is **denitrification**. In this process, some species of microbes break down nitrate and produce nitrogen gas, which they give off into the atmosphere. This process removes nitrogen from the cycle of life and can thus be seen as detrimental to most living things on Earth. Operating through nitrogen fixation, however, microbes bring nitrogen back into the cycle and make it available to the organisms that depend on it for life.

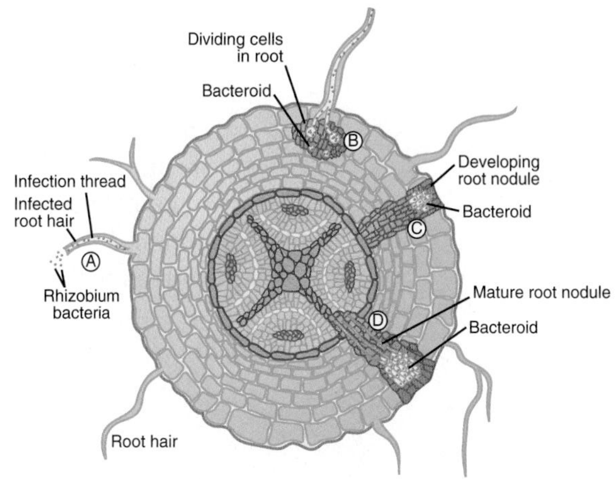

FIGURE 15.3 **A Cross-Section Through a Plant Root.** (**a**) *Rhizobium* cells infect root hairs, triggering (**b**) surrounding root cells to divide. This starts the (**c**) development of a root nodule, which (**d**) then matures and contains the bacteroids.

The importance of nitrogen fixation by symbiotic bacteria has not escaped the attention of biotechnologists. Scientists foresee the day when they can use gene alterations to improve the nitrogen-fixing chemistry in *Rhizobium* and other bacterial genera and increase the efficiency of the bacteria-plant interaction. Agricultural scientists are currently seeking ways to transfer the genes for nitrogen fixation from *Rhizobium* species to bacterial organisms such as *Rhizobium radiobacter* (formerly called *Agrobacterium tumefaciens*). This bacterium inhabits the tissues of a variety of plants such as tobacco, petunias, and tomatoes, and the knowledge base about it and its plant hosts is well developed, as we will discuss later in this chapter. It is conceivable that nitrogen fixation could one day be extended to these plants. Another possibility is to engineer a *Rhizobium* species able to assume a symbiotic relationship with a nonleguminous plant, such as wheat, rice, or corn. Such an advance would go a long way toward solving world hunger problems.

■ *Rhizobium radiobacter*
rī-zō'bē-um
rā-dē-ō-bak'-tėr

■ *Agrobacterium tumefaciens*
ag'rō-bak-ti'rē-urn
tü'me-fa–sh-enz

In the most optimistic of molecular biology circles, some look to the day when the genes for nitrogen fixation can be transferred directly from microbes to animals. Then, the bacterial link in the nitrogen connection could be severed, and animals could extract their nitrogen directly from the atmosphere. The prospect of an animal synthesizing its own amino acids from atmospheric nitrogen fires the imagination of even the most innovative futurists.

Those Remarkable Ruminants

Ruminants are grazing animals such as cattle, sheep, deer, goats, camels, buffaloes, and numerous other "cud-chewing" species. These animals have adapted to a diet of plants, grasses, and other carbohydrate-rich plants; plants whose cells contain much energy-rich starch and whose cell walls consist primarily of cellulose—both polysaccharides being built from chains of glucose molecules. Despite an almost purely carbohydrate diet, ruminants manage to make protein and, consequently, more of themselves. We, in turn, rely on many of these animals for meat and dairy products.

Using a cow as our prototypical ruminant, its secret for converting carbohydrate to protein is intimately bound up with microbes. Although cows can digest starch without trouble, they cannot digest cellulose into its valuable glucose molecules because they lack the enzyme cellulase needed to digest cellulose. (In fact, no vertebrate animal, including humans, can synthesize cellulase or digest cellulose.) In the cow, cellulase is provided by—you guessed it, microbes.

What happens is outlined in FIGURE 15.4a. The cow's large stomach is divided into several compartments, the first and biggest of which is called the **rumen**. The rumen has a constant temperature and is slightly acidic and oxygen-free environment.

It teems with numerous anaerobic bacterial species and other microbes that produce cellulase. The microbial cellulase digests the cellulose in plants, producing the disaccharide cellobiose and, most importantly, the monosaccharide glucose (FIGURE 15.4b). Then, the glucose molecules are fermented by microbes to simple organic acids such as propionic acid and acetic acid along with gases such as methane and carbon dioxide. Some of the organic acids pass through the rumen wall and enter the cow's bloodstream, from where they are transported to its body cells. Here they are key energy sources fueling the cow's metabolism. The methane and carbon dioxide gases from the rumen are eliminated when the cow belches, a process that occurs regularly and vigorously as it reclines under an oak tree in the midday sun.

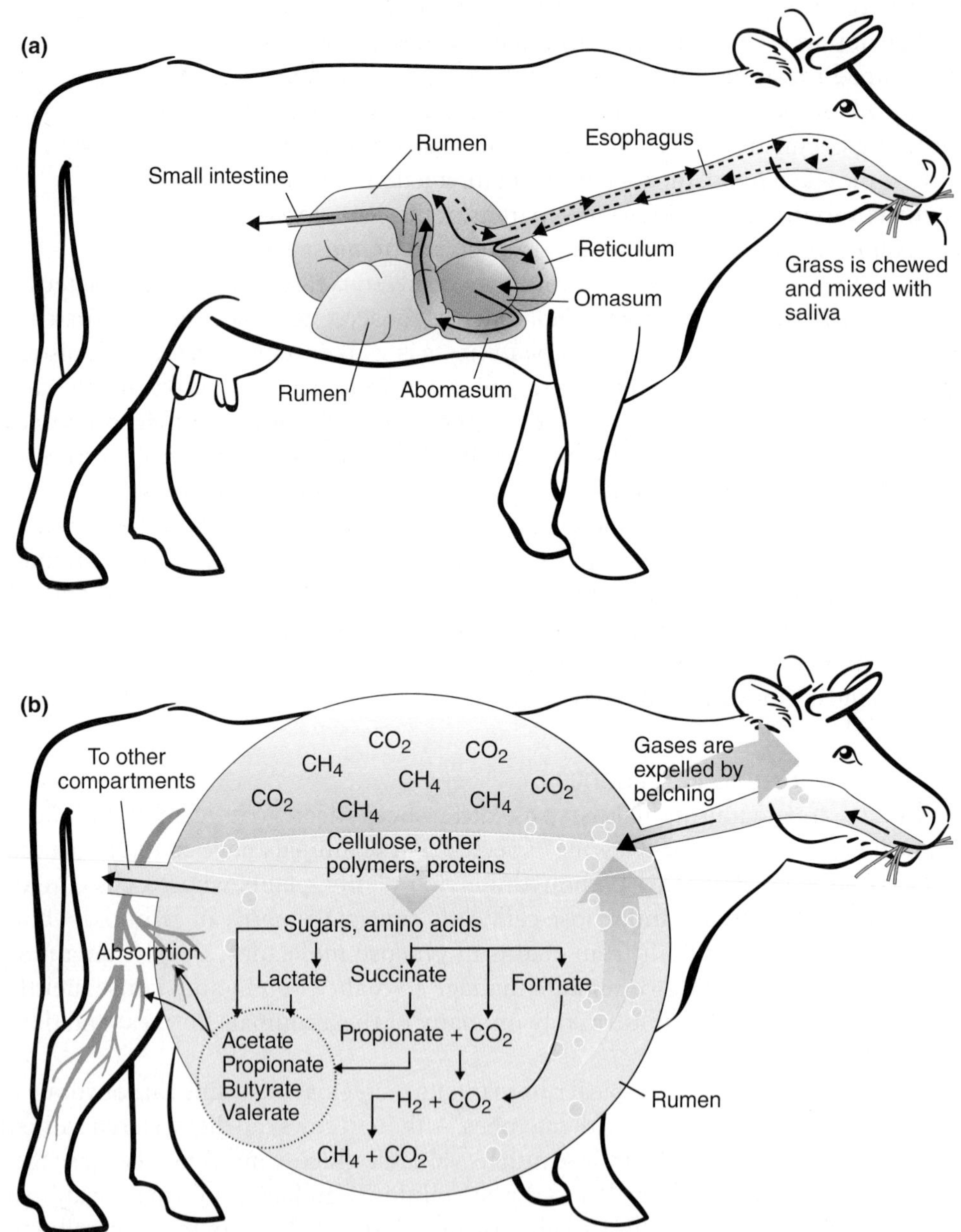

FIGURE 15.4 **Microbes and the Rumen.** (**a**) A schematic diagram of the rumen and other components of the digestive tract of the cow. Arrows show the pathway of food; dashed lines represent the pathway of regurgitated food when the cow chews its cud. Note the large size of the rumen relative to the other compartments of the stomach. (**b**) Some of the chemical reactions taking place in the rumen.

But energy alone does not make cow muscle or dairy products. Quite to the contrary, the energy is used to make more microbes. In fact, while they are performing their biochemical magic on cellulose, the microbes continue to grow and multiply at astoundingly high rates. A single milliliter (20 drops) of rumen fluid may contain up to a trillion (10^{12}) microbes. The rumen is literally a fermentation tank and home to incomprehensible numbers of microbial species. In addition to the organic acids we noted above, the microbes synthesize protein for the cow, using normally toxic urea as a nitrogen source. Even the microbes themselves eventually become a protein source for the cow as they are broken down and their cellular components used to form more cow muscle.

But there is more. As microbes continue to multiply furiously in the rumen, they soon overgrow the space and pass into the second part of the stomach, the reticulum. Here the microbes are squeezed together with undigested plant material to form balls of cud material. At its leisure, the cow regurgitates the cud, mixes it with copious amounts of saliva, and chews it to crush the fibers. Then, the cud is reswallowed and returned to the rumen, where it is redigested by more cellulose-digesting microbes. Next, it passes back into the reticulum for regurgitation and rechewing. Finally, the plant-microbe mass passes into the omasum and the abomasum (together, the "true" stomach of the cow), where protein-digesting enzymes break down the protein to release the amino acids.

From this point on, the physiology of the cow is similar to that of other vertebrates (including humans). Further digestion of protein in the intestines yields amino acids and a bevy of vitamins and other growth factors that are passed into the cow's bloodstream and transported to its muscles and organs. At the fundamental level, the ruminant animal has satisfied its requirement for protein by eating microbes. Humans are next in line for this "microbial steak."

Other ruminants obtain their protein in the same way as the cow. However, nonruminants such as horses, rabbits, and guinea pigs are slightly different. They are also grass-eaters, but they have a side pocket in their gut called the cecum. In this organ, a rich microbial population performs the same task as the microbes in the rumen. Moreover, instead of regurgitating the cud, certain "cecum animals" such as rabbits eat their fecal pellets, thereby giving the food a second pass through the intestine. This activity, called "coprophagy," also gives the animal an opportunity to obtain the vitamins produced by the bacterial cells in the cecum. At one time in its evolution, the human appendix may have been the equivalent of the animal's cecum.

Before we leave our ruminant friends, we note another place where microbes impact on their lives. Cows, goats, and other barnyard animals are fond of silage, a product of microbial fermentation. **Silage** is fermented in the tall cylindrical structure often located adjacent to a barn (FIGURE 15.5). A silo is an enormous fermentation tank into which the farmer loads grasses, grains, legumes, and any other plants available. Under tightly packed, anaerobic conditions, numerous bacterial species consume the plant material and carry on fermentation to reduce the carbohydrates to a more digestible and tastier form. A host of acids and other organic compounds preserve the silage from microbial decay and produce an animal food both economical and pleasing to the palate (the animal palate, that is). Silage also can be used as a biofuel feedstock, a topic discussed in another chapter. On the darker side, the buildup of fermentation gases in the silo can be deadly as **A Closer Look 15.1** explains.

FIGURE 15.5 **Farm Silos.** Silos act as fermentation tanks in which microbes convert plant material into digestible animal feed.

A CLOSER LOOK 15.1

Toxic Atmospheres

Your reading in this text hopefully has made you aware of the variety of products produced by, and speed with which, microbes carry out metabolism. Here is one particularly striking example.

Silage is an important way for farmers to feed cows and sheep during times when the pasture isn't good, such as the dry season. For many farms, silos are used for storing silage. When a silo is filled in late summer and early fall, within 48 hours silage starts to undergo anaerobic fermentation and is essentially complete in about two weeks. Although silage itself poses no danger, the fermentation process being carried out by the microbes produces toxic "silo gases" gases (corn silage forms more silo gas than other crops) that can be a hazard to agricultural workers.

Looking down into a grain silo

Normally, the gases produced remain at low concentrations as long as the hatches and vents of a silo are open. Should silo gases build up, they include carbon monoxide (CO) and nitric oxide (NO) that can react with oxygen (O_2) in the air and form nitrogen dioxide (NO_2), which is toxic and can cause permanent lung damage. Other harmful gases arising from fermentation are methane (CH_4), which is flammable or explosive, ammonia (NH_3), and hydrogen sulfide (H_2S). Oxygen gas can be displaced by the silo gases or depleted by microbial activity.

Should an agricultural worker enter a silo with high concentrations of toxic silo gases or that is oxygen-deficient, the worker may go into respiratory distress, collapse, and die from the gases or be asphyxiated from the lack of oxygen gas. Unfortunately, it is possible to work in the presence of toxic silo gases for some time without ever feeling major discomfort. In fact, victims of silo gas have been known to die many hours later, sometimes in their sleep, from pulmonary edema, which is a buildup of fluid in the lungs.

Therefore, the process of filling and maintaining a silo requires several safety precautions. As mentioned, the vents and hatches should be open to allow the gases to escape. If one must enter the silo, respirators should be worn and ventilator fans operational. Also, farmers can lower the risk of exposure by waiting one month after filling before entering a filled silo.

It is a very sad situation if an individual is harmed or killed by silo gases but it clearly demonstrates the speed with which microbes in mass numbers can produce end products of metabolism.

At the Dairy Plant

The milk from cows provides an excellent medium for the growth of microbes. It is rich in various proteins, including casein, the complex protein that gives milk its white color. It also contains carbohydrates, primarily the disaccharide lactose ("milk sugar"). Moreover, it has a variety of fats, vitamins, and minerals.

Allowed to stand at room temperature, unpasteurized milk rapidly undergoes a natural souring brought about by bacterial fermentation of the lactose and the production of lactic acid. (If excess lactic acid is present, the protein will curdle.) This acidification provides a natural barrier to the growth of other microbes because the acidic environment is generally inhospitable to most pathogens. Although the milk is sour to taste and somewhat offensive to smell, dairy farmers have put this biochemical process to work to produce a variety of fermented milk products. Over the centuries, people have come to accept these dairy products for their resistance to deterioration as well as their interesting and unique flavors.

■ *Lactobacillus bulgaricus* lak-tō-bä-sil'lus bul-gä'ri-kus

One example of a fermented milk product is buttermilk (FIGURE 15.6). Thought to originate in the East European country of Bulgaria, buttermilk is produced by two species of bacteria: *Lactobacillus bulgaricus* (named for Bulgaria) and

Leuconostoc citrovorum. At the dairy plant, these bacterial species are added to skim milk, whereupon the *Lactobacillus* produces lactic acid from the lactose and sours the milk, while the *Leuconostoc* synthesizes polysaccharides to make the milk slightly thick.

Sour cream is produced in the same way as buttermilk, except cream is used instead of skim milk. Biochemists tell us the acid in sour cream retards microbial decay; oblivious to that bit of science, we just know sour cream tastes good on a baked potato.

A very common fermented milk product is yogurt. Also thought to have originated with Bulgarian tribes, yogurt is a popular food today throughout North America and Europe. Among the principal bacterial species used in yogurt production are *Streptococcus thermophilus* and *Lactobacillus bifidus* (other species such as *S. lactis* and *L. acidophilus* can also be used). Milk is first concentrated by adding dried milk protein, then the bacterial cells (the "active cultures") are added by mixing in a sample of previously prepared "mother" yogurt. At a high temperature (about 60°C or 166°F), the streptococci are the first to ferment the lactose. Then, the lactobacilli take over the fermentation and produce more acid and the characteristic yogurt texture and consistency. Although yogurt is really a "spoiled milk" product, we consider it a healthful addition to our diet.

FIGURE 15.6 **Products of Fermentation.** Many dairy products involve microbes and a fermentation process.

Courtesy of Dr. Jeffrey Pommerville

- *Leuconostoc citrovorum* lü-kō-nos'tok sit-rō-vôr'um
- *Streptococcus thermophilus* strep-tō-kok'kus thėr-mo'fil-us
- *Lactobacillus bifidus* lak-tō-bä-sil'lus bī-fid'-us
- *Streptococcus lactis* strep-tō-kok'kus lak'tis
- *Lactobacillus acidophilus* lak-tō-bä-sil'lus a-sid-o'fil-us
- *Saccharomyces kefir* sak-ä-rō-mī'sēs key'-fur

Other fermented dairy products are produced in different parts of the world. They vary according to the source of the milk, temperature of incubation, and species of microbes participating in the fermentation. For example, kefir (first made in the Caucasus Mountains of Ukraine) is produced using lactobacilli, streptococci, and the yeast *Saccharomyces kefir*. Both acid and alcohol are produced during the fermentation, and a unique effervescent quality is present in the sour milk because it contains fermentation gases (much like champagne). A fermentation like this, which produces both acid and alcohol, is referred to as a **mixed fermentation**.

Although we do not often think of butter as a microbial product, microbes contribute to its production. One popular method for preparing butter begins with starter cultures of streptococci and *Leuconostoc* species added to pasteurized sweet cream. The streptococci sour the cream slightly by producing lactic acid, and the *Leuconostoc* species synthesize a substance called **diacetyl**. Diacetyl gives butter its characteristic aroma and taste. Once the reactions have been completed, the slightly sour milk is churned to aggregate the fat globules into butter. Diacetyl also is an ingredient in butter-flavored popcorn and also provides the buttery smell in margarine, candy, and baked goods. It occurs naturally in fermented drinks like beer, and gives some chardonnay wines their buttery taste.

O, O, H_3C, CH_3

Diacetyl

Of Curds and Whey

Legend has it that one day a young Bedouin traveler filled his goatskin pouch with milk and set off on a journey. At midday, he stopped to eat, only to discover the contents of his pouch had turned to large white lumps. Frustrated but hungry, he decided to taste the lumps and found them to be quite good. He washed down the lumps with the leftover fluid and continued on his way...having discovered cheese.

Cheese has historically been a way of maintaining the nutrient quality of milk by preserving the protein for a long period (over the winter, for example, when milking the cows was difficult). Today, cheese is one of the most important products of microbial action on milk.

Cheese is manufactured from milk in three steps: curd formation, curd treatment, and curd ripening. In the initial step, the milk proteins, casein, separate from the milk fluids when the acid produced during microbial fermentation causes the casein to lose its three-dimensional structure and "curdle" out. This results in a sour curd. Alternately, the curd can be produced by altering the casein with the industrial enzyme rennin (which was traditionally obtained from the lining of a calf's stomach but is now produced by microbes). The result is a sweet curd. In both cases, the casein forms an insoluble clot (the curd), plus a butter-yellow liquid called whey. Once the curd has settled as a solid mass, it is separated from the whey by draining. The whey can then be used to make "cheese food" such as ricotta cheese.

Unripened cheese such as cottage cheese or cream cheese is made directly from milk curds. Such cheese varies in fat and calories depending on the starting material. For instance, cottage cheese is produced from skim milk (low fat and calories), while cream cheese is derived from whole milk enriched with cream (high fat and calories). Both types of cheese spoil quickly because they lack salt, which normally inhibits microbial growth, and their pH is close to neutral. Another reason they spoil rapidly is that they have not been preserved by ripening, as we discuss next.

FIGURE 15.7 Various Cheeses Produced by Microbes. Many hard and soft cheeses are the product of bacterial and fungal microbes.

© Agita/ShutterStock, Inc.

In some societies, the ripening of milk curds is considered a form of spoilage; in others, it is the zenith of microbial accomplishments for the palate. The French, for example, relish Camembert cheese, but many Americans consider its smell repulsive. To begin the ripening process, weights are used to put pressure on the curds and make them dense and compact. The curds are then salted, shaped, and inoculated with bacteria and fungi. The temperature and humidity conditions for ripening are controlled carefully, and as the microbes grow, the desired cheese gradually emerges, changing its texture with time and developing its aroma and flavor. Ripened cheeses fall into different categories based on such things as how the curd was formed (sour curd or sweet curd), its texture (hard, semihard, or soft), the ripening microbe (mold or bacteria), the milk source (goat, cow, mare, or ewe), and the country of origin.

Hundreds of different kinds of hard and soft ripened cheeses are available (FIGURE 15.7). For example, hard-ripened cheddar cheese, originally made in Cheddar, England, is produced by heating the curds and pressing them often to remove as much whey as possible (a process called cheddaring). The curds are then wrapped with paraffin or plastic film to prevent surface contamination. Bacterial cells producing lactic acid proliferate in the curd, their enzymes bringing about the chemical changes necessary to give the cheese its cheddarness. Pecorino Romano cheese is Italian cheese made from sheep's milk. In fact, many such cheeses have a unique community of microbes as **A Closer Look 15.2** reports.

■ *Penicillium camemberti* pen-i-sil'lē-um kam-am-bėr'tē

Soft ripened cheeses are typified by Camembert, which is produced by inoculating the curd at its surface with the blue-green mold *Penicillium camemberti*. As the mold grows, its protein-digesting enzymes break down the casein into smaller peptides and amino acids to nourish the mold. In the process, the interior of the cheese gradually changes from a chalky paste to an oozing cream delightful to the palate.

■ *Penicillium roqueforti* pen-i-sil'lē-um rō-kō-fôr'tē

Although many soft ripened cheeses are made around the world, one type remains the province of the French. In southwestern France in a village of the same name, Roquefort cheese is still made as it was centuries ago: Ewe's milk is cured in limestone caves where the temperature and humidity are constant. Spores of the mold *Penicillium roqueforti* fall onto the curds, and as the fungus grows, it seeks cracks in the curd and spreads its hyphae throughout the cheese, giving it the characteristic blue-green veins. The local residents turn out 30 million pounds of Roquefort cheese per year and

A CLOSER LOOK 15.2

Stilton Cheese—Slicing through a Microbial Community

Cheeses represent a unique community of microorganisms and, like human communities, temperaments and personalities may differ within a community. Take Stilton cheese, whose strong aroma and intense flavor some people find delightful and others find nauseating.

Stilton blue cheese (see figure) is made by only six dairies in the English Midlands. It is made by seeding milk curds with *Lactococcus lactis* and the blue mold *Penicillium*. So, how do these microbes give Stilton cheese its aroma and distinct characteristics? Christine Dodd and her colleagues at the University of Nottingham in England decided to find the answer.

Using the latest techniques of DNA sequencing, the microbiologists identified a whole new community of bacterial species within the cheese within nine weeks of ripening. It must be these microbes and *Penicillium* that give Stilton its unique taste and color.

A wedge of Stilton cheese.

The research group sectioned up slabs of cheese and analyzed the microbial populations by DNA sequencing. They discovered there was a distinct distribution of microbial species in different regions of the cheese. Two harmless *Staphylococcus* species were identified on the cheese surface. Mixed *Lactobacillus* species were found in the internal parts of the veins, while much less dense and single species colonies were found in the core. *Lactobacillus plantarum* was detected only underneath the surface (crust), while *Leuconostoc mesenteroides* was evenly spread through all parts of the cheese.

The investigators aren't certain whether these bacterial species arrive on the scene by surviving the milk pasteurization process or by being introduced through equipment or other sources. It is likely each species establishes its regions of growth for ecological reasons; that is, each species finds optimal growth conditions associated with specific regions of the cheese and among the other microbes present there. In the presence of other bacterial species and molds, microbial metabolites probably determine whether a specific species will survive.

The role of each microbial species in Stilton is not known, but their presence or absence may help explain why different batches of the cheese made at the same dairy can have highly different characteristics.

So, does this microbiological knowledge help you better tolerate Stilton cheese? I thought so.

insist that the only cheese to be called Roquefort is the one coming from their town. Gorgonzola, Danablue, and blue cheese are mold-ripened cheeses, but none is Roquefort.

This discussion of cheese closes our explorations of the traditional uses of microbes on the farm. However, microbes have many more agricultural roles. They are the source of novel insecticides; they have made herbicides more effective; they have been used to enhance frost resistance among crop plants; they have been used to create new foods; and they have provided the means to make farm animals more efficient producers of dairy products. We will see how in the next section.

15.2 Biotechnology on the Farm: The Coming of the Transgenic Plant

The techniques of genetic engineering and biotechnology have opened numerous possibilities for improving agriculture and the quality of life, possibilities unimaginable before the breakthroughs in molecular biology in the last half of the twentieth century. Biotechnology has found ways to dramatically increase crop yields, substantially decrease plant disease, and use plants for novel purposes. For example, rotavirus is the cause of severe, often life-threatening diarrhea in young children and infants in many developing parts of the world. In 2013 scientists forced the genes encoding human antibodies against the rotavirus into the chromosomes of rice plants. Simply by eating the rice, the individuals would obtain the antibodies, called "plantibodies."

DNA into Plant Cells

One of the breakthroughs in agricultural technology was the ability to cultivate a whole plant from a single plant cell. Such a cell can be genetically modified by adding the gene or genes of interest and then cultivated in a medium of carefully balanced plant nutrients and hormones. It will develop into a random mass of cells called a **callus**. Some days or weeks later, the forerunners of roots, stems, and leaves appear on the callus. Shortly thereafter, the plantlet is transferred to a container until ready for planting outdoors. The result is a **transgenic** plant; that is, a plant containing a foreign gene deliberately inserted into its genome. All such organisms having been genetically engineered in this way are referred to **genetically modified organisms** (**GMOs**).

Inserting fragments of DNA into plant cells requires the proper technology tools. A valuable method for DNA delivery is the Ti plasmid (short for tumor-inducing plasmid) obtained from the bacterium *R. radiobacter*. This microbe causes plant tumors and a disease called crown gall (FIGURE 15.8a). Crown gall develops when the bacterium releases its Ti plasmid in the plant cell and the plasmid inserts itself into one of the cell's chromosomes. For biotechnology experiments, Ti plasmids are removed from bacterial cells, modified to remove the tumor-inducing genes, and further modified by adding foreign genes (FIGURE 15.8b). Then, the plasmids are inserted into *R. radiobacter* cells,

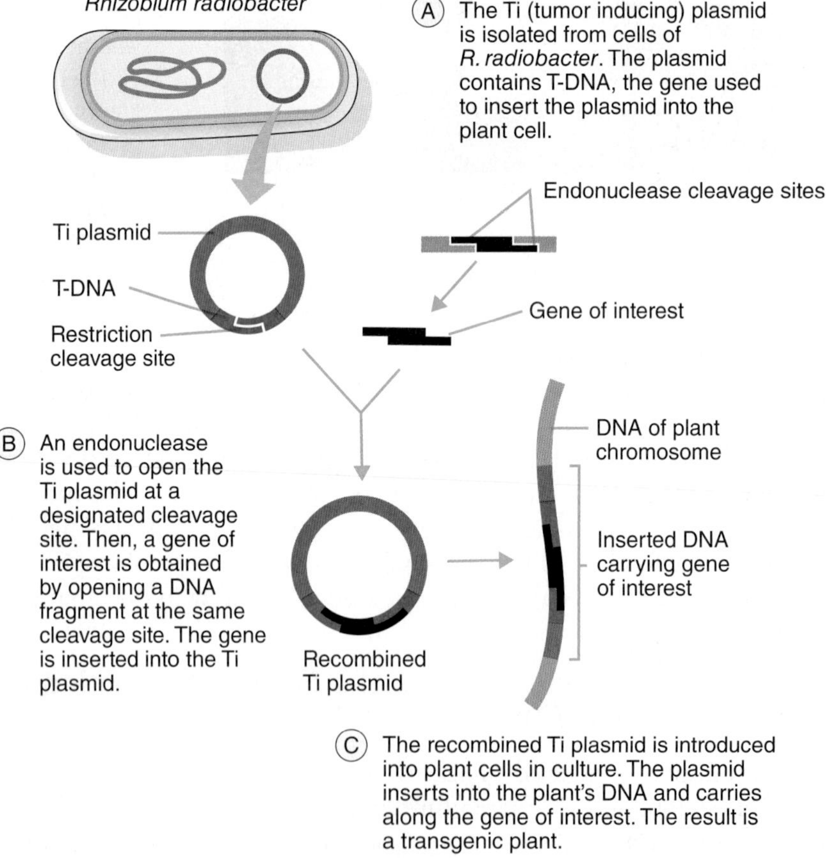

Rhizobium radiobacter
A The Ti (tumor inducing) plasmid is isolated from cells of R. radiobacter. The plasmid contains T-DNA, the gene used to insert the plasmid into the plant cell.
Endonuclease cleavage sites
Ti plasmid
T-DNA
Restriction cleavage site
Gene of interest
DNA of plant chromosome
B An endonuclease is used to open the Ti plasmid at a designated cleavage site. Then, a gene of interest is obtained by opening a DNA fragment at the same cleavage site. The gene is inserted into the Ti plasmid.
Inserted DNA carrying gene of interest
Recombined Ti plasmid
C The recombined Ti plasmid is introduced into plant cells in culture. The plasmid inserts into the plant's DNA and carries along the gene of interest. The result is a transgenic plant.

FIGURE 15.8 ***Rhizobium radiobacter* and the Ti Plasmid.** (**a**) *R. radiobacter* induces tumors in plants and causes a disease called crown gall. A lump of tumor tissue forms at the infection site, as the photograph shows. (**b**) The vehicle for infection is a Ti plasmid carrying DNA. This plasmid is used to carry a foreign gene into plant cells.

(a) Courtesy of Lian Bruno

and the bacterial cells are inoculated into the plant cell. The plasmids then transport the foreign genes into the plant cell chromosome.

Using the Ti plasmid system, plant biotechnologists have designed, developed, and carried out several successful improvements to plants, including resistance to insecticides and herbicides, and improvement of product quality.

Bacterial and Viral Insecticides

To be useful as an insecticide, a microbial toxin should act only on a particular targeted pest and should act rapidly. It should be stable in the environment and easily dispensed, as well as inexpensive to produce. Finding a microbe that produces such a toxin to fit these criteria has been an ongoing challenge for biotechnologists.

In the early 1900s, a scientist named G. S. Berliner found that cells of a newly identified *Bacillus* species could kill moth and butterfly larvae. Berliner named the organism *Bacillus thuringiensis* after Thuringia, the German province where he lived. The bacterial species remained relatively obscure until the 1980s, when scientists discovered that it produces toxic crystalline proteins while forming endospores (FIGURE 15.9a). When the toxic proteins are deposited on leaves, they are ingested by caterpillars. In the caterpillar gut, the protein breaks down the cells of the gut wall causing pores to form in the intestinal membranes. This allows the normal gut bacteria to invade the body cavity, resulting in death of the larvae (FIGURE 15.9b).

larva: A wormlike pre-adult stage of an insect.

Bacillus thuringiensis bä-sil'lus thur-in-jē-en'sis

caterpillar: The larval form of butterflies, moths, and related insects.

Bacillus thuringiensis (commonly shortened to Bt) is harmless to plants and mammals, including humans. It is produced by harvesting the bacterial cells at the onset of spore formation and drying them to make a commercially-available dusting powder. The product is used against tomato hornworms, corn borers, and gypsy moth caterpillars. Its success has encouraged further research and led to the discovery of a new and more powerful strain called *B. thuringiensis* israelensis (commonly known as Bti), first isolated in Israel.

But material sprayed onto plants can soon wash off, so the protective effect of Bt as an insecticide can be limited. Therefore, biotechnologists have sought to provide long-term protection by inserting the toxin-encoding gene directly into plant cells. To accomplish this, they have identified and cloned the gene and spliced it into the Ti plasmid of *R. radiobacter*. Then, they used the Ti plasmid to ferry the toxin gene into plant cells.

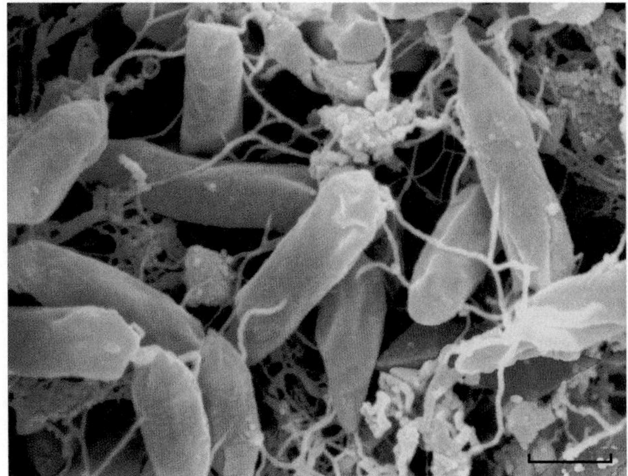

FIGURE 15.9 ***Bacillus thuringiensis* and Its Insecticide.** (**a**) *B. thuringiensis* cells produce crystalline toxins that (**b**) when sprayed on plants kill insect caterpillars that were eating the plant leaves. (Bar = 2 μm.)

The first seeds with inserted Bt genes were put in U.S. soil in 1996, and by 2001, over 35% of all corn planted in the United States consisted of genetically modified corn, popularly known as Bt corn. Losses due to corn borers have dropped dramatically. Genetically modified cotton (Bt cotton) has also been planted on millions of acres in the United States as well as many other countries. Although both crops display resistance to pests, scientists have noted a resistance to Bt toxin emerging in some caterpillars.

Still another approach has involved incorporating insect resistance in corn plants by splicing the Bt toxin gene into corn plants. When the seeds were sown, the toxin-producing plants grow, and when an insect eats the plant, it consumes the toxin and is soon killed. This approach is advantageous because only the insect attacking the plant is subjected to the toxin.

■ *Bacillus sphaericus*
bä-sil'lus sfe'ri-kus

■ *Asticcacaulis excentricus*
as-tik-ä-cäu'lis ek-sen'tri-kus

Another useful species is *Bacillus sphaericus*, a microbe whose toxin kills at least two species of mosquito larvae. To increase the efficiency of the toxin, researchers inserted the gene that encodes the toxin into the bacterium *Asticcacaulis excentricus*. Using *A. excentricus* as a gene carrier has several advantages: It is easy to grow in large quantity; it tolerates sunlight better than *B. sphaericus*; and it floats in water, where mosquitoes feed (the heavier, spore-laden *Bacillus* species sinks). Scientists were delighted to observe insecticidal activity by *A. excentricus* against mosquitoes that transmit such microbial diseases as malaria and viral encephalitis.

■ *Photorhabdus luminescens*
fō-tō-rab'dus lü-mi-nes'senz

Bacteriologists also are investigating the insecticidal potential of the bacterium *Photorhabdus luminescens*. The microbe's toxin, known as Pht, attacks the gut lining of insect larvae (as Bt does), and it is formed as large cytoplasmic crystals (also as Bt is). However, the spectrum of activity of Pht is wider than that of Bt, and it includes numerous species of caterpillars as well as cockroaches. Normally, *P. luminescens* lives in the intestines of soilborne worms called nematodes. The worms invade tissues of insects in the soil, and the toxins produced by *P. luminescens* kill the insects. The toxin-encoding Pht genes have been isolated, and efforts are underway to introduce them to plant cells.

Viruses also show promise as agricultural pest-control agents, in part because they are more selective in their activity than bacterial organisms. Pure preparations of viruses can be used, or biotechnologists can harvest infected insects, grind them up, and use them to disseminate the virus. Among the insects successfully controlled with viruses are the cotton bollworm, the cabbage looper, and the alfalfa caterpillar.

Researchers have also developed an insecticide using a toxin found in the venom of a scorpion. The toxin, which paralyzes the larvae of moths, is encoded by a gene isolated from the cells of the scorpion. To make the insecticide, this gene is attached to a **baculovirus**, a type of DNA virus having a high affinity for insect tissues. After spraying the genetically modified baculoviruses onto the plant foliage, caterpillars consume the virus while eating the foliage. In the caterpillar, the baculovirus DNA is expressed during replication and the scorpion toxin is produced, which kills the caterpillars. Baculovirus-based insecticides are currently being used worldwide.

Resistance to Herbicides

Herbicides are weed-killing chemicals used to clear the land of all plant growth before a field is sown. However, some weed seeds usually remain among the crop seeds. When planted, weeds and crops grow together and the weeds often rob the crop plants of vital nutrients, while crowding them out. It is generally agreed that it would be advantageous to increase the resistance of crop plants to herbicides, so that herbicide could be sprayed on the field during the growing season.

One commonly used herbicide is **glyphosate**, the active ingredient in commercial products such as Roundup®. Glyphosate inhibits the activity of enzymes that synthesize essential amino acids in plant chloroplasts. By a coincidence of nature, *Escherichia coli* cells possess a gene encoding an enzyme with the same synthetic function but more resistant to glyphosate. Biotechnologists have isolated this gene and inserted it into the Ti plasmid of *R. radiobacter*. Then, they used the plasmid to deliver the gene to the cells of tobacco and soybean plants. Here the genes encode the more resistant *E. coli* enzyme and make the plants more resistant to glyphosate than the weeds. Thus, when glyphosate is sprayed on the field, the weed plants die but the crop plants resist the glyphosate and live. In 2012, 93% of the U.S. soybean crop was planted with genetically modified (i.e., "Roundup-resistant") seeds (FIGURE 15.10). By 2011, besides the United States, which is the number one producer of soybeans, Roundup Ready soybeans were also being grown in several of the world's top soybean producing countries (by rank): Brazil (2), Argentina (3), Paraguay (6), Canada (7), Uruguay (8), Bolivia (10), and South Africa (13).

In summary, herbicide-resistant crops have resulted in a 239 million kilogram (527 million pound) increase in herbicide use in the United States between 1996 and 2011, while Bt crops have reduced insecticide applications by 56 million kilograms (123 million pounds).

Pharm Animals

Many of the products of gene-altered microbes have found their way into animals as scientists continue to experiment to improve the quantity and quality of our foods. Most experiments have involved inoculations with hormones produced by gene-altered bacterial species. Animals treated with these hormones and other pharmaceutical products have acquired the catchy name "pharm animals."

Among the first pharm animals was the dairy cow. As early as 1983, the gene for bovine growth hormone (BGH) was isolated and inserted into *Escherichia coli* cells. Soon the cells were secreting the hormone en masse and it was put into commercial production. Injected into beef and dairy cattle, the recombinant BGH (known as recombinant bovine somatotropin or rBST) promotes growth of bone and muscle and increases milk production. Increases in milk production of up to 25% were soon reported, and the FDA has declared that milk produced by hormone-treated cows is

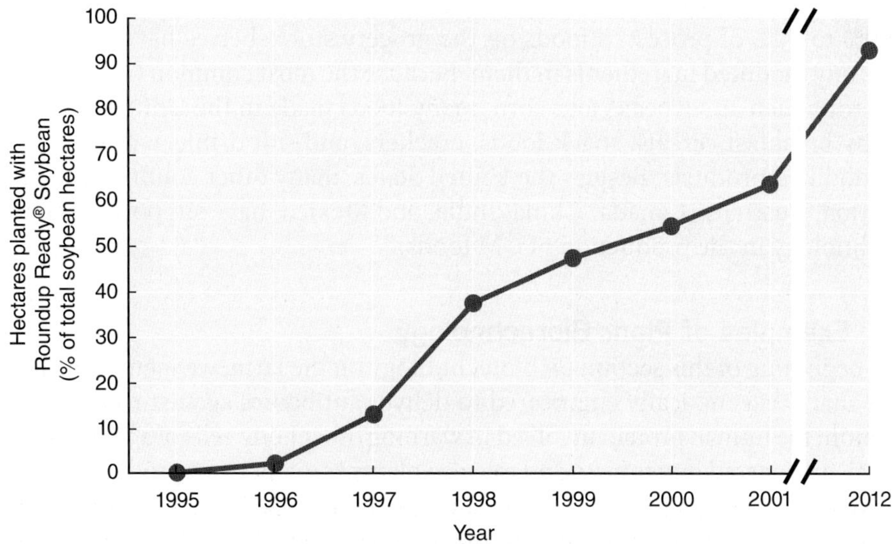

FIGURE 15.10 **Adoption of Roundup Ready® Soybeans in the United States.** Since its introduction in 1996, U.S. farmers have rapidly adopted genetically modified soybean to tolerate glyphosate.

Data are from USDA/Economic Research Service.

safe to drink. However, some people are concerned about rBST and the potential risk of developing cancer. Although the scientific evidence does not support this claim, many dairies have returned to marketing non-rBST milk.

Interesting results have been obtained by researchers attempting to ease wool collection from sheep. Biotechnologists have used a hormone called sheep epidermal growth factor, which is isolated from bacterial cells engineered with the gene to produce the protein. When the hormone is injected into the sheep, it weakens the follicles, causing them to loosen their grip on the wool strands. Several weeks later, the fleece comes off the sheep in a single, unmutilated sheet.

Genetically Modified (GM) Foods

The processes of winemaking, bread making, and food fermentation trace back to antiquity, but there are some imaginative uses of biotechnology in contemporary food production. One of the first **genetically modified (GM) foods** was the FlavrSavr tomato, which was engineered to slow the ripening process while still retaining flavor and color. Unfortunately, this tomato never made it to the market because of numerous technical problems (among them a skeptical public), but the molecular biology used in its development illustrates how innovative thinking can be put to practical use.

Failure of the biotech tomato did not deter scientists. They soon produced a new type of canola oil. Canola oil, extracted from canola seeds, is used in such things as ice cream, detergents, and facial creams. It is also valuable as a type of cooking oil because it contains relatively few saturated fats, an appealing characteristic to nutritionists. Biotechnologists improved canola oil by adding new genes: In the California bay laurel tree, they located a gene encoding lauric acid, and they introduced the gene into canola plant cells. Lauric acid is a fatty acid not normally found in canola oil. It gives added nutritional value to the oil and has the feature of not breaking down easily during cooking, an appealing factor to home cooks and chefs.

Biotechnologists are continuing to work on a host of other GM foods, including: a type of yellow squash that is resistant to two plant viruses (appropriately, the squash is called "Freedom 2") and virus-resistant papayas. Many other products have been developed but have not made it successfully due to consumer concerns over the safety of GM foods. In fact, GM foods are quite prevalent today in the American diet. Experts say 60% to 70% of processed foods on the grocery store shelves have some type of genetically modified ingredients in them. Because the most common GM foods come from crops such as soybeans and corn, many foods made in the United States, such as many breakfast cereals, snack foods, crackers, and soft drinks will have genetically modified products. Besides the United States, many other countries, including Argentina, Australia, Canada, China, India, and Mexico, have supported the use of biotechnology in the production of GM foods.

Other Examples of Plant Biotechnology

At the beginning of this section on biotechnology on the farm, we mentioned a strain of rice that was genetically engineered to deliver antibodies against rotavirus. Also, interferon, the human protein involved in warning other body cells of a virus infection, has been engineered into tomato and tobacco plants as a source for producing the antiviral protein. These are just two more examples of other uses for plant biotechnology.

Today, researchers continue to report encouraging results on the use of transgenic vegetables and fruits as carriers of microbial vaccines. Whimsically called "veggie vaccines," fruit and vegetable carriers are desirable because they are inexpensive to store and are readily accepted in developing countries, where vaccines are needed most.

The vaccines could be produced locally (thus avoiding transportation and refrigeration problems), and vaccine production would go on indefinitely because the plants could be propagated continuously. Moreover, use of the vaccines would not necessitate exposing people to potentially contaminated syringes. And the vaccines would be safer than vaccines containing whole organisms or their parts.

Although we do not yet have a good malaria vaccine, biotechnologists are working on engineering proteins from the causative agent, *Plasmodium*, into the tobacco mosaic virus. The virus would be allowed to infect tobacco plants in which massive amounts of the *Plasmodium* protein would be produced. The plants could then be harvested and the immune-stimulating *Plasmodium* protein made into the vaccine.

Scientists are also studying ways to employ more palatable plants such as potatoes, tomatoes, and bananas as sources of "edible vaccines" (FIGURE 15.11). For example, researchers are exploring the use of potatoes as a vaccine source. In a proof of concept, scientists have inserted a benign segment of toxin genes of pathogenic *E. coli* into potato plants. When the plants grew and chunks of potato were eaten by mice, an antibody response against the toxin was generated. Unfortunately, the potatoes had to be eaten uncooked because cooking reduced the immune response. To resolve this dilemma, researchers need to develop a more heat-stable vaccine form.

FIGURE 15.11 **"Edible Vaccines."** Several fruits and vegetables are being used to develop vaccines that one could eat to obtain immunity to specific infectious diseases.

Courtesy of Dr. Jeffrey Pommerville

Bananas have attracted interest because children like them (children are the main recipients of vaccines); they grow in many tropical countries (where vaccines are essential to public health systems); and they don't have to be cooked before consumption. In fact, scientists have already spliced genes from hepatitis B viruses into banana plant cells and are attempting to induce the plant to produce hepatitis B proteins that will stimulate an antibody response. Ultimately, they hope to engineer a "vaccine banana" where one could be immunized against hepatitis B or any of several other infectious diseases, like gastrointestinal diseases and cholera, by eating a banana.

As expected, using vegetables and fruits as vaccine carriers introduces a unique set of problems that must be addressed: How can researchers ensure the passage of immune-stimulating substances through the acid of the stomach and the enzymes of the small intestine? What methods can be used to encourage passage of the substances from the gastrointestinal tract to the bloodstream, where the immune reaction occurs? Will the vaccine components be strong enough to elicit a protective antibody response? Will vaccine components combine with body proteins and elicit allergic reactions? Despite these obstacles, biotechnologists have great hopes for transgenic vaccines. Unfortunately, many years of laboratory tinkering and many clinical trials will pass before the banana or potato replaces the needle as a vehicle for delivering vaccines. Still, perhaps one day mothers will be telling their children "Eat your vaccines!" rather than "Eat your vegetables!"

A Final Thought

Although still in its infancy, agricultural biotechnology has breathtaking possibilities for improving human health and nutrition. Genetically modified foods stand at the forefront of this revolution. But no revolution is without controversy, and the critics of GM foods are numerous and loud. The American government's position is that genetically engineered crops are safe, they resist pests and disease better, and, as such, more food can be produced to feed starving nations. However, the European Union (EU) has banned all imports into Europe of GM food products. The EU believes there may be a risk that GM foods could harm the public and have negative environmental consequences. They and others point to the potential for allergic reactions in consumers, the possibility of gene transfers in the environment to create "superweeds," and the death of unintended animal victims of insecticides in nature. Another concern of some is the lost profits in the pesticide and herbicide businesses. Unfortunately, GM foods have sometimes been labeled "Frankenfoods" because crop plants have been modified in the laboratory to enhance desired traits, such as resistance to herbicides or improved nutritional content.

Supporters of GM foods respond by saying no food is 100% safe, whether it is genetically modified or not. In fact, most foodborne illnesses today do not come from GM foods. To date, the reports of adverse reactions to a genetically modified food are few and therefore the risks are outweighed by the benefits. They point to a burgeoning world population that must be fed—a population of over 7 billion people in 2013, and an estimated 10 billion by 2050—and suggest biotechnology can dramatically improve productivity where food shortages arise from pest damage and plant disease. They point to the steady decline in the world's arable land, a decline that will accelerate precipitously in the years ahead, and they further suggest biotechnology can raise crop yields in developing countries by over 25%. And they mention other nutritional projects such as "golden rice." Traditional rice lacks sufficient amounts of vitamin A. In areas of the world where rice is the major staple in their diet, vitamin A deficiency is widespread and can lead to vision problems and, in some cases, blindness. Golden rice supplies a precursor of vitamin A as well as increased iron.

But, as with any technology, caution should be taken and, in this specific case, GM foods need to be subjected to strenuous testing before they are widely used in agriculture.

Questions to Consider

1. This chapter relates much about the innovative and imaginative work being done in agricultural biotechnology. Suppose you were the head of a lab and had the choice of pursuing any project you wished. What would that project be? Why would you pursue it? What would be the chances of your success?
2. A carton of yogurt usually carries the message "contains active cultures." However, knowledgeable individuals know that the wording should really be "contains live bacteria." Why do you suppose yogurt manufacturers perpetuate the minor deception, and what can be done to get them to be more transparent?
3. It is intriguing to stop and wonder how microbes came to be where they are. How, for example, did microbes find their way into the rumen of cattle and evolve to become as important as they now are? And what if they had found their way inside some other type of animal? Suppose they had found their way into the human intestine? What answers might you give to these questions?

4. Italians are fond of a dish called pásta e fagiòle. The dish consists of pasta with beans, chick peas, or other legume. Italians know that if they eat enough pásta e fagiòle, they don't have to eat quite so much meat, which is often in limited supply. Microbiologically speaking, why does this idea make sense?
5. Scientists have noted that one of the side effects of the use of Bt toxin has been a reduction in the population of Monarch butterflies. Why do you think this may have happened, and what might you as a concerned citizen do about it?
6. War of the Worlds is a classic novel by H. G. Wells, made into several movies. The story details the invasion of Earth by aliens from the planet Mars. All the might and resources of earthlings are exhausted as the aliens make their way through American cities destroying everything in their path. When all appears lost, the aliens suddenly die and their space vehicles crash to Earth. After a pregnant pause, the narrator solemnly explains, "In the end, the invaders were exposed to 'germs' in the Earth's atmosphere and died." Although War of the Worlds is fiction, nevertheless, a case can be made for microbes contributing mightily to the well-being of society. In what ways has this chapter supported their case?
7. Futurists tell us that a great variety of genetically modified foods will soon be appearing in the supermarkets of the world. If you had your choice, what sort of improvements would you like to see in foods? What foods would you change to improve characteristics such as flavor, shelf life, and nutritional quality? What genetic changes would be necessary to effect these improvements?

Key Terms

Informative facts are necessary for the expression of every concept, and the information for a concept is founded in a set of key terms. The following terms form the basis for the concepts of this chapter. On completing the chapter, you should be able to explain and/or define each one.

anaerobic
bacteroid
baculovirus
callus
denitrification
diacetyl
genetically modified (GM) food
glyphosate
mixed fermentation
nitrification
nitrogenase
nitrogen cycle
nitrogen fixation
rumen
ruminant
silage
symbiotic
transgenic

16 Microbes and the Environment: No Microbes, No Life

Microbes live just about anywhere you want to look on planet Earth. From miles down beneath the surface to miles up in the atmosphere, microbes can be found. They live in Earth's coldest environments in Antarctica to areas around Earth's boiling hot volcanic fields. They thrive on all types of Earth's nutrients, including oil, toxic wastes, and rock. Every time you step on the soil, you step on billions of microbes. But in all these and other Earth environments, microbial metabolism is returning nutrients from dead organisms and nonliving substances to forms used by plants and animals. In turn, plants and animals feed the world. From microbes to man, planet Earth represents a superorganism.

Looking Ahead

It may appear brash to suggest that microbes are among the most important participants in the web of life on Earth, but the pages of this chapter will provide ample opportunities for you to evaluate the validity of that assertion.

On completing this chapter, you should be capable of:

- Identifying the essential roles of microbes in (a) the carbon cycle, (b) the sulfur cycle, (c) the nitrogen cycle, (d) the phosphorus cycle, and (e) the oxygen cycle.
- Describing the three steps of the waste treatment process.
- Identifying the sources of water pollution.
- Explaining the water treatment process.

During the 1960s, the renowned scientist and writer Buckminster Fuller described our planet as "Spaceship Earth." Fuller envisioned Earth as a self-sustaining entity whirling through space, completely isolated from everything but sunlight. He hoped to convey the notion that some of Earth's resources—such as oil and gas—are nonrenewable, but most others are renewable as long as they are recycled.

An essential element of Fuller's vision is that Spaceship Earth depends for life-support on life itself: The living organisms aboard this great planet produce its oxygen, cleanse its air, adjust its gases, transfer its energy, and recycle its waste products—all with great efficiency.

High on the list of key passengers are the microbes. As we will see in this chapter, many important resources of Spaceship Earth could not be recycled without microbial intervention. In fact, without microbes, the plants and animals of Earth would have depleted its resources countless eons ago, and the great experiment of life would have failed miserably. However, life did evolve, which is testament to the power and adaptation of microbes: They have managed to fill every conceivable niche on Earth; they have evolved so they can sustain themselves on anything Earth has to offer; and they have adapted to participate in the intricate web of metabolic activities that permits Spaceship Earth to continue on its long journey through the universe.

16.1 The Cycles of Nature: What Goes Around, Comes Around

Take a pinch of rich soil and hold it in your hand. You are face to face with an estimated billion microbes representing over 10,000 different microbial species. If you were to try cultivating them in the laboratory, perhaps only a few species would grow. The remaining thousands of species would remain unknown until scientists developed the right combination of nutrients and environmental conditions to cultivate them or the molecular techniques to identify them. But our knowledge about them is immaterial to how important they are to us: They recycle our nutrients, refresh our environment, clean our waste, and stand as vital links between life as we know it and a world we cannot begin to imagine. Indeed, because microbes are invisible, their importance in the environment may go unsuspected. Yet, without microbes, life could not continue, as we will see in the pages ahead.

The Earth Ecosystem

Because of their ubiquitous distribution and their diverse chemical activities, microbes provide the basic underpinnings to the cycles of elements on Earth. But they do not operate alone. Rather, they act as part of an **ecosystem**, a community of all the organisms interacting with the nonliving components within a defined space. Certain ecosystems (such as lakes) are fairly small, but other ecosystems (such as the entire North American continent) are enormous. Whatever its size, an ecosystem is a type of "superorganism" having the ability to respond to and modify its environment. Although the lake may be easier to see as an ecosystem, the continent is no less an ecosystem. It has even been suggested that all the microbes on Earth may constitute a single ecosystem—all living in support of one another and the planet (**A Closer Look 16.1**).

Within the ecosystem that is the Earth, the sum total of all living organisms is known as the **biota**. Influenced by physical characteristics such as rain and tides, wastes and sewage, and oxygen and pH, the biota includes unknown numbers of species of microbes as well as hundreds of thousands of species of plants and animals. How these organisms interact with one another and with their physical and chemical environments is the subject matter of **environmental microbiology**.

The physical space or location where a species of microbe lives is called its **habitat**. Habitats may be freshwater lakes, marine environments, or soils. Soil is a complex mixture of mineral particles, along with organic matter from decaying organisms and plants, animals, and microbes. Within this habitat, microbes, primarily bacterial and fungal species, occupy an important niche as the preeminent recyclers of nutrients.

Microbes live together with other organisms in a **community**. The two major elements of a community are energy and chemical substances. Energy enters the ecosystem through sunlight and is used by photosynthetic organisms (cyanobacteria,

A CLOSER LOOK 16.1

"All for One and One for All"

The motto *"All for one and one for all,* united we stand divided we fall" is a line written by Alexandre Dumas in *The Three Musketeers*. It refers to D'Artagnan and his three inseparable friends (the musketeers) who lived life by that motto (see figure). Here's an intriguing thought: Suppose a bacterium was not a single-celled creature, but rather, a subunit of a global organism consisting of the entire bacterial world.

Statue of d'Artagnan and the Three Musketeers at Condom, Midi-Pyrénées, Pyrenees, France.

Such a view may not be quite so preposterous, according to Sorin Sonea of the University of Montreal. Sonea points out the vast majority of bacteria live in communities of mixed strains whose metabolisms complement one another. Such an association may be viewed as a sort of multicellular eukaryote. The communities of bacteria appear to improvise in nature and adapt to changes in the environment; for example, certain strains emerge and others disappear. The difficulty in cultivating various strains in the laboratory is a reflection of their metabolic dependence on one another.

Sonea continues his hypothesis by pointing out that a bacterium can only survive in nature as part of a temporary metabolic association. Such an association makes the bacterium different from the relatively independent eukaryotic cell. Essentially, then, each bacterium is part of a widely dispersed planetary system of subunits, all metabolically linked and all dependent on one another for the common good.

Do you agree with Sonea's hypothesis, or can you find fault with it? What evidence can you offer to support or reject the hypothesis? You might also enjoy investigating the Gaia hypothesis, first stated by James Lovelock. (Gaia is the mythical goddess of Earth.) Lovelock envisions Earth as one "superorganism" and living things as one metabolizing creature that is continually reshaping the landscape to fit its needs. The existence of one "superbacterium" might conceivably fit into this pattern. Certainly, the idea is novel enough to merit more than a passing moment of attention.

algae, and green plants) for the synthesis of organic matter. These organisms, known as **producers**, are then used as nutrients by other organisms called **consumers**. Much of the energy is lost as heat during respiratory processes going on in these organisms, and as primary consumers are in turn eaten by secondary consumers, the energy continues to be lost. Microbes and other feeders, collectively known as **decomposers**, attack the consumers and producers after they die and use the available energy for their metabolism. Eventually, the energy that entered the ecosystem as sunlight is dissipated, primarily as heat energy (FIGURE 16.1).

But the dissipated energy is constantly being replaced by new energy from the Sun, and life will continue as long as the chemical substances are recycled. In the case of carbon, for example, the chemical processes of respiration release carbon from carbohydrates in the form of carbon dioxide gas. The latter can then be used in photosynthesis and recycled back into the ecosystem. Nitrogen, sulfur, and other essential elements are also released during metabolic processes, and thus are made available for reentry into the ecosystem. This is the basic concept underlying the **biogeochemical cycles**; that is, basic elements occurring in living organisms move through the environment by a series of naturally occurring physical, chemical, and biological processes. In the next several pages, we will emphasize the roles that microbes fill and

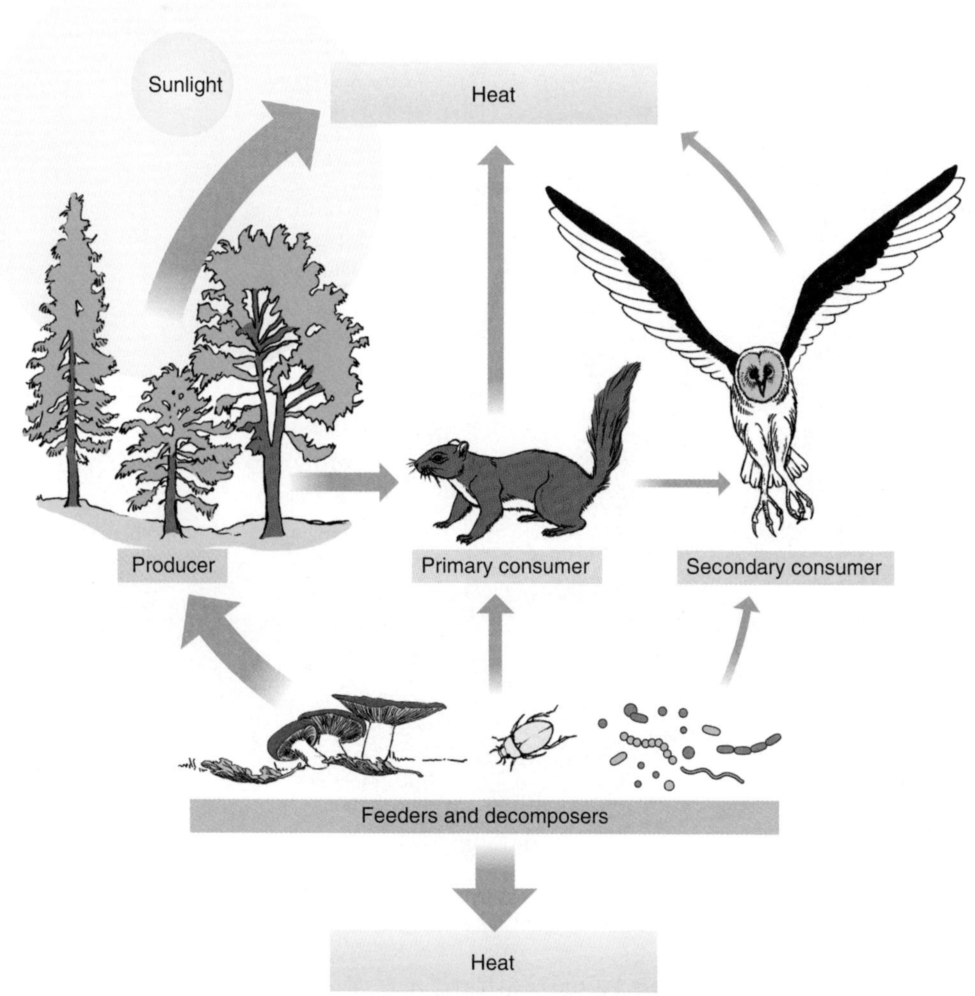

FIGURE 16.1 **The Pathway of Energy in a Forest Community.** This diagram illustrates the entry of energy into the community and its gradual loss during transfers between levels in the ecosystem. The width of the arrow represents the relative quantity of energy lost during each transfer.

reemphasize the insight first enunciated by Louis Pasteur over 100 years ago: "*The influence of the very small is very great, indeed.*"

The Carbon Cycle

Planet Earth is composed of numerous elements and carbon (C) is one of the most important to all life. Because there is a defined amount of carbon on Earth, the **carbon cycle** constantly recycles the carbon between the atmosphere, land, and ocean and encompasses nearly all life. As shown in FIGURE 16.2, photosynthetic organisms take carbon in the form of carbon dioxide and form carbohydrates using the sun's energy and chlorophyll pigments. The vast jungles of the world, the grassy plains of the temperate zones, and all the algae and cyanobacteria in the seas display the results of this process. Photosynthetic organisms, in turn, are consumed by animals, fish, and humans; these creatures use some of the carbohydrates as energy sources and convert the remainder to cell parts. Although some carbon is released by respiration and returns to the atmosphere as carbon dioxide, the major portion of the carbon is returned to the soil when an organism dies.

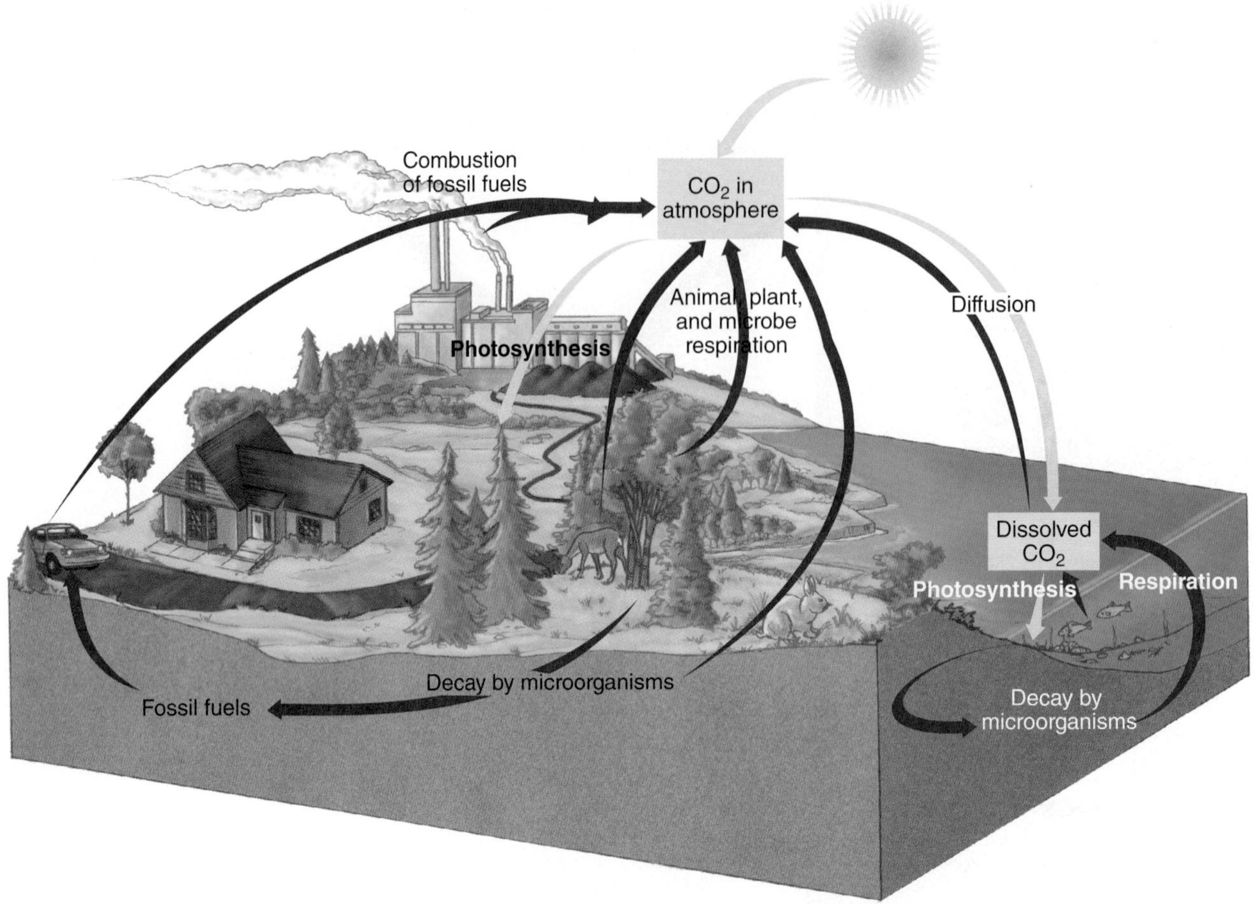

FIGURE 16.2 **A Simplified Carbon Cycle.** Photosynthesis represents the major process by which carbon dioxide is incorporated into organic matter, and respiration accounts for its return to the atmosphere. Microorganisms are crucial to all decay in soil and ocean environments.

Notice in Figure 16.2 that in the soil, microbes exert their influence as the primary decomposers of organic matter. Working in their countless billions, bacteria and fungi consume organic matter and release carbon dioxide for reuse by plants. This activity is carried on as a concerted action by a huge variety of microbes, each with its own nutritional pattern of protein, carbohydrate, or lipid digestion. Aerobic decomposition processes lead to the release of carbon dioxide through the reactions of cellular respiration. A similar process occurs in the oceans, known to marine biologists as "carbon sinks." Here the **phytoplankton** (the free-floating microscopic cyanobacteria and unicellular algae) and the **zooplankton** (the protists and microscopic animals) are important participants in the cycle and bring carbon dioxide back into the system through photosynthesis.

But not all the processes in the carbon cycle are aerobic. To the contrary, chemical decompositions take place in anaerobic environments such as the muddy bottoms of lakes and waterlogged soils, where archaeal species produce methane gas (CH_4). As the CH_4 moves to zones above the sediment that are rich in oxygen, other microbes degrade the CH_4 to carbon dioxide (CO_2) and water (H_2O). Without the action of these species, methane would probably enter the atmosphere and enhance the greenhouse effect to the point that the Earth might become too hot for life to continue.

Hopefully, you are beginning to realize that without the microbes our planet would be a veritable garbage dump of accumulating animal waste, dead plants, and organic debris. Operating in soil and water, microbes break down animal and plant waste and contribute the greatest share of organic matter to the environment. In the soil, much of the organic matter remains unused, in part, because the microbes are killed by the toxic products of the decay. This left-over organic matter then combines with mineral particles to form **humus**, a dark-colored material that retains air and water and is excellent for plant growth.

Microbes accomplish a similar feat under controlled conditions, where they decompose manure and other natural waste materials and convert them to **compost** as crop fertilizers. Although both conventional and organic farms use manure and compost as soil fertilizers, only certified organic farmers are required to have a plan detailing the methods for building soil fertility with these materials. Furthermore, certified organic farmers are prohibited from using raw manure for at least 60 days prior to harvesting crops for human consumption.

Scientists believe coal, being a fossil fuel, was formed 360 to 290 million years ago when plants in marshes and lakes became buried. Under high temperatures and pressure, the microbial decay process was interrupted and the vegetation was transformed into peat and then to coal. Petroleum may have originated in the same way when the organic matter of the phytoplankton and zooplankton in brackish waters sank to the bottom and was incorporated in clay sediments that became sedimentary rocks. This so-called oil shale, again under high pressure and temperature, had the oil (i.e., petroleum) squeezed out into the porous rock.

Again notice in Figure 16.2 that when coal or petroleum is burned, carbon dioxide is released back into the atmosphere. In fact, petroleum buried underground where there is no oxygen does not decay, but when it is brought to the surface, it is used as a food source by aerobic microbes that attack its components and break them down. This microbial growth is an important aspect of petroleum microbiology. For example, jet fuels can support a flourishing microbial population, and the microbes tend to clog fuel strainers and interrupt the flow of fuel to the jet engines, a potentially disastrous situation.

And there is more. Microbes also break down the carbon-based chemicals produced by industrial processes, including herbicides, pesticides, and plastics. Some bacterial species even release the carbon locked up in organic compounds in sedimentary rock called shale, thereby filling another gap in the global cycle of carbon.

The Sulfur Cycle

The **sulfur cycle** may be defined in more specific terms than the carbon cycle. Sulfur (S) is a key constituent of such amino acids as cysteine and methionine, which are important components of proteins. Proteins are deposited in water and soil as living organisms die, and bacterial and fungal species decompose the proteins and break down the sulfur-containing amino acids to yield various molecules, including hydrogen sulfide (H_2S), as shown in FIGURE 16.3. Other bacterial species accomplish this conversion anaerobically (without oxygen) while growing in water-logged soils and muds. The hydrogen sulfide they produce reacts with iron to yield iron disulfide (FeS_2), which is black. Anaerobic muds tend to be black for this reason. In addition, hydrogen sulfide has the odor of rotten eggs, and swamps, marshes, and landfills often give off this odor.

Other anaerobic conversions involve several bacterial genera. Look again at Figure 16.3 and notice that these genera release elemental sulfur (S) from hydrogen

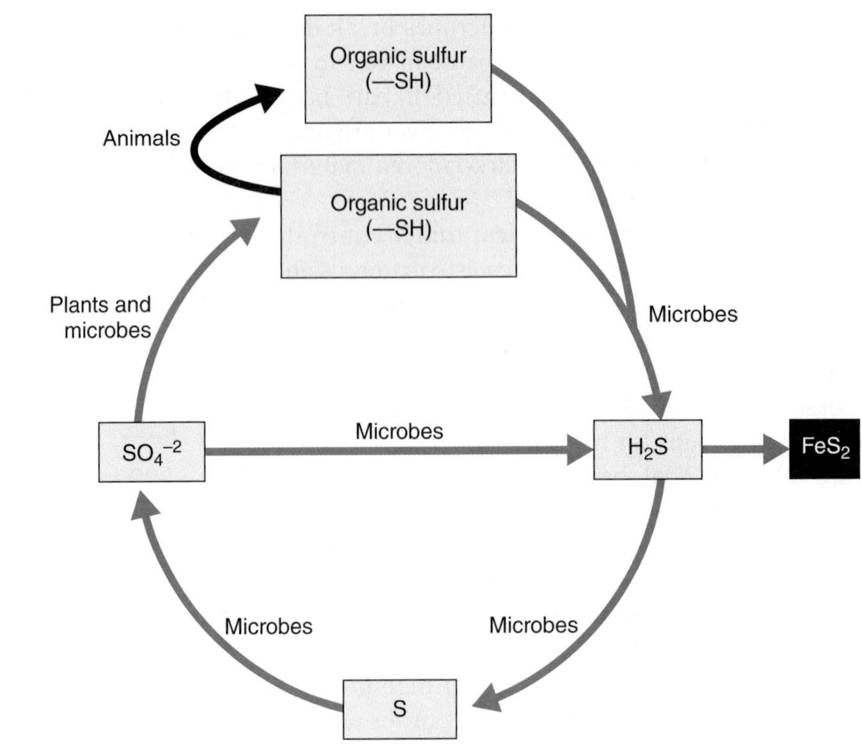

FIGURE 16.3 **A simplified Sulfur Cycle.** Most of the conversions of sulfur are dependent on microbes.

sulfide during their metabolism and convert the sulfur into sulfate (SO_4). The sulfate now is available to plants and microbes, where it is assimilated into the sulfur-containing amino acids. Further assimilation by animals (including humans) completes the cycle.

Other bacteria, the colorless sulfur bacteria, convert hydrogen sulfide into sulfate ions under aerobic conditions and derive energy from this conversion. These sulfur bacteria grow in acid-mine waters, where they break down the iron disulfide in pyrite ("fool's gold") and produce sulfuric acid. Draining from a mine, this sulfuric acid poses a serious environmental problem as an insoluble yellow precipitate that pollutes streams and rivers, making them unsightly and toxic (FIGURE 16.4).

FIGURE 16.4 **Acid Mine Drainage.** The discharge in a stream draining from a mine is extremely acidic (pH 2) and its orangish color is due to iron and copper dissolved in the water.

Courtesy of D. Hardesty/USGS

Sulfur-metabolizing species also have been linked to a deep-sea phenomenon. Oceanographers have found undersea volcanoes associated with a series of rifts (cracks) in the ocean floor. Along the rifts, hydrothermal vents spew black smoke. To their amazement, the area is rich in animal life, particularly giant clams and 2-meter-long tubeworms. It was unclear how these animals could survive in the hellish environment until sulfur-using bacterial species were located in their tissues. Apparently, the bacterial partners metabolize sulfur compounds in the smoke and use the energy they obtain to turn carbon dioxide into organic matter for use by the animals. In return, the tubeworms, clams, and other animals provide a hospitable environment in their specialized organs, where the bacterial species can carry on their biochemical processes while receiving the sulfur and oxygen they need. And all this takes place under extraordinarily high pressure, and where the temperature can be around 120°C (remember, water boils at 100°C).

The Nitrogen Cycle

The **nitrogen cycle** is described at length in another chapter, but we must mention it here because the impact of microbes on the cycling of nitrogen (N) is considerable. Indeed, the cyclic movement of nitrogen is of paramount importance to life on Earth because this element is an essential part of nucleic acids and proteins. Although nitrogen is the most common gas in the Earth's atmosphere, neither animals nor any but a very few species of plants can use it in its gaseous form (N_2). For nitrogen to enter the cycle of life, microbes must intercede (FIGURE 16.5).

After organic waste is deposited in the environment, the process of decay by soil microbes yields various organic nitrogen substances, including a mixture of amino acids. These organic nitrogen sources are converted by some bacterial species into inorganic compounds and ammonia (NH_3). Then, nitrogen recyclers, the so-called nitrifying bacteria, enter the picture. Much of the ammonia is converted to nitrites (NH_2^-) by *Nitrosomonas* species and the nitrites are then converted to nitrates (NO_3^-) by species of *Nitrobacter*.

■ *Nitrosomonas* nī-trō-sō-mō'näs

■ *Nitrobacter* nī-trō-bak'tėr

Nitrate is a crossroads substance: Some is used by plants for their nutritional needs, but most is broken down by various species of cyanobacteria and denitrifying bacteria and liberated to the atmosphere as nitrogen gas. A reverse trip back into living things is an absolute necessity for life as we know it to continue. The process is called **nitrogen fixation**. Once again, microbes, specifically nitrogen-fixing bacteria, play a key role because they possess the enzyme systems to trap atmospheric nitrogen and convert it to compounds useful to plants; that is, ammonia and nitrate. Free-living nitrogen-fixing soil microbes are involved in nitrogen fixation as well as symbiotic species such as *Rhizobium*, which infects and lives within the roots of leguminous plants. In the roots, the bacterial cells fix nitrogen and make nitrogen compounds

■ *Rhizobium* rī-zō'bē-um

■ leguminous plant: A plant that bears seeds in pods, such as peas, alfalfa, peanuts, beans, and soy beans.

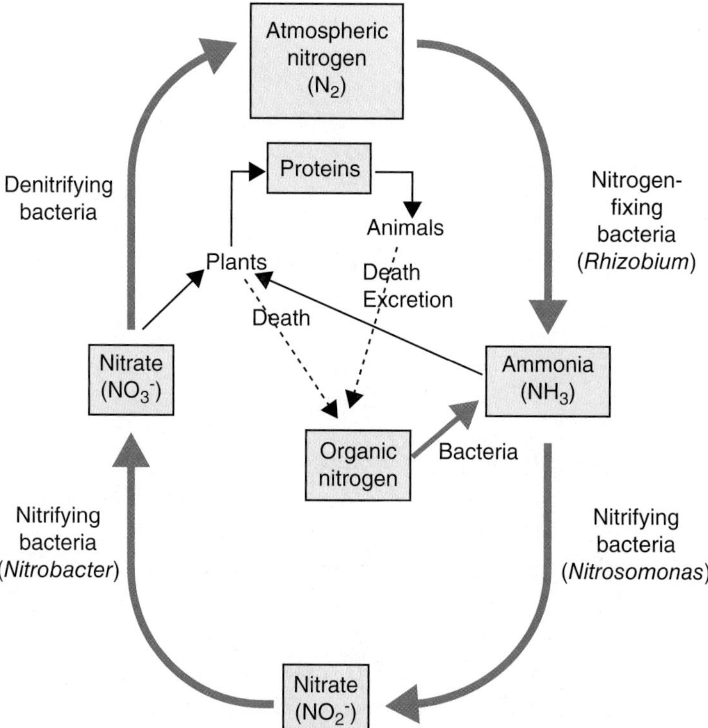

FIGURE 16.5 **A Simplified Nitrogen Cycle.** Plant and animal protein and metabolic wastes (organic nitrogen) are decomposed by certain bacterial species into ammonia. The ammonia may be used by plants, or it may be converted by *Nitrosomonas* and *Nitrobacter* species to nitrate, which also is used by plants. Some nitrate is broken down to atmospheric nitrogen. This nitrogen is returned to the leguminous plants by nitrogen-fixing microorganisms as ammonia.

available to the plant. Without this association, plants would have a difficult time making proteins, although today most crop plants are fertilized with chemical fertilizers containing nitrates.

The importance of microbes in the nitrogen cycle cannot be overstated: Without the vital microbial connections, there would be little opportunity for nitrogen to enter the cycle of life; nor would there be any way for nitrogen to be blended into the nutrients living things must have. There would be few amino acids, few proteins, few enzymes, few structural materials, and few of anything built around nitrogen. In the end, all plant and animal life would disappear.

The Phosphorus Cycle

The element phosphorus (P) has a place in the chemistry of living things in such molecules as nucleic acids, coenzymes, and the all-important energy molecule ATP (adenosine triphosphate). One portion of the **phosphorus cycle** begins when exposed rocks are worn away by rushing water (rain, streams, waterfalls, and so forth) (FIGURE 16.6). The phosphorus enters the sea in the form of dissolved phosphates (PO_4), ready to enter the cycle of life. Microbes come into play right at the outset. Unicellular algae and other microbes take up the phosphorus as they multiply in the waters. These relatively simple producers manufacture the all-important nucleic acids and other phosphorus-rich compounds. The phosphorus compounds are concentrated as the algae are consumed by other microbes, which are then eaten by shellfish and fin fish. These animals feed upon one another and use the phosphorus for organic molecules and such body parts as bones and shells. The death of these animals returns the phosphorus to the sea.

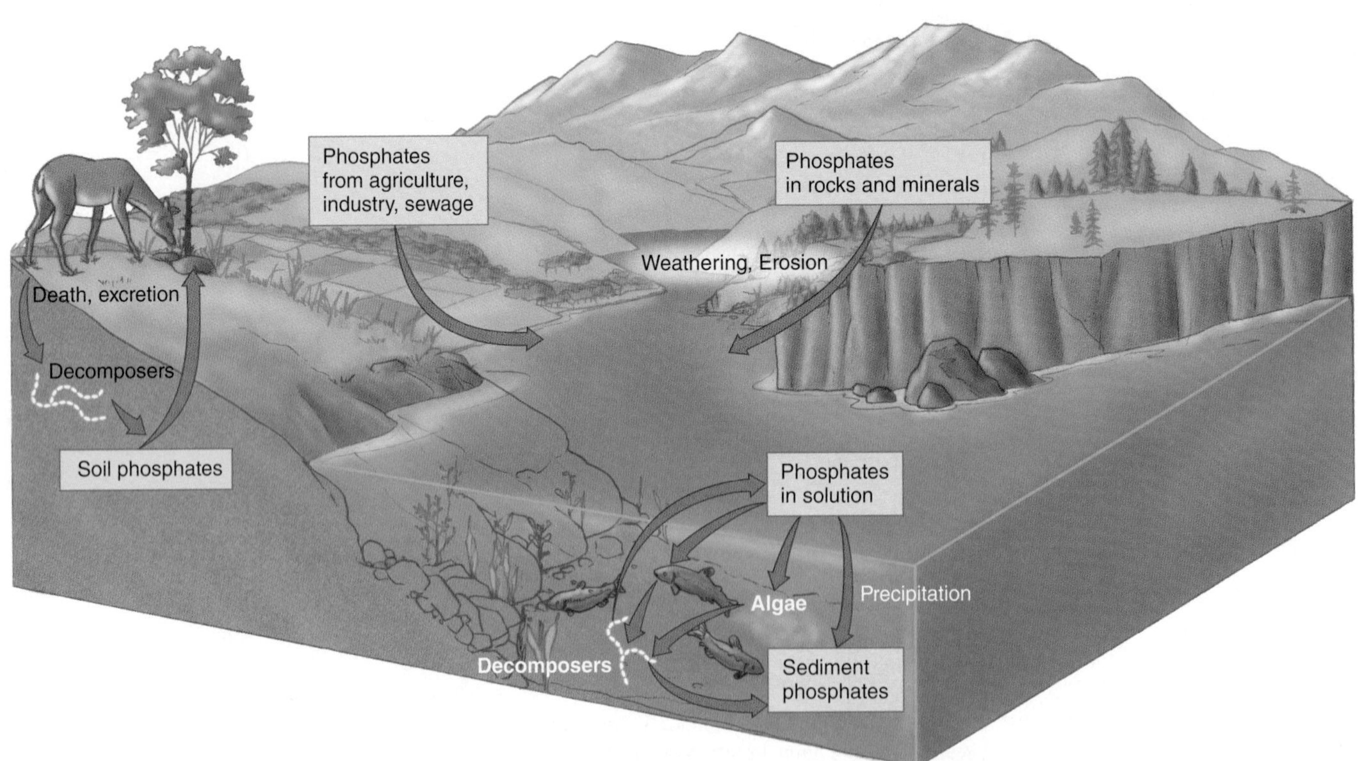

FIGURE 16.6 **The Phosphorus Cycle in Nature.** Phosphorus enters the cycle as phosphate from various points, and waterborne microbes incorporate it into nutrients at the base of the food chain.

Another portion of the cycle begins with phosphorus mined from rocks and used in fertilizer for the soil. Field crops incorporate the phosphorus and the crops are eaten by animals. In time, animal waste and decomposition as well as soil erosion bring the phosphates to the sea, where some solidifies to rock. In this form, it would be unavailable for eons of time. Fortunately, microbes live in the sea. As above, the unicellular algae and other microbes trap much of the phosphates for reuse in the cycle of life. In doing so, they keep the phosphorus cycling and the "wheel of life" turning.

The Oxygen Cycle

To appreciate the **oxygen cycle**, realize the next breath you take probably contains oxygen molecules produced by microbes. Billions of years ago, cyanobacteria started releasing oxygen molecules (O_2) into the atmosphere, and they have been performing that task unceasingly to this moment. Although we often thank green plants for their contribution to our oxygen needs, we should also extend kudos to the cyanobacteria. Cyanobacteria absorb water molecules during photosynthesis, break them down to retrieve the protons, and release the leftover oxygen atoms into the air as oxygen gas. To be sure, plants and algae perform the same type of chemistry, but the amount of oxygen they give off is equal to that of the cyanobacteria.

Animals, plants, and other microbes complete the oxygen cycle during the chemistry of cellular respiration. At the end of the respiratory process (described in another chapter), these organisms use oxygen as a final electron acceptor; they combine it with protons and form molecules of water. Bound in water molecules, the oxygen becomes useless for respiration, and our world would change dramatically if it were not released as the free gas. Not to worry—the plants do their share, but equally important players in the game of photosynthesis are the cyanobacteria. No cyanobacteria would mean much less oxygen gas, which would support fewer plants and animals.

In summary, it should be clear that microbes occupy a prominent position in the elemental cycles of life. To be sure, they keep organic matter fresh and vital as they sustain Spaceship Earth. But the work of microbes does not end here. Microbes also have a key place in preserving Earth's environment. They are the waste digesters, water purifiers, and soil cleansers par excellence. In these respects, they help society solve its age-old problem of how to keep the environment a fit place to live. We rely on microbes even more than we realize, as the next section illustrates.

16.2 Preserving the Environment: Reducing the Human Footprint

The traditional adage is: "*The solution to pollution is dilution.*" This somewhat simplified view recognizes the fact that industrial, agricultural, and human waste can be added to bodies of water and the water will distribute the waste and remain pure. For a small village situated along a quick-flowing river, the adage may be true; but for an industrialized society, the intervention of microbes is a necessary corollary.

Sanitary and Waste Facilities

Sanitation came into full flower in the mid-1800s, as a relatively modern phenomenon. Before that time, living conditions in some Western European and American cities were almost indescribably grim: Garbage and dead animals littered the streets; human feces and sewage stagnated in open sewers; rivers were used for washing, drinking, and excreting; and filth was rampant. Conditions like these fueled the Great Sanitary Movement of the mid-1800s, as we explore in **A Closer Look 16.2**.

A CLOSER LOOK 16.2

The Great Sanitary Movement

In the early 1800s, the steam engine and its product, the Industrial Revolution, brought crowds of rural inhabitants to European cities. To accommodate the rising tide, row houses and apartment blocks were hastily erected, and owners of already crowded houses took in more tenants. Not surprisingly, deaths from typhoid fever, cholera, tuberculosis, dysentery, and other diseases mounted in alarming proportions.

As the death rates rose, a few activists spoke up for reform. Among them was an English lawyer and journalist named Edwin Chadwick (see figure). Chadwick subscribed to the then-novel idea that humans could eliminate many diseases by doing away with filth. In 1842, well before the germ theory of disease was established, Chadwick published a landmark report indicating that poverty-stricken laborers suffered a far higher incidence of disease than those from middle or upper classes. He attributed the difference to the abominable living conditions of workers and declared that most of their diseases were preventable. His report established the basis for the Great Sanitary Movement.

Edwin Chadwick (1800–1890).

Courtesy of the National Library of Medicine

Chadwick was not a medical man, but his ideas captured the imagination of both scientists and social reformers. He proposed that sewers be constructed using smooth ceramic pipes, and enough water be flushed through the system to carry waste to a distant depository. In order to work, the system required the installation of new water and sewer pipes, the development of powerful pumps to bring water into homes, and the elimination of older sewage systems. The cost would be formidable.

Chadwick's vision eventually came to reality, but it might have taken decades longer were it not for an outbreak of cholera. In 1849, cholera broke out in London and terrified so many people that public opinion began to sway in favor of Chadwick's proposal. Another epidemic occurred in 1853, during which an English physician named John Snow proved that water was involved in transmission of the disease. In both outbreaks, the disease reached the affluent as well as the poor, and more than half of the sick perished. Construction of the sewer system began shortly thereafter.

As the last quarter of the 1800s unfolded, Europe's sanitary movement needed a "smoking gun" to hammer home its point. That incident came in 1892 when a devastating epidemic of cholera erupted in Hamburg, Germany. For the most part, Hamburg drew its water directly from the polluted Elbe River. Just west of Hamburg lay Altona, a city where the German government had previously installed a water filtration plant. Altona remained free of cholera. The contrast was further sharpened by a street that divided Hamburg and Altona. On the Hamburg side of the street, multiple cases of cholera broke out; across the street, none occurred. The sanitarians could not have imagined a more clear-cut demonstration of the importance of water purification and sewage treatment.

Systems for the treatment of human waste range from the primitive outhouse, which is nothing more than a hole in the ground, to the sophisticated sewage-treatment facilities used by many large cities. All operate under the same basic principle: Water is separated from the waste, and the solid matter is broken down by microorganisms to simple compounds for return to the soil and water.

In many homes, human waste is emptied into underground **cesspools**, which are concrete cylindrical rings with pores in the wall. Water passes into the soil through these pores and pores on the bottom of the cesspool, where solid waste accumulates. Microorganisms, especially anaerobic bacterial species, digest the solid matter into soluble products that enter the soil and enrich it. Some hardware stores sell enzymes and dried endospores to accelerate digestion.

Some homes have a **septic tank**, an enclosed concrete box for collecting waste from the house. Organic matter accumulates on the bottom of the tank, while water rises to the outlet pipe and flows to a distribution box. The water is then separated into pipes that empty into the surrounding soil. Because digested organic matter is not absorbed into the ground, the septic tank must be pumped out regularly.

The use of sewers can be traced all the way back to Roman times. However, only after the Great Sanitary Movement did sewers and sewage treatment facilities become much more efficient, and during the 1900s, sophisticated facilities were built in most large cities. All operate under the same basic principle: Water is separated from the waste, and the solid matter is broken down by microbes to simple compounds that can safely be returned to the soil and water.

Small towns collect sewage into large ponds called **oxidation lagoons**, where the sewage is left undisturbed for up to 3 months to allow natural digestion of the organic matter to occur. During this time, aerobic bacterial species digest the organic matter in the water, while anaerobic organisms break down the settled material. Under controlled conditions, the waste may be totally converted to simple salts such as carbonates, nitrates, phosphates, and sulfates. At the cycle's conclusion, the bacterial cells die naturally, the water clarifies, and the lagoon may be emptied into a nearby river or stream.

Large municipalities rely on a mechanized sewage-treatment facility to handle domestic wastewater, which contains massive amounts of waste and garbage. The process involves primary, secondary, and sometimes tertiary treatment (FIGURE 16.7).

Primary Waste Treatment. In the first step, a screen is used to remove grit and large insoluble waste. Then, the raw sewage is piped into huge open sedimentation (settling) tanks for organic waste removal. This waste, called **sludge**, is passed into sludge tanks for further treatment. Flocculating materials, such as alum, may be added to the raw sewage to drag microorganisms and debris to the bottom.

Secondary Waste Treatment. The purpose of secondary sewage treatment is to degrade the biological content of the fluid (effluent) coming from the primary treatment. This oxidation process is carried out by aerobic microorganisms as they metabolize the organic matter and produce inorganic end products (mainly CO_2 and H_2O). Two common secondary sewage treatment processes are the activated sludge system and trickling filters. They differ primarily in the way oxygen is supplied to the microorganisms and the rate at which organisms metabolize the organic matter.

In the activated sludge system, an aeration tank is supplied with oxygen so aerobic microorganisms can metabolize the organic matter in the primary effluent. One of the important bacterial species is *Zoogloea ramigera*, a gram-negative rod whose cells form a gelatinous matrix. Following 3 to 8 hours in the system, the active microorganisms (**activated sludge**) are separated from the liquid by sedimentation and the clarified liquid (secondary effluent) is further processed. A portion of the activated sludge is recycled back to the aeration tank to maintain a constant concentration of microbes. The remainder is removed and sent to the anaerobic sludge tank.

■ *Zoogloea ramigera*
zō'ō-glē-ä ram-i-gėr'ä

An alternative oxidation system is the "trickling filter" (also called a "biofilter"), which includes a basin or tower consisting of a bed of stones or gravel, molded plastic, or other porous substances that maintain a high surface-to-volume ratio (FIGURE 16.8). Microorganisms become attached to the bed and form a biofilm (see below). The primary effluent containing the organic matter is sprayed over the bed. As the organic matter in the water percolates through the film, it is rapidly metabolized by the aerobic microbes. Oxygen is normally supplied to the film by the natural flow of air. Because the thickness

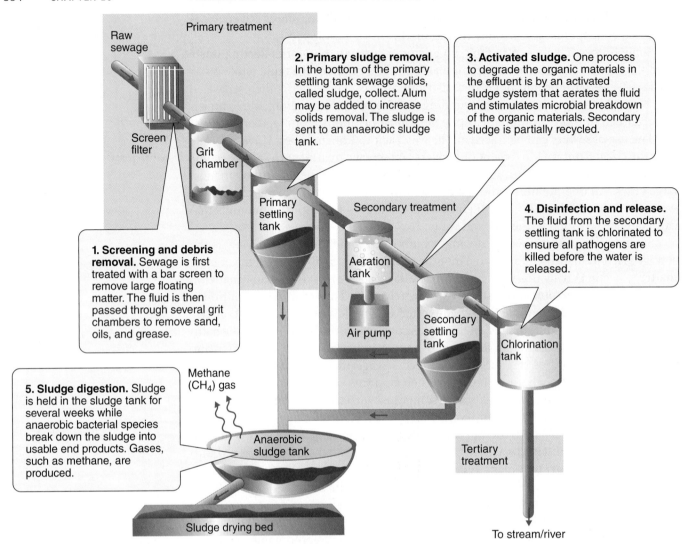

Primary treatment

Raw sewage

Screen filter

Grit chamber

Primary settling tank

1. Screening and debris removal. Sewage is first treated with a bar screen to remove large floating matter. The fluid is then passed through several grit chambers to remove sand, oils, and grease.

2. Primary sludge removal. In the bottom of the primary settling tank sewage solids, called sludge, collect. Alum may be added to increase solids removal. The sludge is sent to an anaerobic sludge tank.

3. Activated sludge. One process to degrade the organic materials in the effluent is by an activated sludge system that aerates the fluid and stimulates microbial breakdown of the organic materials. Secondary sludge is partially recycled.

Secondary treatment

Aeration tank

Air pump

Secondary settling tank

4. Disinfection and release. The fluid from the secondary settling tank is chlorinated to ensure all pathogens are killed before the water is released.

Chlorination tank

Tertiary treatment

To stream/river

5. Sludge digestion. Sludge is held in the sludge tank for several weeks while anaerobic bacterial species break down the sludge into usable end products. Gases, such as methane, are produced.

Methane (CH_4) gas

Anaerobic sludge tank

Sludge drying bed

FIGURE 16.7 **A Sewage-Treatment Facility.** Sewage-treatment facilities use physical, chemical, and biological processes to treat and remove all water contaminants.

FIGURE 16.8 **A Trickling Filter.** In secondary sewage treatment, a trickling filter contains rotating pipes through which sewage waste is sprayed onto a rock or plastic bed of aerobic microbes.

of the biofilm increases as the microbes reproduce, fragments of the biofilm break off the bed and are combined with primary sludge in the anaerobic sludge tank.

The secondary effluent from secondary treatment processes may still contain a few pathogens. Therefore, the water is disinfected, usually by chlorination, before it is released into a stream, river, or ocean.

Finally, the primary and secondary sludge in the sludge tank are digested by anaerobic microbes; thus the sludge tank is very anaerobic. Methane-producing archaeal species predominate and give off methane and carbon dioxide gases. The methane gas may be used to provide the fuel to heat the sludge tank. The digested sludge is dried and used as a soil amendment or placed in landfills.

Tertiary Waste Treatment. Sometimes further treatment of wastewater is needed as not all the water pollutants have been removed. Such tertiary sewage or advanced treatment systems are costly but may be essential for the removal of pesticides, fertilizers, and phosphate, which can pose problems if dumped into a lake or stream.

Biofilms and Bioremediation

Using bacterial species and other microbes to break down waste in sewage treatment has been very effective as you just read. As such, the use of microbes to help alleviate environmental chemical problems seems plausible and has become an immensely appealing idea. Putting microorganisms to work to "clean up" environmental chemical spills and contamination in this manner is a step toward limiting the human footprint in the environment.

For decades, microbiologists have studied free-floating bacterial species growing in laboratory cultures as examples of how bacterial organisms live and behave in nature. In recent years, however, they have come to realize the importance of bacterial species living in biofilms. A **biofilm** is an immobilized population of bacterial species (or other microorganisms) living in a web of tangled polysaccharide fibers adhering to a surface. Examples of environments where biofilms are naturally found include the surfaces of human teeth (dental plaque), dense growths of algae in a stream, slime inside a water pipe, and, as we just saw with sewage treatment, on stones of a trickling filter. Contact with a fluid environment ensures a plentiful supply of nutrients for the microbes of the biofilm.

Among the benefits of biofilms are their uses in **bioremediation**, where bacterial populations are used to stimulate or augment the degradation of toxic wastes and other synthetic products of industry. They also are employed in the oil industry to fill empty spaces after oil has been pumped out. In natural settings, they degrade organic substances, thus retarding pollutant buildup. A few examples show the power of the bacterial biofilm and bioremediation.

Among the first attempts to use bacterial biofilms for bioremediation was after a major oil spill from the tanker Exxon Valdez in 1987 along the Alaska coastline. Previous studies showed that where oil is spilled, there are oil-degrading bacterial species already present, and all technologists had to do was encourage their growth. Thus, after the oil spill occurred, technologists "fertilized" the oil-soaked water with nitrogen sources (e.g., urea), phosphorus compounds, and other mineral nutrients to stimulate the growth of indigenous (naturally occurring) microorganisms into biofilms. Areas treated in this way were cleared of oil significantly faster than non-remediated areas. Indeed, the oil degraded five times faster when microbes were enlisted in the cleanup. Such examples where bioremediation attempts to modify the environment by adding nutrients to activate bacteria is called **biostimulation**.

Bioremediation has also been applied to help eliminate polychlorinated biphenyls (PCBs) from the environment. PCBs were used widely in industrial and electrical machinery before their threat to environmental quality was realized. The contaminants contain numerous chlorine atoms and chlorine-containing groups, and researchers discovered that certain added anaerobic bacterial species can remove those atoms and groups, reducing the contaminants to smaller molecules. Aerobic bacteria then take over and reduce the molecular size still further. Biofilms containing these bacterial species can be very efficient in substantially reducing the contaminants in the environment. When specific bacteria are added to speed up bioremediation, the process is referred to as **bioaugmentation**.

Trichloroethylene (TCE) was once a much-used cleaning agent and solvent. Although TCE causes liver damage and nervous system dysfunction, and is a carcinogen, at the time, scientists did not realize TCE would diffuse through the soil and contaminate underground wells and water reservoirs (aquifers). To combat the problem and degrade the TCE, scientists began using bioaugmentation to exploit the ability of bacterial strains in biofilms to detoxify TCE. As the bacterial cells grow into a biofilm, they eliminate the TCE. The deliberate enhancement of microbial growth yields an environmental cleanup.

From the 1940s through the 1960s, a major component of weaponry was the explosive compound 2,4,6-trinitrotoluene (TNT). This compound, which has contaminated the soil around weapons plants, is poisonous and causes anemia and liver damage in people exposed to TNT. Scientists have found they can reduce the level of contamination by adding molasses to the soil (biostimulation) to encourage bacterial biofilm growth. Researchers have mixed water with TNT-laced soil and added molasses at regular intervals. In a matter of weeks, the TNT concentration plummeted as the TNT was broken town into harmless ethene (C_2H_4). However, to date, biostimulation in contaminated environmental soils has not been as dramatic.

For many years, "haul and bury" was the prevailing method for disposing of synthetic waste, including radioactive waste. As the public has become increasingly intolerant of that approach, the importance of bioremediation is apparent. Technologists are using microbes for their ability to degrade flame-retardants, phenols, chemical warfare agents, radioactive materials, and numerous other waste products of industry. Indeed, one prominent researcher has called bioremediation "*a field with its own mass and momentum*." And the momentum is again to reduce the human footprint on the environment by reducing or eliminating these and other industrial contaminants.

Which brings us to yet another human-produced problem—water pollution.

Water Pollution and Purification

Clean water is essential for bathing, cooking, and cleaning, as well as for industry, irrigation, recreation—and survival. According to the World Health Organization (WHO) and the United Nations Children's Fund (UNICEF), in 2013 more than 780 million people, primarily in the developing world, lacked access to clean water (FIGURE 16.9). As a result, each year more than 3.4 million people die (almost half are children under 5 years of age) from waterborne diseases contracted from unsafe water. Dirty water is the world's biggest health risk and continues to threaten both quality of life and public health for billions of people.

Drinking or **potable** water (*pot* = "drink") by contrast refers to water that is safe to drink. In the United States, a typical family of four uses up to 400 gallons of potable water each day. For most of us, we are thankful to have safe water yet water pollution is still a major issue.

Water pollution is the contamination of lakes, rivers, oceans, and groundwater through human activities. It begins with the addition of sewage and industrial wastes discharged by such direct sources as factories, refineries, waste treatment plants, and paper mills into nearby water supplies. In the United States and other developed countries, these practices are regulated, although pollutants often still find their way into water supplies.

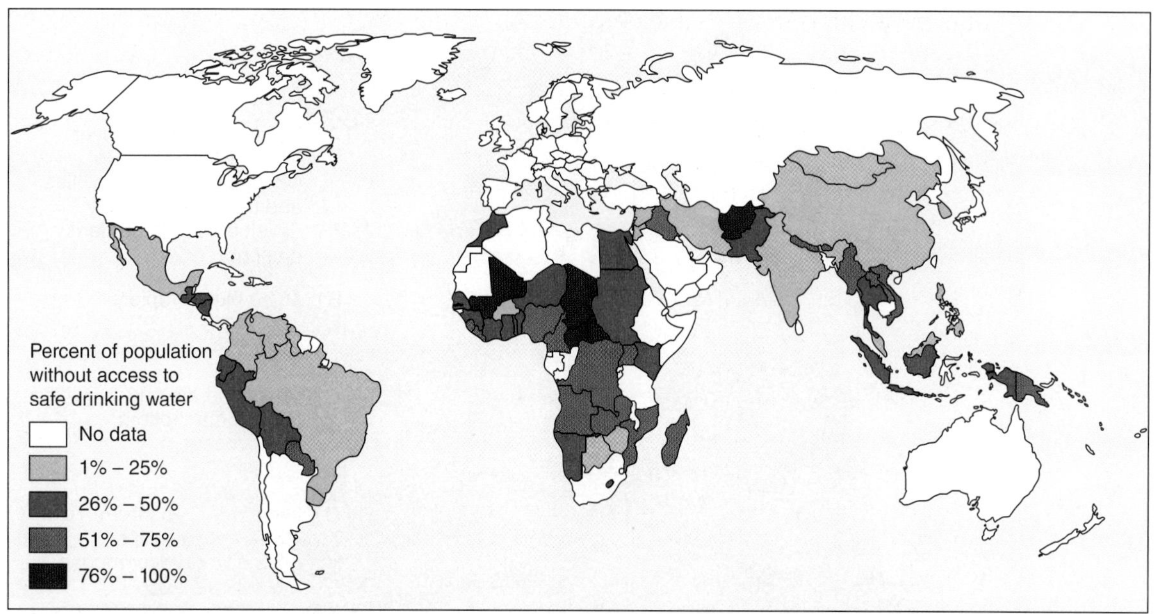

FIGURE 16.9 **Populations Lacking Access to Safe Drinking Water.** In 2012, more than 1 billion people lacked access to clean and sufficient drinking water.

Modified from The World's Water 1998–1999 by Peter H. Gleick. © 1998 Island Press. Reproduced by permission of Island Press, Washington, D.C.

Indirect sources of water pollution include contaminants that enter the water supply from soils/groundwater systems and from the atmosphere via rain water. Soils and groundwater contain the residue of human agricultural practices (e.g., fertilizers and pesticides) and improperly disposed of industrial wastes. Atmospheric contaminants are also derived from human practices (such as gaseous emissions from automobiles and factories).

In many cases, these water supplies contain a nutrient-rich "witch's brew" of pollutants, resulting in polluted water, as FIGURE 16.10 illustrates.

Toxic substances that destroy marine and freshwater species are polluting because the decomposition of aquatic life sparks the bacterial bloom, while yielding horrible odors. As bacterial cells use up the available oxygen, protists, fish, small arthropods, and plants die and accumulate on the bottom. Anaerobic bacterial species thrive in the sediments and produce gases, such as H_2S, that give the water a stench reminiscent of rotten eggs.

An important problem associated with water pollution is the corrosion of water pipelines. This corrosion is often due to hydrogen sulfide that some bacterial species produce from sulfates during the sulfur cycle. Hydrogen sulfide converts the iron in water pipelines into iron disulfide. Anaerobic bacterial species are usually involved, so such corrosion is most obvious on pipes buried in mud.

Water can also be a vehicle for the transport of pathogenic microbes causing human misery. Although the blame generally lies with humans who fail to deal properly with their sewage and waste, natural circumstances sometimes bear the brunt of the blame. Heavy rains, for example, wash pathogens from the soil and into rivers used for drinking water. Whether from human or natural sources, pollution must be detected, and to this end, public health departments apply a series of tests to identify health hazards and sound the alert. We will describe some of these methods next.

Microbe Detection Methods

A major public health concern is the presence of viral, bacterial, and protist pathogens in water supplies. Numerous infectious diseases such as typhoid fever, cholera, and hepatitis A can be transmitted in water. Because it is impossible to test water

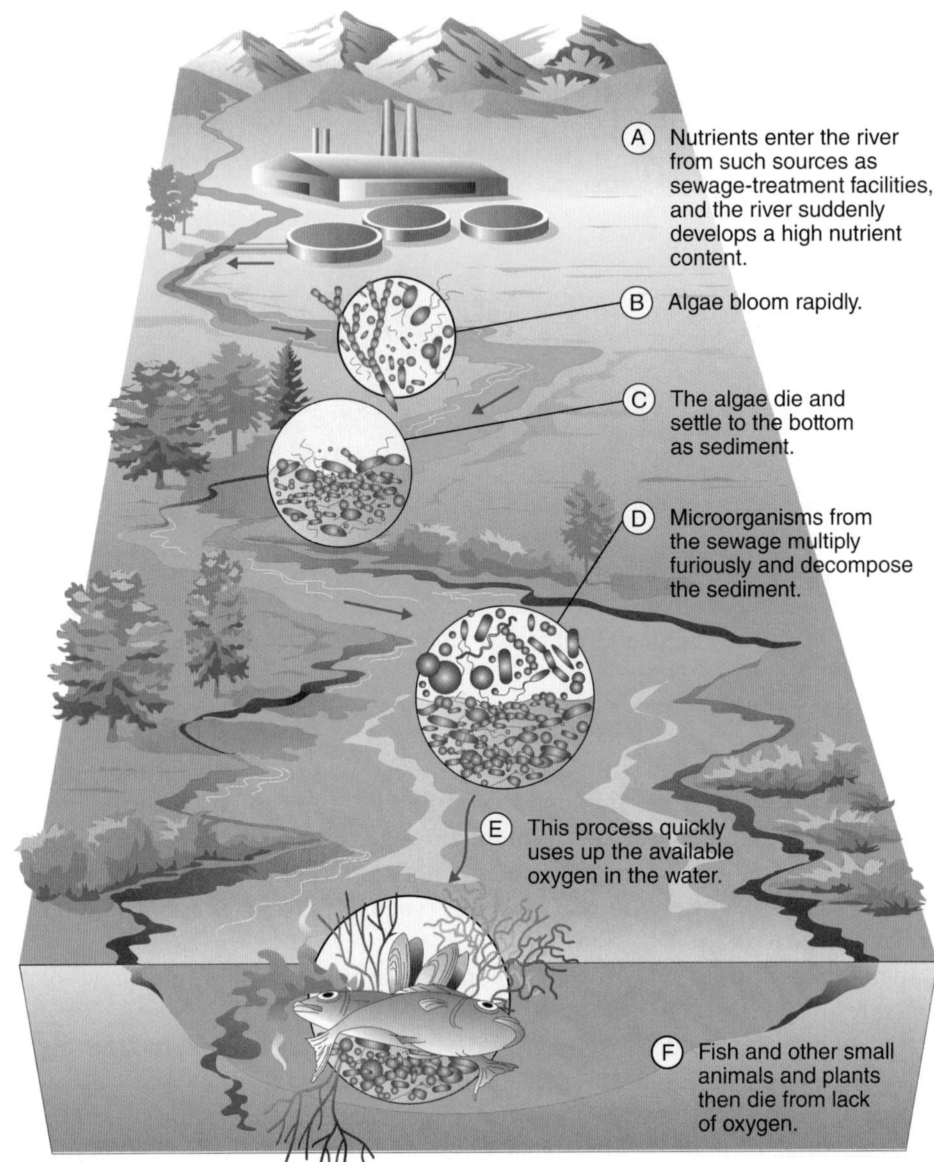

A Nutrients enter the river from such sources as sewage-treatment facilities, and the river suddenly develops a high nutrient content.
B Algae bloom rapidly.
C The algae die and settle to the bottom as sediment.
D Microorganisms from the sewage multiply furiously and decompose the sediment.
E This process quickly uses up the available oxygen in the water.
F Fish and other small animals and plants then die from lack of oxygen.

FIGURE 16.10 **The Death of a River.** The introduction of nutrients can lead to an algal bloom from which other microorganisms derive nutrients. This metabolism can quickly deplete the oxygen in the water, leading to the death of larger organisms, such as fish.

for all pathogens, certain **indicator organisms** are used to determine if water is contaminated. Among the most frequently used indicators are **coliform bacteria**, a group of gram-negative rods normally found in human and animal large intestines. If coliforms are present, then the likelihood is high that the water has been contaminated with intestinal microbes, including intestinal pathogens. Among the well-known coliform bacteria is *Escherichia coli*, found in the intestines of virtually all humans. The species is easy to cultivate in the laboratory, and numerous tests have been devised for its detection.

■ *Escherichia coli*
esh-ėr-ē'kē-ä kō'lī (or kō'lē)

The **membrane filter technique** is a popular laboratory test in water microbiology because it is straightforward and can be used in the field. A water technologist collects a 100 milliliter (mL) sample of water (FIGURE 16.11a). Back in the water testing lab, the water sample is passed through a cellulose-based membrane filter to trap any bacterial cells, and the filter is transferred to a plate of medium able to support bacterial growth. The plate is incubated, and bacterial cells trapped in the filter form

(a)

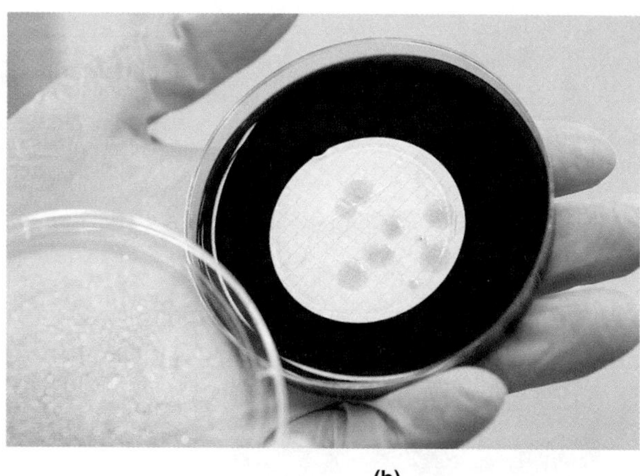

(b)

FIGURE 16.11 **Water Analysis.** (**a**) A water sample is collected and (**b**) potential indicator organisms are identified by the membrane filtration technique.

(a) Courtesy of Scott Bauer/USDA. (b) © Caro/Alamy.

visible colonies on the surface of the filter (FIGURE 16.11b). Assuming each bacterial cell trapped will give rise to a separate colony, the technologist counts the colonies and thereby determines the original number of bacterial cells in the sample of water.

These more traditional detection methods can take several days to complete. However, today the techniques of biotechnology can shorten this time considerably. In one of the most sophisticated methods available, a sample of polluted water is filtered, the bacterial cells are trapped on the filter, and then the cells are broken open to release their DNA. To detect *E. coli*, any DNA present is amplified (described in another chapter) and then, a DNA probe specific for *E. coli* DNA is added. Essentially, the technologist attempts to identify *E. coli* DNA, rather than *E. coli* cells. The process not only saves time, but is extremely sensitive. It has been estimated, for example, that a single *E. coli* cell can be detected in a 100-mL sample of water. In addition, a wealth of other microbes, including many pathogenic species, can be directly identified by DNA probe analysis, thereby eliminating the search for indicator organisms.

■ DNA probe: A known segment of single-strand DNA that is complementary to a desired DNA sequence of a bacterial species.

Ensuring the safety of water supplies for consumers is a high priority of the public health system. Rapid determination of a potential disease health risk permits public health officials to implement measures to benefit the most people. High on the list of these measures is the treatment of water to interrupt the spread of disease.

Water Treatment

It is rare to locate a water source that does not need treatment before consumption. The general rule is that water must be treated to remove potentially harmful microbes and to improve its clarity, odor, and taste.

Three basic steps are included in the preparation of water for drinking: sedimentation, filtration, and chlorination (FIGURE 16.12). As we go through the steps, realize the treatment process does not necessarily produce sterile water, but rather clean, potable water that is free of pathogens.

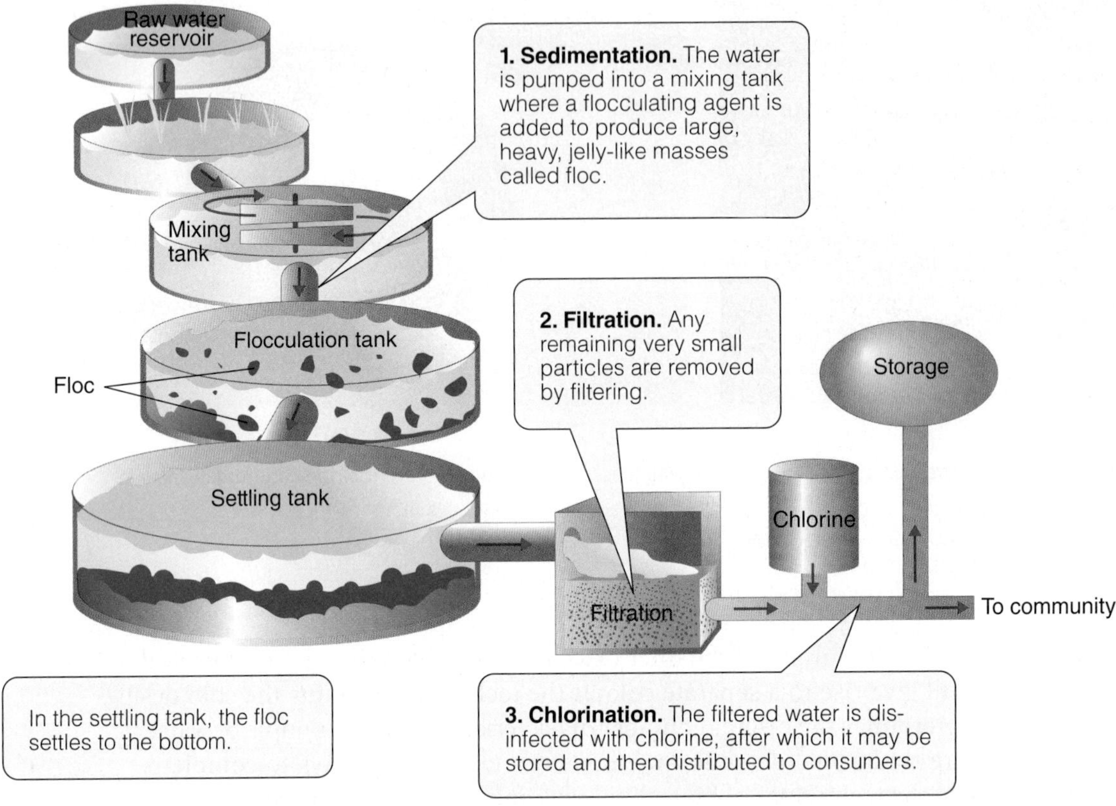

Prior to treatment, large objects are allowed to settle to the bottom of the raw water reservoir and the water may be sprayed to increase the oxygen content.
Raw water reservoir
1. Sedimentation. The water is pumped into a mixing tank where a flocculating agent is added to produce large, heavy, jelly-like masses called floc.
Mixing tank
Flocculation tank
Floc
2. Filtration. Any remaining very small particles are removed by filtering.
Storage
Settling tank
Chlorine
Filtration
To community
In the settling tank, the floc settles to the bottom.
3. Chlorination. The filtered water is disinfected with chlorine, after which it may be stored and then distributed to consumers.

FIGURE 16.12 **Steps in the Purification of Municipal Water Supplies.** Water purification involves sedimentation, filtration, and chlorination.

Sedimentation. **Sedimentation** uses large reservoirs or settling tanks to remove leaves, particles of sand and gravel, and other materials from the soil. Then chemicals, such as aluminum sulfate (alum), are dropped as a powder onto water. The alum sticks together small particles in the water and forms jelly-like masses of congealed material called **flocs**. In the flocculation tank, these masses fall through the water and cling to organic particles and microorganisms, dragging a major portion to the bottom sediment. **A Closer Look 16.3** describes a new, environmentally safe flocculant for developing parts of the world.

Filtration. The second step in water treatment is **filtration**. Although different types of filtering materials are available, most filters use a layer of sand and gravel to trap microorganisms. A slow sand filter, containing fine particles of sand a meter deep, is efficient for smaller-scale operations. Within the sand, a biofilm of microorganisms (bacterial, fungal, and protist cells) acts as a supplementary filter.

A slow sand filter may purify over 3 million gallons of water per acre per day. To clean the filter, the top layer is removed and replaced with fresh sand.

Some water treatment plants use a membrane filter system, where the water flows through submerged hollow fibers with tiny pores. The water molecules can pass through the pores but the larger contaminants stay within the fibers and are filtered

A CLOSER LOOK 16.3

Purifying Water with the "Miracle Tree"

In developed nations, water purification usually uses chemical powders such as aluminum sulfate or iron sulfate for the flocculation process. Many developing nations do not have such chemicals available, nor can they readily afford to purchase them. Could something else be used for the flocculation step in water purification? Yes! It's the "miracle tree."

Scientifically known as *Moringa oleifera,* this tropical tree was given the name "miracle tree" because of the many uses the tree has—including an environmentally friendly way to purify water.

A Moringa tree (*Moringa oleifera*) tree bearing fruit.

M. oleifera survives in arid areas and produces elongated seedpods. The local people use the leaves and roots for food, the wood is used in building, and some parts of the tree are used in traditional medicines. High-grade oil for lubrication and cosmetics is produced from the seeds. However, the seeds also have another use—purifying water. In those parts of Africa and India where the trees grow, people grind up the seeds and add the powder to cloudy water to precipitate all the solid particles. Clean drinking water results.

Ian Marison, research director at École Polytechnique Fédérale de Lausanne in Switzerland, and his colleagues have examined the ground-up seed residue and have isolated a charged peptide. When they add this peptide to cloudy water, within 2 minutes the water goes from cloudy to clear.

Even more exciting is the discovery that the peptide also has bactericidal properties and can disinfect heavily contaminated water. According to Marison, it even works against some drug-resistant strains of *Staphylococcus, Streptococcus,* and *Legionella,* and is also effective against some nonbacterial waterborne pathogens. In fact, Marison's group discovered they could tweak the peptide's structure and increase the antimicrobial effect of the peptide.

This is certainly good news for the developing nations where the trees mainly grow. However, if commercially produced, the plant flocculant also would be useful to developed nations because the traditional chemical flocculants often have safety and environmental concerns associated with them.

As Marison says, "It's a biological, biodegradable, sustainable resource. *Moringa* grows where there is very little water; it grows very, very fast and it costs almost nothing." It truly is a "miracle tree."

out. The filtered water is then pumped on to the next stage of treatment. Both forms of filtration eliminate more than 99% of the microbes and other very small particles from water. Often granular activated carbon, a black, sand-like material, is then used to remove natural organic matter. This step decreases taste and odors in the water.

Chlorination. The final step is **chlorination**. Water is disinfected with chlorine gas, which reacts with any organic matter in water. It is important, therefore, to continue adding chlorine until a residue is present. A residue of 0.2 to 1.0 parts of chlorine per million (ppm) of water often is the standard used in a water storage reservoir. Under these conditions, most remaining microbes die within 30 minutes. If you have a swimming pool at your home or use a public swimming pool, the chlorine level there should be 1 to 3 ppm to make sure any fecal microbes have been eliminated.

Some communities and homeowners also soften water by removing magnesium, calcium, and other salts. Softened water mixes more easily with soap, and soap curds do not form. Water also may be fluoridated to help prevent tooth decay. Scientists believe fluoride strengthens tooth enamel and makes the enamel more resistant to the acid produced by anaerobic bacterial species in the mouth.

In some instances, it is necessary to treat water for drinking on the spot. For example, raw sewage may find its way into water supplies and contaminate the water. Moreover, during drought conditions, sediment from the bottom of reservoirs may be stirred up, bringing bottom-dwelling microbes into the water and making it hazardous to health. Under conditions like these, consumers are generally advised to boil their water for a few minutes before drinking. This treatment kills all microbes with the possible exception of bacterial spores. However, with few exceptions, these spores do not represent a hazard to health. The water is safe to drink, but it is not sterile.

To disinfect clear waters, campers, backpackers, and hikers are advised to use commercially available chlorine or iodine tablets. If these cannot be obtained, the Centers for Disease Control and Prevention (CDC) recommends a half-teaspoon of household chlorine bleach in 2 gallons of water, with 30 minutes contact time before consumption.

As we discuss in two other chapters, polluted water can carry numerous species of microbes associated with infectious disease. Unfortunately, outbreaks of these diseases overshadow the myriad ways in which microbes benefit society through their activities in water, soil, and the environment. Hopefully, this chapter has given you sufficient examples of microbes exhibiting their friendly sides—so much so that they reduce the human footprint of contamination and pollution.

A Final Thought

As we continue our study of the microbes, it is easy to become paranoid about taking a breath of air, bite of food, or drink of water. After all, microbes lurk everywhere. However, we hope this chapter has shown you another side to the microbe story, a side that reflects the positive roles they play in the environment and in our lives.

Just consider, for example, the activity of microbes in the treatment of sewage. Through a complex network of processes, microbes transform the devilish cocktail of sewage into simple compounds that can be handled by the environment. Working with quiet and unceasing competence, they break down the vile mixture of human and animal feces, oily filth from roads, bloody effluent from slaughterhouses, and unimaginably ugly profusion of garbage. This mammoth task is accomplished unfailingly and efficiently by microbes.

Nor does it end here. In the carbon, nitrogen, and sulfur cycles, microbes convert the basic elements of life into usable forms and replenish the soil and oceans to nourish all living things. By far, the great majority of microbes are engaged in socially constructive and beneficial activities.

Questions to Consider

1. The English poet John Donne once wrote: "No man is an island, entire of itself." This statement applies not only to humans, but to all living things in the natural world. What are some roles microbes play in the interrelationships among living things?
2. A student notes in her microbiology class that a particular bacterial species actively dissolves fats, greases, and oils. Her mind stirs, and she wonders whether such an organism could be used to unclog the cesspool that collects waste from her house. What do you think she is considering? Will it work?
3. In the 1970s, a popular bumper sticker read: "Have you thanked a green plant today?" The reference was to photosynthesis taking place in plants. Suppose you saw this bumper sticker: "Have you thanked a microbe today?" What might the owner of the car have in mind?

4. What information might you offer to dispute the following four adages common among campers and hikers? (1) Water in streams is safe to drink if there are no humans or large animals upstream. (2) Melted ice and snow is safer than running water. (3) Water gurgling directly out of the ground or running out from behind rocks is safe to drink. (4) Rapidly moving water is germ-free.
5. The late syndicated columnist Erma Bombeck wrote a humorous book entitled *The Grass Is Always Greener Over the Septic Tank* (an adaptation of the expression "The grass is always greener on the other side of the fence"). Indeed, the grass is often greener over the septic tank. Why is this so? How can you locate your home's cesspool or septic tank in the days following a winter snowfall?
6. Bioremediation holds the key to solving numerous types of environmental problems in the future. Yet the process has been used for generations without people realizing it. How many instances where bioremediation is currently in use can you point out?
7. When sewers were constructed in New York City in the early 1900s, engineers decided to join storm sewers carrying water from the streets together with sanitary sewers bringing waste from the homes. The result was one gigantic sewer system. In retrospect, was this a good idea? Why?
8. The author of a biology textbook writes: "Because the microorganisms are not observed as easily as the plants and animals, we tend to forget about them, or to think only of the harmful ones...and thus overlook the others, many of which are indispensable to our continued existence." How do the carbon, sulfur, nitrogen, and phosphorus cycles support this outlook?

Key Terms

Informative facts are necessary for the expression of every concept, and the information for a concept is founded in a set of key terms. The following terms form the basis for the concepts of this chapter. On completing the chapter, you should be able to explain and/or define each one.

activated sludge
bioaugmentation
biofilm
biogeochemical cycle
bioremediation
biostimulation
biota
carbon cycle
cesspool
chlorination
coliform bacteria
community
compost
consumer
decomposer
ecosystem
environmental microbiology
filtration
floc
habitat
humus
indicator organism
membrane filter technique
nitrogen cycle
nitrogen fixation
oxidation lagoon
oxygen cycle
phosphorus cycle
phytoplankton
potable
primary waste treatment
producer
secondary waste treatment
sedimentation
septic tank
sludge
sulfur cycle
tertiary waste treatment
water pollution
zooplankton

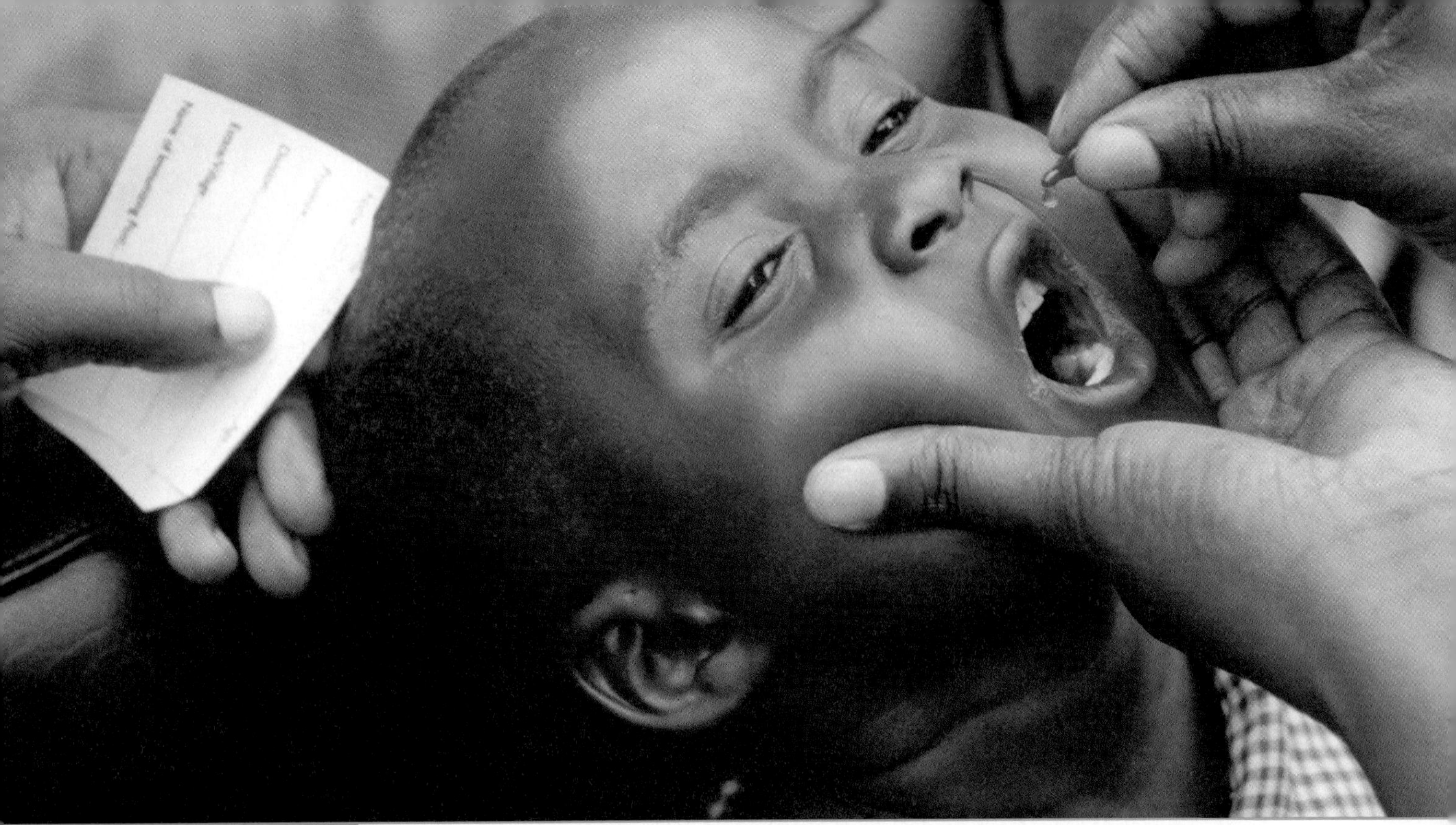

17 Disease and Resistance: The Wars Within

Vaccines provide resistance against infection, such as this oral vaccine against polio. Between 1967 and 1979, a global smallpox vaccination program brought about the eradication of the disease in 1980. Today, through another global vaccination effort, polio is close to eradication. In all cases, vaccination helps bolster immune defenses against potential pathogens.

Looking Ahead

For most of us, our earliest childhood experiences with microbes were related to their ability to cause infectious disease. Fortunately, we survived those early encounters by drawing on a number of resistance processes, as this chapter will illustrate.

On completing this chapter, you should be capable of:

- Discussing the characteristics and origins of the human microbiota.
- Describing pathogenicity and distinguishing between the direct and indirect methods of disease transmission.
- Distinguishing exotoxins from endotoxins.
- Discussing the first and second line immune defenses against disease.
- Describing how the immune system recognizes a specific pathogen.
- Comparing the roles of cell-mediated immunity and antibody-mediated immunity.
- Differentiating active immunity (natural and artificial) from passive immunity (natural and artificial).

The Peloponnesian War (431–404 BC) was a fight for power between Athens (and its empire) and Sparta (with its allies). Sparta had assembled large armies that were very nearly unbeatable. In 430 BC, the huge Spartan threat forced the Athenians to retreat behind the city walls of Athens. The retreat meant throngs of people from the countryside took refuge in the city. Due to the density of people and poor hygiene, Athens became a breeding ground for disease. The result was a devastating epidemic, today called "The Plague of Athens."

In his *History of the Peloponnesian War*, the contemporary Greek historian Thucydides gave an eye-witness account of the coming catastrophe that spared no segment of the population (FIGURE 17.1)

HIPPOCRATE SAUVANT LES ATHÉNIENS DE LA PESTE.

FIGURE 17.1 **The Plague of Athens.** In 430 BC, a disastrous plague hit Athens, Greece, taking the lives of up to 100,000 people.

"The bodies of dying men lay one upon another, and half-dead creatures reeled about in the streets. The catastrophe became so overwhelming that men cared nothing for any rule of religion or law."

Today, we still are not sure what this "plague" was. Several bacterial and viral diseases have been suggested. Whatever the cause, Thucydides remarked:

"As a rule, however, there was no apparent cause, but people in good health were all of the sudden attacked by violent heats in the head, and redness and inflammation in the eyes, the inward parts, such as the throat or tongue, became bloody and emitted an unnatural and foul breath."

As the epidemic spread, Thucydides made an extraordinary observation about resistance to the disease. He wrote:

"Yet it was with those who had recovered from the disease that the sick and dying found most compassion. These knew that it was from experience and now had no fear for themselves, for the same man was never attacked twice, never of the least fatality."

In other words, if one survived the disease, as Thucydides did, that person would never again develop the illness. They were immune.

By the end of the epidemic, historians believe 75,000 to 100,000 people sheltered within Athens' walls died of the disease. The sight of the burning funeral pyres of Athens caused the Spartan army to temporarily withdraw for fear of the disease. But they would be back.

In past centuries, the spread of disease appeared to be wildly erratic. Illnesses would attack some members of a population while leaving others untouched. A disease that for many generations had taken small, steady tolls would suddenly flare up in epidemic proportions. Strange, horrifying plagues, like the Plague of Athens, descended unexpectedly on whole nations.

Scientists now know humans live in a precarious equilibrium with the microbes surrounding them. Most of the time, the relationship is harmonious, because the microbes have no disease potential or because humans have developed resistance to them. However, when human resistance is diminished by a pattern of life that gives microbes the edge, then disease may set in.

In this chapter, we explore the infectious disease process and the mechanisms by which the body responds to disease. We examine the host-microbe relationship and the factors contributing to the establishment of disease, as well as the nonspecific and specific methods by which body resistance develops, with emphasis on the immune system.

17.1 The Host and Microbe: An Intimate Relationship

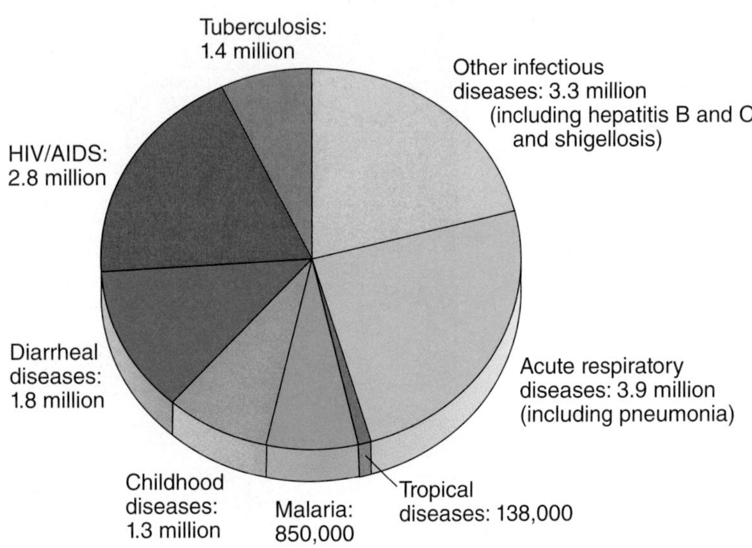

FIGURE 17.2 **Infectious Disease Deaths Worldwide.** This pie chart depicts the leading causes of infectious diseases and the number of worldwide deaths as reported by the World Health Organization. Tropical diseases: African sleeping sickness, Chagas disease, schistosomiasis, leishmaniasis, filariasis, and onchocerciasis. Childhood diseases: diphtheria, measles, pertussis, polio, and tetanus.

In the late 1960s and into the 1970s, the belief was widespread that infectious diseases had been conquered, and the use of antibiotics and vaccines would make the threat of infectious disease of little consequence. However, antibiotic resistance and new emerging viral diseases have thwarted such optimism. In 2013, of the approximately 57 million humans who died worldwide, more than 25% (15 million) died from infectious diseases, making them the second leading cause of death (behind cardiovascular disease) (FIGURE 17.2). In fact, infectious diseases are the leading cause of death in children under 5 years of age.

The Human Microbiota

Infection refers to the entry, establishment, and multiplication of a pathogen in the body, what is referred to as the **host**. A host whose resistance is strong remains healthy, and the pathogens are either driven from or assume a transient relationship with the host. By contrast, if the infection leads to tissue or organ damage or dysfunction, disease develops. The term **disease** therefore refers to any change from the general state of good health. Importantly, disease and infection are not synonymous; a person may be infected without suffering a disease.

Whether host or microbe gets the upper hand often is due in part to the estimated 100 trillion microbes found on and in the human body; a number that dwarfs the 10 trillion human cells composing the body. This remarkable number of microbes (99% bacterial) residing in the body without directly causing disease represents the human **microbiota** and amounts to almost 1.5 kg (3.3 pounds) of human weight. Most of these microbes have established a permanent relationship with various parts of the body. In the large intestine of humans, for example, *Escherichia coli* and the yeast *Candida albicans* are almost always found.

■ *Escherichia coli*
esh-ėr-ē′kē-ä kō′lī (or kō′lē)

■ *Candida albicans*
kan′did-ä al′bi-kanz

Our microbiota resides on or in several body tissues (FIGURE 17.3). These include the skin, the external ears and eyes, and upper respiratory tract. Most of the digestive tract, from oral cavity to rectum, is heavily populated with different communities of microbes as are the urogenital orifices in both males and females. Most other tissues of the body remain sterile; the blood, cerebrospinal fluid, joint fluid, and internal organs, such as the kidneys, liver, muscles, bone, and brain, of a healthy person are free of all microbes and viruses.

One of the important roles our microbiota plays is to protect us from pathogen invasion. By forming a physical barrier over the surface of the respiratory and digestive tracts, pathogens cannot establish a foothold for colonization. In addition, the microbiota successfully competes for the available nutrients and secretes antimicrobial substances to make the environment less hospitable for pathogens. Thus, the relationship between the body and its microbiota is an example of a **symbiosis**, or living together. If the symbiosis is beneficial to both the host and the microbe, the

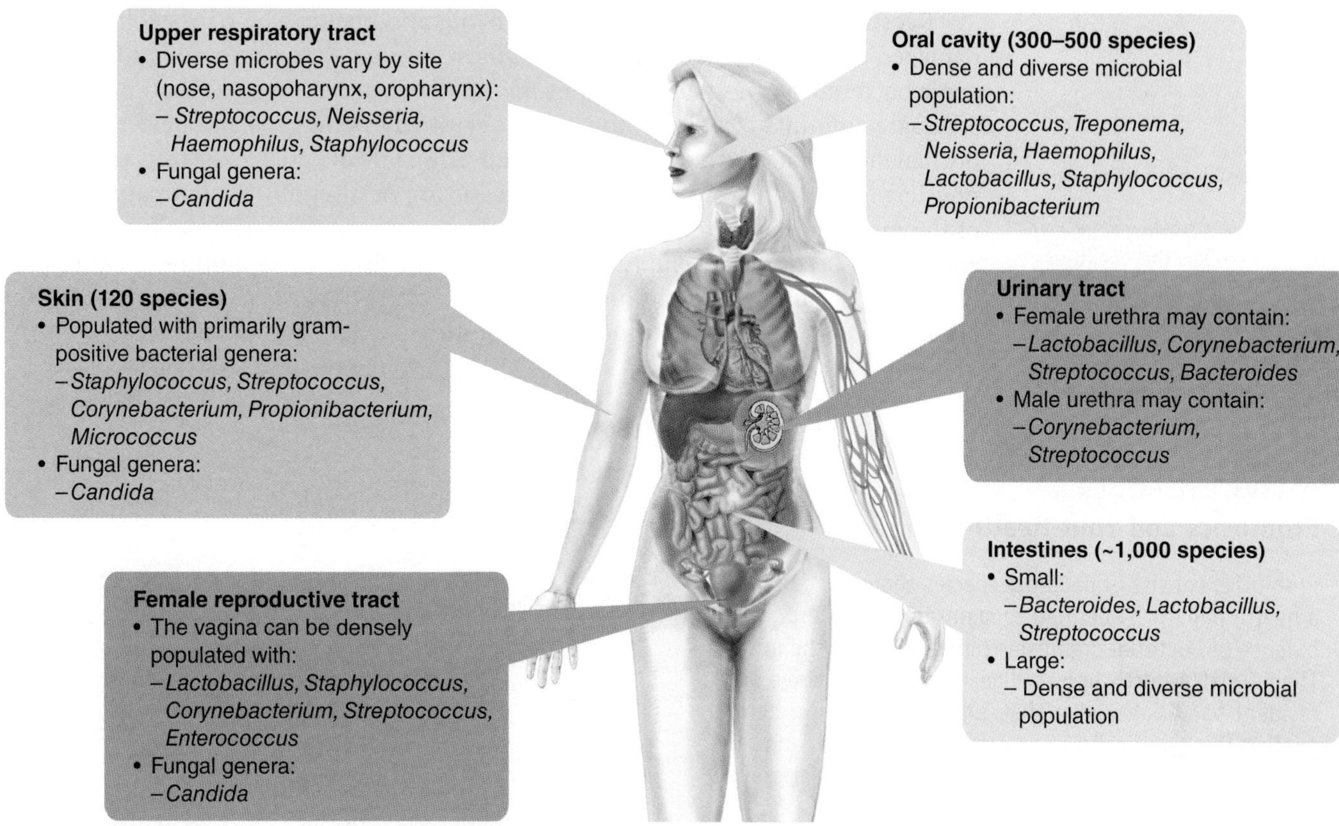

Upper respiratory tract
• Diverse microbes vary by site (nose, nasopoharynx, oropharynx):
– Streptococcus, Neisseria, Haemophilus, Staphylococcus
• Fungal genera:
– Candida
Oral cavity (300–500 species)
• Dense and diverse microbial population:
– Streptococcus, Treponema, Neisseria, Haemophilus, Lactobacillus, Staphylococcus, Propionibacterium
Skin (120 species)
• Populated with primarily gram-positive bacterial genera:
– Staphylococcus, Streptococcus, Corynebacterium, Propionibacterium, Micrococcus
• Fungal genera:
– Candida
Urinary tract
• Female urethra may contain:
– Lactobacillus, Corynebacterium, Streptococcus, Bacteroides
• Male urethra may contain:
– Corynebacterium, Streptococcus
Intestines (~1,000 species)
• Small:
– Bacteroides, Lactobacillus, Streptococcus
• Large:
– Dense and diverse microbial population
Female reproductive tract
• The vagina can be densely populated with:
– Lactobacillus, Staphylococcus, Corynebacterium, Streptococcus, Enterococcus
• Fungal genera:
– Candida

FIGURE 17.3 **A Sampling of the Indigenous Microbiota of the Human Body.** This sampling of microbes represents a few of the thousands of species found on and within some human body systems.

relationship is called **mutualism**. For example, species of *Lactobacillus* live in the female vagina and derive nutrients from the environment while producing acid to prevent the overgrowth of other organisms.

■ *Lactobacillus* lak-tō-bä-sil'lus

A symbiosis also can be beneficial only to the microbe and the host is unaffected, in which case the symbiosis is called **commensalism**. In the gut, *E. coli* is generally presumed to be a commensal, although some evidence exists for mutualism because the bacterial cells produce nonessential amounts of B complex and K vitamins that we can use.

Our appreciation for and understanding of the human microbiota rapidly expanded with the launching in 2007 of the "Human Microbiome Project" (HMP). The HMP, completed in 2012, used genetic and molecular techniques to identify and characterize the DNA composing microbial communities in healthy and diseased humans; this included the microbiota of the oral cavity, skin, and gastrointestinal tract. Researchers also were interested in determining if all humans share a core set of microbes and whether fluctuations in the human microbiota reflect changes in human health.

Although our microbiota is capable of maintaining itself, physical, chemical, or biological injury to the resident microbial communities can sometimes overextend its defensive abilities. For example, taking antibiotics for an infection can destroy and injure the gut microbiota. In some cases, this allows antibiotic-resistant pathogens to overrun the bowel because most resident microbes have been destroyed by the antibiotic. The pathogens now have the competitive edge.

Recent studies suggest that ingesting living microbes can help reestablish normal damaged microbiota. Called **probiotics**, these products are composed of nonpathogenic bacterial species similar to those normally found in the gut. Their presence in adequate amounts is thought to compete with any pathogens and help restore the normal microbiota. Although they are of questionable benefit to a healthy individual, physicians may suggest taking probiotics during or after antibiotic treatment. One of the most familiar probiotics is yogurt. Those containing *Lactobacillus bulgaricus* and *Streptococcus thermophilus* are said to contain "live, active cultures." Other probiotic yogurts may have additional species, especially *Bifidobacterium lactis*. **A Closer Look 17.1** digs into the usefulness of probiotics.

■ *Lactobacillus bulgaricus*
lak-tō-bä-sil′lus bul-gā′ri-kus

■ *Streptococcus thermophilus*
strep-tō-kok′kus thėr-mo′fil-us

■ *Bifidobacterium lactis*
bī-fi-dō-bak-ti′rē-um lak′tis

A CLOSER LOOK 17.1
Probiotics and Your Health

Probiotics consist of bacterial species thought to provide a health benefit by helping to maintain the natural balance of bacterial organisms (microbiota) in the gut. The normal, healthy human digestive tract has some 500 resident bacterial species that, among other roles, defend against the colonization and growth of bacterial and fungal pathogens. The largest group of probiotic bacteria in the intestine is lactic acid bacteria, of which *Lactobacillus acidophilus*, found in yogurt with live cultures, is the best known.

But do the anecdotal reports of yogurt's benefit match up with research studies?

Many people use, and doctors prescribe, probiotics to prevent diarrhea caused by antibiotics. Antibiotics not only kill the "bad" (damaging) microbes causing a disease, but also kill the "good" (beneficial) bacterial species that are part of the normal intestinal microbiota in the human body. A decrease in beneficial bacterial strains, whether a pathogen is present or not, may lead to digestive problems. So, taking probiotics may help reestablish the lost beneficial bacterial strains and thus help prevent diarrhea. And there are many scientific reports supporting the use of probiotics for diarrhea.

Yet other research studies with probiotic yogurts have given mixed results, even within the same study. For example, in 2011, scientists at Washington University of St. Louis wanted to know if probiotic yogurt had a beneficial effect on the gut. Using humans and mice in their studies, they reported that the intake of yogurt supplemented with five bacterial species did not significantly change the composition of the human or mouse gut microbiota, but it did alter gut metabolism, which was reflected by host microbiota responses to the arrival of the new bacterial species. So, perhaps the bacterial strains in a probiotic do not themselves provide protection or alleviate symptoms, but rather rev up the resident microbiota to better perform their role.

Courtesy of Dr. Jeffrey Pommerville

In 2013, another study conducted by Swansea University in the United Kingdom reported that probiotic supplements do not prevent antibiotic-associated diarrhea (AAD). In this study, a large number of patients taking antibiotics were split into two groups, one group given a placebo pill and the other group given a pill containing four bacterial strains typically found in probiotic yogurts. In the placebo group, 10.4% of the patients reported AAD while 10.8% of the patients taking the probiotic pill experienced ADD.

The bottom line is that we still do not know exactly what and how many bacterial strains are needed to confer a positive health outcome. And certainly, the probiotic yogurts on the market today are just an early attempt to define the strains and number required. Certainly, eating a probiotic yogurt cannot hurt, but its true beneficial role in human health remains to be determined.

Origin of the Human Microbiota

As we just learned, humans (and many other animals too) have coevolved with a diverse set of microorganisms. So where did all these microbes making up the human microbiota come from?

An infant's initial exposure to the friendly microbes (bacteria) probably starts in the womb. However, an infant receives copious supplements of microbes as the newborn passes through the birth canal. The birth canal is covered with the mother's resident microbes and these "rub off" onto the newborn at delivery (FIGURE 17.4). Most interestingly, babies born by caesarian section (C-section) can have a markedly different microbiota than those born vaginally because C-section babies are not exposed to the mother's resident microbes but rather to microbes found in the environment and on doctors, nurses, and family members. Additional organisms enter upon first feeding. Again, a very different gut microbiota may develop depending on whether the baby is breast fed or is fed formula.

During the following weeks, additional contact with the mother and other individuals will expose the infant to additional microbes. Besides the gut, the skin will be colonized by many different bacterial and fungal species. The oral cavity and upper respiratory tract will become covered with a diverse group of bacterial species, while the lower urethra will be populated by bacterial and fungal organisms as well as a few potential pathogens.

By 3 years of age, the infant's microbiota is almost adult-like and usually remains fairly stable throughout life, undergoing small changes in response to the internal and external environment of the individual. However, much remains to be learned about our microbiota and the roles it plays in our health and physical wellbeing.

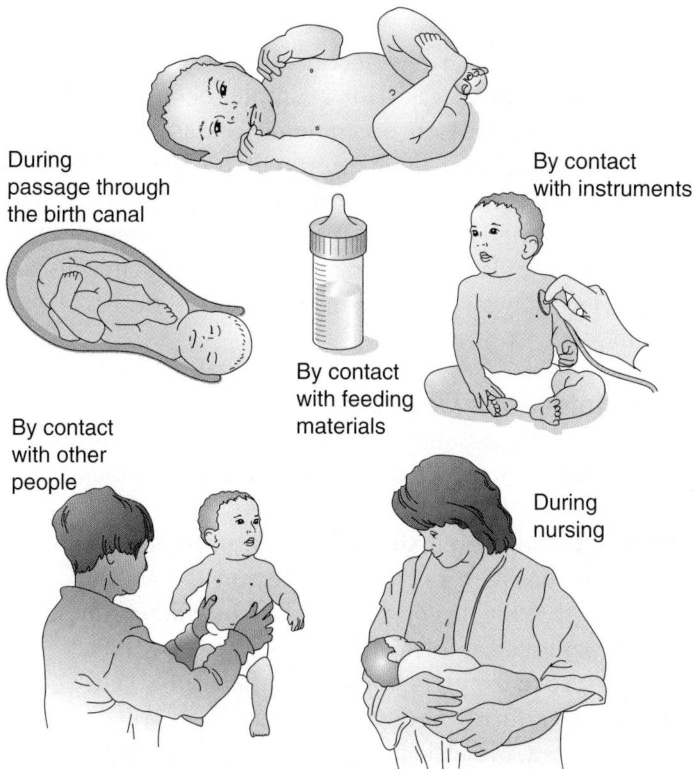

FIGURE 17.4 **Origins of a Newborn's Microbiota.** Barring an infection by the mother, the fetus during the nine-month gestation period remains sterile; there are no microbes associated with its developing body.

17.2 Concepts of Infectious Disease

Besides the mutualism and commensalism relationships with microbes, there can also be a symbiotic relationship called **parasitism**, where a pathogenic microbe causes damage to the host and disease may result.

Pathogenicity and Virulence

Pathogenicity refers to the ability of a microbe to gain entry into the host's tissues and bring about a physiological or anatomical change, resulting in altered health.

Certain pathogens, such as the cholera, plague, and typhoid pathogens, are well known for their ability to cause serious human disease. Others, such as common cold viruses, are usually less serious because they bring about milder illnesses. Still other pathogens are **opportunistic**. These microbes are nonthreatening, often commensal organisms until the normal body defenses are suppressed; then, they seize the "opportunity" to invade the tissues and express their pathogenicity. An example of suppressed defenses is observed in individuals with acquired immune deficiency syndrome (AIDS). These patients are highly susceptible to disease caused by several opportunistic bacterial, fungal, and protist organisms.

Whether a disease is mild or severe depends on the pathogen's ability to do harm to a susceptible host. Thus, the degree of pathogenicity, called **virulence** (*virul* = "poisonous"), depends on the host–microbe interaction. An organism that invariably causes disease, such as the typhoid bacillus, is said to be "highly virulent." By comparison, an organism that sometimes causes disease, such as the yeast *C. albicans*, is labeled "moderately virulent." Certain organisms, described as **avirulent**, are not regarded as disease agents. The lactobacilli and streptococci found in yogurt are examples.

Diseases Within Populations

Infectious diseases may be described according to the level at which they occur in a population. An **endemic** disease, for example, occurs at a low level in a certain geographic area. By comparison, an epidemic disease (or **epidemic**) breaks out in explosive proportions within a population. Often seasonal flu cases result in an epidemic. This should be contrasted with an **outbreak**, which is a more contained epidemic. An abnormally high number of measles cases (above the endemic level) in one American city would be classified as an outbreak. A pandemic disease (**pandemic**) occurs worldwide. Two newsworthy examples here would be AIDS and the 2009 swine flu pandemic.

Transmission of Infectious Diseases

The microbial agents of disease may be transmitted by a broad variety of methods conveniently divided into two general categories: direct methods and indirect methods (FIGURE 17.5).

Direct Methods. Person-to-person transmission implies **direct contact**. Here, physical contact occurs between a susceptible person and another who is infected or has the disease. Hand-shaking, kissing, and sexual intercourse are examples; gonorrhea and genital herpes are examples of diseases spread by the latter. Direct contact may also mean exposure to **droplets**, the tiny particles of mucus and saliva expelled from the respiratory tract during a cough or sneeze (FIGURE 17.6). Diseases spread by droplets include influenza, measles, pertussis (whooping cough), and streptococcal sore throat.

Indirect Methods. Among the **indirect methods** of disease transmission are the consumption of contaminated food or water. Foods are contaminated during

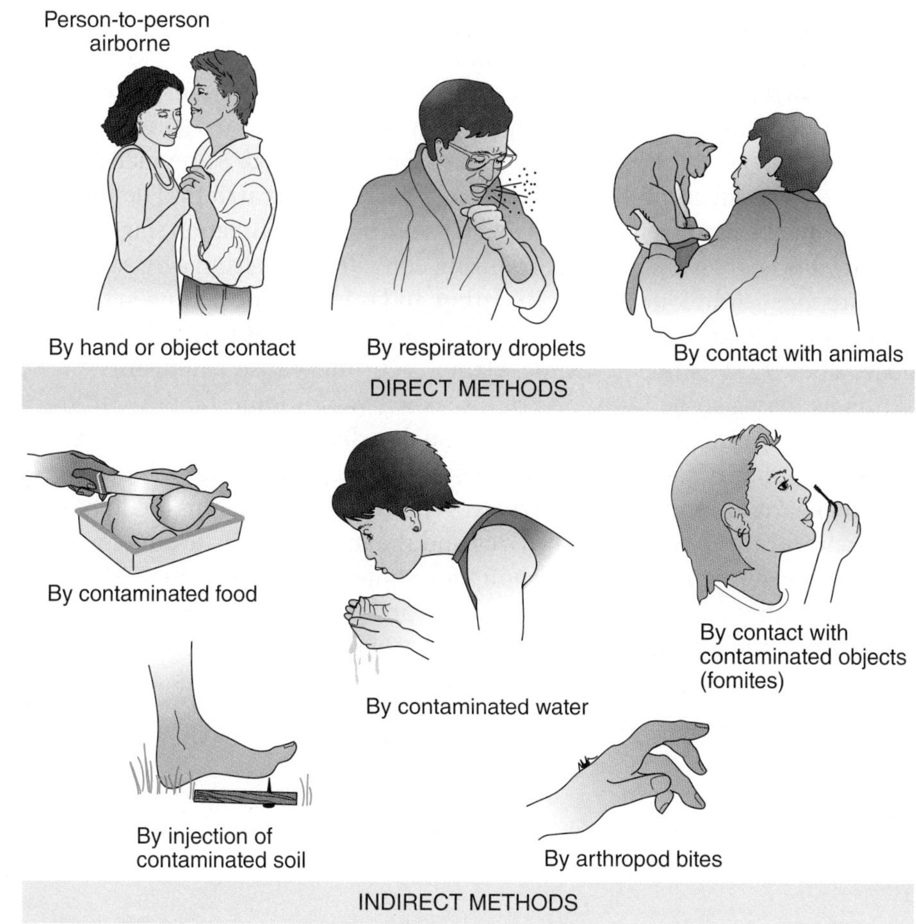

FIGURE 17.5 **Methods of Transmitting Disease through Contact.** Infectious diseases can be transmitted by direct methods and indirect methods.

processing or handling or may be so because they were made from diseased animals. Poultry products, for example, are often a source of salmonellosis because *Salmonella* species frequently infect chickens. Transmission can also occur through **fomites**, inanimate objects that carry disease organisms. For instance, doorknobs or drinking glasses may be contaminated with flu or cold viruses and transmitted to another person when touching the object. This shows the importance of hand washing. Also, contaminated syringes and needles can transport the viruses of hepatitis B and AIDS.

■ *Salmonella*
säl-mōn-el'lä

Arthropods (e.g., insects) are responsible for another indirect method of transmission. In some cases, the arthropod is a **mechanical vector** of disease because it transports microbes on its legs and other body parts. House flies would be examples of mechanical vectors. In other cases, the arthropod itself is diseased and serves as a **biological vector**. The malaria protist and West Nile virus, for instance, infect mosquitoes and accumulate in its salivary glands, from which the protists or viruses can be injected into a human host during the next bite (blood meal).

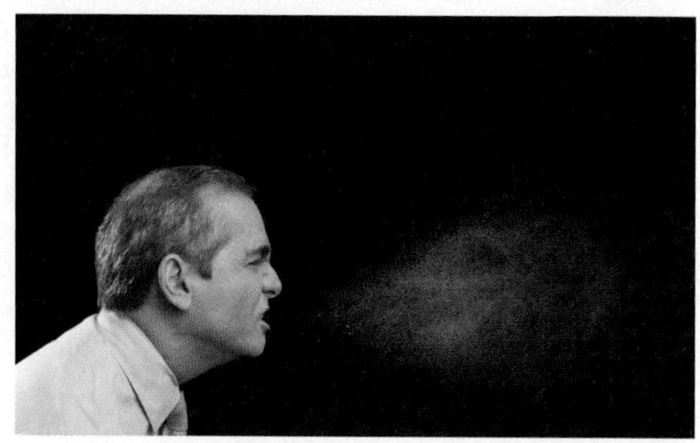

FIGURE 17.6 **Droplet Transmission.** Sneezing or coughing represents a method for airborne transmission of pathogens.

For a disease to perpetuate itself, a habitat in which a pathogen normally resides and multiplies is necessary. These locations are called **reservoirs** and may or may not be the source for human infections. For example, the reservoir for *Clostridium botulinum*, the causative agent for botulism,

■ *Clostridium botulinum*
klôs-tri'dē-um bot-ū-lī 'num

is the soil. However, the source of most human cases of botulism is from improperly canned foods contaminated with *C. botulinum* spores.

Some infectious diseases have human reservoirs. Sexually transmitted diseases, measles, mumps, and many respiratory pathogens are transmitted solely from person to person without any other organism or source being involved. Some such individuals may not show the effects of the illness and are referred to as **carriers**. They are capable of transmitting the pathogen to others, perhaps during the incubation stage of a disease before they know they are infected or during convalescence from a disease. Chronic carriers are those individuals who can harbor and transmit the pathogen for months, perhaps years after an initial infection. Typhoid Mary, whose story is recounted in **A Closer Look 17.2**, is one of the most famous chronic carriers in history.

Animals may also be reservoirs of disease. A rabid domestic dog can transmit the rabies virus to humans through a bite. Such a disease transmitted from animals to humans is called a **zoonosis** (pl. zoonoses). Besides rabies, the plague, West Nile fever, and malaria are zoonoses (the reservoir for the plague bacillus could be rats and the reservoir for the West Nile fever virus and the malaria parasite are mosquitoes).

Soil and water can also be reservoirs for some infectious agents. Many fungal agents, such as those that cause valley fever, live and multiply in the soil. The reservoirs for *Legionella pneumophila*, the pathogen responsible for outbreaks of Legionnaires' disease, are often traced back to water supplies in cooling towers and evaporative condensers.

A CLOSER LOOK 17.2

Typhoid Mary

By 1906, typhoid fever was claiming about 25,000 lives annually in the United States. During the summer of that year, a puzzling outbreak occurred in the town of Oyster Bay on Long Island, New York; one girl died and five others contracted the disease. Eager to find the cause, public health officials hired George Soper, a well-known sanitary engineer from the New York City Health Department. Soper's suspicions centered on Mary Mallon, the seemingly healthy family cook. She had disappeared 3 weeks after the disease surfaced. Soper was familiar with Robert Koch's theory that infections like typhoid fever could be spread by people who harbor the organisms. Quietly he began to search for the woman who would become known as Typhoid Mary.

Soper's investigations led him back over the decade during which Mary Mallon cooked for several households. Twenty-eight cases of typhoid fever occurred in those households, and each time, the cook left soon after the outbreak.

Soper tracked Mary Mallon through a series of leads from domestic agencies and finally came face-to-face with her in March 1907. She had assumed a false name and was now working for a family in which typhoid had broken out. Soper explained his theory that she was a carrier, and pleaded that she be tested for typhoid bacilli. When she refused to cooperate, the police forcibly brought her to a city hospital on an island in the East River off the Bronx shore. Tests showed her stools teemed with typhoid organisms, but fearing her life was in danger, she adamantly refused the gall bladder operation to eliminate the organisms (the causative agent, *Salmonella* Typhi, often colonizes the gallbladder). As news of her imprisonment spread, Mary became a celebrity. Soon public sentiment led to a health department policy deploring the isolation of carriers. She was released in 1910.

But Mary's saga had not ended. In 1915, she turned up again at New York City's Sloane Hospital working as a cook under another new name. Eight people had recently died of typhoid fever, most of them doctors and nurses. Mary was taken back to the island, this time in handcuffs. Still she refused to have her gall bladder removed and vowed never to change her profession. Doctors placed her in isolation in a hospital room while trying to decide what to do. The weeks wore on.

Eventually Mary became less incorrigible and assumed a permanent residence in a cottage on the island. She gradually accepted her fate and began to help out with routine hospital work. However, she was forced to eat in solitude and was allowed few visitors. Mary Mallon died in 1938 at the age of 70 from the effects of a stroke. She was buried without fanfare in a local cemetery.

FIGURE 17.7 summarizes the three elements (host, transmission, and source needed for infection.

The Course of a Disease

In most instances, there is a recognizable pattern in the progress of the disease following the entry of the pathogen into the host. Often these periods are identified by **signs**, which represent evidence of disease detected by an observer (e.g., physician). A low-grade fever or bacterial cells in the blood would be examples. Disease also can be noted by **symptoms**, which represent changes in body function sensed by the patient. Sore throat and headache are examples. Diseases also may be characterized by a specific collection of signs and symptoms called a **syndrome**. AIDS is an example where the individual exhibits over time a typical set of opportunistic infections. Thus, a doctor's **diagnosis** (the identification of a disease or illness) often is based on a patient's signs and symptoms.

Disease progression is distinguished by five stages (FIGURE 17.8).

Incubation Period. The episode of disease begins with an **incubation period**, reflecting the time elapsing between the entry of the pathogen into the host and the appearance of the first symptoms. For example, an incubation period may be as short as 2 to 4 days for the flu; 1 to 2 weeks for measles; or 3 to 6 years for leprosy. Such factors as the number of organisms, their generation time and virulence, and the level of host resistance determine the incubation period's length.

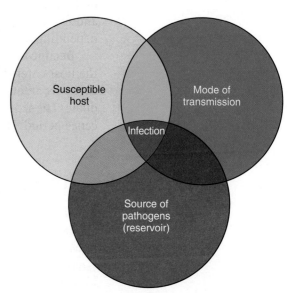

FIGURE 17.7 **Infectious Disease Elements.** Assuming there is a susceptible host, infection requires both a source of the pathogen and a method for transmission.

Prodromal Phase. The next phase in disease is a time of mild signs or symptoms, called the **prodromal phase**. For many diseases including the flu, this period is characterized by indistinct and general symptoms such as headache and muscle aches, which indicate the competition between host and pathogen has begun.

Acute period. In the third stage of a disease, called the **acute period** or **climax**, the signs and symptoms are of greatest intensity. For the flu, patients suffer high fever

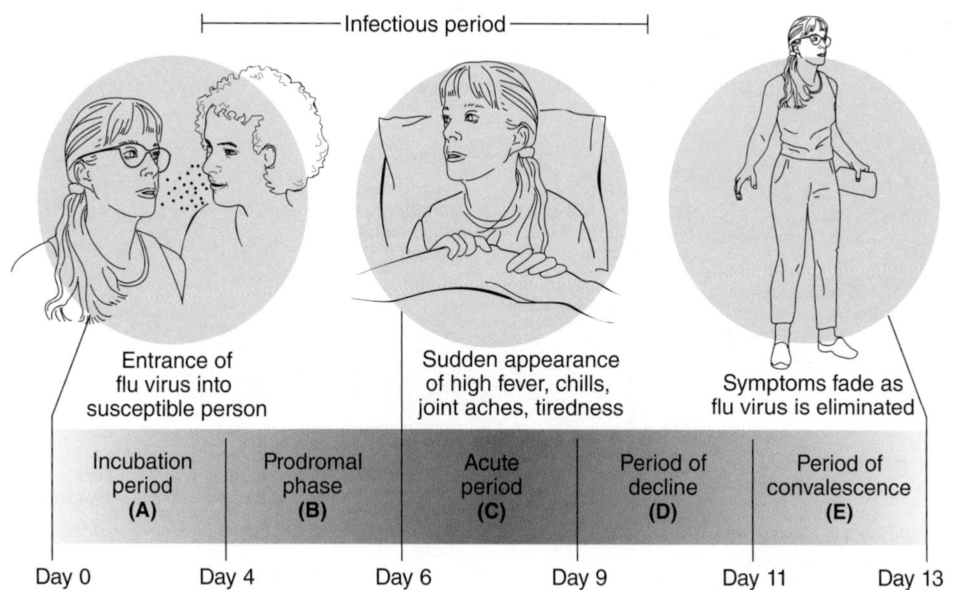

FIGURE 17.8 **The Course of an Infectious Disease.** Most infectious diseases go through a five-stage process. Here, the course of infection is shown for a susceptible person exposed to flu viruses.

and chills, the latter reflecting differences in temperature between the superficial and deep areas of the body. Dry skin and a pale complexion may result from constriction of the skin's blood vessels to conserve heat. A headache, cough, body and joint aches, and loss of appetite are common. The length of this period can be quite variable, depending on the body's response to the pathogen and the virulence of the pathogen.

Decline and Convalescence Periods. As the signs and symptoms begin to subside, the host enters the **decline period**. Sweating may be common as the body releases excessive amounts of heat and the normal skin color soon returns as the blood vessels dilate. The sequence comes to a conclusion after the body passes through a convalescence period. During this time, the body's systems return to normal.

17.3 The Establishment of Disease: Overcoming the Odds

A pathogen must possess unusual abilities if it is to overcome host defenses and bring about the profound changes leading to disease. Before it can manifest these abilities, however, the pathogen must first gain entry into the host in sufficient numbers to establish a population. Next, it must be able to penetrate the tissues and grow at that location. Disease is therefore a complex series of interactions between pathogen and host. FIGURE 17.9 shows a simplified scenario.

Entry and Invasion

The **portal of entry** refers to the site at which the pathogen enters the host. The site varies considerably for different pathogens. These entry routes include: (1) a respiratory portal (inhalation of pathogens in air); (2) a gastrointestinal portal—the **fecal–oral route** (ingestion of food or water containing pathogen-contaminated fecal material); (3) a sexually-transmitted portal; and (4) non-oral portals (e.g., piercing the skin through cuts, animal/insect bites, wounds, injections).

Having reached the appropriate portal of entry, the ability of a pathogen to establish an infection and possible disease usually depends on the **infectious dose**, the number of cells or viruses that must be taken into the body for disease to be established (see Figure 17.9, part A). For example, the consumption of a few thousand typhoid bacilli will probably lead to the disease. By contrast, many millions of cholera bacteria must be ingested if cholera is to be established. One explanation for the difference is the high resistance of typhoid bacilli to the acidic conditions in the stomach, in contrast

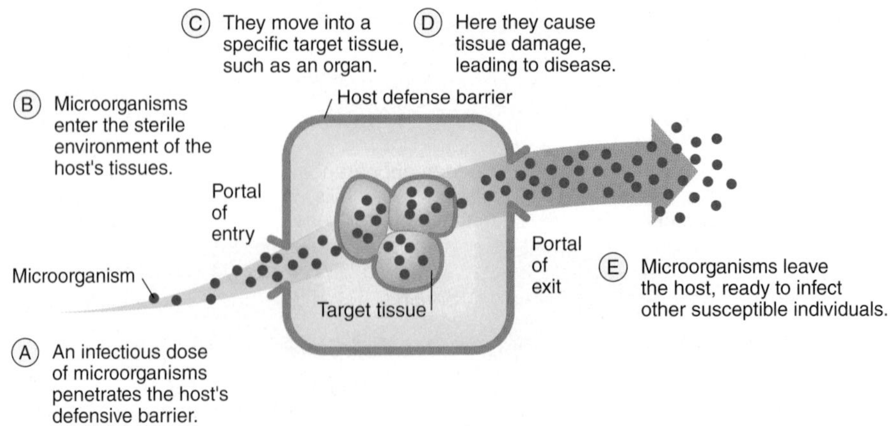

FIGURE 17.9 **The Flow of Events for a Disease.** An infectious dose and portal of entry are required to initiate infection and disease.

to the low resistance of cholera bacilli (most of which thus perish). Also, it may be safe to eat fish taken from water that contains hepatitis A viruses, but eating raw clams from the same water can be dangerous. Clams filter water to obtain nutrients, and in doing so, they concentrate any hepatitis A viruses in the water in their tissues to a concentration representing an infectious dose.

The ability of a pathogen to penetrate tissues and cause structural damage is called **invasiveness** (see Figure 17.9, parts B and C). Many bacterial pathogens have specific genes for invasiveness, which gives them the ability to gain access to body sites unavailable to nonpathogenic organisms. For example, pathogenic *E. coli* 0157:H7 cells can penetrate to sites in the small intestine and urinary system that are not available to nonpathogenic *E. coli* strains present in the small intestine.

The virulence of a pathogen further depends on its ability to produce a series of enzymes that help it to counter body defenses. An example of a bacterial enzyme is **hyaluronidase**. Sometimes called the "spreading factor," this enzyme enhances pathogen penetration through tissues and the pathogen can more easily move to its target site (tissue, organ) in the body and cause disease (see Figure 17.9, part D). A large group of other enzymes also assist in invasiveness.

Bacterial Toxins

Toxins are microbial poisons affecting the establishment and course of disease. Two types of toxins are recognized: exotoxins and endotoxins.

Exotoxins. Some bacterial pathogens have the ability to produce and release **exotoxins**. These are protein molecules that act locally or diffuse to their site of activity in the body. The exotoxin produced by the botulism organism, *C. botulinum*, is among the most lethal toxins known (it is believed 1 pint of the pure toxin would be sufficient to destroy the world's population). In humans, the toxin inhibits the release of a neurotransmitter called acetylcholine at the junction where nerve cells meet muscle cells. This inhibition leads to muscle paralysis.

The body responds to exotoxins by producing antibodies called **antitoxins**. An antitoxin molecule combines with a toxin molecule and neutralizes it. In the laboratory, certain chemicals can be used to alter the toxin and destroy its toxicity without hindering its ability to elicit an immune response in the body. The altered toxin is called a **toxoid**. When toxoid molecules are injected into the body, the immune system responds with antitoxins that circulate and provide a measure of defense against the disease. Toxoids are used as vaccines to protect against diphtheria and tetanus.

Endotoxins. The **endotoxins** are part of the cell wall of gram-negative bacterial species and thus are released only on disintegration of the cell. Endotoxins manifest their presence by certain signs and symptoms, including increased body temperature, substantial body weakness and aches, and general malaise. Damage to the circulatory system and shock may also occur.

■ shock: A state of physiological collapse, marked by a weak pulse, coldness, sweating, and irregular breathing due to too low a blood flow to the brain.

Pathogen Exit

At some point, the pathogen gains the ability to escape the body through some suitable **portal of exit** (see Figure 17.9, part E). Perhaps most common mechanisms for exit are coughing and sneezing, which easily spread nasal secretions, saliva, and sputum as respiratory droplets or aerosols. This is important because easy transmission permits the pathogen to continue its disease-spreading existence in the world. Exit could also be by blood transfusion, insect bite, or released into the environment in urine or feces.

All together then, many pathogens have an arsenal of virulence weapons to give the invaders a competitive advantage over the host, and these advantages must be addressed and counterbalanced if the host is to survive an episode of infectious disease. How the body accomplishes this is explored in the remaining sections of this chapter.

17.4 Nonspecific Resistance to Disease: Natural Born Immunity

Think of the human body as protected by an "immunological umbrella" that can shield us from the torrent of potential microbial pathogens to which we are continually exposed (FIGURE 17.10). As long as the umbrella remains intact we are fairly safe. However, if holes or tears develop, then we run a higher risk of an infection. The skin and its extensions into the body cavities represent part of the umbrella and are a major means of resistance to infection and disease. This resistance is nonspecific, because it exists in all humans and is present from birth. The other half of the umbrella represents a form of resistance that is specific to each invading pathogen and is based on antibodies and immune cells, as we will see later in this chapter.

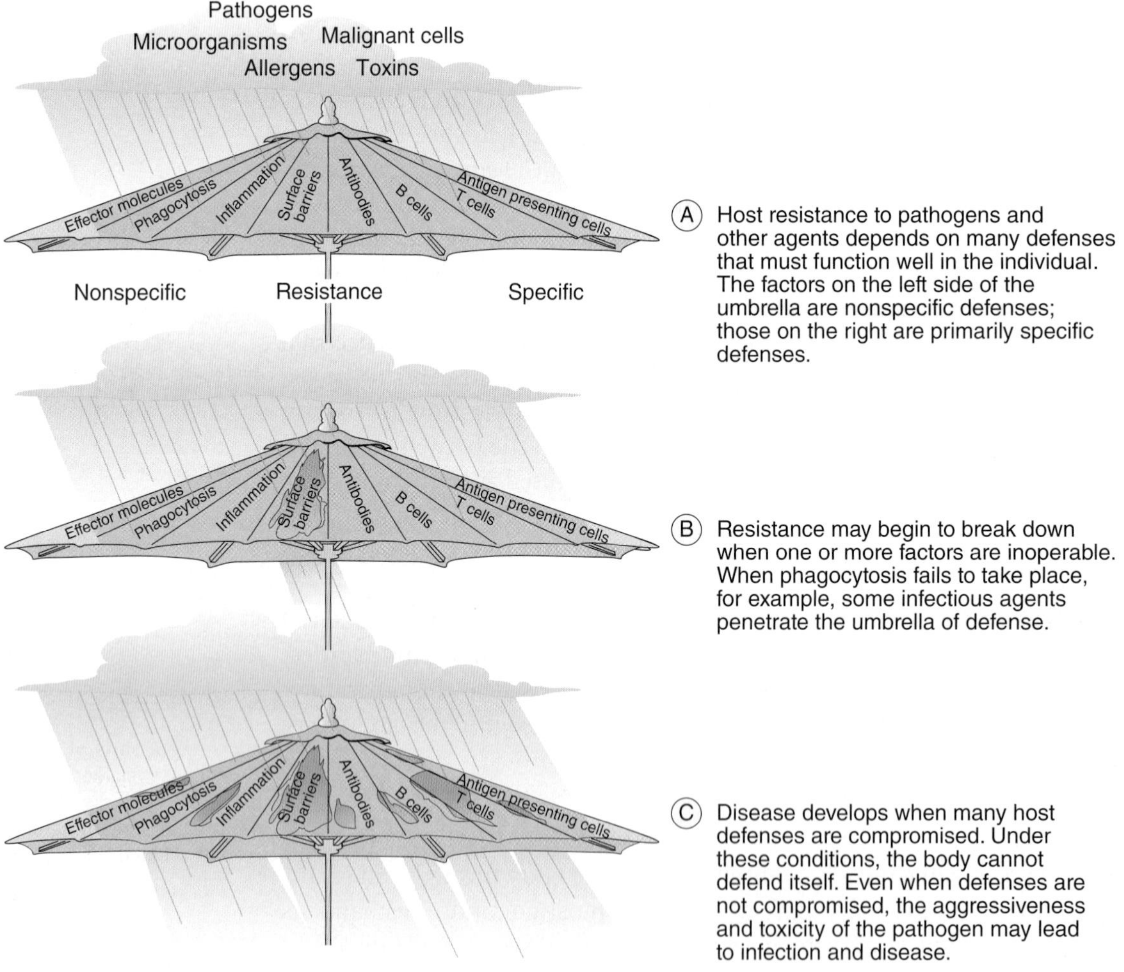

Pathogens
Microorganisms
Malignant cells
Allergens
Toxins
Effector molecules
Phagocytosis
Inflammation
Surface barriers
Antibodies
B cells
T cells
Antigen presenting cells
Nonspecific
Resistance
Specific
(A) Host resistance to pathogens and other agents depends on many defenses that must function well in the individual. The factors on the left side of the umbrella are nonspecific defenses; those on the right are primarily specific defenses.
(B) Resistance may begin to break down when one or more factors are inoperable. When phagocytosis fails to take place, for example, some infectious agents penetrate the umbrella of defense.
(C) Disease develops when many host defenses are compromised. Under these conditions, the body cannot defend itself. Even when defenses are not compromised, the aggressiveness and toxicity of the pathogen may lead to infection and disease.

FIGURE 17.10 **The Relationship Between Host Resistance and Disease.** Host resistance can be likened to a "microbiological umbrella" that forms a barrier or defense against infectious disease.

Mechanical and Chemical Barriers

The intact skin provides an important nonspecific defense against all microbial invaders. The skin itself not only provides mechanical protection, but its cells are constantly being sloughed off, and with them go any attached microbes. In addition, the keratin (the protein in the dead skin cells) is a poor source of nutrients for microbes; the sweat and the fatty acids in sebum (the oily substance in skin glands) contain antimicrobial substances; and, from the viewpoint of a microbe, the low water content of the skin makes the surface a veritable desert. Unless injury to the skin occurs, infectious disease is rare. But penetration of the skin barrier is a fact of everyday life. A cut or abrasion provides a portal of entry into the blood and tissues, and an insect bite acts like a hypodermic needle, permitting any microbes in the infected insect to enter the blood.

The surface of many body cavities, such as the respiratory system and digestive tract, is lined with a **mucous membrane** that provides a physical and chemical barrier to infection. For example, cells of the mucous membranes along the respiratory passageways secrete **mucus**, a viscous substance that traps microbes. In fact, the amount of mucus increases substantially when we are sick with diseases like the cold and flu, as **A Closer Look 17.3** points out. The cilia of other cells nearby then move the mucoid particles along the membranes up to the throat, where they are swallowed into the stomach. Stomach acid destroys the microbes.

Additional chemical barriers exist. Resistance to infection in a woman's vaginal tract is enhanced by the low pH that develops when the resident *Lactobacillus* species synthesize various acids. Indeed, the disappearance of lactobacilli during excessive antibiotic treatment causes the acid to disappear and encourages

A CLOSER LOOK 17.3

Going with the Flow

Every time we get a cold, the flu, or a seasonal allergy, we often end up with the sniffles or a truly raging runny nose. When this happens, it is simply your body's response to what it believes is an infection—or a hypersensitivity to cold temperatures or spicy food. As a defensive barrier against infection, innate immunity uses mucus flow as the best way to wash respiratory pathogens out of the airways.

You always are producing—and swallowing—mucus. Most people are not aware of their mucus production until their body revs up mucus secretion in response to a cold or flu virus, or allergens. So how much mucus is produced? In a healthy individual, glands in the nose and sinuses are continually producing clear and thin mucus—often more than 200 milliliters (about one cup) each day! An individual is not aware of this production because the mucus flows down the throat and is swallowed. Now, if that individual comes down with a cold or the flu, the nasal passages often become congested, forcing the mucus to flow out through the nostrils of the nose. This requires clearing by blowing the nose or (to put it nicely) expectorating from the throat. With a serious cold or flu, the mucus may become thicker and gooier, and have a yellow or green color. The revved up mucus flow in such cases can amount to about 200 milliliters every hour; if you blow your nose 20 times an hour, each blow could amount to anywhere from 2 to 10 milliliters of mucus. If you have watery eyes as well, then that tear liquid can enter the nasal passages and combine with the mucus, producing an even larger "flow per blow."

So, although a runny nose is usually just an annoyance, make sure you drink plenty of water to make up for the lost mucus from that runny nose.

candidiasis ("yeast infection") to develop. Another barrier in the gastrointestinal tract is provided by stomach acid, which has a pH of approximately 2.0. (A cotton handkerchief placed in stomach acid would dissolve in a few moments.) Most microbes are also destroyed in this environment. Another chemical barrier is provided by the enzyme **lysozyme** found in the saliva and tears. Lysozyme disrupts the cell walls of gram-positive bacterial species and causes the cells to swell and burst. The antiviral substance **interferon**, which is produced in response to a viral infection, indirectly interferes with virus replication.

Phagocytosis and Other Factors

Even though the skin and mucous membranes are formidable barriers to infectious agents, these barriers can be breached. Therefore, there needs to be a second line of defense to eliminate any pathogens crossing the skin or mucous membranes. This involves several factors, including phagocytosis, inflammation, and fever.

Phagocytosis. Shortly after the germ theory of disease was verified by Robert Koch, a Russian scientist named Elie Metchnikoff made a chance discovery while observing cells moving freely within transparent starfish larvae. He saw this as a mechanism by which cells could eliminate bacteria. He wrote "*These wandering cells in the body of the starfish larvae, these cells eat food...but they must eat up microbes too! Of course, the wandering cells are what protect the starfish from microbes!* [Therefore] *our wandering cells, the white cells of our blood—they must be what protects us from invading germs.*" This was the beginning of the theory of **phagocytosis**, the process by which certain white bloods cells respond to an infection by attacking and ingesting the invading microbes.

Phagocytosis (literally, "cell-eating") is viewed as a form of nonspecific resistance in the body. The white blood cells involved are called **phagocytes** and they originate in the bone marrow, circulate in the bloodstream for a while, then leave the circulation, and develop in body tissues. Among the most important phagocytes are the highly specialized **macrophages** found in many body tissues and the blood.

When a macrophage encounters a microbe, it carries out the phagocytosis process by enclosing the microbe with a portion of its cell membrane, then infolding the membrane to form an internal sac called a phagosome (FIGURE 17.11). Lysosomes, which contain a large variety of digestive enzymes, fuse with the phagosome and the digestive enzymes destroy and digest the enclosed microbe. Phagocytosis then is a major nonspecific defense mechanism capable of eliminating just about any pathogen.

Inflammation. Inflammation is a nonspecific defensive response that occurs when an injury or traumatic event occurs in human tissues, be it a paper cut to a finger, a burn to the hand, or an infection. The trauma sets into motion a nonspecific process designed to limit the extent of the injury. When close to the skin surface, it is easily identified by its characteristic signs: redness, warmth, swelling, and pain.

For an infection, redness and warmth are due to the dilation (expansion) of the blood vessels bringing an increased flow of blood (and white blood cells) to the infection site. Swelling comes from the accumulation of fluid, the pressure of which on nerve endings accounts for the pain. All these signs work to bring macrophages and other phagocytes quickly to the site of the infection so the cells can begin to phagocytize the invading pathogens. In addition, it sets in motion the events leading to repair of the damaged tissue. In the end, pus, a mixture of blood fluid, dead white blood cells, and dead bacteria, may accumulate at the site before complete repair or a scar is formed.

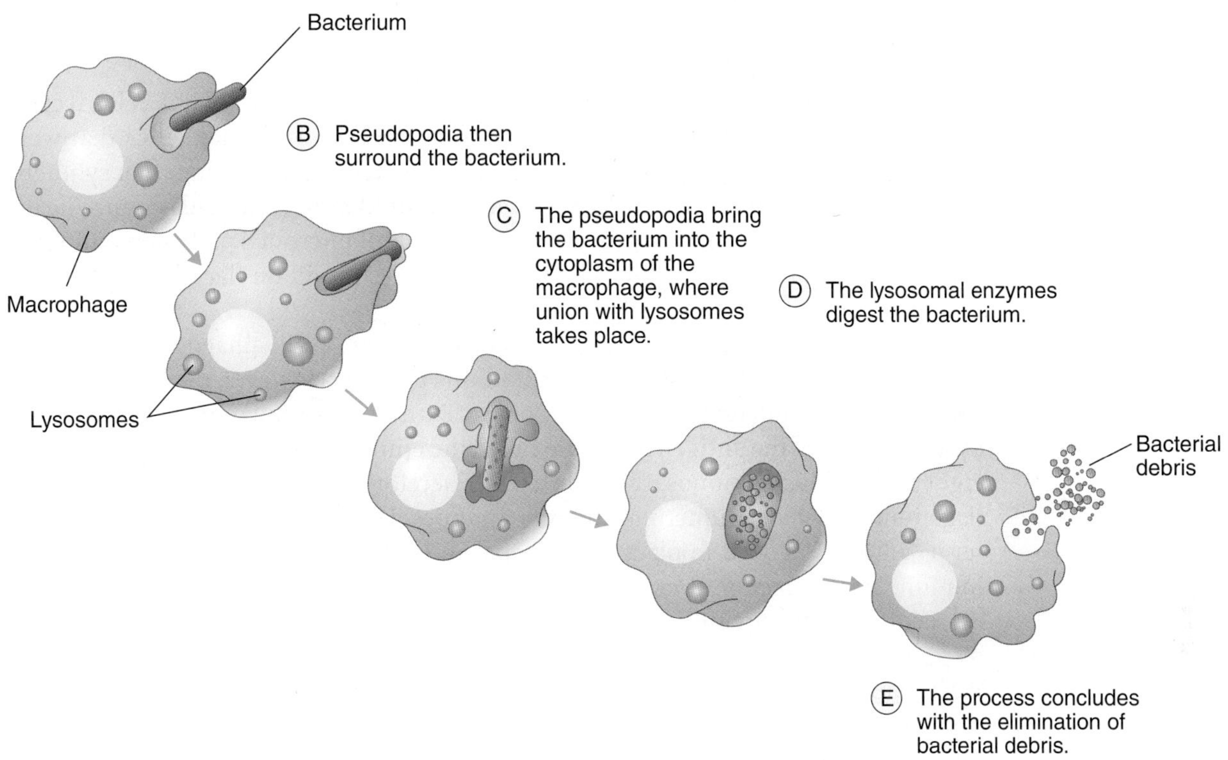

(A) The macrophage attaches to a bacterium.
Bacterium
(B) Pseudopodia then surround the bacterium.
(C) The pseudopodia bring the bacterium into the cytoplasm of the macrophage, where union with lysosomes takes place.
(D) The lysosomal enzymes digest the bacterium.
Macrophage
Lysosomes
Bacterial debris
(E) The process concludes with the elimination of bacterial debris.

FIGURE 17.11 **The Mechanism of Phagocytosis.** The stages of phagocytosis are shown.

Fever. Often accompanying inflammation is **fever**, an abnormally high body temperature. In the case of an infection, fever-producing substances, called **pyrogens** are released in the blood. These include immune proteins produced by macrophages and other white blood cells, and microbial fragments from bacterial cells, viruses, and other microorganisms are released in the blood. The pyrogens move through the blood and affect the anterior hypothalamus, resulting in a dialing up of body temperature above the normal core (oral) temperature of 37°C (98.6°F). As this takes place, cell metabolism increases and blood vessels constrict, thus denying blood to the skin and keeping its heat within the body. Depending on the infection, patients may experience cold skin and chills along with the fever.

■ hypothalamus: The hypothalamus is an area of the brain producing hormones that direct a multitude of important functions, including body temperature, hunger, sleep, and thirst.

A low to moderate fever (up to 38.3°C or 101°F) is a natural defensive response because the elevated temperature inhibits the rapid growth of many microbes, inactivates toxins, encourages rapid tissue repair, and heightens phagocytosis. However, if the fever is prolonged, or rises above 40°C (104°F), convulsions and death may result from metabolic inhibition in the body. Also, infants with a fever above 38°C (100°F), or older children with a fever of 39°C (102°F), may need medical attention.

17.5 Specific Resistance to Disease: Acquiring Immunity

Specific resistance refers to the reactions mounted by the immune system when a specific pathogen has managed to get past the host barriers and nonspecific defenses mentioned above. In these cases, a third defense mechanism comes into play, one that is acquired and aimed solely at that one specific infecting pathogen. It is not something one is born with or can pass on to the next generation. In this last section of the chapter, let's see how this acquired immunity works.

Pathogen Recognition

An **antigen** is any molecule or part of a molecule that is recognized by T cells, B cells, and antibodies. In fact, the term "antigen" was derived from molecules that were anti(body) gen(erating). Most often, antigens are proteins, carbohydrates, or synthetic substances, although in rare cases, they can be of lipid or nucleic acid origin. The list of such antigens is enormously diverse but, in terms of microbial pathogens, antigens include bacterial protein toxins, chemical structures found on flagella and pili, and viral proteins found on the capsid or envelope.

Antigens usually have a large molecular size and are easily phagocytized by macrophages, the necessary first step in the immune process. However, the whole antigen itself does not stimulate the immune response. Rather, the stimulation is accomplished by a small part of the antigen molecule called the **epitope**, which contains about six to eight amino acids or monosaccharide units. Therefore, an antigen, such as a bacterial flagellum, may have numerous different epitopes. However, for simplicity, often the term "antigen" is used in place of "epitope" in conversation or when writing about immune system recognition.

Under normal circumstances, the body's own epitopes do not stimulate an immune response. This failure occurs because these molecular pieces are recognized as "self." What happens is that the immune system suppresses the activity of and/or actually destroys any cells capable of attacking "self." Thus, the immune system develops a tolerance of "self" and remains extremely sensitive in recognizing "nonself" epitopes present on foreign substances like pathogens.

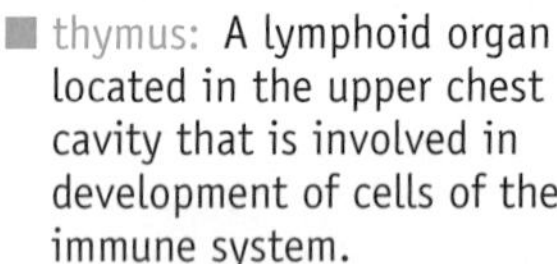

■ thymus: A lymphoid organ located in the upper chest cavity that is involved in development of cells of the immune system.

Cells of the Immune System

The **immune system** is a general term for the complex collection of cells, chemical factors, and processes that provide a specific response to invading pathogens. Besides the phagocytes, the cornerstones of this recognition are a set of white blood cells known as **lymphocytes**. These cells are distributed throughout the body and comprise the organs of the lymphoid system, including the lymph nodes, spleen, tonsils, and adenoids. Lymphocytes are small cells, about 10 to 20 μm in diameter, each with a large nucleus (FIGURE 17.12). Two types of lymphocytes can be distinguished on the basis of developmental history, cellular function, and unique biochemical differences. The two types are **B lymphocytes** (**B cells**) and **T lymphocytes** (**T cells**).

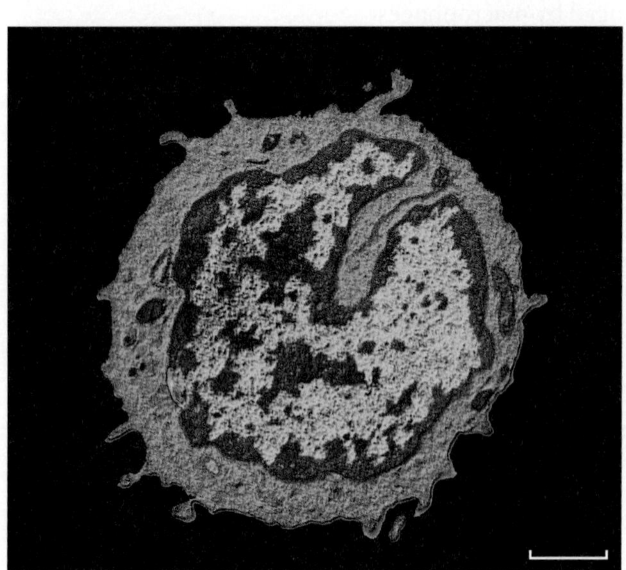

FIGURE 17.12 **A T Lymphocyte.** False-color transmission electron microscope image of a T lymphocyte showing the typical large cell nucleus. (Bar = 2 μm.)

All blood cells, including red blood cells and white blood cells originate from stem cells in the bone marrow (FIGURE 17.13). Some stem cells develop into "myeloid stem cells" that will mature into red blood cells and most of the white blood cells, including the macrophages.

The "lymphoid stem cells" will become lymphocytes of the immune system. Some of these stem cells mature into T cells in the thymus (T for thymus), where they are modified by the addition of surface receptor proteins. They then emerge from the thymus and migrate to secondary lymphoid tissues, such as the lymph nodes, spleen, and tonsils where they are now act as "helper T cells" or "cytotoxic T cells." These cells, as we will see, can interact with "nonself" antigens and are ready to engage in what is called **cell-mediated immunity (CMI)**.

The B cells take a different track. They mature in the bone marrow (B for bone). They then move through the circulation to colonize the secondary lymphoid tissues along with the T cells. B cells are a key cell for antibody production and what is called **antibody-mediated immunity (AMI)**.

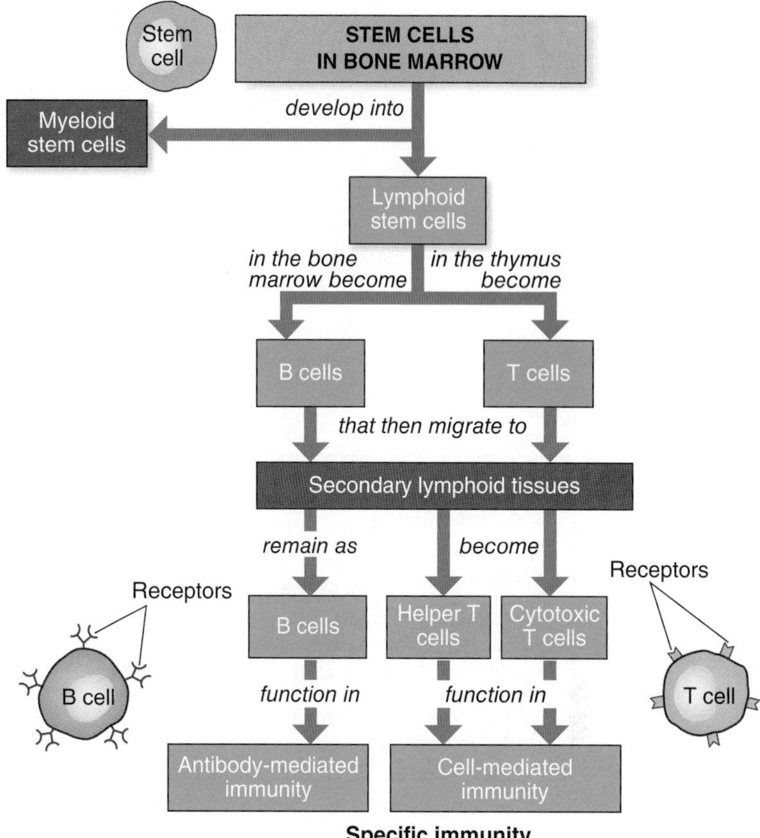

FIGURE 17.13 **The Fate of Lymphoid Stem Cells.** Lymphocytes arise in the bone marrow and mature in the bone marrow (B cells) or thymus (T cells). Insets: a drawing of a B cell and T cells, showing their surface receptors.

Cell-Mediated Immunity

Before we proceed, take a look at the surface receptor proteins in Figure 17.13. These proteins enable B cells and T cells to recognize a specific epitope on an antigen and bind to it.

The CMI response originates with the entry of antigen-containing pathogens into the body and their passage into the tissues. As described above, the pathogens are phagocytized by macrophages and other phagocytic cells, and the antigens are broken down. The macrophages display on their surface digested fragments of the antigen, called "antigen peptides" (FIGURE 17.14). Thus, the macrophages are called **antigen-presenting cells (APC)**.

The APCs travel to an organ of the lymphoid system, such as the lymph nodes, where they "present" the peptide on their surface to the T cells. Thus, the APCs are critical because they are bearing the news that there is an infection and telling the T cells "Hey! There is an active infection occurring in the body. You need to get motivated and get to work to deal with it!"

Within the lymph nodes, the T cells are responsible for aiding and promoting the destruction of antigen and pathogen-infected cells. Here is how they do it.

Death by Cytotoxic T Cell. Many pathogens, including certain bacterial species and viruses, have the ability to enter inside body cells. These infected cells need to be destroyed and that is one of the jobs of the CMI.

For example, when measles viruses enter and infect cells, the infected cells manage to present viral antigen fragments on their exterior surface. This combination is essentially a "red flag" saying, "Help, I've been infected!"

Meanwhile back in the lymph node, an APC mingles among the myriad groups of inactive T cells, searching for a group having the surface receptor recognizing its antigen peptide specific to the measles virus.

Once this matchup has been completed, those and only those T cells are "motivated" and mature into an activated form, called **cytotoxic T cells (CTLs)**, which then divide to produce a population of identical cells called a clone (see Fig. 17.14), all with the capability of recognizing flu virus-infected cells.

The CTLs cells now leave the lymphoid tissue and enter the lymph and blood vessels. They circulate until they encounter their targets, the infected cells displaying the telltale "red flags" on their surface (see Fig. 17.14). Now begins the cell-cell interaction characteristic of CMI. The receptor proteins of each CTL recognize the antigen peptide on the infected cells, triggering the CTLs to release a number of active

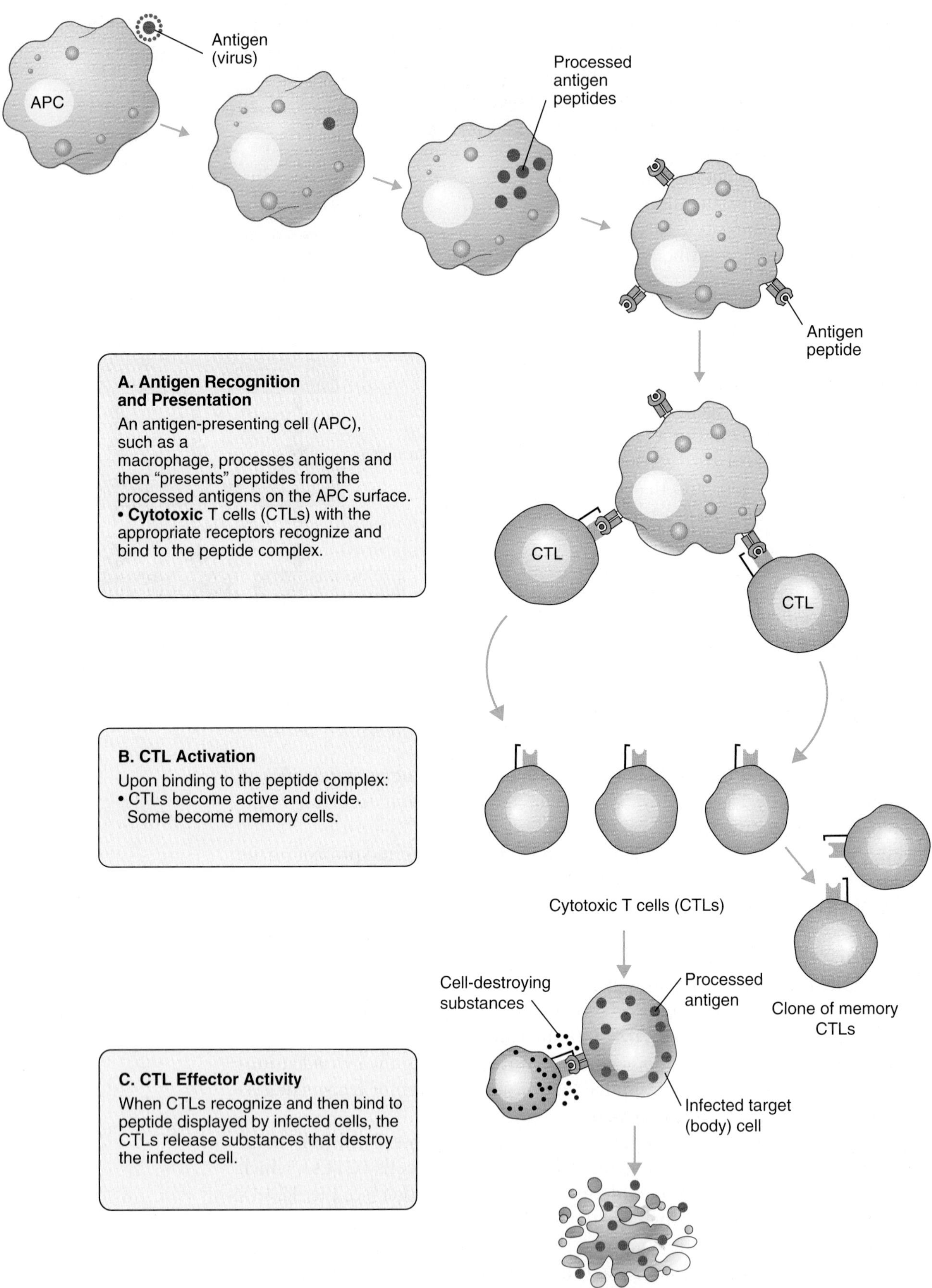

Antigen
(virus)
APC
Processed
antigen
peptides
Antigen
peptide
A. Antigen Recognition
and Presentation
An antigen-presenting cell (APC), such as a macrophage, processes antigens and then "presents" peptides from the processed antigens on the APC surface.
• Cytotoxic T cells (CTLs) with the appropriate receptors recognize and bind to the peptide complex.
CTL
CTL
B. CTL Activation
Upon binding to the peptide complex:
• CTLs become active and divide. Some become memory cells.
Cytotoxic T cells (CTLs)
Cell-destroying
substances
Processed
antigen
Clone of memory
CTLs
C. CTL Effector Activity
When CTLs recognize and then bind to peptide displayed by infected cells, the CTLs release substances that destroy the infected cell.
Infected target
(body) cell
Death of infected cell

FIGURE 17.14 **Cell-Mediated Immunity.** Triggered by antigen peptide presentation, cytotoxic T cells (CTLs) become activated and primed to destroy infected cells.

substances capable of poking holes through the membrane of the infected cells. Small molecules, fluids, and cell structures escape, and death of the infected cell occurs. In this manner, all such infected cells would be destroyed by the population of CTLs.

When the CTLs are dividing in the lymphoid tissues, some cells will mature into **memory T cells** that will provide resistance in the event the same pathogen re-enters the body in the future. Should the pathogen be detected once again in the tissues, the memory T cells will multiply rapidly and immediately set into motion the process of CMI. This is one reason we enjoy long-term immunity to a given disease after having contracted and recovered from that disease (remember what Thucydides remarked at the beginning of this chapter).

Antibody-Mediated Immunity

While CMI is centered in attacking infected cells, AMI attempts to eliminate pathogens in the body fluids. Thus, AMI depends on the activity of **antibodies**, a class of protein molecules circulating in the body's fluids, capable of reacting with toxin molecules in the bloodstream as well as with whole pathogens in the body fluids.

With AMI, the major type of participating immune system cell is the B cell. Recognition by the appropriate B cells occurs when antigens enter a lymphoid organ, bringing the epitopes close to the appropriate B cells where they bind with B cell receptor proteins. After binding, the antigen peptide is internalized and presented on the cell surface (FIGURE 17.15a).

While this process is going on, an APC mingles among the groups of inactive T cells, searching for a group having the surface receptor recognizing its specific antigen peptide.

Besides CTLs, there is another group of T cells, called **helper T cells (HTLs)** capable of recognizing the APC antigen-peptide (FIGURE 17.15b). Once this matchup has been completed, the inactive T cells are "motivated" and mature into activated HTLs, which then divide to produce a population of identical HTLs.

Activation of the B cells begins when HTLs in the lymphoid organ bind to the peptide presented by the B cell (FIGURE 17.15c). That binding stimulates the B cell to divide into a large B cell clone.

The B cells now mature into one of two types of cells. Some become **memory B** cells and like, memory T cells, will be ready for a future exposure to the same pathogen (FIGURE 17.15d). The majority mature into **plasma cells**, which are large, complex cells having no surface protein receptors. Their sole purpose is to produce antibodies. In fact, a plasma cell lives about 4 to 5 days, during which time it produces an incredible 2,000 antibody molecules per second. Realize these antibodies recognize the same epitope on the antigen that originally started the CMI and AMI processes. Thus, once the plasma cells are present, antibodies start flowing into the body fluids. Because of their sheer force of numbers, some of them inevitably meet up with the antigens (i.e., measles viruses), bind to them and mark them for elimination. Essentially, by binding to the pathogen, the antibody "coats" the antigen and acts as a "beacon" that is recognized by phagocytes. These phagocytes then engulf and phagocytize both the antigen and the antibodies. So, in essence, AMI is designed to immobilize and eliminate the antigens in the body fluids.

Antibody Structure. Antibodies are extremely diverse, and viruses or the cells of a single bacterial species may elicit hundreds of different kinds of antibodies able to unite with hundreds of different kinds of viral or bacterial antigens (i.e., epitopes). The term "immunoglobulin" (Ig) is used interchangeably with "antibody" because antibodies exhibit the properties of globulin proteins and are used in the immune response.

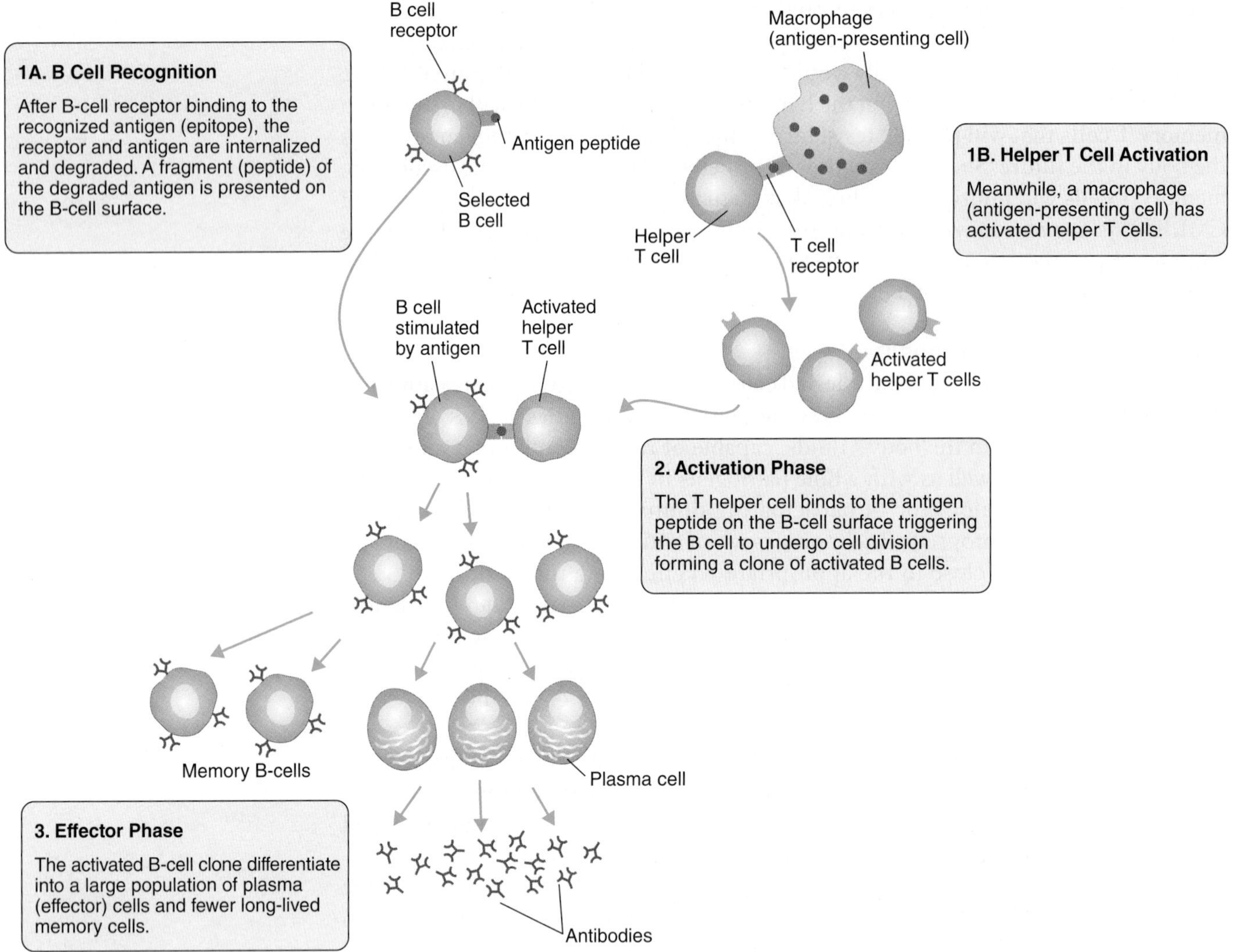

B cell receptor
Macrophage (antigen-presenting cell)
1A. B Cell Recognition
After B-cell receptor binding to the recognized antigen (epitope), the receptor and antigen are internalized and degraded. A fragment (peptide) of the degraded antigen is presented on the B-cell surface.
Antigen peptide
Selected B cell
1B. Helper T Cell Activation
Meanwhile, a macrophage (antigen-presenting cell) has activated helper T cells.
Helper T cell
T cell receptor
B cell stimulated by antigen
Activated helper T cell
Activated helper T cells
2. Activation Phase
The T helper cell binds to the antigen peptide on the B-cell surface triggering the B cell to undergo cell division forming a clone of activated B cells.
Memory B-cells
Plasma cell
3. Effector Phase
The activated B-cell clone differentiate into a large population of plasma (effector) cells and fewer long-lived memory cells.
Antibodies

FIGURE 17.15 **Antibody-Mediated Immunity.** Only those B cells with a surface receptor matching the shape of an epitope will be stimulated. Meanwhile, triggered by antigen peptide presentation, helper T cells become activated and divide. Binding to B cells presenting a peptide-MHC complex, helper T cells stimulate those B cells to divide and mature into plasma cells that secrete antibody.

The basic antibody molecule consists of four polypeptide chains: two larger, identical **heavy** (H) **chains** and two smaller, identical **light** (L) **chains** (FIGURE 17.16a). These chains are joined together by chemical linkages to form a Y-shaped structure. An antibody molecule then has two identical arms, each consisting of a heavy chain and a light chain.

Each polypeptide chain, light or heavy, has both a constant and variable region. The amino acids in the constant region are virtually identical among different types of antibodies. However, the amino acids of the variable region differ among the hundreds of thousands of antibody types. Thus, the variable regions of a light and heavy chain combine to form a highly specific, three-dimensional structure somewhat analogous to the active site of an enzyme. This portion of the antibody molecule, called the **antigen binding site**, is uniquely shaped to combine with an epitope on an antigen. Moreover, because its "arms" are identical, a single antibody molecule has two antigen binding sites, so it can combine with two identical epitopes. These

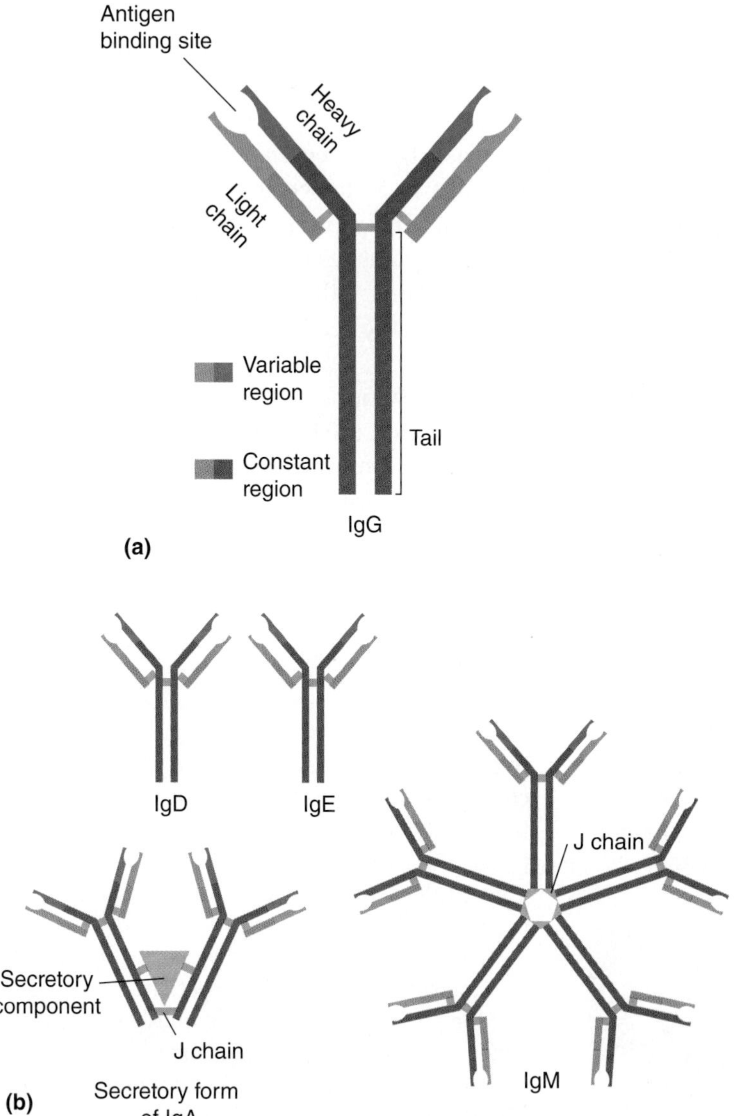

FIGURE 17.16 **Structure and Classes of Antibodies.** (**a**) The basic structure of an antibody consists of two light and two heavy polypeptide chains. The variable domains in each light and heavy chain form a pocket called the antigen-binding site. (**b**) Note the complex structures of IgM (5 antigen-binding sites) and IgA (four antigen-binding sites). IgG, IgE, IgD consist of a single Y-shaped unit.

combinations may lead to a complex of antibody and antigen molecules. The tail region, composed of the two heavy chains, is important for phagocyte recognition.

Antibody Classes. Five classes of antibodies have been identified. Using the abbreviation Ig (for immunoglobulin), the five classes are designated IgG, IgM, IgA, IgE, and IgD.

IgG is the classical "gamma globulin." This type of antibody is the major circulating one, comprising about 80% of the total antibody content in normal **serum**, the fluid portion of the blood. IgG appears about 24 to 48 hours after antigenic stimulation and provides long-term resistance to disease. Booster injections of a vaccine raise the level of this type of antibody considerably. IgG is also the maternal antibody that crosses the placenta and renders immunity to the child until it is fully capable of producing antibodies at about 6 months of age.

The remaining antibody classes are composed of one or more of the Y-shaped immunoglobulin units (FIGURE 17.16b). IgM is the first type of antibody to appear in the circulatory system after B cell stimulation. It is the largest antibody molecule (M stands for macroglobulin) and has 10 identical antigen binding sites. Because of its size, most of it remains in circulation and accounts for 5% to 10% of the antibody in the blood during an infection.

Much of another antibody class, IgA, is secreted and accumulates in body secretions. This antibody, composed of two Y-shaped Ig units and four identical antigen binding sites, provides resistance in the respiratory and gastrointestinal tracts, possibly by inhibiting the attachment of parasites to the tissues. It is also located in tears and saliva and in the colostrum, the first milk secreted by a nursing mother. When consumed by a child during nursing, the secretory antibody provides resistance to gastrointestinal disorders.

IgE plays a major role in allergic reactions by sensitizing cells to certain antigens. IgD antibody is a cell surface receptor on the B cells.

Before concluding the material on immunity, it is worth considering whether your immune system is influenced by "thinking good thoughts." **A Closer Look 17.4** cogitates on this idea.

A CLOSER LOOK 17.4

Can Thinking "Well" Keep You Healthy?

In 2010, psychological scientists at the University of Kentucky and the University of Louisville wanted to see how optimism affects the immune system. Studying law students, they were interested in knowing if student expectations about their future (optimistic or not) was reflected in their immune responses. The 6-month study

© Creatas/Thinkstock

showed an optimistic disposition made no difference in their immune responses. However, as each student had highs and lows in law school, their immune response showed a similar response. In optimistic times, their immune system was quicker to respond to an immunological challenge and in pessimistic times, their immune system was slower to respond to a similar challenge. The scientists concluded that an optimistic outlook may promote a stronger immune system.

The idea that one's mental state can influence the body's susceptibility to, and recovery from, disease has a long history. The Greek physician Galen thought cancer struck more frequently in melancholy women than in cheerful women. During the twentieth century, the concept of mental state and disease was researched more thoroughly, and a firm foundation was established linking the nervous system and the immune system. As a result of these studies, a new field called "psychoneuroimmunology" has emerged.

The outcome of these discoveries is the emergence of a strong correlation between a patient's mental attitude and the progress of disease. Rigorously controlled studies conducted in recent years have suggested that the aggressive determination to conquer a disease can increase the lifespan of those afflicted. Therapies consist of relaxation techniques, as well as using mental imagery suggesting disease organisms are being crushed by the body's immune defenses. Behavioral therapies of this nature can amplify the body's response to disease and accelerate the mobilization of its defenses.

More recently, scientists at Ohio State University concluded that women who practiced yoga regularly had lower levels of immune suppressing chemicals in their blood than those women who did not practice yoga. The benefits of these lower levels was important because when these chemicals are unregulated (e.g., under stress conditions), they have been shown to contribute to the onset of rheumatoid arthritis, inflammatory bowel disease, osteoporosis, multiple sclerosis, and some types of cancer. In the yoga study, after a stressful experience, the women doing yoga showed smaller increases in these immune suppressing chemicals than their nonpractitioners.

Few reputable practitioners of behavioral therapies believe such therapies should replace drug therapy. However, the psychological devastation associated

A CLOSER LOOK 17.4

Can Thinking "Well" Keep You Healthy?

with many diseases, such as AIDS, cannot be denied, and it is this intense stress that the "thinking well" movement attempts to address. Very often, for instance, a person learning he or she is HIV positive goes into severe depression, and because depression can adversely affect the immune system, a double dose of immune suppression ensues. Perhaps by relieving the psychological trauma, the remaining body defenses can adequately handle the virus.

As with any emerging treatment method, there are numerous opponents of behavioral therapies. Some opponents argue that naive patients might abandon conventional therapy; another argument suggests therapists might cause enormous guilt to develop in patients whose will to live cannot overcome their failing health. Proponents counter with a growing body of evidence suggesting patients with strong commitments and a willingness to face challenges—signs of psychological hardiness—have relatively greater numbers of T cells than passive, nonexpressive patients. To date, no study can conclude mood or personality has a life-prolonging effect on immunity. Still, doctors and patients are generally inspired by the possibility of using one's mind to help stave off the effects of infectious disease.

Types of Specific Immunity

The word "immune" (*immuno* = "safe" or "free from") in its most general sense implies a condition under which an individual is protected (safe) from disease. This does not mean, however, that the person is immune to all diseases, but rather to a specific disease or group of diseases.

Two general types of specific immunity are recognized (FIGURE 17.17). The first, called **active immunity**, develops when the immune system responds to antigens and forms antibodies, as we have been discussing in this chapter. Active immunity takes several hours or days to develop fully, but it remains for a long period of time, often throughout a person's life. On the other hand, **passive immunity** develops when antibodies enter the body from an outside source. This second form of immunity comes about immediately when antibodies enter the body, but it lasts only as long as the antibodies are present (several days to a few weeks).

Notice in Figure 17.17 that active and passive immunity are further subdivided into four types. The first is "naturally acquired active immunity." This usually follows a bout of illness. Memory T and B cells residing in the lymphoid tissues remain active for many years and produce IgG immediately if the pathogen later enters the host. "Artificially acquired active immunity," by comparison, develops after an exposure to antigens in a vaccine. The antigens may be toxoids, inactivated viruses, synthetic viral parts, bacterial parts, or other components. Vaccines promote a long-term immune response in the form of memory T or B cells and IgG antibodies. Many vaccines may be effective for a lifetime, but some need to be readministered periodically as booster injections to restimulate the immune response.

The third type of immunity is "naturally acquired passive immunity," also called congenital immunity. This immunity develops when IgG antibodies pass from the mother's bloodstream into the fetal circulation via the placenta and umbilical cord. The antibodies remain with the child for approximately 3 to 6 months after birth and fade as the child's immune system becomes fully functional. Maternal antibodies, predominantly IgA type, also pass to the newborn through the colostrum. Lastly, "artificially acquired passive immunity," the fourth type, arises from the injection of antibody-rich serum into the patient's circulation. This form of therapy is used for serious viral diseases and toxin-related diseases such as botulism and tetanus. The serum injected is often referred to as gamma globulin.

ACTIVE IMMUNITY

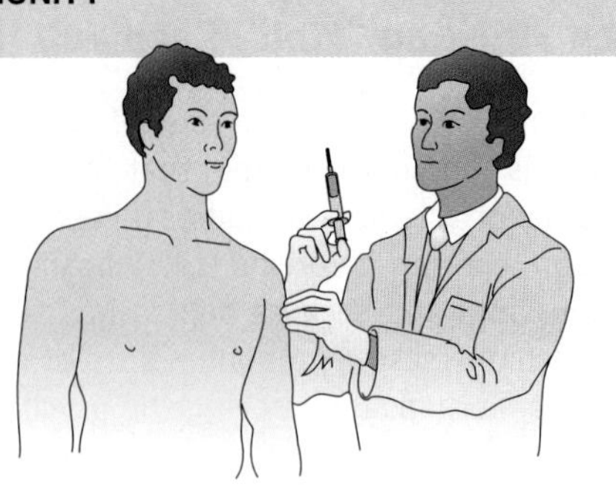

(A) Naturally acquired active immunity arises from an exposure to antigens and often follows a disease.

(B) Artificially acquired active immunity results from a vaccination.

PASSIVE IMMUNITY

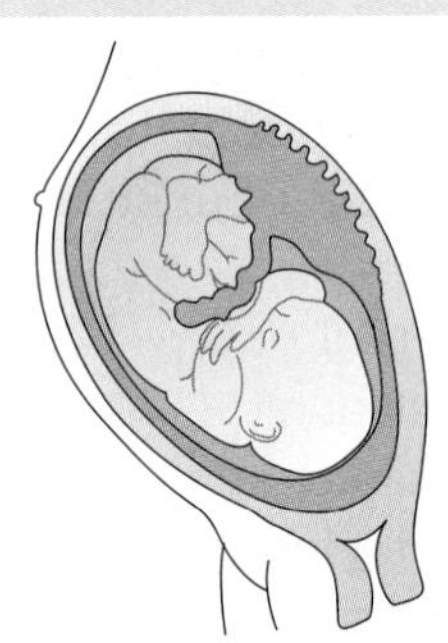

(C) Naturally acquired passive immunity stems from the passage of IgG across the placenta from the maternal to the fetal circulation.

(D) Artificially acquired passive immunity is induced by a transfer of antibodies (antitoxins) taken from the circulation of an animal or another person.

FIGURE 17.17 **Four Ways to Acquire Immunity.** Immunity can be generated in an active or passive form and by natural or artificial means.

A Final Thought

It may have occurred to you that an episode of disease is much like a war. First, the invading microbes must penetrate the natural barriers of the body; then, they must escape the phagocytes and other chemicals constantly patrolling the body's circulation and tissues. Finally, they must elude the antibodies or T cells the body sends out to combat them. How well the body does in this battle will determine whether the individual survives the disease.

Obviously, our bodies do very well, because most of us are in good health today. As it turns out, most infectious organisms are stopped at their point of entry to the body. Cold viruses, for example, get no farther than the upper respiratory tract. Many parasites come into the body by food or water, but they rarely penetrate beyond the intestinal tract. The staphylococci in a wound may cause inflammation at the infection site, but this is usually the extent of the problem.

To be sure, antibiotics and other antimicrobial drugs help in those cases where diseases pose life-threatening situations. In addition, sanitation practices, insect control, care in the preparation of food, and other public health measures prevent microbes from reaching the body in the first place. However, in the final analysis, body defense represents the bottom line in protection against disease, and, as history has shown, it usually works very well. The late Lewis Thomas, an American physician, poet, essayist, and researcher, said it best in his book, *The Lives of a Cell*: "*A microbe that catches a human is in considerably more danger than a human who catches a microbe.*"

Questions to Consider

1. In his classic book *The Mirage of Health* (1959), the French scientist René Dubos develops the idea that health is a balance of physiological processes, a balance that takes into account such things as nutrition and living conditions. (Such a view opposes the more short-sighted approach of locating an infectious agent and developing a cure.) From your experience, describe several other things you would add to Dubos' list of "balancing agents."
2. The transparent windows placed over salad bars are commonly called "sneeze guards" because they help prevent nasal droplets from reaching the salad items. What other suggestions might you make to prevent disease transmission via a salad bar?
3. Several recent reports suggest infectious diseases are still a major public health issue in the United States. Population shifts, modern travel patterns, and microbial evolution are three of the many reasons given for the emergence and reemergence of infectious diseases. What other reasons can you name?
4. An environmental microbiologist has created a stir by maintaining that a plume of water is aerosolized when a toilet is flushed, and that the plume carries bacteria to other items in the bathroom, such as toothbrushes. Assuming this is true, what might be two good practices to follow in the bathroom? Further, suppose you are a microbe hunter assigned to make a list of the ten worst "hot zones" in your home. The title of your top-ten list will be "Germs, Germs Everywhere." What places will make your list, and why?
5. A woman takes an antibiotic to relieve a urinary tract infection caused by *Escherichia coli*. The infection resolves, but two weeks later she develops a "yeast infection" of the vaginal tract caused by *Candida albicans*. What conditions may have caused this to happen?
6. The ancestors of modern humans lived in a sparsely settled world where communicable diseases were probably very rare. Suppose one of those individuals was magically thrust into our present-day world. How do you suppose he or she would fare in relation to infectious disease? What is the immunological basis for your answer?
7. In the book and 1966 classic movie, *Fantastic Voyage,* a group of scientists is miniaturized in a submarine (the Proteus) and sent into the human body to dissolve a blood clot. The odyssey begins when the miniature submarine carrying the scientists is injected into the bloodstream. What perils, immunologically speaking, would you encounter if you were to take such an adventure?

Key Terms

Informative facts are necessary for the expression of every concept, and the information for a concept is founded in a set of key terms. The following terms form the basis for the concepts of this chapter. On completing the chapter, you should be able to explain and/or define each one.

active immunity
acute period (climax)
antibody
antibody-mediated immunity (AMI)
antigen
antigen binding site
antigenic determinant (epitope)
antigen presenting cell (APC)
antitoxin
avirulent
biological vector
B lymphocyte (cell)

carrier
cell-mediated immunity (CMI)
commensalism
cytotoxic T cell (CTL)
diagnosis
direct contact
disease
droplet
endemic
endotoxin
epidemic
epitope
exotoxin
fecal-oral route
fever
fomite
heavy (H) chain
helper T cell (HTL)
host
hyaluronidase
incubation period
indirect method
infection
infectious dose
inflammation
interferon
invasiveness
light (L) chain
lymphocyte
lysozyme
macrophage
mechanical vector
memory B cell
memory T cell
mucus
mucus membrane
microbiota
mutualism
opportunistic
outbreak
pandemic
parasitism
passive immunity
pathogenicity
period of convalescence
period of decline
phagocyte
phagocytosis
plasma cell
portal of entry
portal of exit
probiotic
prodromal phase
pyrogen
reservoir
serum
sign
symbiosis
symptom
syndrome
thymus
T lymphocyte (cell)
toxin
toxoid
virulence
zoonosis

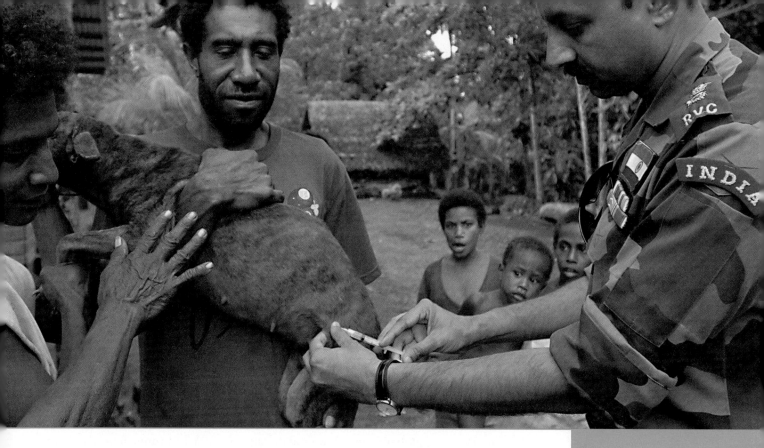

Viral Diseases of Humans: AIDS to Zoster

Looking Ahead

The viral diseases of humans cover a broad spectrum of illnesses and include usually mild diseases such as measles, mumps, and chickenpox as well as very serious maladies as smallpox, rabies, and AIDS. These diseases, and others, will be surveyed in this chapter.

Because this chapter mentions many different viral diseases, the chapter objectives focus on the viral diseases in each section. On completing the chapter, you should be capable of:

- Identifying and describing several viral diseases of the skin.
- Drawing the structure of a type A flu virus, and comparing and contrasting flu symptoms with common cold symptoms.
- Identifying three viral diseases of the nervous system and explaining why they are so dangerous.
- Listing and describing several diseases of the visceral organs, including the blood, liver, and gastrointestinal tract.
- Comparing and contrasting HIV infection and AIDS.

Every September 28 is World Rabies Day, a global health observance aimed at promoting awareness and prevention of human rabies. Why a World Rabies Day? In 2013, rabies took the lives of more than 55,000 people, over 50% being children. In fact, 3.3 billion people are at risk of rabies primarily from bites originating from uncontrolled rabies infections in dogs. Since the 2007 inaugural World Rabies Day, more than 182 million people have been educated about rabies and 7.7 million dogs have been vaccinated, such as this dog in Papua, New Guinea.

Courtesy of Mass Communication Specialist 3rd Class Bryan M. Ilyankoff/U. S. Navy

Every year there are seasonal **influenza** outbreaks ("the flu") that may become epidemic and spread quickly through a large, susceptible population. Occasionally an epidemic becomes a global epidemic called a pandemic. The largest influenza pandemic, called the "Spanish flu," occurred in 1918 to 1919 when a fifth of the world's population was infected by a very virulent form of the flu virus. Estimates suggest between 20 and 50 million people died.

Influenza continues to be an ongoing problem in the twenty-first century. Just look back to 2009. In April of that year, the Centers for Disease Control and Prevention (CDC) identified two children in California who had been infected with a new strain of influenza virus. This strain was soon traced back to the major outbreak of flu in Mexico. By the end of April, cases of this new flu strain, originally called "swine flu", were being reported in cities around the world. It appeared the first flu pandemic in over 40 years was emerging.

As the cases continued to mount up in the spring and early summer, thankfully most infected individuals exhibited only relatively mild symptoms. Since the end of the pandemic in late 2010, the CDC has estimated that somewhere between 150,000 and 500,000 people died worldwide (similar to the seasonal flu). In the United States, there were an estimated 61 million cases, 275,000 hospitalizations, and an estimated 12,500 deaths,which was less than the 19,000 that usually die every year from the seasonal flu (FIGURE 18.1).

There were at least two major take-home lessons from this "mini-flu pandemic." First, in this age of globalization, new viral threats can come from anywhere and spread with the speed of jet travel. Second, we need to be on "virological alert" worldwide for new virus strains. For example, most flu outbreaks and epidemics usually start in Southeast Asia—the swine flu surprised disease experts by coming from Mexico. So, this time we dodged the bullet—next time we may not be so lucky.

Influenza is just one of the present-day viral diseases we will discuss. The discussion will not be traditional though; rather we will present the diseases affecting various body systems in the form of a conversation between you and an infectious disease microbiologist. Let's get started.

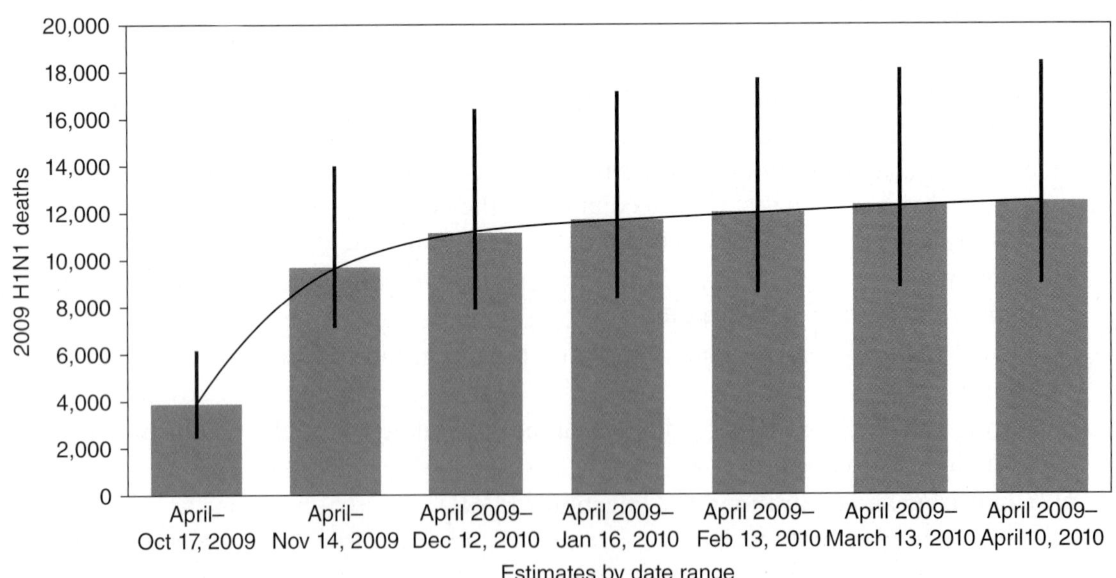

FIGURE 18.1 **Cumulative 2009 H1N1 Deaths in the United States by Date, April 2009–April 2010.** The vertical black lines indicate the range in 2009 H1N1 estimates for each time period.

Reproduced from CDC Estimates of 2009 H1N1 Influenza Cases, Hospitalizations and Deaths in the United States

18.1 Viral Diseases of the Skin: From Mild to Deadly

The viral diseases of the skin are a diverse collection of human maladies. Certain ones, such as herpes simplex, remain epidemic in our time; others, such as measles, mumps, and chickenpox, are under control through effective vaccination programs. These so-called "dermotropic diseases" are generally transmitted by contact with an infected individual, and at least some of the symptoms are seen on the skin tissues. With these diseases, we begin our "microbial conversation."

Herpes Simplex

I have heard there are two types of herpes viruses. What are they? Actually, there are several types of herpes viruses. The two you are most familiar with are **herpes simplex** virus type 1, or HSV-1, and HSV-2. HSV-1 generally infects the facial area around the lips and is the cause of most cold sores or fever blisters as you see in this photo (FIGURE 18.2).

How do you get infected with HSV-1? The first episode is transmitted from another person with an active lesion. Because kissing, or sharing eating utensils or towels, may spread HSV-1, it is important to avoid contact with someone who has active blisters and don't share items that can spread the virus. And wash hands often if you or another person has active blisters because, in some rare cases, HSV-1 can spread to the brain, causing herpes encephalitis or to the eyes, causing herpes keratitis.

Once infected, a person remains infected for life because the virus can lie dormant in nerve cells and emerge again as an active infection. Often recurrence is triggered by some form of stress, such as fatigue, exposure to the sun, or menstruation in women. That is why the malady is called "cold sores" or "fever blisters;" often reactivation comes from the body's stress of having a cold or a fever. Although HSV-1 infections cannot be cured, there are a few antiviral pills or creams that accelerate the healing process and lessen the severity of recurrent episodes.

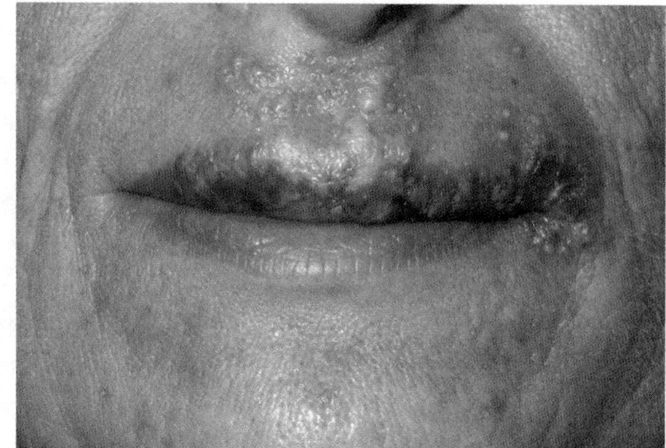

FIGURE 18.2 **Herpes Simplex Viruses Can Cause Cold Sores.** Cold sores (fever blisters) erupt as tender, itchy papules and progress to vesicles that burst, drain, and form scabs. Contact with the sores accounts for spread of the virus.

So, what is HSV-2? HSV-2 is generally responsible for **genital herpes**, a sexually transmitted disease (STD). However, oral sex can spread HSV-1 to the genitals and HSV-2 to the lips.

What's genital herpes like? Genital herpes affects between 10 and 20 million Americans yearly, the majority in people aged 14 to 49 years. Most people are unaware they are infected because there may be few, if any, symptoms. When symptoms do occur, painful blisters on or around the genital area, rectum, or mouth may erupt, crust over, and disappear, usually within about 3 weeks. Again, because of the lifelong infection, genital herpes may reappear, although it may be less severe than the first episode.

Can pregnant women give genital herpes to their newborns? Yes. In a pregnant woman, herpes simplex viruses sometimes pass to the fetus via the placenta and cause **neonatal herpes**, which can produce neurological problems and/or mental impairment in the newborn. Miscarriage and premature birth can also occur. Therefore, it is critical that a pregnant woman tell her doctor if she has ever experienced any symptoms of, been exposed to, or been diagnosed with genital herpes. If so, she may be given antiviral medication to reduce the risk of another episode. Obstetricians recommend women with active genital herpes consider giving birth by cesarean section, which lessens the chance of virus spread to the newborn.

Chickenpox (Varicella)

I am not sure I ever had chickenpox. Is chickenpox easy to catch? Chickenpox, or varicella disease, is a highly communicable disease to people who have never had the disease and have not been vaccinated. The infection is caused by the varicella-zoster virus (VZV), another DNA virus and is usually transmitted by skin contact. Viruses localize in the skin and in nerves close to the skin. As you can see in this photo, the viruses trigger the formation of raised pink bumps that develop into a telltale rash, which is composed of small, teardrop-shaped, fluid-filled blisters (FIGURE 18.3a). The blisters develop over 3 or 4 days in a succession of crops before breaking open to yield highly infectious virus-laden fluid. The blisters eventually form crusts and scabs, and heal over several days. The chickenpox infection usually lasts about 5 to 10 days. The drug acyclovir lessens the symptoms of chickenpox and hastens recovery.

I've heard there is a chickenpox vaccine. How effective is it? Yes. The vaccine for children and adults who have never had chickenpox contains weakened chickenpox viruses to help prevent the disease. This vaccine has been largely responsible for the dramatic decrease in the incidence of chickenpox in the United States. The CDC estimates that the vaccine provides complete protection from the virus for nearly 90% of young children who receive it.

Is chickenpox the same as shingles? Not exactly. **Shingles**, or **herpes zoster**, is an adult disease caused by the same virus that causes chickenpox. Like the herpes simplex viruses, VZV remains dormant in nerve tissue near the spinal cord in an adult who had chickenpox as a child. Years later, the viruses may reactivate and multiply in nerve tissue and travel down the nerves to the skin where, as you see in this photo, they can cause an excruciatingly painful rash often encircling either the left or right side of the torso (FIGURE 18.3b). Many sufferers also experience headaches as well as facial paralysis and sharp "icepick" pains. The condition can occur repeatedly and is linked to emotional and physical stress, as well as to a suppressed immune system or aging.

Antiviral medications can lessen the pain and speed recovery. A shingles vaccine, available for adults over 50 years of age, can reduce the severity of the disease and speed recovery.

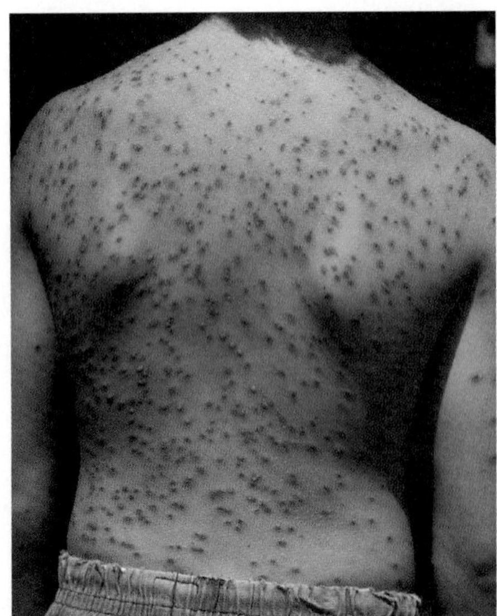

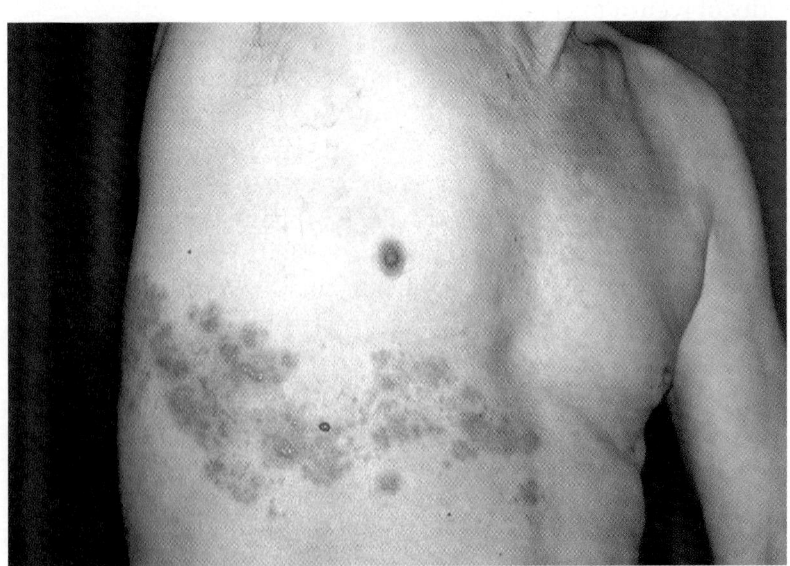

FIGURE 18.3 **The Lesions of Chickenpox and Shingles.**
(**a**) A typical case of chickenpox. The lesions may be seen in various stages, with some in the early stage of development and others in the crust stage. (**b**) Dermal distribution of shingles lesions on the skin of the body trunk. The lesions contain less fluid than in chickenpox and occur in patches as red, raised blotches.

(a) © Gilbert S. Grant/Science Source; (b) © Medical-on-Line/Alamy

Can someone get shingles from a person who has chickenpox? No. If a person who has never had chickenpox comes in contact with a person having shingles, the susceptible person, if infected, will get chickenpox.

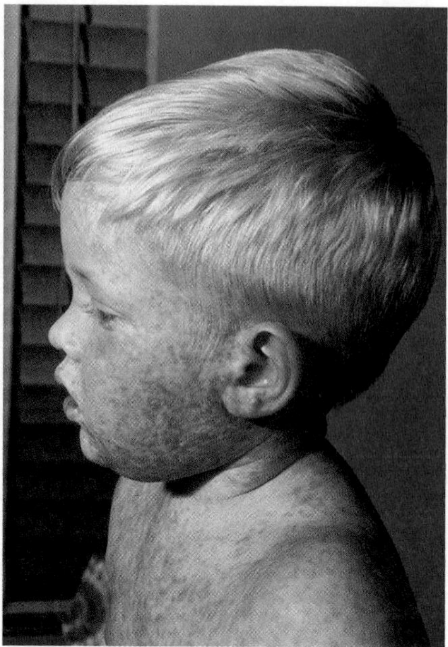

FIGURE 18.4 **The Measles Rash.** A child with measles, showing the typical rash on face and torso.

Courtesy of CDC

Measles (Rubeola)

Speaking of childhood diseases, isn't measles also highly contagious? Measles, also called rubeola, is a highly contagious disease usually transmitted by bits of mucus and saliva, what we call respiratory droplets, coughed out during the early stages of the disease. But its recognizable symptoms appear on the skin, and so we consider it a skin disease. An RNA virus is responsible for measles.

What are the symptoms of measles? Measles symptoms commonly include a hacking cough, sneezing, runny nose, eye redness, sensitivity to light, and a high fever. The characteristic red rash of measles soon appears, beginning as pink-red pimple-like spots and developing into a rash that breaks out at the hairline, covers the face, and spreads to the trunk and extremities. The young boy in this photo displays the classic measles rash (FIGURE 18.4). Within a week, the rash turns brown and fades.

How common is measles? The disease is still common in many parts of Europe, Asia, the Pacific, and Africa where there are 20 million cases and 164,000 deaths each year. Therefore, most cases reported are imported and occur in individuals who have not been vaccinated. Immunization of children in the United States is now mandatory. However, compliance is a thorny issue and can be a social problem because some parents object to some or all immunizations based on religious or philosophical beliefs. In addition, some people incorrectly believe vaccines have serious and dangerous side effects when, in fact, such complications are extremely rare. As a result, since 2013 there has been a steep spike in outbreak cases in these unvaccinated individuals in the United States.

Rubella (German Measles)

So is German measles the same as regular measles? No, the two diseases are caused by different viruses. The rubella virus, like the measles virus, is transmitted through coughing and sneezing. The symptoms of **rubella**, though, are different, usually involving a fever with a variable, pale-pink rash beginning on the face and spreading to the body trunk and extremities. The disease is usually mild and lasts for 2 to 3 days.

I heard that rubella is dangerous to a fetus. Why is that? Rubella can be dangerous when it occurs in a developing fetus. This condition, called **congenital rubella syndrome**, occurs when rubella viruses pass across the woman's placenta. The fetal organs most often affected are the eyes, ears, and cardiovascular organs, and children may be born with cataracts, glaucoma, deafness, heart defects, mental retardation, or liver and spleen damage.

Can rubella be prevented? Yes. The MMR vaccine has brought about a dramatic decline in the disease. In fact, only about 50 cases occur each year in the United States.

Mumps

Speaking of MMR, when I was a kid, I had the mumps and my father and uncles were told to stay away from me until I got over the disease. Why was that? When a person has **mumps**, ducts leading from the parotid glands become obstructed, which retards the flow of saliva and causes the characteristic swelling under one or both ears. The skin overlying the glands becomes taut and shiny, and patients experience pain when the glands are touched.

Mumps is spread by respiratory droplets in coughs and sneezes. Transmission to an adult male can lead to an infection of the testicles and cause a lowering of the sperm count. This condition, called **orchitis**, rarely leads to fertility problems. Still, it can be painful, so to prevent the infection, adult males are told to avoid children with mumps.

How effective is the MMR vaccine for mumps? Quite effective. In 2013, there were less than 450 reported cases in the United States, but almost 700,000 cases globally. Therefore, the risk of exposure to mumps among American travelers can be high, especially for those who are older than 12 months and who do not have evidence of mumps immunity. In 2013, due to waning immunity, reported cases of mumps in the United Kingdom have been rising to levels that are 50% higher than normal.

FIGURE 18.5 Smallpox lesions are raised, fluid-filled vesicles similar to those of chickenpox.

Courtesy of Jean Roy/CDC

Fifth Disease (Erythema Infectiosum)

My little brother had the fifth disease. What the heck is that? In the late 1800s, numbers were assigned to diseases accompanied by skin rashes. Disease I was measles, II was scarlet fever, III was rubella, IV was Duke's disease (also known as roseola), and V was the fifth disease. The agent of the fifth disease is a DNA virus transmitted by respiratory droplets.

Most cases of fifth disease occur in children. The outstanding feature of the illness is a mild rash on the cheeks and ears (hence, another name is slapped-cheek disease) that fades within several days. However, the virus can cause painful or swollen joints in adults.

Smallpox

You mentioned earlier the different pox diseases. What is smallpox? Throughout history, **smallpox**, caused by yet another DNA virus, was a horrible contagious disease that could sometimes be fatal. Those individuals surviving smallpox often had pitted scars or the pox.

The earliest symptoms include fever, body aches, and a pink-red rash that develops into large fluid-filled blisters, as you see in this photo (FIGURE 18.5). Eventually the blisters break open and emit pus.

Are there any smallpox cases today? At present, there is no smallpox in the world. You may remember Edward Jenner, the English physician, who in 1798, discovered he could inoculate people with material from a cowpox lesion to give them immunity to smallpox. However, it was not until 1966 that the World Health Organization (WHO) coordinated a global vaccination campaign to eradicate smallpox. In late 1977, healthcare workers reported isolation of the last smallpox case and the disease was certified as eradicated in 1980.

However, smallpox viruses still remain in two laboratories, one at the CDC in Atlanta, the other in Russia. The WHO has recommended these stocks of smallpox viruses be destroyed, especially because scientists have deciphered the base sequence of the virus' genome. Not all scientists agree. Here is an interesting perspective you might want to read concerning the debate (**A Closer Look** 18.1).

A CLOSER LOOK 18.1
Should We or Shouldn't We?

One of the liveliest global debates in microbiology is whether the last remaining stocks of smallpox viruses in Russia and the United States should be destroyed. Here are some of the arguments.

For Destruction

- People are no longer vaccinated, so if the virus should escape the laboratory, a deadly epidemic could ensue.
- The DNA of the virus has been sequenced and many fragments are available for performing research experiments; therefore, the whole virus is no longer necessary.
- The elimination of the remaining stocks of laboratory virus will eradicate the disease and complete the project.
- No epidemic resulting from the theft or accidental release of the virus can occur if the remaining stocks are destroyed.

A CLOSER LOOK 18.1

Should We or Shouldn't We? (continued)

- If the United States and Russia destroy their smallpox stocks, it will send a message that biological warfare will not be tolerated.

Against Destruction

- Future studies of the virus are impossible without the whole virus. Indeed, certain sequences of the viral genome defy deciphering by current laboratory means. Insights into how the virus causes disease and affects the human immune system cannot be studied without having the genome and whole virus. The virus research may identify better therapeutic options that can be applied to other infectious diseases.
- Mutated viruses could cause smallpox-like diseases, so continued research on smallpox is necessary in order to be prepared.
- No one actually knows where all the smallpox stocks are located. Smallpox virus stocks may be secretly retained in other labs around the world for bioterrorism purposes, so destroying the stocks may create a vulnerability in protecting the public. Smallpox viruses also may remain active in buried corpses.
- Destroying the virus impairs the scientists' right to perform research, and the motivation for destruction is political, not scientific.
- Today, it is possible to create the smallpox virus from scratch. So why bother to destroy it?
- Because the smallpox virus (vaccinia) may have evolved from camelpox, who is to say that such evolution could not happen again from camelpox?

Discussion Point

Now it's your turn. Can you add any insights to either list? Which argument do you prefer?

Note: In 2011, the World Health Assembly of the World Health Organization (WHO) met to again consider the evidence for retention or destruction of the smallpox stocks. The stalemate continues in 2014.

I have heard smallpox viruses could be used by bioterrorists. Is that true? Yes, unfortunately they could. Smallpox is a very dangerous disease, it spreads with relative ease, and as a bioweapon, it could cause a calamity of unprecedented proportions. Smallpox viruses, however, are very difficult to cultivate (as compared to bacterial organisms), and thus are not accessible to most bioterrorist groups. Still, the smallpox virus is of concern as a bioterrorism agent. Here is another short paper concerning bioterrorism that you might want to read (**A Closer Look** 18.2). The prospects are kind of scary.

Warts

I have this small wart between my knuckles. Are warts another viral skin disease? Yes, **warts** are small, usually localized skin growths commonly due to viruses. **Plantar warts** occur on the soles of the feet while common skin warts, such as the one between your knuckles, are most often found on the hands and fingers. **Genital warts** develop on the genital organs and are often transmitted through sexual contact; they are usually moist and pink. An estimated 360,000 Americans are believed to be infected by genital warts each year.

So exactly what causes warts? Warts are caused by the human papilloma viruses (HPV), which are DNA viruses. There are more than 100 such viruses that can enter the body through tiny cuts or breaks on the skin. In most cases, the skin and plantar warts are caused by HPV-1 to HPV-4 and they cause minor, although sometimes unsightly problems. Such warts don't require treatment and they usually disappear over time, though new ones may develop nearby. There are over-the-counter medications available. For stubborn warts or bothersome plantar warts, a doctor may suggest freezing or minor surgery.

Are genital warts also benign? Genital warts are caused by a different group of papilloma viruses. About 90% of genital warts are caused by HPV-6 and HPV-11, so these viruses are considered among the many sexually transmitted viruses and are introduced through oral, vaginal, or anal sex with someone who has genital warts. Although most people who acquire HPV-6 or HPV-11 never develop warts or any

A CLOSER LOOK 18.2

Bioterrorism: What's It All About?

The anthrax attacks that occurred on the East Coast of the United States in October 2001 confirmed what many health and governmental experts had been saying for over 10 years—it is not if bioterrorism would occur but when and where. Bioterrorism represents the intentional or threatened use of primarily microorganisms or their toxins to cause fear in or actually inflict death or disease upon a large population for political, religious, or ideological reasons.

Is Bioterrorism Something New?

Bioterrorism has a long history, beginning with infectious agents that were used for biowarfare. In the United States, during the aftermath of the French and Indian Wars (1754–1763), British forces, under the guise of goodwill, gave smallpox-laden blankets to rebellious tribes sympathetic to the French. The disease decimated the Native Americans, who had never been exposed to the disease before and had no immunity. Between 1937 and 1945, the Japanese established Unit 731 to carry out experiments designed to test the lethality of several microbiological weapons as biowarfare agents on Chinese soldiers and civilians. In all, some 10,000 "subjects" died of bubonic plague, cholera, anthrax, and other diseases. In 1973, after years of their own research on biological weapons, the United States, the Soviet Union, and more than 100 other nations signed the Biological and Toxin Weapons Convention, which prohibited nations from developing, deploying, or stockpiling biological weapons. Unfortunately, the treaty provided no way to monitor compliance. As a result, in the 1980s the Soviet Union developed and stockpiled many microbiological agents, including the smallpox virus, and anthrax and plague bacteria. After the 1991 Gulf War, the United Nations Special Commission (UNSCOM) analysts reported that Iraq had produced 8,000 liters of concentrated anthrax solution and more than 20,000 liters of botulinum toxin solution. In addition, anthrax and botulinum toxin had been loaded into SCUD missiles.

In the United States, several biocrimes have been committed. **Biocrimes** are the intentional introduction of biological agents into food or water, or by injection, to harm or kill groups of individuals. The most well known biocrime occurred in Oregon in 1984 when the Rajneeshee religious cult, in an effort to influence local elections, intentionally contaminated salad bars of several restaurants with the bacterium *Salmonella*. The unsuccessful plan sickened over 750 citizens and hospitalized 40. Whether biocrime or bioterrorism, the 2001 events concerning the anthrax spores mailed to news offices and to two U.S. congressmen only increased our concern over the use of microorganisms or their toxins as bioterror agents.

What Microorganisms Are Considered Bioterror Agents?

A considerable number of human pathogens and toxins have potential as microbiological weapons. These "Tier 1 agents" include bacterial organisms, bacterial toxins, and viruses. The seriousness of the agent depends on the severity of the disease it causes (virulence) and the ease with which it can be disseminated. Tier 1 agents can be spread by aerosol contact, such as anthrax and smallpox, or added to food or water supplies, such as the botulinum toxin (see table below).

Why Use Microorganisms?

Perhaps as many as 12 nations have the capability of producing bioweapons from microorganisms. Such microbiological weapons offer clear advantages to these nations and terrorist organizations in general. Perhaps most important, biological weapons represent "The Poor Nation's Equalizer." Microbiological weapons are cheap to produce compared to chemical and nuclear weapons, and provide those nations with a deterrent every bit as dangerous and deadly as the nuclear weapons possessed by other nations. With biological weapons, you get high impact and the most "bang for the buck." In addition, microorganisms can be deadly in minute amounts to a defenseless (nonimmune) population. They are odorless, colorless, and tasteless, and unlike conventional and nuclear weapons, microbiological weapons do not damage infrastructure, yet they can contaminate such areas for extended periods. Without rapid medical treatment, most of the select agents can produce high numbers of casualties that would overwhelm medical facilities. Lastly, the threatened use of microbiological agents creates panic and anxiety, which often are at the heart of terrorism.

How Would Microbiological Weapons Be Used?

All known microbiological agents (except smallpox) represent organisms naturally found in the environment. For example, the bacterium causing anthrax is found in soils around the world (see figure). Assuming one has the agent, the microorganisms can be grown (cultured) easily in large amounts. However, most of the select agents must be "weaponized"; that is, they must be modified into a form that is deliverable, stable, and has increased infectivity and/or lethality. Nearly all of the microbiological agents in tier 1 are infective as an inhaled aerosol. Weaponization,

A CLOSER LOOK 18.2

Bioterrorism: What's It All About?

therefore, requires the agents be small enough in size so inhalation would bring the organism deep into the respiratory system and prepared so that the particles do not stick together or form clumps. Several of the anthrax letters of October 2001 involved such weaponized spores.

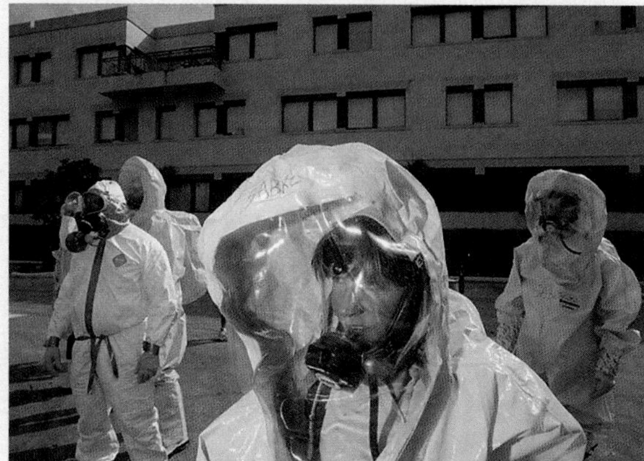

Many bioterrorism agents are airborne and protective suits are required by emergency responders.

Dissemination of biological agents by conventional means would be a difficult task. Aerosol transmission, the most likely form for dissemination, exposes microbiological weapons to environmental conditions to which they are usually very sensitive. Excessive heat, ultraviolet light, and oxidation would limit the potency and persistence of the agent in the environment. Although anthrax spores are relatively resistant to typical environmental conditions, the bacterial cells causing tularemia become ineffective after just a few minutes in sunlight. The possibility also exists that some nations have developed or are developing more lethal bioweapons through genetic engineering and biotechnology. The former Soviet Union may have done so. Commonly used techniques in biotechnology could create new, never before seen bioweapons, making the resulting "designer diseases" true doomsday weapons.

Conclusions

Ken Alibek, a scientist and defector from the Soviet bioweapons program, has suggested the best biodefense is to concentrate on developing appropriate medical defenses that will minimize the impact of bioterrorism agents. If these agents are ineffective, they will cease to be a threat; therefore, the threat of using human pathogens or toxins for bioterrorism, like that for emerging diseases such as 2009 H1N1 flu and West Nile fever, is being addressed by careful monitoring of sudden and unusual disease outbreaks. Extensive research studies are being carried out to determine the effectiveness of various antibiotic treatments and how best to develop effective vaccines or administer antitoxins. To that end, vaccination perhaps offers the best defense. The United States has stated it has stockpiled sufficient smallpox vaccine to vaccinate the entire population if a smallpox outbreak occurred. Other vaccines for other agents are in development.

This primer is not intended to scare or frighten; rather, it is intended to provide an understanding of why -microbiological agents have been developed as weapons for bioterrorism. We cannot control the events that occur in the world, but by understanding bioterrorism, we can control how we should react to those events—should they occur in the future.

TABLE

Table Some Tier 1 Agents and Perceived Risk of Use

Type of Agent	Disease (Microbe Species or Virus Name)	Perceived Risk
Bacterial	Anthrax (*Bacillus anthracis*)	High
	Plague (*Yersinia pestis*)	Moderate
	Tularemia (*Francisella tularensis*)	Moderate
Viral	Smallpox (Variola virus)	Moderate
	Hemorrhagic fevers (Ebola, Marburg virus)	Low
Toxin	Botulinum toxin (*Clostridium botulinum*)	Moderate

other symptoms, if they do occur, they appear as small, flat, flesh-colored bumps or tiny, cauliflower-like bumps anywhere in the genital region or areas around the anus. Visible warts can be removed by freezing or electrosurgical excision, or by using specific chemicals available only to doctors.

Although HPV-6 and HPV-11 are considered to have a low risk of causing **cervical cancer**, HPV-16 accounts for about 50% of cervical cancers, and together with types 18, 31, and 45, account for 80% of cervical cancers. These same viruses can cause penile cancer in males.

I heard there is a vaccine for cervical cancer. Is it effective? Yes, there is a very effective preventative vaccine called Gardasil®. It has been approved for use in females and males 9 to 26 years of age and is aimed at the most prominent types of HPV (6, 11, 16, and 18) causing genital warts and cervical/penile cancer. Importantly, the vaccine prevents infection. It is not a treatment as it does not eliminate the virus in already infected individuals.

18.2 Viral Diseases of the Respiratory Tract: Flus and Colds

A number of viral diseases occur within the human respiratory system. The most common are influenza—the flu—and the common cold that affects millions of Americans annually.

Influenza

Why does the flu always come around every year? Influenza is an acute, contagious disease of the upper and usually the lower respiratory tracts and is transmitted by respiratory droplets. The influenza virus is quite unusual. Notice in this diagram (FIGURE 18.6) that it is composed of eight RNA segments, each associated with protein to form a nucleocapsid. The viral envelope contains a series of protein projections called spikes. The majority of the spikes contain the enzyme **hemagglutinin—H spikes**—that assist the entry of the virus into its host cell. The rest of the spikes contain **neuraminidase—N spikes**—that help the virus be released from the host cell.

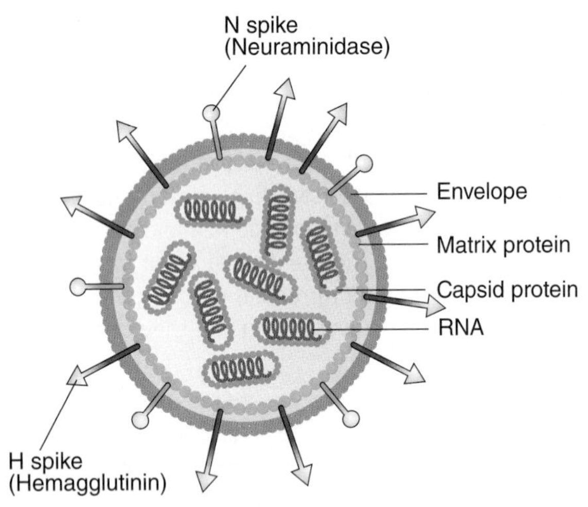

FIGURE 18.6 **The Influenza Virus.** This diagram of the influenza A virus shows its eight segments of RNA and envelope with hemagglutinin and neuraminidase spikes protruding.

Chemical changes occur periodically in these two spike enzymes, thereby yielding new strains of virus. These changes have practical consequences because the antibodies you produced during last year's attack of influenza either partially or completely failed to recognize this year's strain, so individuals will suffer another bout of disease if exposed and not vaccinated. Consequently, we have seasonal flu outbreaks every year.

I've heard of type A and type B flu viruses. Are there differences between the two types? The type A virus can cause outbreaks and epidemics, and does cause most pandemics, while type B is less widespread but can cause more mild seasonal outbreaks or epidemics. The nomenclature for influenza type A viruses is based on the H and N spikes present on the virus. Influenza A is divided into subtypes based on the 17 potential unique H spike proteins, numbered H1 to H17, and 10 potential unique N spike proteins, numbered N1 to N10. The current seasonal flu subtypes are A(H1N1) and A(H3N2). All the H and N spikes on any one virus particle are the same.

I sometimes get the symptoms of the flu and colds confused. Can you remind me of the flu symptoms? The onset of influenza is abrupt, with sudden chills, fatigue, headache, and pain that is most pronounced in the chest, back, and legs. Over a 24-hour period, the body develops a fever and a severe cough. Despite these severe symptoms, influenza is normally short-lived and has a favorable prognosis. However, secondary complications may occur if bacterial pathogens, such as staphylococci, invade the damaged respiratory tissue. These bacterial cells most often cause a form of pneumonia.

Are there any antiviral drugs to treat influenza? Yes, there are prescription drugs. The brand names are Tamiflu® (generic name oseltamivir) that comes in pill form and Relenza® (generic name zanamivir) that is an inhaled powder. Both inhibit the functioning of the neuraminidase enzyme. However, they do not cure but rather, if taken early in the infection, diminish the symptoms and shorten slightly the length of time you are sick.

How about the flu vaccine? I heard you can get the flu from the shot in the arm. Everyone who is at least 6 months of age should get a flu vaccination every fall to prevent the development of serious complications and spreading the virus to high-risk individuals, such as persons over 65 years of age, young children, and others with certain medical conditions such as asthma and diabetes.

Now, the vaccine prepared for the "flu shot" uses inactivated influenza viruses, meaning the vaccine does not contain any "live" influenza viruses (either two type A and one type B virus or two A and two B that match the types predicted for the impending flu season). Because these viruses have been inactivated, you cannot get the flu from the vaccine. But realize there are many other versions of the flu virus in any season so you could still get the flu if you get infected with one of these other nonvaccine strains.

Also, let's say two days ago you were infected with a flu virus (one that is identical to what is in the flu vaccine) and today you decide to get the flu vaccination. It will take at least 10 to 14 days for the vaccine to generate immunity in your body. But you were infected only two days ago, so you will get the flu because your body has not had time to develop immune resistance.

There is also a second type of flu vaccine, one that is given as an inhaled nasal spray. The vaccine contains weakened flu viruses (not "killed"), but it is still extremely unlikely you would get the flu from the vaccine. It is available only for healthy people 2 through 49 years of age who are not pregnant.

I have heard about a dangerous avian or bird flu. What's it all about? Ah yes. Since 2003 a highly pathogenic avian influenza A virus called H5N1 has been circulating primarily Southeast Asia. It is highly contagious among birds, and can be deadly, especially to domestic poultry. Although relatively rare, there have been sporadic human infections from bird to human but not human-to-human. Of the more than 600 human H5N1 cases reported to WHO since November 2003, approximately 60% of those individuals have died. So, it is a virus the world health authorities are keeping a close watch on. In fact, in 2014, there are several avian flu strains (H1N8, H5N8, H7N9, and H10N8) circulating primarily in birds but with some human infections reported in Southeast Asia. These are being closely watched as well.

Severe Acute Respiratory Syndrome (SARS)

I remember hearing about another respiratory disease many years ago that started in China. What was that disease? You are probably thinking of **severe acute respiratory syndrome** (SARS) that was first reported in southeastern China in February 2003. During the 114-day epidemic the WHO reported that more than 8,000 people from 29 countries were infected and 774 died. It demonstrates how fast an emerging disease can spread—and then disappear; it has never reappeared.

This outbreak was an example of a newly emerging disease caused by a previously unknown RNA virus. Bats apparently were the reservoir for the virus. The virus is spread through close person-to-person contact by touching one's eyes, nose, or mouth after contact with the skin of someone with SARS. Spreading also comes from contact with fomites, nonliving objects contaminated (usually through coughing or sneezing) with infectious droplets from a SARS-infected individual.

Many people remain asymptomatic after infection. However, in affected individuals, symptoms include fever, headache, an overall feeling of discomfort, and body aches. After 2 to 7 days, SARS patients may develop a dry cough and have trouble breathing. In those patients, pneumonia usually develops with insufficient oxygen reaching the blood. In 10% to 20% of cases, patients require mechanical ventilation.

Of current concern in 2014 are cases of a SARS-like disease in the Middle East. It is called **Middle East respiratory syndrome** (**MERS**) and the causative agent is a virus somewhat similar to the SARS virus. The disease is spread by close contact with an infected person. More than 820 cases had been reported (about 80% in Saudi Arabia) by mid-2014 and nearly 35% of those patients have died. In early 2014, there were two unlinked cases reported in the United States in individuals who had lived and worked in Saudi Arabia. No spread of those infections occurred. Camels may be the reservoir for the virus.

Respiratory Syncytial (RS) Disease

Is respiratory syncytial disease common? I've just recently heard of it. Since 1985, **respiratory syncytial (RS) disease** has been the most common lower respiratory tract disease affecting infants and children under 2 years of age. Infection takes place in the bronchioles and air sacs of the lungs, and the disease is often described as viral pneumonia. The RS virus, which is an RNA virus, causes the respiratory cells to fuse and form giant cells called **syncytia** (*sing*., syncytium). The syncytia cannot function normally the way the separate cells did. RS disease occasionally breaks out in adults, usually as an upper respiratory disease with influenza-like symptoms.

■ bronchiole: A narrow tube inside the lungs that branches off of the main air passages.

It is worth mentioning an RSV-like illnesses caused by the human metapneumovirus (hMPV). Just about every child in the world has been infected by hMPV by age 10. The RNA virus is responsible for over 10% of lower respiratory tract infections and 15% of common colds in children. It appears to be milder than RS disease and, like RS disease, can be treated with ribavirin.

Rhinovirus Infections

What's going on with a head cold? Upper respiratory infections called **head colds** are caused by the rhinoviruses, a group of over 100 different RNA viruses that take their name from the Greek *rhinos*, meaning "nose" and referring to the infection site. Adults typically suffer two or three colds and children up to six colds a year, usually in the fall and spring.

A head cold involves a regular set of signs and symptoms, what we call a common-cold syndrome. The infection begins with headache, chills, and a dry, scratchy throat. A "runny nose" and obstructed air passageways are the dominant symptoms,

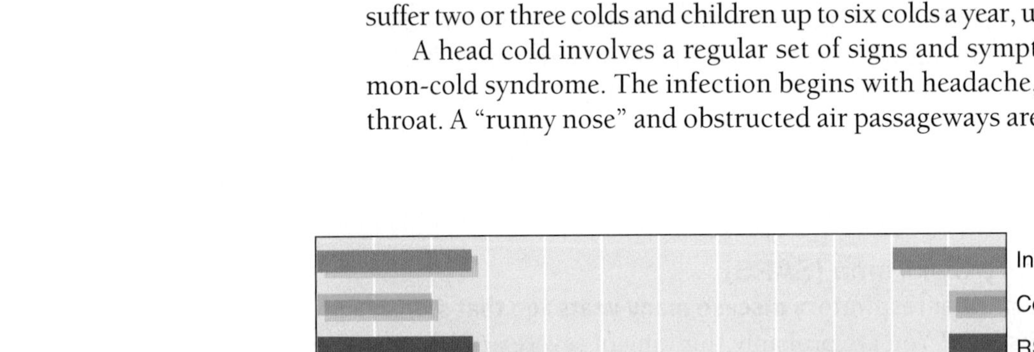

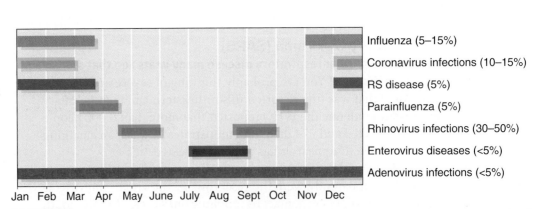

FIGURE 18.7 **The Seasonal Variation of Viral Respiratory Diseases.** This chart shows the seasons associated with various viral diseases of the respiratory tract (and their annual percentage). Enteroviruses cause diseases of the gastrointestinal tract as well as respiratory disorders and are usually acquired from the environment.

but the cough is variable and fever is often absent or slight. The illness usually lasts 7 to 10 days.

Antihistamines can be used to relieve cold symptoms because these are often due to histamines released from damaged respiratory cells. Because so many different viruses are involved, the prospect for developing a vaccine for head colds is not promising.

And here is a chart showing you the other viruses typically causing seasonal respiratory diseases (FIGURE 18.7).

So now I'm confused. How would I know if I have a cold or the flu? Oh, thanks for reminding me. Here is a nice chart listing the similarities and differences between a common cold and the flu (**A Closer Look** 18.3).

A CLOSER LOOK 18.3

Is It a Cold Or the Flu?

Do you know the differences in symptoms between a common cold and the flu? Both are respiratory illnesses but are caused by different viruses. Yet both often have some similar symptoms. In general, the flu has a sudden onset (3–6 hours) while a cold comes on more gradually. Flu symptoms are more severe than cold symptoms (see table) and colds generally do not progress to more serious complications, such as pneumonia and bacterial infections, nor do they usually require hospitalization.

TABLE

Symptoms	Common Cold	Flu
Fever	Uncommon	Common; 38°C to 39°C [occasionally higher, especially in young children (40°C)]; lasts 3 to 4 days
Headache	Uncommon	Common
Chills	Uncommon	Fairly common
General aches and pains	Mild	Common and often severe
Fatigue and weakness	Sometimes	Usual and can last up to 2 to 3 weeks
Extreme exhaustion	Uncommon	Usual and occurs at the beginning of the illness
Stuffy nose	Common	Sometimes
Sneezing	Common	Sometimes
Sore throat	Common	Sometimes
Cough	Mild to moderate hacking, productive cough	Dry, unproductive cough that can become severe

Adapted from National Institute of Allergy and Infectious Diseases website www.niaid.nih.gov/topics/Flu/Pages/coldOrFlu.aspx.

18.3 Viral Diseases of the Nervous System: Some Potential Life-Threatening Outcomes

Some serious viral diseases affect the human nervous system. This fragile system suffers substantial damage when viruses replicate in its tissues. Rabies, a highly fatal and well-known disease, is characteristic of these diseases, as we will see in our continuing conversation.

Rabies

I've heard rabies is a really dangerous disease. Why is it so dangerous? Rabies is notable for having the highest mortality rate of any human disease, once the symptoms have fully materialized. Few people in history have recovered from rabies.

Early signs are abnormal sensations such as tingling, burning, or coldness at the site of the bite. Then, increased muscle tension develops, and the individual becomes alert and aggressive. Soon there is paralysis, especially in the swallowing muscles, and saliva drips from the mouth. Brain degeneration, together with an inability to swallow, increases the violent reaction to the sight, sound, or thought of water. The disease has therefore been called **hydrophobia**—literally, "fear of water." Death follows from respiratory paralysis.

In what animals does rabies occur? Rabies can occur in most warm-blooded animals, including dogs and cats, horses and rats, skunks and bats. The RNA virus enters the tissue through a skin wound contaminated with the saliva, urine, blood, or other fluid from an infected animal. The incubation period may vary from days to years, depending on such things as the amount of virus entering the tissue and the wound's proximity to the central nervous system.

Can anything be done to prevent rabies from developing? Funny you should ask. Once symptoms set in, there is little chance of survival. However, a person who is bitten or has had contact with a rabid animal should be given the rabies vaccine via four or five injections in the arm muscle. Because of the long incubation period of the disease, there is hopefully time for the post-exposure immunization to produce protective antibodies. The number of human cases of rabies in the United States is usually less than five per year, due in large measure to immunizations of domestic animals. Today some 92% of reported rabies cases occur in wild animals (raccoons, skunks, bats, foxes, and rodents). As you learned earlier (see chapter opener), rabies is still a major problem in many developing countries.

Polio

My parents said when they were children polio was a feared disease. How do you get polio and what are the symptoms? Polio is caused by the polio viruses, which are very small RNA-containing viruses that usually enter the body in contaminated water or food.

About 75% of infected people never develop any symptoms while about 20% have minor symptoms such as fever, sore throat, upset stomach, or flu-like symptoms. They do not develop any paralysis or other serious symptoms. Unfortunately, up to 5% develop serious paralytic symptoms due to the damage caused to the nervous system by the viruses. If inflammation or swelling of the meninges occurs, paralysis of the arms, legs, and body trunk may result. In the most severe form of polio, the viruses infect the medulla of the brain. Swallowing is difficult, and paralysis develops in the tongue, facial muscles, and neck. Paralysis of the diaphragm muscle causes labored breathing and may lead to death.

- meninges: The system of membranes that cover the brain and spinal cord.
- medulla: The lower half of the vertebrate brain and continuous with the spinal cord.
- diaphragm: A sheet of internal skeletal muscle that extends across the bottom of the rib cage.

I've heard there are two polio vaccines. What are they made of? The Salk or injected vaccine contains inactivated viruses while the Sabin (oral) vaccine contains attenuated or weakened polio viruses. The Salk vaccine is currently the vaccine of choice in the United States and most countries.

Is it true polio may one day be eradicated from the Earth? The last cases of naturally occurring paralytic polio in the United States were in 1979. As of 2014, Afghanistan, Pakistan, and Nigeria were still reporting endemic cases. Hopefully, worldwide eradication through continued vaccination efforts will soon bring an end to another crippling human disease.

West Nile Virus Infection

I hear about the West Nile virus in the news a lot. How do people contract this infection? **West Nile virus infection**, as the name suggests, is caused by the West Nile virus that is transmitted by mosquitoes. This RNA virus was first appeared in the United States in 1999 and has since spread across the continental United States and Canada.

Most people get infected with West Nile virus by the bite of an infected mosquito that itself became infected when it took a blood meal from an infected bird. Infected mosquitoes spread the virus to humans and other animals, especially horses as shown in this diagram (FIGURE 18.8).

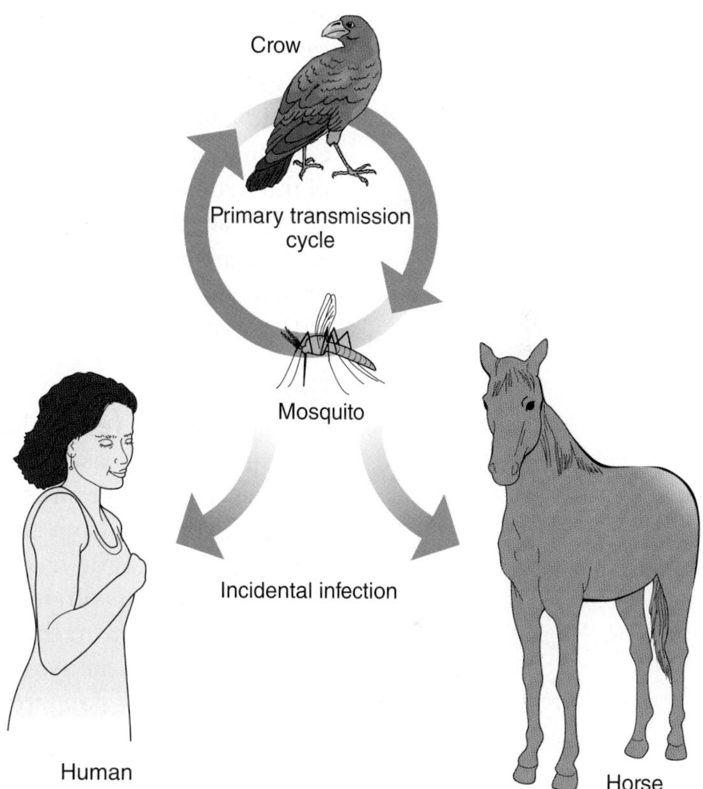

FIGURE 18.8 **The Transmission of the West Nile Virus.** Shown here is the generalized pattern of West Nile virus transmission among various animals, including humans. An incidental infection involves a host that can become infected, but is not required for the survival of the pathogen, in this case survival requires mosquitoes and birds.

Who is at risk of getting infected and what are the symptoms? The risk of infection is highest for people who work outside or participate in outdoor activities as they are at risk of being bit by an infected mosquito. To reduce the risk, people should use insect repellents and wear long sleeves and pants after dusk, which is when mosquitoes are most active. It is also important to make sure there is no standing water around the home where mosquitoes can breed.

The vast majority of people who get infected show no symptoms. However, about 20% of infected individuals develop a febrile illness that includes headache, body aches and joint pains, vomiting, diarrhea, and a rash. Most of these individuals fully recover after an extended period of weakness and fatigue.

About 1% of infected individuals develop a neurological illness involving **encephalitis**, an infection and inflammation of the brain, or **meningitis**, an inflammation of the protective membranes (meninges) covering the brain and spinal cord. Symptoms include a very high fever, a severe headache, disorientation, and possibly a series of convulsions before lapsing into a coma. About 10% of those patients developing neurological symptoms and may die.

What can be done for the more severely ill patients? There are no medications or antiviral drugs to treat the disease, nor are there any vaccines to prevent virus infection. In more severe cases, patients often need to be hospitalized to receive supportive treatment, such as intravenous fluids, pain medication, and nursing care.

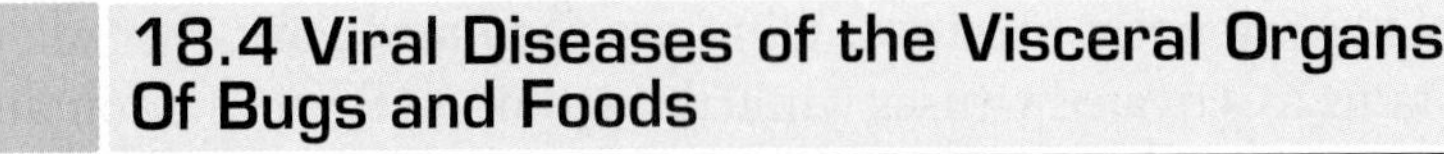

18.4 Viral Diseases of the Visceral Organs: Of Bugs and Foods

Viral diseases of the viscera affect the blood and such organs as the liver, spleen, and small and large intestines. To reach these organs, viruses are generally introduced into the body tissues by insects or by contaminated food and drink, as we will discuss in the following conversation.

Yellow Fever

Why is yellow fever so named? Most people infected with the virus causing **yellow fever** experience a flu-like illness and improve without medical treatment. However, about 15% of infected people develop a more severe form of the disease. In these cases, infection of the liver causes an overflow of bile pigments into the blood, a condition called **jaundice**; the person's complexion becomes yellow (the disease is often called "yellow jack"). And this is the least of their worries. These seriously ill patients experience bleeding gums, bloody stools and vomit, and delirium. Patients may die of internal bleeding as the mortality rates are between 20% and 50%.

Can anything be done for these ill patients? Except for supportive therapy, no treatment exists for yellow fever. However, for people planning on traveling to or living in areas at risk of yellow fever, the disease can be prevented by vaccination with a weakened virus vaccine.

Dengue Fever

Another tropical fever I have heard of is dengue fever. What is it like? **Dengue fever** is a viral disease caused by a DNA virus that multiplies in white blood cells. Like the yellow fever virus, the dengue fever virus also is transmitted by mosquitoes. Following the bite from an infected mosquito, high fever and severe prostration are early signs of dengue fever. These are followed by sharp pains in the muscles and joints, and patients often report sensations that their bones are breaking. The name "dengue" comes from the Swahili word *dinga* that means "cramp-like attack" that describes some of the symptoms.

Is the disease dangerous like yellow fever? As many as 100 million people are infected every year. Death is uncommon, but if another strain of the virus later enters the body, a condition called dengue hemorrhagic fever may occur. Although there is no medical treatment or vaccine for either form of dengue, proper management and supportive care can keep the mortality rate below 1%.

Infectious Mononucleosis

Isn't infectious mononucleosis also called "mono"? Yes, it is. Infectious mononucleosis or "mono" is common in young adults. It is sometimes called the "kissing disease" because it can be spread by contact with saliva.

What kind of disease is mono? Infectious mononucleosis is a blood disease, especially affecting the antibody-producing B cells of the immune system. Enlargement of the lymph nodes ("swollen glands") is accompanied by a sore throat, fever, enlarged spleen, and a high count of damaged lymphocytes. Those who recover usually become carriers for several months and shed the viruses into their saliva.

And this is a viral disease? Yes, the Epstein-Barr virus causes mono. It is one of the most common human viruses and occurs worldwide. In fact, most people become infected with the virus sometime during their lives. There is no specific treatment for mono and there are no antiviral drugs or vaccines available. The virus also has been detected in patients who have Burkitt's lymphoma, a tumor of the connective tissues of the jaw that is prevalent in areas of Africa.

Hepatitis

What exactly is hepatitis? **Hepatitis** is an acute inflammatory disease of the liver caused by any of several very different viruses. The most common types are the hepatitis A virus, hepatitis B virus, and hepatitis C virus.

The hepatitis A virus causes **hepatitis A**, which is most commonly transmitted by contaminated food or water or by infected food handlers. The disease is often transmitted by raw shellfish such as clams and oysters, since these animals filter and concentrate the viruses from contaminated seawater.

Hepatitis B and **hepatitis C** are caused by the hepatitis B and C viruses, respectively. Symptoms include anorexia, nausea, vomiting, and low-grade fever. Discomfort in the abdomen follows, as the liver enlarges. Considerable jaundice usually follows the onset of symptoms and the urine darkens, as well.

I am planning on traveling to a country where there is a fairly high hepatitis A rate. Should I get the vaccine before I go? By all means. There are two vaccines. Both are for individuals over 1 year of age and for those who are at increased risk for infection. In fact, with these vaccines, hepatitis A rates in the United States have dropped 95%.Try to get the shot at least two weeks before you plan on leaving to maximize your body's time to develop immunity.

How is hepatitis A different from hepatitis B and C? All three are liver diseases and the symptoms are similar. But there are important differences. Hepatitis B is caused by a DNA virus, while heptatis C and A are caused by different RNA viruses. Second, transmission of hepatitis B and C is not transmitted through food but rather usually involves contact with an infected body fluid such as blood or semen. This diagram shows the ways hepatitis B can be transmitted (FIGURE 18.9).

Are there any other differences? Yes, there are. Hepatitis B has a longer incubation period than hepatitis A, and the vaccine components are different. For hepatitis B, the vaccine contains only protein fragments from the virus while the hepatitis A vaccine contains

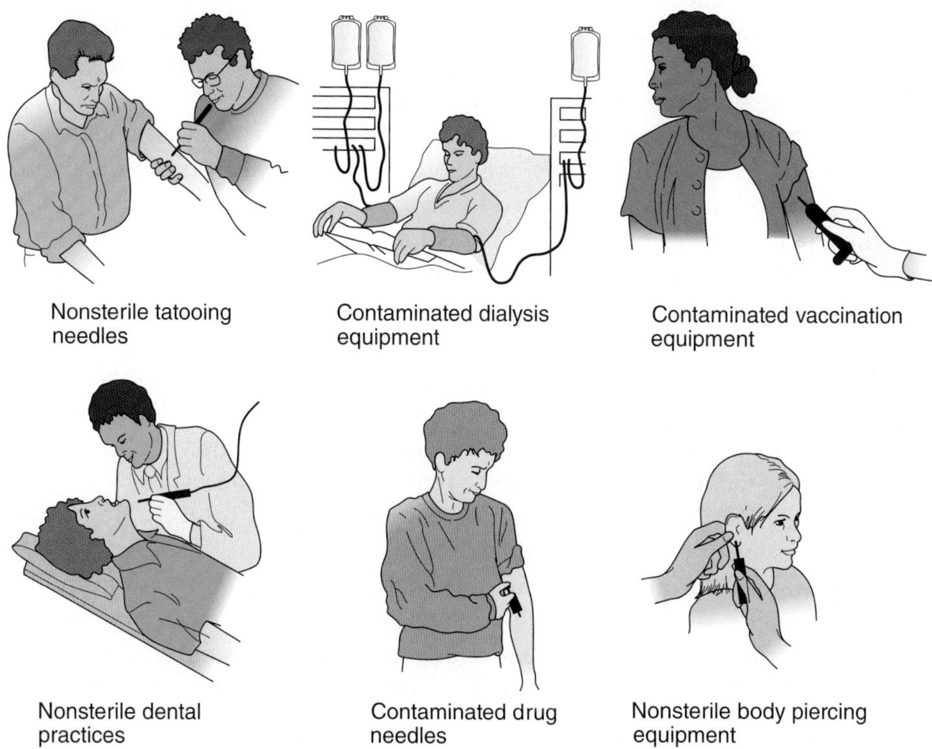

FIGURE 18.9 **Some Methods for the Transmission of Hepatitis B.** Some form of blood contact is involved with all the transmission methods.

inactivated viruses. Also, long-term liver damage may occur as a result of hepatitis B, including liver cancer. Therefore, getting the hepatitis B vaccine can be a good thing. CDC data show that there has been an 82% drop in hepatitis B cases since the introduction of the vaccine in 1991. Yet in 2014, there were still over 2,000 new infections.

Can anything be done if you get hepatitis B? For chronic infections, several antiviral drugs are available and regular monitoring of the patient is necessary to detect if any liver damage has occurred or if liver cancer has developed.

And what about hepatitis C? The one hepatitis virus of most concern is the hepatitis C virus. In fact, HCV infections are the most common type of chronic bloodborne infection in the United States; approximately 3.2 million Americans are chronically infected and there were more than 1,600 reported cases in 2013. HCV is not efficiently transmitted sexually, so as shown in this chart (FIGURE 18.10), most infections come from injecting drugs using "dirty needles."

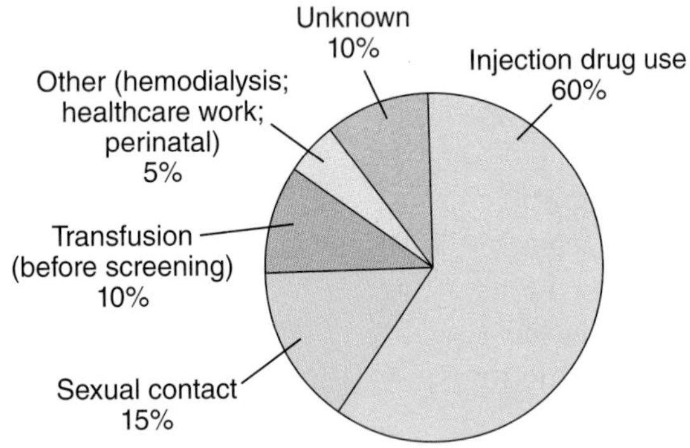

FIGURE 18.10 **Hepatitis C Infections—United States, by Source.** There has not been a reported case of transfusion-related hepatitis C in the United States since 1993.

Data from: CDC

Viral Gastroenteritis

I recently had the stomach flu. Does the flu virus also cause this illness? Stomach flu is actually a form of **viral gastroenteritis**, an inflammation of the stomach and the intestines. It usually has an explosive onset with varying combinations of diarrhea, nausea, vomiting, low-grade fever, cramps, headache, and malaise. It has nothing to do with the influenza viruses. Rather, several other viruses are responsible for viral gastroenteritis.

One is the human rotavirus, an RNA virus. It is transmitted by ingestion of contaminated food or water, or from contaminated surfaces. Infections tend to occur in the cooler months (October–April) in the United States and are often referred to as "winter diarrhea."

gastroenteritis An inflammation of the stomach and the intestines.

The most likely cause of viral gastroenteritis in adults is a norovirus infection. Noroviruses are another RNA virus and are highly contagious. The virus is transmitted primarily through hand contamination with fecal material—we call it the **fecal-oral route**—or by consumption of contaminated food or water.

How common are these illnesses? Public health officials believe rotavirus infections are second in frequency to the common cold. In fact, in developing nations, this form of gastroenteritis is estimated to be the second leading killer of children under the age of five, accounting for 23% of all deaths in this age group. In the United States, the CDC reports that there are up to 70,000 hospitalizations and 20 to 60 childhood deaths each year. The 450,000 childhood deaths worldwide (90% in developing nations) made the development of a safe and effective rotavirus vaccine a high priority objective. And now two such vaccines are available and being made part of each country's national immunization program. In the United Sates, with the use of these vaccines, hospitalizations for severe diarrhea have been reduced 70% to 80% and similar statistics are starting to be reported around the world.

Norovirus infections may lead to dehydration and are responsible for some 21 million cases every year in the United States. Fewer than 100 virus particles can make a person sick. The only treatment for norovirus gastroenteritis is fluid and electrolyte replacement. Washing hands and having safe food and water are important prevention measures. Although there is one or more norovirus outbreaks on cruise ships every year, 70% of infections come from infected food handlers.

Viral Hemorrhagic Fevers

Why were hemorrhagic fevers given that name? Certain viral fevers are accompanied by profuse bleeding or hemorrhaging of the tissues. Physicians report blood spouting from patients' eyes, nose, ears, and gums, and organs turning to liquid. They tell a horror story similar to the one recounted in the 2011 movie *Contagion* where a virus eventually causes a pandemic, leading to panic and chaos among the civilian population and the eventual collapse of society.

I liked the movie. There was an earlier one called Ebola. Was that one based on a real Ebola virus? Well, I don't believe that Hollywood really stuck to the facts, but yes **Ebola virus disease** has occurred sporadically in Africa during the past several decades. As a matter of fact, right now in mid-2014, several west African nations are experiencing the worst ever outbreak of Ebola virus disease. More than 1,700 people have been infected and over 930 have died. Its symptoms are typical of hemorrhagic fever. Luckily, the virus can only be transmitted to an individual who has had direct contact with the blood or bodily fluids of a person that is ill with Ebola virus disease, or though exposure to objects (such as needles) that have been contaminated with infected secretions. Other examples of hemorrhagic fevers are Lassa fever and Marburg virus disease, which have similar symptoms.

What about the disease that broke out in the American Southwest some years ago? Was it a hemorrhagic fever? Yes, it was. In the summer of 1993, a brief epidemic occurred in the southwestern United States. It was named the "Four Corners disease" for the place where four states (Arizona, New Mexico, Colorado, and Utah) come together. The disease is a rapidly developing flu-like illness whose symptoms include hemorrhaging blood and respiratory failure. The disease was formerly called Korean hemorrhagic fever, but it has been renamed **hantavirus pulmonary syndrome** because the viruses responsible for it are a group of RNA viruses called hantaviruses. CDC investigators identified airborne viral particles from the dried urine and feces of rodents (especially deer mice) as the responsible agents.

18.5 HIV Infection and AIDS: A Modern-Day Plague

Speaking of viruses, what is it about HIV that makes it so unique? It has been over 30 years now since what became known as **acquired immunodeficiency syndrome** (**AIDS**) was first reported. The virus is the **human immunodeficiency virus** (**HIV**), which is an RNA virus with an envelope that contains spikes, as this sketch shows (FIGURE 18.11). The virus also carries two copies of an enzyme called **reverse transcriptase**. When the RNA is released in the cytoplasm of a host cell, the enzyme synthesizes a molecule of DNA using RNA as a template. This reversal of the usual mode of genetic information transfer is why the virus is called a **retrovirus**. The synthesized DNA molecule then integrates into the host cell's DNA as a **provirus**, and from that location, it transcribes its genetic message into new HIV particles. The viral particles then "bud" from the host cell and infect other cells. Such an individual now has an HIV infection.

What are the host cells for HIV? The normal host cell for HIV is the **helper T cell**. These cells are essential for the immune system to destroy infected cells and for the production of antibodies. After infection, for a period of time the body keeps up with the virus infection and replaces infected T cells as they are destroyed. Eventually, however, the individual suffers a decline from the normal 1,000 T cells per microliter of blood to less than 200 per microliter. Infections due to opportunistic pathogens follow, which can be severe and even fatal. A summary diagram showing the various diseases and conditions that such a patient might experience is in FIGURE 18.12.

How prevalent is AIDS in the United States and the world? More than 1 million people in the United States are living with an HIV infection and most alarmingly, 20% of those individuals do not know they are infected—that's more than 200,000 individuals! There are more than 47,000 new HIV infections every year. The CDC estimates that approximately 15,000 people with an AIDS diagnosis die each year, and approximately 636,000 people in the United States with an AIDS diagnosis have died since the epidemic began in 1981.

Globally, the Joint United Nations Programme on HIV/AIDS (UNAIDS) reported that in 2013 some 34 million people were infected with HIV with 2.5 million new

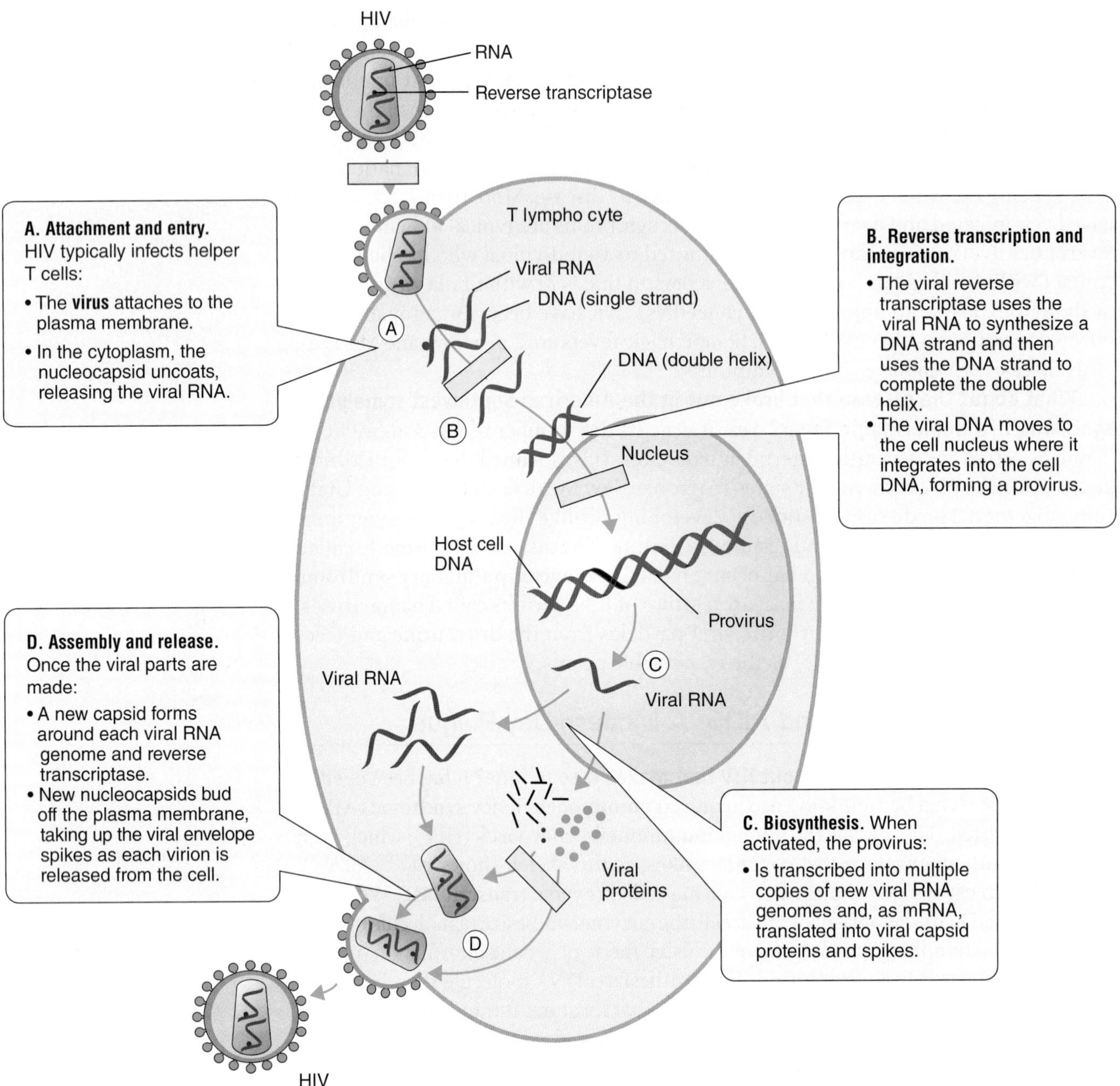

FIGURE 18.11 **The Replication Cycle of the Human Immunodeficiency Virus (HIV).** Replication is dependent on the presence and activity of the reverse transcriptase enzyme. The yellow rectangles represent places in the replication cycle where antiviral drugs block entry or replication of the virus.

infections. Approximately 1.7 million people died from AIDS-related diseases. Both the number of new infections and deaths are dropping from previous years, primarily due to wider treatment.

How is HIV transmitted? Transmission of HIV occurs primarily by infected blood or semen. Intimate sexual contact, including anal intercourse, is a common method of transmission, particularly when rectal tissues bleed and give the virus access to the blood stream. Unprotected vaginal intercourse is also a high-risk sexual activity, especially if lesions, cuts, or abrasions of the vaginal tract exist. Importantly, the use of condoms has been shown to decrease the transmission of HIV significantly. The sharing of blood-contaminated needles by injection drug users also transmits HIV. Moreover, transplacental transfer from mother to child is an acknowledged method of transmission.

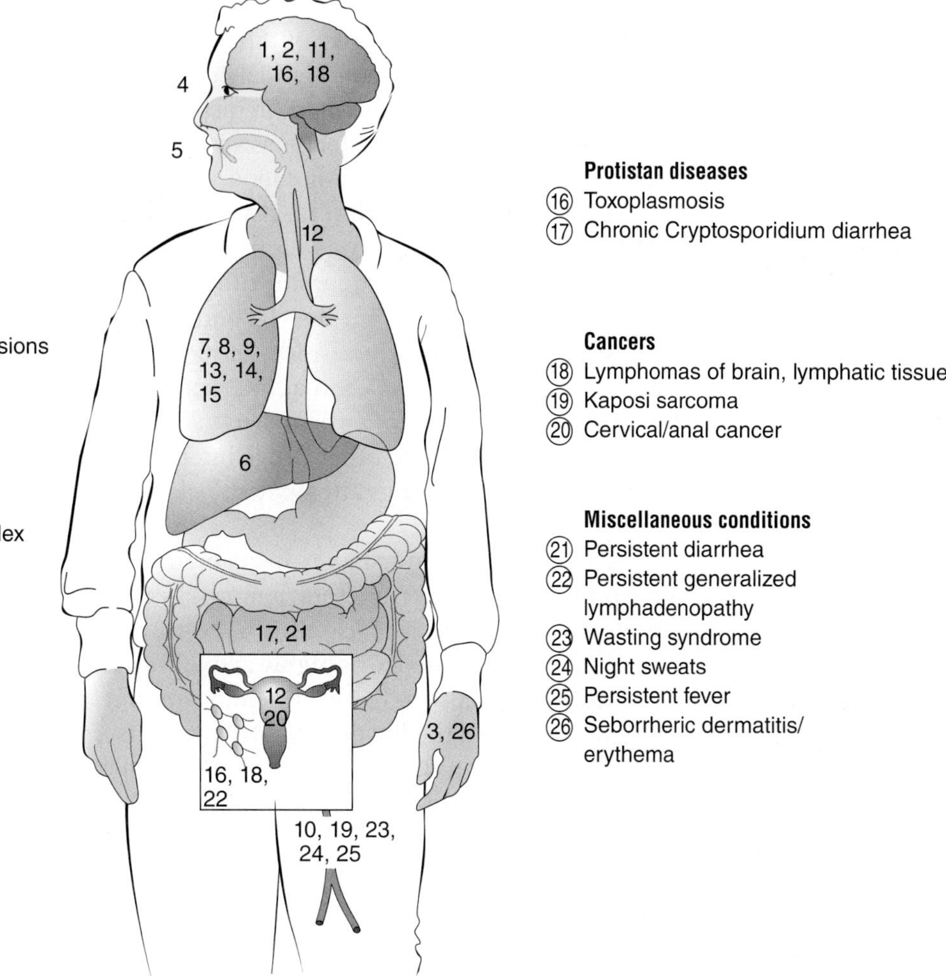

FIGURE 18.12 **Defining Conditions in AIDS Patients.** A variety of opportunistic illnesses can affect the body as a result of infection with HIV. Note the various systems that are affected and the numerous microorganisms involved.

How is AIDS different from an HIV infection? It is important to separate **HIV infection** from the diseases under the umbrella of AIDS. An infected person starts with an **acute HIV infection**. This person may not experience any early symptoms or the individual may suffer flu-like symptoms. In either case, the amount of HIV in the body is increasing and the virus can be easily spread to other individuals during this period.

Most infected people then enter a stage called an **asymptomatic HIV infection**, which can last for many years. Although the virus level stays somewhat stable, HIV is still reproducing at low levels. In fact, if the person is on effective antiviral treatment, the individual has the prospect of living an almost normal life.

However, without treatment, as the years pass the virus starts to get the upper hand and the viral load in the body starts to increase as the T cell population declines. Now the person may start to have unusual illness symptoms.

Once the number of T cells has fallen below 200 per microliter, the person is considered to have AIDS. Often opportunistic illnesses start to occur prior to the 200 level, so that too would be diagnosed as AIDS. With the onslaught of opportunistic diseases and without treatment, an AIDS patient has a life expectancy of three years or less.

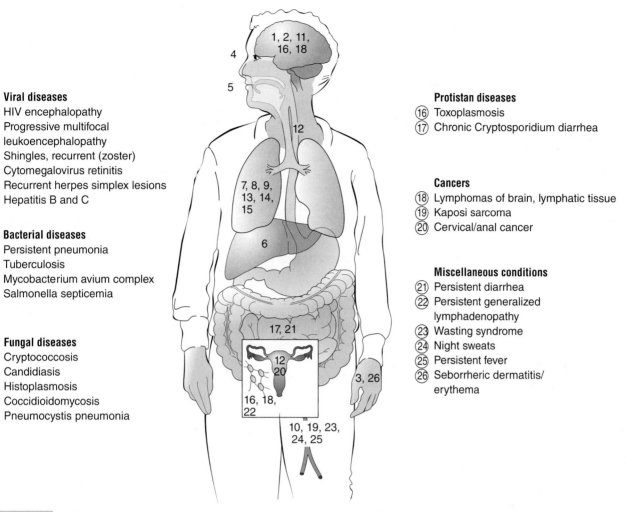

How does a doctor detect if someone has been infected with HIV? There are a number of diagnostic tests a doctor can order to determine whether a person has been infected or not. The simplest involves the detection of HIV antibodies by a blood test. This can be done in the clinic. In addition, testing kits can be purchased at local drugstores or pharmacies—or even on the Internet. A drop of blood is put on a paper disk, and sent to a laboratory. Within a few days, results are available by phoning the lab and giving them the confidential code number that came with the test kit.

Although the blood tests are extremely accurate, if a person does test positive for HIV, the individual should see a doctor as soon as possible to have another test run to detect the presence of actual virus in the blood, and if positive, antiviral treatment can be started.

You mentioned antiviral treatment. Can these drugs actually cure someone of AIDS? First, there is no cure yet for AIDS. Once a person is infected with HIV, the virus remains in that person even with antiviral therapy. Here is an article you might like to read that talks about the HIV hideouts (**A Closer Look** 18.4).

What antiviral drug treatment or antiretroviral therapy does is to control virus levels in the body. Look again at the sketch I showed you earlier (see Figure 18.11) as it shows where some of these drugs act. Notice some interfere with reverse transcriptase activity, while other drugs interfere with the assembly of new viruses. Yet other antiretroviral drugs attempt to block viral entry into host cells or interfere with provirus formation.

HIV can become resistant to many of these drugs when the drug is used singly and for prolonged times. Therefore, a more effective treatment now uses a combination of drugs called a drug "cocktail." When three or more drugs are used together, the combination is referred to as **highly active antiretroviral therapy (HAART)**. Although HAART is not a cure, it has been significant in reducing the risk of HIV transmission and has extended the life of HIV/AIDS patients. In fact, many patients are being told that if they stay on their medications, they can expect a near normal life expectancy.

Will there ever be an AIDS vaccine? An anti-HIV vaccine is being actively researched. Research evidence indicates that inactivated whole viruses may protect susceptible individuals, but there is reluctance to use such vaccines because of the possibility that active HIV may be present. Other vaccines use synthetic viral fragments, such as proteins found in the spikes of HIV. Problems arise because there are no animal models available for testing. Moreover, HIV tends to mutate in the body and avoid the antibody response, and the pool of volunteers for testing candidate

A CLOSER LOOK 18.4

The HIV Hideouts

Even though the levels of HIV in an individual on HAART are extremely low, the person is not cured because antiviral therapy cannot completely eradicate the virus from the body. Therefore, patients who can afford the drugs must keep taking drugs for their entire life. If they stop, the virus will come back with a vengeance.

One place HIV hides out is in memory T cells, a subgroup of T cells that gives us immunity to diseases, such as measles and chickenpox, we have previously experienced in our lives. Many of the infected cells are destroyed by the infection itself or by attack from other immune cells. However, some memory T cells remain inactive and the HIV provirus in such cells remains latent. These "resting T cells" will not be destroyed by antiviral drugs because the drugs only are effective in actively dividing cells. Should the memory T cells reactivate, the provirus also becomes active and starts making more viruses.

Other hideouts also exist. One is the central nervous system (CNS) where HIV-associated dementia arises from the production of neurotoxins released by infected white blood cells in the brain called microglia. Should immune cells called macrophages become infected in some other tissue, they can then cross the blood-brain barrier and take up residence in the CNS, where new viruses can now infect the microglia. Thus, even individuals on antiviral drugs can develop this form of dementia because the blood-brain barrier prevents the entry of potentially effective drugs.

HIV also hides out in other areas where antiviral drugs seem to be ineffective. This includes immune cells in the gastrointestinal and genital tracts. Interestingly, the semen in male HIV patients often contains HIV genomes even though the blood appears to be clear of the virus. If these and other hideouts are to be destroyed, new drugs must be developed that can seek out and destroy the HIV hideouts.

vaccines is limited. Nevertheless, vaccine research continues, and while a vaccine is being developed, emphasis continues to be placed on preventing the disease. Indeed, public health officials emphasize that at the present time, the best vaccine is education.

A Final Thought

In the remote tropics of South America, Africa, and Asia, there lurks a variety of viruses that infect animals but seldom bother humans. Occasionally, however, humans stumble into their paths, with results horrifying enough to mark the annals of medicine. Ebola and Marburg viruses are just two examples. In 2006, the CDC reported 37 cases of a unique hemorrhagic fever in American travelers returning from destinations in the Indian Ocean. Since then the chikungunya virus, an RNA virus carried by mosquitoes, has spread to parts of southern Europe, to the western Pacific, and to the Caribbean. As of mid-2014, more than 600 cases have been reported in the United States and its territories (Puerto Rico and the US Virgin Islands). Some 200 locally-transmitted cases have been reported from Florida, Puerto Rico, and the US Virgin Islands. All other cases were the result of infections contracted in Americans visiting other parts of the world (Caribbean, South America, the Pacific Islands, and Asia). Undoubtedly, these numbers will continue to rise.

As the world shrinks to a global village due to encroachment on wild, tropical areas and to air travel, humans are experiencing and transmitting new and sometimes dangerous microorganisms, especially viruses. Today, it is imperative that health agencies around the world monitor for such outbreaks and be prepared to snuff out any dangerous microbes before they can cause an epidemic or pandemic.

Questions to Consider

1. The CDC reports an outbreak of measles at an international gymnastics competition. A total of 700 athletes and numerous coaches and managers from 51 countries are involved. What steps would you take to avert a disastrous international epidemic and quell the spread of the disease?
2. Thomas Sydenham was a London physician who, in 1661, differentiated measles from scarlet fever, smallpox, and other fevers, and set down the foundations for studying these diseases. How would you go about distinguishing the variety of look-alike skin diseases discussed in this chapter?
3. A child experiences "red bumps" on her face, scalp, and back. Within 24 hours, they have turned to tiny blisters and become cloudy, some developing into sores. Finally, all become brown scabs. New "bumps" keep appearing for several days, and her fever reaches 39°C (102°F) by the fourth day. Then the blisters stop coming and the fever drops. What disease has she experienced? What evidence led you to this conclusion?
4. An airliner bound for Kodiak, Alaska, developed engine trouble and was forced to land. While the airline located and brought in another aircraft, the passengers sat for four hours in the unventilated cabin on the runway. One passenger was coughing heavily. The replacement aircraft arrived and all the passengers made it to their final destination. But within four days, 38 of the 54 passengers on the stranded plane had come down with the flu. What lessons about infectious disease does this incident teach?
5. As a state health inspector, you are suggesting all restaurant workers should be immunized with the hepatitis A vaccine. Do you believe restaurant owners will agree or disagree with your suggestion? Explain.
6. Sicilian barbers are renowned for their skill and dexterity with razors (and sometimes their singing voices). French researchers studied a group of 37 Sicilian barbers and found that 14 had antibodies against hepatitis C, despite never having been sick with the disease. By comparison, when a random group of 50

blood donors was studied, none had the antibodies. Propose an explanation for the high incidence of exposure to hepatitis C among these Sicilian barbers.

7. With many diseases, the immune system overcomes the infectious agent, and the person recovers. With certain diseases, the infectious agent overcomes the products of the immune system, and death follows. Compare this broad overview of disease and resistance to what is taking place with AIDS, and explain why AIDS is probably unlike any other human disease.
8. During a blood donation drive, a young man arrives at the local blood donation bank to donate. The donation goes smoothly and the individual leaves. Because the blood would be used for transfusion purposes, it must be tested for several viruses. When this is done, it is discovered the blood is positive for HIV. Obviously the blood will not be used. However, there is a lively debate as to whether the blood donor should be informed of the positive result. What is your opinion? Explain.

Key Terms

All the names of the viral diseases are amongst the key terms of this chapter. To use the disease names as a study guide, discuss each one for a few moments, indicating what sort of disease it is, where it occurs, whether a treatment or vaccine is available, and other information that is relevant to a concise view of the disease.

adenoid
adenovirus infections
acquired immunodeficiency syndrome (AIDS)
acute HIV infection
asymptomatic HIV infection
cervical cancer
chickenpox (varicella)
congenital rubella syndrome
dengue fever
Ebola virus disease
encephalitis
fecal-oral route
fifth disease (erythema infectiosum)
gastroenteritis
genital herpes
genital wart
hantavirus pulmonary syndrome (HPS)
head cold
helper T cell
hemagglutinin
hepatitis A
hepatitis B
hepatitis C
herpes simplex
highly active antiretroviral therapy (HAART)
HIV infection
human immunodeficiency virus (HIV)
hydrophobia
infectious mononucleosis
influenza
jaundice
measles (rubeola)
MERS
meningitis
mumps
neonatal herpes
neuraminidase
orchitis
planter wart
polio
provirus
rabies
respiratory syncytial (RS) disease
retrovirus
reverse transcriptase
rubella (German measles)
severe acute respiratory syndrome (SARS)
shingles (zoster)
smallpox
syncytium
viral gastroenteritis
viral hemorrhagic fever
warts
West Nile virus infection
yellow fever

Bacterial Diseases of Humans: Slate-Wipers and Current Concerns

19

Arthropods, like fleas, lice, and ticks historically have been pests capable of carrying and passing infectious diseases to other animals and humans. This was true for flea-bitten rats that spread the Black Death (plague), which decimated the population in Europe in the mid-1300s and lice that carried typhus and brought Napoleon's intended conquest of Russia to a halt in 1812. Ticks can be equally dangerous, as this sign warns, on entering a forest trail. They carry and transmit, among other pathogens, a variety of bacterial diseases, including Rocky Mountain spotted fever and Lyme disease.

Looking Ahead

Bacterial diseases are among the most well-known of human afflictions. In some epidemics, such diseases have carved out great swathes through humanity as they coursed over the land. Today, some bacterial diseases continue to pose serious challenges to the public health community. We will see both cases as we pass through the pages of this final chapter.

Because this chapter discusses diverse bacterial diseases, the chapter objectives focus on the bacterial infections in each section. On completing this chapter, you should be capable of:

- Identifying several airborne bacterial diseases, including tuberculosis, and disease prevention strategies.
- Explaining the differences between a bacterial intoxication and a bacterial infection, and identifying several foodborne and waterborne bacterial diseases.
- Comparing and contrasting the soilborne bacterial diseases anthrax and tetanus.
- Identifying several arthropodborne bacterial diseases and the insects that transmit the bacterial pathogen.
- Distinguishing between the three major sexually transmitted diseases.
- Identifying bacterial diseases associated with the skin, the oral cavity, and the urinary tract.
- Explaining what healthcare-associated infections (HAIs) are and identifying several HAIs.

What do Abraham Lincoln, Adolf Hitler, Friedrich Nietzsche, Oscar Wilde, Ludwig van Beethoven, and Vincent van Gogh have in common? Very likely all suffered from syphilis if Deborah Hayden's research is correct—and it more than likely is. In 2003, she wrote a book entitled *Pox: Genius, Madness, and the Mysteries of Syphilis* (New York: Basic Books) in which she looks at 14 eminent figures from the fifteenth to twentieth centuries whose behavior, careers, or personalities were more than likely shaped by this sexually transmitted disease.

Syphilis originally was called the Great Pox to separate it from smallpox. Until the introduction of penicillin in 1943 syphilis was untreatable and caused a chronic and relapsing disease that could disseminate itself throughout the body, only to reappear later as so-called tertiary syphilis. In this most dangerous and terminal form, the disease produces excruciating headaches, gastrointestinal pains, and eventually deafness, blindness, paralysis, and insanity.

Yet sometimes ecstasy and fierce creativity were part of the "symptoms" of tertiary syphilis. As Deborah Hayden says, "one of the 'warning signs' of tertiary syphilis is the sensation of being serenaded by angels." In fact, Danish writer Karen Blixen (Isak Dinesen) who wrote "*Out of Africa*" once said, "*Syphilis sold her soul to the devil for the ability to tell stories*." Deborah Hayden believes it is just such emotions that provided much of the creative spark for many of the notable historical figures she describes in her book. Perhaps the most intriguing is the debated proposal that syphilis may have driven Hitler mad and that he was dying of syphilis when he committed suicide in his Berlin bunker in the final days of World War II.

If Deborah Hayden's arguments are true, it is amazing how a bacterial organism can affect the body and mind, shaping the thoughts of writers and philosophers, the creative genius of artists, composers, and scientists—and yes, the madness of dictators.

In this final chapter of the book, we study some of the major bacterial diseases of humans according to their mode of transmission. Some of the diseases are of historical interest and are currently under control. But just as a garden always faces new onslaughts of weeds and pests, so, too, the human body and society in general are continually confronted with pathogens and newly emerging diseases.

■ *Streptococcus pyogenes* strep-tō-kok′kus pī-äj′en-ēz

19.1 Airborne Bacterial Diseases: Some Major Players

Bacterial diseases of the respiratory tract can be severe. Moreover, the respiratory tract is a portal of entry to the blood, and from there, a disease can spread and affect the more sensitive internal organs. We use the question–answer format in the discussion, so imagine you are conversing with a microbiologist or infectious disease physician.

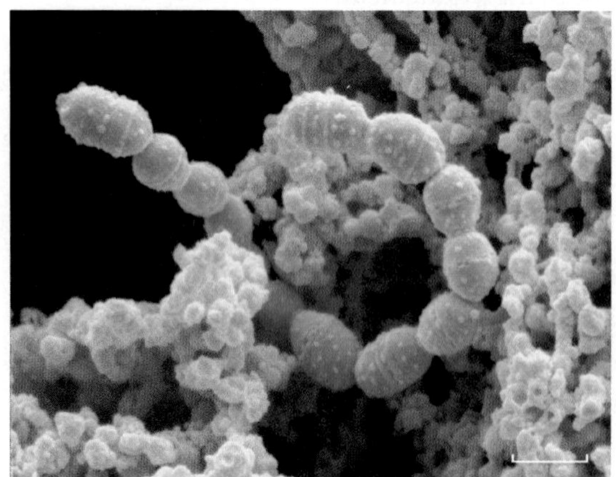

FIGURE 19.1 **Streptococci.** A false-color scanning electron microscope image streptococci growing as long chains. (Bar = 2 µm.)

Streptococcal Diseases

I know about strep throat, but are tooth decay and even yogurt the result of streptococci? They all are in the same genus but they are different species. Streptococci are a large and diverse group of gram-positive bacterial species occurring in chains (FIGURE 19.1). One species, *Streptococcus pyogenes*, causes **streptococcal sore throat**, popularly known as **strep throat**. Patients experience a high fever, coughing, swollen lymph nodes and tonsils, and a fiery red "beefy" appearance to pharyngeal tissues owing to tissue destruction. Scarlet fever is strep throat accompanied by a skin rash.

Another species, *S. mutans*, is involved in tooth decay while the species *S. lactis* is among the important "active cultures" in a cup of yogurt. So you see, the streptococci can be quite dangerous or quite harmless, depending on the species involved.

■ *Streptococcus mutans* strep-tō-kok'kus mū'tans

■ *Streptococcus lactis* strep-tō-kok'kus lak'tis

Just how dangerous is *S. pyogenes*? Although the streptococci usually respond to antibiotics such as penicillin, left untreated streptococcal disease can be dangerous. One potential complication is **rheumatic fever**, which can develop as a complication of inadequately treated strep throat or scarlet fever. The condition is characterized by fever and inflammation of the small blood vessels. The most significant long-range effect is permanent scarring and distortion of the heart valves, a condition called "rheumatic heart disease."

I also read about streptococci causing some "flesh-eating disease." What species causes this and how dangerous is it? The disease you are referring to also is caused by *S. pyogenes* and it is a rare but life-threatening infection. It is not the result of airborne transmission but rather comes from infection of a wound or some other trauma to the skin tissue. But because we are talking about the streptococci, let me mention this disease here.

If a wound or cut becomes infected with *S. pyogenes*, the invasive streptococci infect the fatty tissue, called fascia, lying over the muscles and beneath the skin. The streptococcal toxins produced destroy the fascia so quickly that it appears the pathogen is "eating" the flesh. Tissue destruction causes cell death, which is called necrosis; thus, the real name for the disease is **necrotizing fasciitis**. Here is a photo of such a person whose infection was spreading so fast that extensive tissue had to be removed to stop the infection's progression (FIGURE 19.2).

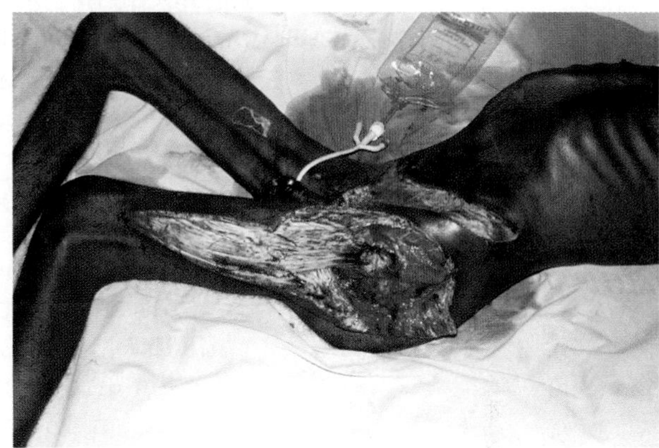

FIGURE 19.2 **Necrotizing Fasciitis.** The extensive removal of infected connective tissue can be seen in the leg of a 15-year-old AIDS patient.

Diphtheria

I never hear of diphtheria in the news. Does this disease still occur? Diphtheria is caused by *Corynebacterium diphtheriae*, a gram-positive rod that is acquired by inhaling respiratory droplets from an infected person. The bacterial cells produce a potent toxin capable of killing the human cells lining the respiratory tract. In complicated cases, there is a buildup of dead tissue, mucous, and fibrous material on the tonsils or pharynx that can block the airway and lead to suffocation.

■ *Corynebacterium diphtheriae* kôr'ē-nē-bak-ti-rē-um dif-thi'rē-ī

The reason you do not hear of the disease is because it no longer occurs in most developed nations thanks to the diphtheria-pertussis-tetanus (DPT) vaccine, which is given to all children in the United States and many other countries.

Pertussis

My father had pertussis as a child, but then it was called "whooping cough." Why is that? Thankfully he survived it, as **pertussis** can be one of the more dangerous diseases of childhood years. After infection by *Bordetella pertussis*, the toxins it produces cause cell dysfunction and mucus accumulation in the airways. This leads to labored breathing. Children then experience violent spells of rapid-fire coughing in one exhalation, followed by a forced inhalation over a partially closed windpipe. The rapid inhalation results in the characteristic "whoop"—and hence the name "whooping cough."

■ *Bordetella pertussis* bor-de-tel'lä pėr-tus'sis

Wasn't there just recently a pertussis outbreak in California? Yes, in 2010. Since the early 2000s, the incidence of pertussis has been on the increase. Every year, globally there are some 16 million cases reported and 195,000 child deaths.

Health experts are not quite sure why there has been this steady increase and the major outbreak in California—and now in 2014, in California and several other states and countries. Perhaps infants and children are not being vaccinated on schedule and not building up protective immunity. Another thought is that immunity in adolescents and adults is waning, so they may get a mild infection and then spread it to unvaccinated children.

Bacterial Meningitis

Neisseria meningitidis nī-se′rē-ä me-nin-ji′tī dis

I heard there was an outbreak of meningitis at Princeton University and the University of California at Santa Barbara in late 2013. What is meningitis? The term **meningitis** refers to several diseases of the **meninges**, the three membranous coverings of the brain and spinal cord. Meningitis patients suffer pounding headaches, neck paralysis, and numbness in the extremities. A particularly dangerous form of meningitis is **meningococcal meningitis**. This disease is caused by *Neisseria meningitidis*, a small gram-negative diplococcus commonly called the "meningococcus." One type of this species was responsible for the outbreaks at Princeton and at UC Santa Barbara.

How do doctors identify the disease and what can be done for patients? Meningococcal meningitis begins as an influenza-like upper respiratory infection. Look at this diagram (FIGURE 19.3).Notice the infection spreads to the bloodstream, where bacterial toxins may overwhelm the body in as little as 2 hours and cause death. This condition is called "meningococcemia." In survivors, a rash also appears on the skin, beginning as bright red patches, which progress to blue-black spots. Paralytic symptoms soon appear.

Therefore, doctors must recognize symptoms early to begin treatment before irreversible nerve damage or death occurs. A principal criterion for diagnosis is the observation of gram-negative diplococci in samples of spinal fluid obtained by a spinal tap. Aggressive treatment with antibiotics is recommended.

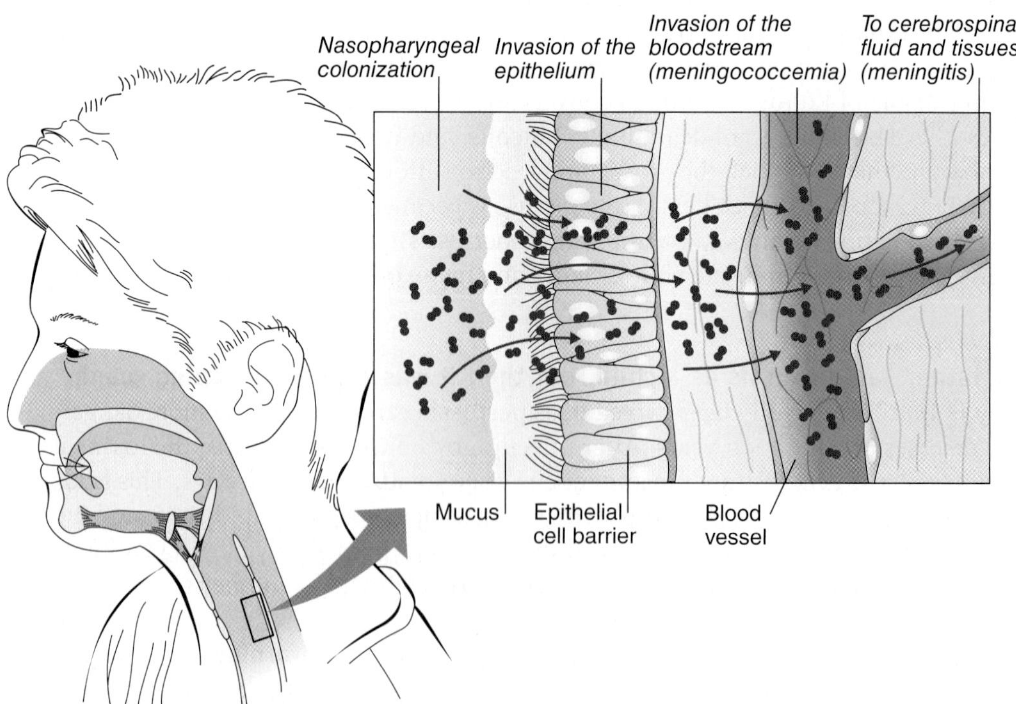

FIGURE 19.3 **Pathogenic Steps Leading to Meningitis.** Bacterial species capable of causing meningitis can colonize the nasopharynx, and then invade the epithelium causing respiratory distress. They then pass into the bloodstream. Finally, the pathogens disseminate to tissues near the spinal cord, causing inflammation and meningitis.

Is this the same type of meningitis that occurs primarily in children? No, it's not. Another form of meningitis, caused by *Haemophilus influenzae* type b, occurs primarily in children between the ages of 6 months and 2 years. Symptoms include stiff neck, severe headache, and other evidence of neurological involvement, such as listlessness, drowsiness, and irritability. There are antibiotics that can be used for treatment.

Haemophilus influenzae hē-mä'fil-us in-flü-en'zī

There's a vaccine for this childhood meningitis, right? Yes, there is, and it's been used quite successfully as there were less than 20 cases reported to the CDC in 2013.

Before I forget, in parts of the developing world meningococcal meningitis occurs in epidemic proportions. According to the World Health Organization (WHO), more than 90% of the meningitis cases, which primarily occurs in infants, children, and young adults, occurs in sub-Saharan Africa's so-called "meningitis belt." The good news is that in 2013 the number of reported cases of meningococcal meningitis in the "belt" reached its lowest level in 10 years thanks to a new meningitis vaccine.

Tuberculosis

Recently, I have been hearing a lot about tuberculosis. Why are health experts so worried? **Tuberculosis (TB)** is caused by *Mycobacterium tuberculosis*. Today, the species represents a re-emerging disease because it has become resistant to so many antibiotics that effective treatment has become a concern.

Mycobacterium tuberculosis mī-kō-bak-ti'rē-um tü-bėr-kū-lō'sis

The rod-shaped cells enter the respiratory tract in droplets. Infected patients experience chronic cough, chest pain, and high fever. In response to the infection, the body responds to the disease by forming a wall of white blood cells and fibrous materials around the organisms, as illustrated here (FIGURE 19.4). As these materials accumulate in the lung, a hard nodule called a **tubercle** arises (hence the name "tuberculosis"). The cells in the tubercles produce no discernible toxins, but growth is so unrelenting that the lung tissues can literally be destroyed.

How do the diagnostic tests for tuberculosis work? A special staining procedure called the **acid-fast test** is an important screening tool used on the patient's sputum. Early detection of tuberculosis is also aided by the tuberculin test, a procedure that begins with the application of a purified protein derivative of *M. tuberculosis* to the skin. If the patient has been exposed to the organism, the skin becomes thick, and a raised, red welt develops within 48 to 72 hours. An X-ray exam can also identify tubercles in the lungs.

Which drugs are used to treat tuberculosis? Physicians have treated tuberculosis patients with several antibiotics. However, as I mentioned, *M. tuberculosis* is now showing strong resistance to these drugs, so patients with "multidrug-resistant tuberculosis" (called MDR-TB) have to be given stronger drugs. Drug therapy is intensive and extends over a period of 6 to 9 months or more. And the scary part, there are now strains of the pathogen that are resistant to almost all drugs used to treat TB. This organism has been a slate-wiper for centuries and it continues to challenge the best medical treatment available today.

I assume a TB vaccine is available to offset this antibiotic resistance? Well, not really. Many potential vaccines are being studied but today there is no effective, long-term TB vaccine for everyone in the population. Short-term protection to TB can be rendered in children with the so-called BCG vaccine. It is not generally recommended for use in the United States because of the low risk of infection with *M. tuberculosis*. Also, the vaccine does not provide life-long immunity, is not effective against adult pulmonary TB, and can produce occasional side effects.

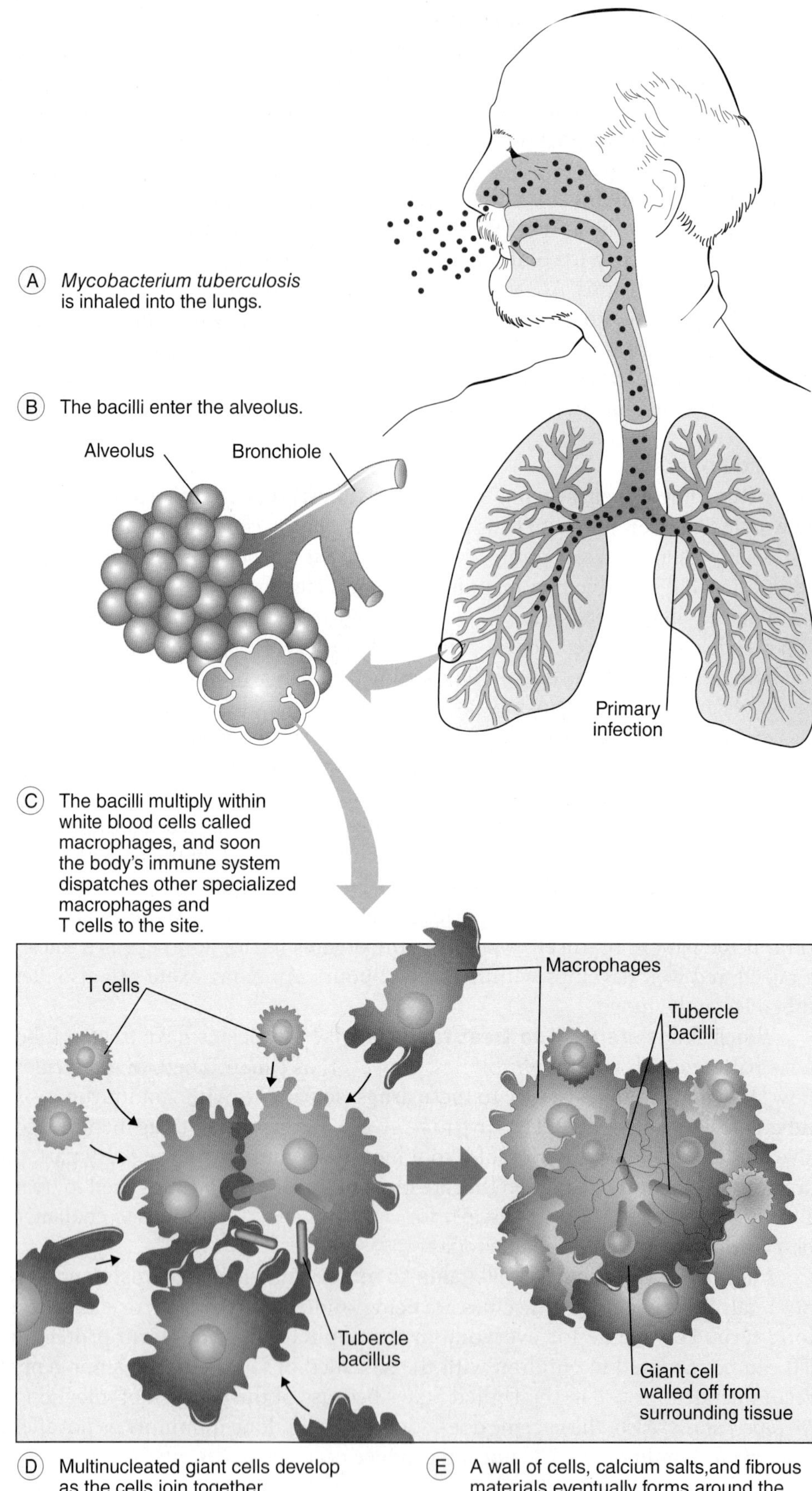

A Mycobacterium tuberculosis is inhaled into the lungs.
B The bacilli enter the alveolus.
Alveolus
Bronchiole
Primary infection
C The bacilli multiply within white blood cells called macrophages, and soon the body's immune system dispatches other specialized macrophages and T cells to the site.
T cells
Macrophages
Tubercle bacilli
Tubercle bacillus
Giant cell walled off from surrounding tissue
D Multinucleated giant cells develop as the cells join together.
E A wall of cells, calcium salts,and fibrous materials eventually forms around the giant cell. This is the tubercle.

FIGURE 19.4 **The Progress of Tuberculosis.** Following invasion of the alveoli, the tubercle bacilli are taken up by macrophages and the immune system attempts to "wall off" the bacilli by forming a tubercle.

Legionnaires' Disease (Legionellosis)

My grandfather contracted Legionnaires' disease several decades ago. He recovered, but what exactly is this disease? In 1976, a respiratory disease occurred at an American Legion convention in Philadelphia. By the time the outbreak was over, 140 conventioneers and 72 other individuals became ill with fever, coughing, and pneumonia. Eventually, 34 individuals died of the disease, which came to be called **Legionnaires' disease**. The bacterial species eventually isolated was named *Legionella pneumophila*.

■ *Legionella pneumophila*
lē-jä-nel′lä nü-mō′fi-lä

The cells of *L. pneumophila* exist where water collects, and the pathogen apparently becomes airborne in wind gusts and breezes. Cooling towers, industrial air-conditioning units, humidifiers, stagnant pools, and puddles of water have been identified as sources of the pathogen. Humans breathe the contaminated droplets into the respiratory tract, and disease develops a few days later. The symptoms of Legionnaires' disease include fever, a dry cough with little sputum, and lung infection. There often are a few outbreaks every year somewhere in the world.

Bacterial Pneumonia

What do doctors mean by "pneumonia," and which bacteria cause it? The term **pneumonia** refers to a microbial disease of the bronchial tubes and lungs. A wide spectrum of organisms, including many different viruses and bacterial species, may cause pneumonia. Over 80% of bacterial cases are due to *Streptococcus pneumoniae*, a gram-positive chain of diplococci traditionally known as the "pneumococcus." The disease is commonly called **pneumococcal pneumonia**. Patients experience high fever, sharp chest pains, difficulty breathing, and rust-colored sputum resulting from blood seeping into the air sacs of the lungs. Antibiotics are used in therapy.

■ *Streptococcus pneumoniae*
strep-tō-kok′kus nu-mō′nē-ī

Is there a vaccine for pneumococcal pneumonia? Currently, there are two pneumococcal vaccines for the American public. One, called PCV13 or Prevnar 13®, is available for people over 50 years of age to help prevent pneumonia. A second vaccine called PPSV or Pneumovax 23®, is also available for use in all adults who are older than 65 years of age and for persons who are 2 years of age and older and at high risk for disease. Here is a short essay on pneumonia and children (**A Closer Look 19.1**).

A CLOSER LOOK 19.1

The Killer of Children

Global Health Magazine recently reported the following: "Chitra Kumal knows the pain of losing a child. When her daughter, Sunita, was 15 months old, she developed a respiratory infection that quickly progressed into pneumonia. With no health facilities in her Nepalese village, Kumal depended on the advice and treatment of a traditional healer or shaman. After just 3 days of fever, fast breathing, and chest indrawing, her only daughter died."

Similar stories are reported everyday around the world. According to the World Health Organization (WHO), pneumonia kills 2 million children under 5 years of age each year—more than AIDS, malaria, and measles combined—accounting for nearly one in five child deaths globally (see figure). However, this number may be an underestimate as nearly half of all pneumonia cases occur in malarious parts of the world where pneumonia often is misdiagnosed as malaria.

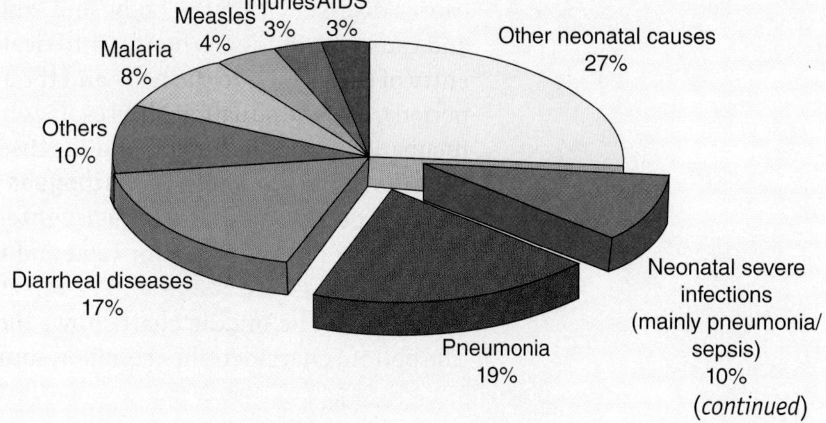

(continued)

A CLOSER LOOK 19.1

The Killer of Children (continued)

The WHO estimates that more than 150 million episodes of pneumonia occur every year among children under 5 in developing countries, accounting for more than 95% of all new cases worldwide, and between 11 and 20 million of these episodes require hospitalization. The highest incidence of pneumonia cases among children under 5 occurs in South Asia and sub-Saharan Africa.

Preventing and treating childhood pneumonia obviously is critical to reducing childhood mortality. However, only about one in four caregivers know the two key symptoms of pneumonia: fast breathing and difficulty breathing (indrawing). If antibiotics were universally available and given to children with pneumonia, around 600,000 lives could be saved each year, which represents only about 25% of the annual cases. Clearly, other control measures are needed.

At the beginning of the twentieth century, pneumonia accounted for 19% of childhood deaths in the United States, a statistic remarkably similar to the rate in developing countries today. Control in the United States was achieved largely without antibiotics and vaccines. Therefore, other control measures and strategies are needed on a global scale.

Key prevention measures include promoting balanced nutrition, reducing environmental air pollution, and increasing immunization rates with vaccines, such as those against *Streptococcus pneumoniae* (pneumococcus) and *Haemophilus influenza* type b (Hib). However, only about 50% of pneumonia cases in Africa and Asia are caused by these two organisms, so other vaccines need to be developed against other bacterial species (and viruses) that cause pneumonia. And of course—hand washing, like in all areas of infectious disease, can play an important role in reducing the incidence of pneumonia.

I hear of pneumonia quite often. Are there other types of pneumonia? Yes. A second type of bacterial pneumonia is termed "primary atypical pneumonia:" "primary" because it occurs in previously healthy individuals, whereas pneumococcal pneumonia usually develops in people who are already ill; "atypical" because the organism differs from the typical pneumococcus and because symptoms are unlike those in pneumococcal disease. The patient experiences fever, fatigue, and a characteristic dry, hacking cough.

■ *Mycoplasma pneumoniae* mī-kō-plaz′mä nu-mō′nē-ī

The agent of primary atypical pneumonia is *Mycoplasma pneumoniae*, one of the smallest bacterial pathogens. Often it is called "walking pneumonia" even though the term has no clinical significance. The disease is rarely fatal.

19.2 Foodborne and Waterborne Bacterial Diseases: Gastroenteritis

In this section, we move to the bacterial diseases transmitted by vehicles: food and water. Most illnesses of the gastrointestinal (GI) tract represent some form of **gastroenteritis**, an inflammation of the stomach and the intestines, usually with vomiting and diarrhea. We will encounter two types of such diseases: "intoxications," where a toxin produced outside the body is consumed in contaminated food or water and then causes disease within the body; and "infections," where microbes grow in the body and cause the illness. When an intoxication occurs, there is a brief time between the entry of the toxin into the body and the appearance of symptoms (i.e., the incubation period), and the situation also resolves in a relatively brief period. For infections, the incubation period is longer, and the disease takes longer to resolve.

What types of bacterial pathogens cause foodborne and waterborne diseases and how do people contract gastroenteritis? We live in a microbial world and there are many "opportunities" for food and water to become contaminated. Just look at these pie charts(FIGURE 19.5). The top chart indicates the major players causing gastroenteritis. The middle chart shows the commodities most associated with illness. The bottom chart identifies common sources of illness. It is most instructive because it shows foodborne microbes can be introduced through the fecal-oral route; from infected humans who handle the food, or by cross-contamination from some other raw product.

As you know, we have quite a diverse intestinal microbiota to help protect us from pathogens. However, in many cases these foodborne and waterborne pathogens either reproduce so quickly and to such high numbers, or generate toxins, that our microbiota may not be able to contain them.

So if our intestinal microbiota gets knocked out, how does it recover? Good question. It might just naturally recover from the survivors present in the intestines. Also, if a person takes probiotics, those good bacteria might help reestablish the normal microbiota. You probably have heard of **probiotics**, the various types of preparations you can buy that contain bacterial species capable of helping maintain the natural balance of bacterial species, the microbiota, in the intestines. Also, there has been much interest in the appendix as a source of microbiota replenishment. Here's the story (**A Closer Look 19.2**)

Anyway, let's discuss some of these foodborne and waterborne organisms and the human diseases they cause.

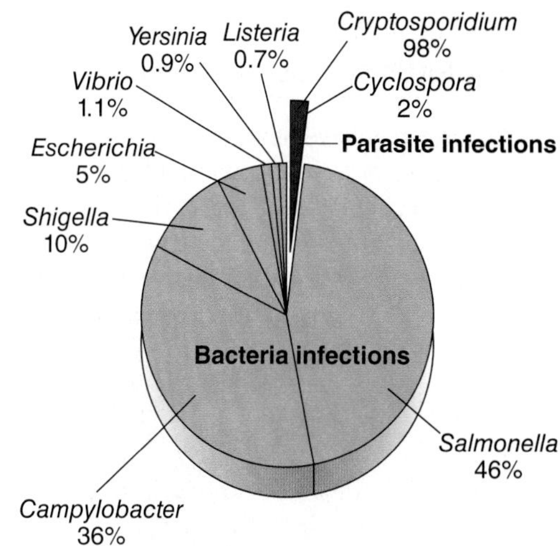

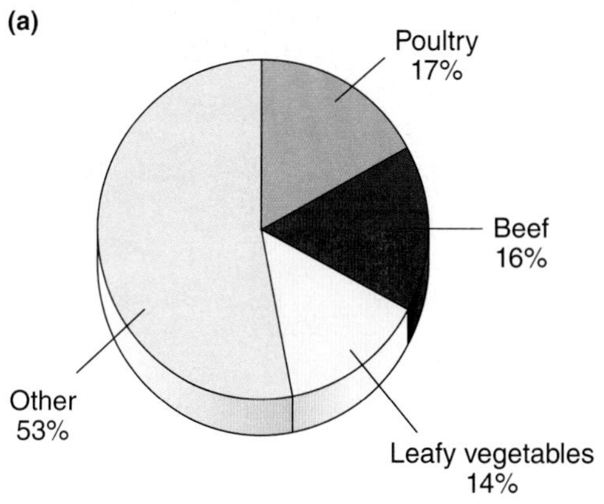

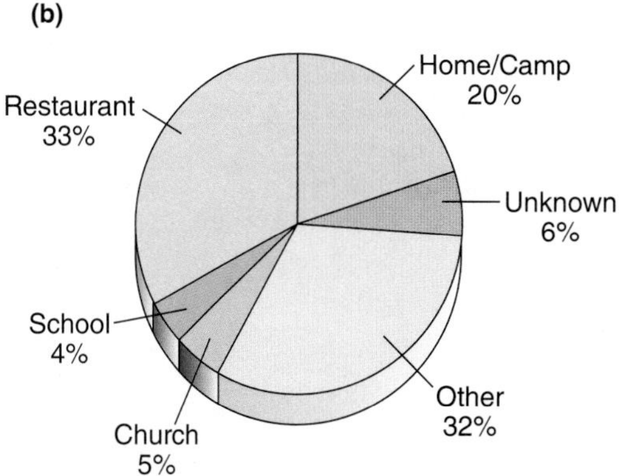

FIGURE 19.5 **Foodborne Illness Surveillance.** (**a**) Among the more than 19,000 lab-identified cases of foodborne illnesses in 2011, bacterial causes predominated. (**b**) Among the foodborne outbreaks, the commodities most associated with illness were poultry, beef, and leafy vegetables. (**c**) The most common source of illness was restaurants.

Data from: CDC.

Botulism

Is botulism a problem to be taken seriously? Absolutely. Of all the foodborne intoxications in humans, none is more dangerous than botulism. The botulism toxin is extremely poisonous. **Botulism** is caused by *Clostridium botulinum*, a spore-forming, gram-positive rod. The spores exist in the soil, and if they inadvertently enter the anaerobic environment of cans or jars, the spores germinate to multiplying bacilli that produce and release the toxin into the food product.

How does one recognize botulism in patients? The symptoms of botulism develop within hours of consuming toxin-contaminated food. Patients suffer blurred vision, slurred speech, difficulty swallowing and chewing, labored breathing, and eventually muscle paralysis. These symptoms are the result of the botulism toxin, which inhibits the release of the neurotransmitter acetylcholine between nerves and muscles. Without acetylcholine, nerve impulses cannot pass into the muscles, and the muscles do not contract. Death follows within a day or two.

Can anything be done for patients? Because botulism is a type of foodborne intoxication, antibiotics are of no value as a treatment. Instead, large doses of special antibodies called **antitoxins** must be administered to neutralize the toxins. Life-support systems such as respirators are also used.

The disease can be avoided by heating foods before eating them because the toxin is destroyed by exposure to temperatures of 90°C (194°F) for 10 minutes.

So is Botox the botulism toxin? Yes, under medical supervision extremely low concentrations of the toxin are used to temporarily remove facial wrinkles and frown lines by paralyzing the muscles. Botox® or Dysport®, as the materials are known, is also used to treat "cross eye" and both have been approved for temporary relief of excessive body sweating and chronic migraine headaches.

■ *Clostridium botulinum*
klôs-tri′dē-um bot-ū-lī′num

A CLOSER LOOK 19.2

The Gut's Microbial Source

You probably have heard people say that if you have a case of diarrhea or even take antibiotics for a bacterial infection, you should eat products like yogurt to replenish the gut microbiota lost through diarrhea or killed by antibiotics. Although this cannot hurt, eating yogurt does not supply nearly all the microorganisms normally colonizing the gut. So, where do all the microbes come from to repopulate the gut? The answer—your appendix!

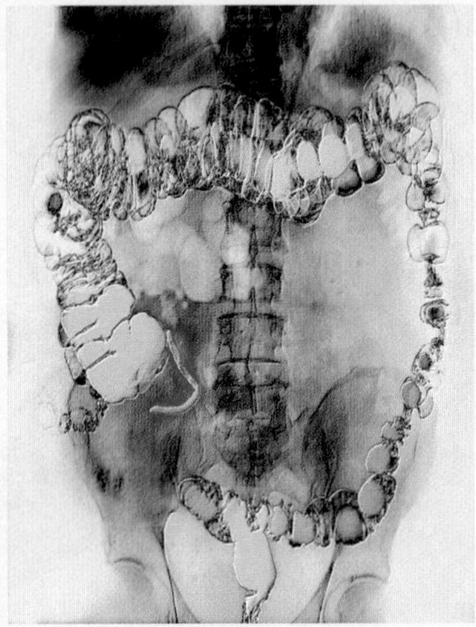

A computer-enhanced image showing the appendix highlighted in red.

The appendix is a slender, hollow, blind-ended pouch projecting from the posterior-medial region of the cecum, near its junction with the small intestine (see figure). For a long time, the appendix was thought to be a vestigial or useless organ that may have had a purpose far back in our evolutionary past. But in 2007, investigators at Duke University Medical Center and Arizona State University suggested the appendix serves as an internal "safe house" for the indigenous microbiota normally living in the gut. The investigators believe the bacterial cells in the appendix form a very well developed biofilm that can survive a bout of diarrhea and emerge afterward to repopulate the gut.

In 2011, another group of investigators at Winthrop University Hospital on Long Island studied a large group of patients with a history of gut infections. The researchers discovered that 45% of patients without an appendix had reoccurring gut infections, while only 18% of those patients with an appendix had a reoccurrence. It appears the appendix can supply "good bacteria" to reestablish a normal gut microbiota.

Staphylococcal Food Poisoning

■ *Staphylococcus aureus* staf-i-lō-kok'kus ô-rē-us

So, what is the most common form of food intoxication? *Staphylococcus aureus* is the most common bacterial species causing foodborne intoxications. This gram-positive sphere is found on the skin and in the noses of up to 25% of healthy people and animals. Therefore, foods can become contaminated with *Staphylococcus* through contact with food workers who carry the pathogen or through contaminated milk and cheeses. Foods containing *S. aureus* toxin lack any unusual taste, odor, or appearance.

Is a toxin involved in staph food poisoning? Like botulism, staphylococcal food poisoning is caused by toxins secreted into foods and consumed by unsuspecting individuals. These individuals will experience abdominal cramps, nausea, vomiting, prostration, and diarrhea as the toxin triggers the release of water. The symptoms last for several hours, and recovery is usually rapid and complete.

■ *Salmonella* säl-mōn-el'lä

How can you tell if a person has staphylococcal food poisoning rather than some other type of food poisoning? One telltale sign of staph food poisoning is the short incubation period, a brief 1 to 6 hours. *Salmonella*, which causes foodborne infections, has a longer incubation period, and the symptoms are different.

Salmonellosis

So how are the symptoms of Salmonella food poisoning different from staph? "Salmonella" as you call it, refers to **salmonellosis**, a foodborne infection. It is caused by ingesting hundreds of *Salmonella* cells, which grow in the body for 1 to 3 days before the patient experiences fever, nausea, vomiting, diarrhea, and severe abdominal cramps. The symptoms may last a week or more, and if dehydration occurs, fluid replacement may be necessary. Look at this cartoon graph from the CDC (FIGURE 19.6). It shows that the number of *Salmonella* outbreaks linked to live poultry has been increasing over the years.

So, it sounds like salmonellosis is a common foodborne infection. Yes, it is. Every year, approximately 42,000 cases of salmonellosis are reported in the United States. However, many cases are fairly mild and would not be reported, so the actual number of total cases is probably well over one million.

As suggested by the graph I showed you (see figure 19.6), poultry products are particularly notorious because *Salmonella* species commonly infect chickens and turkeys. The microbes may be consumed directly from the poultry or in poultry products such as chicken salad or cold cuts made from chicken. Cooking poultry properly and not allowing raw chicken and its fluids to come in contact with raw foods can prevent many salmonellosis outbreaks.

How about eggs? Are they dangerous too? Eggs can be another source of salmonellosis when used in foods such as custard pies, cream cakes, eggnog, ice cream, and mayonnaise. Researchers believe that *Salmonella* species infect the ovary of the hen and pass into the egg before the shell forms. Therefore, contaminated eggs can look perfectly normal and consumers should store eggs in the main compartment of the refrigerator, refrigerate leftover egg dishes quickly in small containers to accelerate cooling, and avoid eating "runny" or undercooked eggs.

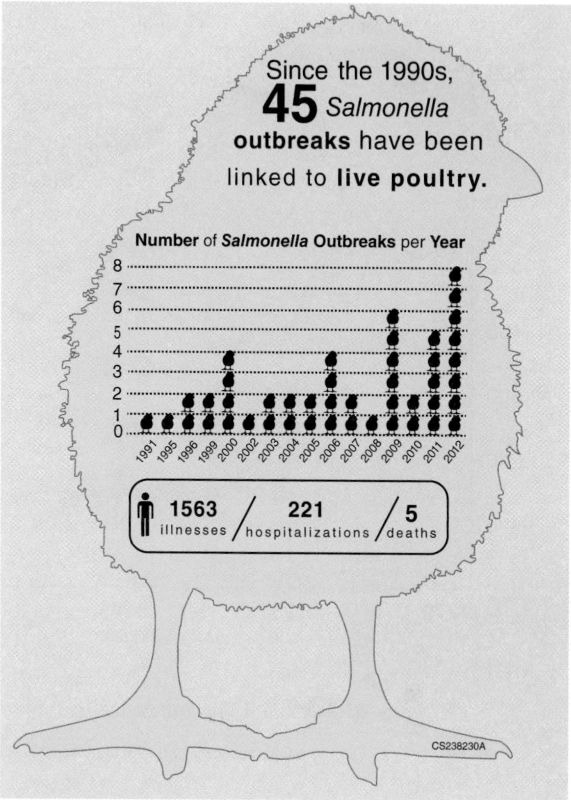

FIGURE 19.6 ***Salmonella* Outbreaks.** The number of *Salmonella* outbreaks linked to live poultry since the 1990s has fluctuated with an increased number in the last few years.

Reproduced from Salmonella/CDC

Typhoid Fever

Is typhoid fever as serious as salmonellosis? **Typhoid fever** is among the classical diseases (one of the "slate-wipers") that have ravaged human populations for generations. *Salmonella enterica* serotype Typhi (we'll call it *S.* Typhi) causes typhoid fever. The CDC estimates there are about 5,700 cases each year in the United States, but about 75% of these were acquired during foreign travel. In fact, globally there are more than 21 million cases every year.

- *Salmonella enterica* (Typhi) säl-mōn-el'lä en-tėr-i'kä (tī'fē)
- serotype: A distinct variation within a species based on immunological characteristics.

What's typhoid fever like? The resistance to environmental conditions outside the body allows *S.* Typhi to remain alive for long periods of time in water, sewage, and certain foods. Thus, it is easily transmitted by the five Fs: flies, food, fingers, feces, and fomites, as seen in this graph (FIGURE 19.7). The food and water could contain fecal material with the pathogen, so transmission often is by the fecal-oral route.

In the small intestine, *S.* Typhi causes deep ulcers and bloody stools. Blood invasion follows, and after a few days, the patient experiences mounting fever, lethargy, and delirium. The abdomen becomes covered with rose spots, an indication that blood is hemorrhaging in the skin.

How is typhoid fever treated? Treatment of typhoid fever is generally successful with antibiotics, but about 5% of recoverers are carriers and continue to harbor and shed the organisms for a year or more. Consequently, people traveling to a country where typhoid fever is common can be vaccinated before traveling. However, the vaccines lose their effectiveness after only a few years.

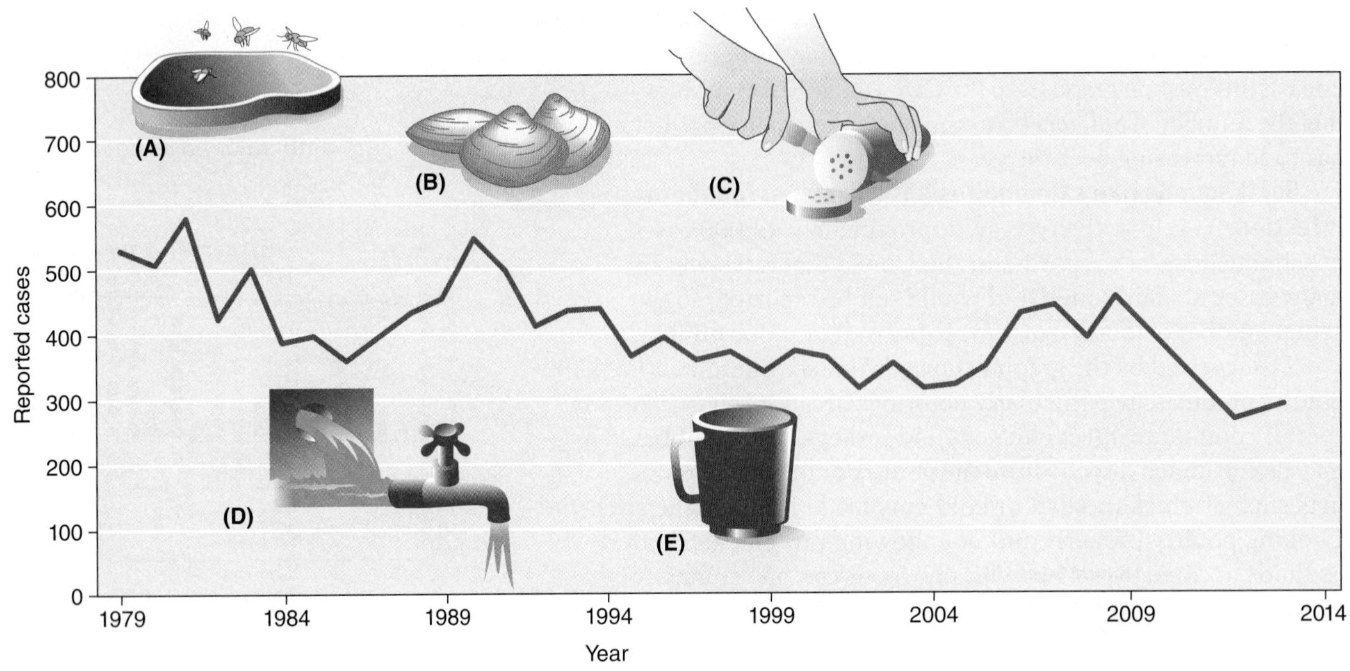

FIGURE 19.7 **The Incidence of Typhoid Fever.** Reported cases of typhoid fever in the United States by year, 1979–2013.

Data from CDC, Summary of Notifiable Diseases, 2013.

Shigellosis

I know someone who had bacterial dysentery. What's that? Bacterial dysentery is technically known as **shigellosis**. It is manifested by waves of intense abdominal cramps and frequent passage of small-volume, bloody, mucoid stools. There usually are around 15,000 cases annually in the United States.

■ *Shigella sonnei*
shi-gel'lä son'nē-ē

Most cases of shigellosis are caused by *Shigella sonnei*. Humans ingest the organisms in contaminated water or foods. *Shigella* bacilli usually penetrate the cells lining the intestines, and, after 2 to 3 days, they produce sufficient toxins to trigger water release. Antibiotics are sometimes effective, but many strains of *Shigella* are becoming resistant to antibiotics.

Cholera

I have read about a cholera outbreak occurring in Haiti. What exactly is cholera? No diarrheal disease can compare with the extensive diarrhea associated with **cholera**. In the most severe cases, a patient infected with *Vibrio cholerae* may lose up to 1 liter of colorless, watery fluid every hour for several hours. The patient's eyes become gray and sink into their orbits; the skin is wrinkled, dry, and cold; muscular cramps occur in the arms and legs; the blood thickens, urine production ceases; and the sluggish blood flow to the brain leads to shock and coma.

■ *Vibrio cholerae*
vib'rē-ō kol'ėr-ī

So, it sounds like cholera is another of those "slate-wipers." Correct? Most definitely. In untreated cases, the mortality rate for cholera may reach 70%.

Even today, the WHO estimates there are more than 100,000 cases and over 1,900 deaths annually. The outbreak in Haiti that began 10 months after the 2010 earthquake continues in 2014 and is considered the worst epidemic of cholera in recent Haitian history. There have been more than 700,000 cases and some 8,500 deaths.

How is cholera treated? The bacilli enter the intestinal tract in contaminated water or food. As they move along the intestinal epithelium, they secrete a toxin that stimulates the unrelenting loss of fluid. Antibiotics kill the bacterial cells, but the key treatment is restoration of the body's water balance with oral or intravenous (IV) rehydration solutions.

E. coli Diarrheas

I thought *Escherichia coli* was a harmless member of our intestines. Does it also cause disease? Besides the harmless strain in your gut, there are other pathogenic strains of *E. coli*. One such strain induces diarrhea in infants when it invades the intestinal lining and produces powerful toxins that cause water loss. Other strains cause **traveler's diarrhea**, a term usually applied to a disease in which a traveler experiences diarrhea within 2 weeks of visiting the affect area; the diarrhea lasts up to 10 days.

■ *Escherichia coli*
esh-ėr-ē′kē-ä kō′lī (or kō′lē)

What about the *E. coli* in foods that I've read about? It is possible to contract a very serious form of hemorrhagic diarrhea due to *E. coli* O157:H7. When confined to the large intestine, the strain causes grossly bloody diarrhea, a complication known as hemorrhagic colitis. When the disease involves the kidneys, it can lead to kidney failure and is called hemolytic uremic syndrome (HUS). Seizures, coma, colon perforation, and liver disorder have been associated with HUS.

How widespread is this 0157:H7 strain? About 200–300 cases of postdiarrheal HUS are reported to the CDC each year, most cases coming from contaminated foods other than ground meat. This includes fresh spinach and romaine lettuce, hazelnuts, cheeses, and even cookie dough. How some of these foods got contaminated is uncertain. The prevailing wisdom from ground meat is that the *E. coli* cells exist in the intestines of cattle but cause no disease in the animals. Contamination during slaughtering brings *E. coli* to beef products, and excretion to the soil accounts for transfer to plants and fruits.

Campylobacteriosis

I recently heard about a death from campylobacteriosis. What is this disease? Today, the infection, called **campylobacteriosis**, is one of the most common causes of diarrheal illness in Americans. Dairy products and water contaminated with *Campylobacter jejuni* are possible sources of infection. However, the cases are sporadic and usually do not occur in the larger outbreaks seen with *E. coli* and *Salmonella*.

■ *Campylobacter jejuni*
kam′pi-lō-bak-tėr jē-jū′nē

What are the symptoms of campylobacteriosis? The symptoms of the disease range from mild diarrhea to severe gastrointestinal distress, with fever, abdominal pains, and bloody stools. Most patients recover in less than a week although the CDC estimates there are about 75 deaths every year from *Campylobacter* infections.

Listeriosis

I heard about an outbreak of listeriosis in cantaloupes a few years ago. What's listeriosis? In the fall of 2011, there was a multistate outbreak of listeriosis linked to cantaloupes. About 150 people were infected and 33 died.

Listeriosis is caused by *Listeria monocytogenes*, a small gram-positive rod. The disease primarily affects older adults, pregnant women, newborns, and individuals with a weakened immune system. It may occur as listeric meningitis, with headaches,

■ *Listeria monocytogenes*
lis-te′rē-ä mo-nō-sī-tô′je-nēz

stiff neck, delirium, and coma. Another form is a blood disease accompanied by high numbers of white blood cells called monocytes. In pregnant women, an infection of the uterus may result in miscarriage.

How does Listeria get into food? *Listeria* species are commonly found in the soil and in the intestines of many animals, so the bacterial cells are transmitted to humans by food contaminated with fecal matter, as well as by the consumption of contaminated animal foods. Cold cuts, as well as soft cheeses (e.g., Brie, Camembert, and feta), have been associated with cases. Antibiotics are effective treatments.

Brucellosis

I recently heard of a brucellosis outbreak in livestock in China. Does this disease also affect humans? Yes, people can contract **brucellosis** from infected animals or contaminated animal products. Animals are most commonly infected with the genus *Brucella* include cattle, goats, sheep, pigs, and dogs.

■ *Brucella*
brü'sel-lä

The major way humans get infected is by eating or drinking unpasteurized or raw dairy products, such as milk and cheeses contaminated with the organism. Patients experience flulike weakness as well as backache, joint pain, and a high fever with drenching sweats in the daytime and low fever with chills in the evening. This fever pattern gives the disease its alternate name of "undulant fever."

Peptic Ulcer Disease

Did I read somewhere that peptic ulcers are now considered an infectious disease? That seems crazy! Crazy as it sounds, it's true. Many cases of **peptic ulcer disease** are caused by the bacterium *Helicobacter pylori*. This gram-negative curved rod is apparently transmitted by the fecal-oral route from person-to-person. Doctors have revolutionized the treatment of ulcers by prescribing antibiotics such as tetracycline. They have achieved cure rates of up to 90%, and relapses are uncommon.

■ *Helicobacter pylori*
hē'lik-ō-bak-tėr pī'lō-rē

But isn't the stomach very acidic? How does the microbe manage to survive? How *H. pylori* survives in the intense acidity of the stomach is interesting. Here, look at this illustration (FIGURE 19.8). The bacterial cells attach to the stomach wall and then secrete an enzyme that digests urea in the area, producing ammonia as an end product. The ammonia neutralizes stomach acid in the vicinity of the infection and the organisms begin their destruction of the tissue, supplemented by the stomach acid.

Infection with *H. pylori* is also the major cause of gastric cancer, which represents the fourth most common cancer and the second leading cause of cancer deaths in the world—about 1 million deaths each year.

Other Intestinal Diseases

What other bacterial pathogens of the intestine should I know about? There are a couple of important ones. *Vibrio parahaemolyticus*, for example, is a gram-negative rod that produces foodborne infections where seafood is the main staple of the diet. *Vibrio vulnificus* occurs naturally in brackish water and seawater, where oysters and clams live and is transmitted to humans by them. *Bacillus cereus* is a gram-positive spore-forming bacillus that causes food poisoning, frequently experienced after consuming cooked rice. And *Yersinia enterocolitica* is a gram-negative rod that causes fever, diarrhea, and abdominal pain.

■ *Vibrio parahaemolyticus*
vib'rē-ō pa-rä-hē-mō-li'ti-kus

■ *Vibrio vulnificus*
vib'rē-ō vul-ni'fi-kus

■ *Bacillus cereus*
bä-sil'lus se'rē-us

■ *Yersinia enterocolitica*
yėr-sin'ē-ä
en'tėr-ō-kōl-it-ik-ä

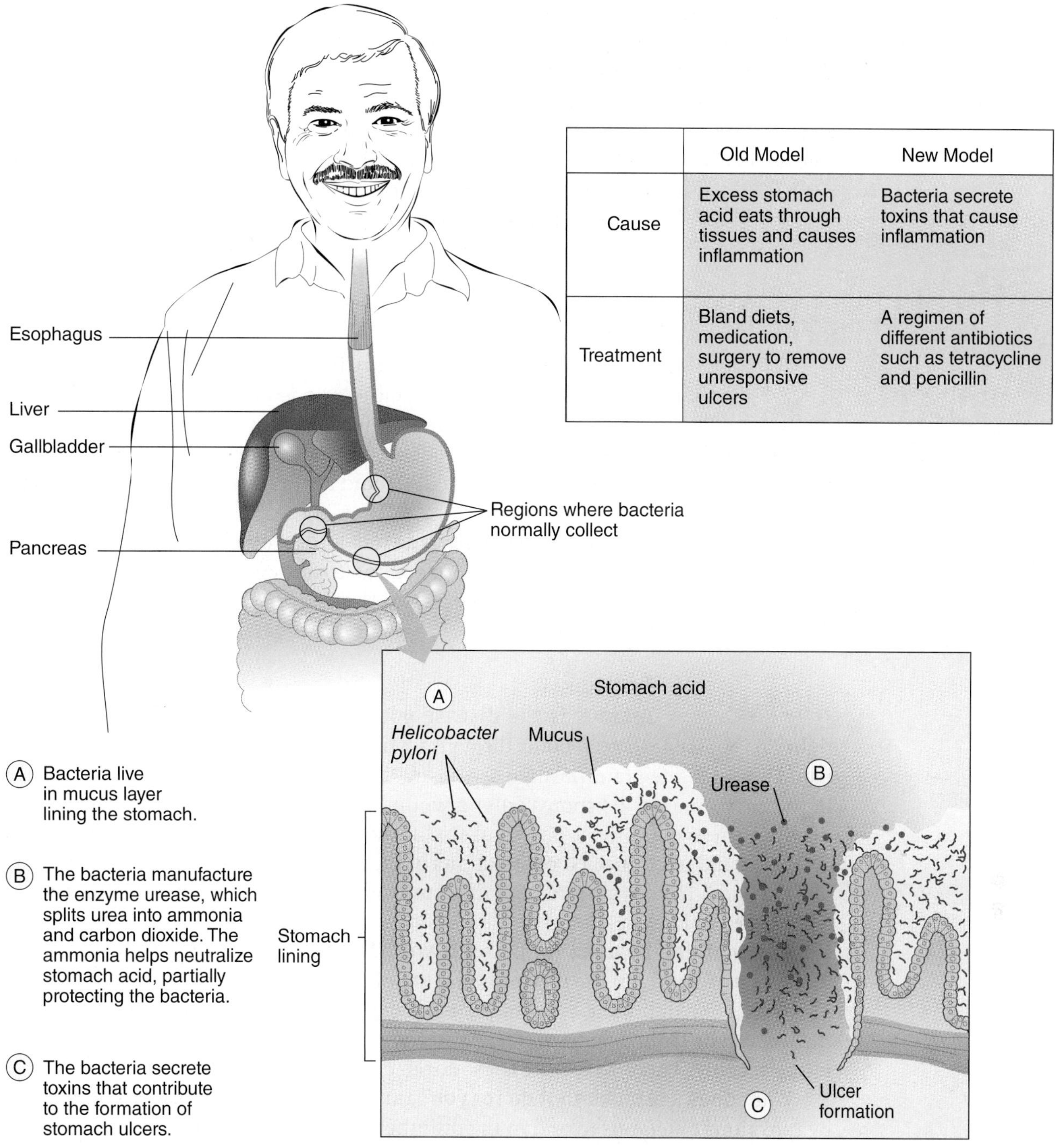

	Old Model	New Model
Cause	Excess stomach acid eats through tissues and causes inflammation	Bacteria secrete toxins that cause inflammation
Treatment	Bland diets, medication, surgery to remove unresponsive ulcers	A regimen of different antibiotics such as tetracycline and penicillin

Esophagus
Liver
Gallbladder
Pancreas
Regions where bacteria normally collect
(A) Bacteria live in mucus layer lining the stomach.
(B) The bacteria manufacture the enzyme urease, which splits urea into ammonia and carbon dioxide. The ammonia helps neutralize stomach acid, partially protecting the bacteria.
(C) The bacteria secrete toxins that contribute to the formation of stomach ulcers.
Stomach lining
(A)
Stomach acid
Helicobacter pylori
Mucus
Urease
(B)
(C)
Ulcer formation

FIGURE 19.8 **The Progression of Gastric Ulcers.** The majority of peptic ulcers are caused by *Helicobacter pylori*. Inset: the old and new models of the cause and treatment of peptic ulcers.

19.3 Soilborne Bacterial Diseases: Endospore Formers

Soilborne diseases are those whose bacterial agents are transferred from the soil to the unsuspecting individual. To remain alive in the soil, the bacterial cells must resist environmental extremes, and often they form endospores, as these three diseases illustrate.

Anthrax

Bacillus anthracis
bä-sil'lus se'rē-us an-thrā'sis

What exactly is anthrax? **Anthrax** is a blood disease that occurs in large animals such as cattle, sheep, goats, and rarely, in humans. The disease is caused by *Bacillus anthracis*, a gram-positive spore-forming rod. Patients inhale the spores or come into contact with them in the air or ingest them in meat, and soon their organs fill with bloody, black, infected fluid. In this photograph, you can see that on the skin there are boil-like lesions covered with a black crust (FIGURE 19.9). Violent dysentery with bloody stools accompanies the intestinal form.

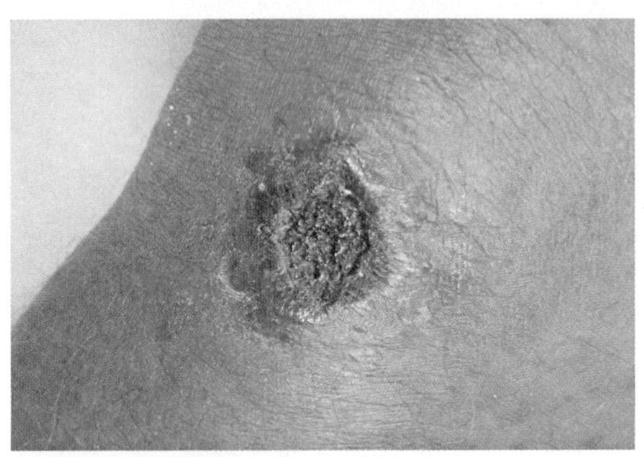

FIGURE 19.9 **Anthrax.** This cutaneous lesion is a result of infection with anthrax bacilli. Lesions like this one develop when anthrax spores contact the skin, germinate to vegetative cells, and multiply.

Courtesy of James H. Steele/CDC

Humans can acquire anthrax by shearing or processing wool from sheep, consuming infected meat, or coming in contact with products that may contain endospores, such as goatskin drums and leather jackets.

Penicillin and other antibiotics are used for therapy. In untreated cases, the mortality rate is more than 80%. There are few, if any cases in the United States because of the testing of imported animal products. Of course, there was an anthrax bioterrorism incident in 2001 and anthrax remains a potential bioterror weapon.

How is anthrax treated and prevented? As mentioned, there are antibiotics used for treatment after exposure or after infection. To be successful though, the illness needs to be identified early.

Tetanus

Clostridium tetani
klôs-tri'dē-um te'tän-ē

Tetanus is the disease you get when you step on a rusty nail, right? It is possible to get **tetanus** through a puncture wound from a soil-contaminated nail. However, the gram-positive organism, *Clostridium tetani*, is found everywhere in the environment. Its spores enter a wound and revert to multiplying bacilli that produce the second most powerful toxin known to science. The toxin provokes sustained and uncontrolled contractions of the muscles, and spasms occur throughout the body. Here is a short story of an unusual way to become exposed to tetanus (**A Closer Look 19.3**).

Can anything be done for patients with tetanus? Tetanus patients are treated with sedatives and muscle relaxants and are placed in quiet, dark rooms. Physicians prescribe penicillin to destroy the bacterial cells and tetanus antitoxin to neutralize the toxin. The United States has had a steady decline in the incidence of tetanus due to the vaccine. There were only a few dozen cases reported by the CDC each year.

What does a tetanus shot do for you? Immunization to tetanus is accomplished by injections of tetanus toxoid in the diphtheria-tetanus-acellular pertussis vaccine. The toxoid induces the immune system to produce protective antibodies called antitoxins. Booster injections of tetanus toxoid in the "tetanus shot" are recommended every 10 years to keep the level of immunity high.

Gas Gangrene

Clostridium perfringens
klôs-tri'dē-um pėr-frin'jens

Is gas gangrene the same as gangrene? No, "gangrene" is self-destruction of the muscle tissues brought about by enzymes from partially repaired tissue. **Gas gangrene** occurs when soilborne *Clostridium perfringens*, another endospore former, invades the dead, anaerobic tissue of a wound. These gram-positive rods ferment the muscle carbohydrates and decompose muscle proteins, producing large amounts of gas that tear the tissue apart. The organism is also a common cause of food poisoning due to the production of a toxin.

A CLOSER LOOK 19.3

Tetanus Outbreak Among Injecting Drug Users

When one thinks about tetanus, the typical image coming into mind is stepping on a rusty nail. Such nails do pose a threat because spores cling to the rough edges of the nail and the nail may cause extensive tissue damage as it penetrates. Actually, most cases in the developed nations of the world are found in older women who become infected by contaminated soil while gardening. However, there are other ways spores can be introduced into the body.

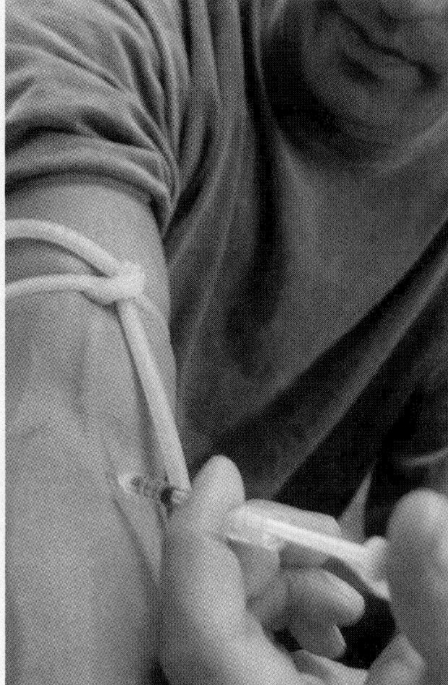

© Oscar Knott/FogStock/Alamy

Between July 2003 and January 2004, 25 cases of tetanus in injecting drug users (IDUs) were reported to public health officials in the United Kingdom. The method of injection was subcutaneous injection of heroin. Thirteen women and 12 men between the ages of 20 and 53 years were identified as having clinical tetanus. Sixteen were hospitalized with severe generalized tetanus. In the end, two patients died. Among the 23 survivors, 2 had mild disease and 21 required intensive treatment for a median of 40 days.

Twenty-two of the 23 IDUs had either not been immunized against tetanus or kept up immunization boosters. The source of the tetanus infection has remained unclear. Because all cases were clustered in a short period, the most likely source was the drug or the adulterant (a substance added to the drug to make it less pure).

Since this cluster in 2003/4, only sporadic cases of tetanus have been reported in IDUs.

What are the symptoms of gas gangrene? The symptoms of gas gangrene include intense pain and swelling at the wound site as well as a foul odor. The site initially turns dull red, then green, and finally blue-black as you can see in this photo FIGURE 19.10. Treatment consists of antibiotic therapy as well as removal of dead skin and tissue. In severe cases, amputation or exposure in a hyperbaric oxygen chamber may be needed to prevent disease spread and death.

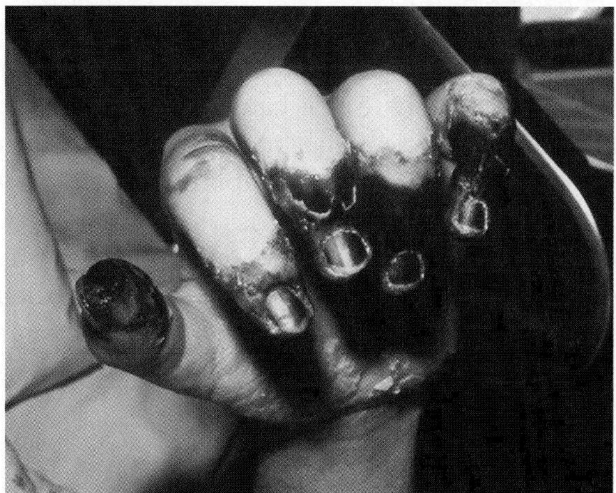

FIGURE 19.10 **Gas Gangrene of the Hand.** A severely infected hand showing gangrene (blackened tissue necrosis). This infection developed from an accident while the patient was scaling fish. Antibiotic drugs may prevent the infection, but in this advanced stage amputation of the hand may be necessary.

Courtesy of Dr. Jack Poland/CDC

19.4 Arthropodborne Bacterial Diseases: The Bugs Bite

Fleas, lice, and ticks are examples of arthropods that transmit diseases to or among humans usually by taking a blood meal from an infected animal or person and themselves becoming infected. Then, these arthropods, called **vectors**, pass the microbes to another individual during the next blood meal. Arthropod-related diseases occur primarily in the blood stream, and they are often accompanied by a high fever and a body rash.

Plague

I've heard plague had a powerful influence on the course of Western civilization. Is that true? Yes, it is. Few diseases have had a more terrifying history than **bubonic plague**, and few can match the array of social, economic, and religious changes the disease has brought about. The pandemic in the 1300s was known as the "Black Death" because of the purplish-black splotches on victims. By some accounts, plague killed an estimated 40 million people in Europe, almost one-third of the population.

Has plague ever occurred in the United States? Yes, indeed. Plague first appeared in San Francisco in 1900, carried by rats on ships from Asia. The disease spread to ground squirrels, prairie dogs, and other wild rodents, and it is now endemic in the southwestern states, where it is commonly called "sylvatic plague."

What's the cause of plague? Plague is caused by the gram-negative rod *Yersinia pestis*. The bacterial cells are transmitted by the rat flea when it takes a blood meal. The bacterial cells localize in the lymph nodes, especially those of the armpits, neck, and groin, where hemorrhaging causes substantial swellings **buboes**—hence the name **bubonic plague**. From there, the microbes spread to the bloodstream, where they cause "septicemic plague" and then to the lungs, where they cause "**pneumonic plague**." There is extensive coughing and hemorrhaging, and many patients suffer cardiovascular collapse. The cells can be spread by respiratory droplets to other people. Mortality rates for pneumonic plague approach 100% unless antibiotic therapy is instituted.

Tularemia

I've never heard of tularemia. What is it? **Tularemia** is a plague-like disease, although it is not as serious as plague. It is caused by *Francisella tularensis*, a small gram-negative rod. The disease occurs in a broad variety of wild animals, especially rodents, and it is particularly prevalent in rabbits (where it is known as "rabbit fever").

How is tularemia spread? Tularemia is among the most transmissible of bacterial diseases. Ticks are important vectors for its spread. Other methods of transmission include contact with an infected animal, consumption of contaminated rabbit meat, splashing bacilli into the eye, and inhaling bacilli. The disease usually resolves on treatment with antibiotics.

Lyme Disease

Does Lyme disease have anything to do with limes? No, it doesn't. **Lyme disease** is named for Old Lyme, Connecticut, the suburban community where the first cluster of cases occurred in 1975. The disease was traced to deer ticks and several years later a spirochete was identified as the causative agent. Researchers named it *Borrelia burgdorferi*. Lyme disease is currently the most commonly reported arthropodborne illness in the United States.

Borrelia burgdorferi
bôr-rel′ē-ä burg-dôr′fėr-ē

What are the symptoms of Lyme disease? For about 20% of people infected, they have nothing more than flu-like symptoms. For many of the rest, the illness starts with a rash at the site of the tick bite. It expands slowly, eventually forming an intense red border and a red center, and it resembles a bull's eye as you can see in this photo (FIGURE 19.11). Left untreated, some cases enter a second stage where the patient experiences pain, swelling, and arthritis in the large joints, especially the knee, shoulder, ankle, and elbow joints. In some individuals a third stage occurs, and the arthritis is complicated by damage to the cardiovascular and nervous systems.

Can the disease be treated with antibiotics? In the rash stage, effective treatment can be rendered with antibiotics. Patients developing neurological or cardiac symptoms may need to be treated with intravenous antibiotics.

Rocky Mountain Spotted Fever

Does Rocky Mountain spotted fever happen only in the Rocky Mountains? Interestingly, very few cases of the disease occur in the Rocky Mountains. Most cases occur in southeastern and Atlantic Coast states where ticks transmit *Rickettsia rickettsii*, the responsible agent.

Rickettsia rickettsii
ri-ket′sē-ä ri-ket′sē-ē

What are the signs and symptoms of the disease? The hallmarks of **Rocky Mountain spotted fever** are a high fever lasting for many days and a skin rash reflecting damage to the small blood vessels. The rash begins as pink spots and progresses to pink-red pimple-like spots that fuse to form a flat, red area on the skin that is covered with small confluent bumps. As you can see in this photo (FIGURE 19.12), the rash generally begins on the palms of the hands and soles of the feet and progressively spreads to the trunk. Antibiotics are therapeutically useful.

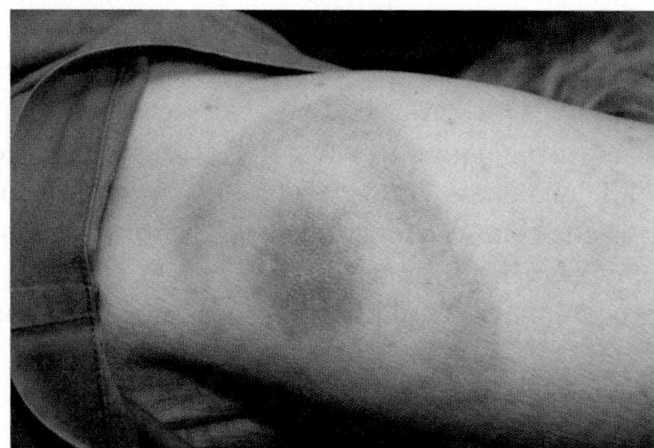

FIGURE 19.11 **The Bull's-Eye Rash.** Lyme disease can start with a rash that consists of a large red patch with an intense red border. It is usually hot to the touch, and it expands with time.

Courtesy of James Gathany/CDC

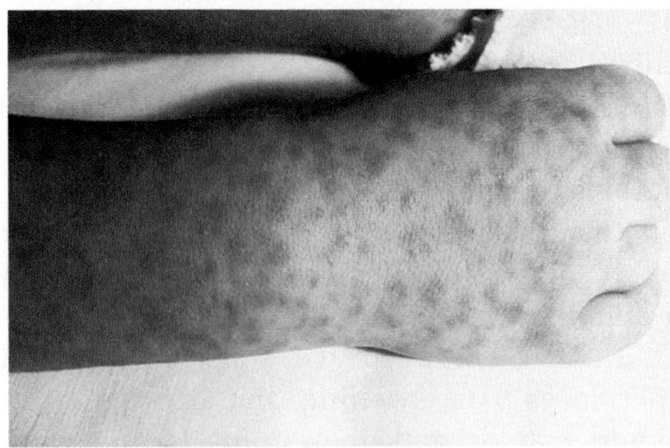

FIGURE 19.12 **Rocky Mountain Spotted Fever.** A child's right hand and wrist displaying the characteristic spotted rash of Rocky Mountain spotted fever.

Courtesy of CDC

19.5 Sexually Transmitted Diseases: A Continuing Health Problem

The **sexually transmitted diseases (STDs)** belong to a broad category of diseases transmitted by contact. The contact in this case is with the reproductive organs. Person-to-person transmission is necessary for bacterial survival because the microbes usually cannot remain alive outside the body tissues. The major STDs continue to be a problem in the United States. As you can see in this bar graph FIGURE 19.13, they make up four of the top ten reported infectious diseases. Incidentally, STDs are often referred to as sexually transmitted infections (STIs) because often there is an infection but no definite disease.

Syphilis

What causes syphilis, and what's the disease like? Over the centuries, Europeans have had to contend with four pox diseases: chickenpox, cowpox, smallpox, and the Great Pox, the latter now known as syphilis. **Syphilis** is caused by *Treponema pallidum*, whose spiral bacterial cells penetrate the skin surface and cause a disease that can progress through three stages: "Primary syphilis," is characterized by a **chancre**, a painless, hard, circular, purplish ulcer often on the genital organs. It persists for 2 to 6 weeks, and then disappears. Several weeks later, the patient experiences "secondary syphilis" with lesions over the entire body surface, fever, rash, and a patchy loss of hair on the head. Recovering patients bear pitted scars from the lesions and they remain "pockmarked." Then comes "tertiary syphilis." Its hallmark is the **gumma**, a soft, gummy granular lesion that weakens the blood vessels, causing them to bulge and burst. In the nervous system, gummas lead to paralysis and insanity. [This is the form that may have affected the bodies and minds of some famous people in history (see Chapter introduction)].

■ *Treponema pallidum*
tre-pō-nē'mä pal'li-dum

Can syphilis be treated? Yes, penicillin is the drug of choice for the primary and secondary stages of the disease, but antibiotics are ineffective in tertiary syphilis. The cornerstone of syphilis control is the identification and treatment of the sexual contacts of patients. Syphilis is also a serious problem in pregnant women because the spirochetes penetrate the placental barrier.

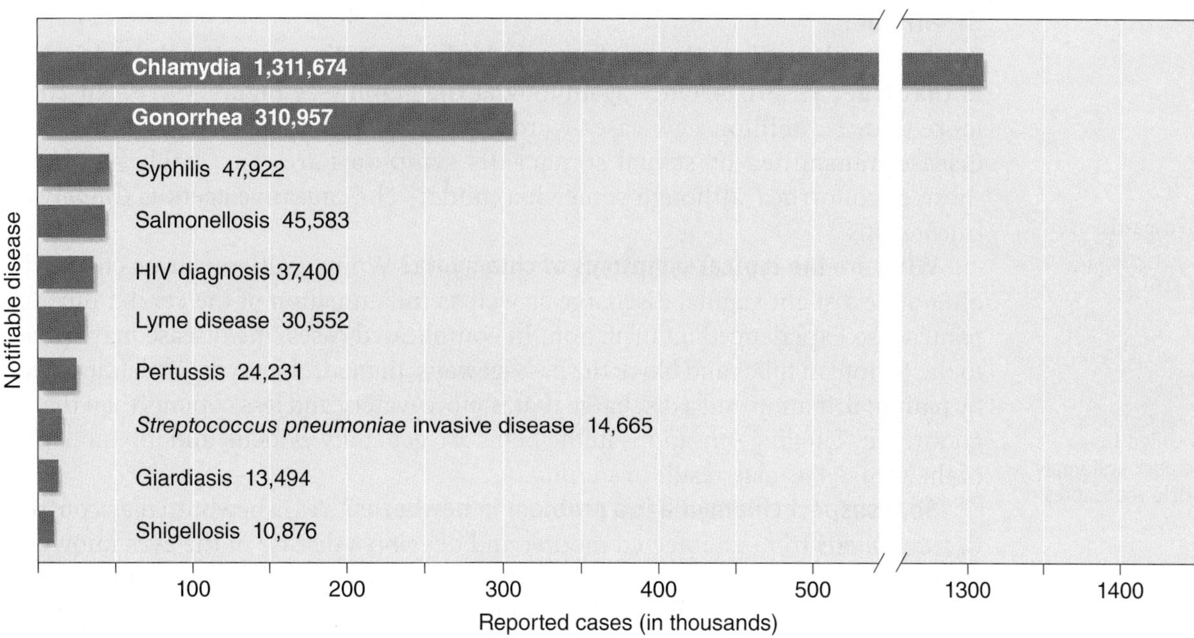

FIGURE 19.13 **Reported Cases of Notifiable Diseases in the United States, 2013.** Four of the top five most reported microbial diseases in the United States are sexually transmitted.

Data from CDC,Summary of Notifiable Diseases, 2013.

Gonorrhea

How common is gonorrhea in the United States? If you look back at the graph (see Figure 19.13), gonorrhea was the second most frequently reported notifiable disease in the United States.

What's the cause of gonorrhea? Gonorrhea is caused by *Neisseria gonorrhoeae*, a small gram-negative diplococcus commonly known as the gonococcus. The great majority of cases of gonorrhea are transmitted by person-to-person contact during sexual intercourse.

■ *Neisseria gonorrhoeae* nī-se'rē-ä go-nôr-rē' ī

Is gonorrhea a deadly disease? Not really. In women, the gonococci invade the cervix and the urethra. Patients often report a discharge, abdominal pain, and a burning sensation on urination. In some women, gonorrhea also spreads to the Fallopian tubes, and these thin passageways become riddled with adhesions, causing a difficult passage for egg cells. In men, gonorrhea occurs primarily in the urethra. Onset is usually accompanied by a tingling sensation in the penis, followed in a few days by pain when urinating. There is also a thin, watery discharge at first, and later a whitened, thick fluid that resembles semen.

■ cervix: The entrance to the womb and leading to the vagina.

■ urethra: The tube in mammals carrying urine from the bladder out of the body.

Can gonorrhea occur in other organs? Gonorrhea does not restrict itself to the urogenital organs. Gonococcal pharyngitis, for example, may develop in the pharynx; in infants born to infected women, gonococci may cause a disease of the eyes called gonococcal ophthalmia. To preclude the blindness that may ensue, most states have laws requiring the eyes of newborns be treated with antibiotics. In adults, gonorrhea therapy consists of antibiotic treatment.

Chlamydia

I've heard chlamydia is the most reported infectious disease in the United States. Is that true? It sure is. Once again look at the graph (see Figure 19.13). In 2013, more than 1.3 million new cases were reported. **Chlamydia** is a gonorrhea-like disease transmitted by sexual contact. Its symptoms are remarkably similar to those of gonorrhea, although somewhat milder. The causative agent is *Chlamydia trachomatis*.

Chlamydia trachomatis kla-mi'dē-ä trä-kō'mä-tis

What are the typical symptoms of chlamydia? Women suffering from chlamydia often note a slight vaginal discharge as well as inflammation of the cervix. Burning pain is also experienced on urination. In complicated cases, the disease may spread to the Fallopian tubes and block the passageways. In men, chlamydia is characterized by painful urination and a discharge that is more watery and less copious than that of gonorrhea. Tingling sensations in the penis are generally evident, and inflammation of the epididymis may result in sterility.

epididymis: A coiled tube attached to the back and upper side of the testicle that stores sperm.

So, I suspect chlamydia is a problem in newborns? Yes, a newborn may contract *C. trachomatis* from an infected mother and develop a disease of the eyes known as chlamydial ophthalmia. Chlamydial pneumonia may also develop in a newborn. Health officials estimate that each year in the United States, over 75,000 newborns suffer chlamydial ophthalmia and 30,000 newborns experience chlamydial pneumonia. The disease can be successfully treated with antibiotics.

19.6 Contact and Miscellaneous Bacterial Diseases: The Need for Skin Contact

Sexually transmitted diseases are a subgroup of a larger set of diseases usually transmitted by contact. Usually, some form of skin contact takes place, as these diseases will illustrate.

Leprosy (Hansen Disease)

I know leprosy was well-known in past centuries. What causes leprosy and how is it spread? For many centuries, leprosy was considered a curse of the damned. It did not kill, but neither did it seem to end. Instead, it lingered for years, causing the tissues to degenerate and deforming the body. In biblical times, the afflicted were required to call out "Unclean! Unclean!" and usually were ostracized from the community. The agent of **leprosy** or Hansen disease is *Mycobacterium leprae*, an acid-fast rod. The cells are spread by multiple skin contacts as well as by droplets from the upper respiratory tract of an infected person. The disease has an unusually long incubation period of 3 to 6 years.

Mycobacterium leprae mī-kō-bak-ti'rē-um lep'rī

Patients with leprosy experience disfiguring of the skin and bones, twisting of the limbs, and curling of the fingers. The largest number of deformities develops from the loss of pain sensation due to nerve damage. A compound known commercially as Dapsone is often used to treat the disease—often with miraculous results as you can see from these photos (FIGURE 19.14).

Staphylococcal Skin Disease

What causes staphylococcal skin disease, and what are the various forms? *Staphylococcus aureus*, the grapelike cluster of gram-positive cocci, is the species usually involved in staphylococcal skin disease. Look at this photo (FIGURE 19.15a).

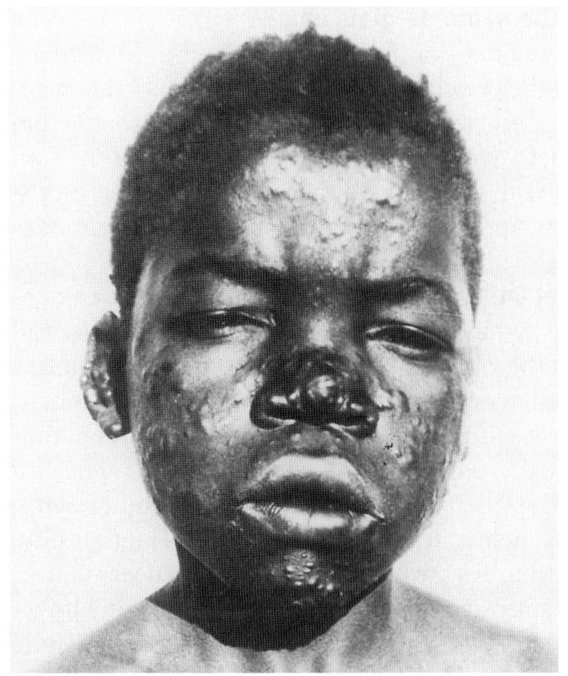

(a)

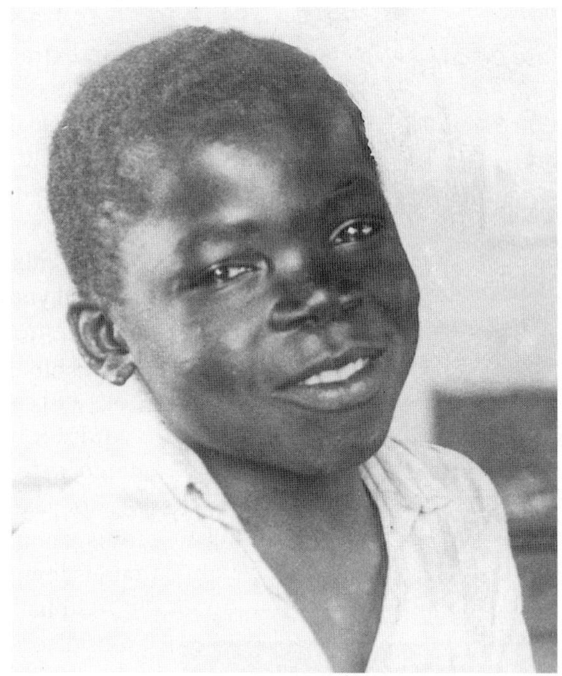

(b)

FIGURE 19.14 **Treating Leprosy.** The young boy with leprosy is pictured (**a**) before treatment with dapsone and (**b**) some months later, after treatment. Note that the lesions of the ear and face and the swellings of the lips and nose have largely disappeared.

The hallmark of disease is the abscess, a circumscribed pus-filled lesion. A boil is a skin abscess while deeper skin abscesses, such as these so-called carbuncles, develop when the staphylococci work their way into the tissues below the skin. In this photo of a patient with **impetigo** (FIGURE 19.15b), the infection is more superficial and involves patches of epidermis just below the outer skin layer.

How is staphylococcal skin disease spread, and can it be treated? Skin contact with an infected individual is the usual mode of transmission. Staphylococcal diseases are commonly treated with penicillin, but resistant strains of *S. aureus*, called **methicillin-resistant *Staphylococcus aureus*** (MRSA), have become a real problem, both in the hospital and the community.

Is toxic shock syndrome a staphylococcal disease? **Toxic shock syndrome** (TSS) is caused by a toxin-producing strain of *S. aureus*. The earliest symptoms of disease include a rapidly rising fever, accompanied by vomiting and watery diarrhea. Patients then experience a sore throat, severe muscle aches, and, as seen in this photo, a sunburn-like rash with peeling of the skin, especially on the soles of the feet and palms of the hands (FIGURE 19.15c). A sudden drop in blood pressure also occurs, leading to shock and heart failure.

Conjunctivitis and Trachoma

Is conjunctivitis an eye disease? Yes, it is. **Conjunctivitis** is a disease of the conjunctiva. When infected, the membrane becomes inflamed, a factor that imparts a brilliant pink color to the white of the eye—hence the name "pink eye." Several microbes cause conjunctivitis, including staphylococci, which are transmitted by face-to-face contact and airborne droplets, as well as by contaminated instruments. The disease normally runs its course in about 2 weeks. Therapy with antibiotics hastens recovery.

■ conjunctiva: The membrane covering the internal part of the eyelid and the front of the eye (cornea).

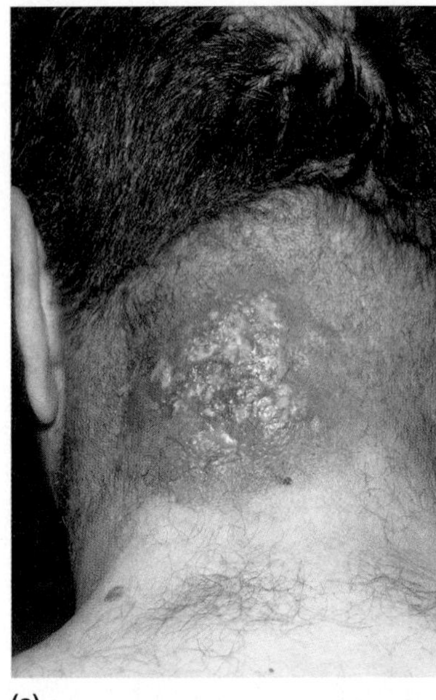
(a)

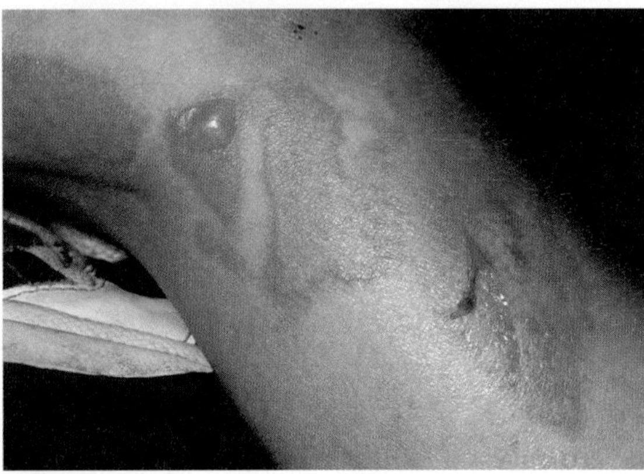
(b)

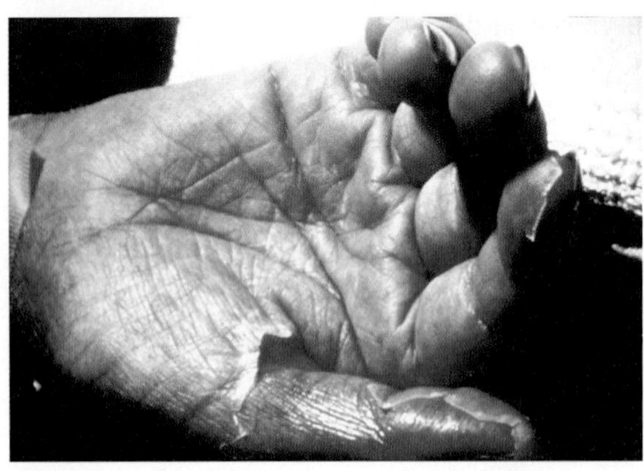
(c)

FIGURE 19.15 **Staphylococcal Skin Diseases.**
(**a**) A severe carbuncle on the back of the head/neck.
(**b**) A patient with impetigo on the cheeks and chin.
(**c**) A patient with toxic shock syndrome.

(a) © Medical-on-Line/Alamy. (b) © SPL/Science Source. (c) Courtesy of CDC.

Is trachoma the same as glaucoma? No, the names only sound similar. Glaucoma is a physiological disease caused by a buildup of fluid in the eye chambers, while **trachoma** is an infectious disease caused by a variant of *Chlamydia trachomatis* that infects the conjunctiva. A series of tiny, pale nodules forms on this membrane, giving it a rough appearance. Fingers, towels, optical instruments, and face-to-face contact are possible modes of transmission.

Can you also get animal diseases from water? Yes. One example is **leptospirosis**, which is the most common and widespread zoonotic disease in the world. It can affect household pets such as dogs and cats as well as rats, mice, and barnyard animals. Humans acquire it by contact with these animals or from water contaminated with their urine. The agent of leptospirosis is *Leptospira interrogans*, a small spirochete. Patients experience flulike symptoms, such as fever, aches, and muscle weakness. The mortality rate is low, and antibiotics are generally used with success.

Globally, leptospirosis is considered to be an emerging infectious disease and has been a concern to adventure travelers. Here, you should read this report from several years ago (**A Closer Look 19.4**).

Dental Diseases

I know it's not a contact disease, but what about dental caries? Scientists estimate that there are between 50 billion and 100 billion bacterial cells in the adult mouth at any time. In order for **dental caries** to develop, three elements must be present: a caries-susceptible tooth with a buildup of plaque; dietary carbohydrates, usually in the form of sucrose (sugar); and acid-producing plaque bacteria. One of the primary bacterial causes of caries is *Streptococcus mutans*. This gram-positive coccus has a high affinity for the smooth surfaces, pits, and fissures of a tooth. The microbes then ferment dietary carbohydrates to lactic acid, along with smaller amounts of acetic acid, formic acid, and butyric acid. The acids dissolve the calcium compounds of the tooth enamel, and protein-digesting enzymes break down any remaining organic materials. Soon dental caries develops.

How can I protect against dental caries? Preventing dental caries has three principal thrusts: protecting the tooth, modifying the diet, and combating the plaque bacteria. Tooth protection may be accomplished by the topical application of fluorides that reduce the solubility of the enamel. Teeth can also be protected by applying polymers to cover pits and fissures in their surfaces, thereby preventing bacterial adhesion. Diet modification requires minimizing sucrose in foods. And of course, normal brushing and flossing of the teeth along with dental cleaning to remove the plaque are foremost.

What about gum disease? Caries is not the only form of dental disease. The teeth are surrounded by tissues that provide essential support. These so-called periodontal tissues may be the site of a **periodontal disease**, which affects some 80% of American adults. Look at these photos. One of the most common forms of early-stage periodontal disease is **gingivitis** (FIGURE 19.16a), which

■ zoonotic: Referring to a disease transmitted from animals to humans.

A CLOSER LOOK 19.4

A Real Eco-Challenge!

It started as a headache on the plane back from Borneo to the States. Within 3 days, Steve went to a hospital emergency room in Los Angeles complaining of fever and chills, muscle aches, vomiting, and nausea.

The Eco-Challenge 2000 in Sabah, Borneo was the site for the annual adventure race. Some 304 participants composing 76 teams from 26 countries competed in the 10-day endurance event, which was designed to push participants and their teams beyond their athletic limits. During Eco-Challenge 2000, teams would kayak on the open ocean, mountain bike into the rainforest, spelunk in hot caves, and swim in local rivers (see figure). Of the 76 teams starting the challenge, 44 teams finished.

A river in Borneo.

About the time Steve arrived at the emergency room, the Centers for Disease Control and Prevention (CDC) in Atlanta received calls from the Idaho Department of Health, the Los Angeles County Department of Health Services, and the GeoSentinel Network (a network of international travel clinics) reporting cases of a febrile illness similar to Steve's.

The CDC quickly carried out a phone questionnaire to 158 participants in the Eco-Challenge. Many reported symptoms similar to Steve's, including chills, fever, headache, diarrhea, and conjunctivitis. Twenty-five respondents had been hospitalized. Within a few days, antibiotic therapy had Steve recovering. In fact, all 135 affected participants completely recovered.

The similar symptoms suggested leptospirosis and laboratory tests either confirmed the presence of *Leptospira* antibodies or positive culture of the organism from serum samples collected from ill participants.

To identify the source and the exposure risk, information was gathered from participants about various portions of the race course. Analysis identified swimming in and kayaking on the Sagama River as the probable source.

Several participants who did not become ill had taken doxycycline as a preventative for malaria and leptospirosis, as race organizers had advised. Unfortunately, Steve had not heeded those words and suffered a real "eco-challenge." Asked if he would participate in the 2001 Eco-Challenge in New Zealand, he said, "Heck yes! It just adds to pushing the limits."

develops when bacterial cells in the plaque multiply and build up between the teeth and gums. Left untreated, gingivitis can progress to what you see in this photo (FIGURE 19.16b), which is **periodontitis**, a serious disease of the soft tissue and bone supporting the teeth. This can result in loosening of the tooth and tooth loss.

Urinary Tract Disease

I believe urinary tract diseases are among the most common diseases in the United States. Is this true? Yes they are. It has been estimated that up to 50% of humans will suffer a **urinary tract infection (UTI)** at some time during their lives. In the United States, the CDC estimates there are about 4 million ambulatory-care visits each year, representing about 1% of all outpatient visits and accounting for about 10 million doctor visits each year. The major bacterial agents causing UTIs are *E. coli* and *Staphylococcus saprophyticus*. Sufferers report abdominal discomfort, burning pain on urination, and frequent urges to urinate. Most infections develop in the bladder, and women are apparently infected more often than men.

Staphylococcus saprophyticus
staf-i-lo-kok'kus sa-pro-fi'-ti-kus

How can UTIs be avoided? Such practices as avoiding tight-fitting clothes and urinating soon after sexual intercourse can reduce the possibility of infection. Some studies have suggested that cranberry juice may inhibit the bacterial cells by increasing the acidity of the urinary tract. Unfortunately, if urinary tract disease is left untreated, the infection can involve the entire bladder or spread to the kidneys.

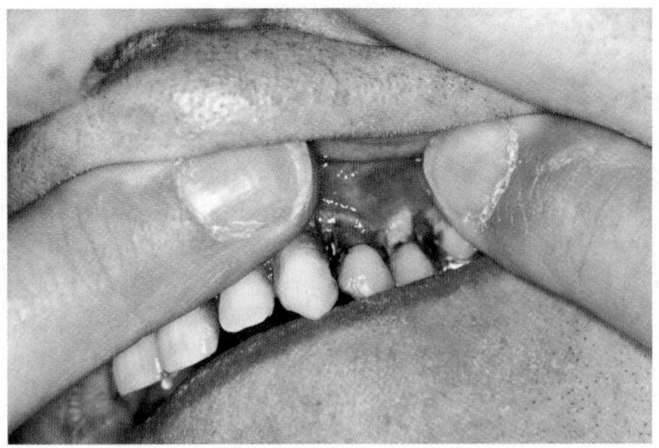

(a)

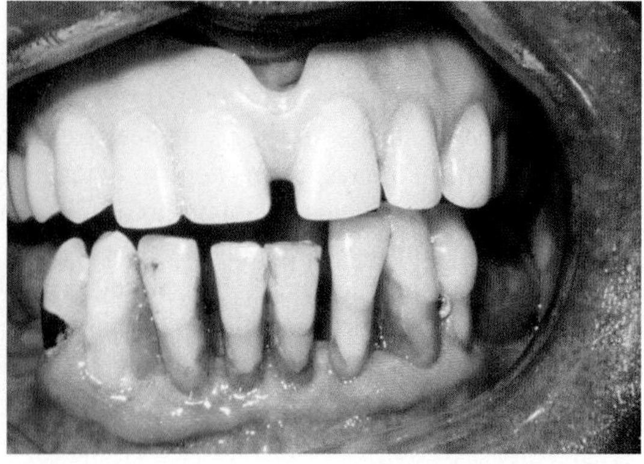

(b)

FIGURE 19.16 **Periodontal Disease.** (**a**) Gingivitis is an inflammation of the gums (gingiva) around the teeth, which is evident in this patient. (**b**) Periodontitis refers to conditions when gum disease affects the structures supporting the teeth. In this example, gum recession and bone loss has occurred, which can lead to loosening of the teeth and even tooth loss.

(a) © Medical-on-Line/Alamy. (b) © SPL/Science Source.

19.7 Healthcare-Associated Infections: Treatment Threats

Healthcare-associated infections (HAIs), also called "nosocomial infections," are infections taking place in the hospital (*noso* = "sickness;" *comi* = "care") or any healthcare setting where patients are receiving treatment for another condition or illness. Sources of the infection include caregivers, hospital staff, and the surfaces and environment of the healthcare facility. Infections also may be associated with devices used in medical procedures, such as catheters or ventilators, and at surgical sites.

Recently, the CDC reported that in the United States alone, healthcare-associated infections account for nearly 1.7 million infections and as many as 99,000 deaths each year. Some estimates of the annual cost in treating nosocomial infections in the United States range from $4.5 billion to $11 billion.

There are many different bacterial organisms capable of causing HAIs; the most common affect the urinary tract or a surgical site, and various pneumonias are also common.

Bacterial Healthcare-Associated Infections

Acinetobacter baumannii a-si-ne′tō-bak-tėr bou-mä′nē-ē

What bacteria can be picked up from and spread in a healthcare facility? One is *Acinetobacter baumannii*, a very short, gram-negative rod commonly found in soil and water. It also can be found on the skin of healthy people, though it rarely causes disease outside of the healthcare setting. This suggests *A. baumannii* targets the weak, infecting primarily only those patients in intensive care units and the very ill, especially those with weakened immune systems, chronic lung disease, or diabetes. HAIs caused by *A. baumannii* can lead to a variety of diseases, including pneumonia, blood infections, and wound infections. *Acinetobacter* is often resistant to antibiotics, and there is no standard treatment or medicine.

A friend of mine is a nurse and she keeps mentioning an organism called *C. diff*. What is she talking about? Yes, *C. diff* or *Clostridium difficile* is a real problem in hospitals and other healthcare facilities and is being reported at historically high levels. The *C. difficile* cells are gram-positive, spore-forming rods that cause severe diarrhea and can lead to proliferative colitis, perforation of the colon, and blood infections. While infection with *C. difficile* can end with a bad bout of diarrhea, its progression to inflammation of the colon may require surgical resection of infected tissue and, in the most serious cases, can lead to death.

resection: The surgical removal of part of an organ, bone, or other body part.

Unlike *Acinetobacter*, C. *diff* can also infect relatively healthy people after a round of antibiotic treatment. The antibiotics that a patient takes to eliminate some infection also can reduce the population of the normal microbiota colonizing the lower gastrointestinal tract. Antibiotic-resistant *C. difficile* survives and can grow in the gut now vacant of the resident bacterial community.

Because they can form spores, *C. diff* can exist on surfaces for years despite dehydration and application of disinfectants. Moreover, as *C. diff* is commonly found wherever people are, the high concentration and traffic of people in hospitals make the hospital setting likely places for *C. difficile* contamination. Combined with the use of antibiotics in healthcare facilities, this explains why C. *difficile* is so dangerous today.

We mentioned MRSA earlier. It too causes HAIs? Definitely. MRSA is resistant to a whole group of antibiotics. And because it causes such a variety of diseases, treatment can be difficult to manage due to antibiotic resistance. Historically, *S. aureus* infections were easily treated with penicillin and its derivatives, such as methicillin. In addition, MRSA infections can easily spread among otherwise healthy people in a community by skin-to-skin contact, so community-associated MRSA is another concern.

I know antibiotic resistance is really a problem today. What are some of the emerging antibiotic-resistant organisms? Probably the most threatening right now are the carbapenem-resistant species. The most noteworthy is *Klebsiella pneumoniae*, a gram-negative bacillus that causes meningitis, infections of the bloodstream, and wound and surgical-site infections. *K. pneumoniae* has been increasingly observed to have developed multidrug resistance, particularly to carbapenems. This is a class of antibiotics that are important in fighting gram-negative bacterial infections. Infection with carbapenem-resistant *Klebsiella pneumoniae* therefore represents a very difficult HAI to treat.

Klebsiella pneumoniae kleb-sē-el'lä nü-mō'ne-ī

So how do we prevent further spread? In most cases, HAIs can be prevented with increased diligence in hand washing and surface disinfection. HAIs, to a great extent, are the result of many people, both patients and healthcare providers, sharing a relatively small space in a setting where people are sick and also taking therapeutic or prevention antibiotics. The presence of multidrug resistant organisms is complicating the problem with HAIs and, in some cases, presenting the patient with a worse illness than that for which treatment was initially sought.

A Final Thought

As we conclude this chapter of *Microbes and Society*, you should take a moment to reflect on the vast body of knowledge you have acquired over the many pages of this book. You have undoubtedly become better aware of the microbes and their profound influence on your life. The last two chapters have focused on the negative aspects, but you will certainly also remember the ways in which they affect your life for the better. You have learned an alphabet of microbes and a language of microbiology that will allow you to keep learning the rest of your life. No longer will you bypass a news headline referring to an emerging disease, a biotechnology breakthrough, or a new drug. Now you can read and digest the contents of the story. Then, in using and sharing your new knowledge, you will become a better citizen. And that's what education is all about. We congratulate you—you are now a citizen microbiologist.

Questions to Consider

1. The CDC estimates 40,000 people in the United States die annually from pneumococcal pneumonia. Despite this high statistic, only 30% of older adults who could benefit from the pneumococcal vaccine are vaccinated (compared to over 50% who receive an influenza vaccine yearly). As an epidemiologist in charge of bringing the pneumonia vaccine to a greater percentage of older Americans, what would you do to convince older adults to be vaccinated?
2. A children's hospital in a major American city reported a dramatic increase in the number of rheumatic fever cases. Doctors were alerted to start monitoring sore throats more carefully. Why do you suppose this prevention method was recommended?
3. The story is told of a doctor in New York City in the early 1800s who was an expert at diagnosing typhoid fever even before the symptoms of disease appeared. He would go up and down the rows of hospital beds, feeling the tongues of patients and announcing which patients were in the early stages of typhoid. Sure enough, a few days later the symptoms would surface. What do you think was the secret to his success?
4. You read in the newspaper that botulism was diagnosed in 11 patrons of a local restaurant. The disease was subsequently traced to mushrooms bottled and preserved in the restaurant. What special cultivation practice enhances the possibility that mushrooms will be infected with the spores *Clostridium botulinum?*
5. A classmate plans to travel to a tropical country for spring break. To prevent traveler's diarrhea, she was told to take 2 ounces or 2 tablets of Pepto-Bismol 4 times a day for 3 weeks before travel begins. Short of turning pink, what better measures can you suggest she use to prevent traveler's diarrhea?
6. In Chapter 9 of the Bible, in the Book of Exodus, the sixth plague of Egypt is described in this way: "Then the Lord said to Moses and Aaron, 'Take a double handful of soot from a furnace, and in the presence of Pharaoh, let Moses scatter it toward the sky. It will then turn into a fine dust over the whole land of Egypt and cause festering boils on man and cattle throughout the land.'" Which bacterial disease is possibly being described?
7. A column by Ann Landers carried the following letter: "I am a 34-year-old married woman who is trying to get pregnant, but it doesn't look promising.... When I was in college, I became sexually active. I slept with more men than I care to admit.... Somewhere in my wild days, I picked up an infection that left me infertile.... The doctor told me I have quite a lot of scar tissue inside my fallopian tubes." The woman, who signed herself "Suffering in St. Louis," went on to implore readers to be careful in their sexual activities. What advice do you think the woman gave to readers?

Key Terms

All the names of the bacterial diseases represent the key terms of this chapter. To use the names as a study guide, talk about each one for a few moments, indicating what sort of disease it is, what causes it, how it is transmitted, where it occurs, whether a treatment or vaccine is available, and other information that is relevant to a concise view of the disease.

acid-fast test
anthrax
antitoxin
bacterial dysentery
botulism
brucellosis
bubo
bubonic plague (Black Death)
campylobacteriosis
chancre

chlamydia
cholera
conjunctivitis
dental caries
gas gangrene
gastroenteritis
gingivitis
gonorrhea
gumma
healthcare-associated infection (HAI)
impetigo
Legionnaires' disease (legionellosis)
leprosy (Hansen disease)
leptospirosis
listeriosis
Lyme disease
meninges
meningitis
meningococcal meningitis
multidrug-resistant *Staphylococcus aureus* (MRSA)
necrotizing fasciitis
peptic ulcer disease
periodontal disease
periodontitis
pertussis
plague
pneumococcal pneumonia
pneumonia
pneumonic plague
probiotic
rheumatic fever
Rocky Mountain spotted fever
salmonellosis
sexually-transmitted disease (STD)
shigellosis
streptococcal sore throat (strep throat)
syphilis
tetanus
toxic shock syndrome (TSS)
trachoma
traveler's diarrhea
tubercle
tuberculosis (TB)
tularemia
typhoid fever
urinary tract infection (UTI)
vector

APPENDIX

Pronouncing Organism Names

Some of the names of microorganisms, which have Latin or Greek roots, can be hard to pronounce. As an aid in pronouncing these names, microorganisms used in this textbook are listed below alphabetically, followed by the pronunciation. The following pronunciation key will aid you in saying these names. The accented syllable (') is placed directly after the syllable being stressed.

Pronunciation Key

a add	ch check	g go	o odd	ou out	u put
ā ace	e end	i it	ō open	sh rush	ü rule
ã care	ē even	ī ice	ô order	th thin	ū use
ä father	ė term	ng ring	oi oil	u up	

Acanthamoeba a-kan-thä-mē'bä
Acetobacter aceti a-sē'tō-bak-tėr a-set'ē
Acinetobacter baumannii a-si-ne'tō-bak-tėr bou-mä'nē-ē
Agaricus bisporis ä-gãr'i-kus bī-spōr'us
Agrobacterium tumefaciens ag'rō-bak-ti'rē-urn tü'me-fāsh-enz
Alcaligenes al'kä-li-gen-ēs
Alternaria äl-tėr-ner'-ēä
Amanita muscaria am-an-ī'tä mus-kãr'e-ä
A. verna vėr-na
Amycolatopsis orientalis am-i-kō-lä'top-sis ôr-ē-en-tä'lis
Anabaena an-a-bē-na
Ashbya gossypii ash'bē-ä gos-sip'ē-ē
Aspergillus favus a-spėr-jil'lus flā'vus
A. niger nī'jėr
A. oryzae ô'ri-zī
Asticcacaulis excentris as-tik-ä-cäu'lis ek-sen'tris
Azotobacter ä-zo'to-bak-tėr
Bacillus anthracis bä-sil'lus an-thrā'sis
B. cereus se'rē-us
B. licheniformis lī-ken-i-fôr'mis
B. polymyxa pol-ē-mik'sä
B. sphaericus sfe'ri-kus
B. subtilis su'til-us
B. thuringiensis thur-in-jē-en'sis
Balantidium coli bal-an-tid-ē-um kō-lī
Batrachochytrium dendrobatidis ba-tra-ko-ki'tre-um den-dro-ba-ti'dis
Bdellovibrio del'ō-vib'rē-ō
Beijerinckia bī-yė-rink'ē-ä
Bifidobacterium lactis bī-fi-dō-bak-ti'rē-um lak'tis
Blastomyces dermatitidis blas-tō-mī'sēz dėr-mä-tit'i-dis
Bordetella bronchiseptica bor-de-tel'lä bron-kē-sep'ti-kä
B. parapertussis pãr'ä-pėr-tus-sis
Bordetella pertussis bor-de-tel'lä pėr-tus'sis
Borrelia burgdorferi bôr-rel'ē-ä burg-dôr'fėr-ē
Botryococcus braunii bo-tre-ō-kok'kus brawn'-ē-ē
Botrytis bo-trī'tis
Bradyrhizobium brad-ē-rī-zo-'bē-um
Brucella brü'sel-lä
Campylobacter jejuni kam'pi-lō-bak-tėr jē-jū'nē
Candida albicans kan'did-ä al'bi-kanz
C. milleri mil'ler-ē
Cephalosporium acremonium sef-ä-lō-spô'rē-um ac-re-mō'nē-um
Chlamydia pneumoniae kla-mi'dē-ä nü-mō'nē-ī
C. trachomatis trä-kō'mä-tis
Chlamydomonas klam-i-dō-mō'äs
Claviceps purpurea kla'vi-seps pür-pü-rē'ä
Clostridium acetobutylicum klôs-tri'dē-um ā-sē-tō- bū-til'i-kum
C. botulinum bot-ū-lī'num
C. difficile dif'fi-sil-ē
C. perfringens pėr-frin'jens
C. tetani te'tän-ē
Coccidioides kok-sid-ē-oi'dēz
Corynebacterium diphtheriae kôr'ē-nē-bak-ti-rē-um dif-thi'rē-ī
Coxiella burnetii käks'ē-el-lä bėr-ne'tē-ē
C. glutamicum glü-tam'-i-kum
Cryptococcus neoformans krip'tō-kok-kus nē-ō-fôr'manz
Cryptonectria parasitica krip-tō-nek'trē-ä pãr-ä-si'ti-kä
Deinococcus radiodurans dī'nō-kok-kus rā-dē-ō-dür'anz
Dunaliella salina dun-al-ē-el' -a sa-lī'-na
Entamoeba histolytica en-tä-mē'bä his-tō-li'ti-kä
Enterobacter en-te-rō-bak'tėr
Enterococcus faecium en-tė-rō-kok'kus fē'sē-um
Epidermophyton ep-ē-der-mō-fī'ton
Erwinia dissolvens ė r-wi'-nē-ä dis-sol-venz
Escherichia coli esh-ėr-ē'kē-ä kō'lī (or kō'lē)
Euglena ū-glē'nä
Flavobacterium flā-vō-bak- tėr'-ē-um
Francisella tularensis fran'sis-el-lä tü'lä-ren-sis
Giardia lamblia jē-är'dē-ä lam'lē-ä
Haemophilus influenzae hē-mä'fil-us in-flü-en'zī
Halobacterium salinarum hā'lo-bak-tėr' ē-um sal-inar'-um
Helicobacter pylori hē'lik-ō-bak-tėr pī'lō-rē
Histoplasma capsulatum his-tō-plaz'mä kap-su-lä'tum
Homo sapiens hō'mō sā'pē-ens
Humulus lupulus hū'mū-lus lū'pū-lus
Klebsiella pneumoniae kleb-sē-el'lä nü-mō'ne-ī

Pronouncing Organism Names

Lactobacillus acidophilus lak-tō-bä-sil'lus a-sid-o'fil-us
L. bifidus bī-fid'-us
L. bulgaricus bul-gā'ri-kus
L. caseii kā'sē-ē
L. sanfranciscensis san-fran-si-sen'-sis
Lactococcus lactis lak-tō-kok'kus lak'tis
Legionella pneumophila lē-jä-nel'lä nü-mō'fi-lä
Leishmania lish'mä-nē-ä
Leptospira interrogans lep-tō-spī'rä in-tėr'rō-ganz
Leuconostoc citrovorum lü-kū-nos'tok sit-rō-vôr'um
Listeria monocytogenes lis-te'rē-ä mo-nō-sī-tô'je-nēz
Micromonospora mī-krō-mō-nos'pōr-ä
Microsporum mī-krō-spô'rum
Mycobacterium leprae mī-kō-bak-ti'rē-um lep'rī
M. tuberculosis tü-bėr-kū-lō'sis
Mycoplasma capricolum mī-kō-plaz'-mä ka-pri-cō'-lum
M. genitalium jen'i-tä-lē-um
M. laboratorium la-bôr-a-tôr'-ēum
M. mycoides mī-coi'-dēz
M. pneumoniae nu-mō'nē-ī
Neisseria gonorrhoeae nī-se'rē-ä go-nôr-rē 'ī
N. meningitidis me-nin ji'ti-dis
Nitrobacter nī-trō-bak'tėr
Nitrosomonas nī-trō-sō-mō'näs
Nostoc nos'tok
Paramecium pãr-ä-mē'sē-um
Penicillium camemberti pen-i-sil'lē-um kam-am-bėr'tē
P. chrysogenum krī-so'gen-um
P. roqueforti rō-kō-fôr'tē
Photorhabdus luminescens fō-tō-rab'dus lü-mi-nes'senz
Phytophthora infestans fī-tof'thô-rä in-fes'tans
Plasmodium plaz-mō'dē-um
Pneumocystis jirovecii nü-mō-sis'tis jėr-ō-vek'ē-ē
Propionibacterium prō-pē-on'ē-bak-ti-rē-um
Proteus prō'tē-us
Pseudomonas aeruginosa sū-dō-mō'näs ā-rü ji-nō'sä
Rhizobium radiobacter rī-zō'bē-um rā-dē-ō-bak'tėr
Rhizopus rī'zo-pus
Rickettsia rickettsii ri-ket'sē-ä ri-ket'sē-ē
Saccharomyces carlsbergensis sak-ä-rō-mī'sēs kä-rls-bėr-gen'sis
S. cerevisiae se-ri-vis'ē-ī
S. ellipsoideus ē-lip-soi'dē-us
S. kefir key'-fur
Salmonella enterica säl-mōn-el'lä en-tėr-i'kä
S. enterica serotype Enteritidis en-tėr-i-tī'dis
S. enterica serotype Typhi tī'fē
S. enterica serotype Typhimurium tī-fi-mur'ē-um
Serratia marcescens ser-rä'tē-ä mär-ses'sens
Shigella dysenteriae shi-gel'lä dis-en-te'rē-ī
S. sonnei son'nē-ē
Staphylococcus aureus staf-i-lō-kok'kus ô-rē-us
S. epidermidis e-pi-der'mi-dis
Streptococcus lactis strep-tō-kok'kus lak'tis
S. mutans mū'tans
S. pneumoniae nü-mō'nē-ī
S. thermophilus thėr-mo'fil-us
Streptomyces cattley strep-tō-mī'sēs kat-tel'-ē
S. coelicolor kō'lē-ku-lėr
S. rimosus ri-mō'sus
S. venezuelae ve-ne-zü-e'lī
Thermotoga maritima thėr'mō-tō-gä mar-i-tē'mä
Thermus aquaticus thėr'-mus a-kwa'-ti-kus
Thiomargarita namibiensis thī'ō-mär-gä-rē-tä na'mi-bē-n-sis
Toxoplasma gondii toks-ō-plaz'mä gon'dē-ē
Treponema pallidum tre-pō-nē'mä pal'li-dum
Trichoderma trik'ō -dėr-ma
Trichomonas vaginalis trik-ō-mōn'äs va-jin-al'is
Trichonympha trik-ō-nimf'ä
Trichophyton trik-ō-fī'ton
Trypanosoma brucei tri-pa'nō-sō-mä brüs'ē
T. cruzi krüz'ē
Vibrio cholerae vib'rē-ō kol'ėr-ī
V. parahaemolyticus pa-rä-hē-mō-li'ti-kus
V. vulnificus vul-ni'fi-kus
Vitis vinifera vi'tis vi-ni'fėr-ä
Volvox vōl'-voks
Vorticella vôr-ti-cel'-lä
Xanthomonas zan'-thō-mō-nas
Yersinia enterocolitica yėr-sin'ē-ä en'tėr-ō-kōl-it-ik-ä
Y. pestis pes'tis
Zoogloea ramigera zō'ō-glē-ä ram-i-gėr'ä

GLOSSARY

Note: This glossary contains concise definitions for microbiological terms and concepts only. Please refer to the index for specific infectious agents, infectious diseases, anatomical terms, taxa, and specific antimicrobial drugs.

A

acellular slime mold A member of the protists that in the vegetative state exists as a multinucleate diploid, acellular ameboid mass called a plasmodium.

acid A substance that releases hydrogen ions (H^+) in solution. *See also* **base**.

acid-fast technique A staining process in which certain bacteria resist decolorization with acid alcohol.

acidophile A microorganism that grows at acidic pHs below 4.

activated sludge Aerated sewage containing microorganisms added to untreated sewage to purify it by accelerating its bacterial decomposition.

active immunity The immune system responds to antigen by producing antibodies and specific lymphocytes.

active site The region of an enzyme where the substrate binds.

acute period (climax) The phase of a disease during which specific symptoms occur and the disease is at its height.

acyclovir A drug used as a topical ointment to treat herpes simplex and injected for herpes encephalitis.

adenosine diphosphate (ADP) A molecule in cells that is the product of ATP hydrolysis.

adenosine triphosphate (ATP) A molecule in cells that provides most of the energy for metabolism.

adenovirus infection A commonly caused illness of the respiratory system, such as a cold.

aerobe (anaerobic) An organism (or referring to an organism) that uses oxygen gas (O_2) for metabolism.

aerobic respiration A set of the metabolic reactions and processes that take place in the cells of organisms to convert biochemical energy from nutrients into cellular energy in the form of ATP.

aflatoxin A toxin produced by *Aspergillus flavus* that is cancer causing in vertebrates.

agar A polysaccharide derived from marine seaweed that is used as a solidifying agent in many microbiological culture media.

alcohol An antiseptic that works by destroying protein, which inhibits the growth of many microorganisms.

alcoholic fermentation A catabolic process that forms ethyl alcohol during the reoxidation of NADH to NAD^+ for reuse in glycolysis to generate ATP.

aldehyde A chemical such as glutaraldehyde or formaldehyde that can be used as a chemical agent to kill or slow the growth of bacteria.

alga (pl. algae) An organism in the kingdom Protista that performs photosynthesis.

algal bloom An excessive growth of algae on or near the surface of water, often the result of an oversupply of nutrients from organic pollution.

amino acid An organic acid containing one or more amino groups; the monomers that build proteins in all living cells.

aminoglycoside An antibiotic that contains amino groups bonded to carbohydrate groups that inhibit protein synthesis; examples are gentamicin, streptomycin, and neomycin.

amoeba A protozoan that undergoes a crawling movement by forming cytoplasmic projections into the environment.

amylase A group of enzymes that break down starch.

anabolism An energy-requiring process involving the synthesis of larger organic compounds from smaller ones. *See also* **catabolism**.

anaerobe An organism that does not require or cannot use oxygen gas (O_2) for metabolism.

anaerobic respiration The production of ATP where the final electron acceptor is an inorganic molecule other than oxygen gas (O_2); examples include nitrate and sulfate.

animalcule A tiny, microscopic organism observed by Leeuwenhoek.

antibiotic A substance naturally produced by bacteria or fungi that inhibits or kills bacteria.

antibody A highly specific protein produced by the body in response to a foreign substance, such as a bacterium or virus, and capable of binding to the substance.

antibody-mediated immunity The form of acquired immunity conferred to an individual through the activity of B cells and the production of antibodies in the body fluids.

anticodon A three-base sequence on the tRNA molecule that binds to the codon on the mRNA molecule during translation.

antigen A chemical substance that stimulates the production of antibodies by the body's immune system; also called immunogen.

antigen binding site The region on an antibody that binds to an antigen.

antigen presenting cell (APC) A macrophage or dendritic cell that exposes antigen peptide fragments on its surface to T cells.

antigenic determinant *See* **epitope**.

antihistamine A drug that blocks cell receptors for histamine, preventing allergic effects such as sneezing and itching.

antimicrobial agent (drug) A chemical that inhibits or kills the growth of microorganisms.

antisepsis The use of chemical methods for eliminating or reducing microorganisms on the skin.

antiseptic A chemical used to reduce or kill pathogenic microorganisms on a living object, such as the surface of the human body.

antiserum (pl. antisera) A blood-derived fluid containing antibodies and used to provide immunity.

antitoxin An antibody produced by the body that circulates in the bloodstream to provide protection against toxins by neutralizing them.

apicomplexan A protozoan containing a number of organelles at one end of the cell that are used for host penetration; no motion is observed in adult forms.

aqueous solution One or more substances dissolved in water.

Archaea The domain of living organisms that excludes the Bacteria and Eukarya.

arthrospore An asexual fungal spore formed by fragmentation of a septate hypha. ascocarp (pl. ascomata) The fruiting body of an ascomycete fungus.

ascospore A sexually produced fungal spore formed by members of the ascomycetes.

ascus (pl. **asci**) A sac-like structure containing ascospores; formed by the ascomycetes.

asexual reproduction The form of reproduction that maintains genetic constancy while increasing cell numbers.

assembly *See* **maturation.**

atom The smallest portion into which an element can be divided and still enter into a chemical reaction.

atomic nucleus The positively charged core of an atom, consisting of protons and neutrons that make up most of the mass.

ATP synthase The enzyme involved in forming ATP during electron transport.

attachment Referring to the association between a virus and its host cell surface.

attenuated Referring to the reduced ability of a bacterium or virus to do damage to the exposed individual.

autoclave An instrument used to sterilize microbiological materials by means of high temperature using steam under pressure.

autotrophic Referring to an organism that uses carbon dioxide (CO_2) as a carbon source; *see also* chemoautotroph.

avirulent Referring to an organism that is not likely to cause disease.

B

bacillus (pl. **bacilli**) (1) Any rod-shaped prokaryotic cell. (2) When referring to the genus *Bacillus*, it refers to an aerobic or facultatively anaerobic, rod-shaped, endospore-producing, gram-positive bacterial cell.

Bacteria The domain of living things that includes all organisms not classified as Archaea or Eukarya.

bacteriocin One of a group of bacterial proteins toxic to other bacterial cells.

bacteriology The scientific study of prokaryotes; originally used to describe the study of bacteria.

bacteriophage (phage) A virus that infects and replicates within bacterial cells.

bacteriostatic Any substance that prevents the growth of bacteria.

bacterium (pl. **bacteria**) A single-celled microorganism lacking a cell nucleus and membrane-enclosed compartments, and often having peptidoglycan in the cell wall.

bacteroid A symbiotic form of the nitrogen-fixing bacteria *Rhizobium*.

baculovirus A group of viruses used as vectors for protein expression in insect and mammalian cells.

barophile A microorganism that lives under conditions of high atmospheric pressure.

base A chemical compound that accepts hydrogen ions (H^+) in solution. *See also* **acid**.

basidiospore A sexually produced fungal spore formed by members of the basidiomycetes.

beer An alcoholic beverage produced by yeast fermentation of carbohydrate.

benign Referring to a tumor that usually is not life-threatening or likely to spread to another part of the body.

beta-lactam nucleus The chemical group central to all penicillin antibiotics.

binary fission An asexual process in prokaryotic cells by which a cell divides to form two new cells while maintaining genetic constancy.

binomial nomenclature The method of nomenclature that uses two names (genus and specific epithet) to refer to organisms.

bioaugmentation The form of bioremediation where bacterial cultures are added to speed up the rate of degradation of a compound.

biofilm A complex community of microorganisms that form a protective and adhesive matrix that attaches to a surface, such as a catheter or industrial pipeline.

biofuel A material that stores potential energy and that is produced from living organisms.

biogeochemical cycle Any of the natural circulation pathways of the essential elements of living matter (i.e. carbon, nitrogen, sulfur, phosphorus, and oxygen).

biological safety cabinet A cabinet or hood used to prevent contamination of biological materials.

biological vector An infected arthropod, such as a mosquito or tick, that transmits disease-causing organisms between hosts. *See also* **mechanical vector**.

bioluminescence The emission of light by a living organism in which the light is produced by an enzyme-catalyzed chemical reaction during which chemical energy is converted to light energy.

bioremediation The use of microorganisms to degrade toxic wastes and other synthetic products of industrial pollution.

biosphere The areas of the Earth that are inhabited by living organisms.

biostimulation The form of bioremediation that modifies the environment to stimulate existing microbes to degrade a compound.

biosynthesis Referring to the manufacture of virus parts during virus replication.

biota The total collection of organisms of a geographic region.

biotechnology The commercial application of genetic engineering using living organisms.

bioterrorism The intentional or threatened use of biological agents to cause fear in or actually inflict death or disease upon a large population.

blanching A process of putting food in boiling water for a few seconds to destroy enzymes.

blastospore A fungal spore formed by budding.

bloom A sudden increase in the number of cells of an organism in an environment.

B lymphocyte (B cell) A white blood cell that matures into memory cells and plasma cells that secrete antibody.

booster shot A repeat dose of a vaccine given some years after the initial course to maintain a high level of immunity.

brandy An alcoholic spirit produced by distilling wine.

bright-field microscope An instrument that magnifies an object by passing visible light directly through the lenses and object. *See also* **light microscope**.

broad spectrum Referring to an antimicrobial drug useful for treating many groups of microorganisms, including gram-positive and gram-negative bacteria.

broth A liquid containing nutrients for the growth of microorganisms.

bubo A swelling of the lymph nodes due to inflammation.

budding An asexual process of reproduction in fungi, in which a new cell forms as a swelling at the border of the parent cell and then breaks free to live independently.

C

callus An unorganized cell mass formed in response to various biotic and abiotic stimuli.

cancer A disease characterized by the radiating spread of malignant cells that reproduce at an uncontrolled rate.

canning A food preservation method in which the food contents are processed and sealed in an airtight container.

capsid The protein coat that encloses the genome of a virus.

capsomere Any of the protein subunits of a capsid.

capsule A layer of polysaccharides and small proteins covalently bound to some prokaryotic cells.

carbohydrate An organic compound consisting of carbon, hydrogen, and oxygen that is an important source of carbon and energy for all organisms; examples include sugars, starch, and cellulose.

carbolic acid *See* **phenol**.

carbon cycle A series of interlinked processes involving carbon compound exchange between living organisms and the nonliving environment.

carbon-fixing reactions The stage of photosynthesis in which electrons and ATP are used to reduce carbon dioxide gas (CO_2) to sugars.

carcinogen A substance capable of causing cancer.

carrier An individual who has recovered from a disease but retains and continues to shed the infectious agent.

carotenoid An orange or yellow pigment in chloroplasts that broadens the spectrum of colors (wavelengths) that can be used for photosynthesis.

casein The major protein in milk.

catabolism An energy-liberating process in which larger organic compounds are broken down into smaller ones. *See also* **anabolism**.

caterpillar The larval form of butterflies, moths, and related insects.

cell envelope The cell wall and cell membrane of a prokaryotic cell.

cell-mediated immunity (CMI) The body's ability to resist infection through the activity of T-lymphocyte recognition of antigen peptides presented on macrophages and dendritic cells and on infected cells.

cell membrane A thin bilayer of phospholipids and proteins that surrounds the prokaryotic cell cytoplasm. *See also* **plasma membrane**.

cell nucleus A membrane-enclosed organelle in eukaryotic cells that contains most of the cell's genetic material, organized as chromosomes (DNA molecules).

cellular respiration The process of converting chemical energy into cellular energy in the form of ATP.

cellular slime mold A member of the protists that exists as amoeboid cells, which aggregate into a multicellular "slug."

cellulase An enzyme produced by fungi, bacteria, and protozoans that is used to degrade and metabolize cellulose.

cellulose A polysaccharide carbohydrate composed of beta-glucose subunits. It forms the primary structural component of cell walls.

cell wall A carbohydrate-containing structure surrounding fungal, algal, and most prokaryotic cells.

central dogma The doctrine that DNA codes for RNA through transcription and RNA is converted to protein through translation.

cesspool Concrete cylindrical rings with pores in the walls that is used to collect human waste.

chancre A painless, circular, purplish, hard ulcer with a raised margin that occurs during primary syphilis.

chemical bond A force between two or more atoms that tends to bind those atoms together.

chemical element Any substance that cannot be broken down into a simpler one by a chemical reaction.

chemical reaction A process that changes the molecular composition of a substance by redistributing atoms or groups of atoms without altering the number of atoms.

chitin A polymer of acetylglucosamine units that provides rigidity to the cell walls of fungi.

chlamydia (pl. **chlamydiae**) A very small, round, pathogenic bacterium visible only with the electron microscope and cultivated within living cells.

chlorination The process of treating water with chlorine to kill harmful organisms.

chlorophyll A green or purple pigment in algae and some bacterial cells that functions in capturing light for photosynthesis.

chloroplast A double membrane-enclosed compartment in algae that contains chlorophyll and other pigments for photosynthesis.

chromosome A structure in the nucleoid or cell nucleus that carries hereditary information in the form of genes.

ciliate A protozoan that moves with the aid of cilia.

cilium (pl. **cilia**) A hairlike projection on some eukaryotic cells that, along with many others, assist in the motion of some protozoa and beat rhythmically to aid the movement of a fluid past a cell of the respiratory epithelial cells in humans.

citric acid cycle A metabolic pathway essential for ATP production during cellular respiration. Also called the Krebs cycle.

class A category of related organisms consisting of one or more orders.

climate change The alteration in global temperatures and rainfall patterns that may occur due to the increased levels of atmospheric greenhouse gases (carbon dioxide, methane) produced by the use of fossil fuels.

clone A population of cells genetically identical to the parent cell.

coccus (pl. **cocci**) A spherical-shaped prokaryotic cell.

codon A three-base sequence on the mRNA molecule that specifies a particular amino acid insertion in a polypeptide.

coenzyme A small, organic molecule that forms the nonprotein part of an enzyme molecule; together they form the active enzyme.

coliform bacterium A gram-negative, non-spore-forming, rod-shaped cell that ferments lactose to acid and gas and usually is found in the human and animal intestine; high numbers in water indicates contamination.

colony A visible mass of microorganisms of one type.

commensalism A close and permanent association between two species of organisms in which one species benefits and the other remains unharmed and unaffected.

commercial sterilization A canning process to eliminate the most resistant bacterial spores.

communicable disease A disease that is readily transmissible between hosts. competence The ability of a cell to take up naked DNA from the environment.

community A group of organisms that live in the same area.

comparative genomics The comparison of DNA sequences between organisms.

compost Organic matter that has been decomposed by microbes and recycled as a fertilizer.

compound A substance made by the combination of two or more different chemical elements.

conidium (pl. **conidia**) An asexually produced fungal spore formed on a supportive structure without an enclosing sac.

conjugation (1) In prokaryotes, a unidirectional transfer of genetic material from a live donor cell into a live recipient cell during a period of cell contact. (2) In the protozoan ciliates, a sexual process involving the reciprocal transfer of micronuclei between cells in contact.

conjugation pilus (pl. **pili**) A hollow projection for DNA transfer between the cytoplasms of donor and recipient bacterial cells.

consolidation Formation of a firm dense mass in the alveoli.

contagious Referring to a disease whose agent passes with particular ease among hosts.

consumer An organism or group of organisms that are the final users of a product.

contractile vacuole A membrane-enclosed structure within a cell's cytoplasm that regulates the water content by absorbing water and then contracting to expel it.

covalent bond A chemical bond formed by the sharing of electrons between atoms or molecules.

critical control point (CCP) In the food processing industry, a place where contamination of the food product could occur.

culture medium A mixture of nutrients in which microorganisms can grow.

cyanobacterium (pl. **cyanobacteria**) An oxygen-producing, pigmented bacterium occurring in unicellular and filamentous forms that carries out photosynthesis.

cyst A dormant and very resistant form of a protozoan and multicellular parasite.

cytochrome A compound containing protein and iron that plays a role as an electron carrier in cellular respiration and photosynthesis; *see also* **electron transport chain**.

cytoplasm The complex of chemicals and structures within a cell; in plant and animal cells excluding the nucleus.

cytotoxic T cell The type of T lymphocyte that searches out and destroys infected cells.

D

dark-field microscopy An optical system on the light microscope that scatters light such that the specimen appears white on a black background.

decline (death) phase The final portion of a bacterial growth curve in which environmental factors overwhelm the population and induce death; also called death phase.

decomposer An organism, such as a bacterium or fungus, that breaks down dead or decaying matter.

dedifferentiation A cellular process in which a cell reverts to an earlier developmental stage.

defective particle A virus particle that does not contain the complete set of viral genes needed for virus replication.

definitive host An organism that harbors the adult, sexually mature form of a parasite.

dehydration synthesis reaction A process of bonding two molecules together by removing the products of water and joining the open bonds.

denaturation A process caused by heat or pH in which proteins lose their function due to changes in their molecular structure.

denitrification The process of reducing nitrate and nitrite into gaseous nitrogen, which is far less accessible to life forms but makes up the bulk of our atmosphere.

deoxyribonucleic acid (DNA) The genetic material of all cells and many viruses.

desiccation The process through which things are made to be extremely dry by removing water.

diacetyl An organic compound having an intensely buttery flavor.

diagnosis The process of identifying a disease, illness, or problem by examining an individual.

diarrhea Excessive loss of fluid from the gastrointestinal tract.

diatom One of a group of microscopic marine algae that performs photosynthesis.

diatomaceous earth Filtering material composed of the remains of diatoms.

differential stain technique A procedure using two dyes to differentiate cells or cellular objects based on their staining; *see also* **simple stain technique**.

dinoflagellate A microscopic photosynthetic marine alga that forms one of the foundations of the food chain in the ocean.

diplococcus (pl. **diplococci**) A pair of spherical-shaped prokaryotic cells.

direct contact The form of disease transmission involving close association between hosts; *see also* **indirect contact**.

disaccharide A sugar formed from two single sugar molecules; examples include sucrose and lactose.

disease Any change from the general state of good health.

disinfectant A chemical used to kill or inhibit pathogenic microorganisms on a lifeless object such as a tabletop.

disinfection The process of killing or inhibiting the growth of pathogens.

DNA *See* **deoxyribonucleic acid**.

DNA polymerase An enzyme that catalyzes DNA replication by combining complementary nucleotides to an existing strand.

DNA probe A known segment of single-strand DNA that is complementary to a desired DNA sequence of a bacterial species.

DNA replication The process of copying the genetic material.

DNase I An enzyme that degrades DNA and is used to treat the buildup of DNA-containing mucus in patients with cystic fibrosis.

DNA vaccine A preparation that consists of a DNA plasmid containing the gene for a pathogen protein.

domain (1) The most inclusive taxonomic level of classification; consists of the Archaea, Bacteria, and Eukarya. (2) A loop of DNA consisting of about 10,000 bases.

double helix The structure of DNA, in which the two complementary strands are connected by hydrogen bonds between complementary nitrogenous bases and wound in opposing spirals.

droplet An airborne particle of mucus and sputum from the respiratory tract that contains disease-causing microorganisms.

dry weight The weight of the materials in a cell after all the water is removed.

E

ecosystem The collection of living and nonliving components and processes that comprise and govern the behavior of some defined subset of the biosphere (the outermost part of the earth's shell, which includes air, land, surface rocks, and water).

electrolyte A mineral or salt in blood and other body fluids that regulates nerve and muscle function.

electron A negatively charged particle with a small mass that moves around the nucleus of an atom.

electron microscope An instrument that uses electrons and a system of electromagnetic lenses to produce a greatly magnified image of an object. *See also* **transmission electron microscope** and **scanning electron microscope**.

electron transport A series of proteins that transfer electrons in cellular respiration to generate ATP.

emerging infectious disease A new disease or changing disease that is seen within a population for the first time; *see also* **reemerging infectious disease**.

encephalitis Inflammation of the tissue of the brain or infection of the brain.

endemic Referring to a constant presence of disease or persistence of an infectious agent at low level in a population.

endocytosis The process by which many eukaryotic cells take up substances, cells, or viruses from the environment.

endophyte A fungus that lives within plants and does not cause any known disease.

endospore An extremely resistant dormant cell produced by some gram-positive bacteria.

endosymbiosis hypothesis This theory explains the origins of organelles in eukaryotic cells as the result of bacteria engulfing other prokaryotic cells through endophagocytosis. These cells with the bacteria trapped inside them entered a symbiotic relationship (symbiosis).

endotoxin A metabolic poison, produced chiefly by gram-negative bacteria that are part of the bacterial cell wall and consequently are released on cell disintegration; composed of lipid–polysaccharide–peptide complexes.

enriched medium A growth medium in which special nutrients must be added to get a species to grow.

enterotoxin A toxin that is active in the gastrointestinal tract of the host.

envelope The flexible membrane of protein and lipid that surrounds many types of viruses.

enveloped virus A virus whose genome and capsid is surrounded by the membrane-like covering.

environmental microbiology The scientific study of microbial processes, microbial communities, and microbial interactions in an environment.

enzyme A reusable protein molecule that brings about a chemical change while itself remaining unchanged.

enzyme-substrate complex The association of an enzyme with its substrate at the active site.

epidemic The occurrence of more cases of a disease than expected in individuals within a geographic area.

epidemiology The scientific study of the source, cause, and transmission of disease within a population.

epitope A section of an antigen molecule that stimulates antibody formation and to which the antibody binds; also called antigenic determinant.

ergotism A disease caused by the transfer of a toxin produced by the fungus *Claviceps purpurea* from rye grain to humans.

Eukarya The taxonomic domain encompassing all eukaryotic organisms.

eukaryotic cell (eukaryote) A cell (organism) containing a cell nucleus with multiple chromosomes, a nuclear envelope, and membrane-bound compartments; *see also* prokaryotic cell.

exoenzyme An enzyme that is secreted by a cell and functions outside of that cell.

exon Any region of a gene that is transcribed and retained in the final messenger RNA product, in contrast to an intron, which is spliced out from the transcribed RNA molecule.

exotoxin A bacterial protein toxin secreted into the environment or body by living bacterial cells.

expiration date The last date that a product, as food, should be used before it is considered spoiled.

extreme acidophile An archaeal organism living at an extremely acidic pH.

extreme halophile An archaeal organism that grows at very high salt concentrations.

extremophile A microorganism that lives in extreme environments, such as high temperature, high acidity, or high salt.

extrinsic factor An environmental characteristic that influences the growth of food microbes. *See also* **intrinsic factor**.

F

facultative Referring to an organism that grows in the presence or absence of oxygen gas (O_2).

FAD *See* flavin adenine dinucleotide.

family A category of related organisms consisting of one or more genera.

fat A type of lipid made up of a glycerol attached to three fatty acids.

fecal-oral route A route of disease transmission, where pathogens in fecal material pass from one person and are introduced into the oral cavity of another person.

fermentation A metabolic process that converts sugar to acid, gases, and/or alcohol.

fermentor (bioreactor) A large fermentation tank for growing microorganisms used in industrial production.

fever An abnormally high body temperature that is usually caused by a bacterial or viral infection.

F factor (plasmid) A plasmid containing genes for plasmid replication and conjugation pilus formation.

filtration A mechanical method to remove microorganisms by passing a liquid or air through a filter.

flagellum (pl. **flagella**) A long, hairlike appendage composed of protein and responsible for motion in microorganisms; found in some bacteria, protozoa, algae, and fungi.

flash pasteurization method A treatment in which milk is heated at 71.6°C for 15 seconds and then cooled rapidly to eliminate harmful bacteria.

flavin adenine dinucleotide (FAD) A coenzyme that functions in electron transfer during oxidative phosphorylation.

floc A jelly-like mass that forms in a liquid and made up of coagulated particles.

fluid mosaic Referring to the model to represent the cell (plasma) membrane where proteins "float" within or on a bilayer of phospholipid.

fluorescence The emission of one color of light after being exposed to light of another wavelength.

fluorescence microscopy An optical system on the light microscope that uses ultraviolet light to excite dye-containing objects to fluoresce.

fomite An inanimate object, such as clothing or a utensil, that carries disease organisms.

food infection Caused by infectious pathogens in the consumed food.

food poisoning Caused by consuming foods that contain toxins.

food spoilage The result of food deteriorating to the point that it is not edible to humans or its quality of edibility becomes reduced.

foraminiferan A shell-containing amoeboid protozoan having a chalky skeleton with windowlike openings between sections of the shell.

freeze drying *See* lyophilization.

fruiting body The general name for a reproductive structure of a fungus from which spores are produced.

Fungi One of the five kingdoms in the Whittaker classification of living organisms; composed of the molds and yeasts.

fungicide Any agent that kills fungi.

fungus (pl. **fungi**) A member of a large group of eukaryotic organisms that includes the yeasts, molds, and mushrooms.

G

gamma globulin A general term for antibody-rich serum.

gastroenteritis An inflammation of the stomach and the intestines, causing vomiting and diarrhea.

gene A segment of a DNA molecule that provides the biochemical information for a function product.

gene expression The processes by which the information in a gene is transcribed and translated into a protein.

generation time The time interval for a cell population to double in number.

genetically modified food A food that has been produced from an organism in which its DNA has been modified through genetic engineering.

genetic code The specific order of nucleotide sequences in DNA or RNA that encode specific amino acids for protein synthesis.

genetic engineering The use of bacterial and microbial genetics to isolate, manipulate, recombine, and express genes.

genetic recombination The process of bringing together different segments of DNA.

genome The complete set of genes in a virus or an organism.

genomics The identification and study of gene sequences in an organism's DNA.

genus (pl. **genera**) A rank in the classification system of organisms composed of one or more species; a collection of genera constitutes a family.

germ theory of disease The principle, formulated by Pasteur and proved by Koch, that microorganisms are responsible for infectious diseases.

gluten A type of protein found in many grains, cereals, and breads.

glycocalyx A viscous polysaccharide material covering many prokaryotic cells to assist in attachment to a surface and impart resistance to desiccation. *See also* **capsule** and **slime layer**.

glycolysis A metabolic pathway in which glucose is broken down into two molecules of pyruvate with a net gain of two ATP molecules.

glyphosphate A broad-spectrum systemic herbicide.

gonococcus A colloquial name for *Neisseria gonorrhoeae*.

gram-negative Referring to a bacterial cell that stains red after Gram staining.

gram-negative cell wall Bacterial cell walls composed of little peptidoglycan and an outer membrane.

gram-positive Referring to a bacterial cell that stains purple after Gram staining.

gram-positive cell wall Bacterial cell walls composed of a thick layer of peptidoglycan and teichoic acid.

Gram stain technique A staining procedure used to identify bacterial cells as gram-positive or gram-negative.

green alga (pl. **algae**) A member of the protists that carries out photosynthesis and may have given rise to green plants.

growth curve The plotted or graphed measurement of the size of a population of bacteria as a function of time.

gumma A soft, granular lesion that forms in the cardiovascular and/or nervous systems during tertiary syphilis.

H

HAART *See* **highly active antiretroviral therapy**.

habitat The place where a particular species lives and grows.

HACCP *See* **Hazard Analysis Critical Control Point**.

halogen A chemical element whose atoms have seven electrons in their outer shell; examples include iodine and chlorine.

halophile An organism that lives in environments with high concentrations of salt.

Hazard Analysis Critical Control Point (HACCP) A set of federally enforced regulations to ensure the dietary safety of seafood, meat, and poultry.

healthcare-associated infection (HAI) An infection resulting from treatment for another condition while in a hospital or healthcare facility.

heavy (H) chain The larger polypeptide in an antibody.

heavy metal A chemical element often toxic to microorganisms; examples include mercury, copper, and silver.

helix (pl. **helices**) A twisted shape such as that seen in a spring, screw, or a spiral staircase.

helper T lymphocyte A T lymphocyte that enhances the activity of B lymphocytes and stimulates destruction of macrophages infected with bacteria.

hemagglutinin (1) An enzyme composing one type of surface spike on influenza viruses that enables the viruses to bind to the host cell. (2) An agent such as a virus or an antibody that causes red blood cells to clump together.

hemoglobin The red oxygen-carrying pigment in erythrocytes.

heredity The passing of genetic traits from parents to offspring.

heterocyst A specialized cell in some cyanobacteria that fix nitrogen gas.

heterotrophic Referring to an organism that requires preformed organic matter for its energy and carbon needs. *See also* chemoheterotroph.

high efficiency particulate air (HEPA) filter A type of air filter that removed particles larger than 0.3 micrometers.

highly active antiretroviral therapy (HAART) The combination of several (typically three or four) antiretroviral drugs for the treatment of infections caused by retroviruses, especially the human immunodeficiency virus.

highly perishable Referring to foods that spoil easily.

holding (batch) method A pasteurization process that exposes a liquid to 63° for 30 min.

horizontal gene transfer (HGT) The movement of genes from one organism to another within the same generation; also called lateral gene transfer.

host An organism infected by a pathogen.

human endogenous retrovirus (HERV) Virus genes that are incorporated into human chromosomes and comprise up to 10% of the human genome.

human genome The complete set a genetic information in a human cell.

Human Genome Project (HGP) An international scientific research project that sequenced the human DNA in a cell, and is identifying and mapping all of the genes.

human growth hormone A small protein hormone that stimulates growth and cell reproduction in humans.

Human Microbiome Project (HMP) An initiative that identified and characterized the microorganisms found in association with both healthy and diseased humans.

human immunodeficiency virus (HIV) The retrovirus that causes acquired immunodeficiency syndrome (AIDS).

human microbiome The community of microorganisms and viruses that normally resides on the surface of the skin and in the mouth, respiratory system, and gastrointestinal and urogenital tracts of the human body. *See also* microbiome.

humus A complex organic substance resulting from the microbial breakdown of plant material.

hyaluronidase An enzyme that digests hyaluronic acid and thereby permits the penetration of pathogens through connective tissue.

hydrogen peroxide An unstable liquid that readily decomposes in water and oxygen gas (O_2).

hydrolysis reaction A process in which a molecule is split into two parts through the interaction of H^+ and $(OH)^-$ of a water molecule.

hydrophilic Referring to a substance that dissolves in or mixes easily with water; *see also* **hydrophobic**.

hydrophobia An abnormal or unnatural dread of water, especially seen with rabies.

hydrophobic Referring to a substance that does not dissolve in or mix easily with water; *see also* **hydrophilic**.

hyperthermophile A prokaryote that has an optimal growth temperature above 80°C.

hypha (pl. **hyphae**) A microscopic filament of cells representing the vegetative portion of a fungus.

I

icosahedron A symmetrical figure composed of 20 triangular faces and 12 points; one of the major shapes of some viral capsids.

idiopathic Referring to a disease or disorder that has no known cause.

IgA (immunoglobulin A) The class of antibodies found in respiratory and gastrointestinal secretions that help neutralize pathogens.

IgD (immunoglobulin D) The class of antibodies found on the surface of B cells that act as receptors for binding antigen.

IgE (immunoglobulin E) The class of antibodies responsible for type I hypersensitivities.

IgG (immunoglobulin G) The class of antibodies abundant in serum that are major disease fighters.

IgM (immunoglobulin M) The first class of antibodies to appear in serum in helping to fight pathogens.

imidazole An antifungal drug that interferes with sterol synthesis in fungal cell membranes; examples are miconazole and ketoconazole.

immunity The body's ability to resist infectious disease through innate and acquired mechanisms.

immunization The process of making an individual resistant to a particular disease by administering a vaccine. *See also* **vaccination**.

immunocompetent The ability of the body to develop an immune response in the presence of a disease-causing agent.

immunoglobulin (Ig) The class of immunological proteins that react with an antigen; an alternate term for antibody.

immunology The scientific study of how the immune system works and responds to nonself agents.

inactivated Referring to a vaccine containing nonreproducing bacterial cells or nonreplicating viruses.

incubation period The time that elapses between the entry of a pathogen into the host and the appearance of signs and symptoms.

indicator organism A microorganism whose presence signals fecal contamination of water.

indirect contact The mode of disease transmission involving nonliving objects. *See also* **direct contact**.

induced mutation A change in the sequence of nucleotide bases in a DNA molecule arising from a mutagenic agent used under controlled laboratory conditions.

industrial fermentation Any large scale industrial process, with or without oxygen gas (O_2), for growing microorganisms. *See also* **fermentation**.

industrial microbiology The field that uses microbes in the manufacturing of food and industrial products, including pharmaceuticals, beverages, and chemicals.

infection The relationship between two organisms and the competition for supremacy that takes place between them.

infectious disease A disorder arising from a pathogen invading a susceptible host and inducing medically significant symptoms.

infectious dose The number of microorganisms needed to bring about infection.

inflammation A nonspecific defensive response to injury; usually characterized by redness, warmth, swelling, and pain.

interferon An antiviral protein produced by body cells on exposure to viruses; triggers the synthesis of antiviral proteins.

intermediate host The host in which the larval or asexual stage of a parasite is found.

intrinsic factor A characteristic of a food product that influences microbial growth. *See also* **extrinsic factor**.

intron A section of DNA that is removed after transcription by splicing the mRNA transcript.

invasiveness The ability of a pathogen to spread from one point to adjacent areas in the host and cause structural damage to those tissues.

ion An electrically charged atom.

ionizing radiation A type of radiation such as gamma rays and X-rays that causes the separation of atoms or a molecule into ions.

isomer A substance that has the same number and types of atoms as another substance but where the atoms are arranged differently.

J

jaundice A condition in which bile seeps into the circulatory system, causing the complexion to have a dull yellow color.

K

kingdom The second highest taxonomic rank below domain Eukarya that is divided into smaller groups called phyla.

Koch's postulates A set of procedures by which a specific organism can be related to a specific disease.

koji The common name for the fungus *Aspergillus oryzae.*

Krebs cycle *See* **citric acid cycle**.

L

lactic acid bacteria A group of Gram-positive, acid-tolerant rods or cocci that are associated by their common metabolic and physiological characteristics.

lactose A milk sugar composed of one molecule of glucose and one molecule of galactose.

lagering The secondary aging of beer.

lag phase A portion of a bacterial growth curve encompassing the first few hours of the population's history when no growth occurs.

larva (pl. **larvae**) A wormlike pre-adult stage of an insect.

latency A condition in which a virus integrates into a host chromosome without immediately causing a disease.

legume (leguminous plant) A plant that bears its seeds in pods.

leukocyte Any of a number of types of white blood cells.

lichen An association between a fungal mycelium and a cyanobacterium or alga.

light (L) chain A smaller polypeptide in an antibody.

light microscope An instrument that uses visible light and a system of glass lenses to produce a magnified image of an object; also called a compound microscope.

lipid A nonpolar organic compound composed of carbon, hydrogen, and oxygen; examples include triglycerides and phospholipids.

lipopolysaccharide (LPS) A molecule composed of lipid and polysaccharide that is found in the outer membrane of the gram-negative cell wall of bacterial cells.

logarithmic (log) phase The portion of a bacterial growth curve during which active growth leads to a rapid rise in cell numbers.

lymph The tissue fluid that contains white blood cells and drains tissue spaces through the lymphatic system.

lymph node A bean-shaped organ located along lymph vessels that is involved in the immune response and contains phagocytes and lymphocytes.

lymphocyte A type of white blood cell that functions in the immune system.

lyophilization (freeze drying) A process in which food or other material is deep frozen, after which its liquid is drawn off by a vacuum; also called freeze-drying.

lysis The rupture of a cell and the loss of cell contents.

lysozyme An enzyme found in tears and saliva that digests the peptidoglycan of gram-positive bacterial cell walls.

M

macronucleus The larger of two nuclei in most ciliates; involved in controlling metabolism. *See also* **micronucleus**.

macrophage A large cell derived from monocytes that is found within various tissues and actively engulfs foreign material, including infecting bacteria and viruses.

malignant Referring to a tumor that invades the tissue around it and may spread to other parts of the body.

malting Referring to the process when barley begins to germinate by being soaked in water to produce simpler carbohydrates.

manufacturer code A system used by manufacturers to identify products quickly.

mechanical vector A living organism, or an object, that transmits disease agents on its surface. *See also* **biological vector**.

membrane filter technique A method to test water quality by identifying any coliform bacteria trapped on a filter.

memory cell A cell derived from B lymphocytes or T lymphocytes that reacts rapidly upon re-exposure to an antigen.

meningitis An inflammation of the membranes surrounding and protecting the brain and spinal cord, often resulting from a bacterial or viral infection.

meningococcus A common name for *Neisseria meningitidis*.

mesophile An organism that grows in temperature ranges of 20°C to 40°C.

messenger RNA (mRNA) An RNA transcript containing the information for synthesizing a specific polypeptide.

metabolic pathway A sequence of linked enzyme-catalyzed reactions in a cell.

metabolism The sum of all biochemical processes taking place in a living cell. *See also* **anabolism** and **catabolism**.

metabolite Any substance produced during metabolism.

metagenome The collective genomes from a population of organisms.

metagenomics The study of genes isolated directly from environmental samples.

metastasize Referring to a tumor that spreads from the site of origin to other tissues in the body.

methanogen An archaeal organism that lives on simple compounds in anaerobic environments and produces methane during its metabolism.

microbe *See* **microorganism**.

microbial forensics The discipline involved with the recognition, identification, and control of a pathogen.

microbial genomics The discipline of sequencing, analyzing, and comparing microbial genomes.

microbial load The total number of bacteria and fungi in a given quantity of water or soil or on the surface of food.

microbiology The scientific study of microscopic organisms and viruses, and their roles in human disease as well as beneficial processes.

microbiome A specific environment characterized by a distinctive microbial community and its collective genetic material.

micrometer (μm) A unit of measurement equivalent to one millionth of a meter; commonly used in measuring the size of microorganisms.

micronucleus The smaller of the two nuclei in most ciliates, which contains genetic material and is involved in sexual reproduction. *See also* **macronucleus**.

microorganism (microbe) A microscopic form of life including bacterial, archaeal, fungal, and protozoal cells.

mitochondrion (pl. **mitochondria**) A double membrane-enclosed compartment in eukaryotic cells that carries out aerobic respiration.

mixed fermentation An anaerobic fermentation where the products are a complex mixture of acids that have numerous applications in biotechnology.

mixotroph An organism that can exist as either an autotroph or heterotroph.

mold A type of fungus that consists of chains of cells and appears as a fuzzy mass in culture.

molecule Two or more atoms held together by a sharing of electrons.

monosaccharide A simple sugar that cannot be broken down into simpler sugars; examples include glucose and fructose.

morphology Refers to the form (shape) and structure of cells and organisms.

mucous membrane A moist lining in the body passages of all mammals that contains mucus-secreting cells and is open directly or indirectly to the external environment.

mucus A sticky secretion of viscous fluid.

mushroom A spore-bearing fruiting body typical of many members of the basidiomycetes.

must The juice resulting from crushing grapes.

mutagen A chemical or physical agent that causes a mutation.

mutant An organism carrying a mutation.

mutation A change in the characteristic of an organism arising from a permanent alteration of a DNA sequence.

mutualism A close and permanent association between two populations of organisms in which both benefit from the association.

mycelium (pl. **mycelia**) A mass of fungal filaments from which most fungi are built.

mycology The scientific study of the fungi.

mycorrhiza (pl. **mycorrhizae**) A close association between a fungus and the roots of many plants.

mycosis A fungal infection of animals, including humans.

N

nanometer (nm) A unit of measurement equivalent to one billionth of a meter; the unit is often used in measuring viruses and the wavelength of energy.

neuraminidase An enzyme composing one type of surface spike of influenza viruses that facilitates viral release from the host cell.

neutraceutical A food or part of a food that may provides medicinal or health benefits, including the prevention and treatment of disease.

neutron An uncharged particle in the atomic nucleus.

nicotinamide adenine dinucleotide (NAD+) A coenzyme that transfers and transports electrons during oxidative phosphorylation and fermentation reactions.

nitrification The biological oxidation of ammonia into nitrates; performed in two steps by two different bacteria collectively known as nitrifying bacteria.

nitrogen cycle The processes that convert nitrogen gas (N_2) to nitrogen-containing substances in soil and living organisms, then reconverted to the gas.

nitrogen fixation The chemical process by which microorganisms convert nitrogen gas (N_2) into ammonia.

nonenveloped virus A virus consisting of only the viral genome and a capsid.

nonperishable Referring to foods that are least likely to spoil.

nucleic acid A high-molecular-weight molecule consisting of nucleotide chains that convey genetic information and are found in all living cells and viruses. *See* DNA and RNA.

nucleobase Any of five nitrogen-containing compounds found in nucleic acids, including adenine, guanine, cytosine, thymine, and uracil.

nucleocapsid The combination of genome and capsid of a virus.

nucleoid The chromosomal region of a prokaryotic cell.

nucleotide A component of a nucleic acid consisting of a carbohydrate molecule, a phosphate group, and a nitrogenous base.

nucleus (pl. **nuclei**) (1) The portion of an atom consisting of protons and neutrons. (2) A membrane-enclosed compartment in eukaryotic cells that contains the chromosomes.

nutrient agar A solidifying agent that contains nutrients for microbial growth.

O

objective lens The lens or mirror in a microscope that receives the first light rays from the object being observed.

obligate intracellular parasite An organism or virus that must get its nutrients from a host cell.

oil A fat that is liquid at room temperature.

oncogene A segment of DNA that can induce uncontrolled growth of a cell if permitted to function.

oncogenic Referring to any agent such as viruses that can cause tumors.

oncology The scientific study of tumors and cancers.

oncovirus A virus capable of causing a tumor or involved with a cancer.

operator A sequence of bases in the DNA to which a repressor protein can bind.

operon The unit of bacterial DNA consisting of a promoter, operator, and a set of structural genes.

opportunistic Referring to pathogens that only cause disease when the person's immune system is weakened.

order A category of related organisms consisting of one or more families.

organelle A specialized compartment in eukaryotic cells that has a particular function.

organic Referring to chemicals that contain carbon atoms.

organic acid A carbon-containing compound with acidic properties.

organic compound A substance characterized by chains or rings of carbon atoms that are linked to atoms of hydrogen and sometimes oxygen, nitrogen, and other elements.

origin of replication The fixed point on a DNA molecule where copying of the molecule starts.

osmosis The net movement of water molecules from an area of high concentration through a semipermeable membrane to a region of lower concentration.

osmotic pressure The force that must be applied to a solution to inhibit the inward movement of water across a membrane.

outbreak A small, localized epidemic.

outer membrane A bilayer membrane forming part of the cell wall of gram-negative bacteria.

oxidation lagoon A large pond in which sewage is allowed to remain undisturbed so that digestion of organic matter can occur.

oxygen cycle The biogeochemical cycle that describes the movement of oxygen through the air, water, and soil.

P

pandemic Refers to a disease occurring over a wide geographic area (worldwide) and affecting a substantial proportion of the global population.

Pap smear A test to detect cancerous or precancerous cells of the cervix, allowing for early detection of cancer.

parasite A type of heterotrophic organism that feeds on live organic matter such as another organism.

parasitism A close association between two organisms in which one (the parasite) feeds on the other (the host) and may cause injury to the host.

passive immunity The temporary immunity that comes from receiving antibodies from another source.

pasteurization A heating process that destroys pathogenic bacteria in a fluid such as milk and lowers the overall number of bacteria in the fluid.

pasteurizing dose The amount of irradiation used to eliminate pathogens.

pathogen A microorganism or virus that causes disease in a host organism.

pathogenicity The ability of a disease-causing agent to gain entry to a host and bring about a physiological or anatomical change interpreted as disease.

pathway engineering The engineering of microbes, such as bacteria or fungi, to produce biochemicals or other products in a more sustainable or useful way.

pellicle A flexible covering layer typical of the protozoan ciliates.

penetration Referring to the entry of a virus and its uncoating in a host cell during replication.

penicillinase An enzyme produced by certain microorganisms that converts penicillin to penicilloic acid and thereby confers resistance against penicillin.

peptide bond A linkage between the amino group on one amino acid and the carboxyl group on another amino acid.

peptidoglycan A complex molecule of the bacterial cell wall composed of alternating units of N-acetylglucosamine and N-acetylmuramic acid cross-linked by short peptides.

period of convalescence The phase of a disease during which the body's systems return to normal.

period of decline The phase of a disease during which symptoms subside.

periplasmic space A metabolic region between the cell membrane and outer membrane of gram-negative cells.

peroxide A compound containing an oxygen–oxygen single bond.

pH A measure of the hydrogen ion (H^+) concentration of an aqueous solution. Solutions with a pH less than 7 are said to be acidic and solutions with a pH greater than 7 are alkaline (basic). Pure water has a pH of 7.

phage *See* **bacteriophage**.

phagocyte A white blood cell capable of engulfing and destroying foreign materials or cells, including bacteria and viruses.

phagocytosis A process by which foreign material or cells are taken into a white blood cell and destroyed.

pharmacopeia A book describing drugs, chemicals, and medicinal preparations.

phenol A chemical compound that has one or more hydroxyl groups attached to a benzene ring, derivatives of which are used as an antiseptic or disinfectant; also called carbolic acid.

phenotype The visible (physical) appearance of an organism resulting from the interaction between its genetic makeup and the environment.

phospholipid A water-insoluble compound containing glycerol, two fatty acids, and a phosphate head group; forms part of the membrane in all cells.

phosphorus cycle The biogeochemical cycle that describes the movement of phosphorus through the water and soil.

photosynthesis A biochemical process in which light energy is converted to chemical energy, which is then used for carbohydrate synthesis.

phylum (pl. **phyla**) A category of organisms consisting of one or more classes.

phytoplankton Microscopic free-floating communities of cyanobacteria and unicellular algae.

pilus (pl., **pili**) A hairlike extension of the plasma membrane found on the surface of many bacteria that is used for cell attachment and anchorage.

plasma The fluid portion of blood remaining after the cells have been removed. *See also* **serum**.

plasma cell An antibody-producing cell derived from B lymphocytes.

plasma membrane The phospholipid bilayer with proteins that surrounds the eukaryotic cell cytoplasm. *See also* **cell membrane**.

plasmid A small, closed-loop molecule of DNA apart from the chromosome that replicates independently and carries nonessential genetic information.

plasmodium The acellular mass of protoplasm consisting of thousands of nuclei typical of the acellular slime molds.

point mutation The replacement of one base in a DNA strand with another base.

polymerase chain reaction (PCR) A technique used to replicate a fragment of DNA many times.

polypeptide A chain of linked amino acids.

polysaccharide A complex carbohydrate made up of sugar molecules linked into a branched or chain structure; examples include starch and cellulose.

polysome A cluster of ribosomes linked by a strand of mRNA and all translating the mRNA.

portal of entry The site at which a pathogen enters the host.

portal of exit The site at which a pathogen leaves the host.

potable Referring to water that is safe to drink because it contains no harmful material or microbes.

primary metabolite A small molecule essential to the survival and growth of an organism; *see also* **secondary metabolite**.

primary structure The sequence of amino acids in a polypeptide.

primary waste treatment The removal (sedimentation) of large solids from the wastewater via physical settling, screening, or filtration.

primordial soup A pond or body of water rich in substances that could provide favorable conditions for the emergence of life.

prion An infectious, self-replicating protein involved in human and animal diseases of the brain.

probiotic Living microbes that help reestablish or maintain the human microbiota of the gut.

prodromal phase The phase of a disease during which general symptoms occur in the body.

producer An organism (autotroph) in an ecosystem that produces biomass from inorganic compounds.

product A substance or substances produced in a chemical reaction.

productive infection The active assembly and maturation of viruses in an animal cell.

prokaryote A microorganism in the domain Bacteria or Archaea composed of single cells having a single chromosome but no cell nucleus or other membrane-bound compartments.

prokaryotic cell Referring to cells or organisms having a single chromosome but no cell nucleus or other membrane-bound compartments. *See also* **eukaryotic cell**.

promoter site The region of a template DNA strand or operon to which RNA polymerase binds.

prophylaxis A measure taken to maintain health and prevent the spread of disease.

protease An enzyme that uses water to hydrolyze and break peptide bonds between amino acids of proteins.

protein A chain or chains of linked amino acids used as a structural material or enzyme in living cells.

protein synthesis The process of forming a polypeptide or protein through a series of chemical reactions involving amino acids.

protist A member of a very large and diverse group of eukaryotic microorganisms that includes the protozoa and single-celled algae.

proton A positively charged particle in the atomic nucleus.

proto-oncogene A region of DNA in the chromosome of human cells; altered by carcinogens into oncogenes that transform cells.

protozoan (pl. **protozoa**) A single-celled eukaryotic organism that lacks a cell wall and usually exhibits chemoheterotrophic metabolism.

provirus The viral DNA that has integrated into a eukaryotic host chromosome and is then passed on from one generation to the next through cell division.

pseudopodium (pl. **pseudopodia**) A projection of the plasma membrane that allows movement in members of the amoebozoans.

psychrophile An organism that lives at cold temperature ranges of 0°C to 20°C.

psychrotroph An organism that lives at cold temperature ranges of 0°C to 30°C.

pure culture An accumulation or colony of microorganisms of one species.

pyrogen A fever-producing substance.

R

radiolarian A single-celled marine organism with a round silica-containing shell that has radiating arms to catch prey.

reactant A substance that interacts with another in a chemical reaction.

recombinant DNA molecule A DNA molecule containing DNA from two different sources.

red tide A brownish-red discoloration in seawater caused by increased numbers of dinoflagellates. *See also* algal bloom.

reemerging infectious disease A disease showing a resurgence in incidence or a spread in its geographical area; *see also* **emerging infectious disease**.

regulatory gene A DNA segment that codes for a repressor protein.

release Referring to the exiting of a virus from a host cell after replication.

rem *See* Roentgen Equivalent in Man.

rennin A protein-digesting enzyme that curdles milk.

replication fork The point where complementary strands of DNA separate and new complementary strands are synthesized.

repressor protein A protein that, when bound to the operator, blocks transcription.

reservoir The location or organism where disease-causing agents exist and maintain their ability for infection.

restriction endonuclease A type of enzyme that splits open a DNA molecule at a specific restricted point; important in genetic engineering techniques.

reverse transcriptase An enzyme that synthesizes a DNA molecule from the code supplied by an RNA molecule.

ribonucleic acid (RNA) The nucleic acid involved in protein synthesis and gene control; also the genetic information in some viruses.

ribosomal RNA (rRNA) An RNA transcript that forms part of the ribosome's structure.

ribosome A cellular component of RNA and protein that participates in protein synthesis.

rickettsia (pl. **rickettsiae**) A very small bacterial cell generally transmitted by arthropods; most rickettsiae are cultivated only within living tissues.

RNA *See* **ribonucleic acid**.

RNA polymerase The enzyme that synthesizes an RNA polynucleotide from a DNA template.

Roentgen Equivalent in Man (rem) A measure of radiation dose related to biological effect.

rumen The first chamber in the digestive tract of ruminant animals.

ruminant Any hooved animal that digests its food in two steps.

S

Sabin vaccine A type of polio vaccine prepared with attenuated viruses and taken orally.

Salk vaccine A type of polio vaccine prepared with inactivated viruses and injected into the body.

sanitation The process of reducing the number of microbes to a safe level.

saturated Referring to a water-insoluble compound that cannot incorporate any additional hydrogen atoms; *see also* **unsaturated**.

sauerkraut Cabbage that has been fermented by various lactic acid bacteria.

scanning electron microscope (SEM) The type of electron microscope that allows electrons to scan across an object, generating a three-dimensional image of the object.

science The organized body of knowledge that is derived from observations and can be verified or tested by further investigation.

secondary metabolite A small molecule not essential to the survival and growth of an organism; *see also* **primary metabolite**.

secondary structure The region of a polypeptide folded into an alpha helix or pleated sheet.

secondary waste treatment The removal of smaller solids and particles remaining in the wastewater through fine filtration aided by the use of membranes or through the use of microbes.

sedimentation *See* **primary waste treatment**.

selective medium A growth medium that contains ingredients to inhibit certain microorganisms while encouraging the growth of others.

semen The thick white fluid containing sperm that a male ejaculates.

semiconservative replication The DNA copying process where each parent (old) strand serves as a template for a new complementary strand.

semiperishable Referring to foods that spoil less quickly.

semisynthetic Referring to a chemical substance synthesized from natural and lab components used to treat disease.

sepsis The growth and spreading of bacteria or their toxins in the blood and tissues.

septicemia A growth and spreading of bacteria in the bloodstream.

septic shock A collapse of the circulatory and respiratory systems caused by an overwhelming immune response.

septic tank An enclosed concrete box that collects waste from the home.

serotype A distinct variation within a species based on immunological characteristics.

serum (pl. **sera**) The fluid portion of the blood consisting of water, minerals, salts, proteins, and other organic sub stances, including antibodies; contains no clotting agents. *See also* **plasma**.

sexually transmitted disease (STD) A disease, such as gonorrhea or chlamydia, that is normally passed from one person to another through sexual activity.

shelf life The length of time that a commodity, such as food, may be stored without becoming unfit for use or consumption.

shock A state of physiological collapse marked by a weak pulse, coldness, sweating, and irregular breathing due to too low a blood flow to the brain.

sign An indication of the presence of a disease, especially one observed by a doctor but not apparent to the patient; *see also* **symptom**.

silage Grass or other green fodder compacted and preserved through fermentation in airtight conditions and used as animal feed in the winter.

simple stain technique The use of a single cationic dye to contrast cells. *See also* **differential stain technique**.

slime layer A thin, loosely bound layer of polysaccharide covering some prokaryotic cells. *See also* **capsule** and **glycocalyx**.

slime mold An amoeboid protist that can, under certain conditions, transform into a multicellular slug and migrate to better growth conditions.

sludge The solids in sewage that separate out during sewage treatment.

solute The substance dissolved in a solution.

solvent The substance doing the dissolving in a solution.

soredium (pl. **soredia**) The disseminated group of fungal and photosynthetic cells formed by a lichen.

soy sauce A condiment made from a fermented paste of boiled soybeans, roasted grain, salt, and *Aspergillus oryzae*.

species The fundamental rank in the classification system of organisms.

specific epithet The second of the two scientific names for a species. *See also* **genus**.

spike A protein projecting from the viral envelope or capsid that aids in attachment and penetration of a host cell.

spirillum (pl. **spirilla**) (1) A bacterial cell shape characterized by twisted or curved rods. (2) A genus of aerobic, helical cells, usually with many flagella.

spirochete A twisted bacterial rod with a flexible cell wall containing axial filaments for motility.

spontaneous generation The doctrine that nonliving matter could spontaneously give rise to living things.

spontaneous mutation A mutation that arises from natural phenomena in the environment.

spore (1) A reproductive structure formed by a fungus. (2) A highly resistant dormant structure formed from vegetative cells in several genera of bacteria, including *Bacillus* and *Clostridium*. *See also* **endospore**.

sporulation The process of spore formation.

staphylococcus (pl. **staphylococci**) (1) An arrangement of bacterial cells characterized by spheres in a grapelike cluster. (2) A genus of facultatively anaerobic, nonmotile, nonspore-forming, gram-positive spheres in clusters.

starch An energy polysaccharide that is built from many glucose molecules.

stationary phase The portion of a bacterial growth curve in which the reproductive and death rates of cells are equal.

sterile Free from living microorganisms, spores, and viruses.

sterilization The removal of all life forms, including bacterial spores.

sterol An organic solid containing several carbon rings with side chains; examples include cholesterol.

sticky end The unpaired ens of a DNA fragment that has been cut with a restriction endonuclease (enzyme).

streptobacillus (pl. **streptobacilli**) (1) A chain of bacterial rods. (2) A genus of facultatively anaerobic, nonmotile, gram-negative rods.

streptococcus (pl. **streptococci**) (1) A chain of bacterial cocci. (2) A genus of facultatively anaerobic, nonmotile, non-spore-forming, gram-positive spheres in chains.

structural gene A segment of a DNA molecule that provides the biochemical information for a polypeptide.

substrate The substance or substances upon which an enzyme acts.

subunit vaccine A vaccine that contains parts of microorganisms, such as capsular polysaccharides or purified pili.

sulfur cycle The processes by which sulfur moves through and is recycled in the environment.

superbug A strain of a bacterial species that has become resistant to multiple antibiotic drugs.

superinfection The overgrowth of susceptible strains by antibiotic resistant ones.

symbiosis (symbiotic) An interrelationship (or referring to an interrelationship) between two populations of organisms where there is a close and permanent association.

symptom An indication of some disease or other disorder that is experienced by the patient. *See also* **sign**.

syncytium (pl. **syncytia**) A giant tissue cell formed by the fusion of cells infected with respiratory syncytial viruses.

syndrome A collection of signs or symptoms that together are characteristic of a disease.

synthetic drug A man-made medicine produced by a pharmaceutical compant or research lab.

T

Taq polymerase a thermostable DNA polymerase used for the polymerase chain reaction (PCR).

taxonomy The science dealing with the systematized arrangements of related living things in categories.

T cell *See* **T lymphocyte**.

teichoic acid A negatively charged polysaccharide in the cell wall of gram-positive bacteria.

tertiary structure The folding of a polypeptide back on itself.

tertiary waste treatment Involves the disinfection of the wastewater through chlorination.

tetrad An arrangement of four bacterial cells in a cube shape.

thermoacidophile An archaeal organism living under high temperature and high acid conditions.

thermoduric Referring to an organism that tolerates the heat of the pasteurization process.

thermophile An organism that lives at high temperature ranges of 40°C to 90°C.

three-domain system The classification scheme placing all living organisms into one of three groups based, in part, on ribosomal RNA sequences.

tincture A low concentration of a chemical dissolved in alcohol.

tissue plasminogen activator A secreted protease (an enzyme that cuts proteins) that cuts plasminogen and thereby converts it to plasmin.

T lymphocyte (T cell) A type of white blood cell that matures in the thymus gland and is associated with cell-mediated immunity.

toxin A poisonous chemical substance produced by an organism.

toxoid A preparation of a microbial toxin that has been rendered harmless by chemical treatment but that is capable of stimulating antibodies; used as vaccines.

toxoid vaccine A vaccine produced from a toxin (poison) that has been made harmless but can still trigger an immune response against the toxin.

transcription The biochemical process in which RNA is synthesized according to a code supplied by the bases of a gene in the DNA molecule.

transduction The transfer of a few bacterial genes from a donor cell to a recipient cell via a bacterial virus.

transfer RNA (tRNA) A molecule of RNA that unites with amino acids and transports them to the ribosome in protein synthesis.

transformation (1) The transfer and integration of DNA fragments from a dead and lysed donor cells to a recipient cell's chromosome. (2) The conversion of a normal cell into a malignant cell due to the action of a carcinogen or virus.

transgenic Referring to an organism containing a gene or genes from another organism.

translation The biochemical process in which the code on the mRNA molecule is translated into a sequence of amino acids in a polypeptide.

transmission electron microscope (TEM) The type of electron microscope that allows electrons to pass through the object, resulting in a detailed view of the object's structure.

transposon A segment of DNA that moves from one site on a DNA molecule to another site, carrying information for protein synthesis.

trichocyst A structure of some ciliate and flagellate protozoans that consists of a cavity that contains long, thin threads that are ejected in response to the correct stimuli.

triclosan A phenol derivative incorporated as an antimicrobial agent into a wide variety of household products.

trophozoite The feeding form of a microorganism, such as a protozoan.

tubercle A hard nodule that develops in tissue infected with *Mycobacterium tuberculosis*.

tuberculin test A procedure performed by applying purified protein derivative from Mycobacterium tuberculosis to the skin and noting if a thickening of the skin with a raised vesicle appears within a few days; used to establish if someone has been exposed to the bacterium.

tumor An abnormal uncontrolled growth of cells that has no physiological function.

tumor suppressor gene A normal gene that inhibits tumor formation.

U

ultra-high temperature (UHT) method A treatment in which milk is heated at 140°C for 3 seconds to destroy harmful bacteria.

ultrastructure The detailed structure of an cell, virus, or other object when viewed with the electron microscope.

ultraviolet (UV) light A type of electromagnetic radiation of short wavelengths that damages DNA.

uncoating Referring to the loss of the viral capsid inside an infected eukaryotic cell.

unicellular alga A member of the protists that exists as a single cell and carries out photosynthesis.

unsaturated Referring to a water-insoluble compound that can incorporate additional hydrogen atoms; *see also* **saturated**.

V

vaccination Inoculation with weakened or dead microbes or viruses in order to generate immunity. *See also* **immunization**.

vaccine A preparation containing weakened or dead microorganisms or viruses, treated toxins, or parts of microorganisms or viruses to stimulate immune resistance.

vector (1) An arthropod that transmits the agents of disease from an infected host to a susceptible host. (2) A plasmid used in genetic engineering to carry a DNA segment into a bacterium or other cell.

venipuncture The piercing of a vein to take blood, to feed somebody intravenously, or to administer a drug.

viable but noncultured (VBNC) Referring to microbes that are alive but not dividing.

vibrio (1) A prokaryotic cell shape occurring as a curved rod. (2) A genus of facultatively anaerobic, gram-negative curved rods with flagella.

vinegar A liquid consisting mainly of acetic acid and water that results from the fermentation of ethanol by acetic acid bacteria.

viroid An infectious RNA segment associated with certain plant diseases.

virology The scientific study of viruses.

virosphere Refers to all places where viruses are found or interact with their hosts.

virulence The degree to which a pathogen is capable of causing a disease.

virulence factor A molecule possessed by a pathogen that increases its ability to invade or cause disease to a host.

virulent Referring to a virus or microorganism that can be extremely damaging when in the host.

virus An infectious agent consisting of DNA or RNA and surrounded by a protein sheath; in some cases, a membranous envelope surrounds the coat.

W

water mold A member of the protists that superficially resembles a fungal mold but is now recognized as having an independent evolutionary history.

water pollution The contamination of water with contaminants.

whey The clear liquid remaining after protein has curdled out of milk.

white blood cell *See* **leukocyte**.

whole agent vaccine A vaccine containing whole bacterial cells, viruses, or toxins.

wine An alcoholic beverage made from fermented grapes or other fruits.

wort A sugary liquid produced from crushed malted grain and water to which is added yeast and hops for the brewing of beer.

Y

yeast (1) A type of unicellular, nonfilamentous fungus that resembles bacterial colonies when grown in culture. (2) A term sometimes used to denote the unicellular form of pathogenic fungi.

Z

zoonotic disease (zoonosis) A disease spread from another animal to humans.

zooplankton Small crustaceans and other animals that inhabit the water column of oceans, seas, and bodies of fresh water.

zoospore An asexual spore that uses a flagellum for locomotion.

zygospore A sexually produced spore formed by members of the Zygomycota.

INDEX

Note: Page numbers followed by *b*, *f*, or *t* indicate materials in boxes, figures, or tables respectively.

B

C

F

G

H

Q

R

S

W

X

Y

Z